The New Uranium Mining Boom

Broder Merkel • Mandy Schipek
Editors

The New Uranium Mining Boom

Challenge and Lessons learned

Springer

Editors
Prof. Dr. Broder Merkel
TU Bergakademie Freiberg
Institute for Geology
Gustav-Zeuner-Strasse 12
09596 Freiberg
Germany
merkel@geo.tu-freiberg.de

Dipl.-Geoökol. Mandy Schipek
TU Bergakademie Freiberg
Institute for Geology
Gustav-Zeuner-Strasse 12
09596 Freiberg
Germany
schipek@geo.tu-freiberg.de

ISBN 978-3-642-22121-7 e-ISBN 978-3-642-22122-4
DOI 10.1007/978-3-642-22122-4
Springer Heidelberg Dordrecht London New York

Library of Congress Control Number: 2011936159

Cover design: deblik, Berlin

Printed on acid-free paper

Springer is part of Springer Science+Business Media (www.springer.com)

Contents

Part 5 Modelling

Preface

Many elements of the periodic system such as arsenic, mercury, lead, etc. have been discussed and evaluated in the scientific community and in the public awareness in contradictory manner during the past 100 years. But uranium and its daughter nuclides and some fission products are probably the most controversially discussed elements in recent days. In 2011, we celebrate the 100th anniversary of the Chemistry Nobel Prize awarded to Marie Curie for the disco-very of radium and polonium that she separated and identified from uranium ore. A few years after the Nobel Price was awarded, Europe was in a rush for radium-rich waters in spas like Joachimsthal and Schlema based on the believe that radium may cure cancer and other diseases. It took a while to find that radon and not radium caused the partly extremely high radioactivity of these waters and that radioactivity is a two-sided sword in terms of health treatment. However, even nowadays radium and radon spas exist in many countries. But nowadays waters with about 1 kBq/l are common while in the very beginning waters with 30 kBq/l up to 160 kBq/l have been reported.

With Hiroshima, Nagasaki and the end of WW II, a new era was initiated characterized by the cold war between east and west and a uranium mining boom to supply both nuclear weapon production and slightly later feeding nuclear power plants. Not before 1963 the first version of the Limited Test Ban Treaty, pledging to refrain from testing nuclear weapons in the atmosphere, underwater, or in outer space, was signed. However, this nuclear weapon tests emitted more than a 100 times the amount of hazardous nuclides into the atmosphere in comparison to the 1986 Tschernobyl accident. Two other major mistakes of human society of these times are related to the way of mining, milling and isotope separating causing severe environmental damages in many countries worldwide on the one side and the missing concept for nuclear waste handling on the other side. The latter has not been solved until recent days.

With the end of the cold war in the late 1980s a new era started. It was characterized by less or no future demand of uranium for weapons, certain conferences on disarmament, and dilution of highly enriched uranium-235 with depleted uranium to produce reactor fuel. The latter action caused a severe drop of uranium price worldwide and in consequence the closure of many uranium mines, uranium processing facilities and as well all plants to recover uranium from phosphate.

However, in 2004 the uranium price continuously started to rise from 10 USD/lb to a peak of 135 USD/lb in 2007 and since then establishing at 50 USD/lb. This was caused by a continuous request of nuclear fuel and construction of new nuclear power plant in particular in Asia (China and India). But as well other newly industrialized countries and developing countries are interested in building nuclear power plants. It is estimated that the about 450 nuclear power plants that exist today will partly be pulled down due to their age of 50 years and more but 450 new nuclear power plants are planned or under construction worldwide. As a consequence uranium mining and milling is booming in many countries with Kazakhstan in 2011 the by far biggest uranium producer followed by Canada, Australia, Namibia, Niger, Russia, Uzbekistan, and the USA.

The accident of the Fukushima nuclear power plant in Japan in March 2011 had no significant impact on the uranium price and the energy policy in most countries. The probably most significant consequence was the hasty decision of the German government to shut down all German nuclear power plants at the latest in 2022, which is an impressive example for the public awareness of accepted risk because in Germany roughly 250.000 persons (8.300 per year) were killed during the last 30 years by traffic accidents; another 180.000 in household accidents, 4.2 million (140.000 per year) due to smoking and 2.3 million (77.000 per year) due to heart attack; but nevertheless the majority of Germans do believe that nuclear power plants are the most severe threat to their health although the number of casualties related to nuclear power plants in German in the last 30 years is probably close to zero.

In contrast to this, the same German government is doing nothing against the permanent poisoning of agricultural soil due to the application of phosphate fertilizers containing uranium, thorium, radium, and polonium although this is out of all reasons because recovering of uranium and other radionuclide is feasible and moreover economical concerning the nowadays uranium prices.

Little to no public awareness is recognized about the fact that solar energy is based on nuclear fusion reactions in the sun and geothermal energy mainly on radioactive decay in crust and mantle. Thus at the end of the days all forms of energy (fossil and renewable) are based on nuclear processes and humans have to deal with this including natural radioactivity in our environment. Besides nuclear power plants, we utilize radionuclides in medical investigations and treatment; and the average dose rates received from this is estimated to be at least one third of the total dose rate of 2.4 mS/y. Many human activities, in particular oil and gas production and combustion of coal, lead to man-made enrichment of Naturally Occurring Radioactive Material (NORM) or distribution in the environment as by the already mentioned case of phosphate fertilizers. Thus, radioactivity is a phenomenon we have to consider and accept. And research related to occurrence of radioactive nuclides and radiochemistry-controlled processes is of great importance. A need in additional research can be stated in rather different fields such as clean-up of contaminated sites, distribution of radionuclides by natural and man-made processes, and radiation from outer space and the sun. The international conference Uranium Mining and Hydrogeology VI in Freiberg is an excellent platform to discuss at least part of these research needs, technologies that can be applied, and legal aspects in a globalizing world.

Freiberg, September 2011

Broder J. Merkel
Technische Universität
Bergakademie Freiberg (TUBAF)

Part 0
Plenary

Marie Curie and the Discovery of Radium

Fernando P. Carvalho

Abstract. Marie Curie gave outstanding contributions to science and society that were recognized still in her lifetime. In particular, the discovery of radium completely changed the therapeutic methods for treatment of cancer and other diseases, and allowed the development of radiotherapy and nuclear medicine. Radium was also used in many non-medical applications. Radium applications fostered the growth of uranium mining industry during the first half of 20th century. During the second half of the past century, with developments of artificial radionuclides production and particle physics, radium was gradually replaced by shorter-lived radionuclides and electron and photon beams in cancer therapy. In the 70s and 80s most radium sources in cancer hospitals were replaced while in non-medical applications radium had been substituted already. Notwithstanding, the avenue for medical use of radioactivity and radionuclides opened with Marie Curie discoveries and radium applications still goes on. This avenue is currently pursued in curietherapy and nuclear medicine.

Introduction

This year one completes the 100th anniversary of the Chemistry Nobel Prize awarded to Marie Curie in 1911 for the discovery of radium and polonium, two radioactive elements she identified and separated from uranium ore. These discoveries were made based on measurements of ionizing radiation emitted by the ore and they steered a fantastic number of scientific discoveries made during the first

Fernando P. Carvalho
Nuclear and Technological Institute (ITN) Department of Radiological Safety and Protection E.N. 10, 2686-953 Sacavém, Portugal
E-mail: carvalho@itn.pt

B. Merkel, M. Schipek, *The New Uranium Mining Boom*
DOI 10.1007/978-3-642-22122-4 © Springer 2011

half of 20th century. Likewise, they paved the way to numerous radioactivity applications, particularly in medicine (NN 2011, Fevrier 2011).

In all the research she made, Marie Curie revealed a curious mind, open to novelty, while keeping an unfailing application of the scientific method in hypothesizing, planning experiments, interpreting results and reporting the findings. Furthermore, she combined that with perseverance, hard work, and thoroughness in the laboratory experiments, while fighting for research funding and teaching in the University. In those days these were not activities of easy access for a woman.

Marie Curie was also a woman sensitive to the problems of her time and, beyond her work for radium application in medicine, she was active in societal matters providing her direct contribution. Just two examples are given here. One, during the First World War, when she installed X-rays apparatus on cars, called the "little curies" and worked as volunteer radiographer in the screening of wounded to support medical surgery near the war front line. Other, when after the war she acted as one of the founding members of the Commission for the Intellectual Cooperation of the Society of Nations, which is an ancestor of UNESCO (NN 2011, Fevrier 2011, Curie E. 1938).

The history of scientific contributions given by Marie Curie has common roots with the uranium mining history. It is therefore appropriate to remind and celebrate her scientific discoveries here.

Marie Curie (1867–1934)

Borne Manya (Marie) Sklodowska, in Warsow, Poland, she was the youngest of the five children of a couple of school professors. In 1891 she travelled to Paris, joining her sister Bronya who had concluded studies in Medicine. In Paris, Marie Sklodowska followed courses in the Faculté de Sciences de Paris, and obtained her diploma in Physics in 1893, and in mathematics in the following year. Marie met Pierre Curie, Professor and researcher at the Paris University, in 1894 when she started research for her doctoral thesis in his laboratory. They married in 1895 and had two children, Eve and Irene. In 1898, with Pierre Curie, she discovered two new elements, polonium and radium. Pierre passed away in 1906, victim of an accident. She continued research isolating polonium and radium to fully demonstrate the existence of the elements discovered through their radioactive emissions. She was awarded two Nobel Prizes, one in 1903 together with Pierre Curie and Henri Becquerel for the discovery of radioactivity, and other in 1911 for the discovery of polonium and radium (Fig. 1). In 1910 she published her "Traité sur la Radioactivité". In 1914 the Institute du Radium de Paris was built, but with the start of the First World War, the opening was postponed. She helped, together with her daughter Irene as radiologists at the war front. After the war, she resumed her research on radioactive elements in the Institute du Radium. She visited the USA in two occasions, in 1921 and 1929, and received each time a donation of 1 g radium. The first she used in the Radium Institute in Paris and the second she

Fig. 1 Marie Curie aged 44, the year she received the Nobel Prize of Chemistry

offered to the Radium Institute in Warsaw. Marie Curie deceased in 1934 victim of leukemia caused by the exposure to ionizing radiation for many years. Her body ashes, as well as those of Pierre Curie, were transferred in 1995 to the Pantheon, in Paris (NN 2011, Fevrier 2011, Curie E. 1938).

The Discovery of Radium

Near the end of the 19th century, in 1895, Conrad W. Roentgen discovered the X-rays, the new type of radiation that allowed seeing internal structures of opaque bodies. This was followed by the discovery of the "uranic rays" made by Henri Becquerel in 1898, during his research on the phosphorescence of uranium salts.

When Marie Sklodowska, a young Polish lady that had completed the Degrees in Physics and Mathematics at the Sorbonne University in Paris, arrived in 1894 in Pierre Curie's laboratory at the Faculté de Sciences looking for a doctoral research subject, she considered first to work on X-rays. However, although very recent this subject had been already thoroughly investigated in comparison with the "uranic rays". Until late 1896 about 1000 articles and 50 books had been published already on Roentgen radiation (X-rays), while the Becquerel "uranic rays" were still mysterious and only 20 articles had been published on this subject. Furthermore, a new instrument recently invented by Pierre Curie and his brother Jacques, the Curie's electrometer that allowed much more precise measurements of the air ionization by radiations, might have contributed to Marie Sklodowska's selection of "uranic rays" as the subject for her research (NN 2011).

She started work in Pierre's laboratory in Rue Lhomond, Paris, using pitchblende uranium ore from the Joachimstal mine in Poland, at that time ruled by Austria. She tried to identify which substances and minerals besides pitchblende could also emit ionizing radiation. For this purpose she tested all compounds and minerals she could find in the Sorbonne University and noted that all uranium salts emitted radiation. She observed the same emission also with thorium salts, not being aware that in Germany another scientist had discovered the radiation emitted by thorium a few weeks before.

She focused her attention on the pitchblende from Joachimstal mine and noted that this rock emitted about 2.5 times more radiation than explained by the rock uranium content. She wrote on her laboratory note book: "I searched whether other substances, besides uranium compounds, could be susceptible of ionizing the air and allowing for electric conductivity". She found that all uranium salts and natural uranium phosphate were even more active than metallic uranium and about this she wrote: "This is an outstanding fact and makes one to think that these ores may contain another element much more active than uranium" Curie M. (1921).

Using another supply of uranium ore she initiated a lengthy and laborious chemical procedure to extract the mysterious radioactive element. Her husband Pierre Curie joined her in this research and they looked also into the collaboration of a chemist, Gustav Bémont, Professor at the École de Physique et Chimie de la Ville de Paris. For this work they could only use a shed in the university yard. There Marie treated the ore and, following Fresenius chemistry methodology, she separated several salt fractions and measured the radioactivity in each fraction (NN 2011, Fevrier 2011).

In the bismuth fraction, she measured a strong radioactivity emission, and they wrote about this in her article to the Académie des Sciences: "We believe that the substance extracted from the pitchblende contains an unknown metal with analytical properties close to those of bismuth. If the existence of this metal is confirmed we propose to name it Polonium, according to the name of the country of origin of one of us". The symbol Po is written for the first time in the laboratory note book on 13 July 1893 by the hand of Pierre Curie. The article published in the Comptes-rendus de l' Académie des Sciences announces the discovery of a new element, more radioactive than uranium, but not seen as yet. The word radioactive was used for the first time in this article (Curie & Curie 1898).

The research by the Curies on the radioactivity of uranium ore was not over yet. In the physical and chemical separation procedure applied to treat the uranium ore, they observed also high radiation emission in another chemical fraction containing barium. They hypothesized that eventually other substance with a chemical behavior close to barium could be present. They planned to check this hypothesis in a three step approach. Firstly, they checked that normal barium was not radioactive. Next, they found out that a radioactive substance could be concentrated through fractional crystallization starting with the radioactive barium chloride fraction obtained from the pitchblende. At last, they pursued this separation procedure until obtaining a chloride salt with 900 times more radioactivity than metallic uranium (Curie 1921).

The emission spectroscopy analysis of this salt revealed that the lines observed did not match any known element and their intensity increased along with radioactivity increase, i.e., with the purification of the chloride, while barium lines decreased in intensity. The Curies wrote: "There is a strong reason to believe that the substance obtained contains a new element. We propose to name it as Radium. This new radioactive substance obtained probably still contains a large amount of barium mixed therein, but radium radioactivity seems enormous" (Curie et al. 1898). In their laboratory notebook the word Radium, followed by a question mark, was first annotated on the 18 November 1898.

The acceptance of this discovery was dependent upon separation, purification and chemical characterization of the element with confirmation by spectroscopic emission that the element emission spectrum was different of known elements.

Marie Curie initiated then a lengthy procedure to isolate radium from uranium ore tailings from the processing of Joachimstal mine ore, given by Austria. Eventually she succeeded to obtain a precipitate of radium chloride of few milligrams obtained from one ton of ore.

Fig. 2 Marie Curie, Pierre Curie (*at the center*) and their laboratory technician (*at left*) in the laboratory of rue Lhomond, Faculté de Sciences, Paris. On the bench, the Curie's electrometer used to measure ionization of the air by radioactivity

On 21 July 1902, Marie Curie using a sample containing 0.129 g of barium-radium chloride, which could contain probably only 1:1000 parts of radium in barium, made the first determination of the atomic mass of this new element, and obtained 223.3. Later she would correct this to 225 ± 1 and to 226.4 (actual value 226.0254). That chloride sample contained for the first time radium in visible amounts, and was 1 million times more radioactive than uranium (NN 2011).

In November 1903, Marie Curie presented her Doctoral Thesis, entitled "Recherches sur les Substances Radioactives" at the Sorbonne University, Paris. The same year the Swedish Academy awarded her, together with Pierre Curie and Henri Becquerel, the Nobel Prize of Physics for the discovery of radioactivity.

In 1911 again the Swedish Academy of Sciences awarded to Marie Curie the Nobel Prize, this time of Chemistry, for her discovery and characterization of the new elements radium and polonium (Curie 1911).

Following the adoption of the curie (in memory of Pierre Curie) as the unit of radioactivity, and upon request by the International Radium Standards Committee in 1911, Marie Curie prepared a radium standard with 21.99 mg of pure radium chloride in a sealed glass tube. The secondary certified radium standards for other countries were prepared and tested against this primary standard, now deposited in the Bureau International de Poids et Mesures in Sevres, near Paris (Curie 1912).

Radium Applications and the Dawn of Radiotherapy

Few years only after Becquerel's discovery of the "uranic rays", Ernest Rutherford and Frederick Soddy demonstrated that the radiation emitted by radioactive substances was indeed composed of three distinct types of radiation. These radiations behaved differently when under an electromagnet, and had very different energy. More of 80% of the emitted radiation was due to alpha particles with little penetrating power; beta radiation was also made of particles with electric charge but much smaller than alpha particles and with much higher energy and able to penetrate to a certain depth in the tissues. The third type of radiation were gamma rays (first identified by P. Villard in 1900), which accounted only for 1% of radiation with high energy, short wavelength and strong penetrating power (NN 2011, Mould R F (2007).

The idea of using radium in medicine to destroy tumors dates back to 1900, to the skin burns observed by two German scientists, Friedrich Walkoff and Friedrich Giesel, and to the incident of Henry Becquerel who carried a tube with a small amount of radium in the pocket of his waistcoat during 14 days. He developed a skin burn, and mentioned that to Pierre Curie. Pierre Curie amazed with the radium effect decided to confirm it and applied a small source of radium against his arm for 10 hours, obtaining a localized burn as well. That gave him the idea that radium could be used in medical work. Meanwhile, Becquerel's burn was serious and he went to see a dermatologist at the Hospital St Louis, Paris. The dermatologist, Dr. Ernest Besnier, noted that the Becquerel's radium burn was similar to X-ray burns and immediately thought that radium could be used in medical therapy such as the X-rays. He persuaded the Curies to lend a small amount of radium to a hospital colleague, Dr. Henri Danlos, who successfully used the radioactive material to treat lupus and other dermatological diseases and published his work in 1901 (Danlos & Bloch 1901).

Across the Atlantic, the first use of radium in therapy seems to have occurred by the same time, by the hands of a physicist, Francis Williams, in Boston. He was aware of the successful use of X-rays in the treatment of lupus and speculated that the radiation emission of the recently discovered radium could be used in treating this disease. In late 1900s, Francis Rollins seem to have prepared about 500 mg of radium chloride and placed them in a sealed capsule, and passed it over to his brother-in-law, Dr. William Rollins, for use in therapeutics (Mould 2007).

In his discussion of the first 42 cases treated with radiation from the encapsulated radium, William Rollins compared the results with the previous use of X-ay in similar therapy. He wrote: "The comparison at the present time is greatly to the advantage of radium.(…) When radium is employed for healing purposes no cumbersome apparatus is necessary: radium is portable and always ready for use. Further, the dose from radium is uniform; the strength of the output does not vary, so that the dose depends entirely on the length of exposure and the distance of the radium from the part to be treated. Radium may be applied to parts that are not readily accessible to the X-rays, such as the mouth or vagina. Furthermore, the healing action of the radium is more prompt". He also wrote about the severe burns that radium could cause in health tissues and the need for protection (Mould 2007, Danlos & Bloch 1901).

Since the very beginning, many medical doctors used radium in the treatment of tumors and malignant neoplasms, including for example the work in the Gussenbauer Clinic of Vienna (1902), the work at the Biological Laboratory of Radium in Paris by Louis-Frederick Wickham (1905); the work by Robert Abbe (1904) and William Morton (1914) in New York amongst many others, all testing improvements and innovation in radium therapy (Abbé R 1904, Eisenberg 1992).

Foundation of the Radium Institute in Paris

Encouraged by the successful therapy applications of radium, particularly those of Henri Danlos in Paris, the foundation of the Radium Institute was agreed in 1909 between the Pasteur Institute and the University of Paris. The Radium Institute was set up to further develop the research on radioactivity started by the Curies and to foster their application in medicine and biology. The Radium Institute would house two laboratories, the Curie Pavilion for research on physics and chemistry of radioactive elements under the direction of Marie Curie, and the Pasteur Pavilion for research on radium applications in biology and medicine and treatment of patients with Claudius Regaud, a biologist from Lyon, as director Fevrier (2011).

The construction of the Radium Institute started in 1911 and it was nearly ready in 1914 when the First World War started. The opening of the Institute was postponed and took place only after the war, becoming fully operational in 1919. The Radium Institute was renamed Institute Curie in 1932 and was leaded by Marie Curie until she died in 1934 Fevrier (2011).

The medical treatments achieved using radium inspired the foundation of radium institutes in many other countries. The treatments performed with radium rapidly increased in number and the demand for this element increased as well.

The Radium Mining Industry

Following the pioneer work carried by doctors in Paris and Vienna immediately after the discovery of radium, the use of this radioactive element in therapy rapidly expanded. In Europe, around 1913 the use of radium therapy was already widespread, but still little used in USA. Most American radiologists performed therapy with X-rays. This was due to the availability of the newly invented high energy X-rays and to short radium supply in USA (Mould 2007).

During the first two decades of the 20th century the radium commercially available was mainly produced from the Joachimstal mine, in Bohemia, and some small mines in France, Portugal, and a few other countries. Joachimstal in 1913 contributed alone to about 80% of the world radium supply. Nevertheless, the amounts produced were modest, rendering radium salts price astronomically high.

The search for other sources led to discovery of radium in carnotite, in Colorado, USA. The Standard Chemical Company, founded by Joseph Flannery of Pittsburgh, purchased the Colorado mines in 1911 and began to supply radium in 1913. This company published from 1913 to 1925, a bulletin entitled Radium, to publicize the radium production but where most of the knowledge and applications of this element were abstracted as well (Mould 2007).

In 1910 in USA was founded also the National Radium Institute that jointly with the US Bureau of Mines set up a radium recovery plant in Denver. The production by this plant substantially decreased the cost of radium and made it more available in the USA. By 1922, some 80% of the world's supply of radium was produced already in the USA.

A few years later, a turning point in radium industry took place with the discovery of the richer uranium ores of the Chinkolobwo mine in the Haut Katanga Province, in Belgian Congo. Large amounts of radium from Congo ores were produced in Oolen, Belgium, and this nearly closed radium production in American and European mines (Mould 2007).

In the beginning of last century most radium supply was from the mines of Joachimstal in Poland and nearby, that produced 100 g radium from 1899 to 1940. The Colorado, New Mexico and Utah mines in USA, produced about 200 g between 1900 and 1926. Port Radium in Canada, produced 60 g between 1933 and 1937, while Haut Katanga mines in Belgium Congo, produced 50–100 g per year from 1922 to 1940 (total between 900 and 1800 g). Radium productions in other countries were of a few grams or even less (Mould 2007).

In the beginning of the 20th century, radium price was at $ 180,000 per gram (about 1.8 million Euros today). In 1921, when a subscription was made in the USA to offer 1 g radium to Marie Curie, this amount did cost $ 100,000 (about 1 million Euros today) and it was produced from the uranium mines in Colorado and Utah. Later, with the discovery of uranium in Katanga, the prices dropped to $ 70,000 per gram. After the Second World War, other mines were exploited and together with the decreasing use of radium in medicine and industry the price of 1 g dropped to $ 25,000 in the fifties (Fevrier 2011, Mould 2007).

The availability of radium from the 1920s and the decrease in market price definitively enabled the adoption of radium therapy in many medical centers. This allowed for the development of therapeutic methods and increased reporting of successful healings in many countries.

At the same time, the abundance of radium and popular believes that this element could in small amounts be miraculously beneficial to health, lead to the invention of a high number of side applications marketed for public use. In that time were commonly advertised and used beauty products such as face powders and skin creams, hygiene products such as toothpaste, and radioactive water to the benefit of stomach. All these products contained added radium and were available for unrestricted use (Mould 2007).

Other radium applications were invented, such as the use in phosphorescence quadrants of wrist watches, alarm clocks, night reading instruments, including industrial, aviation and military instruments. NN 2011 (Fevrier 2011, Mould 2007).

The Onset of Radium in Medicine and Other Applications

The noxious effects of radiation emission from radium were noticed early, and Marie Curie herself noticed also how easily instruments and clothes became contaminated with radium in the laboratory. She encouraged the use of caution to her collaborators and students. However, the radiation protection as a science was not born yet, and sadly she was victimized by the element she discovered and handled for so many years.

The radiation protection knowledge was progressively accumulated and led to improved protection measures and regulations today used worldwide in research laboratories, hospitals, mining and chemical industry, and nuclear power plants Fevrier (2011).

Radiation protection considerations and the carcinogenic potential of prolonged exposure to radium led to abandon the use of radium in applications for public use, such as luminescent dials. In medicine, with the discovery of artificial radioactivity, fission products such as cobalt-60 and cesium-137 with much shorter half-lives than radium and high energetic radiation replaced radium allowing for safer applications in teletherapy. In curietherapie other radionuclides, such as iridium-192, gradually replaced the radium needles. After several incidents with lost radium needles, an international program was implemented to recuperate these needles and dispose them safely. Production of radium sources for use in medicine ended in the 1970s (Mould 2007, Eisenberg 1992).

The improved knowledge on biological effects of ionizing radiations and approval of international basic safety standards against the harmful effects of ionizing radiation, led also to improved mining and milling waste management and to the concept of environmental remediation of former radium and uranium producing facilities (IAEA 2005).

The life cycle of radium applications was than completed.

Medical applications of radium were developed for about 70 years, many patients underwent radiotherapy, many successful healings were achieved and human lives extended. During those years many technologies and new sciences started with radium use. Although artificial radioisotopes have already replaced radium to increase the benefits of ionizing radiation and reduce the radiological and environmental risks, and new technologies such as photon and electron beams produced by linear accelerators, and nowadays using proton emission technology (PET), have in turn taken their place, the way was opened with the discovery of radium (IAEA (2009).

Radiation therapy and nuclear medicine continue developing today, making use of much shorter lived radioisotopes, and using as radionuclide carriers organic compounds tailored to specifically bind to certain cells and tissues. The outcomes are much improved radio diagnostic and less invasive therapy techniques along with better targeted cancer treatments.

It shall be remembered that this path was opened also by the Curie family with the discovery of artificial radioactivity by Frederic and Irene Joliot-Curie, Nobel Prize of Chemistry in 1935.

What Future?

During the first half of the 20th century, radium was used in exciting applications, which drove the uranium mining to grow. Today, radium has no major applications and it is not the aim for mining industry. Nevertheless, the mining of radioactive ores and uranium for energy production continues.

The radium applications particularly in medicine, devised by Marie Curie and other scientists of her time, were continued and refined generally through the use of radionuclides with shorter half-lives. The curietherapy is still a method currently applied in medicine to treat cancer tumors, with recent developments in the treatment of prostate and gynecological cancers.

For a while at least, radium is left now in peace by human kind. In nature radium continuously provides a significant contribution to the maintenance of mild temperatures that allowed and sustain human and non-human life on Earth. Nevertheless, applications of radium isotopes are not over. Naturally-occurring radium isotopes are currently used as natural tracers in hydrogeology and in the investigation of groundwater discharges into the oceans. Most likely, other applications or uses of radium isotopes will be invented.

References

Abbé R (1904): The subtle power of radium. Med Record, New York 66:321.

Curie E. (1938): Madame Curie. Editions Gallimard, Paris. New-edition 2010.

Curie M. (1921): The discovery of radium. Address by Madame M. Curie at Vassar College 14 May 1921. Ellen S. Richards Monographs, No 2, Vassar College.

Curie P. & Curie M. (1898): Sur une substance nouvelle radio-active contenue dans la pechblende. Comptes rendus de l'Académie des Sciences. 127:175–178.

Curie P., Curie M, Bemont G. (1898): Sur une nouvelle substance fortement radio-active contenue dans la pechblende. Comptes rendus de l'Académie des Sciences. 127:1215–1218.

Curie M. (1911): Radium and the new concepts of chemistry. Nobel Conference Lecture, Stockholm. 11 1911.

Curie M. (1912): Les mésures en radioactivité et l'etalon du radium. J. Phys 5th Series 2:795–798.

Danlos H & Bloch P (1901): Note sur le traitement du lupus erythemateux par des applications du radium. Ann Dermatol and Syphilog: 986–988.

Eisenberg R(1992): Radiology: an illustrated history. Mosby Year Book, New York.

Fevrier (2011): L' héritage Marie Curie. Les dossiers de la Recherche, N 42, Fevrier 2011, Paris.

IAEA (2005): Environmental Contamination from Uranium Production facilities and their Remediation. Proceedings of an International Workshop, Lisbon, 11–13 February 2004. International Atomic Energy Agency, Vienna, 2005.

IAEA (2009): Clinical Translation of Radiolabelled Monoclonal Antibodies and Peptides. IAEA Human Health Series No. 8. International Atomic Energy Agency, Vienna, 2009.

Mould R F (2007): Radium History Mosaic. Nowotwory Journal Oncology (Warsaw), Supp. 4, Vol 57.

NN (2011): Chemistry International, Vol 33, no 1. (www.iupac.org/publications/ci/index.html).

Part 1
Renaissance of Uranium Mining

License Procedure for Uranium Recovery from Talvivaara Deposit, Eastern Finland

Ari Luukkonen, Esko Ruokola

Abstract. Since 2007, Talvivaara Mining Company has held permits to exploit black schist hosted Ni-Zn-Cu-Co sulfide deposit that is one of the largest in Europe. In 2010, the company conducted an evaluation on the environmental impacts of U production. The company will need a license to produce U concentrate. The application was submitted to Finnish Government in April 2010. The licensing process is expected to be accomplished during 2011. Commissioning of the U recovery facility is subject to Radiation and Nuclear Safety Authority approval.

Geology of Talvivaara

Since 1930s, the geology of Talvivaara area has been studied in several occasions. Talvivaara Ni-Zn-Cu-Co mineralization was discovered in the 1970s. Talvivaara area is located in Early Proterozoic Kainuu schist belt that is dominated by quartzite, back schist, and mica schist rocks. The western border of the Kainuu schist belt can be genetically related to black schists of the famous Outokumpu Province. Especially in eastern Finland, the black schist formations are very thick compared for example to younger 'Kupferschiefer' black schists in Central Europe. Talvivaara drillings have penetrated even 400 meters thick beds of black schist while the thickness of 'Kupferschiefer' is around one meter (Loukola-Ruskeeniemi 1990).

It has been long known that many heavy metals (e.g. lanthanides, Y, Ta, and U) have a tendency to slight enrichment in organic fraction of carboniferous sedi-

Ari Luukkonen
Radiation and Nuclear Safety Authority, Laippatie 4, FI-00880 Helsinki, Finland

Esko Ruokola
Radiation and Nuclear Safety Authority, Laippatie 4, FI-00880 Helsinki, Finland

B. Merkel, M. Schipek, *The New Uranium Mining Boom*
DOI 10.1007/978-3-642-22122-4 © Springer 2011

ments (Vine and Tourtelot 1970). It was already relatively early indicated that e.g. the average U concentration of black schists of Outokumpu province is around 50 ppm (Peltola 1968, 1960). According to more recent studies, Talvivaara black schists contain on the average 19 ppm U (Loukola-Ruskeeniemi and Heino 1996). However, it was concluded in 1991 that U concentrations in Finnish black schists are so low that they do not solely indicate ore potential (Loukola-Ruskeeniemi 1991a). In Talvivaara U occurs as thucholite (U bearing pyrobitumen) that is a hydrocarbon mineraloid. Usually thucholite has uraninite (UO_2) core that is surrounded with hydrocarbons.

Geochemical and mineralogical evidences and general relation to ophiolites indicate that Kainuu and Outokumpu black schists are of marine origin, and are related to hydrothermal exhalative vents that have supported anaerobic bacterial biota. Similarly to present day mid-ocean ridge conditions, bacterial life was based on sulfur utilizing chemosynthesis in warm strongly reducing conditions. The black schists adjacent to sulfide ore deposits have likely originated in such an environment (Loukola-Ruskeeniemi 1988, 1991b). Uranium has possibly bound itself to organic matter via metabolic and/or chemical processes. Strongly reducing conditions may have enhanced U precipitation from seawater to carboniferous sediments. Hydrothermal fluids may have also added metals to black shale sediments later during diagenesis and/or metamorphosis.

Products of Talvivaara

Profitable base metals (Ni, Zn, Cu, Co) in Talvivaara are located in several sulfide phases. Main ore minerals are pyrrhotite (FeS), pyrite (FeS_2), sphalerite ((Fe,Zn)S), pentlandite ($(Fe,Ni,Co)_9S_8$), and alabandite (MnS). Nickel is concentrated especially to pentlandite and pyrrhotite, and with minor extent to pyrite. Currently, Talvivaara Mining Company estimates that Talvivaara contains some 1200 million tons of ore (measured or indicated) that would support an anticipated production for at 46 years. In 2007, Talvivaara Mining Company received its environmental permits for base metal production and in 2008 it got its licenses for extensive usage and storage of hazardous chemicals. Mining operations at the site started in spring 2008. Profitable base metals are precipitated as sulfides from leaching solutions. In 2010, Talvivaara Mining Company produced about 10,000 tons and 25,000 tons Ni and Zn, respectively (Talvivaara Mining Company 2010). The annual base metal production is still in ramp-up stage.

According to the mining company, bioheap leaching method for base metals makes also U extraction profitable as a by-product. The company justifies U recovery as follows: 1) Removal of U from mine tailings is environmentally more sustainable, 2) the production of U will not increase excavations in the mine, 3) utilization of ore, as well as possible, is in accordance with the Mining Act 4) the increase in usage of hazardous chemicals is small compared to the total amounts used, 5) produced Ni concentrates need not to be U decontaminated in the smelting plant, and 6) self-sufficiency of Finland in respect of U would be better. The uranium end-product from Talvivaara mine would be yellowcake that contains mostly (ca. 80%) uranium peroxide ($UO_4{\cdot}xH_2O$), and at lesser extent also other uranium oxides (e.g. UO_2, UO_3, and U_3O_8). Talvivaara Mining Company plans to produce annually approximately 350 tons of U, or 410 tons of yellowcake. The mining company has preliminary agreed with Cameco Corporation on complete delivery of produced yellowcake to Canada for further upgrading (Talvivaara Mining Company 2010).

License Process for Uranium Recovery

The nuclear license procedure related to Talvivaara Mining Company U recovery application is illustrated in Fig. 1. In addition to the U recovery license, the company will need updates of licenses not shown in Fig. 1. The company will need an update of the environmental license pursuant to Environmental Protection Act. Update of this license will be issued by Kainuu Centre for Economic Development, Transport and the Environment (Kainuu ELY). The company will need a permit for handling and storing hazardous chemicals necessary in the U leaching process. This permit will be issued by Finnish Safety and Chemicals Agency (TUKES). New buildings at the site will need a construction license that will be issued by Sotkamo Municipality. The license for U uranium recovery will be a separate license in addition to the existing mining license. Commissioning of the U recovery facility is subject to STUK's safety judgment and approval.

Environmental Impact Assessment

Talvivaara Mining Company started its application procedure by launching environmental impact assessment studies on the possible U recovery (Fig. 1). Results of these studies were published in 2010, and Kainuu ELY (contact authority) began the evaluation of environmental impact assessment. Environmental impact assessment is based on law. Its purpose is to estimate possible impacts and to help the planning of facilities so that negative impacts can be minimized, but also to increase information delivery to citizens and to give opportunities to citizens to influence to the facility plans. The applicant is obliged to give opportunities to interested citizens and groups of people to express opinions on plans, if they feel that planned actions have influence to their living. Hearings were arranged as public hearings and briefings. The essential goal of environmental assessment is to produce information for the license process and finally for the decision making.

The assessment report describes possible impacts related to possible indoor U recovery and the alternative of no recovery. The important part of the report is to consider the potential consequences of U recovery to ground sediments, ground-

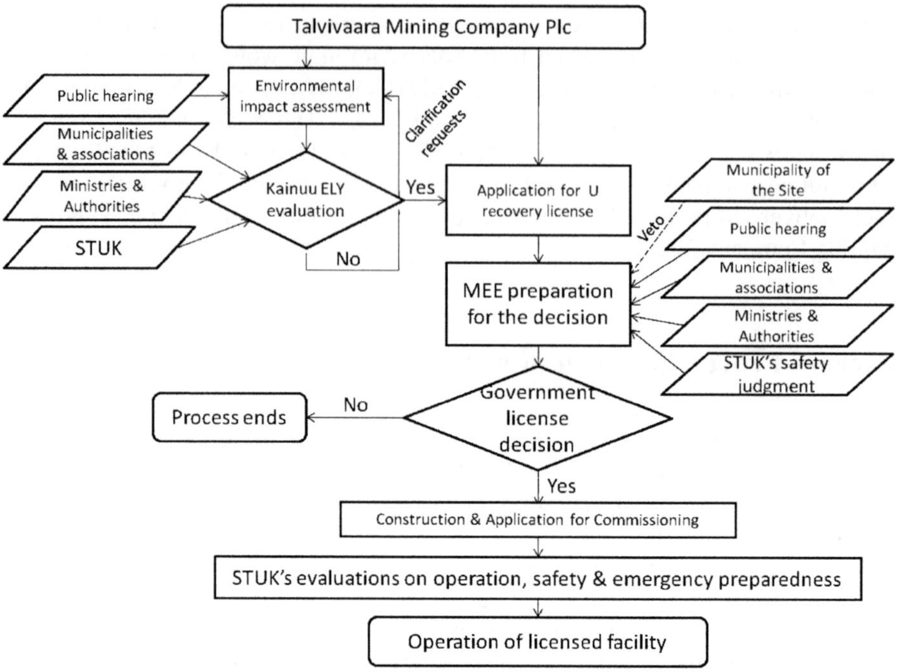

Fig. 1 The license process related to Talvivaara U recovery application. The license for U recovery is issued by Finnish Government. The illustration excludes procedures related to the environmental license updates, permits related to handling and storing hazardous chemicals, and construction licenses related to new buildings. Kainuu ELY is acronym for Kainuu Centre for Economic Development, Transport and the Environment. MEE is acronym for Ministry of Employment and the Economy. STUK is acronym for Radiation and Nuclear Safety Authority

and surface waters, biota (e.g. animals, plants, fungus, and micro-organisms), diversity of nature, and air quality. The consequences to human health, habitability, community structure, landscape, and cultural heritage need to be covered. The report also illustrates chemicals and their amounts to be used in the recovery, handling of waste waters, and purification equipments for gases (e.g. hydrogen sulfides, Rn). Also other waste material flows than water and rock, transports of ore concentrates, recognition of safety risks, emergency preparedness, and monitoring the environment need detailed descriptions.

However, environmental impact assessment is not directly assigned either to the update processes of environmental license or to U recovery license. The adequacy and extent of assessment program and reporting is evaluated by Kainuu ELY that writes the official statement on the environmental impact assessment. This statement can then be attached to the license applications. Before making the official statement Kainuu ELY asked expert opinions from several authorities. Radiation and Nuclear Safety Authority (STUK) expressed its views on e.g. operational radiation and transport safety, and on long-term radiation safety of waste rock and mine tailings. Opinions were requested also form many other authorities, e.g. from MEE, Geological Survey of Finland, Ministry of Social Affairs and Health, Kainuu Regional Authority, several municipalities (Sotkamo, Paltamo, Valtimo, and Rautavaara), and Kajaani town.

Also local associations were entitled to express their opinions and follow-up the assessment work. Associations included for example the local office of Finnish Association for Nature Conservation in Kainuu, local village, hunting and fishing associations, Vuokatti Holiday Centre, and Kainuu Rescue Department. The company also informed on the progress of the assessment work in its web-site and mailed progress updates to interested parties.

Application of Uranium Recovery

In April 2010, Talvivaara Mining Company submitted application to produce U concentrate from Talvivaara deposit. The application was based on the demand declared in Nuclear Energy Act (990/1987), and was supplemented with details defined in Nuclear Energy Degree (161/1988). According to the degree, the application must contain 1) a description of the land use rights and a copy of the mining license (referring to Mining Act (503/1965)), 2) a geological site description including details on U content, 3) a description of planning details for the mine and enrichment plants and their immediate vicinity, 4) a description of the planned enrichment method, 5) an outline of the radiation protection arrangements, 6) a description of environmental impacts caused by the planned actions and a criteria how potential damages are minimized, 7) a description of the applicant's financial prerequisites and the economical viability of the project, 8) an outline of the ownership and occupation of the facilities, 9) a description of the quality and quantity of the ores, nuclear materials and nuclear waste that will be produced, handled and stored, 10) a description of transportation arrangements, 11) a description of plans and available methods for arranging nuclear waste management, including the decommissioning or demolition of the facilities, 12) a description of the radiation protection expertise and operating organization, and 13) any other information considered necessary by the authorities.

MEE makes the preparations for the license decision (Fig. 1). In October 2010, MEE requested Talvivaara Mining Company to supplement its original application with 1) land register certificate of the location of U recovery plant, 2) more detailed description of the planned enrichment method, 3) an outline of radiation protection arrangements, and 4) a description of available radiation protection expertise. Like in the previous evaluation stage expert and everyman opinions were requested from authorities and associations. However, this time the request was broader than in the environmental impact assessment stage. Opinions were requested from 13 municipalities and towns, 17 authorities, and 18 associations. Among the authorities, STUK has an emphasized expert role because STUK produces safety judgment evaluation on the license application for the usage of MEE. Also several ministries are able to make strong arguments to license preparations. Civil associations that had an opportunity to influence the license preparation can be divided environment protection organizations (e.g. the Finnish Association for Nature Conservation, WWF, Greenpeace), interest groups of industry and commerce (e.g. employer associations, labor unions), and local associations (e.g. local village, hunting, and fishing associations, and local nature conservation associations).

At the moment, Mining Act of Finland is subject to modernization. The current act was made in 1965 and has been amended several times since then. The new act is expected to come in operation later this year and it will emphasize more the rights of local citizens in the mining license decisions. In future this means that local municipality may use veto in licensing decisions for U production (Fig. 1). However, in the present procedure Finnish Government has full power in the U recovery decision.

Commissioning of the Recovery Plant

The U recovery license procedure is expected to be accomplished during 2011. If Finnish Government grants the U recovery license for Talvivaara Mining Company, then commissioning of the recovery facilities is subject to STUK's approval. STUK regulatory control will then focus on occupational and environmental radiation protection, safety of transport of uranium concentrate, management of residues from U recovery, and nuclear safeguards.

References

Loukola-Ruskeeniemi K (1991a) Suomen proterotsooisten mustaliuskeiden uraanipitoisuudesta (in Finnish). Geol. Surv. Finl. M19/3344/-91/1/30

Loukola-Ruskeeniemi K (1991b) Geochemical evidence for the hydrothermal origin of sulphur, base metals and gold in Proterozoic metamorphosed black shales, Kainuu and Outokumpu areas, Finland. Mineralium Deposita 26:152–164

Loukola-Ruskeeniemi K (1990) Karelidien mustaliuskeiden hiili- ja rikkipitoisuudet kerrostumisympäristön kuvastajina (in Finnish with English summary). Geologi 42:95–101

Loukola-Ruskeeniemi K (1988) Early Proterozoic metamorphosed black shales in the Kainuu schist belt and in the Outokumpu region. Geol. Surv. Finl., Special Paper 10:103–106

Loukola-Ruskeeniemi K, Heino T (1996) Geochemistry and genesis of the black shale-hosted Ni-Cu-Zn deposit at Talvivaara, Finland. Econ. Geol. 91:80–110

Peltola E (1968) On some geochemical features in the black schists of the Outokumpu area, Finland. Bull. Geol. Soc. Finland 40:39–50

Peltola E (1960) On the black schists in the Outokumpu region in Eastern Finland. Bull. Comm. geol. Finlande 192

Talvivaara Mining Company Plc (2010) Annual results review 2010

Vine J, Tourtelot E (1970) Geochemistry of black shale deposits – A summary report. Econ. Geol. 65:253–272

New Uranium ISR Satellites at Beverley North, South Australia

Horst Märten, Richard Phillips, Peter Woods

Abstract. Recently discovered uranium deposits, including Pepegoona and Pannikan, have been developed as in-situ recovery (ISR) satellite mines to the Beverley plant. Both deposits are located within the Eocene Eyre formation of the Frome Embayment. The paper characterizes mineralogy, hydrogeology and geochemistry briefly and describes ISR satellite operation in combination with central processing, environmental management and specific requirements of the approval process including mine closure concepts. After performing a field leach trial in 2010, the ISR satellite wellfields and plants were commissioned in 2011.

Introduction

Heathgate Resources Pty Ltd (Heathgate) operates the technologically advanced in-situ recovery (ISR) Beverley and Beverley North uranium mines, located about 550 km north of Adelaide, South Australia, on the arid plains between the northern Flinders Ranges and Lake Frome. The original Beverley uranium mine has operated since 2000 under close environmental scrutiny (Jeuken et al. 2008, Woods 2011, Märten et al. in press). Since the discovery (Curtis et al. 1990) and mining of the original Beverley uranium resources, additional uranium deposits have been found in the area and mining is extending over a larger area. In 2009 Heathgate's

Horst Märten
Heathgate Resources Pty. Ltd., Level 4, 25 Grenfell Street, Adelaide, S.A. 5000, Australia
UIT GmbH Dresden, Zum Windkanal 21, 01109 Dresden, Germany

Richard Phillips
Heathgate Resources Pty. Ltd., Level 4, 25 Grenfell Street, Adelaide, S.A. 5000, Australia

Peter Woods
Heathgate Resources Pty. Ltd., Level 4, 25 Grenfell Street, Adelaide, S.A. 5000, Australia

B. Merkel, M. Schipek, *The New Uranium Mining Boom*
DOI 10.1007/978-3-642-22122-4 © Springer 2011

exploration group discovered the first of several new deposits to the north of Beverley. The Pepegoona and Pannikan deposits are the first ISR uranium satellite mines in Australia and form the nuclei of the Beverley North Project.

Mineralization at original Beverley is hosted within a series of palaeochannels ~100–140 m below the surface within unconsolidated Beverley sands, silts and clays of the Miocene Namba Formation. Ore primarily occurs as coffinite and uraninite associated with organic-rich layers of the palaeochannel sediments (Curtis et al. 1990). By contrast, at Beverley North mineralization occurs in the Eocene Eyre Formation (Stoian 2010) as will be described below.

Geological Setting and Characteristics

A schematic stratigraphic cross-section of the region indicating known uranium deposits is shown in Fig. 1.

The geological setting of the Beverley deposits is in Tertiary sediments of the Frome Embayment of the Callabonna sub-basin of the Lake Eyre Basin. These sediments cover an area of approximately 25.000 km^2 between the Mount Painter Inlier in the north-west, the Olary block to the south and Broken Hill block to the east. Basement to this sub basin may be variably Cretaceous Palaeozoic, Adelaidean or Mid to Lower Proterozoic, generally following that progression from north-west to south-east. The sub basin comprises an almost flat lying sequence reaching 300 m maximum thickness.

In its southern part, Lower Tertiary (Eocene) Eyre formation sediments are restricted to incised palaeovalley infill. Overall gentle palaeovalley gradients are to the north where skeletal valley fill sediments grade into widespread blanket sands overlying the Cretaceous Frome Embayment.

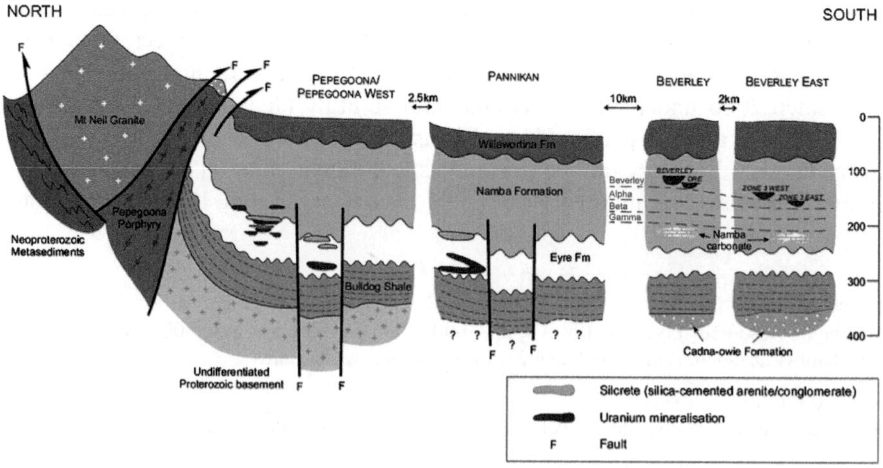

Fig. 1 Schematic stratigraphic cross-section including Beverley North/Beverley deposits

Mesoproterozoic, uranium rich granites within infringing or underlying terrains are considered the source for the uranium mineralization.

Interpretation of seismic data has previously identified the major faults as west-dipping structures that extend well down into Proterozoic basement rocks and show west over east compression. The structural regime was established in the early Palaeozoic and has periodically re-activated. The faults generally extend steeply to within 100 m of surface; some are expressed at the surface.

Uranium mineralization at the currently known Beverley North deposits is interpreted to be in a series of stacked rollfronts. At the first-discovered deposits of Pepegoona (including its extension, Pepegoona West) and Pannikan mineralization occurs between approximately 200 and 260 m below the ground surface.

The main reactive materials within and in the vicinity of the mineralized zones of Beverley North include the major reductants pyrite and carbonaceous material as well as calcite (neutralizing), kaolinite and other clay minerals (partly neutralizing, cation exchange capacity). The overall abundance of both organic carbon and calcareous minerals is below the empirical limit for acid ISR technology (<3% and <2% CO_2 equivalent, respectively, cf. IAEA 2001).

The hydrogeology of the Beverley North area (Heathgate 2010) consists of a series of Tertiary and Mesozoic sedimentary aquifers and aquitards overlying a highly variable bedrock aquifer. The deposits are sandwiched between the overlying Namba Formation and the underlying Bulldog Shale (Fig. 1). The Eyre formation is dominated by sandy lithologies, with thin shale beds in the middle and lower parts of the sequence and silcrete-like layers in the upper parts. Hydraulic transmissivity in the zone of interest (Eyre formation) varies from 70 to 240 m^2/d

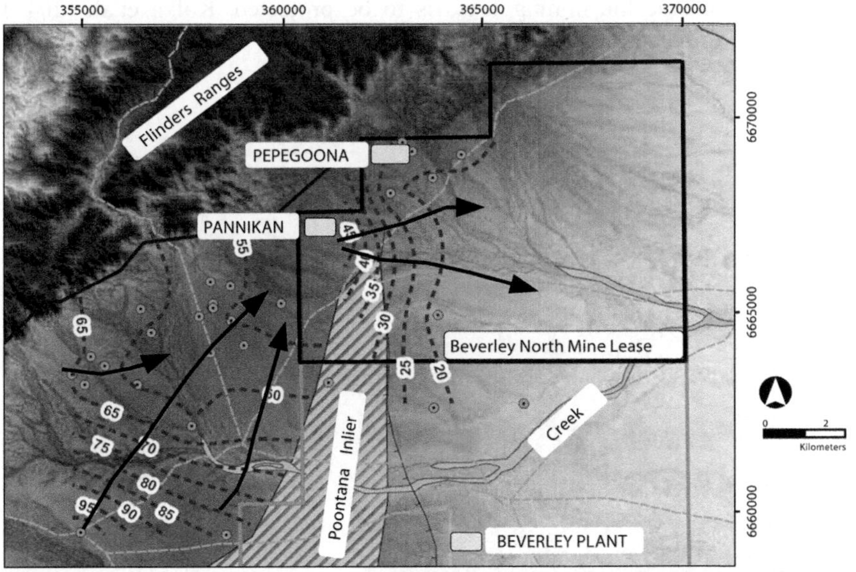

Fig. 2 Beverley North map and aquifer conditions within Eyre Formation (*dashed lines* – water level in Eyre formation, *arrows* – interpreted groundwater flow)

(from pump tests) corresponding to groundwater flow velocities in the order of 10 to 25 m/a.

ISR Satellite Mines: Permitting and Operation

The operations of Beverley are subject to stringent environmental regulations. The mines operate in accordance with the environmental standards and procedures initiated through a comprehensive environmental impact assessment process which incorporates input from State and Commonwealth Governments as well as from the broader community through consultation. This process is ongoing as continual improvement is sought be all parties. As well as the original Beverley EIS (Heathgate 1998), thorough reviews and updating of environmental and radiation protection aspects, and public consultation arrangements, were undertaken as part of the Public Environmental Report (PER) processes for the extension of mining at Beverley North (Heathgate 2010). At the same time significant review and associated updates to its Radiation and Radioactive Waste Management Plans was undertaken (Kutty et al. 2010). Operational documents, namely the current Mining and Rehabilitation Programs (MARPs) and annual environmental reports for the Beverley and Beverley North mines are public documents.

Permitting also required consideration of future rehabilitation and mine closure. At Beverley North, as at Beverley, the area will be returned to low-intensity pastoral use. Although the mined aquifer does not meet regulatory guidelines for potable, irrigation or stock water use, the surrounding aquifers and the Eyre Formation aquifer outside the mining lease is to be protected. Kalka et al. (2011) describe the hydrogeochemical modeling undertaken as part of the permitting

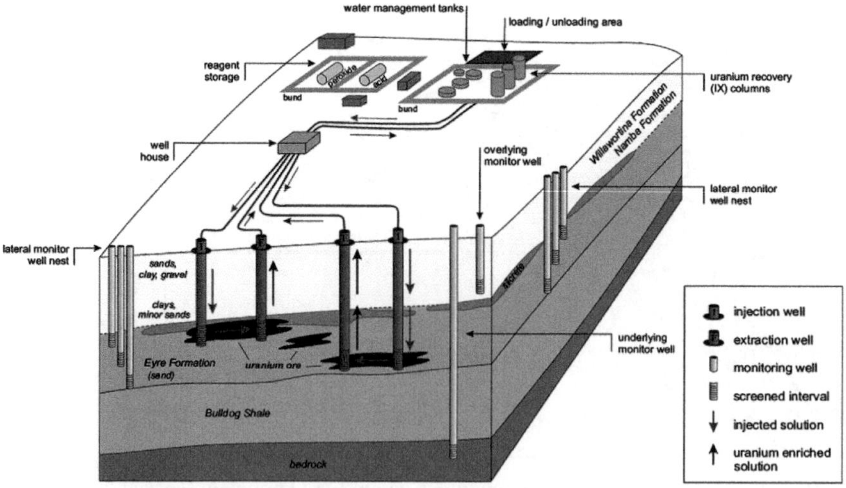

Fig. 3 Schematic of ISR operation at Beverley North

process and Douglas et al. (2011) one of the optional supplementary groundwater treatments that could be considered if required in the future.

Typically, ISR satellites are standalone operations of wellfields in combination with an ion-exchange (IX) facility to capture the uranium out of the mining solution cycle. An ISR operation is schematically shown in Fig. 3. The major components of a satellite ISR operation include:

1. Wellfield consisting of injection and extraction wells both designed with regard to position and screened interval to enable the most efficient flow of the mining solution through the ore body and to selectively mobilize uranium by oxidation [U(IV) minerals → soluble U(VI) uranyl complexes].
2. System of lateral monitoring wells (screened intervals at various depths in the mining aquifer) together with underlying and overlying monitoring wells within adjacent formations. Monitoring wells are used control the mining solution within a well-defined area and to avoid any migration of mining solution to the outside (monitoring data to be assessed in combination with a hydrological model of the mining aquifer).
3. Wellhouses with distributors to the injection wells and collectors to combine the uranium enriched solution from the extractors towards the IX capture facility. Wellhouses are equipped with bag filter banks for both extraction and injection solution as well as electrical systems (in particular for the extraction pumps), measuring gauges, data acquisition system and sampling points.
4. IX facility (within a bund) for uranium capture consisting of 3 or 5 columns operated in merry-go-round (pair(s) of lead and tail columns and one column idle for resin transfer), see Fig. 4.

Fig. 4 Pepegoona ISR satellite with IX columns on the right

5. Water management tanks including barren mining solution tank (de-gassing), resin transfer infrastructure etc.
6. Bund with storage tanks for leaching chemicals (sulfuric acid and hydrogen peroxide) together with dosage systems to injection pipeline.
7. Loading/de-loading area (for both chemicals and resin).
8. Diesel generator and diesel tank for standalone power supply.
9. Miscellaneous facilities including electrical and digital control systems, workplace safety facilities, mini-laboratory, etc.

The new satellite plants at Pepegoona and Pannikan undertake the first part of in-situ recovery mining, absorbing mobilized uranium onto ion-exchange resin beads. The loaded resin is then trucked to the Beverley plant where the processing (elution, i.e. de-loading of uranium from the resin, and further down-stream processing and packing) is completed, and the regenerated resin trucked back to Beverley North sites for re-use.

After a successful field-leach trial in a few wellfield patterns at Pepegoona performed in 2010, the full-scale operation of the Pepegoona satellite plant (also operating in-situ recovery from the adjacent Pepegoona West deposit) commenced at the beginning of 2011, followed by the commissioning of the second Pannikan satellite in July 2011. These two ISR satellites are operated in parallel with the ongoing recovery from remaining wellfields close to the Beverley plant by using pipelines to pump the mining solution to the plant and – after capture of uranium – back to the wellfields (ISR cycle).

As for the original Beverley operation, the ISR chemistry applied to Beverley North is moderately acidic (sulfuric) in combination with the addition of hydrogen peroxide as oxidant. Compared with the natural geochemistry within the Namba formation at Beverley, the Beverley North deposits are hosted in the more reducing Eyre formation (determined by the abundance of both pyrite and carbonaceous material). This has required some adjustments of ISR chemistry in the first phase of wellfield operation (increased acidification, whereas the H_2O_2 dosage is adjusted under the condition that practically all Fe in the mining solution is oxidized to ferric). The stronger reducing conditions in the mineralized zones at Beverley North have caused retarded uranium recovery reaching its maximum about 2–3 months after wellfield startup, whereas most of the historical Beverley wellfields showed a maximum recovery within about 0.5 to 1.5 months, followed by the typical decline. The Beverley North geochemistry is further analyzed in comparison with a reactive geochemical model in order to optimize ISR performance in the new areas.

Due to the stronger reducing conditions in the Eyre formation it can be expected that the natural attenuation (NA) after ISR operation will be more efficient than originally assumed. The computer simulation of NA within post-mining scenarios (together with enhanced effects by additional measures – ENA) performed by the use of a reactive-transport model (recently published by Kalka et al. 2011) has demonstrated the efficient restoration of the Beverley North aquifer towards reducing chemical conditions in combination with the compensation of the introduced acid potential by neutralizing minerals in a reasonable time frame (to be shortened by ENA). From the real wellfield performance it can be concluded that characteristic (E)NA decline parameters will be significantly shorter than predicted by the model so far. Further work to improve the model predictions is in progress.

Summary and Conclusions

Heathgate has developed two new ISR satellite operations at Beverley North. After performing a field leach trial in 2010, the ISR satellite wellfields and plants were commissioned in 2011.

The Beverley North orebodies discovered to date are each significant. Economically, each deposit provides a vital ongoing source of production for Heathgate's operations at the Beverley Uranium Mine processing facility. Further, the new deposits and two satellite processing plants constitute the first instance of Satellite Mining for ISR uranium mining in Australia. They also open up additional exploration targets for Heathgate and for the uranium mining industry in South Australia generally.

The mineralogical and geochemical conditions of the Pepegoona and Pannikan deposits within the Eocene Eyre formation differ from the historical Beverley deposits within the Miocene Namba formation considerably. The strongly reducing chemical milieu within and around the new deposits had to be compensated by adjusting the ISR chemistry on the one side, but are favorable with respect to post-mining restoration of the aquifer by natural attenuation effects.

The future validation of the hydrogeological and hydrochemical (E)NA modeling based on real-world data will be a valuable opportunity to refine the conceptual and numerical modeling tools, as well as providing important evidence of successful closure of the Beverley North mining areas. It will also be used to optimize the ISR operation by balancing the mobilization of uranium against interfering leaching effects.

Acknowledgements The geological setting described here is a summary of the work of many geologists and support staff. Similarly, permitting, design, construction, commissioning and now operating of the Beverley North satellite mines and ongoing further processing at the original Beverley treatment plant is the result of the work of Heathgate's team and subcontractors too numerous to acknowledge individually.

References

Curtis J, Brunt D, Binks P (1990) Tertiary Palaeochannel Uranium Deposits of South Australia. Geology of the Mineral Deposits of Australia and Papua New Guinea (Ed. F.E. Hughes). AusIMM. Melbourne: 1631–1636

Douglas G, Wendling L, Usher K, Woods P (2011) Neutralisation and Trace Element Removal from Beverley In-Situ Recovery Uranium Mine Barren Lixiviant via Hydrotalcite Formation. These proceedings

Heathgate (1998) Beverley Uranium Mine: Environmental Impact Statement – Main Report. Heathgate Resources. June 1998. Adelaide

Heathgate (2010) Beverley North Project Mining Lease Proposal and Draft Public Environment Report. Prepared for Heathgate Resources Pty Ltd by URS Australia Pty Ltd. 8 April 2010

IAEA (2001) Manual of acid in situ leach uranium mining technology. IAEA-TECDOC-1239

Jeuken B, Märten H, Phillips R (2008) Uranium ISL Operation and Water Management under the Arid Climate Conditions at Beverley, Australia. Mine Water and the Environment. Proceedings 10th IMWA Congress, June 2–5 2008, Carlovy Vary, Czech Republic. Rapantova N, Hrkal Z, Eds. Esmedia DTP: 487–490

Kalka H, Märten H, Woods P (2011) ISR Mine Closure Concepts. Nachhaltigkeit und Langzeitaspekte bei der Sanierung von Uranbergbau- und Aufbereitungsstandorten. Proceedings des Internationalen Bergbausymposiums WISSYM_2011, Ronneburg, May 25–27 2011. Wismut GmbH: 201–215

Kutty S, Woods P, Dayal E, Jagger A. (2010) Keeping Radiation Management at Beverley Uranium Mine at Best Practice. Plans, Responses and Outcomes. 35th Australasian Radiation Protection Society Conference, Adelaide, Oct 18–20 2010

Märten, H, Phillips R, Woods P (in press) Environmental Management of the Beverley Uranium Mine. IAEA UMREG Meeting "Low environmental impact uranium mining and remediation: 15 years of multinational experience through Uranium Mine Remediation Exchange Group", Bruge, Sep 05 2007, IAEA-TECDOC-Number to be assigned

Stoian, LM (2010) Palynology of Mesozoic and Cenozoic sediments of the Eromanga and Lake Eyre basins: results from recent drilling in the northwest Frome Embayment. MESA Journal 57: 27–35

Woods P (2011) Sustainability Aspects of the Beverley Uranium Mines. AusIMM Bulletin, Melbourne, June 2011, no.3: 30–36

Waste Water Treatment of $CO_2 + O_2$ in-situ Leaching Uranium

Lechang Xu, Naizhong Liu, Guofu Zhang

Abstract. An in-situ leaching uranium mine locates in Northern China. The mine applies processes of $CO_2 + O_2$ leaching uranium of ore-bearing aquifer, The mine reduces consumption of the industrial reagent and water and generation and discharge of waste water as much as possible by comprehensive waste water treatment technology with process water recycle, reverse osmosis and natural evaporation. Process water of the mine that can be recycled and reused include barren fluid, solution washing loaded resin, solution regenerating barren resin, precipitating mother solution and filtered liquor of yellow cake.

Introduction

In-situ leaching uranium includes acid and alkaline leaching. Acid in-situ leaching has been developing for 20 years and now produces 200 t U/a. Alkaline in-situ leaching experiment has carried out for about 10 years. However, commercial alkaline in-situ leaching uranium mine was not built before 2010.

Alkaline in-situ leaching uranium has obvious environmental advantages over traditional mining/milling and acid in-situ leaching. But the control and treatment of liquid effluents from uranium milling operations have assumed greater impor-

Lechang Xu
Beijing Institute of Chemical Engineering and Metallurgy, CNNC,
P.O. Box 234, Beijing 101149, China

Naizhong Liu
Beijing Institute of Chemical Engineering and Metallurgy, CNNC,
P.O. Box 234, Beijing 101149, China

Guofu Zhang
China Uranium Co. Ltd, Beijing 10082, China

B. Merkel, M. Schipek, *The New Uranium Mining Boom*
DOI 10.1007/978-3-642-22122-4 © Springer 2011

tance because of increased environmental awareness and more stringent regulatory requirements. Conservation of water has also placed more pressure on mill operators to recycle water, which both minimizes usage and limits quantities requiring disposal.

Site Characterization

An in-situ leaching uranium mine locates in Northern China. The topography of the mine site is flat and spacious (Fig. 1). The precipitation of the site is 200 ~ 577 mm and evaporation is 1467 ~ 2242 mm. There no surface water system such as lake and river near the site. The uranium ore body occurs in Yaojia formation of Cretaceous in 212–315 m depth. Yaojia formation is a 21–50 m thickness confined sandstone aquifer overlying 4–12 m mudstone and underlying 0.5–2 m mudstone. Chemical composition in groundwater of the ore-bearing aquifer is simple. Chemical type of the groundwater is mainly HCO_3–Na and HCO_3·Cl–Na, TDS 3.5–5.7 g/L, groundwater temperature 15–16 °C, pH 7.2–8.4, HCO_3^- 1120–2520 mg/L, Cl^- 287–647 mg/L, SO_4^{2-} 176–682 g/L, Fe^{2+}/Fe^{3+} 0.5–1.5, $O_2 < 2$ mg/L and U 0.443–1.80 mg/L.

Fig. 1 Topography of the site

In-Situ Leaching Uranium Process

The mine applies processes of $CO_2 + O_2$ leaching uranium of the ore-bearing aquifer, ion-exchange recovering uranium of lixiviate, NaCl + $NaHCO_3$ eluting uranium of ion-exchange resin and NaOH precipitating uranium of the eluate.

Wellfield is one of the key part of the process. The wellfield areas have been divided into mining units for scheduling development. The wellfield of the first developed area consists of 5 mining units with 246 injection wells, 105 recovery wells and 15 monitoring wells which is about 50–100 m away from the nearest mining unit boundary (Fig. 2).

Injection and recovery wells are cased and cemented to isolate the open hole or screened ore bearing interval from all other aquifers. The specification of injection and recovery well is $\Phi 148 \times 14$ mm and $\Phi 100 \times 10$ mm, respectively.

Wellfields were arranged as seven-spot pattern, which is 6 injection wells arranged in a hexagon with the recovery or extraction well in the centre (Fig. 3). The distance between injection and recovery well is 35 m. All wells, including monitor wells, were connected via buried pipelines to a wellfield "headerhouse" where injection and recovery well flow meters, pressure gauges, flow controls, oxygen mixers, and sample ports were located.

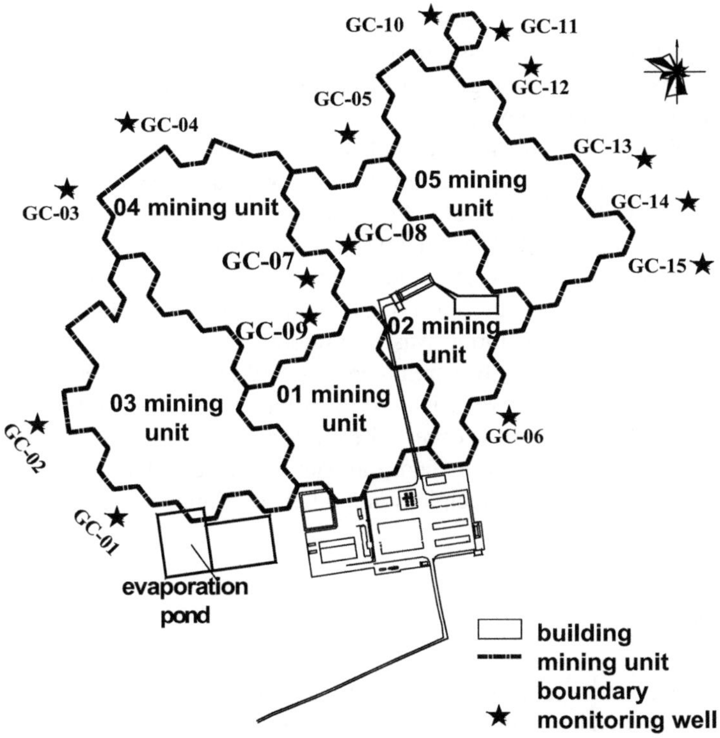

Fig. 2 Mining units and monitoring wells of the first developed area

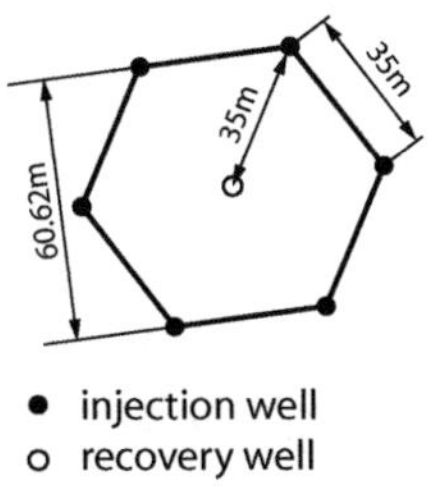

Fig. 3 Wellfield pattern

Waste Water Treatment

The mine reduces consumption of the industrial reagent and water and generation and discharge of waste water as much as possible by comprehensive waste water treatment technology of process water recycle, reverse osmosis and natural evaporation.

Water Recycle and Reuse

During production operations, about 99.5% of the produced water is re-fortified with oxygen and carbon dioxide gas and re-injected into the wellfield. A bleed of 0.5% is taken to maintain a hydrological cone of depression to ensure the leach solutions are contained within the wellfield.

Process water of the mine that can be recycled and reused include barren fluid, solution washing loaded resin, waste solution regenerating barren resin, precipitating mother solution and filtered liquor of yellow cake.

99.5% of barren fluid is recycled to make up of leachant, wash loaded and barren resin and regenerate barren resin. The solutions from washing loading resin are also recycled to make up of leachant. 90% of precipitating mother solution is recycled as eluant. The resin regeneration waste solutions generate 25% salt water and 75% fresh water by reverse osmosis treatment. 52% of the fresh water is recycled as leachant, 35% of the fresh water is used to irrigate local grass and 13% of the fresh water is used to wash yellow cake. Filtered liquor of yellow cake is recycled as yellow cake precipitation. In the end, radioactive waste water of the mine is only from small barren fluid and precipitation mother solution.

Reverse Osmosis

The resin regeneration waste solutions containing higher chloride and other chemical constituents are treated by reverse osmosis as mentioned above.

The reverse osmosis facility is given in Fig. 4.

Technical parameters of the reverse osmosis treatment technology are given in Table 1.

Nature Evaporation

Salt water from osmosis, waste solution from washing barren resin and 10% of precipitating mother solution are pumped in evaporation pond and be treated by nature evaporation (Fig. 5).

Fig. 4 Reverse osmosis facility

Fig. 5 Evaporation pond and evaporating waste water

Table 1 Parameters of reverse osmosis

Parameter	Index
Treatment capacity	40 m^3/h
Chloride concentration of feed solution	$\leq$2.0 g/L
RO Recovery	75%
Typical Salt Rejection	90%–95%
Operation mode	Intermittence

The evaporation pond can be preventing from seepage by setting 2 layer of 0.3 mm thick HDPE and 1 layer of 600 mm thick compacted clay.

The evaporation pond has a total capacity of 23,855 m^3 with the total area of 16,570 m^2 and a depth of 1 m.

Groundwater Quality and Waste Water Treatment Efficiency

Tables 2 and 3 present groundwater quality and reverse osmosis treatment efficiency.

Table 2 Groundwater quality

Monitoring point	pH	U (mg/L)	^{226}Ra (Bq/L)	As (μg/L)	Se (μg/L)	F^- (mg/L)	Cl^- (mg/L)	SO_4^{2-} (mg/L)
GC-01	8.82	0.438	0.053	14.2	6.22	4.45	174	239
GC-02	7.80	0.503	0.176	19.8	7.25	4.70	279	478
GC-03	7.80	0.288	0.091	8.62	4.0	1.71	232	233
GC-04	8.4	0.569	0.117	20.4	<2.0	4.13	473	538
GC-05	8.1	0.320	0.168	15.9	<2.0	2.82	493	603
GC-06	7.9	0.229	0.296	24.0	7.46	1.47	324	381
GC-07	8.0	0.458	0.117	4.02	<2.0	1.67	129	44.8
GC-08	7.95	0.200	0.349	16.5	2.10	1.18	516	560
GC-09	7.86	0.495	0.016	4.26	11.3	<1.0	1380	1410
GC-10	7.90	0.483	0.344	18.5	9.62	<1.0	826	1010
GC-11	8.05	0.164	0.314	15.8	7.26	1.51	681	712
GC-12	8.12	0.183	0.477	87.5	17.8	1.20	770	859
GC-13	8.18	0.105	0.170	21.0	2.04	<1.0	646	745
Upper aquifer	7.97	0.004	0.005	3.42	4.72	<1.0	7.17	<2.0
Criteria [a]	6.5–8.5	–	–	≤50	≤10	≤1.0	≤250	≤250

[a] Criteria is taken from Ⅲ class of GB/T 14848-93.

Table 3 Contaminants removal efficiency of reverse osmosis

Monitoring point	U (mg/L)	^{226}Ra (Bq/L)	As (μg/L)	Se (μg/L)	F^- (mg/L)	Cl^- (mg/L)	SO_4^{2-} (mg/L)
Raw water	0.062	0.671	4.84	329	1.99	1080	97.0
Salt water	0.102	2.42	26.1	1200	8.07	4490	403
Fresh water	0.047	0.093	<2.0	1.62	<1.0	39.3	<2.0
Removal efficiency (%)	24.19	86.14	>58.68	99.51	>49.75	96.36	>97.94
Criteria [a]	0.05	1.1	50	20	2	250	

[a] U and ^{226}Ra are taken from GB23727-2009; As, Se, F^- and Cl^- are taken from GB5084-2005.

It is known from Table 2 that the groundwater quality of the site is more remarkable spacious variation and has slightly higher background. However, most of the chemical constituents in groundwater meet Ⅲ class water quality of "Quality standard for ground water" (GB/T 14848-93).

Table 2 shows that reverse osmosis has very higher treatment efficiency. U and ^{226}Ra concentration of fresh water from reverse osmosis meet "regulations for radiation and environment protection in uranium mining and milling" (GB 23727-2009); As, Se, F^- and Cl^- meet "Standards for irrigation water quality" (GB5084-2005).

Conclusions

Alkaline in-situ leaching uranium has more obvious environmental advantages over traditional uranium mining/milling and acid in-situ leaching. But the mine is the first commercial alkaline in-situ leaching uranium mine in China and its waste water treatment has still assumed greater importance. The comprehensive waste water treatment technology with process water recycle, reverse osmosis and natural evaporation is very effective technology. It reduces not only consumption of the industrial reagent and water, but also generation and discharge of waste water as much as possible.

References

GB/T 14848-93 Quality standard for ground water. China standards press: Beijing, 1993.

GB 23727-2009 Regulations for radiation and environment protection in uranium mining and milling. China standards press: Beijing, 2009.

GB5084-2005 Standards for irrigation water quality. China standards press: Beijing, 2005.

Radiation Protection and Environmental Safety Surveillance in Uranium Mining and Ore Processing in India

A.H. Khan, V.D. Puranik

Abstract. Radiation protection and environmental safety surveillance has been an integral part of the entire uranium mining and ore processing industry in India since the first uranium mine and the mill were commissioned at Jaduguda in 1967. Workplace and environmental monitoring for gamma radiation, radon, airborne long-lived alpha activity and evaluation of workers' exposure are undertaken in all facilities. Treated tailings are safely confined in engineered containment facilities and regularly monitored to ensure regulatory compliance. Many monitoring devices such as scintillation cell, low level radon detection system, continuous radon monitor, Rn working level meter and SSNTD/TLD based personal dosimeters have been indigenously developed, calibrated and used for monitoring of workplace, personnel and the environment including dwellings. Regular up-gradation of mining and ore processing operations including ventilation system has helped in keeping workers' exposures well below the prescribed limit. Treated effluents are mostly recycled for use in plants. All releases to the environment are monitored. Used up part of the tailings pond is being provided with appropriate soil and vegetation cover to reduce gamma radiation and radon flux to near natural background levels. Radiation dose to mine workers has shown a downward trend averaging around 5 $mSv.y^{-1}$ during the past several years. The indigenously developed instruments and monitoring methodologies are described and workplace and environmental monitoring data for over a decade are summarized in this paper. Role of

---

A.H. Khan
Raja Ramanna Fellow, Environmental Assessment Division, Bhabha Atomic Research Centre, Trombay, Mumbai – 400085, India
E-mail: ahkhan1214@gmail.com, ahkhan@barc.gov.in

V.D. Puranik
Head, Environmental Assessment Division, Bhabha Atomic Research Centre, Trombay, Mumbai – 400085, India
E-mail: puranik@barc.gov.in

B. Merkel, M. Schipek, *The New Uranium Mining Boom*
DOI 10.1007/978-3-642-22122-4 © Springer 2011

vegetation and plants in the long-term stabilization, confinement and remediation of uranium tailings is discussed. Impact of the operations on the public domain is found to be negligible.

Introduction

Several low grade uranium ore deposits are identified in different parts of India. Uranium mining and ore processing commenced at Jaduguda in eastern India in mid 1960s. Subsequently, other mines were developed and are in operation at Bhatin, Narwapahar, Turamdih, Banduhurang and Bagjata in the same region with an addition large mill at Turamdih commissioned in 2007. Now, another mine and an ore processing mill are being developed at Tumallapalle in southern India. Proposals for development of other deposits are underway. The low grade of ore requires large tonnages to be mined simultaneously from several workings which, in turn, require relatively large workforce. Similarly, the processing plant and tailings management also need significant man power. Workers are exposed to radon and its progeny along with low levels of gamma radiation and airborne ore dust of low specific activity. The workplace and environmental radiation monitoring and evaluation of dose to workers are regularly carried out in all the facilities by an independent Environmental Survey Laboratory and Health Physics Unit established at site by the Bhabha Atomic Research Centre (Beri 1998; Bhasin 1998a, b, 1999; Gupta 2004). A brief description of the monitoring methodologies, indigenously developed instruments and dose evaluation with a summary of results is presented in this paper.

Materials and Methods

Internal exposure of workers is evaluated by monitoring radon at the workplace and occupancy period using an equilibrium factor. The external radiation exposure is evaluated either by area monitoring and occupancy period or by deploying personal thermo-luminescent dosimeters (TLDs).

Development of Monitoring Devices

The main radiation dose in mines arises from inhalation of radon with its progeny. Radon in mines is monitored using pre-evacuated scintillation cells while in lower concentration areas like the mill, tailings site and the environment a low level radon detection system (LLRDS) or a solid state nuclear track detector based radon dosimeter is used. Often, radon and the potential alpha energy concentration of its progeny are measured simultaneously to evaluate the equilibrium factor, F.

Scintillation Cell Technique

The indigenously developed and calibrated scintillation cell technique is one of the simplest and very convenient for collection of large number of air sample from the mine workings. It is a 150 ml metallic cylindrical cell coated inside with ZnS(Ag) with a transparent window which is optically coupled to a photomultiplier tube and counting system as shown in Fig. 1.

Pre-filtered air is introduced into the evacuated cell which is later counted for alpha activity over a known period, typically for about 600 s after ~180 min post sampling, when the progeny attain equilibrium with radon in the cell. Radon concentration is estimated using the relation (Raghavayya 1981),

$$C_{Rn} = \frac{0.0697 \times C}{E.V.\exp(-\lambda t).[1-\exp(-\lambda T)]}$$

where

C_{Rn} – radon concentration, Bq/m^{-3}
C – net count rate (cps)
E – efficiency of counting (fraction)
V – volume of the sampler (m^{-3})
λ – decay constant of radon (s^{-1})
t – time delay post-sampling
T – counting period (s)

Fig. 1 Radon sampling and counting system

Low Level Radon Detection System (LLRDS)

This indigenously developed system is based on electro-deposition of freshly formed positively charged (~90%) ^{218}Po atoms on a negatively charged plate for a predetermined collection and alpha counting period (Srivastava et al, 1984). The system consists of a 5 liter cylindrical aluminum chamber with a ~5 cm dia Al disk on the top lid. Filtered air is introduced in the system. A negative voltage of –800 V supplied to the metallic disc for a period of about 90 min and the alpha activity deposited on the plate is counted ideally from 1 to 75 min post collection. The radon concentration is estimated from the relation,

$$C_{Rn} = \frac{C}{E.V.Z.0.9(1-e^{(0.042H-4.31)})}$$

where

C – net counts observed
E – efficiency of the counter
V – volume of the chamber (m^3)
H – relative humidity in the sample (%)
Z – alpha emission factor

$$Z = \sum_{i=1}^{3} K_i \left(1-e^{-\lambda_i t}\right)\left(e^{-\lambda_i T_1} - e^{-\lambda_i T_2}\right)$$

$K_1 = 277$ s, $K_2 = 982$ s, $K_3 = -5599$ s
λ_i – decay constant of the radon daughters (s^{-1})

This system is also used for measurement of radon in the exhaled breath of workers to evaluate internally deposited radium inhaled in the form of ore dust (Srivastava 1983).

Continuous Radon Monitor

A continuous radon monitoring (CRM) system has been developed to monitor the diurnal variations in radon concentrations near the tailings facility and in the environment (Khan et al. 1989). The CRM is a 600 ml perforated cylindrical chamber covered with suitable foam and provided with a negatively charged plate at the bottom and a ZnS (Ag) scintillator at the top optically coupled to photomultiplier tube and counting assembly. Air with radon diffuses into the chamber and freshly formed, positively charged radon daughters are deposited on the negatively charged plate at the bottom. Size of the chamber is such that only alpha particles from radon reach the scintillator. A portable microprocessor attached to the counter receives the pulse out put from the counter for preset time and stores for subsequent display. The system has been modified for continuous use for several days (Ashokkumar 2008). A data transmission system can be attached for on-line measurement of radon.

Radon Progeny Working Level Meter

A radon progeny working level monitor has also been developed for use in mines. It comprises of a small portable air sampler, an alpha counter and a microprocessor housed in an aluminum casing. The sampling and counting are programmed and the microprocessor attached to the unit receives the pulses from the counter and displays radon progeny WL (Khan et al. 1989). The system can measure radon progeny down to 10 mWL.

SSNTD Based Radon Evaluation

For evaluation of very low levels of radon in the environment or in dwellings and for dosimetry of mine workers a system for integrated measurement of radon using SSNTD has been developed and calibrated. The system is shown in Fig. 2.

The device comprises of a 60 ml cylindrical aluminum chamber covered with a permeable membrane which allows only radon to diffuse in, due to its relatively longer half life, and serves as a barrier for Rn-219, Rn-220 and aerosols. A 1.8 cm × 3 cm SSNTD film is placed between two TLD chips mounted on a disc at

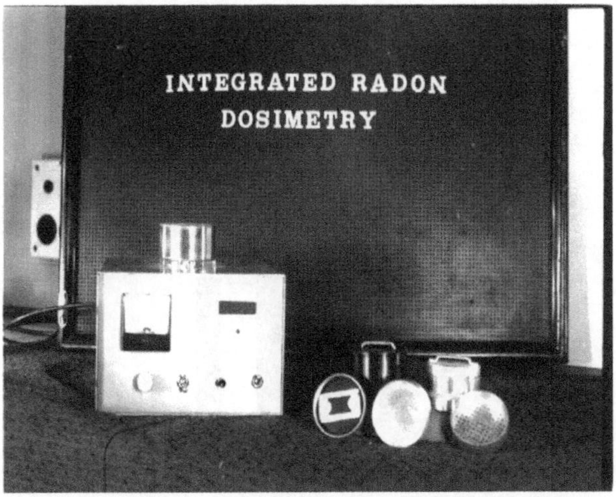

Fig. 2 SSNTD and TLD based Rn and γ dosimeter

the base of the chamber. The SSNTD film and the TLD in the chamber are replaced every two months. The TLDs are processed to give cumulative gamma radiation exposure. The SSNTD film is etched with 10% sodium hydroxide solution at a temperature of 60 °C for 90 min. The tracks are electronically counted which are correlated to radon exposure using a calibration system (Jha and Raghavayya 1983; Raghavayya et al 1986). The cumulative radon exposure is obtained from the track density as,

$$E\left(Bq.h.l^{-1}\right) = 0.554\,T\,,$$

where E is the cumulative exposure, T is track density per cm^2, h is the exposure period in hours and 0.554 is the calibration factor between the track density and radon concentration ($Bq.l^{-1}$).

Average radon concentration for the exposure period can be obtained from the relation,

$$C\left(Bq.l^{-1}\right) = 0.554\,T/h\,.$$

Dose Evaluation

Average Radon Concentrations

The ventilation system in all the underground mines is regularly reviewed and upgraded to keep radon and progeny concentrations well within the regulatory limits. Average radon concentration trend in the Jaduguda uranium mine during the past decade is shown in Fig. 3.

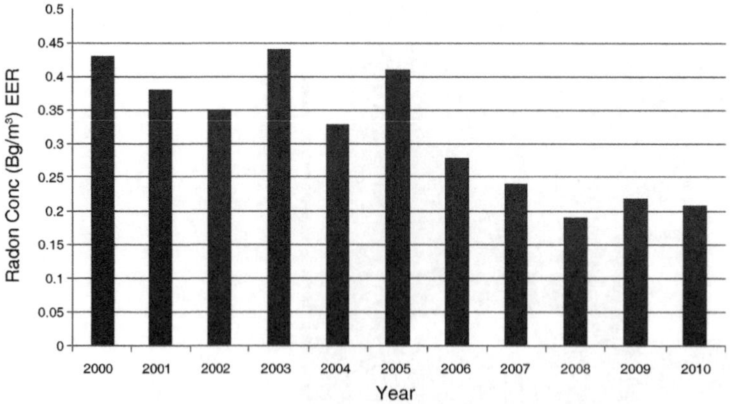

Fig. 3 Average radon concentrations in Jaduguda mine (2000–2010)

Dose Evaluation for Mine Workers

For dose evaluation, radon progeny concentration (WL) is computed from the radon monitoring data and the average equilibrium factor 'F' for different locations. From the progeny concentrations (WL), occupancy period and gamma dose rate obtained from the TLD, the annual effective dose is given as (Khan et al. 2009),

$$H(mSv) = \frac{K.A.\sum_{i}^{n} WL_i T_i}{170} + \sum_{i}^{n} G_i T_i$$

where

H = effective dose, mSv
K = conversion factor (1 WLM = 5 mSv)
WL_i = radon daughter conc. at ith location (WL)
T_i = time spent at ith location (h)
n = no. of locations
A = annual attendance (d/y)
G_i = gamma dose rate at ith location.

The average internal and external doses to workers from all the five underground and one opencast mine during the period of 2006 to 2010 are shown in Fig. 4.

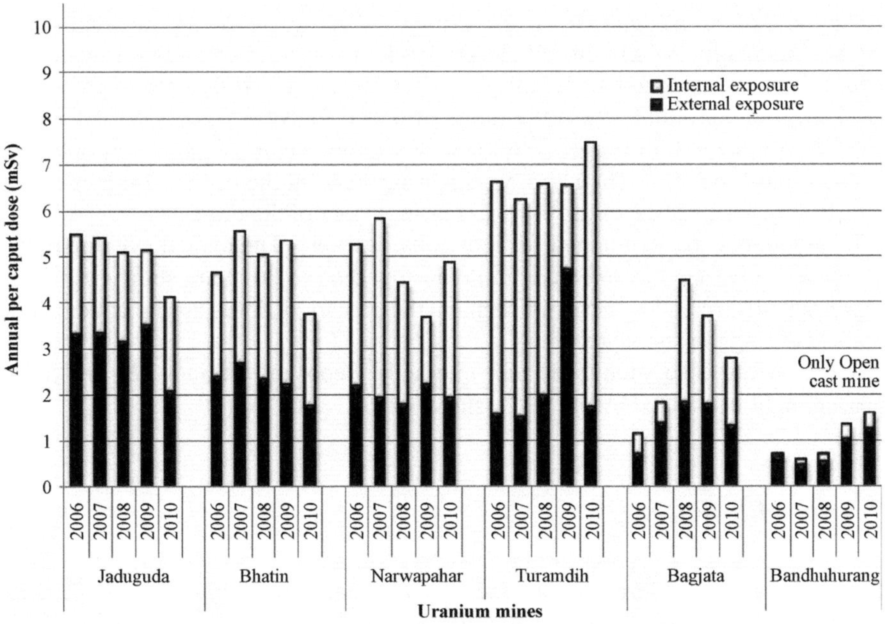

Fig. 4 Average doses to workers in different uranium mines (2006–2010)

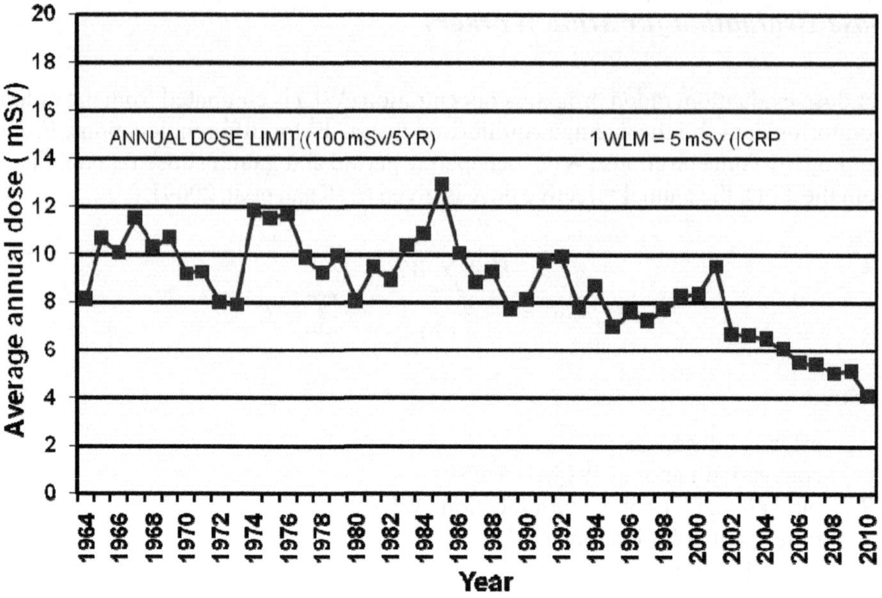

Fig. 5 Average dose to Jaduguda mine workers from during 1964–2010

The underground mines at Jaduguda, Bhatin and Narwapahar have been operating for over 30 to 40 years and their ventilation systems are well developed and stabilized. The Turamdih and Bagjata mines are in the process of development where ventilation systems are in the process of up-gradation by opening up of fresh air routes for radon control. Hence, while the average radiation doses in the older mines are around 4 to 5.5 $mSv \cdot y^{-1}$ showing a gradual downward trend, the doses in Turamdih mine are in the range of 6 to 6.5 $mSv \cdot y^{-1}$. Improvement in the ventilation system is in progress to bring down the average doses to an optimum level around 5 $mSv \cdot y^{-1}$. The slightly increasing trend in the opencast mines is due to relatively better grade of ore with increasing depth of the mine.

Dose records are maintained for individual workers. The annual average radiation doses to workers in Jaduguda mine during 1965 to 2010 are shown in Fig. 5. A consistently reducing trend is seen in the radiation doses to Jaduguda mine workers during the last 9 years.

Workers are also monitored for internally deposited radium-226 by radon measurement in exhaled breath (Srivastava, 1983).

Dose to Mill Workers

Radiation, radon and airborne radioactivity are regularly monitored in mill. Dose to the mill workers is relatively small as most of the operations are carried out in buildings with adequate natural ventilation augmented by well designed ventilation systems and appropriate enclosures. The average dose to mill workers during the past five years has been around 1.7 $mSv{\cdot}y^{-1}$ (Khan et al 2009). Bioassay of mill workers is also carried out, especially for workers in the final product section. These are well within the appropriate limit.

Environmental Monitoring

Environmental monitoring around the mining facility, especially around the tailings, is carried out regularly. Gamma radiation and radon are monitored using TLDs and LLRDs, respectively. Sampling and analysis of effluents and the ground and surface waters are carried out ensure that operations comply with regulatory norms and exposure of members of the public is kept well with in the prescribed limits. Analysis of soil, vegetation, plants and local foodstuff is also undertaken periodically.

The radon levels at 1 m above the tailings pond surface and the tailings embankment are typically around 30–35 $Bq{\cdot}m^{-3}$ and 20 $Bq{\cdot}m^{-3}$, respectively, which reduces to the natural background level of about 10–15 $Bq{\cdot}m^{-3}$ within a short distance. Similarly, the gamma radiation on the tailings surface vary from 1.50 to 3.00 $\mu Gy{\cdot}h^{-1}$, which reduce to around 0.40 $\mu Gy{\cdot}h^{-1}$ at the embankment and reach the natural background level of 0.10 to 0.15 $\mu Gy{\cdot}h^{-1}$ with in 20 to 30 m from the embankment (Khan et al 2000, Jha et al, 2001, Kumar et al, 2006).

Radon and radiation levels are also monitored in the dwellings in the region. The radon levels vary from 25 to 185 $Bq{\cdot}m^{-3}$ and average around 100 $Bq{\cdot}m^{-3}$. The gamma radiation levels vary from 0.07 to 0.20 $\mu Gy{\cdot}h^{-1}$, averaging around 0.13 $\mu Gy{\cdot}h^{-1}$. These levels are similar to those reported from many regions in India and abroad with temperate climate (Khan and Puranik 2010).

Tailings Management and Remediation

The uranium mill tailings management system comprises of treatment of barren liquor from the ion exchange columns in uranium mill with lime stone slurry followed by addition of lime slurry to raise the pH to 10–10.5 and mixing with the barren cake slurry, obtained after filtration of the dissolved uranium. The pH is maintained above 9.5 to precipitate the residual uranium, radium, other radionuclide and chemical pollutants. The treated slurry is classified into coarse and fine fractions. The coarse material, forming nearly 50% of the total mass, is backfilled in mines. The fine tailings are pumped to an engineered tailings pond for permanent containment.

At present there are three valley-dam type of tailings ponds at Jaduguda. The first and second tailings ponds of about 33 and 14 hectare (ha) surface area, respectively, located adjacent to each are now nearly filled up. The third stage of the tailings pond having an area of about 30 ha is currently in use. The underlying soil and the bedrock of these tailings ponds have very low permeability.

The effluents from the tailings pond are clarified and a major portion is recycled for use in the milling process. The rest is treated first with $BaCl_2$ and then with lime slurry to precipitate the radioactive and chemical pollutants, especially ^{226}Ra and Mn. It is clarified and the settled sludge carrying the $Ba(Ra)SO_4$ and $Mn(OH)_2$ precipitates is returned to the tailings pond with the main tailings and the clear effluent is discharged to environment after monitoring.

Remediation of the used up part of the mill tailings is an important aspect of radiation and environmental safety in uranium industry.

As an interim measure, vegetation cover of grass and shrubs, not grazed by cattle, such as *Saccharum spontaneum*, *Typha latifolia, Ipomoea carnia* and *lantana* are grown over the used up portion of the first two tailings ponds. Some plants pick up radioactivity and trace metals from tailings or the cover material which may have a small potential of reaching cattle and the human food chain. Hence, uptake of radionuclide by these plants and shrubs including the *cynodon dactylon* grass has been studied. The *cynodon dactylon* shows highest affinity for Mn and Zn which are present in the tailings. It is also the only component of vegetation that may be grazed by cattle. All other species mentioned above showed soil to plant transfer factor in the range of 10^{-3} to 10^{-2}. *Typha latifolia* showed relatively higher uptake (Basu et al. 2000). Roots of vetiver grass (*vetiveria zyzanoides*) are found to have good soil binding properties but laboratory experiments show a high uptake for uranium from the tailings (Venu Babu 2007). Hence, this grass may be a good candidate for phytoremediation.

On an experimental basis, soil cover has been provided on one of the tailings ponds to reduce radiation levels and radon emanation rate to near natural background levels. After studying its effectiveness other used up parts may be provided with an appropriate soil cover. Studies are under way to provide vegetation cover of such grass and shrubs which have low pick up of radionuclides or heavy metals, have shallow root penetration and are not grazed by cattle.

Acknowledgement The support provided by "DAE-BRNS Raja Ramanna Fellowship" is gratefully acknowledged. Thanks are due to Dr. A. K. Ghosh, Director, HS&EG, and BARC for the support provided. Cooperation received from Uranium Corporation of India Ltd is acknowledged.

References

Ashokkumar, P., Raman, A., Srivastava, V.S., Patnaik, R.L., Babu, D.A.R. and Sharma, D.N. (2008), Radon measurement studies with indigenously developed continuous radon monitor (CRM), Sixteenth national Symposium on Environment, Guru Jambeshwar University of Science and Technology, Hissar, Haryana, July 16–18, 2008.

Basu, S.K., Jha, V.N., and Khan, A.H., (2000), Uptake of radionuclides and trace metals by plants growing on or near uranium tailings, Ninth National Symposium on Environment, Bangalore University, India, June 5–7, 2000, pp. 138–141.

Beri, K.K., (1998), Uranium ore processing and environmental implications; Metal, Materials and Processes, Vol. 10, No.1, pp. 99–108.

Bhasin, J.L., (1998), Mining and milling of uranium ore: Indian scenario; IAEA-Techdoc-1244 (2001), pp. 189–200.

Bhasin, J.L., (1998), Uranium exploration and mining; Nu-Power, vol. 12, No.1.

Bhasin, J.L., (1999), Radiation Protection in uranium mining and milling, Proceedings of 24th Annual Conference of the Indian Association of Radiation Protection, Kakrapar, India, January 20–22, 1999.

Gupta, R. and Siddique, S., (2003), Management of the tailings and liquid efflux-ents in uranium mining and milling operations; Radiation Protection and Environment, vol. 26, No.3 & 4, 2003, pp. 506–515.

Gupta, R. (2004), Uranium mining and environment management – The Indian scenario, Proceedings of Thirteenth National Symposium on Environment, June 5–7, 2004, North Eastern Hill University, Shillong, pp. 1–5.

Jha, G. and Raghavayya, M., (1983), Development of a passive radon dosimeter, 5th National Symposium on Radiation Physics, Calcutta, November 1983.

Jha, S., Khan, A.H. and Mishra, U.C. (2001), A study of the technologically modified sources of ^{222}Rn and its environmental impact in an Indian U mineralized belt, J Environ. Radioact. 53, 183–197 (2001).

Khan, A.H., Raghavayya, M., Sharma, D.N., Raman, N., Jethmalani, P., Krishnamachari G. And Iyer, M.R., (1999), Development of a continuous radon monitor and instant working level meter, Bull. of Rad. Prot., IARP, vol.12, No.3, pp. 56–61 (1999).

Khan, A.H., Basu, S.K., Jha, V.N., Jha, S. and Kumar, R. (2000), Assessment of environmental impact of mining and processing of uranium ore at Jaduguda, India, The Uranium Production Cycle and the Environment, IAEA International Symposium on uranium Production Cycle and the Environment Vienna, Austria, October 2–6, 2000 (C&S Papers Series, 10/P.

Khan, A.H., Sahoo, S.K. and Puranik, V.D., 'Radiation Protection of workers in mining and processing of uranium ore, Radiation Protection and Environment, Vol. 26, No. 3 & 4, 2003 (India).

Khan, A.H., and Puranik, V.D., (2010), Radon in the environment and Dwellings in a uranium mining area in eastern India: An overview, Radiation Protection Dosimetry, Special Issue on Radon2010, April, 6 2011, (ncr061).

Kumar, R., Patnaik, R.L., Shukla, A.K. and Khan, A.H., (2004), Indoor radon levels and gamma radiation in uranium mineralized belt around Jaduguda, Jharkhand State, Environmental Geochemistry, Vol. 9, No.1, 2006, pp. 84–87.

Raghavayya, M., (1981), An inexpensive Radon scintillation cell. Health Phys. 40, 894 (1981).

Raghavayya, M., Khan, A.H. and Jha, G., 'Dose estimate for Jaduguda mine workers', Bulletin of Radiation Protection, Vol. 8 (1986), pp. 69–73.

Srivastava, G.K., (1983), Estimation of radium body burden of uranium miners by measuring radon in breath using electrostatic technique, M. Sc. Thesis, University of Bombay, 1983.

Srivastava, G.K., Raghavayya, M., Khan, A.H. and Kotrappa, P., (1984), A low level radon detection system, Hlth. Physics J., 46 (1), 1984, pp. 225–228.

Venubabu, P. (2007), Study of vetiver for uptake of uranium from tailings, Unpublished personal communication, BARC, Mumbai, 2007.

Estimates of Effective Doses Among Czech Uranium Miners

Ladislav Tomasek, Jiri Hulka, Petr Rulik, Helena Mala, Irena Malatova, Vera Beckova

Abstract. Radiation doses in Czech uranium mines are estimated using measurements of physical and chemical characteristics of the mining aerosol. The main parameters were: size, chemical solubility in lung fluid, and amount of Rn gas emanating from uranium particles. Based on these parameters, committed effective doses from long lived radionuclides in uranium dust were calculated using the IMBA software. In conditions at mine Rozna in 2000–2009, mean annual effective doses are 1.9 mSv from long lived radionuclides, 4.1 mSv from radon and its progeny and 2.2 mSv from external gamma radiation.

Introduction

Radiogenic risk in uranium mines is usually linked to radon exposure, particularly in earlier periods of mining, when ventilation in mines was limited. At present,

Ladislav Tomasek
National Radiation Protection Institute, Bartoskova 28, Prague, Czech Republic

Jiri Hulka
National Radiation Protection Institute, Bartoskova 28, Prague, Czech Republic

Petr Rulik
National Radiation Protection Institute, Bartoskova 28, Prague, Czech Republic

Helena Mala
National Radiation Protection Institute, Bartoskova 28, Prague, Czech Republic

Irena Malatova
National Radiation Protection Institute, Bartoskova 28, Prague, Czech Republic

Vera Beckova
National Radiation Protection Institute, Bartoskova 28, Prague, Czech Republic

B. Merkel, M. Schipek, *The New Uranium Mining Boom*
DOI 10.1007/978-3-642-22122-4

radon levels are 1–2 orders of magnitude lower. Therefore, exposures to other radiation components – external gamma and long lived radionuclides in mining aerosol are of increased importance.

Czech uranium mining started on industrial base in the 1890s. It is estimated that the total production has been 110.000 t of uranium and the uranium industry has employed nearly 100.000 underground workers.

The aim of this report is to estimate size distribution of aerosol particles containing uranium and its decay products in uranium mine Rožná I in dependence on conditions in the mine and mill.

Methods

The estimation of physical parameters of mine aerosol was based on samplings and measurements in mine Rozna I, where the ore is mined from the depth up to 1200 m. Five and six stage cascade impactors were repeatedly placed at three different sites where higher dust content was expected – at stopes, at crushing plants, and at the end of the chute where carts are loaded by the ore. Gamma spectrometric analyses were performed using HPGe detectors.

The radiation dose from the mine aerosol depends on several physical and chemical parameters. The physical ones include isotopic composition, including proportion of radon activity escaping from particles, which determines how the gross alpha activity is distributed into radionuclides of the uranium series. Estimates of this proportions in terms of activities of Rn decay products (Bi-214, Pb-214) and Ra-226 were based on gamma spectrometry as above. The particle size in terms of activity median aerodynamic diameter (AMAD) for each working site were estimated assuming log-normal distribution of activities corresponding to stages of cascade impactors (Hinds 1999), including geometric standard deviations (GSD).

Chemical parameters of particles, particularly the solubility of particles in lung fluid, were estimated by own measurements for U-234 and U-238 (Beckova and Malatova 2008) and parameters for other radionuclides were taken from (Duport et al. 1991).

For conversion of unit intakes into effective doses, we used software IMBA (Integrated Modules for Bioassay Analysis), which implements the biokinetic and dosimetric models currently recommended by the International Commission on Radiological Protection (ICRP) and enables the users to specify their own parameter values to the customized internal dose calculations (IMBA 2005).

Results

Samplings in mines resulted in a total of 14 data sets. The character of activity distribution obtained from these samplings is similar. The highest fraction of activities is in the two last categories (4.2–10.2 and > 10.2 μm) and lowest fraction in category 0.39–0.69 μm. Surprisingly, the activity in the very low category (< 0.39 μm) is higher then in next categories suggesting bi-modal distribution. For these reasons, analyses were conducted ignoring the first category (representing about 20% of activities) and parameters (AMAD and GSD) were estimated from such limited data. Particle size for different working sites in terms of AMAD were in the range 2.0–9.2 μm with the mean of 7.3 μm and GSD were in the range 2.1–6.5 with the mean 3.2. These results generally correspond to data in workplaces (Dorrian and Bailey 1995).

The fraction of escaped radon in different samplings varied from 32% to 63%. The mean value was 43%. Based of this observation, the isotopic composition of U-238 series can is summarized in Table 1. The weights in this table indicate fractions of alpha radionuclides, both long lived (LL) and short lived (SL) in the gross alpha activity.

Effective doses from unit intake of radionuclides were calculated for long lived radionuclides. These calculations were performed using the IMBA software (IMBA 2005) with user defined absorption parameters (Table 2) similarly as in (Marsh et al. 2011). Results of these calculations in terms of conversion coefficients are given in Table 3 containing calculations for different particle sizes (AMAD < 0.4 μm and AMAD = 5, 10, and 7 μm). The resulting conversion of 3.4 mSv/kBq corresponding to AMAD = 7 μm was used to illustrate mean annual

Table 1 Isotopic composition and numbers of alpha emitters assuming 43% radon escape for U-238 and U-235 decay series

	Emitter type [a]	Weight [b]
U-238	LL	1
U-234	LL	1
Th-230	LL	1
Ra-226	LL	1
Rn-222	SL	0.57
Po-218	SL	0.57
Po-214	SL	0.57
Po-210	LL	0.57
Total		6.28
U-235	LL	0.0526
Pa-231	LL	0.0526
Th-227	LL	0.0526
Ra-223	LL	0.0526
Rn-219	SL	0.0526
Po-215	SL	0.0526
Bi-211	SL	0.0526
Total		0.3682

[a] Emitter type: long lived (LL), short lived (SL)

[b] Assuming fraction $^{235}U/^{238}U = 0.0526$ and 43% Rn escape, i.e., $n = 6.65$ alpha emitters

Table 2 Absorption parameters of long lived radionuclides

Nuclide	f_r^a	$s_r\ (d^{-1})^b$	$s_s\ (d^{-1})^c$	f_1^d
^{224}Ra, ^{226}Ra, ^{228}Ra	0.11	7.32	0.000412	0.2
^{210}Pb	0.26	3.91	0.00101	0.2
^{228}Th, ^{230}Th, ^{232}Th	0.14	4.56	0.000683	0.0005
^{234}U, ^{235}U, ^{238}U	0.22	0.78	0.00144	0.2
^{231}Pa	0.18	4.1	0.000886	0.0005
^{227}Ac	0.18	4.1	0.000886	0.0005
^{210}Po	0.18	4.1	0.000886	0.1

[a] Rapid fraction
[b] Solubility of rapid component
[c] Solubility of slow component
[d] Fraction of activity absorbed in blood from intestines

Table 3 Estimates of conversion coefficients (mSv/kBq) for different long lived radionuclides and particle size (AMAD) assuming 43% radon escape and resulting conversion for U-238 and U-235 series

AMAD	<0.4 μm	5 μm	10 μm	7 μm	5 μm [a]	10 μm [a]	7 μm [a]
U-238	5.84	3.09	1.87	2.50	3.72	2.75	3.25
U-234	6.79	3.98	2.41	3.23	4.64	3.38	4.04
Th-230	8.78	4.96	3.06	4.04	5.83	4.34	5.09
Ra-226	10.6	6.18	3.92	5.09	7.18	5.38	6.31
Po-210	5.15	3.32	1.80	2.60	3.78	2.56	3.20
Pb-210	0.76	1.10	0.94	1.02	1.03	0.90	0.97
U-235	6.19	3.41	2.07	2.76	4.05	2.98	3.53
Pa-231	10.3	5.82	3.56	4.73	6.84	5.03	5.96
Ac-227	36.2	24.5	15.0	19.9	27.1	19.5	23.5
weighted sum [b]	38.09	22.46	13.88	18.33	26.07	19.20	22.76
resulting conversion	5.8	3.4	2.1	2.8	3.9	2.9	3.4

[a] Correction for 20% fraction of fine particles (<0.4 μm)
[b] Assuming fraction $^{235}U/^{238}U = 0.0526$ and 43% Rn escape, i.e., $n = 6.61$ alpha emitters

effective doses at mine Rozna in the period 2000–2009, which is 1.9 mSv. For comparison, the mean effective annual dose from radon and its decay products is 4.1 mSv using the latest recommendation of ICRP on radon (ICRP 2009) and the mean annual doses from external radiation is 2.2 mSv.

Acknowledgement The authors thank the staff of the National Institute for Nuclear, Chemical, and Biological Protection for carrying out the sampling and the staff of mine Rozna, who made the realization of the project possible. The research was supported by the State Office for Nuclear Safety (Project No VZ60022490).

References

Bečková V, Malátová I (2008) Dissolution behaviour of 238U, 234U and 230Th deposited on filters from personal dosemeters. Radiation Protection Dosimetry 129:469–472.

Duport P, Robertson R, Ho K Horvath F (1991) Flow-through dissolution of uranium-thorium ore dust, uranium concentrate, uranium dioxide, and thorium alloy in simulated lung fluid. Radiat. Prot. Dosim. 38:121–133.

Dorrian MD, Bailey MR (1995) Particle size distribution of radioactive aerosols measured in workplaces. Radiation Protection Dosimetry 60:119–133.

Hinds WC (1999) Aerosol Technology: properties, behavior and measurement of airborne particles. John Wiley & Sons, Inc.

IMBA (2005) User Manual for IMBA Professional Plus (Version 4.0), © HPA, 2005 (*http://www.imbaprofessional.com/about/description*).

ICRP (2009) International Commission on Radiological Protection. Statement on Radon. (*http://www.icrp.org/docs/ICRP_Statement_on_Radon(November_2009).pdf*)

Marsh JW, Blanchardon E, Gregoratto D, Hofmann W, Karcher K, Nosske D, Tomasek L (2011) Dosimetric calculations for uranium miners for epidemiological studies. Radiation Protection Dosimetry, in print.

Evaluation of Transboundary Impact of Toxic Metals of Uranium Mine Mailoo-Suu (Kyrgyzstan)

V.B. Aparin, J.P. Voronova, S.K. Smirnova

Abstract. Uranium mine Mailoo-Suu is located within the valley of the same name river in Kyrgyzstan territory. Tailings impoundments and heaps of Mailoo-Suu are the sources of contamination of surface waters with radionuclides and heavy metals – selenium, arsenic, manganese, copper. In bottom sediments heightened contents of uranium, selenium, chromium, nickel, cobalt are established. In plants and human biosubstrata (nails, hair) are established high contents of selenium, uranium, molybdenum, nickel, zinc.

Introduction

The frontier between Uzbekistan and Kyrgyzstan lies through piedmont part of the Ferghana intermountain depression, located within Tiyn Shan mountain System.

Uranium deposit Mailoo-Suu is located within the valley of the same name river (Fig. 1) in Kyrgyzstan territory at a distance of 30 km from the frontier of Uzbekistan. Here as a result of working of the Western metallurgical plant which during 22 years (1946–1968) has been intensively treated of uranium ores of the deposit, as well as ores from the Eastern Germany, Czechoslovakia, Bulgaria and China, radioactive wastes have been accumulated. They are concentrated within 23 tailings impoundments and 13 heaps of non-commercial ores. 16 tailings im-

V.B. Aparin
Department of Geoecology, "Institute of Hydrogeology, Engineering Geology and Geoecology" State corporation, Republic of Uzbekistan

J.P. Voronova
Department of Geoecology, "Institute of Hydrogeology, Engineering Geology and Geoecology" State corporation, Republic of Uzbekistan

S.K. Smirnova
Institute of Geology and Geophysics, Academy of Science of Uzbekistan

B. Merkel, M. Schipek, *The New Uranium Mining Boom*
DOI 10.1007/978-3-642-22122-4 © Springer 2011

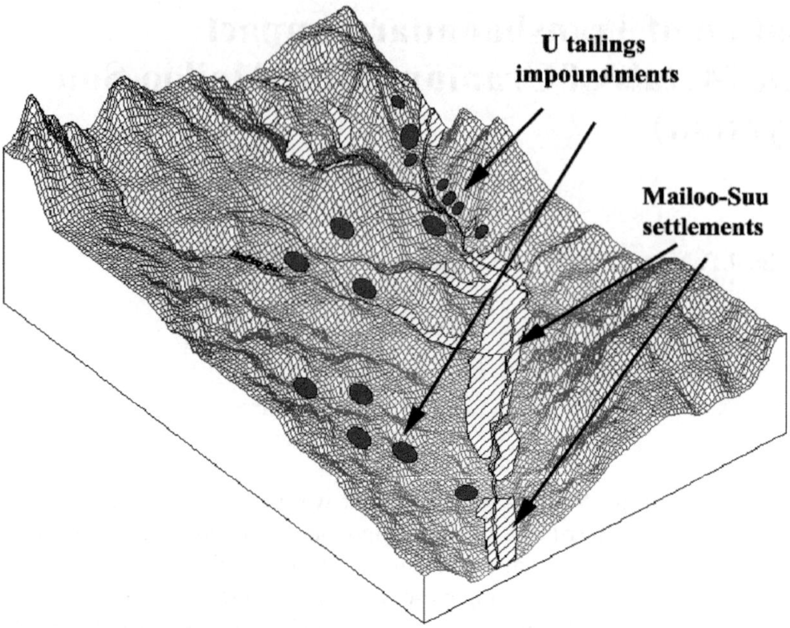

Fig. 1 Location of Mailoo-Suu settlements and tailings impoundments within relief

poundments out of 23 were build according to projects, but within the other 7 accumulation of tails took place without order. Total volume of radioactive tails is near 2 million m^3 or more than 4 million t.

As a result of investigation of tailings impoundments # 3,7,10 it was established that buried radioactive tails are thin crushed limestones in paste condition. Chemical composition of wastes is corresponded to initial content of uranium ore. Besides radionuclides there are established heightened concentrations of lead, molybdenum, vanadium. As well as in tails material toxic chemical reagents, using for extraction and enrichment of uranium oxide, are contained. Among them there are: sulfuric acid, manganese oxides, ammonium sulfates, arsenic, etc. The main radionuclide is radium created maximum equivalent dose (MED) some thousand microroentgen/hour (mcR/h). Besides radium within tailings impoundments have been concentrated significant quantities of thorium, the most part of it is represented with its isotope – ionium (Th-230), as well as uranium remnants. Radioactive wastes are concentrated within 13 non-cultivated heaps and are non-commercial ores with radioactivity up to the first hundreds mcR/h.

Materials and Methods

Field works are necessary to collect samples for this study. Sampling of rocks, dumps, soils, waters and plants have been carried out in four above mentioned locations. It is designed to assess the spatial distribution of heavy metals in the area.

Rock and ore samples have been crushed to <2 mm in a jaw crusher, quartered and then pulverized using "Pulverizette 6" grinding mill. Dry plant samples and soil samples have been disaggregated, sieved and pulverized. Selected ore, tailings and soils samples have been subjected to density separation using heavy liquids. High density fraction have been made into polished mounts.

Preliminary study of sample compositions (major and some trace elements) have been carried out using XRF spectrometer (Oxford Instruments ED-2000Rh). Major minerals have been identified using X-ray diffractometer (DRON-2). Powdered solid samples have been digested with appropriate acids (HNO3/HClO4; aqua regial) using microwave digestion system (Paar Physica "Multiwave"). Some of quartered samples have been leached with deionized water (> 18 MOhm) to determine water-soluble elements. Resulted solutions as well as water samples have been analyzed for radioactive and heavy metals by ICP-MS (Perkin Elmer-Sciex ELAN-6000).

Mineral forms of heavy and radioactive metals have been studied in polished mounts using electron microprobe (JEOL JXA-8800R, 3 wavelength spectrometers and 1 energy-depressive spectrometer Link ISIS-300).

Results and Discussion

Tails composition corresponds to composition of ore which was treated at that moment. Analysis of tails of tailings impoundment #3 (Table 1) shows that carbonates and silicates are prevailed, as well as following metal concentrations: uranium – up to 0,1–0,15%; selenium – 26–36 mg/kg; copper – 0,11–0,18%; manganese – 0,5–1,0%; cobalt – 20–34 mg/kg. Total α-activity of tails is ranged from 67 up to 113 thousand Bq/kg, that can be regarded to radioactive wastes according classification.

Using electron microprobe analyses a number of minerals were studied in tailings: uraninite, orthobrannerite, selenian pyrite, uranotallite, nisbite, cerussite, etc.

Table 1 Data of chemical sampling of tails in tailings impoundment #3

Area	Chemical composition of tails, %	Content of toxic metals in tails, mg/kg
Tailing impoundment #3, depth 6m	CaO-35.1, CO_2-22.8, SiO-25.8, Al_2O_3-4.7, MgO-3.07, Fe_2O_3-3,28, Na_2O-0.82, K_2O-1.07, TiO_2-0.35, P_2O_5-0.06, SO_3-1.19, Cl-0.01	U-1500, Th-3.64, Mn-10285, Cu-1879, As-52.6, Se-26.45, Pb-44.2, Co-33.47, Ni-44.9, Zn-99, Sr-293, Mo-4.42, Cd-1.11
Tailings impoundment #3, depth 12	CaO-42.4, CO_2-31.0, SiO-15.8, Al_2O_3-2.6, MgO-2.75, Fe_2O_3-2.36, Na_2O-0.47, K_2O-0.67, TiO_2-0.19, P_2O_5-0.04, SO_3-0.69, Cl-0.01	U-1000, Th-3.07, Mn-5696, Cu- 1059, As-43.8, Se-35.71, Pb-26, Co-19.93, Ni-49, Zn-62.4, Sr-291, Mo-3.09, Cd-0.7

Dissolution of these minerals by percolated water is a reason for contamination of adjacent areas.

Tailing impoundments of uranium deposit Mailoo-Suu are buried with neutral soils of 15–25 cm thickness. On surface of impoundment gamma-radiation changes from background values up to anomalous ones (100–500 mcR/h). Heightened values were observed locally in setting of impoundments as well. As a result there was established that on the surface of soil cover of tailings impoundments take place contaminations with chromium and especially selenium. Contents of selenium are from 2 to 20 mg/kg. This contamination are fixed within all impoundments. For some of them weak contamination with nickel, cobalt, lead, zinc, cadmium is observed. These elements are not characterized for ores of Mailoo-Suu deposit and to all appearance they were connected with imported ores from Germany and Czechoslovakia.

In marginal parts and borders of impoundments anomalous concentrations of selenium, chromium and for some of them, uranium, molybdenum, manganese, lead, zinc are observed as well. Migration of significant amounts of heavy metals into soil cover is connected with two manner of migration: mechanical (owing to destruction and removal tails from impoundments) and migration of elements in soluble (ionic) form.

Thus, top-soil of Mailoo-Suu tailings impoundments are contaminated with radionuclides and heavy metals – selenium, chromium, uranium, molybdenum, manganese, lead, zinc.

Analyses of water samples from Mailoo-Suu river was shown that in limits of studied region river waters are fresh, weak alkaline, with normal gustatory sense. Their mineralization is rather increased from the north to the south (from 0.19 to 0.26 g/l), composition as a rule, is stable (bicarbonate, calcic-sodic). Hardness of water is changed from 2.5 to 3.3 mg/cq.

Microcomponent content of river water (Table 2) is characterized mainly by low concentrations of elements. Uranium content in water is increased downstream from 0.2 up to 6–11 mcg/l, content of radium – from $5*10^{-13}$ to $1*10^{-12}$ Ci/l, selenium – from 1 to 6 mcg/l. Below the Mailoo-Suu settlement contents of lead, zinc, molybdenum are increased sharply. This is connected with technogenic pollution stipulated by electric bulb' plant activity.

Ground waters are pinched out down the river bed of Mailoo-Suu and in foots of tailings impoundments. In such a case waters are especially polluted. In ground

Table 2 Microcomponent content of Mailoo-Suu river water

Point	Cr mcg/l	Mn mg/l	Co mcg/l	Ni mcg/l	Cu mcg/l	Zn mcg/l	As mcg/l	Se mcg/l	Mo mcg/l	Sb mcg/l	Pb mcg/l	U mcg/l
M1	1.6	0.23	<0.2	0.41	0.31	1.05	0.92	1.03	0.35	0.15	<0.2	0.21
M16	2.4	8.44	0.19	1.64	1.36	3.41	1.08	3.27	0.91	0.15	<0.2	0.50
M58	6.0	6.59	0.15	2.25	2.07	0.25	2.45	5.93	1.1	0.41	<0.2	3.1
M60	4.3	5.04	0.15	3.39	1.73	259.6	1.26	2.81	17.4	0.32	28.9	5.8
M61	4.2	7.87	0.17	2.43	1.32	243.5	1.27	2.37	20.0	0.32	27.4	3.4

waters pinched out from under tailings impoundment #3 (M 42), uranium content is 1648 mcg/l, from under impoundment #5 (M 18) – 1938 mcg/l, from under impoundment #7 (M 48) – 131.4. Besides uranium high concentrations of selenium, manganese, arsenic and elevated contents of molybdenum and copper are established (Table 3).

For investigation the dominating plants were selected – absinthe, camel bur. Surface part of the plant (leaves) was sampled, then after drying plant samples have been disaggregated, pulverized and dissolved with acids, using microwave digestion system. Plants have analyzed by IPS-MS (Perkin-Elmer Sciex ELAN 6000).

Investigations show that concentrations of elements are quite different (Table 4). Thus, for uranium deposit Mailoo-Suu it was observed high contents of Se, U, Mo, Cu, Ni, Zn in plants. The Chatkal-Kurama region and the in first place mining and metallurgical plants are characterized by heightened contents of selenium in top soil. Plants growing close to tailings impoundments and heaps as a rule are enriched by selenium. Especially accumulation of this element takes place in plants with deeply rooted. This is in the first place a Camel bur is accumulated selenium up to 216 mg/kg. Absinth growing in tailings impoundments, heaps and adjoined areas is accumulated selenium as well (from 0.1 to 1 and more mg/kg). Only in Mailoo-Suu content of selenium in absinth was reached 32 mg/kg. Element contents are depended on content of selenium in top soil and underlying rocks.

Table 3 Microcomponent content of ground waters of uranium tailings impoundments of Mailoo-Suu deposit

Point	Cr mcg/l	Mn mcg/l	Co mcg/l	Ni mcg/l	Cu mcg/l	Zn mcg/l	As mcg/l	Se mcg/l	Mo mcg/l	Sb mcg/l	Pb mcg/l	U mcg/l
M18	4.66	0.23	0.13	3.82	2.34	3.96	133.31	97.9	155.4	0.44	3.34	1938
M42	4.23	1.42	0.08	2.89	14.12	20.92	13.07	13.4	67.2	1.87	0.26	1648
M48	5.18	314.09	0.38	15.72	2.44	3.28	2.31	2.63	92.3	0.45	<0.2	131

Table 4 Microcomponent content of plants in zone of technogenic impact of mines and metallurgical plants in Mailoo-Suu

Point	Species	Cd	As	Se	Pb	Zn	Co	Ni	Cu	Mo	Mn	U
M7	Absinth	0.13	1.27	32	1.1	36	0.3	4.3	16	1.8	42	0.20
M21	Caemal bur	0.08	5.73	45	2.3	71	0.2	1.3	25	14.7	89	8.62
M33	Absinth	0.16	1.41	0.3	2.6	55	0.7	14.7	21	1.4	94	0.16
M34	Caemal bur	0.03	0.84	2	1.0	28	0.7	4.6	23	1.1	30	2.67
M47	Caemal bur	0.04	7.21	216	0.7	74	0.2	2.5	12	4.7	64	8.06
M49	Absinth	0.14	1.40	16	1.2	50	0.4	6.6	17	1.1	48	0.83
M55	Caemal bur	0.04	1.32	34	0.7	75	0.2	6.4	12	5.7	40	9.12
M1	Absinth	0.49	0.43	0.1	1.9	69	1.4	16.0	31	0.7	99	0.15

Thus, it is established that owing to migrate properties some toxic metals of mineral deposits transfer into plants and reach in some cases accumulations. For uranium tailings impoundment Mailoo-Suu the most danger are high contents of selenium, uranium, molybdenum, nickel, zinc.

Contents of heavy metals in varies human organs ad tissues studied by many researchers. Especial attention are paid to those accessible for study object as nails and hair. However contents of elements not always are connected with contents of these ones in environment. Different methods and analysis also have been made difficulties in comparison of data. Last achievements is mass-spectrometer instruments permit us to give answers on many questions of metals occurrence in human organism in connection with contents within environment.

The most important question for estimation of metals contents in human nails and hair is conception of "background level of metal", which is of region significance. The Tables 5 and 6 show average of metals in human nails and hair from mining and metallurgical plants of Ferghana valley and comparative published data. It was established that there are differences in element concentrations within biosubstrates in various deposits and some common tends to high concentrations.

In the first place siderophile group of elements attracts attention. On the background of high contents of chromium, iron it observed low contents of cobalt. Cobalt is a biological active elements and always is present in animals' organisms and in plants. Low contents of chromium, iron and cobalt cause anemia. During field works there was inquest of inhabitants about state of health. In all settlements anemia is the most widely – distributed disease.

Analysis of metals contents in different age groups shows that for children and young people (into 35–40) it is observed tends to increase of metals contents in organism in comparison with other age groups this tendency by low metals contents in organisms doesn't display clear. However increasing of concentrations is

Table 5 Microcomponent content in nails of peoples inhabitants in zone of technogenic impact of mines and metallurgical plants in Mailoo-Suu

Birthday	Cr	Co	Ni	Cu	Zn	As	Se	Mo	Pb	U
1926	2	0.09	0.6	5.8	44	<0.01	0.11	0.09	1.81	0.02
1951	57	0.04	2.4	1.7	10	<0.01	0.22	0.1	0.27	0.01
1962	82	0.05	3.7	2.4	11	<0.01	0.34	0.09	0.33	0.01
1963	87	0.17	8	3.8	18	0.14	0.22	0.59	0.99	0.04
1964	109	0.09	6.3	4.6	14	<0.01	0.52	0.27	0.84	0.01
1964	34	0.03	3.2	1.1	8	<0.01	0.14	0.05	0.17	0.01
1965	255	0.26	13	8	23	<0.01	0.89	0.5	1.5	0.04
1967	50	0.03	4.1	1.9	10	<0.01	0.2	0.07	0.55	0.01
1973	1	0.12	2.2	13.5	55	3.67	1.2	0.38	2.15	0.02
1976	85	0.08	4.7	3.3	12	<0.01	0.4	0.14	7.83	0.01
1977	20	0.77	4.1	20.7	235	419.2	1.85	0.99	43.2	0.15
1981	530	0.17	20	11.7	26	<0.01	1.14	0.44	1.59	0.02
1982	3	0.1	1	8.4	149	43.7	0.39	0.17	3.33	0.02

Table 6 Microcomponent content in hair of peoples inhabitants in zone of technogenic impact of mines and metallurgical plants in Mailoo-Suu

Birthday	Cr	Co	Ni	Cu	Zn	As	Se	Mo	Pb	U
1926	2.18	0.11	1.05	18.1	86	0	0.46	0.18	10.19	0.07
1951	6.08	0.01	0.7	0.76	9.7	0.01	0.04	0.02	0.11	<0.01
1959	13.5	0.01	0.65	0.97	8.1	<0.01	0.06	0.03	0.16	<0.01
1962	3.03	0.01	0.21	0.72	11.1	<0.01	0.06	0.01	0.02	<0.01
1963	2.26	0.01	0.13	0.46	9.9	<0.01	0.04	0.01	0.03	<0.01
1964	23.1	0.02	1.34	1.36	13	<0.01	0.12	0.03	4.84	<0.01
1964	2.99	0.01	0.12	0.62	9.99	<0.01	0.04	<0.01	0.01	<0.01
1964	8.27	0.01	0.35	0.73	7.61	<0.01	0.04	0.06	0.26	<0.01
1965	0.17	0.01	0.01	0.32	4.99	<0.01	0.03	<0.01	0.01	<0.01
1966	6.03	0.01	0.39	0.72	5.08	<0.01	0.03	0.01	0.23	<0.01
1967	0.78	0.01	0.13	0.58	6.82	<0.01	0.04	0.01	0.11	<0.01
1973	4.48	0.14	1.12	10.3	124	1.62	0.48	0.49	1.21	0.02
1976	11.1	0.02	0.63	1.18	7.9	<0.01	0.09	0.03	0.45	<0.01
1977	2.43	0.09	1.01	11.6	61.7	0.8	0.29	0.27	14.6	0.08
1981	5.9	0.01	0.41	1.25	11.91	<0.01	0.02	0.02	0.24	<0.01
1981	6.11	0.01	0.45	1.14	8.24	<0.01	0.05	0.01	0.18	<0.01
1982	0.55	0.1	1.6	44.4	99.8	0.8	0.67	0.67	52.9	0.15

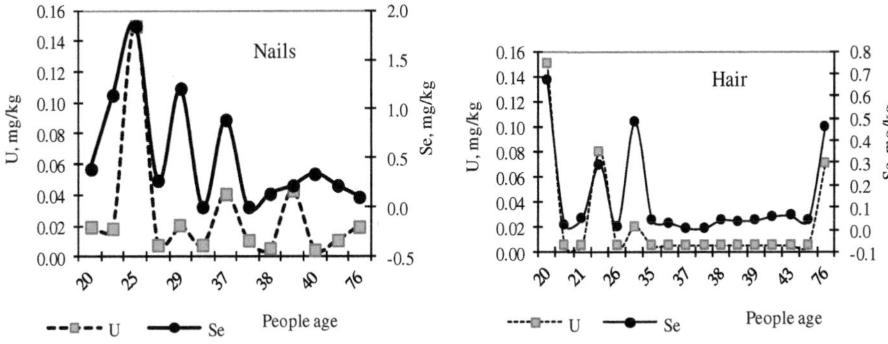

Fig. 2 Uranium and selenium contents in nails and hair of peoples inhabitants of mining areas

distinct connected with age differentiation. This dependence is shown in plots (Fig. 2). It is evident that accumulation of elements in young organism is going more intensive owing to active metabolic processes.

Conclusion

Realized geoecological investigations of transborder territories of Uzbekistan and Kyrgyzstan give the opportunity to estimate of the ecological situation in Ferghana valley at present and to determine geological risk and real danger in a case of state of emergency stipulated by destruction of wastes produced by the mining and processing plants in uranium, lead and gold deposits.

Environmental impact of mining and processing activity is studied on the basis of chemical and geochemical investigations provided valuable information on concentrations of pollutants in soils, surficial flows, groundwaters, plants and human biosubstrata (nails, hair).

Present situation and estimation of mining and industrial wastes, tailings impoundments, dams is analyzed.

Main elements-pollutants influenced on all natural units are established for every object. For uranic object Mailoo-Suu the main pollutants are Se, U, Cr, As.

References

Aitmatova J. I., Aparin V.B. (2003) Tailing dumps of radioactive wastes and their influence on environmental components on the territory of uranium mines of Mailuu-Suu and Charkesar. J Science and New technologies 3:71–83.

Uranium in Water and Sediments of the Mulde River

Stefan Bister, Torben Lüllau, Florian Koenn, Maruta Bunka, Jonny Birkhan, Beate Riebe, Rolf Michel

Abstract. Pollution of a river is determined by the contamination of the water and the corresponding sediments. This paper presents the comparison of the variation of the activity per unit mass of uranium-238 in water and sediments along the course of the Mulde River. The Mulde River drains the former Saxonian uranium mining and milling areas. The uranium activities per unit mass of the sediments show more variability, but they generally reveal the same variation along the course of the Mulde River as uranium activities of the water.

Stefan Bister
Institut für Radioökologie und Strahlenschutz, Leibniz Universität Hannover, Herrenhäuser Str. 2, 30419 Hannover, Germany

Torben Lüllau
Institut für Radioökologie und Strahlenschutz, Leibniz Universität Hannover, Herrenhäuser Str. 2, 30419 Hannover, Germany

Florian Koenn
Fachbereich Chemie und Biotechnologie, Fachhochschule Aachen, Campus Jülich, Ginsterweg 1, 52428 Jülich, Germany

Maruta Bunka
Institut für Radioökologie und Strahlenschutz, Leibniz Universität Hannover, Herrenhäuser Str. 2, 30419 Hannover, Germany

Jonny Birkhan
Institut für Radioökologie und Strahlenschutz, Leibniz Universität Hannover, Herrenhäuser Str. 2, 30419 Hannover, Germany

Beate Riebe
Institut für Radioökologie und Strahlenschutz, Leibniz Universität Hannover, Herrenhäuser Str. 2, 30419 Hannover, Germany

Rolf Michel
Institut für Radioökologie und Strahlenschutz, Leibniz Universität Hannover, Herrenhäuser Str. 2, 30419 Hannover, Germany

B. Merkel, M. Schipek, *The New Uranium Mining Boom*
DOI 10.1007/978-3-642-22122-4 © Springer 2011

Introduction

At the time of the Warsaw Pact, the former German Democratic Republic (GDR) was the third largest producer of uranium in the world and the most important supplier of uranium for the USSR. The Zwickauer Mulde River in Saxony and the Weiße Elster River in Thuringia are the most important river systems draining the uranium mining and milling dominated areas of the western Ore Mountains and its foreland, resulting in accordingly high heavy metal loads. Thus they are of particular interest for radiation protection and radioecology. Today this area is subject to a remediation project which, due to its large scale, can be characterized as pioneer work.

The Mulde River is a left side tributary of the Elbe River and mainly situated in Saxony. The river system consists of three main rivers: Zwickauer Mulde, Freiberger Mulde and Vereinigte Mulde. The Muldenberg Reservoir in the western Ore Mountains is deemed to be the source of the Zwickauer Mulde River. Eight kilometers downstream of the dam, which is used as a drinking water reservoir, the river already receives contaminated water from the former mine Schneckenstein. On its way through the Ore Mountains the river flows through the uranium mining area Aue and Bad Schlema. Downriver of Zwickau the Zwickauer Mulde River passes Crossen and its former uranium ore processing plants and merges with the Freiberger Mulde River at river kilometer 163. The source of the Freiberger Mulde River is located in the Czech Republic, about 5 km from the German border on the ridge of the eastern Ore Mountains. It drains mining areas, however there is no uranium mining in its catchments area. The Vereinigte Mulde River, which is often simply called Mulde River, originates from the merging of the two frontal flows Freiberger Mulde and Zwickauer Mulde and flows into the Elbe River north of Dessau at river kilometer 307. At high water-levels the Mulde River is deemed to be the fastest flowing river in Central Europe. Due to its origin in a high-mineral and high-ore affected area it is the most significant reason for heavy metal entering into the Elbe River and thus into the North Sea.

This research project was established to quantify the long-term effect of the former uranium mining and milling activities by investigating the uranium content of the water and sediment of the Mulde River. It is part of a work package dealing with transport and availability of uranium and its decay products in the Mulde floodplains, which in turn is part of a joint project on radionuclides in the environment and their transport to man via food chains, supported by the German Federal Ministry for Education and Research (BMBF).

The condition of a river is defined by its water and sediments. The interaction between these compartments is characterized by a continuous mass transfer, caused by solution- and precipitation reactions, by desorption and adsorption as well as by suspended particles. This particularly applies to recent (fresh) surface sediments, which are in close contact with the flowing water, and thus primarily are affected by different exchange processes. Although surface sediments are subject to continuous dislocation by flowing water, they can be regarded as quasi-stationary in comparison to the river water. The river acts like a chromatographic system, in which the water body performs the task of the mobile phase, and the sediment forms the stationary phase. As the river water reacts more rapidly to changes than the river sediment, the water rather reflects the present condition of the river, while the surface sediments give information about changes in the more recent past. Furthermore, undisturbed stratified subsurface sediments can serve as environmental archives.

Sampling and Analysis

A total of 26 water samples and more than 36 sediment samples were collected in April, May and October 2008. 19 water samples and 29 sediment samples were collected along the course of the Zwickauer Mulde River and the Vereinigte Mulde River, respectively. Water and sediment from the Freiberger Mulde River, as well as from a small influent of the Zwickauer Mulde River were used as reference samples. Only water samples and the corresponding sediment samples were considered. Additionally, two water samples from the headwaters of the Zwickauer Mulde River were analyzed. Figure 1 shows the Mulde River system and the sampling locations.

Generally, four liters of water were collected at each sampling location. Uranium was concentrated by coprecipitation with ferric hydroxide and separated by solid-phase chromatography using UTEVA (purchased from Eichrom). Following

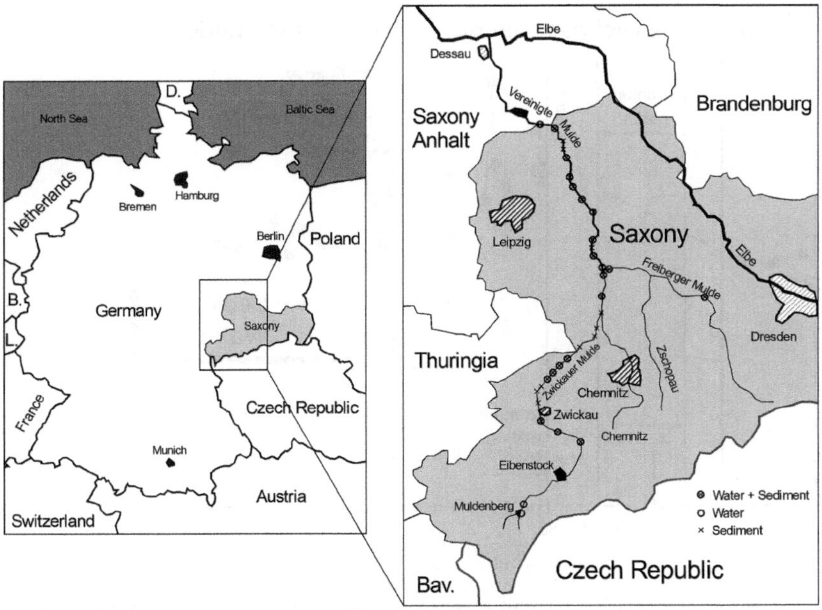

Fig. 1 The Mulde River system with the sampling locations

this procedure the sample was transferred to a stainless steel planchet via electro-deposition and measured by Alpha- spectrometry. A description of this method is given in detail in a previous paper [1]. Sampling and preparation of the sediment samples followed the measuring instructions of the German Federal Ministry for the Environment, Nature Conservation and Nuclear Safety (BMU) [3]. Sampling of surface sediment was performed using a dipper. Two liters of wet sediment were taken from the river bed, preferably in the middle of the river. The samples were dried, ground, passed through a 2 mm sieve, and measured using Gamma-spectrometry [2].

Results and Discussion

Table 1 summarizes the measured activities per unit mass of uranium-238. Figure 2 shows the variation of the mass-related activity values of uranium-238 in water and sediment plotted against the chainage of the Zwickauer and Vereinigte Mulde River. There is no official chainage for the Mulde River system. Thus, the dam of the Muldenberg Reservoir (Gauss–Krüger coordinates: 4528788, 5586248) was defined as the start chainage. The uncertainties are given as combined A and B type standard uncertainty of the measurement according to GUM.

Results reveal that the mass-related activity level of the sediment samples is more than 1000 times higher compared to the corresponding water values. The earth's primary uranium deposits consist of virtually insoluble uranium oxides

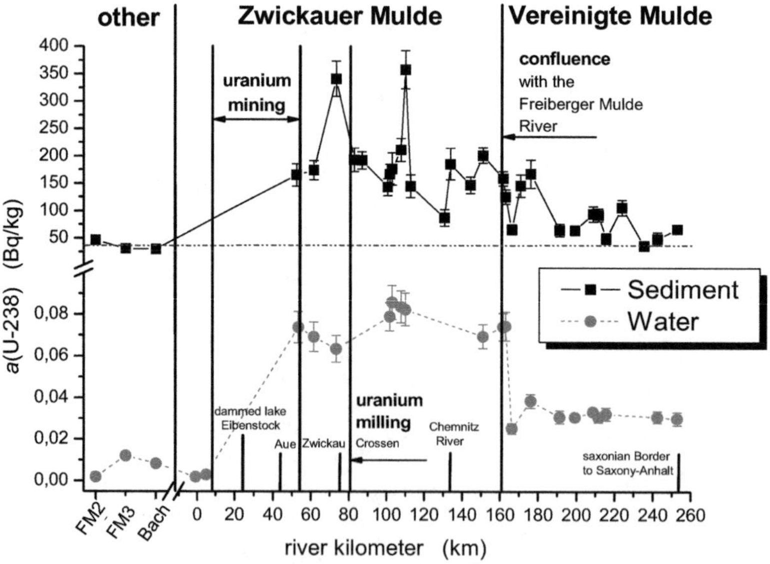

Fig. 2 Variation of the activities per unit mass of uranium-238 in water and sediment of the Mulde River plotted against river kilometers

(UO_2 up to U_3O_8), which have to be oxidized to the uranyl cation (UO_2^{2+}) in atmospheric oxygen to become soluble. The uranyl cation is indeed well soluble, but it is subjected to adsorption and coprecipitation processes. This behavior favors a shift to the solid phase (sediment), and subsequently enhances particulate transport of uranium.

Table 1 Activities per unit mass (*a*) and related uncertainties (u(*a*)) for uranium-238 in water and sediments of the Mulde River

river	river kilometer (km)	Sediment U-238 (Bq/kg)		Water U-238 (mBq/kg)	
		a	u(*a*)	*a*	u(*a*)
other	FM2	46,4	7,5	2,0	0,2
	FM3	31,2	7,0	12,2	1,2
	Bach	29,8	4,9	8,4	1,0
Zwickauer Mulde River	-1			2,0	0,2
	5			3,0	0,4
	52	165,6	20,4	73,9	7,5
	62	174,0	17,5	69,3	7,1
	73	341,2	32,1	63,5	6,4
	83	193,1	21,5		
	87	192,2	15,9		
	101	143,5	15,3		
	102	167,2	17,9	79,0	6,8
	103	176,4	29,5	85,9	8,1
	108	211,5	20,9	83,7	7,7
	110	358,2	34,8	82,4	7,8
	113	145,2	20,6		
	131	87,8	15,2		
	134	185,7	28,8		
	144	147,6	15,3		
	151	201,3	14,1	69,5	5,8
	162	159,4	13,2	74,0	6,6
Vereinigte Mulde River	163	125,8	13,1	74,4	6,9
	166	66,2	9,3	25,2	2,6
	171	146,0	20,3		
	176	167,9	25,5	38,6	3,3
	191	65,2	11,4	30,9	3,1
	199	65,1	8,8	30,8	1,4
	209	94,7	13,6	33,4	1,5
	212	93,2	11,6	30,8	2,6
	216	50,1	9,8	32,4	3,1
	224	106,4	14,1		
	235	36,8	8,9		
	242	49,7	11,6	30,9	2,9
	253	67,0	7,7	30,1	3,2

Furthermore, the measured values for the activity per unit mass in water shows less variability than the values determined for the sediment samples. This finding was due, as flowing water is considerably more affected by mixing and homogenizing processes than the 'quasistatical' sediments. Sedimentation conditions can vary on a small scale, and accordingly the measured activities exhibited a higher level of variability. The bond of heavy metals to sediments depending on the size of the particles, is called particle size effect. Adsorption reactions are surface reactions, and the decrease of the particle size results in a considerable increase of the reactive surface. This interrelationship is employed, e.g., when using active carbon as filter material. Investigations within the scope of a monitoring program on sediments and alluvial soils [4] showed that uranium accumulates in sediments with a high amount of fine particles. The enrichment of uranium-238 in fine-grained sediments (silty sand) was about three times higher regarding the arithmetic mean as compared to sediments from coarse material (sandy gravel). The particle size of the deposited sediments depends, in turn, on the flow velocity of the river. This means that the uranium enriched fine-grained sediments preferably settle in river reaches of low flow velocity. Generally, river bank areas are places of low flow velocity with the plant cover causing additional deceleration. In this connection, it has to be considered that sediments often contain a certain amount of organic matter, which on one hand 'dilutes' the sediment and so reduces the activity per unit mass. But on the other hand, decomposition of organic matter is an oxygen-consuming process, which promotes the reductive precipitation of uranium, and thus results in an increase of the activity per unit mass. River bank areas were only sampled, if no other sampling location was available. Furthermore, sediments can be 'diluted' by mixing with material introduced by less contaminated influents, as well as by erosion of farmland soils caused by flood events.

In Fig. 2, the first three samples of sediment and the first five samples of water, respectively, are used as reference samples as they are unaffected by uranium mining activities, and thus represent the natural background content of uranium in this area. Samples FM2 and FM3 originate from the Freiberger Mulde River. Sample FM2 was taken near the town of Nossen from the middle reaches of the Freiberger Mulde River (Gauss–Krüger coordinates: 4593475, 5655187). FM3 was sampled close to the small village of Sermuth about 1.5 km in front of the confluence of the Freiberger Mulde River with the Zwickauer Mulde River (Gauss–Krüger coordinates: 4556998, 5669149). The third sample (Bach) originates from a small influent of the Zwickauer Mulde River at river kilometer 110, which drains the farmland on the flood plains. The sampling was carried out close to the inflow into the Zwickauer Mulde River (Gauss–Krüger coordinates: 4544984, 5640405). The next two water samples (−1 km and 5 km) originate from the uncontaminated headwaters of the Zwickauer Mulde River. The first one was collected from the Muldenberg Reservoir, which is used as a drinking water reservoir. Downstream of river kilometer 8, the Zwickauer Mulde River receives contaminated water for the first time from an influent coming from the former mine Schneckenstein. Using these reference values results in an arithmetic mean of 5.5 mBq/kg for the natural uranium background in the water. For sediments, an arithmetic mean of 35.8 Bq/kg for the natural uranium background was obtained, which is inserted in Fig. 2 as an auxiliary line.

The uranium activity of the two environmental mediums increases significantly after the Mulde River has passed the uranium mining area. The values of the activity per unit mass in the water reveal a slight decrease as the river approaches the town of Zwickau. This effect can be explained by dilution with less contaminated water from tributaries. Downstream of Zwickau the activity per unit mass of uranium in the water increases again. This increase results from the effluents of the former uranium milling industry site near Crossen. In the further course of the Zwickauer Mulde River, the activity remains approximately constant up to the confluence with the Freiberger Mulde River. Due to measurement uncertainties, the values are in agreement with a slight, but not significant decrease of the activity per unit mass corresponding to dilution caused by tributaries, especially by the Chemnitz River, which is the largest tributary of the Zwickauer Mulde River. Due to the strong deviations, this trend could not be observed for the sediment samples. With three exceptions, the uranium activities remain largely constant along the course of the Zwickauer Mulde River. The strong increase of the uranium activity at kilometer 73 can by explained by the course of the river. Here, the Zwickauer Mulde River leaves the mountain range of the Ore Mountains, which comprises a significant decrease in slope and flow velocity. Thus, a huge proportion of the particles carried away from the Ore Mountains is deposited in this reach. This is reflected by the high uranium activity here. At river kilometer 110 the Mulde River is calm and deep, and flows very slowly. Very fine-grained sediment was collected near the river bank, which could explain the increase of the activity per unit mass. In contrast, the sample collected at river kilometer 131 exhibits a relatively low uranium activity. Here, the river is also quite calm. The river bank being part of a pasture shows impact of cattle treading. Soil erosion of the bank caused by grazing lifestock may be responsible for a ‘dilution’ of the river sediment, and subsequently for a decrease in measured activity value.

A considerable decrease in the activity per unit mass of uranium-238 in the water becomes evident as a result of the confluence of the Zwickauer Mulde River and the uncontaminated Freiberger Mulde River due to the strong dilution effect. The variation of the first three measured values reflects inhomogeneous and slow mixing of waters of the two rivers. Samples were collected at both sides of the Vereinigte Mulde River. The average uranium activity decreases from 75.5 mBq/kg in the Zwickauer Mulde River (arithmetic mean from km 52 to km 163) to 31.5 mBq/kg in the Vereinigte Mulde River (arithmetic mean from km 166 to km 253). There is a similar situation for the sediment samples, although the decrease of the uranium activity is considerably less pronounced. Nevertheless, it is clearly discernable. The average activity per unit mass of uranium-238 decreases from 186.8 Bq/kg in the Zwickauer Mulde River (arithmetic mean from km 52 to km 163) to 126.0 Bq/kg in the Vereinigte Mulde River (arithmetic mean from km 166 to km 253). The mixing area extending to kilometer 191 is represented by five samples. The effect of the sampling on different sides of the river is clearly illustrated by the analogue behavior of water and sediment sampled in parallel.

From river kilometer 191 on, the uranium activity of water and sediment remains nearly constant along the course of the Vereinigte Mulde River, even though the values of the sediment samples show high variability. The uranium activity of the water of the Vereinigte Mulde River is still considerably higher than the activity of uncontaminated samples, whereas the uranium activities of some of the sediment samples reach the level of the uncontaminated samples.

Conclusions

As expected, the activities per unit mass of uranium-238 in sediment samples are much higher than in water samples. The variability of the uranium activity along the course of the river is more pronounced for the sediments. This reflects the strong dependence on local sedimentation conditions. In general, the activities per unit mass of uranium-238 of water and sediment show the same variation along the Mulde River. A strong increase of the uranium activities can be observed downstream of the former uranium mining area, as compared with natural background values, represented by samples from the Freiberger Mulde River, by a small influent of the Zwickauer Mulde River, and by water from the headwaters of the Zwickauer Mulde River, respectively. The arithmetic means of the activity per unit mass of the water increase from 5.5 to 75.5 mBq/kg, while the arithmetic means of the activity per unit mass of the sediments increase from 35.8 to 186.8 Bq/kg in the course of the Zwickauer Mulde River. After the confluence with the uncontaminated Freiberger Mulde River, a decrease of the activities per unit mass of uranium-238 in water and sediment can be observed. The arithmetic mean of the activity per unit mass of the water decreases from 75.5 mBq/kg in the Zwickauer Mulde River to 31.5 mBq/kg in the Vereinigte Mulde River, and the arithmetic mean of the activity per unit mass of the sediment decreases from 186.8 to 126.0 Bq/kg, respectively. The water of the two rivers needs about 30 km for thoroughly mixing.

Acknowledgment This research project is supported by the German Federal Ministry for Education and Research (BMBF) under contract number 02NUK002D.

References

1. Bister S., Koenn F., Bunka M., Birkhan J., Lüllau T., Riebe B., Michel R. (2010) Uranium in the water of the Mulde River. J Radioanal Nucl Chem 286: 367–372
2. Lüllau T. (2009) Gammaspektrometrische Untersuchungen von anthropogen beeinflussten Umweltproben. diploma thesis, download: http://www.zsr.uni-hannover.de/arbeiten/diplluel.pdf
3. Mundschenk H. (2006) Meßanleitung für die Überwachung der Radioaktivität in Sedimenten. Messanleitungen für die Überwachung der Radioaktivität in der Umwelt und zur Erfassung radioaktiver Emissionen aus kerntechnischen Anlagen, German Federal Environment Ministry (BMU)
4. Gesellschaft für Anlagen und Reaktorsicherheit mbH (GRS) and Beak Consultants GmbH (1996) Radionuklidbelastung von Sedimenten und Auenböden – Datenerfassung, Erstauswertung, Ergebnisdarstellung, Freiberg

Assessment of Background Uranium Concentration in Groundwater Around a Proposed Mining Area

K. Brindha, L. Elango, R.N. Nair

Abstract. This study was carried out in and around a proposed mining area (Lambapur and Peddagattu) located in Andhra Pradesh, India with the aim of assessing the spatio-temporal variation and the impact of rainfall on uranium concentration in groundwater. The concentration of uranium in groundwater varied from 0.2 to 77.4 ppb. There was significant increase in uranium concentration in groundwater after rainfall which was due to the leaching of uranium ions present in the unsaturated zone along with rainwater. The reason for the presence of uranium in groundwater of this region is highly localized that depend on the lithology inherent to this region.

Introduction

In most parts of India, groundwater forms the only reliable source of drinking water. Groundwater contains a number of dissolved constituents as a result of chemical and biochemical interaction between groundwater and the geological materials inherent to the region. Also contribution from atmosphere and surface water bodies may play a role in deciding the chemical composition of groundwater in a region. Increase in the concentration of several ions in groundwater such as fluoride, nitrate, arsenic, total hardness and some toxic metal ions have been no-

---

K. Brindha
Department of Geology, Anna University, Chennai-600 025, India

L. Elango
Department of Geology, Anna University, Chennai-600 025, India

R.N. Nair
Environmental Assessment Division, Bhabha Atomic Research Centre, Mumbai-400 085, India

B. Merkel, M. Schipek, *The New Uranium Mining Boom*
DOI 10.1007/978-3-642-22122-4 © Springer 2011

ticed in groundwater in several parts of India. The presence of high concentration of certain ions in groundwater may also be due to the presence of huge quantity of ore deposits of a particular element.

Due to the increasing demand for radioactive minerals, exploration for radioactive ores is being carried out in several parts of the world. Uranium is one of the radioactive mineral which is of great importance. After mining and milling of uranium bearing rocks, the waste called as uranium tailings are stored by ponding on ground surface. Contamination of groundwater due to uranium around mines has been studied worldwide (Gómez et al., 2006; Neves and Matias, 2008; Leijnse et al., 2001).

In India uranium mining is being carried out in Jaduguda and Turamdih. Uranium reserves have been found in Lambapur, Peddagattu and Thumnapalli of Andhra Pradesh; Western Kasi Hills district of Meghalaya and Gogi of Karnataka. It is planned to carry out mining of uraninite from Lambapur and Peddagattu area of Andhra Pradesh, India in near future. In view of this, it is important that the groundwater quality be assessed before the commencement of mining in order to indentify if there is an impact of mining activity on groundwater quality in future. Preliminary investigation has been carried out on uranium concentration in groundwater by Brindha et al (2009). The objective of this study is to assess the influence of rainfall on the spatial and temporal variation in uranium concentration of groundwater in the proposed mining area of Lambapur and Peddagattu, Andhra Pradesh, southern India.

Study Area

The study area is located at a distance of 135 km ESE of Hyderabad and forms a part of Nalgonda district, Andhra Pradesh, India (Fig. 1). It covers the proposed Lambapur and Peddagattu uranium mining areas and Seripalli tailings pond area. The area was demarcated with well defined watershed boundary covering an area of about 750 sq km. The southeastern side of the study area is surrounded by the Nagarjuna Sagar reservoir and the southern side of the area is bounded by Pedda Vagu river. The northern side is bounded by Gudipalli Vagu river. This area experiences arid to semi-arid climate. The study area goes through hot climate during the summer (April–June) with a temperature ranging from 30° to 46.5°C and in winter (November–January) it varies between 17 and 38°C. The average annual rainfall in this area is about 1000 mm falling mostly during the southwest monsoon (June–September). There are four rain guage stations in this area. The recorded rainfall in two of these stations is shown in Fig. 2. The surface runoff has resulted into the development of dentritic to subdentritic drainage pattern in this area. The elevation of the study area varies from a maximum of 348 m in the northwestern side to a minimum of 169 m in northeastern side. Agriculture is the main land use in this area. Paddy, pulses and cotton are widely grown.

Geology

Granitic rock forms the basement of this region, which is traversed by numerous dolerite dykes and quartz veins (Fig. 1). Most part of the investigated area has exposures of granitic rocks belonging to late Archaen. Medium to coarse grained granites are present in this area. The Srisailam formation, the youngest member of the Cuddapah supergroup uncomformably overlies the basement granite with a distinct unconformity. The Srisailam metamorphic formation is exposed in the southeastern part of the study area. The sediments of Srisailam formation are mainly arenaceous and include pebbly-gritty quartzite shale with dolomitic limestone, intercalated sequence of shale-quartzite and massive quartzite. The litho units of this formation are dipping at an angle ranging from 3 to 5° towards SE. The generalized stratigraphic sequence of this area is given in Table 1 (after GSI 1995). The uranium deposit occurs adjacent to the unconformity between basement granite and the overlying Proterozoic Srisailam quartzite in the north western margin of the Cuddapah basin.

Groundwater occurs in unconfined condition in this area mostly in the weathered rocks. Groundwater head fluctuates between 0 and 15 m bgl. Groundwater is used for various purposes from these bore wells and dug wells. The dug wells of this area are shallow with depth ranging from 1.5 to 20 m and their diameter is from 2 to 5 m. The bore wells are more than 10 m deep and are generally of 15 cm diameter.

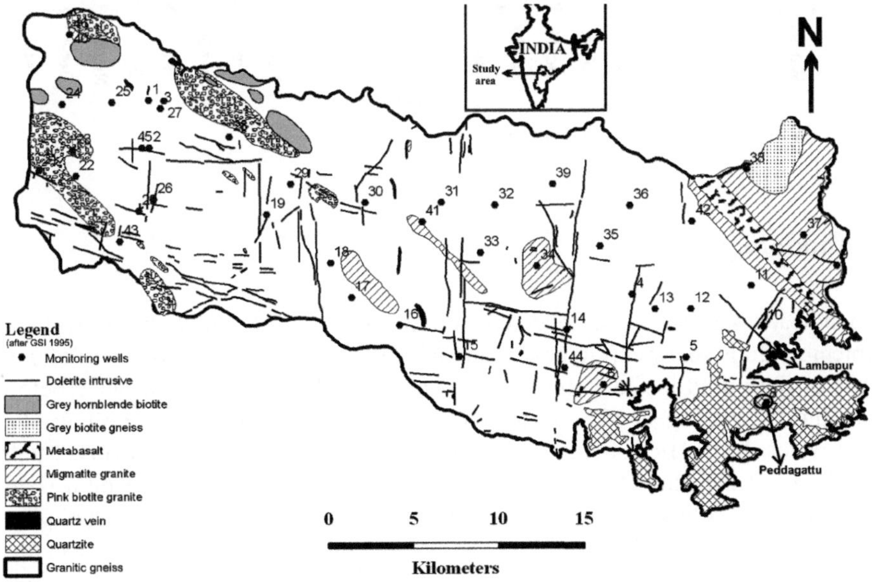

Fig. 1 Geology of the study area with monitoring wells

Table 1 General stratigraphic sequence (after GSI 1995)

Cuddapah supergroup Srisailam formation	Massive quartzite Upper shale Quartzite with shale intercalation Lower shale with limestone intercalation Pebbly and gritty quartzite/arenite Uranium mineralized region
-----------------------------Unconformity----------	(**Uranium mineralized region**)-----------
Late Archean/Lower Proterozoic	Granite/granitic gneiss intrusion of basic dykes and vein quartz

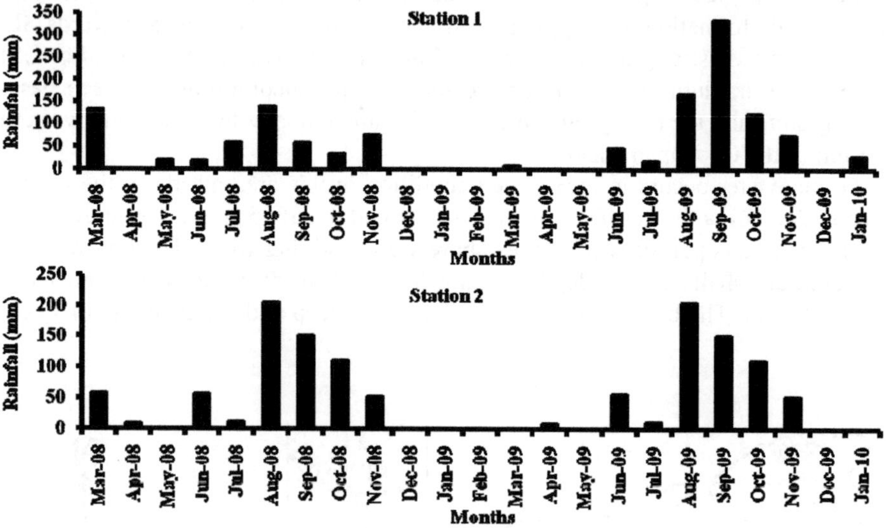

Fig. 2 Rainfall (mm) at different stations

Methodology

Groundwater samples were collected from 36 wells in July 2009 and 44 wells in September 2009 representing before and after rainfall. Polythene bottles of 500 ml capacity were used for collecting groundwater samples. They were cleaned before sampling by soaking in 10% nitric acid for a day followed by washing with distilled water. In the field before collecting the sample, the bottle was rinsed with the filtrate of the sample. Groundwater level was measured in-situ using a water level recorder (Solinst 101). After collection, the samples were kept in a cool place away from sunlight. The concentration of uranium in the groundwater samples was measured using laser fluorimeter at Health Physics Unit, Nuclear Fuel Complex, Hyderabad, India. Blanks and standards were run at regular intervals during the analysis.

Results and Discussion

Uranium concentration in groundwater ranged from 0.2 to 77.4 ppb. Statistical summary of uranium concentration before and after rainfall is given in Table 2.

As groundwater is mostly used for drinking in this area without proper treatment, it is essential to monitor its suitability based on uranium concentration. Maximum permissible limit for uranium in drinking water is 30 ppb as per USEPA (2003) standards. Of the 80 groundwater samples, 19% were above 30 ppb thus not fit for drinking.

Table 2 Uranium concentration in groundwater

Month	Number of samples	Min (ppb)	Max (ppb)	Mean (ppb)	% of samples above 30 ppb of uranium
July 2009	36	0.2	65.1	12	14
September 2009	44	1.4	77.4	19	23
Total	80	0.2	77.4	15.8	19

Temporal Variation

The temporal variation in groundwater level and uranium concentration was assessed (Fig. 3). Obviously there was increase in groundwater level (m bgl) in all the dug wells after rainfall. The bore wells were not considered as their exact level could not be measured. The fall or rise in uranium concentration in groundwater after rainfall is also shown in Fig. 3. This shows that uranium concentration in groundwater has increased significantly in most of the wells before and after rainfall. This is because the uranium in the weathered zone percolates with the rainwater and reaches the groundwater table thus increasing uranium concentration after rains. Also the number of groundwater samples exceeding 30 ppb has increased from July 2009 to September 2009 (Table 2).

Spatial Variation

Figure 4 shows the spatial variation in uranium concentration in groundwater. Uranium concentration in groundwater is higher in the northern and south eastern part of the study area. It is comparatively less and within limits in the proposed mining area. This may be due to difference in the intensity of weathering of uranium rich rocks in this area and also the wells in these locations penetrate only the Cuddapah formation.

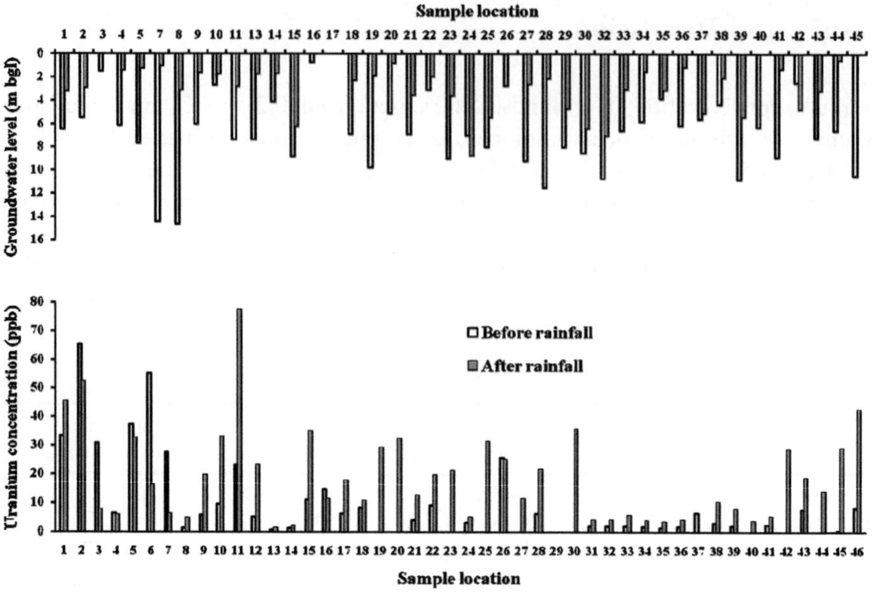

Fig. 3 Comparison of groundwater level (m bgl) and uranium concentration in groundwater (ppb) in monitoring wells before and after rainfall

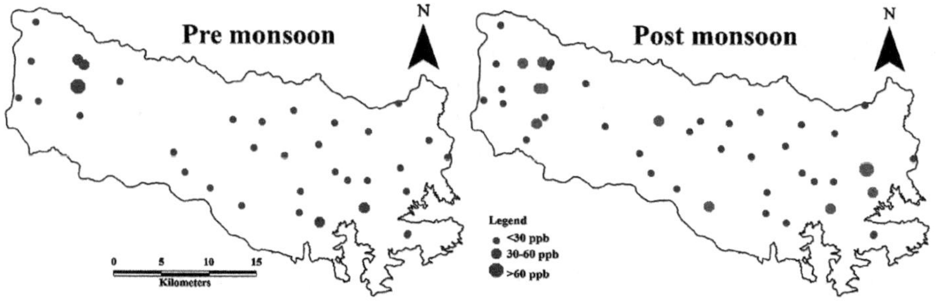

Fig. 4 Spatial variation in uranium concentration (ppb) in groundwater

Source

Previous studies have shown that the granitic rocks of this region contain higher concentration of uranium. Shrivastava et al. (1992) reported that the granitic rocks of Nalgonda district contain uranium in the range of 10.2 to 116 ppm with an average of 35 ppm. The weathering of this granitic rocks and rock water interaction are the reason behind the increased uranium concentration in groundwater in parts of this area. Sinha et al. (1995) reported that uraninite, pitchblende, kasolite and uranophane are the main uranium minerals present in the deposits of Lambapur and Peddagattu. Lambapur mineralisation occurs in the surface or near surface whereas it occurs below 50 m in Peddagattu (Sinha 1996). This may be the reason for high uranium in groundwater in the south eastern part where the mineral deposits are present.

Conclusion

The spatial and temporal variation of uranium concentration in groundwater in a proposed uranium mining area in southern India was assessed. Groundwater had uranium concentration ranging between 0.2 and 77.4 ppb. Uranium concentration was above USEPA maximum permissible limit for drinking (30 ppb) in 19% of the wells. The concentration of uranium in groundwater increased after rainfall depicting the leaching of uranium salts from the unsaturated zone during rainfall recharge. Spatially uranium concentration was high in the northern and south eastern parts of the area. This is because of the presence of uranium rich granitic rocks spread throughout this area. Depending on the intensity of weathering, the concentration of uranium in groundwater varies. The concentration of uranium is comparatively less in the proposed mining area. It is necessary to decrease the uranium concentration in groundwater as it being used for drinking purposes by adopting remedial measures such as rainwater harvesting.

Acknowledgement We would like to thank the Board of Research in Nuclear Sciences, Department of Atomic Energy, Government of India for the financial support (Grant No. 2007/36/35). First two authors thank Dr. G.K. Shrivastava, Health Physics Unit, NFC, Hyderabad for providing the analytical facilities. First two authors also thank the DST-FIST (Department of Science and Technology-Funds for Improvement in Science and Technology) (Grant No. SR/FST/ESI-106/2010) and UGC-SAP (University Grants Commission-Special Assistance Programme) (Grant No. UGC GRS II F.550/10/DRS/2007(SAP-1)) for their financial support which helped in creating facilities to carry out part of this work.

References

Brindha K, Murugan R, Rajesh R, Elango L, Prasad KV, Shrivastava GK, Manikandan ST (2009) Status of uranium occurrence in groundwater of Devarkonda and Peddagattu region of Nalgonda district, Andhra Pradesh, India. 8th IAHS Scientific Assembly and Joint IAHS & IAH International Convention "Water: A Vital Resource Under Stress – How Science Can Help", Hyderabad, India.

Gómez P, Garralón A, Buil B, Turrero MJ, Sánchez, de la Cruz B (2006) Modelling of geochemical processes related to uranium mobilization in the groundwater of a uranium mine. Science of the Total Environment 366: 295–309.

GSI (Geological Survey of India) (1995) Geology and minerals map of Nalgonda district Andhra Pradesh, India.

Leijnse A, van de Weerd H, Hassanizadeh SM (2001) Modelling uranium transport in Koongarra, Australia: the effect of a moving weathering zone. Mathematical Geology 33 (1); 1–29.

Neves O, Matias MJ (2008) Assessment of groundwater quality and contamination problems ascribed to an abandoned uranium mine (Cunha Baixa region, Central Portugal). Environmental Geology 53: 1799–1810.

Shrivastava VK, Parthasarathy TN, Sinha KK (1992) Geochemical study of uraniferous granites from Lambapur area Nalgonda District Andhra Pradesh, India. Exploration and Research for Atomic Minerals 5: 41–52.

Sinha RM, Shrivastava VK, Sarma GVG and Parthasarathy TN (1995) Geological favourability for unconformity-related Uranium deposits in northern parts of the Cuddapah Basin: evidences from Lambapur Uranium occurrences, Andhra Pradesh, India. Exploration and Research for Atomic Minerals 8: 111–126.

Sinha RM (1996) Lambapur uranium deposit and its significance on the uranium potential of Srisailam sub-basin, Cuddapah basin, Andhra Pradesh Exploration for atomic minerals, Part III, AMD Training Course Handbook, 3: 26–42.

USEPA (2003) United States Environmental Protection Agency current drinking water standards Ground water and drinking water protection agency: Report prepared by Wade Miller Associates 1–12.

Geochemistry of Radionuclides in Groundwaters at the Former Uranium and Radium Mining Region of Sabugal, Portugal

Fernando P. Carvalho, João M. Oliveira, W. Eberhard Falck

Abstract. Recent concerns over potential radiological hazards posed by the uranium mining and milling wastes have led to assess the radioactivity in groundwaters of County of Sabugal, central region of Portugal. Waters from the Bica Mine were the highest concentrations measured in waters from this region. Water from irrigation wells generally showed low concentrations with the exception of several wells near the Bica Mine. Waters from public drinking water supplies in the villages and towns contained ^{238}U, ^{226}Ra, ^{230}Th, ^{210}Po and ^{232}Th in concentrations below values recommended for drinking water. Only one local spring exceeded the limit for alpha radioactivity. Overall water resources were not significantly contaminated by the historic uranium mining activities. Nevertheless, mine waters from Bica Mine still require treatment to prevent dispersal of acid mine drainage and radionuclides into aquifers. Other mines nearby, with near neutral pH and anoxic waters show low dissolved concentrations of uranium.

Introduction

The uranium resources of Sabugal in the Beira–Baixa province of Portugal have been known since the early 20th century. These deposits occur as secondary miner-

---

Fernando P. Carvalho
Nuclear and Technological Institute, Department of Radiological Protection and Nuclear Safety, E.N. 10, 2686-953 Sacavém, Portugal
E-mail: carvalho@itn.pt

João M. Oliveira
Nuclear and Technological Institute, Department of Radiological Protection and Nuclear Safety, E.N. 10, 2686-953 Sacavém, Portugal

W. Eberhard Falck
Université de Versailles St. Quentin-en-Yvelines, Laboratoire REEDS, Rambouillet, France

B. Merkel, M. Schipek, *The New Uranium Mining Boom*
DOI 10.1007/978-3-642-22122-4

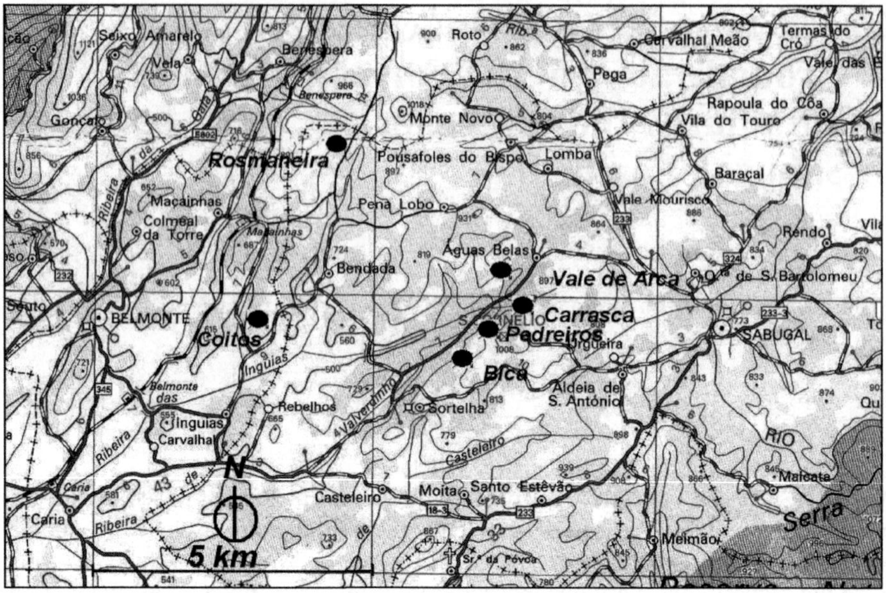

Fig. 1 Old radium-uranium mines (*solid circles*) in the Sabugal region, Portugal

alization of uranium (gummite, sabugalite etc.) formed as vein deposits in granites. The Rosmaneira mine was exploited as early as 1912 for radium, whereas at the others uranium production only began after 1945. The main uranium mine was Bica mine that operated until the late 1980s as an underground mine (Fig. 1). In the last years of production also sulfuric acid was injected into the rock for in situ extraction of low grade uranium ores. After its closure, the mine was let to flood naturally, which led to the generation of acid mine drainage (AMD). The AMD is now pumped to the surface, neutralized and Ra is co-precipitated by $BaCl_2$ addition. The water level in the underground mine fluctuates with the seasonal rainfall, and with water pumping for neutralization, which has caused AMD to migrate into the surrounding aquifers. The AMD also promotes dissolution of U-series elements. Now, nearly 20 years after the closure of the Bica mine, the resulting radioactivity in nearby groundwaters and in drinking water supplies was assessed.

Materials and Methods

Groundwater samples were collected from wells in the vicinity of the six uranium mines in the Sabugal-Belmonte region, namely the mines of Bica, Pedreiros, Carrasca, Vale de Arca, Rosmaneira and Coitos (Fig. 1), as direct sampling of mine water was possible only in the Bica and Coitos mines. Drinking waters from public water supplies in towns and villages relying on local groundwaters, were sampled at the tap (Fig. 1). Sampling and sample preparation procedures are described elsewhere (Carvalho and Oliveira, 2009; Carvalho et al., 2009a, b). Radionuclide analyses were performed on water samples, suspended matter, soils, and mining waste materials by wet radiochemistry and alpha spectrometry (Oliveira and Carvalho, 2007). In order to get a better understanding of the uranium migration behavior, some simple modeling runs using PHREEQC (Parkhurst and Appelo, 1999) were performed.

Results and Discussion

Water from the Bica mine, collected from the underground mine gallery, showed very a low pH-value, of 3.05 (Table 1), an effect of the acid *in situ* leaching that together with oxic conditions promotes further U dissolution. Th, Ra and Po also present in the ore are also partially dissolved at low pH-values, which facilitate their migration into adjacent aquifers (Table 2). In other mines no acid *in situ* leaching was employed and the slight acidity observed is due to pyrite oxidation. Figure 2 shows the calculated uranium species distribution for a number of selected samples (cf. Tables 1–3).

Radionuclide mobilization was clearly observed in the irrigation wells at farms in the valley below the Bica mine to a distance of about 1 km. Mobilization was much lower at the other uranium mines, which all had a higher pH. At the Carrasca mine, with a pH of 6.45 and a low Eh (–95 mV) waters contained little U, consistent with the lower solubility of U(IV). Conversely, a much higher Po concentration than in other mines was measured in these waters (Tables 1 and 2).

For the calculation of inorganic uranium species distributions for selected samples equilibrium with atmospheric CO_2 was assumed, while maintaining the measured pH values. It is not known, whether any organic ligands, such as humic or fulvic acids, were present. The dominating species are the uranyl ion and uranyl-hydroxo species, with the exception of samples 20.4 and 21.4 that seem to have elevated phosphate concentrations derived from agriculture (Fig. 2). Most of the species are uncharged, which tends to promote their migration.

The particulate phase accounts for a significant fraction of total radioactivity in unfiltered water (Tables 2 and 3). In well-water the average fractions were 29% (range 0.2–66%) for U, 32% (0.8–67%) for ^{226}Ra, 54% (12–92%) for ^{210}Pb, and 71% (35–91%) for ^{210}Po. Thus, U and Ra were mostly dissolved. There is a clear trend that with increasing pH more U is bound onto particles (Fig. 3). The increasing predominance of hydro-species as pH-values become higher (Fig. 2) might indicate some colloid formation. The aqueous chemistry of Th is complicated by the fact that it forms poly-nuclear complexes and colloids, which is not sufficiently reflected in the currently available thermodynamic databases.

Table 1 Water samples collected in the Sabugal region (2002)

Area	Sampling station	Latitude (N)	Longitude (W)	Altitude (m)	pH	Eh (mV)	Suspended matter (mg L^{-1} dry wt)	Obs
Bica Mine	Mine water #10	40° 20.484’	07° 12.841’	631	3.05	437	37	Collected through main shaft
	Spring source #11	40° 20.492’	07° 12.796’	660	5.24	398	1.2	Mountain. Above the mine altitude
	Neutraliz pond #12	40° 20.557’	07° 12.875’	614	6.40	273	6.0	Treated water
	Well #14	40° 20.466’	07°13.064’	607	4.39	401	1.7	Farm in the valley
	Well #16	40° 20.403’	07° 13.063’	586	4.78	400	2.0	Farm in the valley
	Well #17B	40° 20.581’	07° 12.932’	600	5.28	489	4.4	Farm in the valley
	Fountain #23B	40° 20.005’	07° 13.766’	566	5.22	475	1.5	Azenha village
Pedreiros Mine	Well #7	40° 20.648’	07° 12.540’	653	4.75	453	0.4	Farm near the mine
	Well #8	40° 20.651’	07° 12.468’	650	4.39	345	6.8	Farm near the mine
Carrasca Mine	Well #18.10	40° 21.095’	07° 11.894’	650	5.96	385	4.7	Farm near the mine
	Mine water #18.12	40° 21.095’	07° 11.856’	671	6.45	–95	13.8	Mine overflow
V. Arca Mine	Well #19.15	40°22.041’	07° 11.146’	609	5.22	485	4.8	Farm in the valley, below mine
	Well #19.16	40°22.030’	07° 11.189’	619	5.75	405	3.7	Farm in the valley, below mine
	Well #19.17	40° 21.867’	07° 11.269’	612	5.54	456	3.6	Farm in the valley, below mine
	Well #20.1	40° 21.427’	07° 11.842’	548	5.67	363	6.1	Farm in the valley, below mine
	Well #20.4	40° 21.548’	07° 11.530’	583	5.60	461	4.0	Farm in the valley, below mine
	Well #24.3	40° 21.028’	07° 13.329’	515	6.24	438	2.6	Caldeirinhas village, in the valley
Rosmaneira Mine	Well #21.4	40° 24.226’	07° 15.325’	679	5.78	388	4.0	Farm near the mine
	Well #21.6	40° 24.295’	07° 15.347’	695	4.90	486	1.0	Farm near the mine
	Well #21.13	40° 24.427’	07° 15.029’	666	5.28	466	15	Farm near the mine
	Well #21.18	40° 24.495’	07° 15.069’	661	4.79	491	6.7	Farm near the mine
Coitos Mine	Mine water #22.1	40° 21.755’	07° 16.201’	522	5.51	314	16.9	Collected through main shaft
	Well #22.13	40° 21.766’	07° 15.998’	502	5.45	393	13.6	Farm near the mine

Table 2 Radionuclide concentrations (mBq $L^{-1} \pm 1$ SD) in groundwater's dissolved phase in the Sabugal region

Area	Sampling station	^{238}U	^{235}U	^{234}U	^{230}Th	^{226}Ra	^{210}Pb	^{210}Po	^{232}Th
Bica Mine	Mine water #10	$(4.4 \pm 0.2) \times 10^3$	169 ± 11	$(4.2 \pm 0.2) \times 10^3$	361 ± 60	$(1.5 \pm 0.2) \times 10^3$	$(1.82 \pm 0.05) \times 10^3$	226 ± 10	78.6 ± 14.0
	Spring source #11	20.7 ± 0.9	0.92 ± 0.09	22.5 ± 0.9	–	12.2 ± 0.8	83.8 ± 3.2	10.6 ± 0.8	.
	Neutraliz pond #12	$(2.5 \pm 0.1) \times 10^3$	103 ± 8	$(2.4 \pm 0.1) \times 10^3$	4.6 ± 1.1	69.2 ± 5.6	14.9 ± 0.9	22.8 ± 1.5	0.242 ± 0.242
	Well #14	234 ± 7	9.5 ± 0.4	236 ± 8	1.5 ± 0.1	43.9 ± 8.2	67.0 ± 1.4	10.1 ± 0.4	0.039 ± 0.024
	Well #16	33.3 ± 1.1	1.7 ± 0.1	34.8 ± 1.2	1.2 ± 0.1	17.1 ± 1.2	31.0 ± 0.9	8.8 ± 0.3	0.113 ± 0.044
	Well #17B	57.6 ± 2.7	2.6 ± 0.2	58.8 ± 2.8	–	11.6 ± 0.8	47.8 ± 2.4	1.26 ± 0.06	.
	Fountain #23.B	34.8 ± 1.2	1.6 ± 0.1	34.1 ± 1.2	0.595 ± 0.111	13.0 ± 0.8	170 ± 2	2.3 ± 0.1	0.128 ± 0.065
Pedreiros Mine	Well #7	35.3 ± 1.3	1.4 ± 0.1	35.6 ± 1.3	–	11.0 ± 1.0	72.6 ± 3.8	8.1 ± 0.4	.
	Well #8	17.6 ± 0.7	0.9 ± 0.1	17.9 ± 0.8	4.4 ± 0.6	17.7 ± 3.4	50.0 ± 2.3	12.3 ± 0.5	0.499 ± 0.166
Carrasca Mine	Well #18.10	51.6 ± 1.5	2.1 ± 0.1	52.6 ± 1.6	2.4 ± 0.2	71.1 ± 2.5	407 ± 12	4.8 ± 0.2	0.492 ± 0.065
	Mine water #18.12	123.3 ± 4.2	5.3 ± 0.3	125.3 ± 4.3	6.9 ± 0.5	116.4 ± 5.2	158.2 ± 5.1	93.6 ± 3.0	0.592 ± 0.077
V.Arca Mine	Well #19.15	43.9 ± 1.7	1.6 ± 0.2	44.5 ± 1.7	4.9 ± 0.4	15.7 ± 0.9	11.4 ± 0.3	42.8 ± 1.7	0.571 ± 0.093
	Well #19.16	35.0 ± 1.0	1.65 ± 0.06	34.6 ± 1.0	3.9 ± 0.4	9.1 ± 0.6	30.1 ± 1.6	13.9 ± 0.6	0.581 ± 0.152
	Well #20.1	41.0 ± 1.4	2.0 ± 0.1	55.1 ± 1.9	10.4 ± 0.5	66.9 ± 3.0	50.9 ± 2.0	16.5 ± 0.7	0.104 ± 0.021
	Well #20.4	33.8 ± 1.0	1.8 ± 0.1	34.3 ± 1.0	4.0 ± 0.5	14.2 ± 0.7	74.8 ± 3.8	16.1 ± 0.6	0.980 ± 0.209
	Well #24.3	19.5 ± 0.5	0.90 ± 0.05	18.9 ± 0.5	1.2 ± 0.1	12.9 ± 1.0	47.7 ± 1.9	6.3 ± 0.3	0.070 ± 0.019
Rosmaneira Mine	Well #21.4	45.1 ± 1.3	1.9 ± 0.1	42.7 ± 1.2	3.5 ± 0.2	92.7 ± 5.5	50.2 ± 2.4	24.5 ± 0.8	0.067 ± 0.018
	Well #21.6	27.0 ± 0.8	1.06 ± 0.06	25.8 ± 0.7	–	39.4 ± 2.3	.	17.6 ± 0.9	.
	Well #21.13	30.6 ± 1.0	1.40 ± 0.06	31.3 ± 1.0	0.646 ± 0.050	15.7 ± 1.3	.	43.8 ± 2.2	0.058 ± 0.020
	Well #21.18	22.1 ± 0.8	0.99 ± 0.09	22.7 ± 0.8	1.1 ± 0.1	128.2 ± 8.7	.	34.3 ± 1.2	0.035 ± 0.035
Coitos Mine	Mine water #22.1	100.2 ± 3.1	4.3 ± 0.4	101.3 ± 3.2	–	259.2 ± 12.7	156.6 ± 6.5	41.6 ± 2.1	.
	Well #22.13	12.7 ± 0.5	0.50 ± 0.06	12.1 ± 0.4	.	24.0 ± 1.0	.	18.3 ± 0.7	.

Table 3 Radionuclide concentrations (mBq $L^{-1} \pm 1$ SD) in groundwater's particulate phase in the Sabugal region

Area	Sampling station	^{238}U	^{235}U	^{234}U	^{230}Th	^{226}Ra	^{210}Pb	^{210}Po	^{232}Th
Bica Mine	Mine water #10	10.4 ± 0.4	0.47 ± 0.05	10.0 ± 0.4	16.1 ± 1.1	11.1 ± 0.9	266 ± 6	937 ± 71	0.07 ± 0.03
	Spring source #11	1.8 ± 0.1	0.11 ± 0.02	2.0 ± 0.1	1.2 ± 0.1	1.8 ± 0.2	21.7 ± 0.8	5.7 ± 0.3	0.10 ± 0.03
	Neutraliz pond #12	22.5 ± 0.9	1.0 ± 0.1	21.0 ± 0.9	3.2 ± 0.3	20.9 ± 3.2	14.3 ± 0.6	38.3 ± 1.5	0.16 ± 0.04
	Well #14	14.2 ± 0.5	0.69 ± 0.07	14.7 ± 0.6	2.0 ± 0.1	4.1 ± 0.3	–	5.0 ± 0.2	0.09 ± 0.02
	Well #16	25.3 ± 0.7	1.16 ± 0.06	25.2 ± 0.7	3.4 ± 0.2	4.8 ± 0.3	24.3 ± 0.9	10.8 ± 0.5	0.25 ± 0.04
	Well #17B	17.4 ± 0.6	0.88 ± 0.08	17.2 ± 0.6	9.0 ± 0.7	5.8 ± 0.5	43.6 ± 1.6	13.8 ± 0.6	0.42 ± 0.07
	Fountain #23.B	13.2 ± 0.4	0.63 ± 0.04	13.1 ± 0.4	3.4 ± 1.4	9.8 ± 0.5	74.3 ± 2.3	20.6 ± 0.8	0.7 ± 0.6
Pedreiros Mine	Well #7	1.56 ± 0.07	0.065 ± 0.012	1.53 ± 0.07	2.6 ± 0.2	3.2 ± 0.3	10.3 ± 0.3	6.4 ± 0.3	0.07 ± 0.02
	Well #8	6.6 ± 0.2	0.37 ± 0.04	6.6 ± 0.2	6.5 ± 0.6	5.7 ± 0.6	50.7 ± 1.2	22.8 ± 1.0	0.37 ± 0.09
Carrasca Mine	Well #18.10	18.8 ± 0.5	0.81 ± 0.04	19.0 ± 0.5	4.8 ± 0.4	13.0 ± 0.9	338 ± 10	29.6 ± 1.1	0.23 ± 0.04
	Mine water #18.12	206 ± 12	7.1 ± 0.6	206 ± 12	324 ± 18	81 ± 2	880 ± 25	414 ± 23	0.44 ± 0.06
Vale de Arca Mine	Well #19.15	20.8 ± 0.5	0.79 ± 0.03	21.4 ± 0.5	31.8 ± 1.9	20.8 ± 1.5	158 ± 5	103 ± 4	0.34 ± 0.05
	Well #19.16	25.9 ± 0.7	1.19 ± 0.06	26.6 ± 0.7	25.9 ± 2.8	18.3 ± 1.0	51.9 ± 1.8	50.2 ± 1.9	0.11 ± 0.06
	Well #19.17	15.7 ± 0.6	0.67 ± 0.06	15.9 ± 0.6	6.4 ± 1.2	7.3 ± 0.5	41.0 ± 1.5	24.6 ± 0.9	0.08 ± 0.08
	Well #20.1	20.1 ± 0.6	0.76 ± 0.06	26.2 ± 0.8	12.3 ± 0.9	47.0 ± 2.9	378 ± 11	56.3 ± 2.1	0.19 ± 0.04
	Well #20.4	11.6 ± 0.5	0.58 ± 0.07	11.6 ± 0.5	4.8 ± 0.4	6.0 ± 0.4	48.0 ± 1.6	28.3 ± 1.1	0.35 ± 0.06
	Well #24.3	5.8 ± 0.2	0.25 ± 0.03	6.3 ± 0.2	–	7.2 ± 0.4	22.0 ± 0.8	12.1 ± 0.5	–
Rosmaneira Mine	Well #21.4	30.0 ± 1.0	1.3 ± 0.1	28.5 ± 0.9	5.6 ± 0.3	52.8 ± 3.3	58.8 ± 1.8	37.7 ± 1.4	0.29 ± 0.04
	Well #21.6	7.4 ± 0.4	0.38 ± 0.07	7.4 ± 0.4	3.8 ± 0.2	7.9 ± 0.6	18.2 ± 0.7	10.0 ± 0.4	0.25 ± 0.03
	Well #21.13	4.7 ± 0.2	0.18 ± 0.02	5.0 ± 0.2	3.8 ± 0.2	6.4 ± 0.4	–	23.3 ± 1.1	0.29 ± 0.04
	Well #21.18	13.0 ± 0.3	0.51 ± 0.03	13.0 ± 0.3	3.8 ± 0.2	12.5 ± 0.5	125 ± 4	45.6 ± 1.8	0.37 ± 0.05
Coitos Mine	Mine water #22.1	58.0 ± 3.2	2.2 ± 0.3	56.2 ± 3.2	37.4 ± 2.0	172 ± 8	958 ± 27	168 ± 14	0.37 ± 0.03
	Well #22.13	20.9 ± 0.6	0.98 ± 0.06	21.2 ± 0.7	5.9 ± 0.3	22.4 ± 1.6	79.3 ± 2.4	38.0 ± 1.4	0.57 ± 0.04

Table 4 Radionuclide concentrations (kBq kg^{-1} ± 1 SD) in the fine grain fraction (<63μm) of soils and mining waste in the Sabugal region

Site	^{238}U	^{235}U	^{234}U	^{230}Th	^{226}Ra	^{210}Po	^{232}Th
Soil, Caldeirinha village	0.34±0.02	0.018±0.002	0.39±0.02	0.82±0.05	1.4±0.1	0.68±0.03	0.19±0.01
Soil, Quarta-Feira village	0.35±0.01	0.016±0.002	0.37±0.01	0.47±0.03	1.10±0.08	0.47±0.02	0.098±0.007
Soil near Bica mine (#18)	0.39±0.02	0.018±0.002	0.40±0.02	0.65±0.06	1.8±0.2	0.70±0.03	0.21±0.02
Bica mine (waste #1)	10.7±0.4	0.48±0.03	11.4±0.4	30±2	50±4	29±1	0.18±0.02
Rosmaneira mine (waste)	29±2	1.18±0.08	30±2	67±4	48±2	50±2	0.079±0.008

Table 5 Radionuclide concentrations (mBq L^{-1} ± 1 SD) in drinking water from public supplies in towns and villages of Sabugal region

	^{238}U	^{235}U	^{234}U	^{230}Th	^{226}Ra	^{210}Pb	^{210}Po	^{232}Th
Sabugal	13.6±0.5	0.59±0.06	12.9±0.5	–	0.64±0.09	37.9±1.1	1.04±0.03	–
Casteleiro	25.2±0.8	1.21±0.08	24.3±0.8	0.891±0.083	9.0±0.7	82.2±2.7	0.74±0.05	0.160±0.028
Pousafoles	21.3±0.7	0.92±0.07	18.5±0.6	1.1±0.1	22.3±2.0	148±5	6.0±0.4	0.116±0.024
Águas Belas	13.5±0.5	0.62±0.06	13.3±0.5	0.360±0.050	47.8±2.1	196±6	2.7±0.1	0.094±0.021
Belmonte	6.1±0.3	0.28±0.04	6.0±0.3	0.572±0.070	3.7±0.4	24.5±1.0	2.0±0.1	0.168±0.040
Inguias	3.3±0.1	0.18±0.02	3.2±0.1	–	7.2±0.5	6.6±0.3	6.7±0.2	–
Sortelha	1.8±0.1	0.10±0.02	2.0±0.1	–	13.4±0.8	14.3±0.8	26.6±1.1	–
Benespera	22.6±0.8	0.89±0.07	24.6±0.8	–	10.3±0.8	36.2±1.4	12.8±0.6	–
Quarta-Feira	42.1±2.3	1.8±0.2	35.4±2.0	–	10.6±1.3	99.7±4.9	7.5±0.3	–
Azenha	230.8±9.6	9.2±0.5	282.2±11.8	–	107.2±5.4	469±16	19.7±1.2	–
Caldeirinhas	44.9±2.5	1.7±0.2	41.1±2.3	1.2±0.1	9.2±0.6	67.8±2.3	38.2±1.5	0.112±0.036

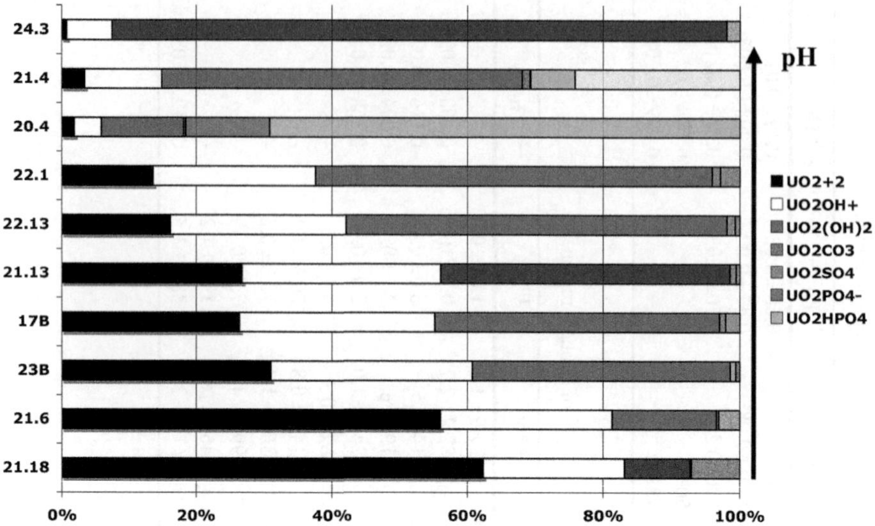

Fig. 2 Uranium species distribution in selected samples calculated using PHREEQC

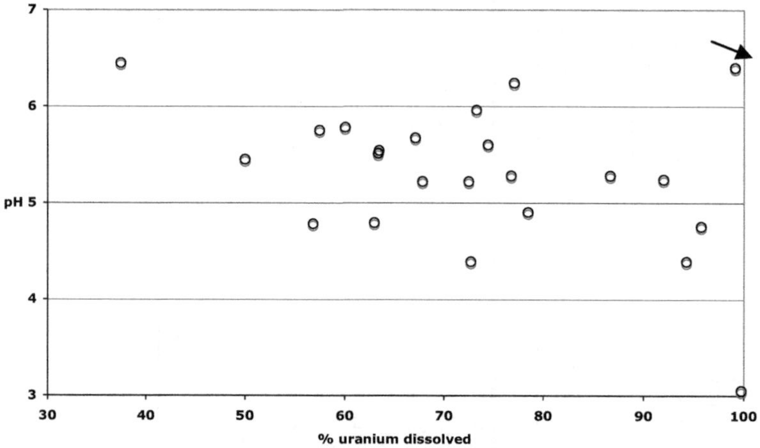

Fig. 3 Percentage of dissolved uranium of total uranium vs. pH. The arrow indicates the supernatant from Bica Mine neutralization pond

The AMD neutralization sludge from Bica are collected in decantation ponds and stockpiled in the open nearby. Sludge and mining waste piles were not covered so far. Determination of radionuclides in nearby stream showed that the treated discharge water still contains 57% of the U (Fig. 3), 1% of the ^{230}Th, 5% of the ^{226}Ra, 0.8% of the ^{210}Pb and about 10% of the ^{210}Po initially dissolved in mine water. This mine water treatment, although effective to neutralize the acid, is not so effective for the removal of U.

Another potential source of radionuclide groundwater contamination is the near-surface disposal of mining waste at several mines. Concentrations in the fine fraction of these materials as well as in agriculture soils from the region are shown in Table 4. Mining residues disposed of near the Bica and Rosmaneira mines show concentrations about three orders of magnitude above that in the soils of the Sabugal region. Continued weathering may release these radionuclides and cause their migration into underlying groundwaters.

The analyzed tap waters (Table 5) contain always higher concentrations of U series than Th and Ac series radionuclides, reflecting their relative distribution in the source rock. However, these waters generally meet the requirements of the European Drinking Water Directive (CEU 1998) and the WHO drinking water guideline (WHO 2006), i.e., 1 Bq l^{-1} for beta and 0.5 Bq l^{-1} for alpha radioactivity. The water from the public water supply of Azenha village is the only exception and this could be due to infiltration of Bica mine water into groundwater in the valley.

Remediation of the AMD in the Bica mine may be improved by inducing reduction of U(VI) with organic matter or Fe (II) injected into the flooded mine, or by bioremediation with reducing bacteria (e.g., Renshaw et al. 2009). However, any such routes need to be preceded by a more detailed analysis of the hydraulic and the hydrogeologic system.

Conclusions

In the Sabugal region of Portugal the legacy of past radium and uranium mining activities, although discontinued about two decades ago, is still very much present and may require continued intervention for years to come. However, most of the old mines in this region have mine water with no noticeable acidity and the near neutral water pH coincides with low radioactivity concentrations. The main exception is the Bica mine, where acid *in situ* uranium was carried out. Residual acid is still dissolving radionuclides that then migrate into the surrounding groundwater. The treatment of AMD seems justified and will continue until a more satisfactory water pH is obtained. However, the current neutralization procedure effectively raises the pH, but is not effective in removing the uranium.

In consequence, several irrigation wells near the Bica mine showed enhanced radionuclide concentrations and the water is not suitable for human and animal consumption. Preventing radionuclide leaching from the respective waste heaps by e.g. capping may make the waters useable eventually. Waters from other mines in this region do not need treatment and can be used for irrigation. Other groundwater-based drinking water resources in the Sabugal-Belmonte region are generally not contaminated and meet the EU radiological standards.

However, in the long run, mining wastes and sludge from mine water treatment, especially at the Bica and Rosmaneira mines, should be capped in order to stop their weathering and ensuing leaching of radionuclides into the environment.

References

Carvalho F.P., Oliveira J.M. (2007). Alpha emitters from uranium mining in the environment. J Radioanal Nucl Chem 274:167–174.

Carvalho F.P., Oliveira J.M. (2009). Performance of alpha spectrometry in the analysis of uranium isotopes in environmental and nuclear materials. J Radioanal Nucl Chem 281:591–596.

Carvalho F.P., Oliveira J.M., Faria I. (2009a). Alpha Emitting Radionuclides in Drainage from Quinta do Bispo and Cunha Baixa Uranium Mines (Portugal) and Associated Radiotoxicological Risk. Bull Environ Contamin Toxic 83:668–673.

Carvalho F.P., Oliveira J.M., Malta M. (2009b). Analyses of radionuclides in soil, water and agriculture products near the Urgeiriça uranium mine in Portugal. J Radioanal Nucl Chem 281:479–484.

Council of the European Union (1998). Council Directive 98/83/EC of 3 November 1998 on the quality of water intended for human consumption. OJ L 330, 05/12/1998 P. 0032–0054.

JEN (1968). Uranium and Nuclear Raw Materials in Portugal. Junta de Energia Nuclear. Lisboa.

Oliveira J.M., Carvalho F.P. (2006). A Sequential Extraction Procedure for Determination of Uranium, Thorium, Radium, Lead and Polonium Radionuclides by Alpha Spectrometry in Environmental Samples. (Proc. 15th Radiochemical Conference). Czech J. Physics 56 (Suppl. D):545–555.

Povinec P.P. et al. (2007). Reference material for radionuclides in sediment, IAEA-384 (Fangataufa Lagoon sediment). J Radioanal Nucl Chem 273:383–393.

Parkhurst, D.L., Appelo, C.A.J. (1999). User's guide to PHREEQC – A computer program for speciation, batch-reaction, one-dimensional transport, and inverse geochemical calculations. – U.S. Geological Survey Water-Resources Investigations Report 99-4259.

Pham M.K. et al. (2006). Certified reference material for radionuclides in fish flesh sample IAEA-414 (mixed fish from the Irish Sea and North Sea). Appl Radiat Isot 64: 1253–1259.

Renshaw J.C., Butchins L.J.C., Livens F.R., May I., Charnock J.M., Lloyd J.R. (2005). Bioreduction of Uranium: Environmental Implications of a Pentavalent Intermediate. Environ Sci Technol 39:5657–5660.

WHO (2006). Guidelines for Drinking Water Quality. 3rd Edition. World Health Organization, Geneva.

Simulation of Radionuclide Transport and Fate in Surface Waters in the Vicinity of a Past Uranium Processing Plant

Maria de Lurdes Dinis, António Fiúza

Abstract. This paper describes an application of a transport and fate model for radionuclides in surface waters. The generic approach was based on the IAEA models used for assessing the impact of discharges of radioactive substances into the environment. The case study refers to a discharge of an effluent with high radium concentration into a stream that in turn is affluent into a river. Several exposure points were considered along the minor watercourse strategically defined by the measured data. Radium concentrations were estimated for these points permitting to establish a relationship between concentrations and the distance to the discharge location. The concentration in the river sediments was also estimated for the same exposure points.

Introduction

Uranium mining and processing generate effluents which contain radioactive and non-radioactive elements as well as chemical compounds that, if not properly treated or contained, can represent a serious environmental impact.

Liquid effluents are generated at all stages of uranium production that manipulates water and chemicals. They contain radioactive elements, such as uranium and

Maria de Lurdes Dinis
Geo-Environment and Resources Research Center (CIGAR), Engineering Faculty, Porto University, Rua Dr. Roberto Frias, 4200-465, Porto, Portugal
E-mail: mldinis@fe.up.pt

António Fiúza
Geo-Environment and Resources Research Center (CIGAR), Engineering Faculty, Porto University, Rua Dr. Roberto Frias, 4200-465, Porto, Portugal
E-mail: afiuza@fe.up.pt

B. Merkel, M. Schipek, *The New Uranium Mining Boom*
DOI 10.1007/978-3-642-22122-4 © Springer 2011

radium, and non-radioactive elements such as arsenic, nickel, molybdenum, manganese, magnesium, selenium, fluorides and sulfates. In addition, acid mine drainage may also be generated due to ore leaching and mineralized waste rock.

Effluents treatment and control is a long-term task that must continue after uranium production ceases. It also goes through the decommissioning and rehabilitation of the site as well as through the monitoring stage after reclamation operations are complete. Generated effluent at any stage must be treated before being discharged into any watercourse.

Active effluent treatment is preferably used during uranium production and passive treatments during reclamation and long-term monitoring. Nevertheless both treatments may be used during operations and decommissioning.

One of the important active effluent treatments consists in neutralization with lime by precipitation of most contaminants. In addition barium chloride is added to remove the radium present resulting in high barium-radium sulfate sludge. The resulting effluent from this sludge sedimentation may be then discharged in the environment.

Radioactive elements remaining in the effluent that is discharged into a watercourse are subject to physical and chemical processes that affect their transport from the sources point and concentration during their movement along the watercourse. Among these processes we may refer advection, dispersion and diffusion for the physical processes and adsorption, precipitation and ion exchange for the chemical processes.

This work describes an application of the radionuclides transport model in surface water. The radionuclides concentrations in water and sediments are estimated considering routine discharges from the effluent treatment by barium chloride into different types of water bodies. The radionuclides discharge into a river was considered for applying the model to a specific case study. The generic approach was based on the IAEA models used in assessing the impact of discharges of radioactive substances into the environment (IAEA 2001).

Methods and Materials

Description of the Model – Radionuclide Concentration in the Water

Radionuclides transport in watercourses may be affected by transport processes such as advection, hydrodynamic dispersion and molecular diffusion. On the other hand, many chemical and biological reactions may take place, such as attenuation, retardation, ionization or dissolution, changing radionuclides fate.

Radionuclides transport and fate in superficial waters may be predicted, with more or less detail, by a simple mass balance with a multidimensional numerical solution. In a generic way, radionuclide transport and fate in surface water may be described by the advection-diffusion equation in a three dimensional form.

$$\frac{\partial C}{\partial t}+U\frac{\partial C}{\partial x}+V\frac{\partial C}{\partial y}+W\frac{\partial C_w}{\partial z}=\varepsilon_x\frac{\partial^2 C}{\partial x^2}+\varepsilon_y\frac{\partial^2 C}{\partial y^2}+\varepsilon_z\frac{\partial^2 C}{\partial z^2}-\lambda_i C+S \quad (1)$$

In this generic equation *C* is the concentration of the radionuclide transported in superficial water: *U*, *V* and *W* are the pore velocity in the directions x, y and z, respectively; *S* represents the radionuclide addition or subtraction and λ is the radionuclide decay constant. This generic governing equation was simplified to obtain appropriate solutions according to the water body considered, calculating the radionuclide concentration at defined distances from the source; it is implicitly admitted that potential receptors are located at those points.

The methodology adopted in IAEA models is based on analytical solutions to advection-diffusion equations describing radionuclide transport in surface waters with steady state uniform flow conditions (IAEA 2001).

Radionuclide concentrations in water and sediment may be calculated for specific locations where members of a critical group may be exposed through contact with the water resulting from drinking, fishing, irrigation or swimming and still through sediment usage for agricultural activities. These exposure points are defined by a specific distance to the source and are incorporated as an input to the model.

For different sources of radionuclides discharges, radionuclide concentration should be calculated separately for each radionuclide at the same location. The individual results may then be summed for cumulative radionuclide concentration at the considered location (IAEA 2001).

Radionuclide concentration in water may be calculated at a location along the river bank where the discharge takes place or along the opposite bank.

Considering radionuclide concentration along the opposite river bank as the discharge, radionuclide transport occurs at least to a distance of half of river width to reach the exposure location. In this situation, radionuclide concentration can be calculated by (IAEA 2001):

$$C_w=\frac{Q}{q}e^{\left(-\frac{\lambda \cdot x}{U}\right)}=C_t, \quad (2)$$

where C_w (Bq/m^3) is the radionuclide concentration in the water, Q (Bq/s) is radionuclide annual average discharge rate, q (m^3/s) is the mean river flow, λ (s^{-1}) is the radioactive decay constant, x (m) is the distance between the receptor and the discharge point and U is the river velocity (m/s).

Considering radionuclide concentration along the same river bank as the discharge, two situations may occur depending on exposure location. If the exposure location is very near to the discharge point ($x \leq 7D$, D is the river depth) and before the vertical mixing can occur, this means that radionuclide concentration in water is undiluted and consequently is the same as the concentration at the discharge point (IAEA 2001):

$$C_w = C_0, \quad x \leq 7 \cdot D, \tag{3}$$

where C_0 (Bq/m^3) is the radionuclide concentration in the discharged effluent and D (m) is the river depth.

If the exposure is considered to take place at a location after vertical mixing occurs ($x > 7D$), the concentration must be corrected for partial lateral mixing with a correction factor P_r, which depends only on site characteristics (IAEA 2001):

$$C_w = C_t \cdot P_r, \quad x > 7 \cdot D \tag{4}$$

$$P_r = \frac{1}{0,142 \cdot \pi} \exp\left(\frac{1,5 \cdot D \cdot x}{B^2}\right) k_0 \left[\frac{1,5 \cdot D \cdot x}{B^2}\right], \tag{5}$$

where P_r (unitless) is the correction factor for partial lateral mixing, B (m) is the river width and k_0 (unitless) is the modified Bessel function of second kind of zeroth order.

Description of the Model – Radionuclide Concentration in the Sediments

Three types of calculations should be considered for radionuclide concentration in sediments: i) dissolved in water; ii) suspended sediments and iii) bottom sediments (IAEA 2001).

The dissolved radionuclide concentration in water (C_{ws}, Bq/m^3) may be estimated by using the distribution coefficient for the respective radionuclide, K_d (L/kg), which expresses the exchange of radionuclides between the dissolved and sediment sorbed phases (IAEA 2001):

$$C_{ws} = \frac{C_w}{1 + 0,001 \cdot K_d \cdot S_s}, \tag{6}$$

where S_s is the suspended sediment concentration (kg/m^3 or g/l) and 0.001 is the conversion factor from l/kg to m^3/kg. For the suspended sediments the radionuclide concentration (C_{sw}, Bq/kg) may be estimated by (IAEA 2001):

$$C_{sw} = \frac{0,001 \cdot K_d \cdot C_w}{1 + 0,001 \cdot K_d \cdot S_s} = 0,001 \cdot K_d \cdot C_{ws}. \tag{7}$$

And the radionuclide concentration in bottom sediments (C_{sb}, Bq/kg) can be given by (IAEA 2001):

$$C_{sb} = \frac{(0,1) \cdot (1) \cdot K_d \cdot C_w}{1 + 0,001 \cdot K_d \cdot S_s} \cdot \frac{1 - e^{(-\lambda \cdot T_e)}}{\lambda \cdot T_e} = 0,1 \cdot C_{sw} \frac{1 - e^{(-\lambda \cdot T_e)}}{\lambda \cdot T_e} \tag{8}$$

where T_e is the effective accumulation time (s). The IAEA (2001) recommends a default value of $3,15 \times 10^7$ s (equivalent to 1 year).

The methodology for estimating radionuclide concentration in water (Bq/m^3) and in sediments (Bq/kg) resulting from routine radionuclide discharges in surface water was implemented in Matlab. The particular case of routine discharges into rivers was adopted (IAEA 2001); (USNRC 1977, 1978). The outputs of model are the radionuclide concentration in water and in sediments in the same river bank and on the opposite bank relatively to the discharge at specific exposure locations.

Study Area – Scenario Description

The radionuclides transport model was applied to a specific case study: the mining area of a former uranium mine located at Urgeiriça, near the city of Viseu (North-East of Portugal). The Urgeiriça mine was active almost during one century, from 1913 to 2001. The activities of this mine, in addition to the exploitation of others uranium mines in the same region, led to the accumulation of large amounts of solid wastes (tailings) in the vicinity of the uranium processing plant. In particular, a tailings pile with an estimated volume of 1.39 million m^3 with a high radioactive content. This tailings pile was the main concern of the national plan for the rehabilitation of former uranium mining sites due to either its volume either to its radioactive content.

The contaminated waters from this mine and from the percolation through the mill tailings used to be collected and neutralized before discharge into a stream near the contaminated site, Pantanha stream, which is also the main watercourse that drains the mining area. This stream is also a tributary of a major river – the Mondego river.

The liquid effluents from the uranium chemical treatment at Urgeiriça site have also been treated and discharged into Pantanha watercourse, for many years. For these reasons, the model simulation was done for the discharge of an effluent with a high radium concentration into that particular stream. Also, even after treatment, effluents discharges do not always occur properly and may originate radionuclides concentrations in the environment above the permissible legal limits.

Two different situations were considered in the model simulation: a worst scenario case by considering the discharge of untreated effluent with an annual average concentration of 960 Bq/m^3 (Pereira et al. 2004) and the discharge of a neutralized effluent with an average concentration of 271 Bq/m^3 (Dinis and Fiúza 2008). The effluent results from a mixture of water from the mine with acid infiltrations from the tailings pile.

The model simulation was done for several exposure points located along the Pantanha stream. These points, upstream and downstream of the discharge point, were strategically defined by the available surface water measurements carried out by the former Uranium National Company (ENU) (Fig. 1). Radium concentration in superficial water, as well as in sediments, was estimated for these same exposure points to determine the relation between the diluted concentrations and the distance to the discharge location.

Monitoring and exposure points C_1, C_2, C_3 and C_4 are located at Pantanha stream, C_5 is located at another stream (Corujeira), C_6 and C_7 are located at Mondego river. Distances between the considered monitoring points are presented in Table 1.

For the model simulation ^{226}Ra is to be released to Pantanha stream at a release rate of $8{,}13 \times 10^7$ Bq/a with an effluent discharge of $0{,}01\,m^3/s$. Dissolved sediments, suspended sediments and bottom sediments were also considered to be affected by the effluent discharge.

The nearest potentially exposed receptor may be located at 150 (C_2) or 1650 (C_3) or 2750 (C_4) or 3200 m (CP) from the discharge point (DP) on the same side

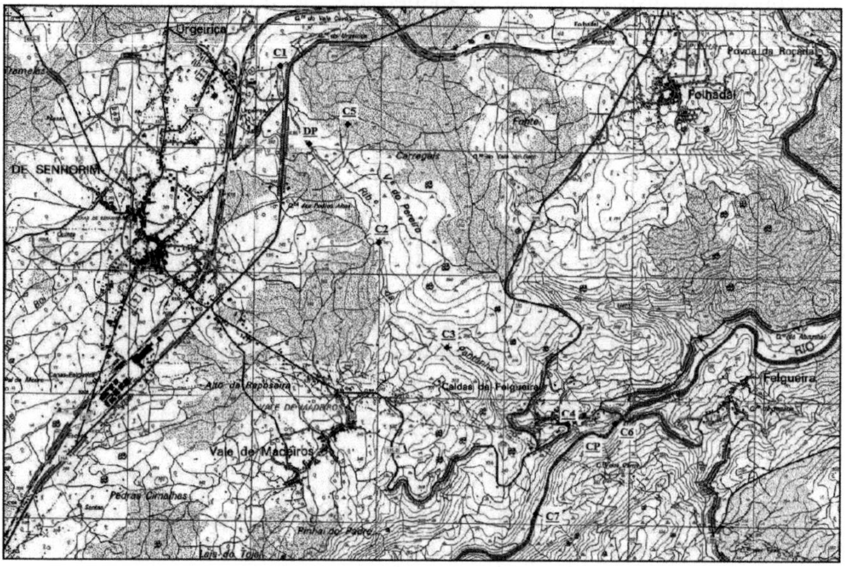

Fig. 1 Monitoring points along Pantanha stream (1:25,000). DP: discharge point; CP: confluent point between Pantanha stream and Mondego river

Table 1 Distance between the considered monitoring points (Dinis 2007)

Monitoring point	Distance (m)
$C_1 - DP$	950
$DP - C_2$	150
$C_1 - C_2$	1100
$C_2 - C_3$	1500
$C_3 - C_4$	1100
$C_4 - CP$	450
$CP - C_6$	200
$CP - C_7$	800

river bank of the discharge. Eventually, exposure may also occur at the opposite side of the river bank, but it isn't considered in the presented simulation. The river conditions were stated according to IAEA (2001) models.

Results and Discussion

The results from the model simulation for radium concentration in the surface water of Pantanha stream are presented in Table 2 (Dinis 2007). The exposure points C_6 and C_7 are located at Mondego river. Depth, width and flow rate for this river were recalculated according to IAEA models (IAEA 2001).

Model results for radium concentration in sediments originated by the same effluents discharges are presented in Tables 3 and 4 according to discharge radium concentration (Dinis 2007).

In general, model results suggest a clear decrease in radium concentration downstream with the distance to the discharge point. Radium concentration in sediments follows the same variation pattern.

Table 2 Radium concentration in surface water for the exposure points considered

^{226}Ra	DP	C_2	C_3	C_4	CP	C_6	C_7
Effluent (Bq/m^3)	960	23.08	8.36	5.46	5.46	0	0.005
Effluent (Bq/m^3)	271	6.51	2.36	1.54	1.54	0	0.001

Table 3 Radium concentration in sediments for the exposure points considered (radium discharge of 960 Bq/m^3)

Radium discharge 960 Bq/m^3	C_{sb} (Bq/kg)	C_{ws} (Bq/m^3)	C_{sw} (Bq/kg)
C_2	1.10	21.98	10.99
C_3	0.40	7.96	3.98
C_4	0.26	5.20	2.60
PC	0.26	5.20	2.60
C_7	2.7×10^{-4}	0.0054	0.0027

Table 4 Radium concentration in sediments for the considered exposure points (radium discharge of 271 Bq/m^3)

Radium discharge 271 Bq/m^3	C_{sb} (Bq/kg)	C_{ws} (Bq/m^3)	C_{sw} (Bq/kg)
C_2	0.31	6.20	3.10
C_3	0.112	2.25	1.12
C_4	0.073	1.47	0.73
PC	0.073	1.47	0.73
C_7	7.68×10^{-5}	0.0015	7.68×10^{-4}

At the confluence point (CP) between Pantanha stream and Mondego river, radium concentration is 5.46 or 1.54 Bq/m^3 in consequence of the effluent discharge (960 or 271 Bq/m^3 for radium concentration, respectively).

Upstream exposure points (such as C_6) won't be affected by the effluent discharge but for downstream exposure points (such as C_7) radium concentration was estimated in 0.005 and 0.001 (Bq/m^3) according to discharge conditions.

Nevertheless, due to Mondego river characteristics (width and depth) radium concentration in water and, consequently in sediments, will turn to near background values with the distance.

In what concerns to Pantanha stream, it was also observed a huge fall down for radium concentration in surface water a few meters after the discharge point (Fig. 2). This corresponds to the turning point where the vertical complete mixing is achieved and radium dilution occurs. From this point forward radium concentration gradually decreases until a constant value. This is due to the fact that partial mixing coefficient, P_r, approaches unity as the downstream distances increases; this coefficient should be greater than or equal to unity (IAEA 2001).

Radium concentration in sediments follows the same variation pattern.

The results from the model referring to the exposure points C_2, C_3, C_4 and C_7 were compared with ENU measurements carried out during the year of 1997 (ENU 1998). These measurements are presented in Table 5.

From the analysis of the ^{226}Ra measurements in the exposition points it is possible to observe that the highest values both in water and in sediments do not cor-

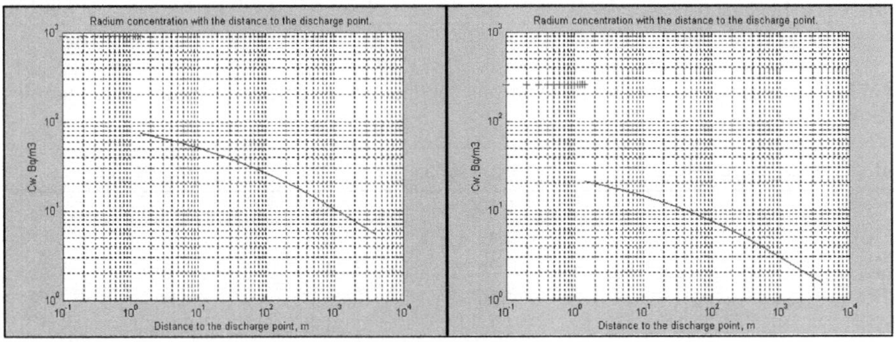

Fig. 2 Radium concentration variation at Pantanha stream with the distance to the discharge point (effluent discharge: 960 and 271 Bq/m^3, respectively)

Table 5 Radium concentration in surface water of Pantanha stream, Bq/m^3 (ENU 1998)

ID	Jan.	Feb.	Mar.	Apr.	May	Jun.	Aug.	Sept.	Oct.	Nov.	Dec.	Mean
C2	150	80	100	100	70	90	210	100	110	70	60	103.6
C3	140	50	70	70	150	80	80	120	140	50	90	94.5
C4	80	50	40	60	60	120	210	160	350	80	60	115.5
C7	–	–	110	110	–	70	–	140	–	–	–	106.7

respond to the considered discharge point but at a distance comprised between 1650 (C_3) and 2750 m (C_4) distant from the discharge point. The radium concentration continues to decrease with the distance afterwards. This suggests that probably there are other contamination sources besides the discharged effluent; the model considers that the contaminated effluent is the only source of contamination. Nevertheless, surface runoff and seepage of rainwater flowing through the tailings pile or through contaminated nearby soils may represent a significant contribution to radium concentrations in surface water and sediments. This can be stressed by the fact that the superficial water measurements performed by ENU are always much higher than those obtained in the model outputs and the model simulation was carried out considering only the contribution from the discharged effluent.

Sediments measurements carried out by Pereira et al. (2004) also showed that the highest values for sediments concentration do not occur in the vicinity of the discharge location but at a distance comprised between 1500 and 2000 m from it. These anomalous concentrations may also result from others sources than the discharged effluent such as the irrigation of agricultural soils with contaminated water beside the direct relation with the surface water concentration.

Conclusions

This work presents the results of an application of the radionuclides transport model in surface water based on the IAEA models for assessing the impact of discharges radioactive substances into the environment (IAEA 2001). The radionuclides discharge into a river was considered for applying the model to a specific case study. Radium concentrations in surface water and in sediments were calculated for different exposure points.

The observed pattern variation for radium concentration could be explained by the presence of others contamination sources besides the discharged effluent, as mentioned before. Nevertheless, model results are below the measured values at Pantanha stream and Mondego river, for the considered exposure points.

The discharge effluent rate adopted in the model simulation refers to the year of 2000 while the experimental data refers to the year of 1997. Urgeiriça mine extraction stopped in 1992 but the processing plant was used to process uranium from others mines, until 2001. Presumably in 1997, and before that, the effluent rate discharge has been much higher than the adopted value; between 1951 and 1991 the processing plant was producing an annual average of 125,000 t of uranium concentrate. This could explain the lower values for radium concentration in water achieved from the model simulation.

Sediments measurements for the same year and exposure points weren't available for comparison but taking in consideration the sediments concentration direct dependency on water concentration we conclude that the model results would be lower than the measured values.

Most of the processes inherent to the model transport are highly site specific, such as the annual river flow rate, flow depth and river velocity, which were estimated based on generic hydrological studies due to unavailable site specific data. Model structure also has some limitations such as assuming that river cross section and flow velocity do not change with distance or time, which is not always true.

Uncertainty studies weren't carried out but it is clear that uncertainty will result at least from the adopted assumptions and calculated hydrological river characteristics. Also a great error may arise from sediment adsorption coefficient, K_d, and suspended sediment concentration, S_s; site specific values for these parameters should be obtained to reduce the error.

One may conclude that model simulation was limited mainly by considering the effluent discharge as the only contamination source. In addition, the lack of updated data for effluent discharges rate and site specific data for the most sensitive parameters contributed for the observed differences.

References

Dinis M.L. (2007) Phenomenological Models for the Inter-Compartmental Distribution of Radioactive Substances, Doctoral Degree in Environmental Engineering, Porto, FEUP.

Dinis M.L. and Fiúza A. (2008) Integrated Methodology for the Environmental Risk Assessment of an Abandoned Uranium Mining Site, In: Uranium, Mining and Hydrogeology, pp. 163–176, ISBN: 978-3-540-87745-5, Springer-Verlag, Berlin.

ENU (1998) Empresa Nacional de Urânio Boletins de Análises ENU, Urgeiriça, Nelas.

EXMIN (2003) Estudo Director de Áreas de Minérios Radioactivos – 2.ª fase. Companhia de Indústria e Serviços Mineiros e Ambientais, SA.

IAEA International Atomic Energy Agency (2001) Generic Models for Use in Assessing the Impact of Discharges of Radioactive Substances to the Environment, Safety Reports Series N.º 19, IAEA: Vienna, Austria.

Pereira A.J.S.C., Neves L.J.P.F., Dias J.M.M., Campos A.B.A. and Barbosa S. (2004) Evaluation of the radiological hazards from uranium mining and milling wastes (Urgeiriça, Central Portugal). Proceedings do 11th International Congress of the International Radiation Protection Association, 10 p.

USNRC United Sates Nuclear Regulatory Commission (1977) Estimating Aquatic Dispersion in Effluents from Accidental and Routine Releases for the Purpose of Implementing, Appendix I, ISNRC Guide I.113, Revision I, Office of Standards Development, USNRC, Washington, DC.

USNRC United Sates Nuclear Regulatory Commission (1978) Liquid Pathway Generic Study, Impacts of Accidental Radionuclide Releases to the Hydrosphere from Floating and Land-based nuclear Power Plants, Rep. NUREG-0440, USNRC, Washington, DC.

Neutralisation and Trace Element Removal from Beverley in-situ Recovery Uranium Mine Barren Lixiviant via Hydrotalcite Formation

Grant Douglas, Laura Wendling, Kayley Usher, Peter Woods

Abstract. An assessment of hydrotalcite formation to neutralize acidity and remove trace elements was undertaken using barren lixiviant from Heathgate Resources' Beverley in situ recovery uranium mine in South Australia. Batch-scale studies demonstrated proof of concept for neutralization of acidity using $MgO + NaAlO_2$ with concomitant removal of a range of trace elements. The hydrotalcite formed during neutralization, hosted a range of potential contaminants including substantial uranium (~ 1% U) and rare earth elements (~2% REE).

Introduction

Beverley Uranium Mine

The Beverley in situ recovery (ISR) uranium mine in South Australia, owned and operated by Heathgate Resources Pty Ltd (Heathgate), is the only operational ISR uranium mine in Australia. The Beverley deposit is located ~550 km north of

---

Grant Douglas
CSIRO Land and Water, Private Bag No. 5, Wembley, WA, 6913, Australia

Laura Wendling
CSIRO Land and Water, Private Bag No. 5, Wembley, WA, 6913, Australia

Kayley Usher
CSIRO Land and Water, Private Bag No. 5, Wembley, WA, 6913, Australia

Peter Woods
Heathgate Resources Pty Ltd, Suite 1, Level 4, 25 Grenfell Street, Adelaide, SA, 5000, Australia

B. Merkel, M. Schipek, *The New Uranium Mining Boom*
DOI 10.1007/978-3-642-22122-4 © Springer 2011

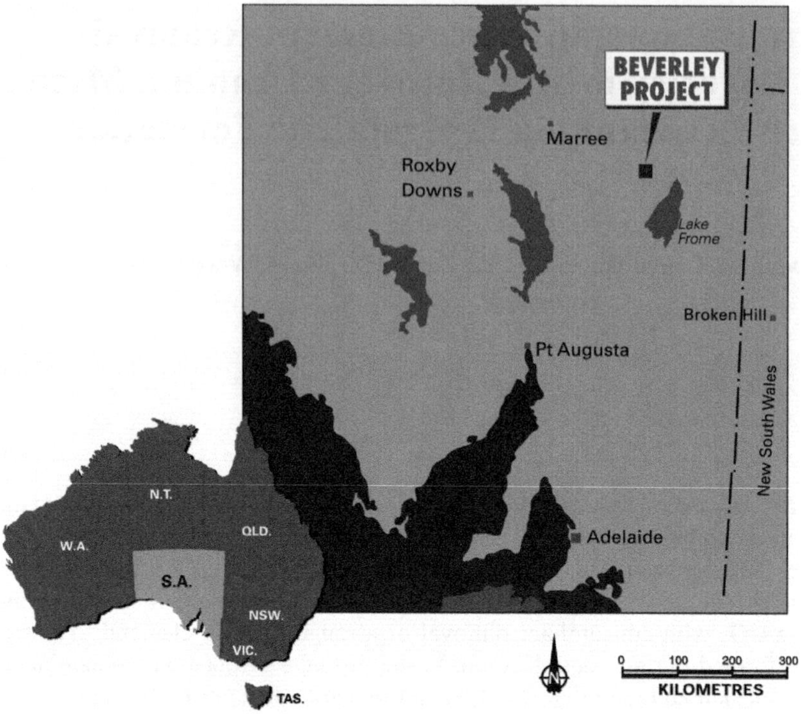

Fig. 1 Location of Beverley in situ recovery (ISR) uranium mine in South Australia

Adelaide (Fig. 1). Uranium mineralisation was first discovered in 1969. In 1990, Heathgate acquired the 21,000 t deposit and commenced ISR mining in 2000.

Mineralisation is hosted within a series of palaeochannels ~ 100–140 m below the surface within unconsolidated Beverley sands, silts and clays of the Tertiary (Miocene) Namba Formation. Ore primarily occurs as coffinite and uraninite associated with organic-rich layers of the palaeochannel sediments. Other element enrichments include Th, Zn, Mo, V and As (Curtis et al. 1990; Ruperto and de Caritat 2006; Hou et al. 2007). Palaeochannel groundwater salinity varies from 3000–15,000 mg/l with aquifer waters of an Na-HCO3-Cl-SO_4 type (Table 1).

Uranium Extraction and Recovery

Aquifer pH is reduced from ~ 7.0–8.0 to 1.5–2.5 using H_2SO_4, with H_2O_2 used as an oxidant to assist in U solubilization. Uranium within pregnant lixiviant is pumped to the surface and recovered. Post U extraction, the lixiviant is refortified and re-injected to assist in further ore extraction. After cessation of mining within a mineralized aquifer, under current arrangements barren lixiviant remains in the mined-out aquifer to undergo natural attenuation. Typical lixiviant compositions are given in Table 1.

Table 1 Typical groundwater and average extraction and injection lixiviant compositions (mg/L) from Beverley ISR U mine (Heathgate data) neutralized lixiviant (Lix + MgO + $NaAlO_2$) and post-Beverley aquifer sediment column (this study)

Analyte	Ground-water	Lixiviant Injection	Lixiviant Extraction	Barren lixiviant	Lix + MgO + $NaAlO_2$	Post-column
pH	7.3	2.1	2.3	1.7	6.9	7.3
EC	831	1324	1199	2470	1900	1980
Al	0.22	36.7	36.9	200	0.094	0.011
As	0.002	0.04	0.05	<0.05	<0.05	0.026
Ba	53	37	39	0.035	0.065	0.055
Ca	370	546	547	638	705	657
Cd	0.2	41	42	0.16	0.078	0.014
Cl	1988	1895	1975	3700	4220	4580
Co	0.11	4.06	4.18	6.7	1.7	0.7
Cr	0.02	0.03	0.58	0.18	0.003	0.002
Cu	0.03	0.12	0.03	0.47	0.005	0.008
Fe	0.74	26.1	25.9	79	0.013	<0.005
HCO_3	n/a	<1	<1	<1	67	165
K	42	59	59	117	109	121
La	n/a	n/a	n/a	0.79	0.13	0.001
Mg	198	303	303	454	1570	1580
Mn	0.19	0.52	0.55	14	11	9.9
Na	1188	1423	1427	1710	2300	2680
Ni	0.004	1.99	2.04	3.7	1.2	0.4
Pb	0.04	0.1	0.15	0.16	<0.0005	<0.0020
S	508	1378	1303	2300	2500	2600
Sc	n/a	n/a	n/a	0.53	<0.0025	<0.0025
Se	0.001	0.26	0.25	<0.05	0.06	0.09
Si	48	125	126	300	37	42
Th	n/a	n/a	n/a	1.1	0.0051	0.0024
Ti	n/a	n/a	n/a	0.028	<0.002	<0.002
U	0.41	15	211	19	1.3	0.7
V	0.001	0.76	0.78	0.9	<0.005	<0.005
Zn	0.29	0.17	0.25	4.2	0.44	3

Hydrotalcite Formation

Hydrotalcites (HT) are a class of naturally-occurring and synthetic materials characterized by positively charged mixed metal hydroxide layers separated by interlayers containing water molecules and exchangeable anions (Cavani et al. 1991; Shin et al. 1996; Ulibarri et al. 2001). Hydrotalcites form via co-precipitation of divalent and trivalent metal cation solutions at moderate to high pH with the general formula: $M_{(1-x)}{}^{2+}M_x{}^{3+}(OH)_2A^{n-} \cdot yH_2O$, where M^{2+} and M^{3+} are divalent and trivalent metal ions, respectively (Taylor 1984), x is the proportion of trivalent metal ion, y is interlayer water and A^{n-} the anion. In Beverley barren lixiviant, Mg^{2+}, Al^{3+}, and SO_4^{2-} predominate giving the formula: $Mg_6Al_2(OH)_{16}SO_4{\cdot}nH_2O$.

Hydrotalcites typically contain Mg^{2+} and Al^{3+}, in the molar ratio of 2 : 1 to 3 : 1, but other cations including Ni, Zn, and La may occur (Behrens et al. 2010; Vucelic et al. 1997) via co-precipitation and ion-exchange both during and after formation (Cavani et al. 1991; Miyata 1983; Seida and Nakano 2000, 2002). Simultaneous removal of a suite of inorganic contaminants (to form a polymetallic HT) is highly advantageous in treatment of contaminated waters.

Addition of Mg and/or Al and an alkalinity source are required to optimize conditions for HT formation in barren lixiviant (Douglas et al. 2010). Alkalinity, Mg and Al may be obtained from a variety of sources including sodium aluminate ($NaAlO_2$), other Al salts, Bayer liquor from alumina refining, sodium hydroxide (NaOH) and calcined magnesia (MgO). Formation of HT in the specific context of Beverley ISR barren lixiviant may also have other advantages, including:

- broad spectrum contaminant (common within U mineralisation) removal;
- rapid formation and dewatering, resulting in a high solids concentration in addition to flocculation of colloidal/material from suspension;
- formation of stable HT precipitates above a pH of ca. 4–5 with stability increasing with pH;
- removal of Mg^{2+} and substantial SO_4^{2-} while only adding Na^+; and
- generation of neutralized solute (mostly Na-SO_4) potentially amenable to RO/electrodialysis of a concentrated solution producing H_2SO_4 and NaOH for use in lixiviant production/barren lixiviant neutralization, respectively.

Advantageously, the formation of HT in Beverley ISR Mine barren lixiviant is favored by high Mg concentrations in the aquifer augmented by Mg and Al liberated during ISR (Table 1). With a Mg : Al mol ratio of ~ 1.6, Beverley ISR barren lixiviant is potentially suitable for HT formation after addition of further Mg and Al (to increase the mass of HT precipitated) and a source of alkalinity. Importantly, low Fe mitigates formation of unstable green rusts (Genin et al. 2001).

It was postulated that the formation of HT of suitable stoichiometry, (Mg^{2+} : Al^{3+} molar ratio of ~ 2 : 1 to 3 : 1), from the Mg, Al-rich Beverley barren lixiviant could be facilitated via MgO + $NaAlO_2$ addition. The addition of these reagents serves two simultaneous purposes: adjustment of the Mg : Al molar ratio to a desired range without Fe addition, and as a source of alkalinity to increase pH to induce the formation of HT as a sink for a suite of contaminants including U.

Methods

Beverley ISR Mine Batch Neutralisation Experiments

A 2 l sample of barren lixiviant was rapidly stirred in an acid-washed flask. Both MgO and $NaAlO_2$ were rapidly added as dry powders. A precipitate formed immediately and was stirred for ~ 0.5 h. The resultant suspension was covered and aged at 50°C in a oven for 7 days, separated, washed with deionized water, and oven-dried at 50°C. Part of the neutralized supernatant was passed through a 1 m long column containing unmineralized Beverley aquifer sands. A precipitate subsample was ground in a mortar and pestle and calcined in a Pt crucible at 650°C for 30 min and re-weighed for mass loss. Dried and calcined precipitate mineralogy was analyzed using X-ray diffraction (XRD) and major and trace elements using X-ray fluorescence (XRF). Supernatant solutions were analyzed for major and trace elements via ICP-MS/OES using standard methods.

Scanning electron microscopy of HT was performed on a Zeiss 1555 VP-FESEM equipped with an energy dispersive spectrometer (EDS). For imaging, dried samples were placed onto Al stubs and Au-coated. For microprobe analysis, dried samples were carbon-coated with EDS spectra from single points used to determine elemental composition.

Results and Discussion

Beverley ISR Mine Batch Neutralisation Experiments

Relative to the initial barren lixiviant pH of 1.7, $MgO + NaAlO_2$ addition resulted in a final pH of 6.9 (Table 1). with a concomitant reduction of dissolved Al concentrations (200 to <0.1 mg/L). Considerable Mg (454 to 1570 mg/L) was present as MgO was added in excess of stoichiometric requirements. Reductions in lixiviant major and trace elements concentrations included Al, Ca, Cd, Co, Fe, La, Ni, Th, and U from (19 to 1.3 mg/L). Most trace element concentrations were further reduced, and pH increased to 7.3 after passage of neutralized solute through a column of unmineralized Beverley aquifer sediments (Table 1). The resultant solute chemistry is similar to the original groundwater + $Mg\text{-}Na\text{-}SO_4$ (Table 1).

Geochemical modeling (PHREEQC, Parkhurst 1995) indicated after addition of $MgO + NaAlO_2$, HT at pH 6.9 was ~ 10^{23} times oversaturated with respect to the solution. Mineralogical (XRD) analysis confirmed HT as the principal mineral.

The major and trace element composition of the HT precipitate formed from the addition of $MgO + NaAlO_2$ to Beverley ISR mine barren lixiviant and after calcination at 650°C is dominated by MgO and Al_2O_3 that collectively constitute ~ 45 and 59%, respectively (Table 2). The 30% mass increase is consistent with the conversion of HT to spinel (confirmed by mineralogical analysis). Sulfate, (as SO_3) was the major solution anion, and hence predominant HT interlayer anion constituting ~ 10% in the HT and 14% after calcination to spinel.

Dissolved U from the barren lixiviant (19 mg/L) was concentrated within the HT (7162 μg/g, 0.85% eU_3O_8) and calcined HT (9778 μg/g, or 1.15% eU_3O_8), comparable to primary ore grade (Table 2). Rare earth elements (REE) were also concentrated; the sum of REE (La + Ce + Nd + Sm + Yb) was 5,867 and 8233 μg/g, re-

Table 2 Major element oxide (%) and trace element (µg/g) geochemistry of Beverley ISR mine barren lixiviant (lix) precipitates

Element/oxide	Lix + MgO + $NaAlO_2$	Calcined Lix + MgO + $NaAlO_2$
SiO_2	7.45	9.91
TiO_2	<0.002	0.01
Al_2O_3	25.8	34.14
Fe_2O_3	3.86	4.99
MnO	0.19	0.26
MgO	18.79	24.77
CaO	1.36	2.03
Na_2O	1.31	1.71
K_2O	0.06	0.09
SO_3	10.16	13.91
As	22	29
Ce	1896	2708
Co	2432	3125
Cr	96	116
Cu	179	218
La	317	448
Mo	18	24
Nd	2068	2894
Ni	1175	1527
Pb	51	14
Sc	290	405
Se	10	9
Sm	946	1308
Th	431	569
U	7162	9778
V	470	647
Y	4731	6403
Yb	640	875
Zn	1859	1717

spectively, in the HT and calcined HT. Estimated total REE concentrations based on interpolation were 9000 and 12,500 µg/g, respectively. Notable was the enrichment of heavy REE over light REE. Heavy REE such as Yb and Y, a surrogate mid-HREE, had concentrations 4731 and 6403 µg/g in the HT and calcined HT (spinel), respectively. Scandium, frequently found in association with REE, also had elevated concentrations of 290 and 405 µg/g, respectively.

In addition to U and REE, many trace elements (Co, Ni, Zn) exceeded 1000 µg/g in both the HT and the calcined derivative, with other elements (Cu, Cr, Th) exceeding 100 µg/g. Given the enrichments of U, Th and trace elements, it is likely that U-Th series radionuclides were also concentrated within the HT.

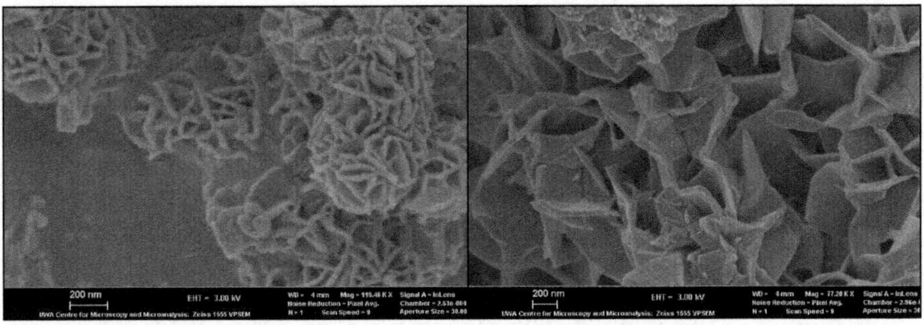

Fig. 2 Hydrotalcite (HT) botryoids and detail of face to edge and edge to edge HT aggregates. Scale bars are 200 nm.

Analysis by SEM revealed flat, hexagonal crystals arranged as edge-to-edge aggregates within botryoidal masses (Fig. 2). Element mapping revealed similar distributions of Mg-Al with an Mg : Al molar ratio of ~ 1. While lower than that of HT, abundant Si (~ 7%) and a Si : Al molar ratio of ~ 1 suggests co-precipitation of an amorphous phase with other cation substitution also decreasing the Mg : Al ratio. Element mapping of U (~ 1%) indicated that it was present in a diffuse distribution within HT rather than as a discrete secondary uraniferous precipitate.

Batch experiment results have demonstrated that a range of major and trace elements are removed during HT precipitation and confirms aspects of the patents of Douglas (2006, 2009), that the combined addition of $MgO + NaAlO_2$ act concurrently as a pH neutralizing agent and as building blocks for HT formation. A HT precipitation reaction, reflecting the barren lixiviant composition is:

$$\begin{aligned}&6MgSO_4 + 2NaAl(OH)_4 + MgO_{(ex)} + 8NaOH \rightarrow \\ &Mg_6Al_2(OH)_{16}SO_4 \cdot nH_2O + 5Na_2SO_4 + MgO_{(ex)}\end{aligned}, \quad (1)$$

where $MgO_{(ex)}$ reflects excess MgO added as a Mg source and neutralizing agent. Post-neutralization solutes were predominantly Na-SO_4 reflecting the presence of SO_4 from the acidic barren lixiviant and Na derived from $NaAlO_2$ addition.

Calcination resulted in the dehydration and re-crystallization of HT to spinel ($MgAl_2O_4$). Conversion to spinel occurs by the following reaction where the Mg not utilized in the spinel is converted to $MgSO_4$ or MgO:

$$\begin{aligned}&Mg_6Al_2(OH)_{16}SO_4 \cdot nH_2O \rightarrow \\ &MgAl_2O_4 + 8H_2O + 4MgSO_4 + MgSO_4 + nH_2O\end{aligned}. \quad (2)$$

Calcination of HT resulted in a mass loss of ~30% due to dehydration. The MgO formed during the calcination is present in a 4 : 1 mol ratio, constituting a large buffering capacity in the event of low pH conditions. Alternatively, HT addition of silica as an interlayer anion results in a chlorite-like composition that could potentially be used as a "natural" repository. Formation of an Na_2SO_4- rich solution after precipitation also enables the possible use of electrolysis to produce H_2SO_4 for lixiviant generation, and NaOH for neutralization.

Conclusions

Batch-scale studies have demonstrated proof of concept for the neutralization of acidity and removal of contaminants from Beverley ISR mine barren lixiviant using $MgO + NaAlO_2$. Hydrotalcite formed during neutralization hosts a range of contaminants, notably ~1% U and ~2% REE indicating a potential to both remediate barren lixiviant prior to aquifer re-injection and to offset remediation costs. Importantly, the ionic composition of the neutralized barren lixiviant is similar to existing groundwater allowing for direct disposal. Hydrotalcite formed during neutralization may be further stabilized via calcination and/or silicification producing minerals potentially amenable for inclusion in a long-term waste repository.

In the Beverley context, the HT-based remediation technology is considered a potential groundwater treatment, if required, for the future closure of its Beverley North operations. Whilst Beverley's ionic composition would mean easier application of this technology there, potential exists to apply it to other mines. This remediation technology, after scale-up and performance validation, allows for the prospect of a fully integrated ISR mining, processing and lixiviant remediation strategy consistent with stringent environmental and mine closure standards.

References

Behrens M, Kasatkin I, Kuhl S, Weinberg, G (2010) Phase-Pure Cu, Zn, Al Hydrotalcite-like Materials as Precursors for Copper rich $Cu/ZnO/Al_2O_3$ Cat, Chem Mater 22:386–397

Cavani F, Trifiro F, Vaccari A (1991) HT-type anionic clays: preparation, properties and applications. Catal Today 11:173–301

Curtis J, Brunt D, Binks P (1990) Tertiary Palaeochannel Uranium Deposits of South Australia. In: Geology of the Mineral Deposits of Australia and Papua New Guinea (Ed. FE Hughes) pp. 1631–1636, AusIMM, Melbourne

Douglas GB (2006) Provisional Patent Application. Remediation of groundwater

Douglas GB (2009) Provisional Patent Application. Treatment and Remediation of Natural and Wastewater

Douglas GB, Wendling LA, Pleysier R, Trefry MG (2010) Hydrotalcite formation for contaminant removal from Ranger mine process water. Mine Water Env 29:108–115

Genin J-M, Refait P, Bourrie G, Abdelmoula M, Trolard F (2001) Structure and stability of the Fe(II)-Fe(III) green rust "fougerite" mineral and its potential for reducing pollutants in soil solutions. Appl Geochem 16:559–570

Hou B, Fabris AJ, Keeling JL, Fairclough MC (2007) Cainozoic palaeochannel-hosted uranium and current exploration methods, South Australia. MESA J 46:34–39

Miyata S (1983) Anion-exchange properties of HT-like compounds. Clays Clay Min 31:305–311

Parkhurst DL (1995) Users guide to PHREEQC – A computer program for speciation, reaction-path, advective transport, and inverse geochemical calculations. USGS Wat Res Inv Rpt 95-4277

Ruperto L, de Catitat P (2006) Geological review of the southern Curnamona region. CRC LEME Open File Report 183. 24 pp

Seida Y, Nakano Y (2000) Removal of humic substances by layered double hydroxide containing iron. Wat Res 34:1487–1494

Seida Y, Nakano Y (2002) Removal of phosphate by layered double hydroxides containing iron. Wat Res 36:1306–1312
Shin HS, Kim MJ, Nam SY, Moon H-C (1996) Phosphorus removal by HT compounds (HTLcs). Wat Sci Tech 34:161–168
Taylor RM (1984) The rapid formation of crystalline double hydroxy salts and other compounds by controlled hydrolysis. Clay Min 19:591–603
Ulibarri MA, Pavlovic I, Barriga C, Hermosín MC and Cornejo J (2001) Adsorption of anionic species on hydrotalcite-like compounds: effect of interlayer anion and crystallinity. Appl Clay Sci 18:17–27
Vucelic M, Jones W, Moggridge GD (1997) Cation ordering in synthetic layered double hydroxides. Clays Clay Min 45:803–813

Concentration of U and Th in the Bloedkoppie Granite, Namibia

Fred Kamona

Abstract. The leucocratic Bloedkoppie Granite is generally considered to be the source rock of most of the uranium concentrated in the nearby Langer Heinrich calcrete-hosted uranium deposit. Ground gamma-ray radiometric measurements for K, U and Th along two survey profiles within the Bloedkoppie Granite showed that the intrusive body is uranium-rich with 15 ppm U and 33 ppm Th. Pegmatite dykes within the granite contain relatively higher U concentrations (23 ppm) compared to the granite (10 ppm).

Introduction

Numerous granites associated with uranium mineralization in Namibia are concentrated in the Central Zone of the Pan-African Damaran Orogen (Jacob et al. 1986). These granites include leucocratic alkali granites with potentially economic primary uranium mineralization at the Rössing, Valencia, Ida Dome and Goanikontes deposits as well as those considered to be probable source rocks of uranium for the various secondary calcrete-hosted uranium occurrences in the Namib Desert (Hambleton-Jones 1984, Fig. 1).

Langer Heinrich and Trekkopje are the two major secondary uranium deposits in Namibia. Uranium production at Langer Heinrich started in 2006 where 12.9 Mt ore of and 11.3 Mt of waste have been mined with an ore feed of 850 ppm U_3O_8 (Langer Heinrich 2010, personal communication). The Trekkopje deposit is currently under development and mining is scheduled to start in 2012. The Bloedkoppie and Spitzkoppe granites have been considered as the probable sources of ura-

Fred Kamona
University of Namibia, Windhoek, Namibia

B. Merkel, M. Schipek, *The New Uranium Mining Boom*
DOI 10.1007/978-3-642-22122-4

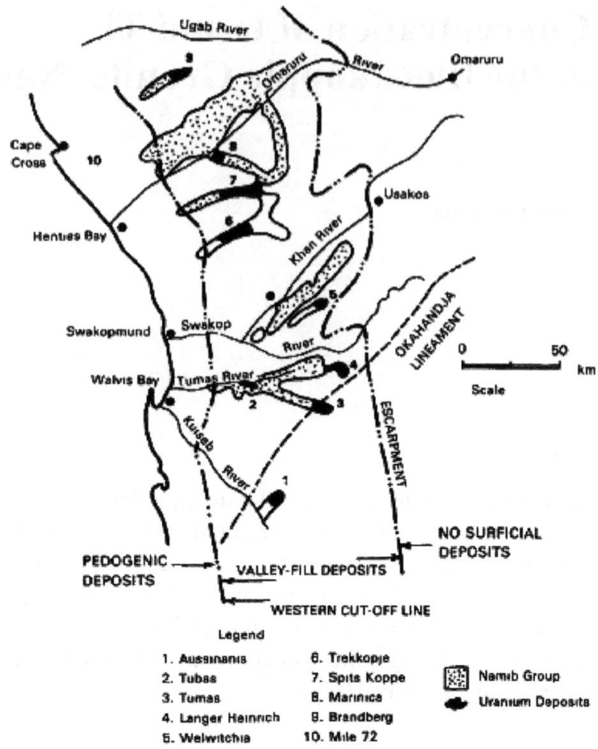

Fig. 1 Distribution of calcrete-hosted carnotite deposits in central Namibia (from Hambleton-Jones 1984)

nium at Langer Heinrich and Trekkopje, respectively (Hambleton-Jones 1976, 1984; Jacob et al. 1986).

The Bloedkoppie Granite is the main granite underlying the catchment area of a 15 km long Upper Cretaceous palaeochannel containing near surface mineralization, 1 to 30 m thick and 50 to 1100 m wide depending on the width of the palaeovalley (Langer Heinrich 2010, personal communication). Carnotite ($K_2(UO_2)_2(VO_4)_2 \cdot 3(H_2O)$) is the main uranium mineral which lines cavities, fracture planes and also occurs as grain coatings and disseminations in the calcretized sediments. According to Hambleton-Jones (1984), carnotite precipitation occurred during the Upper Tertiary in zones of high porosity within the fluviatile sediments of the Langer Heinrich Formation.

Methodology

A field radioactive survey using a portable gamma spectrometer was conducted in order to establish the distribution of uranium in the Bloedkoppie Granite and assess the potential use of ground radiometric surveys in identifying uranium bearing granites in the Damaran Orogen. The radiometric survey was made with a portable gamma ray spectrometer (GF Instruments model) with Thallium-activated sodium iodide NaI(Tl) crystals as detectors capable of recording the full gamma ray spectrum as well as sum 512 channels over broad energy windows for the in-situ estimation of K, U and Th radioelement concentrations. Automatic spectrum stabilization was by a low-energy peak from a reference isotope (^{137}Cs at 0.662 MeV). Measurement of the concentrations of K, U and Th were made for periods of two minutes at each station along two selected profiles within the Bloedkoppie Granite over distances of 110 and 300 m, respectively.

The Bloedkoppie Granite

The Bloedkoppie Granite is a syn- to post-tectonic leucocratic alkali granite intrusive into biotite-muscovite schist, calc-silicate and marble units of the Tinkas Formation of the Damaran Supergroup (Miller 2008). The granite is composed mainly of microcline, quartz, biotite and plagioclase with accessory zircon and varies in texture from medium- to coarse-grained. It is intruded by numerous quartz-feldspar pegmatite veins with minor to accessory garnet, magnetite and tourmaline (schorl). Recent age determinations based on U-Pb single zircon geochronological dating indicates an emplacement age of 554 ± 25 Ma for the granite (Kamona, unpublished data).

Profile 1

A field gamma-ray radiometric survey to determine the concentration of K, U and Th in the Bloedkoppie Granite was initially conducted along a NNW-WSW 110 m profile in a part of the granite body with well exposed pegmatite veins varying in thickness from 1 to 290 cm and generally trending at a bearing of 110°. The results of the survey are summarized in Table 1 for both granite and pegmatite, granite (excluding pegmatite) and pegmatite (excluding granite) respectively.

From the data it is observed that the radiometric concentrations of K and Th along Profile 1 in the granite and pegmatite veins are not significantly different (Table 1). However, the average uranium concentration in the pegmatite veins is nearly three times higher than in the granite, indicating a preferential enrichment of uranium in late pegmatitic fluids associated with the formation of pegmatite veins.

The variations in the concentrations of K, U and Th along Profile 1 are shown in Fig. 2 which is a profile that includes both granite and pegmatite rocks encoun-

Table 1 Average values of K, U, Th and Th/U in Profile 1

Rock Type	K [%]	U [ppm]	Th [ppm]	Th/U	n
Granite and Pegmatite	4.8 ± 0.1	15.5 ± 1.6	31.6 ± 1.3	4.5 ± 1.4	58
Granite	4.9 ± 0.1	8.5 ± 0.7	33.0 ± 1.7	5.2 ± 1.7	33
Pegmatite	4.7 ± 0.2	24.8 ± 2.8	29.6 ± 2.2	3.6 ± 2.2	25

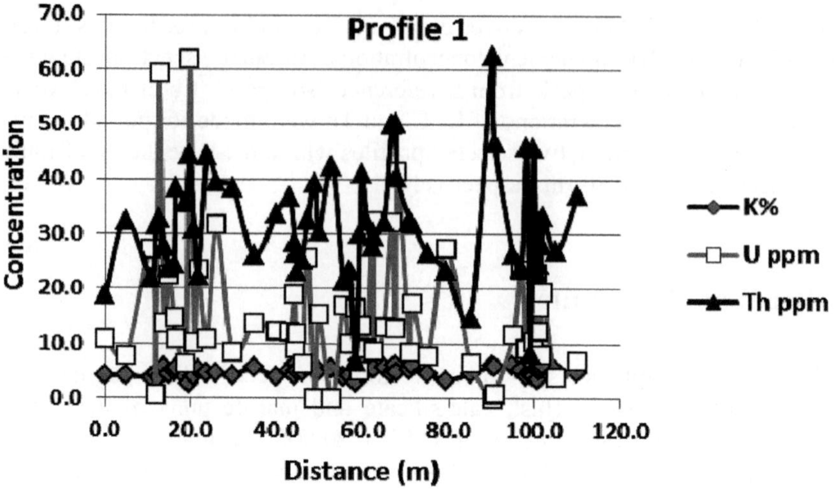

Fig. 2 Variation in concentrations of K, U and Th in the Bloedkoppie Granite along Profile 1

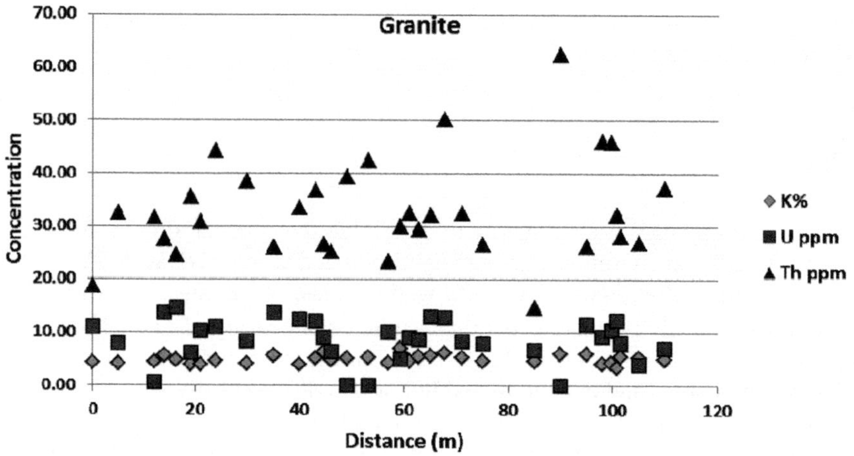

Fig. 3 Variation in concentrations of K, U and Th in granite along Profile 1

tered along the survey. Whereas both U and Th exhibit wide ranges in their respective concentrations (0–61.9 ppm U and 6.8–62.5 ppm Th), the concentration of K is fairly constant in both granite and pegmatite along the profile.

The concentrations of the radionuclides in the two rock types (granite and pegmatite) along the survey profile are markedly different, with relatively lower concentration of uranium in the granite which varies from 0 to 14.7 ppm U (Fig. 3) in comparison to the pegmatite veins with 0.8–61.9 ppm U (Fig. 4). As a consequence of the low uranium concentration, the relative concentration of Th with respect to U in the granite is consistently higher than in the pegmatites as indicated by the Th/U ratios of 5.2 and 3.6 in the granite and pegmatite veins, respectively (Table 1).

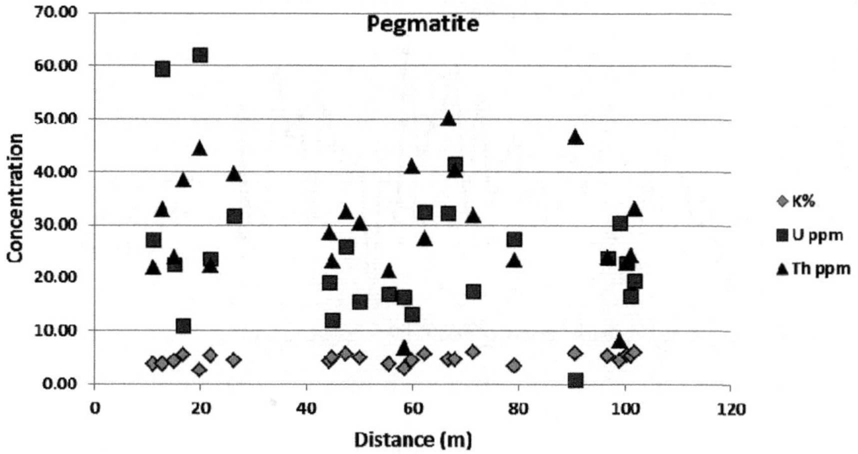

Fig. 4 Variation in concentrations of K, U and Th in pegmatite along Profile 1

Profile 2

The second radiometric survey was conducted along a 250 m long N-S profile in the Bloedkoppie Granite with fewer and thinner (2–40 cm), randomly oriented pegmatite veins within a generally light pink, medium- to coarse-grained biotiterich granite. The radiometric concentrations of K and Th along Profile 2 in the granite and pegmatite veins are not significantly different (Table 2). However, the average uranium concentration in the pegmatite veins is nearly twice that in the granite, indicating a preferential enrichment of uranium in late pegmatitic fluids associated with the formation of pegmatite veins. It is also apparent that the U and Th values along this profile vary widely (0.5 to 46.4 ppm U and 12.9 to 60.8 ppm Th), whereas the K concentrations are much less variable (Fig. 5).

Separation of the data with respect to rock type confirms the relative enrichment of uranium in pegmatite with 12 to 46.4 ppm U (Fig. 6) rather than in granite which contains from 0.5 to 30.6 ppm U (Fig. 7) as observed in Profile 1 above. In this case the granite has average values of 11.2 ppm U compared to 21.9 ppm U in the pegmatite veins with corresponding Th/U ratios of 7.5 and 1.8, respectively (Table 2). In contrast the average concentration of Th and K are quite similar in both rock types (33.3 ppm Th and 4.9% K in granite compared to 34.6 ppm Th and 4.1% K in pegmatites).

Table 2 Average values of K, U, Th and Th/U in Profile 2

Rock Type	K [%]	U [ppm]	Th [ppm]	Th/U	n
Granite and Pegmatite	4.7 ± 0.1	13.9 ± 1.9	33.7 ± 1.6	6.1 ± 2.2	52
Granite	4.9 ± 0.2	11.2 ± 1.0	33.3 ± 1.8	7.5 ± 2.9	39
Pegmatite	4.1 ± 0.2	21.9 ± 2.6	34.6 ± 3.4	1.8 ± 0.3	13

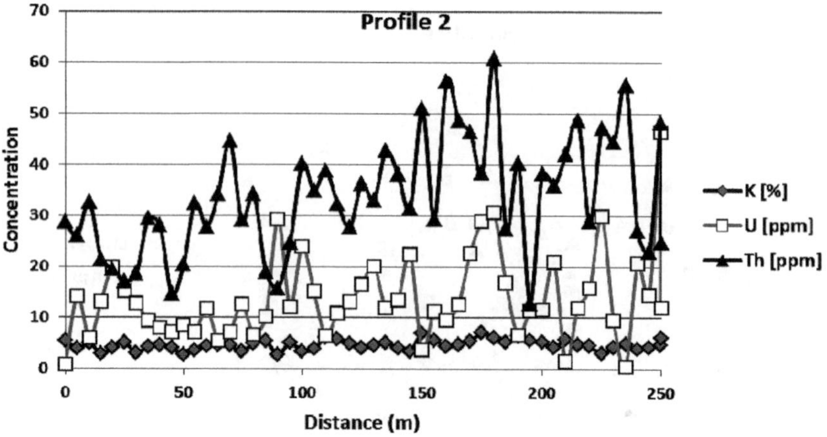

Fig. 5 Variation in concentrations of K, U and Th in the Bloedkoppie Granite along Profile 2

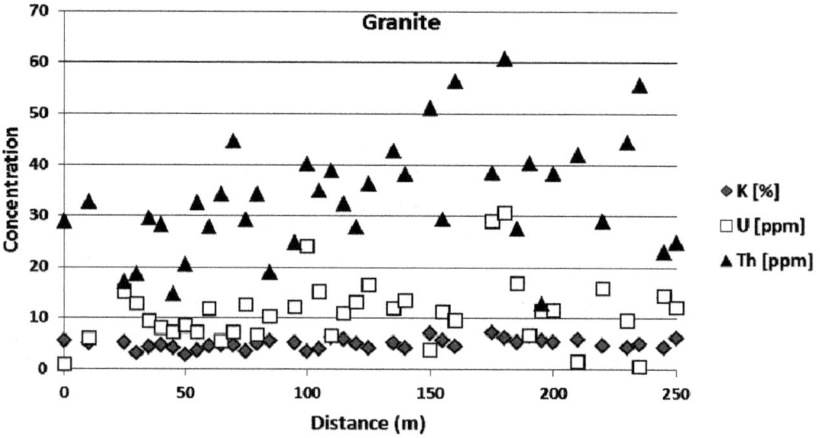

Fig. 6 Variation in concentrations of K, U and Th in granite along Profile 2

Discussion

The radiometric surveys have demonstrated that the leucocratic Bloedkoppie Granite is a likely source of uranium for the Langer Heinrich calcrete-hosted deposit due to its relatively high uranium content (15 ppm U) which is typical of uranium-rich granites with 10 to 20 ppm U (e.g. Lehman 2008). Granite emplacement occurred at 554 ± 25 Ma as a result of partial melting of crustal rocks during the Damaran Orogeny. According to Jacob et al. (1986), the uranium was concentrated in granitic melts during high-grade metamorphism of the basement gneisses of the Paleoproterozoic Abbabis Complex (which contain 3 to 5 ppm U_3O_8) and metaquartzite of the overlying Neoproterozoic Nosib Group.

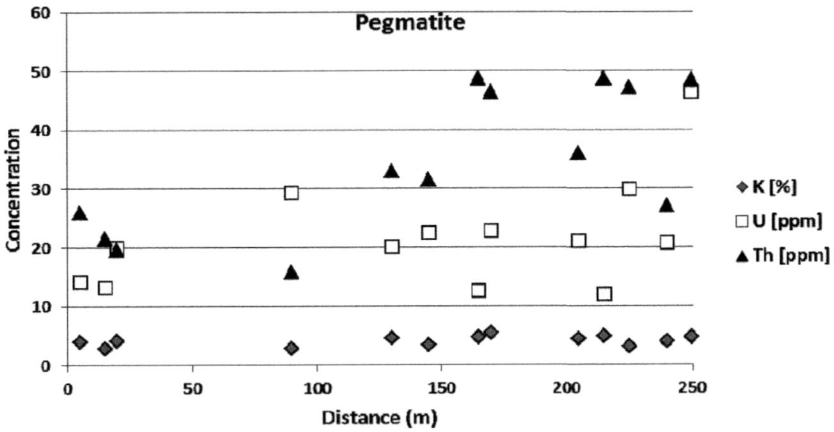

Fig. 7 Variation in concentrations of K, U and Th in pegmatite along Profile 2

Concentration of U and Th was due to normal fractionation processes resulting in enrichment of these large highly charged cations in granitic melts (Jacob et al. 1986; Cuney 2008). However, most of the uranium was concentrated in residual pegmatitic melts from which intrusive pegmatite veins formed within the Bloedkoppie Granite as indicated by the higher uranium content of the pegmatite veins (23 ppm U) in comparison to the granite (10 ppm U).

Weathering of the Bloedkoppie Granite under semi-arid to arid conditions resulted in the leaching and transportation of highly soluble hexavalent uranium (U^{6+}) as a uranyl carbonate complex in warm meteoric waters and its fixation as U^{4+} together with potassium and vanadium due mainly to evapotranspiration (Lehman 2008) to form carnotite as void fillings, fracture coatings and as disseminated cement in porous valley-fill sediments in the Upper Tertiary (5–2 Ma) at Langer Heinrich (Hambleton-Jones 1984).

Conclusion

The sensitivity of modern spectrometers is capable of discriminating between uraniferous rock types such as granites and pegmatites as observed at the Bloedkoppie Granite where pegmatite veins contain up to three times more uranium in comparison to the leucocratic granite. Ground radiometric surveys are therefore very useful in uranium exploration to identify anomalous areas with primary mineralization as well as potential source rocks of secondary surficial uranium mineralization such as that found at Langer Heinrich.

References

Cuney, M. (2008). The extreme diversity of uranium deposits. Mineralium Deposita, *www.springerlink.com*, accessed March 2010.

Hambleton-Jones, B.B. (1976). The geology and geochemistry of some epigenetic uranium deposits near the Swakop River, South-West Africa. Ph.D. Thesis, University of Pretoria, 306 pp.

Hambleton-Jones, B.B. (1984). Surficial uranium deposits in Namibia. In: TOEN, P.D. (ed) *Surficial Uranium Deposits*. IAEA-Tecdoc-322, Vienna, 205–216.

Jacob, R.E., Corner, B., Brynard, H.J. (1986). The regional and structural setting of the uraniferous granitic provinces of southern Africa, 1807–1818. In: Anhaeusser, C.R., and Maske, S. (eds.) Mineral Deposits of Southern Africa, II. Geological Society of South Africa, 2335 pp.

Lehman, B. (2008). Uranium ore deposits. *www.advanced-mining.com*, Issue 2, accessed February 2010.

Concentration Dynamics and Speciation of Uranium in a Boreal Forest Creek – Six Years of Weekly Observations

Stefan Karlsson, Bert Allard

Abstract. The concentrations of major components as well as uranium and some transition metals (Cu, Zn, Cd, Pb) and DOC were measured in a small creek within a boreal catchment during a six years time period. Variations up to four orders were found during the entire period. There was an evident correlation between uranium concentration with concentrations of Fe and Al, however not with DOC (in the absence of carbonate, average pH of 5.5). It is evident that environmental quality monitoring in small catchments must include the highly variable conditions leading to large variations in concentrations and speciation.

Introduction

Although the hydrogeochemistry of uranium, as well as of transition metals, in the environment is well known in general, there are only few studies of the these elements at natural background levels in boreal catchments. Background levels of uranium are generally substantial in Sweden, reflecting the contributions from the granitic bedrock (some 5 mg/kg of uranium), as well as from Cambrian shales in some regions (up to some 300 mg/kg of uranium). In general U(IV) is sparingly soluble, while U(VI) is mobile and without obvious solubility constraints.

Stefan Karlsson
Man-Technology-Environment Research Centre, School of Science and Technology, Örebro University, SE-701 82 Sweden

Bert Allard
Man-Technology-Environment Research Centre, School of Science and Technology, Örebro University, SE-701 82 Sweden

B. Merkel, M. Schipek, *The New Uranium Mining Boom*
DOI 10.1007/978-3-642-22122-4 © Springer 2011

Recent studies have shown that interactions with colloidal matter in groundwater might decrease the mobility of U(VI)-species and increase the mobility of U(IV)-species (Zänker et al. 2006). Identical migration times in sand columns, except for the impact of colloid filtration, were demonstrated for both oxidation states in the presence of humic acid (Mibus et al. 2007). This would be of importance in many till environments considering the high stability constants for U(VI) humate complexes (Szabó et al. 2006). Also the stability of iron-U(VI) surface complexes is high, and the effect of sulfate as well as silicate at environmental levels is minor. Field studies have identified two dominating size classes of uranium carrier phases in soil (Claveranne-Lamolère et al. 2009), one corresponding to high molecular-weight humic acids and the other one to aggregates of mineral/organic origin. It should also be noted that interactions with humic substances in the presence of Fe(III) and Fe(II) might stabilize U(IV) (Ulrich et al. 2006; Jang et al. 2008). It is essential to include hydrological as well as chemical mechanism in soils and sediments when analyzing the transport of uranium and other elements at background levels in catchments where the conditions inevitably will vary. In general, the transport will be governed by chemical processes in solution and on surfaces, as well as by the physical transport of dissolved species and suspended particulate carrier substances.

The mobility of uranium and of some transition metals at trace levels, in a small boreal catchment is described in this study, focusing on the importance of the variations of physical as well as hydrochemical conditions over time.

Materials and Methods

The Field Site

The catchment, that has an area of some 4 km^2, is located in Dyltabruk, some 200 km west of Stockholm, Sweden. The parent bedrock underneath the rather coarse till soil is granite/gneiss with lenses of calcareous minerals, notably limestone, and clays. Some 80% of the catchment has quite dense tree stands of coniferous (80%) and deciduous (20%) species. The lower vegetation is dominated by grass and shrubs (*Vaccinium spp*). The remaining part of the catchment contains bogs and mires. Annual precipitation is some 650 mm, and the related evapotranspiration is some 200–250 mm. Typically, winter begins in early November and lasts until early April, when the snow pack melts in a few weeks.

The Listrebäcken creek is a typical surface water pathway representative of forested catchments in this part of Sweden. It is 0.5–1.5 m wide, has a depth of 0.1–1 m and it is to some extent resulting from drainage of the surrounding forest soil. The sampling site was some 2.3 km downstream in the catchment.

Sampling and Analysis

Samples were collected and water levels measured weekly although with higher frequency during a few rain storms as well as during snow melt episodes. The weekly samples were taken at one specific site, but the whole creek was also sampled at least once a year. Samples for metal analysis were taken in polypropylene test tubes (Sarstedt; 15 ml for metal analysis, 50 ml for other parameters). Samples for metal analysis were acidified to 1% HNO_3 with sub-boiled distilled acid.

Electrical conductivity and pH were recorded with conventional probes. Ion chromatography using a Dionex AS-12 column with bicarbonate/carbonate buffer as mobile phase was used for the determination of inorganic anions in filtered samples. The dissolved carbon content (filtration through 0.20 um membranes) was quantified as dissolved organic carbon (DOC) and inorganic carbon (IC) with a Shimadzu TOC-V CPH analyzer. The content of low molecular weight organic acids (LMWOA) was measured in the filtrate by capillary electrophoresis (Dahlén et al. 2000). The molecular weight distribution of humic substances was estimated from gel permeation chromatography in a separate filtered sample.

Metal analysis was made by ICP-MS (Agilent 4500, Agilent 7500) on the acidified original samples as well as on those resulting after different treatments.

Results and Discussion

General Hydrochemistry

During the period of study the precipitation has deviated slightly from the Swedish reference period (Fig. 1). Autumns and springs have slightly higher precipitation while summers and winter were dryer. No impact could be seen in the creek because of the start of the time series. The system is highly variable, with water levels ranging from a few mm during summer droughts to 60 cm at snow melt, as shown by the variation of the general hydrochemical parameters and principal metal concentrations (Tables 1 and 2).

According to Tables 1 and 2 (e.g. mean values) the water is low in total salts (ions), high in DOC and slightly acidic, as expected in this kind of catchment dominated by acidic organic material (humic). Generally, concentrations are high at low water levels and vice versa. Low water level would indicate a larger contribution from the groundwater. There is also a pronounced seasonality of organic matter as well as conservative and reactive elements (c.f. Fig. 2a, b).

There are periods that are highly different from the general pattern. Individual rain storms with high intensities regularly result in resuspension and influx of soil material. This is evident for Al and Fe where the particulate fraction periodically increases from 20 to more than 95% (not illustrated) whereas only minute changes of particulate Mn were observed. Geochemical modeling showed that unfiltered samples were close to equilibrium or oversaturated with respect to $Al(OH)_3$(am), $Fe(OH)_3$(am) and FeOOH(s).

The DOC has similar concentration dynamics and is one of the constituents that showed a tendency of increasing concentrations over the six years time period. This becomes more evident when the time series for the DOC/Cl ratio is plotted (Fig. 3). The Cl was conservative and did not show any significant concentration changes during the period of measurement, except for its seasonality.

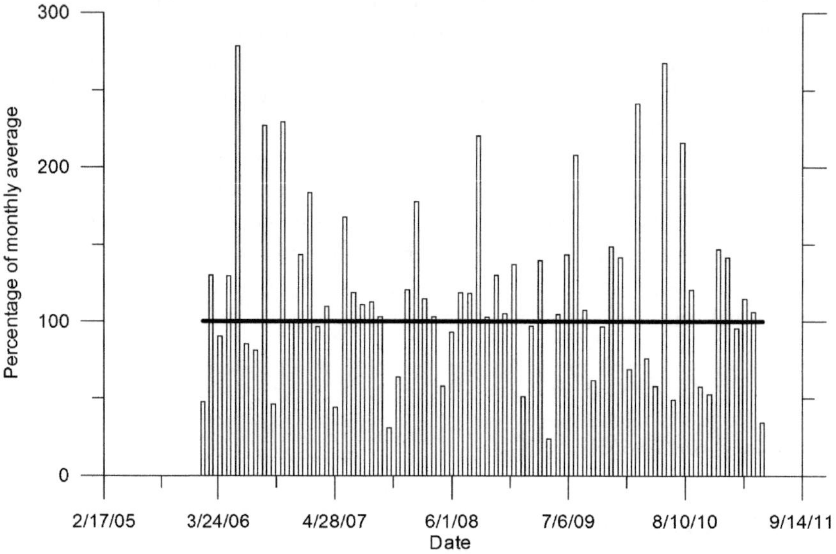

Fig. 1 Percentage of monthly precipitation in relation to the Swedish reference period

Table 1 Summary of general hydrochemical parameters (n = 249)

	El. Cond (uS/cm)	pH	DOC (mg/l)	F^- (mg/l)	Cl^- (mg/l)	SO_4^{2-} (mg/l)	NO_3^- (mg/l)
Min	28.7	4.47	0.1	0.02	1.20	0.67	0.10
Max	114.6	6.85	90.9	1.22	15.3	114.9	14.4
Mean	46.0	5.48	30.3	0.33	3.43	4.60	1.51
S.D.	12.21	0.45	13.2	13.22	1.41	7.97	2.15

Table 2 Summary of principal metal concentrations (mg/l) (n = 249)

	Ca	Mg	Al	Fe	Mn
Min	1.92	0.69	0.26	0.15	0.001
Max	14.86	5.17	4.40	16.28	0.670
Mean	4.03	1.40	0.91	2.47	0.056
S.D.	1.29	0.45	0.60	2.19	0.068

The quality of the DOC has two dominating features, according to the gel permeation chromatography. Fulvic acids (mw 400–2000 D) are always present and correlates (r^2 0.89) with concentrations of DOC. During late autumn and occasionally at intense rainstorms humic acids (mw > 10,000 D) were found.

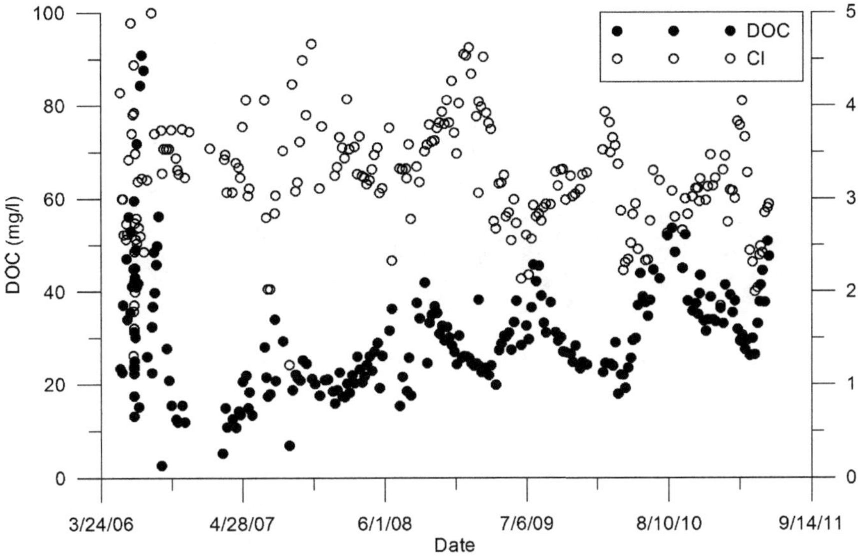

Fig. 2a Time series of DOC (mg/l) and Cl^- (mg/l)

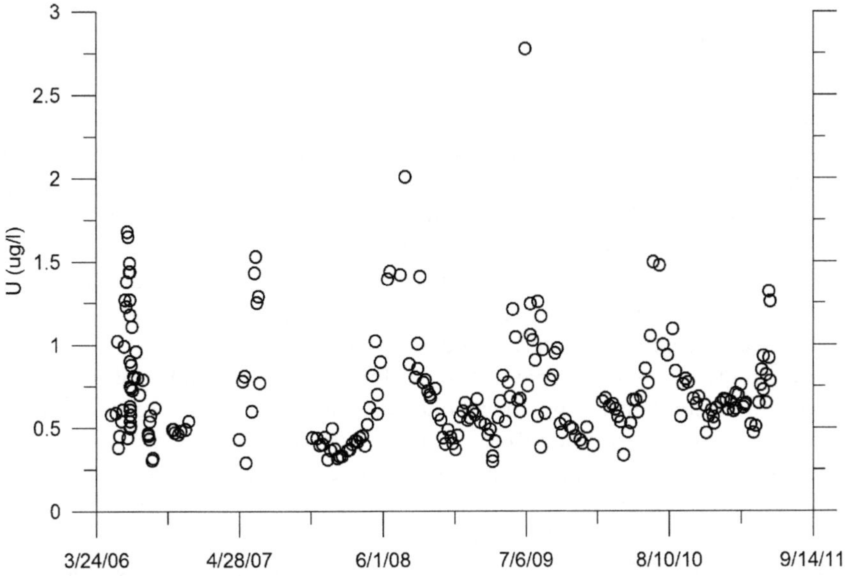

Fig. 2b Time series of uranium (ug/l)

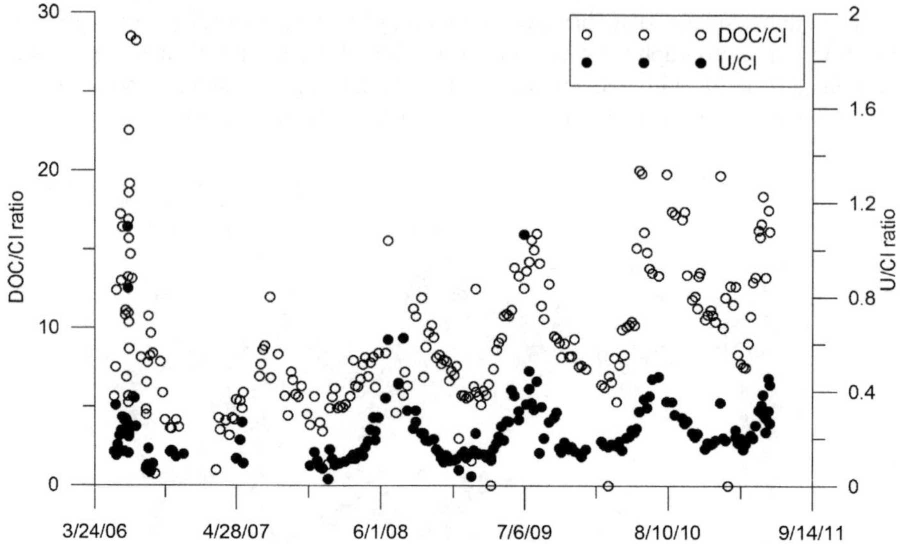

Fig. 3 Time series of ratios for DOC/Cl and U/Cl

The LMWOA, which represent a minor fraction of the total DOC (generally less than 5%), is dominated by formate, acetate, citrate and oxalate although lactate was occasionally found during late winters. Their concentrations seem to be more related to temperature than the hydrodynamics.

Trace Elements

The concentrations of trace metals are generally low in the present creek, Table 3. Data from "extreme" conditions are excluded in this compilation in order to better account for ordinary conditions. Maximum concentrations at some periods with intense rain storms and high fluxes of solid matter were as high as 92 µg/l of U, as well as 25, 130 and 90 µg/l of Cu, Zn and Pb, respectively, in the aqueous phase. The high uranium concentration is not expected in surface waters but may occasionally be found in wells in the region. It should also be noted that at DOC concentrations was as high as 460 mg/l at one occasion.

No stoichiometric solid phases were found for Cu, Zn, Cd and Pb, according to the geochemical modeling why their solid/solution partitioning would be controlled by adsorption. Increasing ratios with Cl were observed for Cu, Zn and Pb why the concentrations of these elements are increasing in the creek water. The reason is unclear but correlation coefficients between their Cl-ratios and DOC are in the range of 0.10 to 0.20 while they are 0.44 to 0.60 for Fe and Al. Thus it, appears as if the two latter elements have some impact, or a corresponding dynamic behavior.

Table 3 Summary of principal metal concentrations (mg/l) (n = 249)

	U	Cu	Zn	Cd	Pb
Min	0.029	0.34	2.55	0.00002	0.10
Max	2.770	9.21	58.91	0.150	11.38
Mean	0.724	1.72	7.92	0.033	1.49
S.D.	0.345	1.29	5.96	0.039	1.56

The concentration changes of uranium are very similar to the transition metals, i.e. a pronounced seasonality, a U/Cl ratio that increases with time and total concentrations below those of likely stoichiometric solids (Fig. 2). The uranium concentrations are independent of pH. Geochemical modeling using the NO_3^-/NO_2^- and SO_4^{2-}/S_{tot} ratios as estimators for the Eh regime suggested a clear dominance of U(VI).

The uranium concentrations are highly correlated with those of Al (r^2 0.70) and Fe (r^2 0.89), just as for the transition metals, which would indicate similar sources or redistribution to those potential carrier phases rather than to DOC (r^2 0.32). This is partly supported by analysis of uranium in a few GPC separations where less than 20% co-eluted with the fulvic acid fraction. The particulate fraction of uranium was generally high, with a maximum of 72% associated with particles.

Conclusions

The observations regarding uranium – concentrations and mobility, can be summarized:

- The concentrations in the creek varies by more than three orders of magnitude – from 0.03 to 2.8 μg/l, with some extreme levels at storm events, above 90 μg/l
- There is no significant correlation with DOC-levels
- There is no significant correlation with pH (and not with carbonate; average pH of 5.5)
- There is a significant correlation with Al and Fe, indicating that these element may serve as carriers, (or that U, Fe and Al come from the same source)

The study demonstrates the importance to jointly consider physical properties of the catchment and its hydrogeochemistry in the analysis and interpretation of concentration and distribution data. The seasonal concentration changes are pronounced for most parameters and during the six years of the study the concentrations of U, as well as of Cu, Pb, Zn and of DOC, show increasing tendencies in relation to the more conservative Cl^-. The reason for this is not obvious. The increase in concentrations of DOC may be more related to physical changes such as water level changes in the system rather than chemical changes. No systematic changes of pH were observed. It is clear, however, that particles with high contents of Al and Fe may serve as the major carrier phase, possibly enhanced by particulate organic material.

Acknowledgements The authors express their gratitude to the landowner, Dyltabruk Förvalning AB, for access to the site and to Ms C Kullberg for compiling climatic data.

References

Claveranne-Lamolère C, Lespes G, Dubascoux S, Aupais J, Pointurier F, Potin-Gautier M (2009) Colloidal transport in soil: Size fractionation and characterization by field-flow fractionation – multi detection. J Chromatogr A 1216:9113–9119

Dahlén J, Hagberg J, Karlsson S (2000) Analysis of low molecular weight organic acids in water with capillary zone electrophoresis employing indirect photometric detection. Fres. J. Anal. Chem. 366:488–493

Jang J-H, Dempsey B A, Burgos W D (2008) Reduction of U(VI) by Fe(II) in the presence of hydrous ferric oxide and hematite: Effects of solid transformation, surface coverage, and humic acids. Water Res. 42:2269–2277

Mibus J, Sachs S, Pfingsten W, Nebelung C, Bernhard G (2007) Migration of uranium(IV)/(VI) in the presence of humic acids in quartz sand: A laboratory column study. J. Contam. Hydrol. 89:199–217

Szabó Z, Toraishi T, Vallet V, Grenthe I (2006) Solution chemistry of actinides: Thermodynamics, structure and reaction mechanisms. Coord. Chem. Rev. 250:784–815

Ulrich K-U, Rossberg A, Foerstendorf H, Zänker H, Scheinost A C (2006) Molecular characterization of uranium(VI) sorption complexes on iron(III)-rich acid mine water colloids. Geochim. Cosmochim. Acta 70:5469–5487

Zänker H, Ulrich K-U, Operl K, Brendler V (2006) Influence of colloids on uranium transport in nuclear waste repositories and abandoned uranium mines – A critical comparison. Goldschmidt Conference Abstracts A731

Uranium Pollution of Grand Water in Karakalpakstan, Uzbekistan

Yoshiko Kawabata, Masaaki Yamada, Onwona-Agyman Siaw, Aparin Vyacheslav, Berdiyar Jollibekov, Masahiro Nagai, Yukio Katayama

Abstract. We investigated uranium concentrations in ground water in Karakalpakstan, Uzbekistan. In the rural area of Karakalpakstan, main drinking water are ground water. In the Half of sampling points for drinking water, uranium concentrations exceeded the WHO (2008) guideline level for drinking water. Since uranium is a suspected carcinogen that can also have a toxic effect on kidneys. However, WHO addressing only the chemical aspects of uranium prescribed that uranium concentrations in drinking water. The effect of uranium exposure from drinking water on people in these areas is significant.

Yoshiko Kawabata
Tokyo University of Agriculture and Technology, 3-5-8, Harumi-cho, Fuchu, Tokyo, 183-8509, Japan

Masaaki Yamada
Tokyo University of Agriculture and Technology, 3-5-8, Harumi-cho, Fuchu, Tokyo, 183-8509, Japan

Onwona-Agyman Siaw
Tokyo University of Agriculture and Technology, 3-5-8, Harumi-cho, Fuchu, Tokyo, 183-8509, Japan

Aparin Vyacheslav
Ministry of Geology, 4a, Navoi, Str., Tashkent, 700000, Republic of Uzbekistan

Berdiyar Jollibekov
Rural Development Group Environment & Regional Development Department, Oriental Consulting Co. LTD, Uzbekistan

Masahiro Nagai
The University of human environments, Faculty of human environment, 6-2 Kamisanonmatsu, Motojukucho, Okazaki-shi, Aichi 444-3505, Japan

Yukio Katayama
The University of human environments, Faculty of human environment, 6-2 Kamisanonmatsu, Motojukucho, Okazaki-shi, Aichi 444-3505, Japan

B. Merkel, M. Schipek, *The New Uranium Mining Boom*
DOI 10.1007/978-3-642-22122-4 © Springer 2011

Introduction

Uzbekistan is located along the ancient Silk Road of Central Asia, has been one of the centers of civilization since ancient times. On the other hand, this area is well known as mineral producers, and served as a supplier of minerals including uranium for the Soviet Union since the 1940s. This area, however took over the negative heritage of environmental pollution when the Union collapsed and was nationalized.

Access to safe drinking water is essential to health, component of effective policy for health protection, and development issue at a national regional and local. Uranium has negative effects on the human body, both as a carcinogen and as a kidney toxin. WHO, addressing only the chemical aspects of uranium, prescribed those uranium concentrations in drinking water should be less than 15 g/l (WHO 2008). Uranium is a naturally occurring radioactive metal. Its two main applications are nuclear bomb production and nuclear electricity generation. Uranium is widely distributed in the earth's crust, but is concentrated in certain rock formations. Most of the uranium for nuclear arms produced in the Soviet Union during the Cold War came from Central Asia.

In the Central Asia, the large scale irrigated agriculture in arid land started in 1950s, especially in the Aral Sea basin. Many villages were built in arid area and peoples are made rice, cotton, wheat, and vegetables in this area. They are drinking ground water or river water. The gradual climate change over the centuries was accelerated by the Aral Sea ecological disaster of the late 20th century in Aral Sea basin, especially Karakalpakstan. By determining EC values, Papa et al. (2004) conducted research on the effect of salinization caused by large-scale irrigation. Crosa et al. (2006) analyzed pesticides in the Amu–Darya basin and reported about identifying compounds with a high risk of contamination.

We researched the uranium concentration of drinking water in central part of Uzbekistan in 2000 (Kawabata et al. 2004). These results were exceeded the guideline level for drinking water prescribed by WHO (1998). Republic of Karakalpakstan is located lowest part of Amu-Darya basin. The gradual climate change over the centuries was accelerated by the Aral Sea ecological disaster of the late 20th century in Aral Sea basin, Karakalpakstan. Therefore, we investigated water pollution in this area.

Material and Method

The sampling sites are shown in Fig. 1. The samples of drinking water were collected 11 points. Water samples from agriculture land were collected 22 points in Karakalpakstan.

The position of each sampling spot was determined by using a e Trex Legend portable Global Positioning System (GARMIN Ltd, Japan). pH, electronic con-

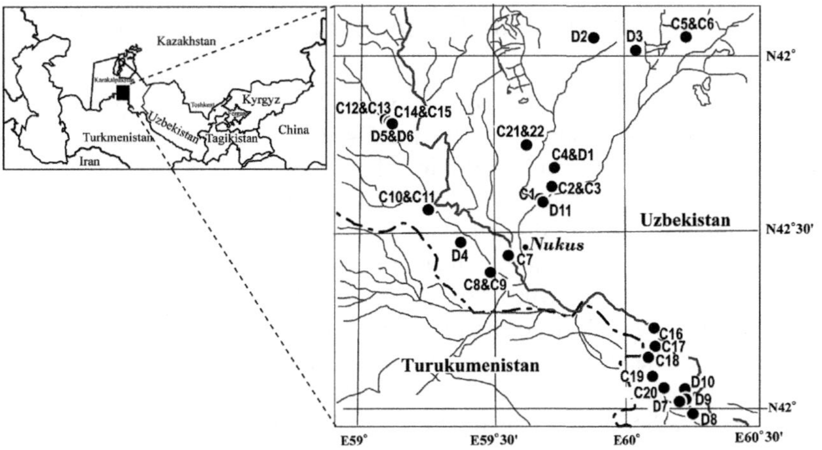

Fig. 1 Location of water sampling sites

ductivity (EC), dissolved oxygen (DO), water and temperature were measured at the sampling site using a U-20 Horiba Multiple water quality monitoring system (Horiba, Ltd., Japan). Water for uranium was filtered through a 0.45 m Millipore filter (Millipore Corporation) and collected in a polypropylene bottle.

Major Ions Measurements of Water Samples and Phosphate Fertilizer

Na^+, Mg^{2+}, Ca^{2+}, Cl^- and SO_4^{2-} concentrations were determined using an LCA-10 A ion chromatographic analyzer (Shimadzu Corporation, Japan). Standard solution was prepared from a Wakenyaku standard solution for ion chromatography in accord with the makers instructions.

Uranium Measurements of Water Samples

Uranium and strontium concentrations of water were determined by a Hewlett Packard Company (Agilent Technologies, Japan) inductively coupled plasma mass spectrometer (ICP-MS). Standard solution was prepared from Spex XSTC-469 (SPEX CertiPrep Inc.). The internal standard is Rh. Finally, accuracy and precision of the analytical methods, used here were tested using Spex XSTC-469 (SPEX CertiPrep Inc.) and shown to be satisfactory with 1–5% error.

Radioactivity Measurements of Phosphate Fertilizer and Soil

Uranium isotopes from phosphate fertilizer were determined by α-spectrometry (Seiko EG&G Model 7800) after radiochemical separation. Each samples treated with a mixture of HNO_3, HF, and $HClO_4$ in a Teflon beaker (100ml). The residue remaining after the acid treatment was fused with Na_2CO_3 in a platinum crucible and the melt was dissolved in diluted HCl. The solutions were combined and a mixture of ^{232}U as yield traces in the chemical separation was added to the solution. Samples were heated in a 100°C oven for about 30 min in order to dry them, and then 10 ml HCl was added to dissolve the residue. The solution was passed through an anion exchange resin in the Cl^- form (Dowex 1 x 8 of 100–200 mesh, column: 8 mm diameter; 5 cm length). Uranium adsorbed strongly on the column. The column was washed with a small amount of 8M HNO_3 to remove adsorbed iron, and then by sufficient 10M HCl to remove some of the other elements. Uranium was eluted from the resin with 2M HCl and the solution was evaporated to dryness. Purified uranium was electrodeposited onto a polished stainless steel disc (Talvitie 1972) and the isotope amount was measured with a surface barrier Si detector (Tennelec TC256 spectrometer coupled to a 1 K channel pulse height analyzer, Seiko EG & G Model 7800) for 1 or 2 days were recorded for uranium.

Results and Discussion

Drinking Water

Sampling points and Uranium, strontium, and major ion concentrations (Na^+, Ca^{2+}, Mg^{2+}, Cl^-, and SO_4^{2-}) of drinking water are shown in Table 1.

Uranium concentrations in drinking water at D3, D8, D9, and D10 exceeded the guideline level for drinking water prescribed by WHO (2008). In particular, uranium concentration in the water sample from D9 was 1.8 times higher than the prescribed WHO value. Kichikkangly village, which is close to D9, had a uranium concentration in its drinking water of 41.92 μg/L (Kawabata et al. 2003). It is possible that a uranium pollutant source is near this area; therefore, it is important that the average uranium concentrations in this area and the pollutant sources be investigated in more detail.

WHO (2008) did not detail a health-based guideline value for major ion concentrations. In Japan, the water quality standards for tap water include limits for Na^+, Ca^{2+} and Mg^{2+} concentrations that affect the acceptability of drinking water. The concentrations of Na^+ at D1, D2, D3, D4, D5, D8, D9, D10, and D11 were higher than those allowed by the water quality standard for tap water in Japan. Further, Ca^{2+} and Mg^{2+} concentrations at D8, D10 and D11 were also higher than those allowed in Japan. However, correlations were not possible between the concentrations of the uranium and Japanese standards.

Table 1 Sampling site description and ion concentrations in drinking water

	N	E	U (μg/l)	Sr^{2+} (mg/l)	Na^{+} (meq/l)	Mg^{2+} (meq/l)	Ca^{2+} (meq/l)	Cl^{-} (meq/l)	SO_4^{2-} (meq/l)
D1	42°41'14,7"	59°43'49,3"	0.01	0.411	33.29	0.59	1.13	14.58	14.00
D2	43°03'04,7"	59°51'55,5"	14.80	3.075	12.83	6.91	10.39	8.27	13.49
D3	43°01'49,3"	60°01'29,4"	16.04	4.190	19.92	11.49	15.92	16.73	23.29
D4	42°28'06,4"	59°21'15,5"	6.39		9.09	6.29	10.43	7.51	11.94
D5	42°49'05,9"	59°05'28,2"	0.78	1.550	11.75	6.30	9.58	8.85	13.08
D6	42°49'13,4"	59°05'33,2"	0.15	1.524	5.12	5.21	10.99	6.28	8.79
D7	42°01'24,2"	60°11'42,7"	4.83	4.635	11.83	7.83	13.38	11.83	14.56
D8	41°59'16,9"	60°14'33,1"	15.80	4.473	11.76	8.80	9.31	9.43	12.12
D9	42°02'52,1"	60°12'27,7"	27.63	3.449	13.04	7.90	13.57	10.11	14.62
D10	42°03'06,0"	60°12'39,5"	4.93	1.373	7.08	3.73	7.79	5.94	7.34
D11	42°35'47,3"	59°40'02,7"	20.53	3.062	18.08	10.50	14.78	16.81	20.31
WHO			15.00	n.d.	200.00	n.d.	n.d.	250.00	500.00

Though WHO (2008) did not detail health-based guideline values for strontium concentrations, the average strontium concentrations were also high in the samples sites; investigations into the sources of strontium pollutants are needed.

Canal Water

Sampling points and uranium, strontium, and major ion concentrations (Na^{+}, Ca^{2+}, Mg^{2+}, Cl^{-}, and SO_4^{2-}) of canal water are shown in Table 2.

The uranium concentrations in canal water from all sampling points were lower than those prescribed by WHO (2008) in the *guideline level for drinking water*. In the season that we samples water, rice fields were being flooded with water and likewise cotton fields had water added every day; this might have greatly increased the flow of water and be a reason for the low uranium concentrations in the samples.

The concentrations of Sodium in discharged water in this area were 1 to 11 times higher than in the irrigation water. Likewise, Mg^{2+} concentrations were 1 to 9, Ca^{2+} concentrations were 0 to 4, Cl^{-} concentrations were 1 to 12, and SO_4^{2-} concentrations were 1 to 7 times higher. Each ion showed a different tendency; especially, Ca^{2+} concentration in drainage water at C6 is lower than in irrigation water at C5. Therefore, the mechanism of ion transfer in agriculture lands in this area needs to be investigated in more detail.

Table 2 Sampling site description and ion concentrations in canal water

	N	E	U (mg/L)	Sr^{2+}	Na^{+} (meq/L)	Mg^{2+}	Ca^{2+}	Cl^{-}	SO_4^{2-}
C1	42°35'13,8"	59°40'15,6"	2.78	0.784	4.41	2.42	3.84	3.29	4.85
C2	42°38'35,2"	59°42'37,9"	2.55	0.735	3.86	2.08	3.56	2.80	4.14
C3	42°38'36,5"	59°42'37,6"	3.20	0.834	43.96	2.44	3.88	3.36	4.94
C4	42°41'08,0"	59°43'43,2"	2.04	0.594	3.03	1.69	3.17	2.21	3.26
C5	43°03'03,7"	60°13'47,1"	2.48	0.744	3.44	1.68	27.17	2.32	3.27
C6	43°03'03,2"	60°13'44,0"	1.63	1.411	3.48	2.04	3.62	2.59	3.77
C7	42°26'34,7"	59°32'56,7"	2.03	1.254	4.11	2.27	3.87	3.08	4.58
C8	42°23'03,2"	59°28'12,2"	1.96	1.067	7.88	2.36	3.60	3.12	4.59
C9	42°23'03,5"	59°28'10,7"	11.28	4.640	41.90	20.33	10.61	38.07	30.63
C10	42°33'57,3"	59°14'12,1"	1.78	0.953	4.29	2.20	4.19	3.29	4.99
C11	42°34'03,3"	59°14'16,5"	9.38	4.423	24.54	16.62	12.36	20.84	26.44
C12	42°49'41,5"	59°04'49,5"	1.61	0.832	3.87	2.08	3.65	2.92	4.41
C13	42°49'42,0"	59°04'49,6"	1.59	0.000	3.77	1.99	3.81	2.90	4.35
C14	42°49'49,7"	59°05'24,2"	1.54	0.828	3.95	2.15	4.43	3.01	4.47
C15	42°49'49,3"	59°05'23,8"	1.63	0.969	5.05	3.08	4.68	3.85	5.92
C16	42°13'15,0"	60°06'27,7"	2.55	0.886	2.63	1.29	2.06	1.92	2.60
C17	42°11'28,4"	60°05'21,6"	2.60	0.885	2.45	1.33	2.19	1.81	2.52
C18	42°08'25,7"	60°04'05,8"	5.28	2.542	9.11	6.21	5.30	9.55	8.55
C19	42°05'08,2"	60°05'52,5"	2.42	0.830	2.30	0.86	2.12	1.71	0.00
C20	42°03'34,1"	60°08'27,8"	6.60	2.198	15.26	9.51	8.01	12.95	15.55
C21	42°45'34,1"	59°37'45,5"	3.73	1.062	3.95	2.34	3.68	3.21	4.92
C22	42°45'35,7"	59°37'43,5"	11.31	3.810	30.15	13.70	13.49	24.19	26.27

Phosphate Fertilizer

Phosphate Fertilizer

Uranium concentrations in phosphate fertilizer are shown in Table 3. Uranium concentrations were the same as in phosphate fertilizers made from volcanic rock (Komura 1985). The uranium concentrations of phosphate fertilizers made from sediment rocks were lower than that from volcanic rocks. In these areas, phosphate fertilizer appears to have been made from volcanic rocks. In this area, 150 kg of phosphate fertilizer is used for 1 ha of rice field; the calculation rates for uranium concentrations in 1 ha of rice field per year of phosphate fertilizer application are shown in Table 3.

ICP-MS samples were dissolved only for just the water analyses; however, uranium concentrations for ICP-MS samples were only about 10% less than the samples for α-spectrometry. This indicates that the uranium in phosphate fertilizer can be dissolved and so enter water. Thus the results here suggest that the uranium in phosphate fertilizer used on agricultural lands dissolved in water and ran into the

Table 3 Uranium Concentration in Phosphorus Fertilizer

Sample	U concentration by ICP-MS (mg/kg)	U concentration by a-spectrometry (mg/kg)	U concentration per 1 ha rice field per 1 year (g)	
Central part of Uzbekistan	19.8	n.d.	3.0	
East part of Uzbekistan	1.0	n.d.	0.1	
C18	68.5	74.4	10.3	(11.2)*
D10	25.4	29.2	3.8	(4.4)*

* These were calculated using the data of a-spectrometry.

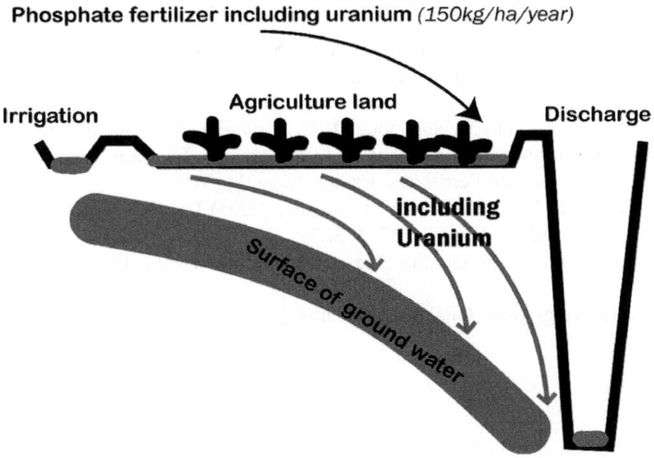

Fig. 2 Schematic view of Mechanism of uranium pollution in agriculture land

groundwater passing through that agriculture land. Irrigation systems in these areas have drains deeper than 5 m to protect against salinization; thus it would be easy for water carried in these drains to mix with groundwater (Fig. 2). Thus the results here suggest that the uranium passes into the groundwater through that agriculture land by the mechanism that is seen in Uzbekistan. This clearly would have an effect on local inhabitants because ground water is used as their drinking water.

Conclusions

This study showed that it is possible that the cause of the high uranium concentration in ground water in the areas studied is use of phosphate fertilizers on agricultural land. In the eastern and central area of the Republic of Uzbekistan, phosphate fertilizers with low uranium concentrations are using for farming. A solution to the uranium concentration problem in the Karakalpakstan area might be for phosphate fertilizer used hereto be altered to the same type as is used in the east of the Republic of Uzbekistan.

Acknowledgments This work was supported by Grant-in-Aid for Science Research, the Ministry of Education, Culture, Sports, Science and Technology of Japan, 2010 (Monbusho International Scientific Research Program No.20405004, represented by Kawabata, Y.).

References

E. Papa, S. Castiglioni, P. Gramatica, V. Nikolayenko, O. Kayumov, and D. Calamari (2004) Screening the leaching tendency of pesticides applied in the Amu Darya Basin (Uzbekistan). Water Res. 38, 3485–3494E.

G. Crosa, F. Stafan, C. Bianchi, and A. Fumagalli (2006a) Water Security in Uzbekistan: Implication of Return Waters in the Amu Darya Water Quality. Environ Sci Pollut Res, 13 (1) 37–42.

K. Komura, M. Yanagizawa, J. Sakurai, and M. Sakanoue (1985) Uranium, Thorium and Potassium Contents and Radioactive Equilibrium State of the Uranium and Thorium Series Nuclides in Phosphate Rocks and Phosphate Fertilizers, Radiosotopes 34, 529–536.

Y. Kawabata, M. Yamamo, K. Shiraishi, S. Ko, and Y. Katayama (2004) Uranium Pollution in Republic of Uzbekistan, J Arid Land Stud 13 (4) 227–233.

WHO (1998) Guidelines for Drinking-water Quality 2nd edition, World Health Organization, Geneva, 554 p.

WHO (2008) Guidelines for Dinking-water Quality, 3rd edition, World Health Organization, Geneva, 668 p.

Environmental Impact of the Kadji-Sai Uranium Tailing Site, Kyrgyzstan

Zheenbek Kulenbekov, Broder J. Merkel

Abstract. Radioactive waste and tailings of the former Kadji-Sai uranium mine in north eastern Kyrgyzstan and treatment plant are a severe environmental concern. In 2009 and 2010 hydrochemical in-situ measurements were carried out and laboratory investigations of surface and ground water were performed. Concentrations of uranium in natural springs, wells, and lake water were found to be above guideline values of WHO. The average gamma radiation at the area of interest is 2.6 mSv/y, however, at one point of the tailing site a gamma dose of 29.9 mSv/y was read.

Introduction

From 1907 to 1913 820,000 kg uranium and radium ores was mined at the Teo-Moyunsky underground mine. Most of it was shipped to St. Petersburg. Industrial exploitation of the large uranium deposit Mayly Suu and others started in 1945. At the beginning of 1950 uranium mining started in Kadjy-Say where uranium was recovered from uranium rich coal by deep mining.

As a result of the uranium mining activities in the entire Kyrgyzstan 30 tailings covering a total area of 6,500,000 m^2 and containing 50,000,000 m^3 tailings are located in different areas of Kyrgyzstan. In total 25 dump sites with a total area of 230,000 m^2 and 4,000,000 m^3 of rocks and tailings were created. (Torgoev et al. 2002). Distribution of uranium mining, milling and dump sites are given in Fig. 1.

Zheenbek Kulenbekov
TU Bergakademie Freiberg, Gustav-Zeuner-Str. 12, 09596 Freiberg/Sa, Germany

Broder J. Merkel
TU Bergakademie Freiberg, Gustav-Zeuner-Str. 12, 09596 Freiberg/Sa, Germany

B. Merkel, M. Schipek, *The New Uranium Mining Boom*
DOI 10.1007/978-3-642-22122-4 © Springer 2011

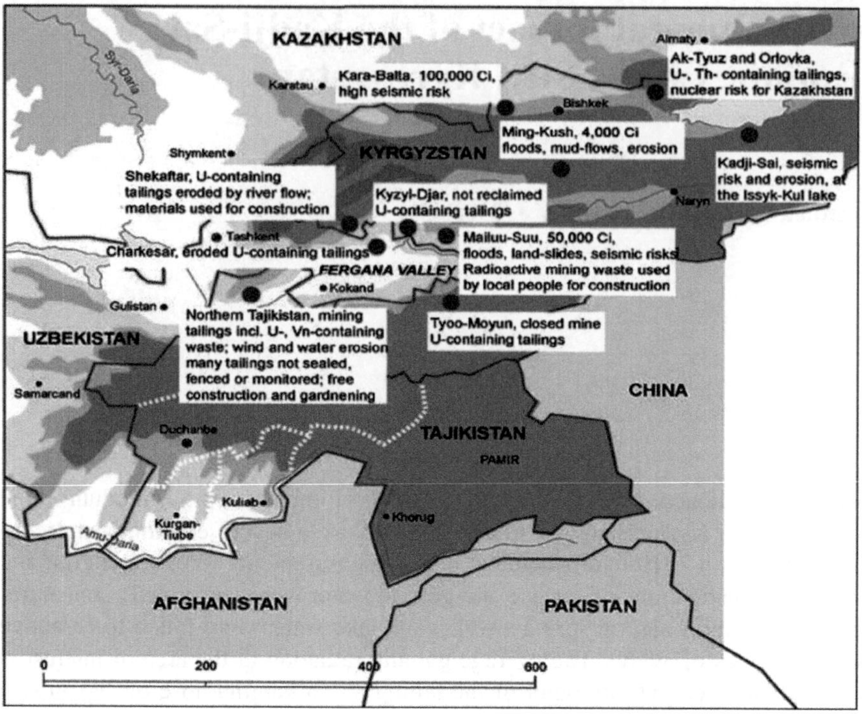

Fig. 1 Map of Uranium tailing storage sites in Kyrgyzstan and brief description of problems related with the tailing sites (UNEP, GRID ARENDAL 1997)

Because Kyrgyzstan is a mountainous area land slides, mud flows and erosion in general is a severe threat in particular for heaps, waste rock piles and tailings sites. High seismic activities and neo-tectonic movements are an additional risk which has to be considered. Therefore risk assessment is an important issue for Kyrgyzstan and can not be performed in the same way as in countries with a rather flat and moderate morphology.

Regional Setting

Issyk-Kul Lake, in northeastern Kyrgyzstan is one of the largest mountainous lake in the world. More than 100 rivers flow into the lake and is fed as well by springs, including many hot springs. Issyk-Kul Lake is a terminal lake with no current outlet, although some hydrologists speculate about a subsurface outflow into the Chu River (Romanovsky 2002). The area around Lake Issyk-Kul is a rather densely populated area, a favorite tourist destination and as well characterized by agricultural activities. However, radioactive waste and tailings of the former

Kadji-Sai uranium mine and treatment plant are a severe environmental threat. The study area is located in an ecologically adverse zone because the inhabitants of this region are exposed directly or indirectly to the uranium tailing (Tynybekov et al. 2008). The uranium tailing is located 2 km from Issyk-Kul Lake and Kadji-Sai village with a population of five thousand is 3 km west of the study area. Livestock are grazing around and over the tailing site because the protection fences around the tailing site was destroyed by local people.

The region is characterized by semiarid climate with annual precipitation of 196 to 240 mm, a rather high evaporation rate and thus only minor groundwater recharge rates. Groundwater is therefore a rather vulnerable and precious good in Kyrgyzstan.

Two perennial small rivers courses are passing through the former mining and milling area. The distance of the talweg of the valleys is as close as 30–35 m from the surface of tailing storage and 15–20 m from the foot hill of process coal ash.

The natural tailing storage relief of the tailing storage is impacted by man maid activity; during coal mining and the tailing storage operation. The tailing storage region is limited from the east by “East canyon”, from the west by “West canyon”. It merges with “Djilubulak-Sai” ravine approximately 450 m to the north of “West canyon” area.

Presently Kadji-Sai tailing and the catchment pools are endangered by natural erosion processes and anthropogenic impact as well. Persistent erosion of the tailings with each flood flow and the risk of potential contamination of the Issyk-Kul lake remain.

Methods

In 2009 and 2010 hydrochemical investigations were carried out. In-situ measurements (pH, redox potential, electrical conductivity, temperature) were performed and surface and ground water samples for major, minor and trace elements including uranium and thorium were taken. Equivalent dose rate of gamma radiation was measured in situ by a gamma ray detector (DKS-96).

Four springs, two streams, two wells, tap-water and lake water in the area of interest (Fig. 2) were investigated. Temperature, pH, redox potential and EC were measured in situ using portable devices (WTW: pH320, LF320, MultilineP4, Germany). From each sampling point 100 ml were collected in a pre-cleaned polythene bottles for major cations and anions analysis. Additional 25 ml were filtered with 200 nm membrane filter (Membrex 25CA) and acidified with ultra pure HNO_3 (65%) for ICP-MS analysis in pre-cleaned PE bottles. Cations and anions were analyzed by ion chromatography (Metrohm: 881CompactICpro AnionMCS-2.881.0030 and 850Professional IC 2.850.1010). Trace elements were performed by ICP-MS (Thermo Scientific XSERIES 2). Total inorganic carbon TIC was determined by LiquiTOC (Elementar Analysensysteme GMBH). The distribution of major species was calculated by PHREEQC using nea_2007 database.

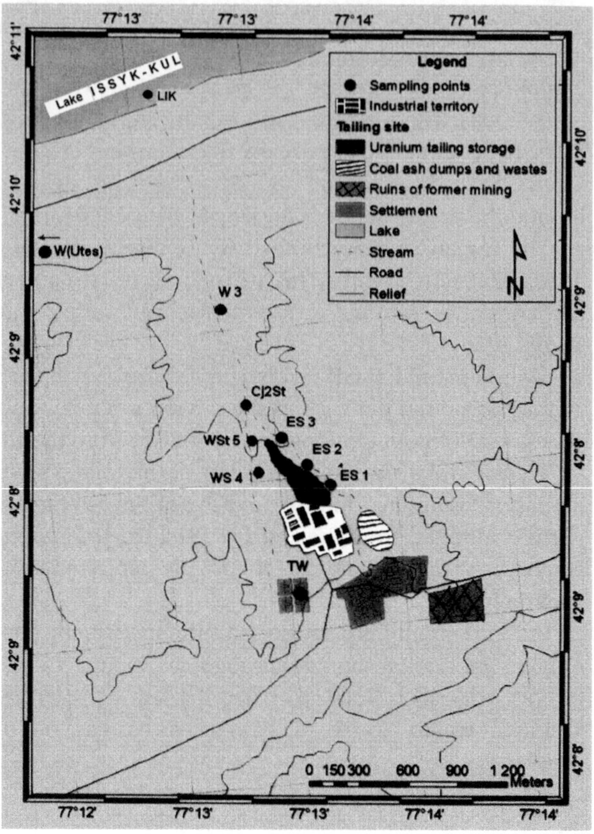

Fig. 2 Map of uranium tailing site and water sampling points, (Datum: WGS-84)

Results and Discussions

Gamma radiation in the vicinity of the tailing site varied between 0.2 and 0.4 μSv/h (1.75 to 4.28 mSv/y) with an average of 0.3 μSv/h (2.6 mSv/y). The average is thus rather close to worldwide natural background radiation. The highest value at one point of the tailings was 3.42 μSv/h (30.0 mSv/y) where technological coal ashes and equipment are deposited.

All water samples taken were not according to the recommendation of the WHO drinking water guideline (World Health Organization 2003). This is true in particular for the uranium concentration which exceeded the recommended value of 15 μg/l (Table 1).

Surface and groundwater is circum neutral and at the edge to oxidizing conditions. However, TDS and EC are rather high which is mainly due to elevated calcium and sodium concentrations on the one and elevated sulfate and chloride concentrations on the other side (Tables 1 and 2).

Table 1 In situ chemical parameters, TDS and uranium concentrations

Code	Samples name	Location	T °C	pH	eH mV	Pe	EC μS/cm	TDS mg/l	U μg/l
ES1	East canyon, spring 1	42° 09′ 31.07″ 77° 13′ 50.5″	21.3	7.95	402	6.8	5470	3829	170.1
ES2	East canyon, spring 2	42° 08′ 11.8″ 77° 11′ 25.9″	16.7	7.17	339	5.7	3550	2485	288.3
ES3	East canyon, spring 3	42° 09′ 18″ 77° 13′ 0.3.9″	30	8.13	293	5	2080	1456	163.2
WS4	West canyon, spring 4	42° 09′ 25.0″ 77° 12′ 58.1″	23.5	8.13	326	5.5	1795	1557	86
WSt5	West canyon, S5 stream	42° 09′ 17.2″ 77° 12′ 58.2″	19.7	7.39	297	5	1772	1240	99.2
Cj2St	Conjunction of 2 streams	42° 09′ 31.9″ 77° 13′ 7.0″	23.8	8.23	367	6.2	2040	1428	167.3
W3	Well water 3*	42° 09′ 21.9″ 77° 13′ 17.0″	15	7.14	374	7.1	1700	1190	46
W(Utes)	Utes (well water)	42° 09′ 49.6″ 77° 11′ 6.1″	17.8	7.56	430	7.3	1611	1128	107.1
LIK	Lake Issk-Kul	42° 10′ 18.2″ 77° 12′ 38.7″	19.2	8.3	371	6.2	8400	5880	64.92
TW	Tap water	42° 09′ 13.7″ 77° 13′ 12.9″	17.9	8.21	373	6.8	660	462	26.2

* Well 3 – data of Ministry of Emergency Situation of Kyrgyz Republic

Uranium forms complexes with sulfate and phosphate as well as with carbonate and hydroxide ions, which increase the total solubility of uranium (Langmuir 1978). However these two Ca-UO_2-CO_3 complexes are not contained in the nea_2007 databases therefore the reaction and log-K values were added to calculate the uranium species.

Hexavalent uranium is highly soluble because it forms stable complexes such as $UO_2(CO_3)_3^{4-}$. As a result of complex formation uranium is a ubiquitous element. This is especially true for oxygen-rich surface waters and shallow groundwater (Merkel et al. 1998). In areas affected by uranium solution mining using sulfuric acid, UO_2SO_4(aq) will be an important species. In alkaline waters, carbonate complexes dominate. Bernhard et al. (1998) studied uranium speciation in water from uranium mining districts in Germany (Saxony) using laser spectroscopy and found that $Ca_2UO_2(CO_3)_3$(aq) was the dominant species in neutral pH carbonate- and Ca-rich mine waters; $UO_2(CO_3)_3^{4-}$ was the dominant aqueous species in basic (pH = 9.8), carbonate-rich, Ca-poor mine waters; and UO_2SO_4(aq) dominated in acidic (pH = 2.6), sulfate-rich mine waters. In the Helmsdorf tailing, uranyl carbonate complexes, mainly $UO_2(CO_3)_3^{4-}$ and $UO_2(CO_3)_2^{2-}$, are the main solution species, Bernhard et al. (1998).

Table 2 Results of Ion Chromatography and TIC determination compared to EPA recommended drinking water MCL's

	East canyon, spring 1	East canyon, spring 2	East canyon, spring 3	West canyon, spring 4	West canyon, stream 5	Conjunction of 2 streams	Well 3**	Utes (well water)	Lake Issyk-Kul	Tab water	EPA
Li	0.30	0.29	0.18	0.12	0.21	0.19	nd*	0.061	nd*	0.016	***
Ca	332	214	74	88.6	77.4	92.3	72	96	110	32.6	***
Mg	55	41.2	19.6	18.3	13.4	20.8	12	20.2	274	18.6	***
Na	963	279.6	210	128	213.4	266	553	225	1436	40.5	***
K	11.7	9.42	4.24	3.1	4.19	5.07	4.72	3.04	66.4	3.8	***
NH_4	0.16	0.19	0.21	3.46	5.16	0.073	0.3	0.011	0.5	0.07	0.5
F	3.0	2.38	5.0	3.27	4.3	nd*	nd*	2.1	11.9	1.59	4.0
Br	0.099	0.072	0.054	0.047	0.044	0.108	nd*	0.170	1.07	0.04	***
Cl	717	378	276	185.5	226.3	218	543	207	1584	93.5	250
SO_4	1643	609	194	165.5	247	461	561	314	875	60.4	250
NO_3	4.39	0.44	4.49	14.4	11	9.08	3.28	30.6	0.63	5.0	43
PO_4	*nd	11.7	4.56	4.1	3.7	0.001	nd*	nd	23.4	0.056	0.3
HCO_3	231	101	63	68	76	64	108	95	57.3	23	***

* nd – not detectable
** Well3 – data of Ministry of Emergency of Kyrgyz Republic
*** No drinking water standard has been established

Uranium speciation in spring waters (ES1, ES2, ES3 and WS4) as well as in surface water (WSt5 and Cj2St) from the uranium tailing site in Kyrgyzstan (Kadji-Sai) is similar to Helmsdorf tailing waters, Germany: $UO_2(CO_3)_3^{4-}$ is the dominant species in neutral to slightly basic pH (7.17–8.13), carbonate is elevated but Ca is lower in this tailing waters (ES1: 97%, ES2: 52%, ES3: 82%, WS4: 83% respectively and Cj2St: 83%). Around 17% of major species like $UO_2(CO3)_2^{2-}$ in down streams sampling point (WSt5) and 83% of $UO_2(CO3)_3^{4-}$ are dominant as well. The situation regarding distribution of species in groundwater is a bit different. 54% of major species such as $UO_2(CO_3)_3^{4-}$ is dominant in groundwater (W3) which is located on the detritus cone of two streams, adjacent to the tailing dump. About 70% of species such as $UO_2(CO_3)_3^{4-}$ is dominant in well water W(Utes) that is located in 2.5 km to north-western side of tailing dump. 96% of major species such as $UO_2(CO_3)_3^{4-}$ is dominant in Lake Issyk-Kul which is located in 2 km to the north from the tailing dump. In uranium mining areas in Cunha Baixa, Portugal, the uranium speciation calculations showed that in local tap water the $UO_2(OH)_3^-$ was the predominant species (>99%) whereas $UO_2(HPO_4)_2^{2-}$ (>99%) was modeled as the dominant species in the non-contaminated water (Neves et al. 2008). In local tab water in Kadji-Sai tailing site $UO_2(CO_3)_3^{4-}$ is dominant with 57% of $UO_2(CO_3)_3^{4-}$ and 16% is $UO_2(CO_3)_2^{2-}$.

Uranium speciation modeling of a water sample from east canyon (ES2) adjacent to the uranium tailing dump is plotted in Figs. 3 and 4. The theoretical mobility of uranium is depending on the speciation of uranium. UO_2^{2+} is forming two species with CO_3^{-2} and Ca^{2*}: one complex is negative charged, the other is zero

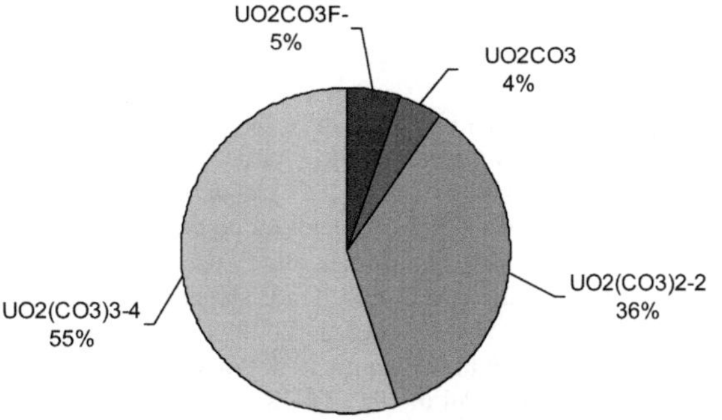

Fig. 3 Calculation of uranium speciation without $Ca\text{-}UO_2\text{-}CO_3$ complexes

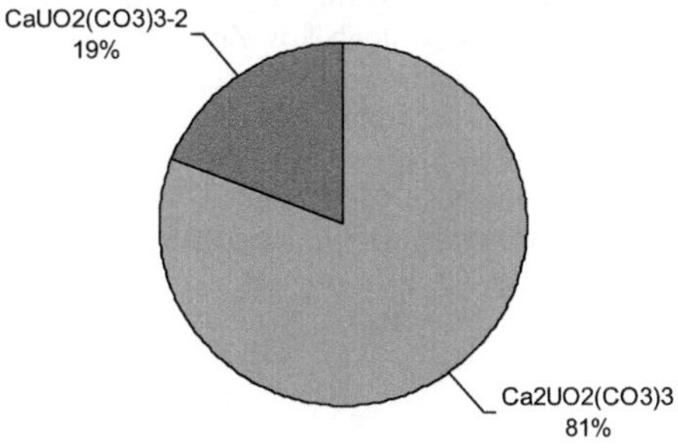

Fig. 4 Calculation of uranium speciation including $Ca\text{-}UO_2\text{-}CO_3$ complexes

charged. The zero charged complexes are assumed to be the most mobile and thus enhancing the transport of uranium in the subsurface, because no sorption occurs (Figs. 3 and 4). Calculation of uranium speciation including positively charged species Ca-UO_2-CO_3 complexes shows a rather different speciation pattern than those without this species.

Conclusion

The unfenced Kadji-Sai Uranium tailing site, located on the south part of Lake Issyk-Kul, Kyrgyzstan is a threat on the shoreline of the terminal lake. The lake water has elevated uranium concentrations (65 µg/L) well above natural background. The source of the elevated uranium concentration of the lake water is unknown. The highest uranium concentration of spring water was found to be 288 µg/L, which can be a natural background value but as well due to the impact of the tailing site. The average equivalent dose rate of gamma radiation around the tailing site is rather close to worldwide natural background radiation. However, the highest value at one point of the tailings with 3.42 µSv/h (30.0 mSv/y) shows that the capping of the tailing is either incomplete or violated by erosion. Erosion is probably the biggest risk for this site due to it position between two perennial creeks.

References

Bernhard G, Geipel G, Brendler V, Nitsche H (1998) Uranium speciation in waters of different uranium mining areas. J Alloys and Compounds: 201–205

Garrison Investigative Board (1977) Water quality report (Appendix A). Garrison: 389–403

Langmuir D (1978) Uranium-mineral equilibrium at low temperatures with applications to sedimentary ore deposits. Geochemica Cosmochemica Acta 42: 547–569

Merkel B & Sperling B (1998): Hydrogeochemische Stoffsysteme II. DVWK Schriftenreihe: 117

Neves O, Abreu M, Vicente E (2008) Uptake of Uranium by Lettuce (*Lactuca sativa* L.) in Natural Uranium Contaminated Soils in Order to Assess Chemical Risk for Consumers. Water Air Soil Pollut 195: 73–78

Romanovsky V (2002) Water level variations and water balance of the Lake Issyk-Kul. Klerkx J, Imanackunov B, Lake Issyk-Kul: Its Natural Environment. NATO Science Series: 45–57

Tynybekov AK, Kulenbekov ZhE, Aliev MS (2008) Radiological investigation and ecological risk of southern coastal section of Lake Issyk-Kul. Fifth International Conference Uranium Mining and Hydrogeology (UMH V) Freiberg, Springer, Germany: 477–486

Torgoev I, Aleshin Y, Havenith HB (2002) Impact of Uranium Mining and Processing on the Environment of Mountainous Areas of Kyrgyzstan. Uranium in the Aquatic Environment. Springer-Verlag Berlin Heidelberg: 93–98

World Health Organization (2003) Background document for development of WHO Guidelines for Drinking-water Quality

Assessment of Distribution Coefficients (K_d) of Radionuclides of the Uranium-Thorium Chain in the Uranium Manufacturing Tailing Dumps

Valentyn Protsak, Valery Kasparov, Igor Maloshtan, Sviatoslav Levchuk, Vasyl Yoschenko, Irina Kalyabina, Olga Marinich

Abstract. One of the problems related to maintaining the ecologically safe state of tailing dumps of uranium manufacturing is the radionuclides migration from the dumps with groundwaters. A convenient way for evaluation of this process is application of the semiquantitative models of sorption interaction, which are based on utilization of distribution coefficient (K_d) describing the radionuclides concentrations in solid phase and soil solution after reaching the equilibrium state in the system. Aim of this study was the experimental assessment of the radionuclides K_d in five tailing dumps of Prydniprovsky Chemical Plant (PCP).

Valentyn Protsak
Ukrainian Institute of Agricultural Radiology of NUBiP of Ukraine

Valery Kasparov
Ukrainian Institute of Agricultural Radiology of NUBiP of Ukraine

Igor Maloshtan
Ukrainian Institute of Agricultural Radiology of NUBiP of Ukraine

Sviatoslav Levchuk
Ukrainian Institute of Agricultural Radiology of NUBiP of Ukraine

Vasyl Yoschenko
Ukrainian Institute of Agricultural Radiology of NUBiP of Ukraine

Irina Kalyabina
Institute of geochemistry of the Environment of NAS and MES of Ukraine

Olga Marinich
Institute of geochemistry of the Environment of NAS and MES of Ukraine

B. Merkel, M. Schipek, *The New Uranium Mining Boom*
DOI 10.1007/978-3-642-22122-4

Site Description

PCP is the first Soviet enterprise for processing the uranium ore. It was built in 1947 in suburb of Dniprodzerjisk in Ukraine. Up to 65% of all uranium ore of the Soviet Union was processed at PCP. Table 1 presents the brief information about the tailing dumps where the studies were carried out.

Table 1 Brief information about the PCP tailing dumps

Name	Area, ha	Mass of waste	Period of operation
Western	4	770 thousand t	1949–1954
Central Ravine	2,4	220 thousand t	1950–1954
Southeastern	3,6	330 thousand t	1956–1990
Dniprovs'ke	73	12 million t	1954–1968
Sukhachivs'ke (I section)	150	19 million t	1968–1983
Sukhachivs'ke (II section)	70	150 thousand t	Since 1983

Instruments and Methods

The radionuclides K_d were determined by batch-method (EPA 402-R-99-004 A 1999; A 2004). Samples of sludge were picked at various horizons of the tailing dumps in the process of auger drilling. Sorption equilibrium of radionuclides between water solution and solid phase of sludge was studied at the phase ratio of 5 : 1. The suspensions were incubated for 24 h at 25°C with a periodical mixing and then filtered through Millipore 0.22 μm filters. Samples of water extracts after injection of the tracers and stable carriers were concentrated by means of evaporation till the wet salts state. Activities of uranium and thorium were measured by means of α-spectrometry after their ion-exchange extraction. Sample for α-spectrometry were prepared by the electrolytic deposition of the extracted fractions of uranium and thorium from the solution of $(NH_4)_2SO_4$ at pH of 2.3 onto the plates of stainless steel.

^{210}Pb was absorbed by the anion-exchange resin from 2M solution of HCl, and after desorption it was deposited in the form of chromate. Radiometric determination of ^{210}Pb was based on measurement of its progeny, ^{210}Bi. ^{226}Ra activities were measured by means of γ-spectrometry of the dry residues obtained by evaporation of aliquots of the extracts.

γ-spectrometric measurements were carried out at the HpGe detector 7229P (Canberra). α-spectrometric measurements were performed at EG&G ORTEC OCTETE PC with the silicon detectors of BU-017-450-100 ULTRA series.

K_d of the radionuclides are derived as

$$K_d^i = \frac{A^i_{\mathrm{sludge}}}{A^i_{\mathrm{water}}}\frac{V}{m}, \quad \mathrm{L\,kg^{-1}},$$

where A^{i}_{sludge} and A^{i}_{water} denote the ith radionuclide specific activities in the solid phase of sludge after extraction, Bq kg^{-1}, and in the solution, Bq L^{-1}, respectively, V is the volume of the water solution, and m is the dry weight of the solid phase.

For estimation of the uranium forms in the tailing dumps the method of thermodynamical modeling was applied using GEM software (http://gems.web.psi.ch/overview.html; Karpov et al. 2001).

Results and Discussion

Obtained values of the radionuclides distribution coefficients and pH of the water extracts of the charge mixtures in the tailing dumps profiles are presented in Table 2. Analysis of the radionuclides K_d values in the PCP tailing dumps conditions shows that migration with the convective water flux of ^{238}U must be up to three orders of magnitude more intensive than that of ^{230}Th. The lowest K_d values of ^{238}U and ^{230}Th and the highest of ^{226}Ra were measured in the tailing dump Central Ravine, where the media is acid (pH = 2.6 – 3.4). Therefore one may expect one-two order of magnitude more intensive water migration of uranium and thorium in

Table 2 Values of pH and K_d (mean ± STD) of radionuclides in profiles of the PCP tailing dumps

Tailing dump	h, m	pH	K_d, L kg^{-1} ^{238}U	^{230}Th	^{226}Ra	^{210}Pb
Dniprovs'ke	7.25	7.76	37 ± 12	58300 ± 13400		
	8.25	7.86	55 ± 10	106500 ± 24500	290 ± 90	
	10.75	7.26	149 ± 41	135700 ± 27200	750 ± 220	
	11.25	7.77	49 ± 23	145000 ± 34800		
Dniprovs'ke	8.25	6.02	82 ± 48	188500 ± 43300		
	9.75	7.35	550 ± 220	413300 ± 86800		
	12.25	7.62	304 ± 93	327300 ± 72000		
Sukhachivs'ke	2.5	7.62	140 ± 69	3200 ± 800		
	12.5	7.93	227 ± 61	8200 ± 1600		
	19.5	7.83	123 ± 28	6250 ± 1440	409 ± 123	
	22.5	7.84	77 ± 25	4290 ± 1030		
	23.5	8.07	316 ± 88	9090 ± 2360		
Southeastern	7.25	8.58	490 ± 13	33060 ± 8260	7.7 ± 2.6	
	8.75	8.52	250 ± 39	25000 ± 6000		
	11.75	8.78	77 ± 21	31300 ± 6300	19.1 ± 5.8	
	16.75	8.96	214 ± 39	45600 ± 9200		
Western	6.75	9.45	18.6 ± 4.5	1030 ± 260		
	10.25	9.33	29 ± 12	13380 ± 3080		
	12.25	8.87	32 ± 15	16470 ± 3850		
Central Ravine	3.75	2.62	26 ± 4	32 ± 7	1910 ± 470	
	9.25	3.35	19.1 ± 4.5	890 ± 180	2540 ± 610	260 ± 90
	16.25	2.91	5.4 ± 1.3	290 ± 60	1450 ± 430	112 ± 41

this dump as compared to the other ones, and less intensive migration of radium. K_d values of ^{238}U and ^{230}Th in the samples collected beneath the tailing dumps were several times higher than in the dumps. One may suppose the lower intensity of migration of these radionuclides in the bedrock as compared to that in the tailing dump profiles.

The radionuclides K_d in the tailing dumps depend on the pH of media. Figure 1 depicts the dependencies of logarithm of the distribution coefficients, $\log(K_d)$, on pH of water extracts of the charge mixtures for ^{238}U, ^{230}Th and ^{226}Ra. These dependences can be satisfactory approximated by the polynomial functions with the corresponding parameters. K_d values obtained at certain pH are in the good accordance with those earlier reported (Keum et al. 2002; Vandenhove et al. 2009).

The well-known feature of actinides is their ability to be present in the tailing dump material in several oxidization forms at the same time (Meinrath et al. 2003; Wall and Krumholz 2006). This feature is caused by the dynamics of the processes of hydrolysis, complexing and formation of the new solid phases and, consequently, influences absorption and migration of the radionuclides.

Forms of uranium in the waste of PCP were estimated for the tailing dumps Western and Central Ravine. Media of the dumps are characterized by the alkaline and acid reactions, respectively. Modeling was performed for the system Si – Al – Fe – K – Na – Ca – Mg – Cl – C – S – U – O – H. Model system included aqueous and gaseous phases and mineral phases that were likely to occur in conditions of the tailing dumps according to the results of the phase X-ray analysis of the collected samples (Table 3).

Minerals are considered as the simple phases (minerals of the permanent composition) and phases of the solid solutions (minerals of the changeable composition). Uranium concentrations were derived from its average concentrations in the tailing dumps (Investigation 2001). pH was modified by means of changing concentration of Cl^- in the model system. Calculations were done for the standard conditions, i.e. $T=25°C$ and $P=1$ bar. Taking into account that the oxidization level of uranium substantially depends on the redox conditions, uranium forms in the tailing dump Western were estimated at various values of Eh. Eh values were modified by variation of Fe(II)/Fe(III) ratio in the material composition.

Results of the model calculations show that the major minerals in the tailing dump Western at $pH=9.8$ and $pH=8.5$ are plagioclase and quartz. To a large degree, iron oxides and hydroxides are presented (up to 13% in total), and, to a less extent, calcite, clay minerals and mica.

Obtained results correlate with the results of the performed phase X-ray analysis of the collected samples. Thus, for the mentioned values of pH and in the Eh range from –369 to –175 mV a major fraction of uranium in the tailing dump Western is presented in the mineral sparingly soluble form of coffinite ($USiO_4$) (Fig. 2), and up to 15% of the total amount of uranium is presented in the aqueous solution. It agrees well with the results of specification of the mobile forms of uranium in the tailing dump Western (up to 4% in aqueous extracts and up to 27% in acetate extracts) and with the results of determination of the dissolved uranium in water of technogenic water-bearing horizon of the tailing dump Western (up to 12% (Investigation 2001)).

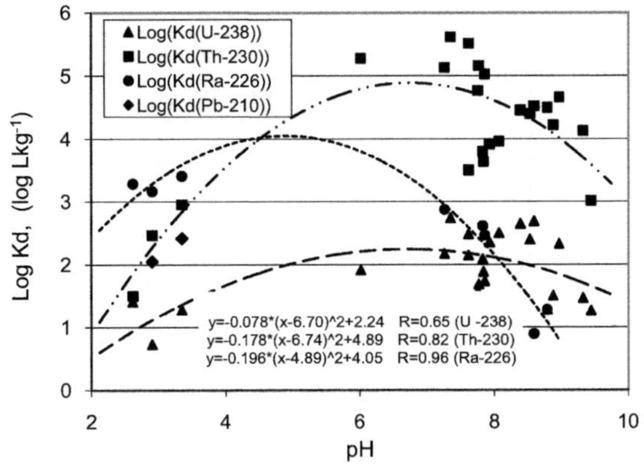

Fig. 1 $\log(K_d)$ of the radionuclides in the PCP tailing dumps as a function of pH

Table 3 Phases included to the model system

Phase	Description	Single Minerals
Aqueous solution	Aqueous compounds of Si, A1, Fe, K, Na, Ca, Mg, Cl, C, S та U	Gibbsite
		Kaolinite
Gaseous phase	Mixture of CO_2, CH_4, H_2, N_2, O_2, H_2S	Aragonite
		Calcite
Solid solutions		Dolomite
Plagioclase	Anorthite	Portlandite
	Albite	Gypsum
Chlorite	$Fe_{4,4}Al_{3,2}Si_{2,4}O_{10}(OH)_8$	Siderite
	$Fe_{4,4}Fe_{1,6}Al_{1,6}O_{10}(OH)_8$	Hematite
	$Mg_4Al_4Si_2O_{10}(OH)_8$	Magnetite
	$Mg_5Al_2Si_3O_{10}(OH)_8$	Ferrihydrite
	$Mg_4Fe_2Al_2Si_2O_{10}(OH)_8$	Goethite
	$Mg_5FeAlSi_3O_{10}(OH)_8$	Pyrite
Illite	$Al_2Si_4O_{10}(OH)_2$	Microcline
	$Ca_{0,5}Al_3Si_3O_{10}(OH)_2$	Muscovite
	$KAl_3Si_3O_{10}(OH)_2$	Magnesite
	$KFe_3AlSi_4O_{10}(OH)_2$	Quartz
	$KMg_3AlSi_4O_{10}(OH)_2$	Uranium minerals
	$NaAl_3Si_3O_{10}(OH)_2$	Coffinite $USiO_4$
Smectite (montmorillonite)	$Al_2Si_4O_{10}(OH)_2$	Shoepite $UO_2(OH)_2$
	$Ca_{0,5}Al_3Si_3O_{10}(OH)_2$	$UO_2(s)$
	$Fe_2Si_4O_{10}(OH)_2$	Rutherfordine UO_2CO_3
	$Fe_3Si_4O_{10}(OH)_2$	
	$KAl_3Si_3O_{10}(OH)_2$	
	$Mg_3Si_4O_{10}(OH)_2$	
	$Mg_3Si_4O_{10}(OH)_2$	

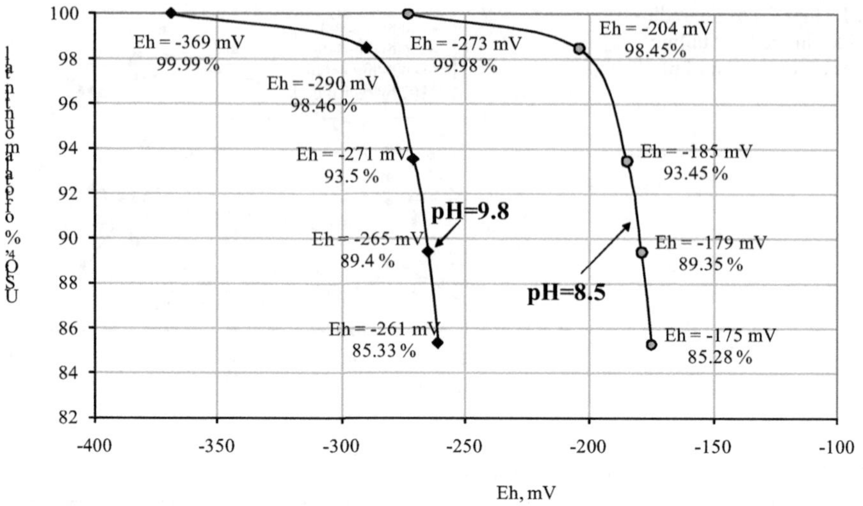

Fig. 2 Formation of the sparingly soluble uranium compounds in material of the tailing dump Western depending on pH and Eh

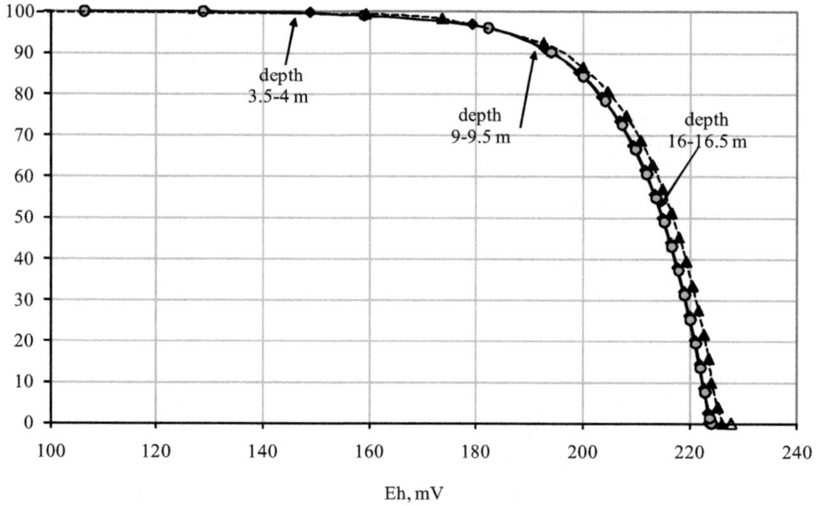

Fig. 3 Formation of the sparingly soluble uranium compounds in material of the tailing dump Central Ravine

It should be noted that formation of coffinite in these conditions significantly depends on Eh. At fixed pH the coffinite fraction of uranium decreases from ~100% to ~85% with increase of Eh by ~100 mV. According to the calculations, soluble uranium in the tailing dump Western is almost exclusively presented in the oxidization degree of +6 in the forms of carbonate complexes ($UO_2(CO_3)_3^{4-}$ – up to 99.9%, and $UO_2(CO_3)_2^{2-}$ – up to 0.1%). Total content of U(IV) and U(V) in solution does not exceed 0.2% in the considered ranges of pH and Eh values.

Modeling of uranium forms in the tailing dump Central Ravine shows that the major minerals in the upper horizons are quartz and gypsum. Kaolinite contents increases with depth up to 6%. In this tailing dump formation of coffinite begins at Eh > 223 mV and sharply increases with increasing of Eh reaching more than 80% of total amount of uranium in $USiO_4$ form at Eh > 200 mV. Obtained data correlate with the results of specification of the uranium mobile forms in the samples from Central Ravine (up to 11.75% in aqueous and up to 5% in acetate extracts). Experimental fraction of the mobile forms of uranium corresponds to Eh ~ 220 mV. It may be noted that the curves of formation of coffinite are practically the same for various depths in the tailing dump (Fig. 3). Also, the calculations show that almost all uranium in the aqueous solution is characterized by the oxidization degree of +6 as UO_2^{2+} (>65%) and chloride (>8%) and sulfate (>6%) complexes. Total content of U(V) in the aqueous solution does not exceed 0.2% in the considered ranges of pH and Eh values.

Therefore, the modeling results allow inferring that

- taking into account the high content of potential minerals-absorbents (iron oxides/hydroxides up to 13% and clay minerals up to 10%) in the tailing dump Western, the possible mechanism of the postfixation of uranium is absorption of U(VI) by iron oxides/hydroxides and clay minerals,
- taking into account that potential minerals-absorbents (iron oxides/hydroxides and clay minerals) in the tailing dump Central Ravine are practically absent, the postfixation of uranium in the upper horizons of the dump is unlikely, and its mobility is determined by the dissolution rate of the residual amounts of uranium, which had not been extracted during the technological process. In the lower horizons of the dump the postfixation of uranium is possible due to its absorption by kaolinite.

The presented results are preliminary, because the calculations were based on the averaged chemical compositions of the mineral part of the tails and averaged water compositions of the technogenic water-bearing horizons of the tailing dumps. Besides, the modeling method implied the thermodynamic equilibrium in the system, and the obtained results should be interpreted as the qualitative assessment of the principal processes determining distribution of the forms of uranium and its mobility in conditions of the PCP tailing dumps.

Conclusions

The distribution coefficients of the uranium-thorium chain radionuclides were measured for five tailing dumps of Prydinprovsky Chemical Plant. Dependence of K_d on pH was determined.

Method of thermodynamical modeling was applied for assessment of the forms of uranium in the tailing dump.

Obtained results are important for prognosis of the radiological situation at the studied territory, for modeling the processes of the radionuclides water migration and for elaboration of the strategy in concern of the tailing dumps.

The study was supported by SE Barrier.

References

ASTM Designation: D 4646-03 (2004) Standard test method for 24-h batch-type measurement of contaminant sorption by soils and sediments

EPA 402-R-99-004 A (1999) Understanding variation in partition coefficient, K_d, values. Volume 1: The K_d model, methods of measurement, and application of geochemical reaction codes

http://gems.web.psi.ch/overview.html

Investigation and development of the RAW storage technology at the SE Barrier (2001) Analysis of safety of maintaining activities for storage of RAW. Tailing dump Western. Book #1. Evaluation of conditions of storage of RAW

Karpov IK, Chudnenko KV, Kulik DA, Avchenko OV, Bychinski VA (2001) Minimization of Gibbs free energy in geochemical systems by convex programing. Geochemistry Int 39: 1108–1119

Keum DK, Choi BJ, Baik MH, Hahn PS (2002) Uranium(VI) adsorption and transport in crushed granite. Environ Eng Res 7: 103–111

Meinrath A, Schneider P, Meinrath G (2003) Uranium ores and depleted uranium in the environment, with a reference to uranium in the biosphere from the Erzgebirge/Sachsen, Germany. J Env Radioactivity 64: 175–193

Vandenhove H, Gil-Garcia C, Rigol A, Vidal M (2009) New best estimates for radionuclide solid-liquid distribution coefficients in soils. Part 2. Naturally occurring radionuclides. J Env Radioactivity 100: 697–703

Wall JD, Krumholz LR (2006) Uranium reduction. Annu Rev Microbiol 60: 149–166

How Much Uranium Can Be Left at Former U Mining Sites? The Need for a Complex Assessment Framework

P. Schmidt, E. Kreyßig, W. Löbner

Abstract. Uranium mining and milling legacies are sources of releases of uranium into the environment. At the WISMUT sites in Saxony and Thuringia, in Eastern Germany, as a result of remedial actions considerable amounts of uranium have been already removed from soil and water, and releases into the environment have been continuously reduced during the last two decades under the WISMUT project. However, there are sites where complete removal is impossible and releases of uranium into the environment continue. In terms of optimization this raises the question of how much uranium in the environment is acceptable. The paper discusses a number of approaches to assess the radiological, ecological and toxicological risks of uranium at former uranium mining and milling sites.

Problem Definition

The quantity of uranium contained in waste rock piles, tailings management areas, and on plant areas at uranium mining and milling sites of WISMUT at the time

P. Schmidt
Wismut GmbH, Chemnitz, Germany

E. Kreyßig
Wismut GmbH, Chemnitz, Germany

W. Löbner
Wismut GmbH, Chemnitz, Germany

B. Merkel, M. Schipek, *The New Uranium Mining Boom*
DOI 10.1007/978-3-642-22122-4 © Springer 2011

when uranium production was terminated in 1991 was in the order of 30,000 t[1]. Transfer of uranium from mining wastes into the environment was predominantly water-borne. Due to the remedial progress achieved (covering of waste rock piles and of tailings management areas, controlled mine flooding associated with water treatment) the total load of uranium released to the environment has significantly decreased by now. However, officially approved uranium emission into the environment continues because complete removal of uranium from water is not feasible within reasonable technological and spending limits.

While the controlled discharge of uranium to receiving streams totaled ca. 27 t in 1991, it was down to only 4.4 t in 2010. Assessment of diffuse release of uranium from mining wastes into receiving streams is fraught with great uncertainties. Estimates for the Schlema site, for example, where some in 2010 still some 2.7 t of uranium were discharged in controlled manner into the Zwickauer Mulde river, are that diffuse discharge to the Mulde amounted to an additional quantity of 20% of the controlled discharge. Downstream of the Schlema as well as of the Crossen site, which is the last location downstream of uranium mining legacies to impact the Zwickauer Mulde river, an average uranium concentration of about 8 µg/l was established in 2010.

Waste rock piles relocated at the Ronneburg site had a total volume of 125 million m^3. As the worked-out open pit mine had a residual volume of only 84 million m^3, the excess mine waste was dumped on top of the backfilled open pit to form what is known in Wismut parlance as "fill body" or "dump body". The total uranium content of the fill body thus created may be put at 6000 t (assumption: 30 ppm uranium content, density 1.6 t/m^3). Waste rock pile footprints left behind after waste rock relocation covered an area of 400 ha. Some of these footprint areas have been reclaimed in the meantime. Given that uranium-laden seepage from the waste rock piles has infiltrated into the subsoil for decades, there is still relatively much uranium contained in the pore water of the footprint grounds, which is not immediately identifiable on the basis of the specific activity in the solid matter. In the presence of an appropriate chemical environment, this uranium content may be washed out and released over extended periods of time. In the surface layers of reclaimed areas, the uranium contained in pore water may be incorporated by plants and animals.

In summary, uranium will still be present in elevated concentrations in the environment of reclaimed mine land even after the completion of essential remedial operations. From this arises the question whether to what extent radiation protection issues do adequately take the environmental impact by uranium sufficiently into account in terms of environmental conservation and how much uranium the small-sized receiving streams (both in water and in their sediments) and soils as part of human, animal and plant habitats can tolerate.

---

[1] Assumption: mean uranium concentration in 311 million m^3 waste rock pile material: 30 ppm (ca. 0.4 Bq/g U-238); mean uranium concentration in 160 million m^3 processing residues: 80 ppm (ca. 1 Bq/g U-238).

Uranium Background Levels

Since uranium occurs naturally as trace element in all types of rock, soil, and water, knowledge of typical background levels is crucial to the assessment of its ecological relevance. General background levels for soils and water are given in (BfS 2010). Utermann and Miller (Utermann and Mille 2008) provide a more thorough analysis of background levels in soils in Germany. Uranium levels typically are in the range of <0.5 mg/kg (for sandy soils) to >3 mg/kg (soils derived from magmatic and metamorphous rocks). No significant differences were noted between topsoil and subsoil. The 90th percentile was indicated as >6 mg/kg. Mean uranium concentration in seawater is about 3 µg/l, and in German streams concentrations vary from 1 to 3 µg/l (Birke et al 2009). A concentration range of 1 µg/l to >100 µg/l is given for unimpacted groundwater in Germany (Merkel 2002). In principle, at former sites of uranium mining more elevated background levels must be expected as compared to mean background levels for soil and water applicable to Germany.

Radiological Assessment of Uranium in Soil and Water

In the context of the assessment conducted on remedial operations of Wismut GmbH the radiological assessment of uranium contained in soil and water was performed on the basis of an exposure pathway analysis in accordance with the German Calculation Guide Mining (BfS 2010) and comparison of calculated annual dose with the guidance level of 1 mSv/a.

Experience of radiological assessment shows that due to relatively low dose coefficients for the ingestion of the nuclides U-238, U-234 and U-235 for use scenarios where no drinking water consumption is being assumed the effective doses to the population turn out to be relatively low and often do not justify remedial measures to be taken. Agricultural use of the soil involving the exposure scenario „consumption of locally produced foodstuff" is the most sensitive type of use in terms of radiological assessment. When the findings of the assessment of the exposure to humans are measured against the guidance level of the mining-induced dose of 1 mSv/a, then uranium levels significantly >16 mg/kg (dry matter), – which is equivalent to 0.2 Bq/g of U-238 –, may be formally accepted. The use of surface or groundwater for drinking water purposes constitutes the most sensitive type of use in terms of assessing the uranium contamination in water. Only uranium concentrations in excess of 240 µg/l uranium (assumption: U-238 and U-234 in radioactive equilibrium) will result in an effective dose of 0.1 mSv/a to an adult, provided that he or she consumes 350 liters of such water annually (BfS 2010). The limiting dose level of 0.1 mSv per year is stipulated in the Drinking Water Ordinance (BMU 2001), – this ordinance, however, proceeds on the assumption of an adult consumption of 730 l annually.

Table 1 lists annual age-depending doses for various exposure pathways. The effective doses were formally calculated using (BfS 2010). Formal here means that all potential exposure pathways are being considered with regard to the use of water with an uranium content of 240 µg/l as well as agricultural use of soils with a uranium concentration of 16 mg/kg dry matter. The calculations assumed a radioactive equilibrium U-238 and U-234 as well as a natural isotopes ratio U-235/ U-238.

Values in Table 1 attest that uranium itself has only low radiotoxicity. The common practice in radiological protection of exposure analysis including dose assessment does not embrace the ecotoxicological effect of the heavy metal uranium

Table 1 Uranium-induced effective doses for various exposure scenarios acc. to BfS (2010)

Use of water (240 µg/l uranium)	Effective dose [mSv/a] for different age groups		
	< 1 a	1–2 a	2–7 a
Drinking water consumption	0.117	0.075	0.050
(consumption rate)	*(55 l/a)*	*(100 l/a)*	*(100 l/a)*
Preparation of breast milk substitute	0.247	–	–
Consumption of agricultural products (50% locally produced*) surface irrigation, livestock watering	0.215	0.145	0.156
Fish consumption	0.001	0.002	0.002
Agricultural use of soil contaminated with 16 mg/kg U ***	< 1 a	1–2 a	2–7 a
Consumption of agricultural products (50% locally produced), grown above ground	0.017	0.011	0.012
Use of water (240 µg/l uranium)	Effective dose [mSv/a] for different age groups		
	7–12 a	12–17 a	> 17 a
Drinking water consumption	0.064	0.064	0.100
(consumption rate)	*(150 l/a)*	*(150 l/a)*	*(350 l/a)***
Preparation of breast milk substitute	–	–	–
Consumption of agricultural products (50% locally produced*) surface irrigation, livestock watering	0.150	0.150	0.097
Fish consumption	0.002	0.002	0.002
Agricultural use of soil contaminated with 16 mg/kg U ***	Effective dose [mSv/a] for different age groups		
	7–12 a	12–17 a	> 17 a
Consumption of agricultural products (50% locally produced), grown above ground	0.012	0.012	0.008

* Exposure pathways: plant-human transfer; plant-animal-meat-human transfer; plant-animal-milk/milk products-human transfer; plant/animal-mother's milk-baby transfer.

** In contrast to the Drinking Water Ordinance (730 l annual consumption), the calculation guide mining assumes adult drinking water consumption of 1 l/day.

*** With respect to the photon radiation emanating from them, uranium nuclides are radiologically irrelevant, which is why the exposure pathway "external radiation" is not to be considered.

Ecotoxicological Derivation of Uranium Threshold Values

In recent years, great efforts have been made on a worldwide scale to further expand current scientific knowledge in the field of ecological hazard assessment for the benefit of environmental protection. The ecotoxicological assessment approach consists in analyzing the potential impacts of contaminants on the ecosystem components with a view to fostering environmental protection decisions. What matters is the protection of population growth and reproduction, it is not directly about mortality.

In Canada, threshold values (Predicted No-Effect Concentrations – PNECs) were derived for the chemical toxicity of uranium with regard to flora and fauna (non-human biota) (Sheppard et al. 2005). Values predicted that way for soil, water, and sediments are compiled in Table 2. These are not limits, but values of a preventive nature in the sense of sustainable handling of nature. In Germany, an analogous approach was used to develop the concept of thresholds of insignificance ("Geringfügigkeitsschwelle").

The comparison of the PNEC values in Table 2 with background figures typically found in Germany demonstrates that elevated uranium concentrations in soil do not really impair the growth of terrestrial plants and the development of populations of soil biota. In contrast to that, in water invertebrates and fresh water plants react very sensitive to even slightly elevated uranium concentrations, with PNEC's in the order of background concentrations. For fresh water fishes, depending on the water hardness, uranium concentrations well above background concentrations are of low significance.

In List II of families and groups of substances contained in daughter directive to 2006/11/EG (Dangerous Substances Directive), the EU Water Framework Directive (WFD) identifies uranium as an element for which measures should be taken to reduce discharges into the environment. During the transposition process in Germany, the Federal Ministry of the Environment has encouraged the estab-

Table 2 PNCC values for uranium derived from ecotoxicological assessments (Sheppard et al. 2005)

Environmental medium	Protected natural resource (non-human biota)	PNEC
Soil	Terrestrial plants	250 mg(U)/kg dry matter
Soil	Other soil biota	100 mg(U)/kg dry matter
Water	Fresh water plants	0.005 mg(U)/l water
Water	Invertebrates	0.005 mg(U)/l water
Water	Benthos	100 mg(U)/kg dry sediment
Water, < 10 mg $CaCO_3$/l, very soft water	Fresh water fish	0.4 mg(U)/l water
Water, 10–100 mg $CaCO_3$/l, soft water	Fresh water fish	2.8 mg(U)/l water
Water, > 100 mg $CaCO_3$/l, hard water	Fresh water fish	23 mg(U)/l water
Soil, water	Mammals	0.1 mg(U)/kg(BW)/d

lishment of environmental quality standards (EQS) for the protection of the aquatic biota which would also include uranium (Nendza 2003). In analogy to the investigations performed in Canada, PNEC values were derived. With specific focus on metals in water, an Added-Risk-Approach was implemented by which the natural background loads of water bodies were taken into account. At the same time, safety factors were brought to bear in order to extrapolate from No Observed Effect Concentration (NOEC) values to PNEC values. A level of uranium concentration in water of $C_{Background} + 0.15$ µg/l was derived for the protected natural resource aquatic biocoenosis, where the NOEL value for the cladoceran community (*Ceriodaphnia dubia*) tipped the balance in favor of the low PNEC value. The values of 0.5 and 0.66 mg/kg uranium, respectively, each in addition to the natural background were derived for suspended solids and sediments.

Use of such extremely low values was subject to controversial discussion in preparation for the Surface Water Bodies Ordinance to be enacted in Germany, and they were not applied thereafter, particularly since the proposed environmental quality standards for suspended solids and sediment are significantly lower than the range of natural background values. Instead the (soon to be enacted) Surface Water Bodies Ordinance stipulates an environmental quality standard of 2 µg/l for the assessment of the ecological status of a water body. On the one hand, it is based on the methodology to determine NOEC values in accordance with Annex V, item 1.2.6, WFD involving a two-stage relevance test procedure: a) determination of the substance's relevant impact on the aquatic environment and b) derivation of an environmental quality standard based on measured data. For uranium, a NOEC value of 5.5 µg/l was established, which subsequently was consolidated by an additional factor of safety. Furthermore, when determining the EQS, the natural uranium background level of 2.2 µg/l was taken account of as the 90th percentile for all of Germany and established as the quasi EQS.

Assessment of the Chemo-Toxic Impact of Uranium

The chemo-toxic impact of uranium on mammals and human beings is by nature a very complex process. Toxicological data describing the impact are given as a dose, in units of mg (uranium) per kg of body weight. In risk assessments, the quantity TDI (Tolerable Daily Intake) is compared with the real intake. According to the World Health Organization WHO the TDI for uranium is 0.6 µg per kg body weight and day. Taking into account a daily consumption rate of 2 l for drinking water, for a person of 60 kg body weight a still acceptable concentration of uranium in this water of 15 µg/l can be derived (WHO 2005). This WHO reference level has been used for many years in Germany to assess risks arising from uranium in drinking and mineral water.

It should be added that neither epidemiological studies nor animal experiments have clearly indicated that the oral intake of uranium via drinking water cause cancer (WHO 2005). The dominating adverse effect on mammals and human

Table 3 Canadian guidelines SQG for uranium in soil according to (CCME 2007)

	Land use			
	Agricultural	Residential/parkland	Commercial	Industrial
SQG [mg(U)/kg]	23	23	33	300

beings is the disturbance of the renal function. The TDI reference value given by the WHO has therefore always been considered as a preliminary value. The revised German Drinking Water Ordinance specifies now a value 10 µg/l of uranium (BMU 2001), even if this Ordinance is for the time being not yet fully in force. There can be no doubt that it will be adopted. The value in turn is being justified by the chemo-toxicological effect of uranium, in particular with regard to the renal function

In Canada, both eco-toxicological effects of uranium on flora and fauna as well as the protection of human health (chemo-toxicity) were taken in account when in 2007 guidelines for uranium depending on land use were issued (CCME 2007). The guidelines are listed in Table 3.

The figures in Table 2 demonstrate again that uranium concentrations well above the natural background are acceptable. The level of uranium is governed by the type of land use.

The currently applicable German Federal Soil Protection and Contaminated Sites Ordinance (BMU 2009) as well as the Federal Soil Protection Act (BMU 2004) do not specify test values for uranium which, if exceeded, would require case-by-case investigations on pathways of effects to be considered. In recent publications approaches are described to develop land-use specific test values (Gellermann et al. 2010).

Aspects of Weighing the Fate of Residual Contamination – a Case Example

Decisions concerning the remediation of mining legacies were or are made on a site-specific basis in a complex weighing process with due regard to the situation at the site and the potential reuse. Basically, decision-making is done with due regard to the legal framework on the one hand (legislation on water, soil, radiological protection, nature conservation), but also in consideration of highly site and object-specific facts such as:

- local lithological, hydrogeological, and hydrological conditions,
- technical options to establish effective spread barriers,
- possibilities to reduce contaminant release or immobilize contaminants,
- present and potential future scenarios of use, and
- stakeholder interests, on the other.

This reveals that complete removal of environmental risks emanating from contaminants (e.g. from uranium in soil and water) cannot and must not be the goal of remedial measures. Left behind are residues, residual contaminations, and emissions which for long periods of time will continue to cause environmental impacts by uranium. At the same time, compliance with preset threshold or guidance levels cannot in each and every case be reached immediately after completion of physical remedial work. That makes it imperative for the process of assessing environmental impacts, and those of uranium in soil and water in particular, to identify a tolerable range of levels that will also be acceptable to the general public under the aspect of sustainability. The same applies to accepting the accompanying effects such as extremely fast growing volumes of wastes that have to be disposed of and permanently monitored in the event of highly intensified uranium removal from effluents, but also accepting the financial burden to society as a whole.

In order to achieve a well optimized and sustainable rehabilitation decision, environmental quality standards should not be raised to the level of dogma, but rather include more thoroughly all factors involved as well as aspects of "natural attenuation" during post-remedial phases. In accordance with the EU WFD, German legislation is definitely taking such aspects into account by allowing for respective exemptions.

Case Study Crossen

The case study of the Crossen site exemplifies a number of aspects considered in weighting the fate of residual contaminations. Completed in 2008, remediation of the Crossen plant area was unprecedented in its kind under the WISMUT environmental rehabilitation project (see Fig. 1).

Crossen had been the second-largest chemical uranium processing facility in Europe. Between 1950 and 1989, a total of 77,000 t of uranium were produced. The prevailing intricate contamination of soil and of near-surface groundwater (amongst other things caused by infiltration of uranium and other heavy metals,

Fig. 1 Crossen plant area after termination of uranium ore processing in 1991 as well as after completion of remedial work in 2008

with the concurrence of processing chemicals) as well as the plant location right on the bank of the Zwickauer Mulde river were the reasons why the decision-making process came to the conclusion that complete removal of the contamination would be hardly feasible and disproportionately expensive. The optimized remedial solution entailed excavating the contaminated material down to the top of the groundwater level as well as excavating in addition partial areas down to below the groundwater table. This was done with a view to reducing to a tolerable level the source of further contaminant releases from the residual contamination left in place.

The optimization process took local hydrological and geochemical conditions of uranium mobilization into account. Contaminated materials totaling 412,000 m³ were removed during the remediation works – a volume of surplus excavated material of 34% compared to original work plans. The entire former plant area was capped with inert soil material and the site was blended into the floodplain landscape of the Zwickauer Mulde river.

The concluding evaluation of the remedial performance came to the result that the total quantity of uranium left behind on the former plant area after surface excavation amounts up to a maximum of 30 t. Nonetheless, no significant increase is ascertained in uranium concentrations in the water of the Zwickauer Mulde river stemming from residual contamination on the former plant area. However, the residual contamination releases uranium via the water pathway, albeit to a reduced extent. Predicted uranium loads vary from 150 to 300 kg/a. As a result of ongoing washing processes, mean uranium concentrations downstream of the site will diminish according to predictions. Therefore, groundwater at the site will be subject to restrictions of use and require surveillance under the environmental monitoring program both in the mid and long term.

Do We Have Sufficient Knowledge of Uranium in the Environment?

To date, the international discussion on the protection of the environment is driven by the rapid development of computer codes and evaluation methods. In this process it becomes evident that significant data gaps and variabilities remain to be overcome. As a consequence, it appears that the current state of knowledge is not yet sufficient to determine internationally recognized limit values for the protection of the environment against uranium. Limit values for uranium as discussed and model-evaluated thus far are to be seen as screening or benchmarking values and they are suited to indicate a hazard. But they do not constitute a basis for legally substantiated decision-making.

Complex processes of uranium migration, sorption, and bioaccumulation as they occur in nature are, as a rule, not yet completely implementable by ecological modeling. These approaches are based on generalized transfer factors which fail to sufficiently take account of forms of chemical bounding and of bioavailability as they are encountered in the natural environment. Thus far, data gaps hamper efforts to fully take into account the variability of background levels for site-specific geological conditions in areas of ore deposits. Therefore, studies have to be performed to investigate the impact of elevated uranium concentrations stemming from left behind "residual contaminations". In doing so it might turn out that nature is well capable of tolerating higher levels of uranium without damage as compared to what is currently derived from unreliable information or is sometimes assumed for reason of prevention. The reclaimed uranium mining legacies containing residual contaminations which will be "reconquered" by nature before long constitute suitable investigation sites to fathom the correlation of uranium contained in water and soil with animals and plants and to specify more precisely conclusions with respect to protect the environment against radioactive and chemo-toxicological contaminants.

References

BfS (2010) Berechnungsgrundlagen zur Ermittlung der Strahlenexposition infolge bergbaubedingter Umweltradioaktivität (Berechnungsgrundlagen-Bergbau), Bundesamt für Strahlenschutz, BfS-SW-07/10

Birke M, Rauch U, Lorenz H (2009) Uranium in stream and mineral water of the Federal Republic of Germany, Environ. Geochemistry and Health 31: 693–706

BMU (2001) Verordnung über die Qualität von Wasser für den menschlichen Gebrauch (Trinkwasserverordnung – TrinkwV). BGBl.: 959

BMU (2004) Bundesbodenschutzgesetz, in der Fassung vom 9. Dez. 2004. BGBl. I: 3214

BMU (2009) Bundes-Bodenschutz- und Altlastenverordnung, in der Fassung vom 31. Juli 2009. BGBl. I: 2585

CCME (2007) Canadian Soil Quality Guidelines for Uranium: Environmental and Human Health. Canadian Council, Ministry of Environment

Gellermann R, Günther P, Evers B (2010) Beurteilung von Bodenkontaminationen mit Radioaktivität im Gebiet Hannover-List nach Maßstäben und Ansätzen der Bundesbodenschutzverordnung. Umweltwissenschaften und Schadstoffforschung 22: 116–122

Merkel B J (2002) Uran im Trinkwasser, Institut für Geologie TU Bergakademie Freiberg. see: http://www.geo.tu-freiberg.de/merkel/uran_index.htm

Nendza M (2003) Entwicklung von Umweltqualitätsnormen zum Schutz aquatischer Biota in Oberflächengewässern. BMU Umweltforschungsvorhaben 202 24 276

Sheppard S C, Sheppard M I, Gallernad M, Sanipelli B (2005) Derivation of ecotoxicity thresholds for uranium. J Environ Rad 79: 55–83

Utermann J, Fuchs M (2008) Uranium in German Soils. In: De Kok L, Schnug E (ed), Load And Fate of Fertilizer-derived Uranium. Backhuys Pub. Leiden, Netherlands: 33–45

WHO (2005) Uranium in Drinking-water, Background doc. for Development of Guidelines for Drinking-water Quality, WHO/SDE/WSH/3.04/118.
see: http://www.who.int/wter_sani-tation_health/

Health Hazards and Environmental Issues at the Uranium Mine Near Tatanagar, India

S.K. Sharma

Abstract. The problem is uranium has always been a highly dangerous material that poisons people and the environment. Uranium mining near Tatanagar in the state of Jharkhand in India, is no exception it. The waste from this process is packed into drums and sent again to uranium mine to be dumped in the tailing dams. The contents of these dams are highly radioactive, even though uranium has been extracted but the remaining 'uranium tailing" material remains active in the atmosphere for many years. This poisonous nuclear waste has caused unimaginable health damage to the local people, animals and the environment.

Introduction

In India, progress and pollution come as a package deal. Though, nuclear energy is the cleanest source of power, no soot, smoke or green house gases. And, if the safety norms are strictly adhered to, there will be insignificant fractional increase of back-ground radiation either in the operating area or outside the power plants. Located near Tatanagar in the Sighbhum district of Jharkhand State, the uranium for the country's nuclear program is mined here from three underground mines up to a depth of 700 m below the earth's surface. The nuclear waste which is called "tailings" or "uranium tailing" comes from the mines and mills as a waste after the uranium ore is mined and processed for purification and it is then dumped in the tailing ponds. The poisonous nuclear waste has the potential to cause unimaginable damage to the surroundings specially where the water table is low in the area and surface water flows though it. The devastating effects caused to local people

---

S.K. Sharma
Department of Environmental Education, Carman School, Dehradun, India

B. Merkel, M. Schipek, *The New Uranium Mining Boom*
DOI 10.1007/978-3-642-22122-4

due to uranium mining and the interaction of tailings with groundwater at uranium mine near Tatanager in Jharkhand State has been investigated where local people are fighting this invisible enemy. The Uranium ore (Uraninite) is mined near Tatanagar for retrieving the rare uranium.

Brief Geology and Geothermal History of the Area

The Son-Narmada-Tapti lineament zone is a fault bound mega lineament belt in Central part of the country (VII), with a large number of hot springs manifestations. Temperature gradients in the 40 to 120°C/km range and heat flow values from 70 to 300 mW/m^2 have been recorded at several locations in the Son-Narmada-Tapti province (Joga Rao and Rao 1987). Godavari (VIII) and Mahanadi (IX) valleys are fault bound grabens with post-Gondwana and possibly late Tertiary/Quarternary reactivations. Moderate temperature gradients of 39 ± 10°C/km and heat flow values of 80 ± 21mw/m^2 have been recorded in the Godavari Valley. Isolated warm springs are found in Southern India (X). A systematic study and geothermal exploration is required in this province.

The exposed rocks in Southern Bihar show two facies, an un-metamorphosed one in the south and a metamorphosed one in the north, separated by a major thrust zone of 2 to 5 km wide and is bordered on its north by well marked shearing. The rocks within the thrust zone are metamorphosed shales and sandy shales with subordinate metavolcanics such as chlorite schist, amphibole schists etc. Near the northern border of the zone are bands of quartz-mica-schists with tourmaline and kyanite. It is postulated that the mineralization took place in three stages. The earliest was the formation apatite – magnetite lenses, followed by the bands of chlorite and amphibole-schists. Uranium mineralization is noted in the form of disseminated uranite, tobernite and autunite along the chlorite schist. Workable grade of uranium bearing ore, amounting to a few million tons are present in the uranium mines neat Tatanagar in Jharkhand. The tailing piles at the mine were leached by water from the nearby geothermal springs present in Hazaribagh area in Jharkhand where few radioactive hot springs are reported.

There are many thermal springs near Tatanager area, in southern Bihar indicated by "triangles" at Jharia and Surajkund. in the following Fig. 1 (Ravishankar et al. 1992).

Data Collection at Tailing Pond

The chemical composition and radium concentration of thermal water has been collected from one of the tailing ponds, Fig. 2, near Tatanagar and has been summarized in Table 1.

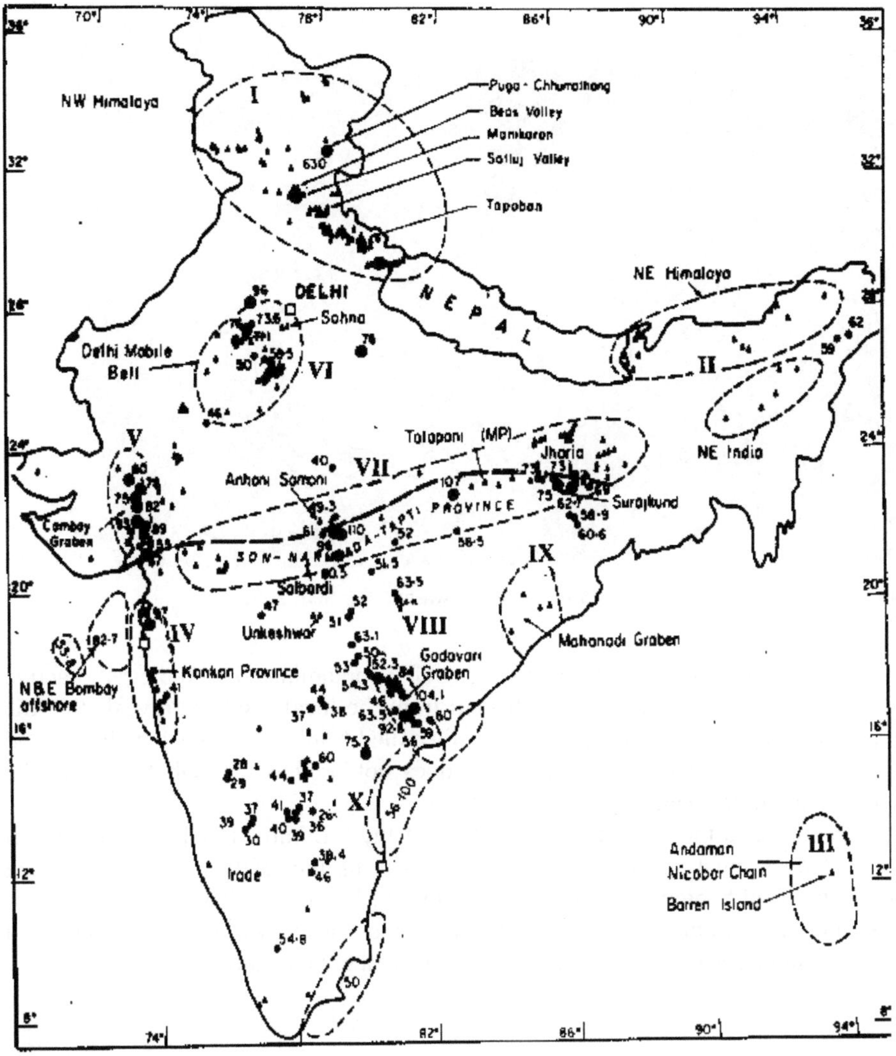

Fig. 1 Geothermal provinces of India and regional flow patterns (Data source: Ravi Shanker et al. 1992)

Table 1 Chemical composition and radium concentration of thermal water

pH	>9
TDS	520 ppm
F	21 ppm
Ca	2.8 ppm
Mg	0.3 ppm
Na	150 ppm
K	6.5 ppm
Cl	90 ppm
SiO_2	110 ppm
Radium	7 to 9 ppm

Fig. 2 One of the tailing ponds to store nuclear waste (after SOE 2001)

It is observed that the surface water is being contaminated with the radio-active waste sediments/dust. The data collected from the local residents reveals that the water in wells and small tributaries (locally called "nallas") has become black, and sometimes the water tastes salty, soap does not produce enough foam and the cloths remain unclean.

According to the Uranium Corporation of India Limited (UCIL) policy which is the Government agency responsible for ensuring the safety of uranium mining sites, the local habitants and the environment in India, all villages within 5 km of the mine and tailing ponds should be evacuated but the settlement is found well within 100 to 200 m thus putting the residents at a great health risk.

Interpretation

It is observed that the contents of the tailing ponds are highly radio-active, even though uranium has been extracted but the ponds/tailings remains radio-active for a very long time. Radio-active tailings have permeated the groundwater and contaminated surface water sources. The thermal spring water coming in contact with these tailing also becomes radio-active and the concentration of radium of the order of 7 to 9 ppm in these waters becomes an invisible enemy. Local habitants suffer from fatigue, lack of appetite, respiratory ailment, infant mortality, skeletal deformities such as fused fingers, skin disease, leukemia, thalassemia and even Parkinson's disease have been reported (Bhatia 2001).

The radium values in thermal waters varying from 7 to 9 ppm from the tailing pond can be compared with that of Karia of Portugal.

Conclusions

When the contaminated was seen to be reaching the external natural environment, the mining authorities decided to collect and pump the water for neutralization. The study described here aimed to investigate possibility that infiltrated rainwater and natural runoff are reaching the tailing pond material where they form new groundwater in contact with the tailing materials. Radium salts have clearly been dissolved and, in the future this contaminated groundwater will reach the natural hydrological cycle. The purpose of this study was to evaluate the interaction between the tailing materials and the hydrological cycle. The tailings were leached by the groundwater, thermal springs or the rainwater and the outflows of water was found to be seriously contaminated by high levels of radium, of the order of 7 to 9 ppm. The local people are poor in the region and are entirely dependant on this radio-active contaminated water available to them for agriculture and their day to day domestic activities. They have less awareness of radiation and its ill-effects and thus, are subjected to variety of diseases.

Recommendations

The high potential for pollution by dissolution of toxic salts inside the tailing ponds suggests that planning for environmental control should include the best possible dealing of substratum before tailing are deposited. Moreover, no settlement be permitted within five kilometers of the uranium mine to safeguard the health of local habitants.

Acknowledgements The author thanks to the team of the Geological Survey of India for fruitful discussions and valuable suggestions to complete this write up.

References

Bhatia, B (2001) Fighting and invisible enemy. The Hindu Survey of the Environment: 129–135 pp

Joga Rao, MV, Rao, AP (1987) Results of Geophysical Survey in the Tattapani Hot Spring Area, Distt. Surguja, Madhya Pradesh, *GSI Records*, Vol. 115, Part 6, 66.83 pp

Prasad, JM (1996) Geothermal energy resources of Bihar, Geothermal Energy in India, GSI. Spl. Pub. No 45: 99–117 pp

Ravishankar Guha SK, Seth NN, Muthuraman K, Pitale UL, and Sinha RK (1991) Geothermal Atlas of India. Special Publication No. 19, Geol. Sur. Ind: 144 pp

Cadmium and Uranium in German and Brazilian Phosphorous Fertilizers

Geerd A. Smidt, Franziska C. Landes, Leandro Machado de Carvalho, Andrea Koschinsky, Ewald Schnug

Abstract. A collection of phosphorous (P) fertilizers used in Germany (N = 75) and southern Brazil (N = 39) was analyzed for cadmium (Cd) and uranium (U). Both collections show high mean concentrations of Cd (12.0 and 18.6 mg/kg respectively) and U (61.3 and 70.16 mg/kg respectively), while maximum concentrations of 56 mg Cd/kg and up to 200 mg U/kg were found. Currently, up to 42 t Cd and 228 t U are distributed annually on German and 611 t Cd and 1614 t on Brazilian agricultural soils by mineral P fertilizers.

Geerd A. Smidt
Jacobs University Bremen, Integrated Environmental Studies Program, Campus Ring 1, D-28759 Bremen
E-mail: g.smidt@jacobs-university.de

Franziska C. Landes
Jacobs University Bremen, Integrated Environmental Studies Program, Campus Ring 1, D-28759 Bremen

Andrea Koschinsky
Jacobs University Bremen, Integrated Environmental Studies Program, Campus Ring 1, D-28759 Bremen

Leandro Machado de Carvalho
Institute for Analytical Chemistry, Federal University of Santa Maria (UFSM), Campus Universitário, CP 5051, Santa Maria (RS), Brazil

Ewald Schnug
Technical University Braunschweig, Department of Life Sciences, Pockelsstr. 14, D-38106 Braunschweig
E-mail: e.schnug@tu-braunschweig.de

B. Merkel, M. Schipek, *The New Uranium Mining Boom*
DOI 10.1007/978-3-642-22122-4 © Springer 2011

Introduction

Uranium (U) and cadmium (Cd) are natural heavy metals, which may impair biological systems by radioactivity (U) and chemical toxicity (U and Cd). The production of phosphorus fertilizers requires raw phosphate rock material, which originates to about 87% from marine sedimentary sources (Kratz et al. 2008). Uranium, Cd and many other trace elements (Cd, Zn, Pb, Cu, Ni, Co, Sn, Cr, Hg, As, Sb, Th, Ra, Sr, Y, Rb, Mo, Se and many Rare Earth Elements) are enriched in these phosphates during deposition and diagenesis. Phosphate (P) rocks contain concentrations in a range of 23–220 mg/kg in general (Kratz and Schnug 2006) and 30–130 mg/kg U in Brazilian P rocks (Saueia et al. 2005; da Conceicao et al. 2006; Santos et al. 2006). The pure raw phosphate is very water insoluble and for fertilization requirements an industrial processing is needed. Phosphoric acid is produced by dissolving the P-rich apatite minerals with H_2SO_4, with phosphogypsum as a by-product. During this process the radionuclides U and Th and many heavy metals become enriched in the P fertilizer to about 150% of the original value of the rock phosphate (Sattouf 2007). Since all uranium isotopes form very stable phosphate compounds, U concentrations are enhanced in the phosphoric acid, while radium isotopes are enriched in the by-product phosphogypsum. This by-product has no application in Germany, and the stockpiles of phosphogypsum are a radiological environmental concern, while in Brazil it is used for liming. The last step to produce a multi-component fertilizer involves the reaction of phosphoric acid with ammonia to produce monoammoniumphosphate (MAP) and diammoniumphosphate (DAP), and with potassium chloride (KCl) to produce PK or NPK fertilizers. The production of singlesuperphosphate (SSP) and triplesuperphosphate (TSP) involves the reaction of phosphoric acid with phosphate rock which yields in a higher concentrated P fertilizer (Saueia et al. 2005; Kratz and Schnug 2006).

Several studies have indicated that the U content of P fertilizers increases with an increasing P_2O_5 content of the product (Spalding and Sackett 1972; Barisic et al. 1992). In P fertilizers U and Cd are abundant in higher concentrations favorably in phosphate fertilizers of sedimentary origin with concentrations of 0.1–362 mg/kg U and 0.1–60 mg/kg Cd (Birke et al. 2009; Aydin et al. 2010). Despite the voluntary self-restriction of the industry (IVA 2010) to substitute sedimentary with magmatic rock phosphates at the end of 1984 to decrease the heavy metal loads to arable soils, there is no trend of decreasing U concentrations. In fact, the import of phosphates with low Cd and U contents decreased by 33% from 1997 to 2007, while the import of U rich phosphates (100–150 mg U/kg) from P rock deposits in Israel increased by 50% (Heffer and Maene 2008). Also Sattouf et al. (2008) determined lower U concentrations in fertilizers with mineral P components produced in 1974–1984 than from 1995–2005 by trend (Sattouf et al. 2008). As U and Cd contents are highly correlated in fertilizers due to their high affinity to phosphates, it can be concluded that the Cd content also did not decrease after 1984. Contrary to Cd, no legal labeling obligation or threshold limits exist for U concentrations in fertilizers, neither in Brazil nor in Germany.

Under oxidizing conditions U(VI) forms complexes with carbonates, which are highly mobile in arable soils. The uncontrolled application of fertilizer-derived U might play a bigger role for the hazard of groundwater bodies in areas of intense agricultural production than the co-application of Cd, due to the high solubility of U and higher concentrations in fertilizers compared to Cd.

This study aims to compare the concentrations of Cd and U in German and Brazilian P-containing fertilizers as well as to estimate and evaluate possible amounts and consequences of the high loads of these toxic elements to the agroecosystems.

Cadmium and Uranium in Soils

Naturally contents of U in soils range from <0.4 up to >10 mg U/kg and <0.5 up to >3 mg Cd/kg (Scheffer and Schachtschabel 2002). Utermann and Fuchs (2008) mapped the U content in German top- and subsoils. They found the tendency of higher U content in topsoils compared to subsoils. The difference of the median values in forest and arable soil is 0.15 mg/kg U. This study agrees with the results of Salminen et al. (2005), who determined an enrichment factor of 1.2 for northern German soils. Such enrichment was also found by Huhle et al. (2008), who detected a difference of 0.2 mg/kg U between arable and forest soils. The U content of topsoils is by trend higher than that of subsoils as is demonstrated by a general map of U in German soils (Utermann and Fuchs 2008). There is a striking difference of 0.15 mg/kg U between the median value for arable and forest soils. This value is in close agreement with the results of other research groups, in particular for Northern Germany (Huhle et al. 2008; Rogasik et al. 2008).

In contrast to other environmentally relevant elements, for instance Zn and Cd, U complexes are highly mobile in agricultural soils under oxic conditions and especially in the presence of carbonate ions (Utermann and Fuchs 2008; Salminen et al. 2005; Sheppard et al. 2005; Zheng et al. 2003). On average four times more U (0.06–3.9 μg/l) was found in percolation water under arable soils than under forest soils in Germany (0.006–0.5 μg/l) (Huhle et al. 2008). Cadmium, in contrast, was found to be on average four times more mobile under sandy forest soil (0.36–0.58 μg/l in percolation water) in comparison to sandy arable soil (0.09–0.13 μg/l in percolation water) (Duijnisveld et al. 2008). These results confirm the higher mobility of fertilizer-derived U in arable soils in comparison to Cd.

Accumulation of Fertilizer-Derived Cd and U in Soils

The annual U load varies between 2.8 g U/ha if P is applied exclusively by organic fertilizers at rates following "Good Agricultural Practices" (GAP) (22 kg P/ha*yr) and 15 g U/ha if P is applied in mineral form according to GAP (Kratz et al. 2008). Rogasik et al. (2008) studied the influence of long-term P fertilization on the accumulation of U in soils at 7 sites in Germany. In all experiments P rates correlated with the increase of the U content in topsoils. The mean accumulation rate was 3.7 µg/kg*yr U with a range of variation from 1 to 15 µg/kg*yr U. The latter value is in close agreement with the rate of 14.5 µg/kg*yr U that has been calculated for the long-term field experiments in Johnston Castle/Ireland (Tunney et al. 2009). In contrast, accumulation rates reported for New Zealand and Australia are distinctly higher with 19–37 µg/kg*yr (Taylor and Kim 2008; Lottermoser 2009). The reason for this appear to be lower U leaching rates, while higher P rates or higher U concentrations in the fertilizer products proved to be no causes. A calculated accumulation rate of 6 µg/kg*yr U and a mean increase of the U content in soils of 0.32 mg/kg U is to be expected when estimated on the basis of U loads in the P balance. Rogasik et al. (2008) found, however, an average increase of 0.13–0.20 mg/kg U in different long-term field experiments in Germany if P was applied by reason of crop demand. It is suggested that U losses from soils by leaching are the reason for this persistent difference throughout this and other studies on U in soils (Utermann and Fuchs 2008; Huhle et al. 2008).

It is estimated that a total amount of up to 13,333 t U and 4305 t Cd (Kratz et al. 2011) has been applied with mineral fertilizers by German agriculture from 1951 to 2005. This equals a cumulative load of up to 1 kg U/ha and 0.30 kg Cd/ha (Kratz et al. 2011) on agricultural land. For comparison at the nuclear waste storage salt mine "Asse II" 102 t U are deposited in total (www.atommuell-endlager.de).

Material and Methods

Phosphorous fertilizers traded in Germany (N = 78) and Brazil (N = 10) were analyzed for the element composition in the laboratories of the Institute of Plant Nutrition and Soil Science of the Federal Agricultural Research Center (FAL-PB, now the Institute for Crop and Soil Science of the Federal Research Centre for Cultivated Plants (JKI-PB)) and in the Geochemistry laboratory of the Jacobs University Bremen by common standard procedures for fertilizer analysis. For origin and analysis of the fertilizers from Germany detailed information are provided by Kratz et al. (2011) and Smidt (in prep.) The Brazilian phosphate fertilizer samples collected in southern Brazil were obtained from the laboratory LACHEM, Federal University of Santa Maria, RS. Both laboratories employed common standard procedures for fertilizer analysis.

Results and Discussion

Cadmium and Uranium in Fertilizers

The collections of Brazilian and German fertilizers analyzed in this study show Cd concentrations in a range of <0.2–>100 mg Cd/kg and U concentrations in a range of <1–>200 mg U/kg (Tables 1, 2). The average phosphorous concentrations are 30.1% in Brazilian and 22.8% in German samples, with maximum concentrations up to 52.5 and 38.0%, respectively. The Brazilian fertilizers show higher average and maximum concentrations for Cd (18.6 and 56.8 mg/kg, respectively) in comparison to the German fertilizers (12 and 34.8 mg/kg, respectively). The Brazilian fertilizer collection was found to have a higher average U concentration of 70.6 mg/kg and maximum concentration of 200 mg/kg, whereas the German fertilizer collection had a comparable maximum U concentration of 206 mg/kg U and average U concentration of 61.3 mg/kg. The determination of rare earth elements and yttrium (La, Ce, Pr, Nd, Sm, Eu, Gd, Tb, Dy, Y, Ho, Er, Tm, Yb, Lu) (REY) and the normalization of the REY concentrations against Post Achaean Australian Shale (PAAS, Taylor and McLennan 1985) enabled to identify the marine sedimentary or the igneous deposit source of the P-rock used for the production of the P-fertilizers in Brazil. Of our 40 fertilizer samples, 4 could be identified as igneous P-rock derived, while the remaining 36 were of marine sedimentary origin (Smidt in prep.). This distribution percentage is consistent with the value of 87% given in literature for the origin of P-fertilizers from marine sedimentary P deposits.

In 96 samples taken for fertilizer quality control by the German Fertiliser Ordinance (DüMV) of federal states in Germany in the framework of a joint project, the following results were obtained: 19% of all samples showed a P_2O_5-content of $<5\%$ and out of these samples 11% had a Cd content of >1 mg/kg Cd which are obliged to be labeled and 6% contained more Cd than the limit value of 1.5 mg Cd/kg (DüMV 2008). The remaining 81% of samples ($N=78$) had P_2O_5-contents of $>5\%$ (Table 1), with 23% out of these below the obligation to label threshold of 20 mg Cd per kg P_2O_5 and 48% above the limit value of 50 mg Cd per kg P_2O_5 (DüMV 2008). The collection of Brazilian fertilizers analyses in this study contain to 98% P_2O_5 higher than 5 and 33% of this group show Cd values below 20 mg Cd/kg P_2O_5. 44% could not be traded on German market, as they were above the limit value of 50 mg Cd/kg P_2O_5. In Brazilian legislation P fertilizers can contain 4 mg Cd/kg P_2O_5 in Brazil with the highest concentration limit of 57 mg Cd/kg product (Ministry of Agriculture Brazil 2006). Considering these values, only 4 fertilizers or 10% of the analyzed Brazilian fertilizers showed Cd concentrations above the maximum limit concentration of 4 mg Cd per kilogram P_2O_5 content. The main results of the fertilizer analyses are summarized in Table 1 and 2 for samples with a P_2O_5-content of $>5\%$.

A comparison of these values with U concentrations reported in literature for same fertilizer products traded in Germany (Kratz and Schnug 2006) provided the following results (values for German products in brackets): Triple-Super 219 (106), NP-fertilizers 95 (17), PK-fertilizers 92 (82), NPK-fertilizers 39 (5.9) mg U/kg and soft rock phosphates 55 (65). The results of this study indicate that the U concentrations in new fertilizer products traded in Germany presented were on an average 26% higher than those reported in literature.

A significant correlation was found between Cd and U content per kg P_2O_5, but the constant term of the regression is high with values of 86 and 106 so that in principle a Cd-free product will have a U content which is by far higher than suggested limit values.

Table 1 P_2O_5, Cd and U concentration in fertilizers with a P_2O_5-content of >5% which have been traded in Germany in 2007 ($n=78$) and weighted mean values provided by Dittrich und Klose (2008)

	P_2O_5 (%)	Cd (mg/kg)	U (mg/kg)	mg Cd/kg P_2O_5	mg U/kg P_2O_5
Mean[a]	22.8	12.0	61.3	47.0	283
Median	17.0	7.40	39.8	50.0	264
Minimum	5.00	0.11	0.73	0.24	6.39
Maximum	38.0	34.8	206	107	1713
Percentile 25	10.8	2.89	11.7	18.0	79.8
Percentile 50	17.0	7.40	39.8	49.9	264
Percentile 75	40.0	20.2	87.4	67.1	402
Weighted mean values from Dittrich and Klose (2008) ($n=193$)[b]	25.8	9.40	63.3	37.0	245

[a] Note: Out of 78 samples 29% had a Cd content above the obligation to label and 48% exceeded the limit value.
[b] Note: Out of 193 samples 41% had a Cd content above the obligation to label and 17% exceeded the limit value.

Table 2 P_2O_5, Cd and U concentration in fertilizers with a P_2O_5-content of >5% which have been traded in Brazil in 2007 and 2008 ($n=39$) and weighted mean values taken from da Conceicao et al. (2006)

	P_2O_5 (%)	Cd (mg/kg)	U (mg/kg)	mg Cd/kg P_2O_5	mg U/kg P_2O_5
Mean[a, b]	30.1	18.6	70.6	61.5	248
Median	21.5	10.8	59.9	52.0	283
Minimum	14.2	1.30	0.30	2.80	1.40
Maximum	52.5	56.8	200	189	498
Percentile 25	17.8	7.4	25.2	40.0	113
Percentile 50	21.7	10.8	59.9	52.0	283.1
Percentile 75	46.0	28.3	98.2	76.4	257
Weighted mean values from da Conceicao (2006) ($n=6$)	42.7	2.17	65.2	6.64	190

[a] Note: Out of 39 samples, 4 had a Cd content above the Brazilian limit value for P-containing fertilizers of 4 mg Cd per percent of P_2O_5.
[b] Note: Out of 39 samples 23% had a Cd content above the obligation to label and 44% exceeded the limit value.

Fertilizer-Derived Cd and U Loads to German and Brazilian Soils

An actual conservative estimate delivered an average total amount of about 114 t U and 22 t Cd (Kratz et al. 2011) that are distributed annually on agricultural land in Germany by mineral phosphorus fertilizers. Under the assumption of "Good Agricultural Practices (GAP)" with fertilization rates of 22 kg P/ha, an average annual load of 1.4 g Cd/ha and 8.2 g U/ha is calculated (Kratz et al. 2011).

The average phosphorous fertilization in Brazil is 48 kg P_2O_5 per hectare (FAO 2004), which is in the same range as the amount of 50.4 kg P/ha used under the assumption of GAP in Germany. In Brazil, the amount of 2.9 g Cd/ha and 11.8 gU/ha is applied annually with 48 kg P_2O_5/ha, which contain 61.5 mg Cd/kg P_2O_5 and 248.2 mg U/kg P_2O_5 in average (Table 2). Maximum loads of 9.1 g Cd/ha and 23.9 g U/ha are potentially applied annually, assuming maximum concentrations of 189 mg Cd/kg P_2O_5 and 498 mg U/P_2O_5 (Table 2) in fertilizers.

In the central western areas of Brazil (Cerrado) the application of P_2O_5 is on average 76 kg/ha*yr, causing an average co-application of 4.7 (*14.3*) g Cd/ha*yr and 18.7 (37.9) g U/ha*yr (maximum loads). The soil in the Cerrado is extremely P deficient and special crops like potato are intensively fertilized using 433 kg P2O5/ha*yr, causing a potential load of 26.6 (81.68) g Cd/ha*yr and 106.6 (215.7) g U/ha*yr (maximum loads).

In 2009, a total of 3.24 million t of P_2O_5 were applied to arable land in Brazil (Gasques et al. 2009), so the calculated total loads of cadmium and uranium to Brazilian soils were in average 199 t Cd and 798 t U with a maximum 611 t Cd and 1615 t U. It is expected that the total usage of P_2O_5 will be increased to an amount of 5.36 million t in the year 2020 (Gasques et al. 2009). As a result the Cd and U loads would increase to an average of 330 t Cd and 1319 t U with a maximum of 1011 t Cd and 2671 t U.

Conclusions

As intensive agriculture in Germany and Brazil requires phosphorous fertilization to maintain soil fertility and hence high quality food products, 50 kg P_2O_5 per hectare are fertilized in average in both countries annually. Our analyses on a representative selection of German and Brazilian phosphorous fertilizers has shown that due to the high Cd and U contents in the products, the calculated loads of Cd and U on arable soils indicate a significant contamination risk of the agroecosystems with these toxic heavy metals. The increasing worlds demand for food and energy plants as sugar cane provides a big economical opportunity for Brazil. To supply the worlds hunger for proteins and bio-ethanol intensification of cropping production and hence intensification of fertilization is expected. While the hazard of Cd mobility and availability can be considered as moderate, the risk of U mobilization from arable soils under oxic conditions in the presence of carbonate ions is high in Germany. Brazilian soils are often very acidic and carbonate free, so the mobilization potential of U is lower than that of Cd. Nevertheless, as recent research indicates leaching of U from arable soils and presence of fertilizer-derived U in ground- and drinking water (Smidt et al. 2011), it is suggested that the uncontrolled loading of the toxic and radioactive heavy metal to soils should be regulated by state authorities, as it is done for cadmium.

References

Atommüll-Endlager.de (http://www.atommuell-endlager.de/index.php?option=com_content&task=view&id=15&Itemid=1)

Aydin I, Aydin F, Saydut A, Bakirdere G, Hamamci C (2010) Hazardous metal geochemistry of sedimentary phosphate rock used for fertilizer (Mazıdag, SE Anatolia, Turkey) Microchem. J. 96:247–251

Barisic D, Lulic S, Miletic P (1992) Radium and uranium in phosphate fertilizers and their impact on the radioactivity of waters, Wat. Res. 26:607–611

Baumann N, Arnold T, Read D (2008) Uranium ammunition in soil. In: de Kok, L. J. and Schnug, E. (eds.). Loads and Fate of fertilizer-derived uranium. Backhuys Publishers, Leiden. ISBN/EAN 978-90-5782-193-6:73–78

Birke M, Rauch U (2008) Uranium in stream water of Germany. In: de Kok, L. J. and Schnug, E. (eds.). Loads and Fate of fertilizer-derived uranium. Backhuys Publishers, Leiden. ISBN/EAN 978-90-5782-193-6:79–91

da Conceicao FT, Bonotto DM (2006) Radionuclides, heavy metals and fluorine incidence at Tapira phosphate rocks, Brazil, and their industrial (by) products. Environ. Poll. 139:232–243

Dittrich B, Klose R (2008) Bestimmung und Bewertung von Schwermetallen in Düngemitteln, Bodenhilfsstoffen und Kultursubstraten. Schriftenr. Sächs. Landesanst. Landwirtsch. 3/2008

Duijnisveld WHM, Godbersen L, Dilling J, Gäbler H-E, Utermann J, Klump J, Scheeder G (2009) UBA-Forschungsbericht FKZ 20472264. Datenanhang.

DüMV – Düngemittelverordnung (2008) Verordnung über das Inverkehrbringen von Düngemitteln, Bodenhilfsstoffen, Kultursubstraten und Pflanzenhilfsmitteln (Düngemittelverordnung – DüMV)

FAO – Food and Agriculture Organization of the United Nations (2004) Fertilizer use by crop in Brazil. First version, published by FAO, Rome

Gasques JG, Bastos ET, Silva LF da (2009) Projeções do Agronegócio – Brasil 2008/09 a 2018/2019. Ministério da Agricultura, Pecuária e Abastecimento – Assessoria de Gestão Estratégica. Brasília, DF

Heffer P, Maene L (2008) International Fertilizer Manufacturers Association, Paris, personal communication

Huhle B, Kummer S, Merkel B (2008) Mobility of uranium from phosphate fertilizers in sandy soils. In: de Kok, L. J., Schnug, E. (eds.). Loads and Fate of fertilizer-derived uranium. Backhuys Publishers, Leiden. ISBN/EAN 978-90-5782-193-6:47–56

IVA – Industrieverband Agrar (2010) Industrieverband Agrar für europaweiten Grenzwert Cadmium in phosphathaltigen Mineraldüngern. http://www.iva.de/themen-und-positionen-des-iva/industrieverband-agrar-fuer-europaweiten-grenzwert-cadmium-phospha-thal

Kabata-Pendias A, Mukherjee AB (2007) Trace Elements from Soil to Human. ISBN-10 3-540-32713-4 Springer Berlin Heidelberg New York

Kratz S, Schnug E (2006) Rock phosphates and P fertilizers as sources of U contamination in agricultural soils. In: Merkel, B.J., Hasche-Berger, A. (eds.). Uranium in the environment. Springer, Berlin Heidelberg: 57–68

Kratz S, Knappe F, Rogasik J, Schnug E (2008) Uranium balances in agroecosystems. In: de Kok, L.J., Schnug, E. (eds.). Loads and Fate of fertilizer-derived uranium. Backhuys Publishers, Leiden. ISBN/EAN 978-90-5782-193-6:179–190

Kratz S, Godlinski F, Schnug E (2011) Heavy metal loads to agricultural soils from the application of commercial phosphorus fertilizers in Germany and their contribution to background concentration in soils. In: Merkel, B. (ed.). Uranium Mining and Hydrogeology VI

Lottermoser B (2009) Trace metal enrichment in sugarcane soils due to the long-term application of fertilisers, North Queensland, Australia: geochemical and Pb, Sr, and U isotopic compositions. Austr. J. Soil Res. 47:1–10

Ministry of Agriculture Brazil (2006) Instrução Normativa N° 27, de 05 de Junho de 2006

Read D, Trueman E, Arnold T, Baumann N (2008) The fate of uranium in phosphate-rich soils. In: de Kok, L. J. and Schnug, E. (eds.). Loads and Fate of fertilizer-derived uranium. Backhuys Publishers, Leiden. ISBN/EAN 978-90-5782-193-6:65–72

Rogasik J, Kratz S, Funder U, Panten K, Barkusky D, Baumecker M, Gutser R, Lausen P, Scherer HW (2008) Uranium in soils of German long-term fertilizer experiments. In: de Kok, L.J., Schnug, E. (eds.). Loads and Fate of fertilizer-derived uranium. Backhuys Publishers, Leiden. ISBN/EAN 978-90-5782-193-6:135–146

Salminen R (2005) Geochemical Atlas of Europe. Part 1: Background Information, Methodology and Maps, Geological Survey of Finland, Espoo. ISBN 951-890-921-3

Santos AJG, Mazzilli BP, Favaro DIT, Silva PSC (2006) Partitioning of radionuclides and trace elements in phosphogypsum and its source materials based on sequential extraction methods. J. Environ. Radioactiv. 87:52–61

Sattouf M (2007) Identifying the origin of rock phosphates and phosphorus fertilizers using isotope ratio techniques and heavy metal patterns, Landbauforsch. Völkenrode, Agric. Res. 311:1–78

Sattouf M, Kratz S, Diemer K, Fleckenstein J, Rienitz D, Schiel D, Schnug E (2008) Significance of uranium and strontium isotope ratios for retracing the fate of uranium during the processing of phosphate fertilizers from rock phosphates. In: de Kok L.J., Schnug E. (eds.). Loads and Fate of fertilizer-derived uranium. Backhuys Publishers, Leiden. ISBN/EAN 978-90-5782-193-6:65–72

Saueia CH, Mazzilli BP, Favaro DIT (2005) Natural radioactivity in phosphate rock, phosphogypsum and phosphate fertilizers in Brazil. J. Radioanal. Nucl. Chem. 264:445–448

Scheffer F, Schachtschabel P (2002) Lehrbuch der Bodenkunde, 15th Ed. Spektrum Akademischer Verlag, Heidelberg, Berlin, Germany

Sheppard SC, Sheppard MI, Gallerand MO, Sanipelli B (2005) Derivation of ecotoxicity thresholds for uranium. J. Env. Radio. 79: 55–83

Smidt GA, Hassoun R, Birke M, Erdinger L, Schäf M, Knolle F, Schnug E (2011) Uranium in German tap and groundwater – occurrence and origins. In: Merkel, B. (ed.). Uranium Mining and Hydrogeology VI

Smidt G (in prep.) Mobility of fertiliser-derived uranium in arable soils and its contribution to uranium concentrations in ground and tap water. Dissertation thesis Jacobs University Bremen

Spalding RF, Sackett WM (1972) Uranium in runoff from gulf of Mexico distributive province – anomalous concentrations. Science 175:629–631

Taylor SR, McLennan SM (1985) The Continental Crust: its Composition and Evolution he Continental Crust. Oxford, London, Edinburgh, Boston, Palo Alto, Melbourne: Black well Scientific. 312 pp.

Taylor M, Kim N (2008) The fate of uranium contaminants of phosphate fertilizers. In: de Kok L.J, Schnug E. (eds.). (2008) Loads and Fate of fertilizer-derived uranium. Backhuys Publishers, Leiden. ISBN/EAN 978-90-5782-193-6:147–156

Tunney H, Stojanovic M, Mrdakovic Popic J, McGrath D, Zhang C (2009) Relationship of soil phosphorus with uranium in grassland mineral soils in Ireland using soils from a long term phosphorus experiment and a National soil database. J. Soil Sci. Plant. Nutr. 172: 346–352

Utermann J, Fuchs M (2008) Uranium in German soils. In: de Kok L.J., Schnug, E. (eds.). (2008) Loads and Fate of fertilizer-derived uranium. Backhuys Publishers, Leiden. ISBN/EAN 978-90-5782-193-6:33–46

Zheng ZP, Tokunaga TK, Wan JM (2003) Influence of calcium carbonate on U(VI) sorption to soils. Env. Sci.Tech. 37:5603–5608

Radiological and Hydrochemical Investigation of Underground Water of Issyk-Kul Region

Azamat Tynybekov

Abstract. Issyk-Kul basin is closed lake with an area along watersheds 22,080 km^2. In terms of geological this is synclinal structure in eastern part of plicate area of north Tien-Shan, which is characterized with difficult many stores building with predominance in mixture of abyssal and metamorphic rocks of Riphean-low Paleozoic age. In terms of hydrogeological this is a difficult inter-mountain artesian basin Mesozoic age in the central part basin, which is symmetrical framed from south and north of Teskei and Kungei hydro-geological massifs. Displaying the newest tectonics essentially affect the hydrogeological situation of region. Feature of that is increasing the amplitude of newest deformation from west to east and from north to south. Radiological and hydrochemical characterization of underground waters in quaternary deposits of Issyk-Kul lake coastal zone was analyzed.

Introduction

Issyk-Kul basin is closed lake with area along watersheds 22,080 km^2. In terms of geological this is synclinal structure in eastern part of plicate area of north Tien-Shan, which is characterized with difficult many stores building with predominance in mixture of abyssal and metamorphic rocks of Riphean-low Paleozoic age.

In terms of hydro geological this is a difficult intermountain artesian basin Mesozoic age in the central part basin, which is symmetrical framed from south and north of Teskei and Kungei hydro-geological massifs. Displaying the newest

Azamat Tynybekov
Kyrgyż Russia Slavonic University, Kyrgyzstan
E-mail: atynybekov@gmail.com

B. Merkel, M. Schipek, *The New Uranium Mining Boom*
DOI 10.1007/978-3-642-22122-4

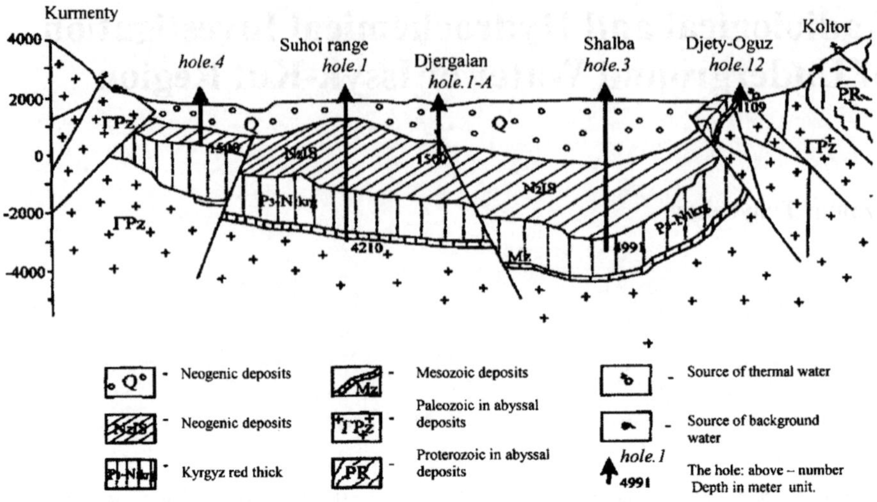

Fig. 1 The scheme in principle of depth hydro-geological building on costal zone of Issyk-Kul lake

tectonics essentially affect the hydro-geological situation of region. Feature of that is increasing the amplitude of newest deformation from west to east and from north to south. The scheme in principle of depth hydrogeological building on costal zone of Issyk-Kul lake (Fig. 1).

As other regions the resources saving factors of underground water on costal zone of Issyk-Kul lake combine into natural and man-caused complexes (Mangeldin 1999).

Among the naturals key role play:

- Physics-geographical complex of factors (relief, climate, hydrology);
- Geological complex of factors (mixture of reservoir sediments, tectonic structure and neotectonics);
- Hydro-geological complex of factors (Source of feeding of underground water, transit conditions and unloading them, hydro-geological parameters of water-bearing stratum).

The key of man-caused factors the following:

- Water economics situation;
- Geology- technical exploitation condition and drawdown of underground water resources;
- The source of pollution (character, working regime, volume and mixture of fault and so on).

In the condition of Issyk-Kul artesian basin displayed all listed factors, but role of part of them displayed different, which give the different result both for the units of resources magnitude and for the waters quality including the radioactivity (Fig. 2).

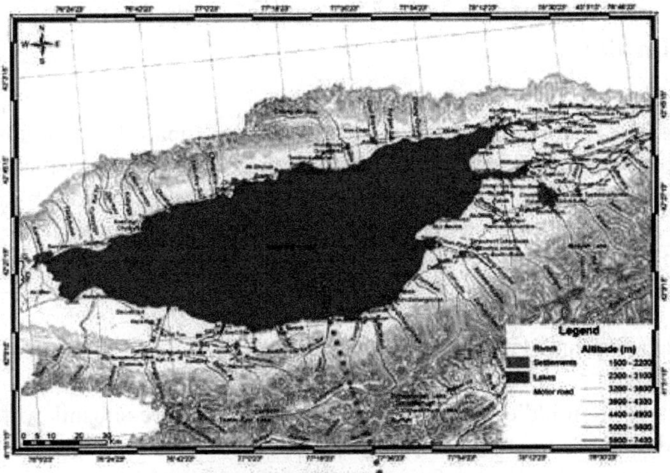

Picture 2. Issyk-Kul lake

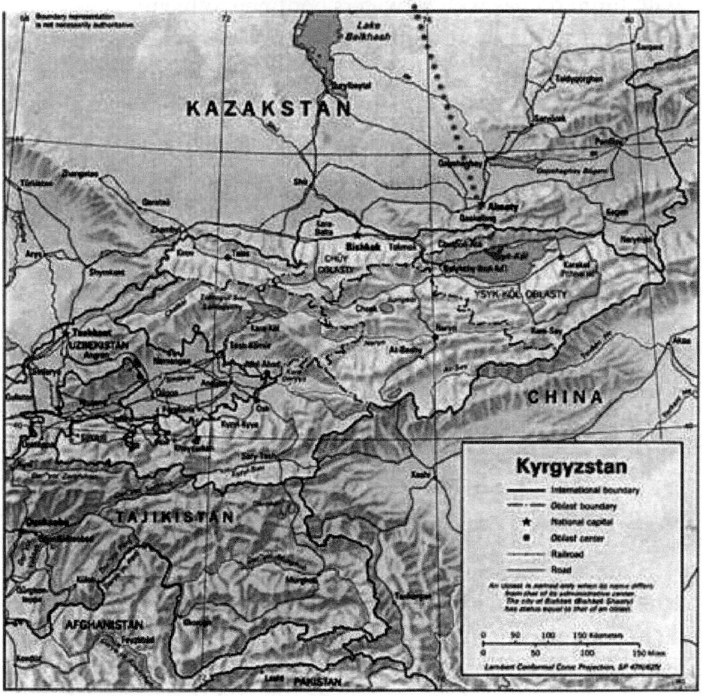

Fig. 2 Costal zone of Lake Issyk-Kul

Short Characterization of the Basic Water-Bearing Horizons and Region Complex

On costal zone of Issyk-Kul lake has not got practically (apart from drained) water-bearing sediments even most waterproof from them – Mesozoic, which represented primary siltstones and carbonaceous shale's consisted strata of coal, gravelite, sandstones watered by system of fissuring coarse-grained cut intervals.

For the characterization of underground water accumulation in some coarsening stratigraphic scheme distinguish the following:

1. underground water of upper structure floor – subsoil and pressure waters in quaternary deposits;
2. underground water of middle structure floor – pressure, pore-subsoil waters rarely and mainly porous fractured waters of Issyk-Kul Neogene suite and Kyrgyz red color thickness of Miocene-Oligocene; Paleogene of Mesozoic rock has been developed with limitation as well as waters confined to them;
3. underground water of lower structure floor – in the base of artesian basin and outcrops in hydro-geological massif of before Age of Reptiles rocks; these are groundwater zone of open fissuring and pressure interstitial-vein waters of depth circulation of tectonic fissuring zone.

The Underground Water in Quaternary Deposits

The quaternary deposits involved region represent mainly alluvial and lake sediments within artesian basin as well as alluvial and glacial formations on hydrogeological massif. Prevailing their feeding-infiltration and providing with atmospheric precipitation, river waters as well as spray waters on lowland part of basin. On the segregated parts the feeding of underground waters of quaternary deposits assisted with overflow of interstitial-vein waters base in concealed focus of their unloading both on mountains and sub-mountain even flat part of territory.

In case of infiltration feeding of underground waters of quaternary sediments is faintly alkali or close to neutrals (for magnitude pH – 7.1 … 7.4), hydro-carbonate calcium or sulfate-hydro-carbonate, magnesium- calcium with mineralization from less then 0.1 g/dm^3 until 0.5 g/dm^3 and on common case with less mineralization then top absolute mark outcrops on the ground.

Given below the distinctive formulas of salt composition involved waters from springs around Issyk-Kul and different move of them as well.

Probe 1

Volley of Turasu river, spring from upper-quaternary moraine, absolute height of abruption – 2640 meters.

$$M_{0.08}\frac{HCO_3 62SO_4 26}{Ca58Mg34} \qquad t = 1.2°C$$

Probe 2

Volley of Arashan river, Holocene alluvial boulder-pebbles, absolute mark of abruption – 2450 m.

$$M_{0.10}\frac{HCO_3 84SO_4 11}{Ca73Mg16} \qquad t = 2.4°C$$

Probe 3

Volley of Kalmaksu river, Holocene alluvial pebbles, absolute mark – 2180 m.

$$M_{0.12}\frac{HCO_3 79SO_4 12}{Ca77Mg13} \qquad t = 7.5°C$$

Probe 4

Volley of Dgergalan river, upper-quaternary alluvial pebbles, absolute mark – 1625 m.

$$M_{0.08}\frac{HCO_3 71SO_4 12}{Ca73Mg18} \qquad t = 8.4°C$$

Probe 5

Volley of Akterek river, poorly cemented clumpy boulder-pebbles of Sharpyldak suite of Neogene-lower-quaternary age, zone of tectonic transgression, absolute mark – 1645 m.

The quaternary deposits on the costal of Issyk-Kul lake influenced by limited impact of factors encountered displaying water and other composition – with intensive anthropogenic pollution as well as sulfate-chloride-sodium with heightened content of carbon dioxide and pH – lower 7 in recent lake deposits of costal zone, but before reduced reasons we drop their consideration.

Radiological Characterization of Underground Waters of Issyk-Kul Lake Costal Zone

Radioactivity of region underground waters as other place formed under influence three group of reasons: firstly natural processes, secondly man-caused with incursion into natural course of redistribution the radioactive elements on changing the structure and composition of underground waters flow under mining (deposits of radioactive ores, underground waters, mountain road heading) or other changing of territory water balance (filtration of west tailings, channels, stream flow redistribution and other.), finally changing the nature conditional of underground waters radioactivity be able to take place in case of nuclear explosions and working the nuclear energy enterprises, which lead to advent of in natural waters more two dozens with half-disintegration from several times to hundred thousand and million years [3, 4].

The question of underground waters radioactivity of the costal of Issyk-Kul lake conditioned with radionuclide, in our view it requires independent consideration and with applied relation to that it is not actuarial: in this region the nuclear energy enterprises absence, nuclear tests have not been conducted and influence of explosions Lobnor, Semipalatinsk and other testing areas should assessing on higher regional level then we consider the region (with area of two tens km^2 on common background area affected nuclear tests).

There are all bases to guess that exploitation the deposits of radioactive ores near Kadji-Sai village in the South costal of Issyk-Kul lake with no doubt has changed the radiological situation in zone of influence its mining and obviously has been told upon character of radioactivity of affected its natural waters, but since special, public works to discover this phenomenon has not been conducted up to date so the range of questions connected with this issue excluded from our consideration.

Underground waters natural radioactivity of the territory characterized with three components – uranium, radium and radon. Common quantity of analyses of their underground waters content for uranium and radon form more hundreds identifications, for radium and uranium isotope – tens one, which fulfilled basically under the projects with many searching the deposits of radioactive ores for assessing the quality and conditions to form fresh and thermo-mineral waters.

In case of no additional instructions, uranium has been defined with standard method that has sensitivity $1*10^{-7}$ g/dm^3 in Kyrgyz Republic Central Laboratory of Geological Committee (Department of geology). Isotope content of uranium and radium (method of emanation) has been defined in Radiological Laboratory of Physics and Math Institute of Republic Science Academy. Radon has been defined with electrometer CG-11M or Emanometer EM-6 by supporting of executors of field works. In first and second cases reliability of results assessing as higher level and in third case the results assessing as representative, but no excluded the content of errors due to subjective factors.

Radiological Characterization of Underground Waters in Quaternary Deposits of Issyk-Kul Lake Costal Zone

In quaternary deposits of Issyk-Kul lake costal zone for radioactivity components have been tested more than hundred objects of underground waters (springs, draw-wells, holes) at the same time radiological anomalies have not been discovered. Anomaly concentration of radon were fixed in Zhety-Oguz, Turasu villages, but their connection with fractured-vein waters no doubt, which were discharging through thickness of quaternary deposits from base rocks, so these anomaly should considerate in content of underground waters characterization of Paleozoic formations.

Typical for underground waters in quaternary deposits the uranium and radon concentrations adduced in table (since in practice of field work one water probe has been picked out usually at once for uranium, fluorine and boron these components adduced in Table 1 too).

As we see in table the average content of uranium closer at $1 * 10^{-6}$ g/dm^3, radon at 0.6 nCi/dm^3 and there is not essential correlation between adduced indexes of underground waters in table.

Taking into consideration the macro-component content and pH – (7.1 ... 7.4), we can guess that probable form of uranium migration in this condition is ionic as UO_2^{2+} and $UO_2(OH)^+$.

Table 1 Underground waters content of uranium, fluorine, boron and radon in quaternary deposits of Issyk-Kul lake costal zone

Probe No.	Water mineralization g/dm^3	Content, mg/dm^3			Content: Radon eman/l
		Fluorine	H_2B0_4	Uranium 10^{-3}	nCi/dm^3
1	0.08	0.4	0.1	1.3	4.4/0.44
2	0.10	0.48	0.34	0.69	7.4/0.74
3	0.12	0.4	0.1	0.83	9.0/0.90
4	0.32	0.8	0.1	1.1	5.5/0.55
5	0.26	3.5	0.1	4.4	11.5/1.15

Table 2 Concentrations in underground waters in quaternary deposits of Issyk-Kul lake costal zone

Radon	(nCi/dm^3)	From 0.7 to 1.2$\pm$0.01
Radium – 226	(10^{-13} Ci/dm^3)	From 3 to 5$\pm$1
Uranium – 238	(10^{-6} g/dm^3)	From 3 to 17$\pm$2
Uranium – 234 / Uranium – 238		From 1.25 to 1.30$\pm$0.03

Parallel definition of uranium, radium and radon adduced in several holes, underground waters in quaternary deposits of Issyk-Kul lake costal zone (in region of Kursk-Boztery) are uncovered.

Properly of these data and on this region content of radioactive elements hesitate near (background) average value for Issyk-Kul lake costal zone (Tynybekov and Matychenkov 1999).

Uranium isotopes correlation 234/238 present some increasing lighter isotope over balanced content and in defined level confirm actually that uranium lixiviation does not come from uranium minerals (Ferronskii 1975). Exactly adduced isotope correlation it is typical for underground waters in coarse Neozoic deposit of Fergana artesian basin (Sultanhodjaev et al. 1972). The assessment for underground waters of sub-mountain trains about at the same value intervals of isotope correlations or several more magnitudes are passage the waters of tectonic fissures and loose deposits, which let the authors divide flow of infiltration formation and flows, which supplying entry in some level from tectonic fissures.

References

Ferronskii V.I. (1975). The natural isotopes of hydrosphere, M: Nedra.

Mangeldin P.S. (1991). The fresh water resources of intermountain cavity within Tien-Shan, Ilim, Bishkek: 148.

Sultanhodjaev A.N., Azizov G.U., Latipov S.U. and others. (1972). The experience assessment of underground waters age of artesian basins of contorted-mountain regions, Fan: 136.

Tynybekov A.K., Matychenkov V.E. (1999). Radiological characteristics of underground waters in quaternary deposits of adjoint Issyk-Kul artesian basin and its mountain frame. J Environment and People Health. A7: 78–86.

Interaction of *Chlorella vulgaris* and *Schizophyllum commune* with U(VI)

M. Vogel, A. Günther, M. Gube, J. Raff, E. Kothe, G. Bernhard

Abstract. U(VI) interaction with biomass was studied using the alga *Chlorella vulgaris* and the fungus *Schizophyllum commune.* Both organisms bind significant amounts of U(VI) dependent on its concentration and pH. Interestingly, living *Chlorella* cells are able to mobilize bound uranium while *Schizophyllum* cells accumulate large amounts thereof. The molecular structure of formed uranyl-bio-mass-complexes was investigated by different spectroscopic techniques and demonstrated an involvement of carboxylic and organic/inorganic phosphate groups with varying contributions dependent on biomass, uranium concentration and pH.

M. Vogel
Helmholtz-Zentrum Dresden-Rossendorf, Institute of Radiochemistry,
P.O. Box 510119, D-01314 Dresden, Germany

A. Günther
Helmholtz-Zentrum Dresden-Rossendorf, Institute of Radiochemistry,
P.O. Box 510119, D-01314 Dresden, Germany

M. Gube
Friedrich-Schiller-Universität Jena, Institute of Microbiology,
Neugasse 25, D-07743 Jena, Germany

J. Raff
Helmholtz-Zentrum Dresden-Rossendorf, Institute of Radiochemistry,
P.O. Box 510119, D-01314 Dresden, Germany

E. Kothe
Friedrich-Schiller-Universität Jena, Institute of Microbiology,
Neugasse 25, D-07743 Jena, Germany

G. Bernhard
Helmholtz-Zentrum Dresden-Rossendorf, Institute of Radiochemistry,
P.O. Box 510119, D-01314 Dresden, Germany

B. Merkel, M. Schipek, *The New Uranium Mining Boom*
DOI 10.1007/978-3-642-22122-4

Introduction

Uranium is a radioactive highly toxic heavy metal which is released into the environment from geogenic deposits and from mining and milling areas by weathering and anthropogenic activities. A comprehensive elucidation of the behavior of uranium in geo- and biosphere is necessary for a reliable risk assessment of radionuclide migration in the environment.

Microorganism, like algae, fungi and bacteria, are widespread in nature and play an important role in biogeochemical cycling of elements. Because of their ubiquitous occurrence and the ability to fix or to mobilize radionuclides the influence of microorganisms on the migration behavior of uranium is of special interest e.g. determining effective and economical remediation strategies for contaminated soil and water. Additionally, the risk of a radionuclide transfer along the food chain by microbial bound actinides could be estimated.

In this study, the unicellular green algae *Chlorella vulgaris* and the wood-rotting basidiomycete *Schizophyllum commune* were used as model organisms living in aquatic and soil habitats to investigate the interaction with U(VI). Both organisms are known to posses the capability of binding heavy metals and also uranium (Gadd 1988; Merten et al. 2004). The aim of this study was to characterize the sorption of U(VI) by living organisms and to investigate the molecular structure of formed complexes. In contrast to previous investigations, the present study with *Chlorella* focuses on the biosorption of uranium(VI) especially at environmentally relevant uranium concentrations in relation to their metabolic activity. In comparison to the algae the binding capacity of *Schizophyllum* for U(VI) was investigated over the naturally occurring pH range from 4 to 7. The formed uranyl complexes on the biomasses in dependence of the used organism and U(VI) concentration were spectroscopically characterized by time-resolved laser-induced fluorescence spectroscopy (TRLFS), extended X-ray absorption fine structure (EXAFS) spectroscopy and in case of algal biomass also by attenuated total reflection Fourier transform infrared (ATR-FTIR) spectroscopy.

Material and Methods

Test Organisms and Culture Conditions

Chlorella vulgaris and *Schizophyllum commune* were purchased from the SAG Culture Collection (Göttingen, Germany) and Friedrich Schiller University Jena (Germany). All experiments regarding culturing and incubation of the test organisms were performed under sterile conditions. All chemicals were used in analytical grade.

Chlorella cells were grown in liquid algal full medium in bioreactors under light and continuous air supply (for details of cultivation see Vogel et al. (2010)).

Cells for the following experiments were harvested by centrifugation (10,000 × g, 8°C) after the batch culture reached the stationary phase. Cell pellets were then washed with algal mineral medium (Vogel et al. 2010).

Shizophyllum was cultivated in Erlenmeyer flasks with liquid medium containing tryptone (2 g/l), yeast extract (2 g/l), glucose (20 g/l), tryptophane (1 g/l), $MgSO_4 \times 7\,H_2O$ (0.5 g/l), KH_2PO_4 (0.5 g/l) and K_2HPO_4 (1 g/l). After 7 days of cultivation the biomass was harvested by filtration and washed with fungal mineral medium ($MgSO_4$, $FeCl_3$, K_2HPO_4, KH_2PO_4) which was used also for the sorption experiments.

Sorption Experiments

Sorption experiments with algal biomass were carried out in algal mineral medium at uranium concentrations of 5 µM to 1000 µM at different pH values in the range of 4.4 to 7. The uranium concentrations ≥ 100 µM were applied only at pH 4.4 to avoid uranium precipitation. The concentration of algal biomass during the experiment was 0.76 g algae dry weight/l. The incubation of algal cells with the uranium containing medium was carried out under culturing conditions as described in the previous section to keep algal cells metabolic active. Biomass was harvested immediately after addition of uranium, after 48 h and after 96 h by centrifugation (5 min, 10,000 × g).

For the sorption experiments with the fungal biomass (0.3 g dry weight/l) fungal mineral medium containing uranium in a concentration range of 2 µM to 400 µM at pH 4, 5, 6 and 7 was used. After incubation for 24 h biomass was separated from the medium by filtration.

After washing with 0.1 M $NaClO_4$ biomass was used for spectroscopic investigations. The uranium concentration in initial and final solutions was quantified by inductively coupled plasma mass spectrometry (ELAN 9000, Perkin Elmer). The amount of biomass-bound uranium was calculated from the difference between the amount of applied uranium and the amount remaining in the supernatant after sorption experiments. The sorbed uranium was normalized to dry biomass.

Time-Resolved Laser-Induced Fluorescence Spectroscopy (TRLFS)

Luminescence spectra were obtained using a Nd-YAG laser (Minilite, Continuum) with laser pulses at 266 nm and a beam energy of about 250 µJ. The emitted luminescence light was detected using a spectrograph (iHR 550, HORIBA Jobin Yvon) and a ICCD camera (HORIBA Jobin Yvon). TRLFS spectra were recorded from 371.4 nm to 664.3 nm by accumulating 100–200 laser pulses using a gate time of 2 µs. Computer control was ensured by the software Labspec 5. The obtained data were processed by using Origin 7.5 (OriginLab Corporation, Northampton, MA, USA) including the PeakFit module 4.0. For the TRLFS measurements living and dead *Chlorella* biomass was used incubated mineral medium with 5 and 100 μM uranium at pH 4.4 for 30 min. Fungal biomass was incubated in mineral medium with 2.3 to 410 μM uranium at pH 5 for 24 h. All biomass samples were washed with 0.1 M $NaClO_4$ to remove unbound uranyl species.

Results and Discussion

Uranium Speciation in Mineral Media

The uranium speciation in the media influences its bioavailability and by this the sorption behavior and interaction of uranium with the organisms. Uranium speciation depends on uranium concentration, pH value and ionic strength in the media. For both algal and fungal mineral medium an example for the speciation of uranium in dependence of pH value is shown in Fig. 1.

Figure 1 shows that at pH values around 4 the speciation in both cases is dominated by the fee uranyl cation. With increasing pH the uranyl hydroxides and uranyl carbonates become the main species in the mineral medium. Changes in the uranium concentration in the experiment (not shown) influenced the speciation only slightly.

Sorption of Uranium by Chlorella and Schizophyllum

The sorption experiments showed that *Chlorella* and *Schizophyllum* biomass binds significant amounts of U(VI) dependent on uranium concentration and pH in the range from 4 to 7. In Fig. 2 a comparison of the sorption capacity of both biomasses

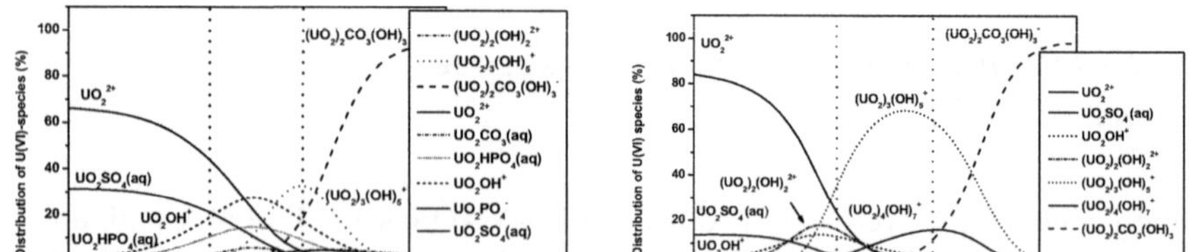

Fig. 1 Distribution of main uranyl species in algal (*left*) and fungal (*right*) mineral medium in dependence of pH value with uranium concentrations of 5 and 70 μM. Speciation calculated with the computer program EQ3/6 (Wolery 1992) using the NEA database (Guillaumont et al. 2003)

is shown as a function of the initial uranium concentration. In case of *Schizophyllum* also the pH was varied. At uranium concentrations up to 200 µM the fungal biomass shows comparable binding capacities of around 130 mg U/g dry weight independent of pH value. At the highest tested uranium concentration of 400 µM clearly higher binding capacities ranging from 230 to 280 mg U/g dry weight at pH 5 and 6 for still living *Schizophyllum* biomass were obtained. In contrast to that, *Chlorella* binds at a uranium concentration of 550 µM only 75 mg U/g dry weight at pH 4.4. This is attributed to the more than two times higher biomass concentration compared to the experiment with fungi. Nevertheless, the fungal biomass shows better sorption capacities as a doubling of the uranium concentration to 1000 µM did not increase the amount of algal bound uranium in the same way (Fig. 2). Furthermore, *Chlorella* cells die at concentrations ≥ 100 µM within 48 h.

The behavior of living algal biomass in the presence of U(VI) was investigated in further experiments with 5 µM uranium in a pH range of 4.4 to 7.0 (Fig. 3). Under this condition living algae firstly bind almost all uranium within 5 min, but then again mobilize up to 80% of the bound uranium during ongoing incubation. As the algae were alive during the entire experiment and such a mobilization of uranium did not occur if dead biomass was used (Fig. 3) the mobilization is ascribed to a release of metabolism related substances. As potential leachates for

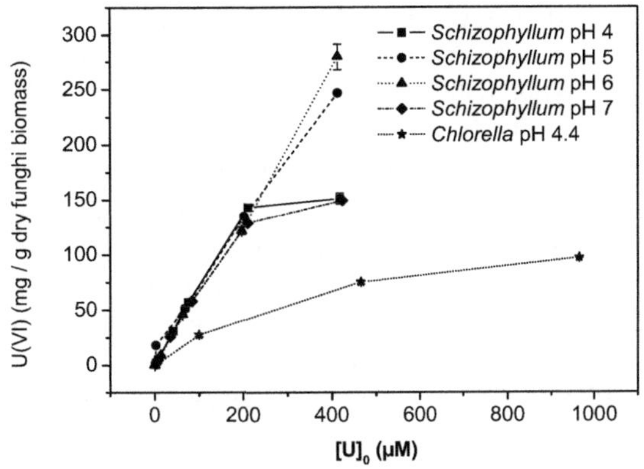

Fig. 2 U(VI) binding capacities of *Schizophyllum* from pH 4 to 7 and *Chlorella* at pH 4.4

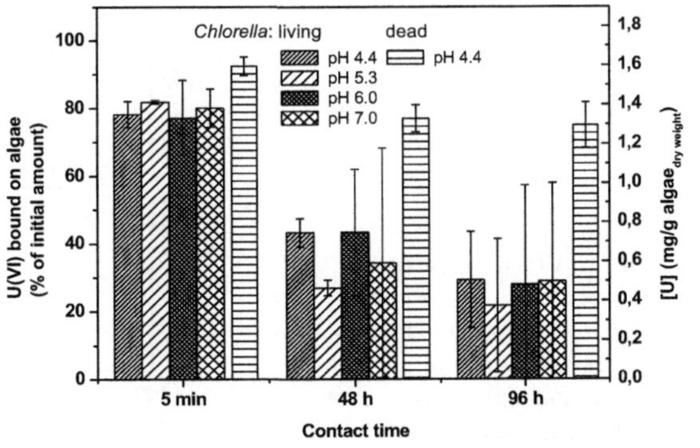

Fig. 3 Uranium bound by living (pH 4.4 to 7.0) and dead algae (pH 4.4) at $[U]_0 = 5$ µM

algal-bound uranium oxalate, citrate and ATP were tested and found to be able to mobilize 55–87% of the algal-bound uranium within 24 h (Vogel et al. 2010).

Spectroscopic Characterization of Formed U(VI)-Biomass-Complexes

As both *Chlorella* and *Schizophyllum* sorbed significant amounts of U(VI) spectroscopic techniques were used to characterize the formed complexes between U(VI) and the biomass. With TRLFS luminescence spectra of the U(VI)-biomass-complexes were measured in dependence of pH value and applied initial uranium concentration. For both biomasses luminescence spectra were obtained which showed shifts in the main emission bands compared to those spectra of uranyl species in the initial media. A change of pH value between 4.4–7.0 at a uranium concentration of 5 μM had no influence on the obtained U(VI)-*Chlorella* spectra. At 70 μM uranium the luminescence spectra of *Schizophyllum* biomass showed also no difference in the pH range of 4 to 7.

In Fig. 4 U(VI)-luminescence spectra of *Chlorella* at pH 4.4 and *Schizophyllum* biomass at pH 5 in dependence of applied uranium concentration and in case of algae dependent on cell status are shown.

The different cell status of the algae were investigated because sorption experiments with uranium concentrations ≥ 100 μM showed that living *Chlorella* cells died off within an incubation time of 48 h. The results show that the cell status of *Chlorella* has an influence on the spectra and emission bands of the formed U(VI)-algae-complexes. The luminescence spectra of dead cells are shifted to higher wavelength compared to those spectra of living cells (Fig. 4). With increasing uranium concentration this spectral shift to higher wavelength becomes even more clearly. A comparison of the obtained emission bands of the U(VI)-complexes with living and dead algae (Table 1) with emission bands of different model uranyl complexes (Table 2) suggests different functional groups

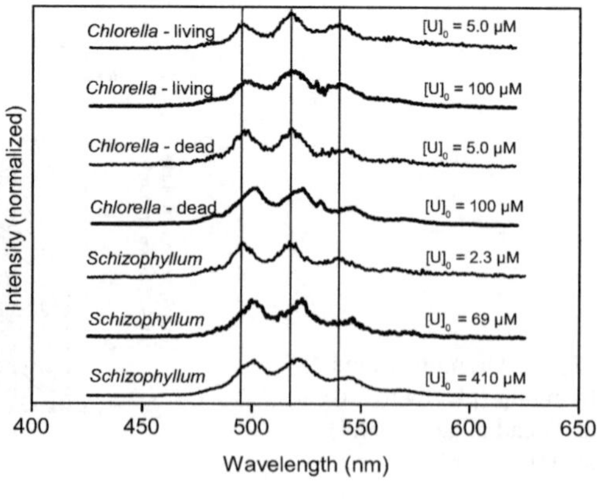

Fig. 4 U(VI)-luminescence spectra of *Chlorella* at pH 4.4 and *Schizophyllum* at pH 5 in dependence of uranium concentration and in case of algae dependent on cell status

are responsible for the uranium binding. Living *Chlorella* cells bind U(VI) via carboxylic and organic phosphate groups. The shift to higher wavelengths makes the emission bands of U(VI)-complexes with dead cells more comparable to those of U(VI)-complexes with inorganic phosphates. This means, for dead algae not only carboxylic and organic phosphate groups are involved in U(VI) binding but also inorganic phosphates. The latter complexation becomes even more important with increasing uranium concentration.

For fungal biomass increasing uranium concentration in the initial solution causes a shift of the luminescence emission signals of U(VI)-fungi species to higher wavelengths (Fig. 4). This means, different luminescent species with varying amounts were formed. The emission bands of U(VI)-fungi species (Table 1) are comparable with bands of several uranyl phosphates (Table 2). The positions of the bands at low uranium concentration suggest a U(VI) binding mainly by organic phosphate groups, whereas, at higher uranium concentrations the binding by inorganic phosphate groups becomes more important. This can be explained by a progressive biomineralization process.

The involvement of different functional groups in the interaction of U(VI) with *C. vulgaris* and *S. commune* was confirmed by EXAFS investigations and additionally by ATR-FTIR spectroscopy in case of algal cells (data not shown).

Table 1 Luminescence emission bands of uranyl species on/in *C. vulgaris* (C_L = living, C_D = dead) at pH 4.4 and *S. commune* (S_L = living) at pH 5 in dependence of uranium concentration

Sample	$[U]_0$ (μM)	Emission bands (nm)						
C_L	5.0		480.5	496.8	517.6	539.9	565.4	
C_L	100.0	465.7	481.4	497.8	518.1	540.4	563.8	590.8
C_D	5.0	463.0	481.6	497.9	518.9	541.6	566.0	591.7
C_D	100.0	463.5	483.5	500.1	521.1	543.7	568.2	596.0
S_L	2.3		480.7	496.2	517.0	539.3	564.4	603.3
S_L	69.0		486.2	500.6	521.1	543.7	569.1	594.3
S_L	410.0		487.1	500.1	520.6	543.5	567.6	597.5

Table 2 Luminescence emission bands of different uranyl model compounds. PThr = o-phospho-L-threonine

Sample	Emission bands (nm)						
UO_2^{2+} [a]		471.1	487.9	509.1	532.3	557.8	587.5
UO_2(acetate)[a]	462.9	478.8	494.6	514.3	535.9	562.0	
UO_2(acetate)/UO_2(PThr)[a]	462.8	481.7	496.7	517.1	539.9	563.5	594.0
UO_2(fructose-6-phosphate)[b]		478.9	497.1	519.0	543.3	568.9	598.0
$Ca(UO_2)_2(PO_4)_2 \cdot 10\ H_2O$[c]		488.6	504.0	524.2	548.0	573.9	602.4

[a] Vogel et al. 2010
[b] Koban et al. 2004 (2004)
[c] Geipel et al. 2000 (2000)

Conclusions

The obtained results from the sorption experiments and TRLFS investigations show that the interaction of U(VI) with *Chlorella* and *Schizophyllum* differs. The uranium binding capacity of the alga is much lower than that one of the fungus. Bioaccumulation by metabolism dependent transport was not observed for the algal cells whereas it is suggested for the fungus. The cell activity of *Chlorella* had an influence on the biosorption of uranium. In presence of viable cells permanent binding of uranium to the cell was prevented most likely by secreted algal metabolites. The formed U(VI)-complexes on the biomasses differ in dependence on the organism and uranium concentration. For living algae mainly carboxylic and organic phosphate groups are involved in the binding of uranium, whereas in dead algal biomass additionally inorganic phosphates are relevant. In contrast to that, in fungus only phosphate groups are responsible for the uranium coordination. At high uranium concentrations the role of inorganic phosphates for the U(VI) coordination became more important in the fungus and dead algal biomass.

Acknowledgement This work was funded by BMBF under contract number 025851.

References

Gadd, G. M. (1988). Accumulation of metals by microorganisms and algae. In Biotechnology (ed. h. J. Rehm), p. 401–433, VCH, Weinheim.

Geipel, G., Bernhard, G., Rutsch, M., Brendler, V., Nitsche, H. (2000). Spectroscopic properties of uranium(VI) minerals studied by time-resolved laser-induced fluorescence spectroscopy (TRLFS). Radiochim. Acta 88, 757–762.

Guillaumont, R., Fanghänel, T., Fuger, J., Grenthe, I., Neck, V., Palmer, N. A., Rand, M. H. (2003). Update on the chemical thermodynamics of uranium, neptunium, plutonium, americum and technetium. Elsevier, Amsterdam.

Koban, A., Geipel, G., Rossberg, A., Bernhard, G. (2004). Uranium(VI) complexes with sugar phosphates in aqueous solution. Radiochim. Acta 92, 903–908.

Merten, D., Kothe, E., Büchel, G. (2004). Studies on Microbial Heavy Metal Retention from Uranium Mine Drainage Water with Special Emphasis on Rare Earth Elements. Mine Water and the Environment 23, 34–43.

Vogel, M., Günther, A., Rossberg, A., Li, B., Bernhard, G., Raff, J. (2010). Biosorption of U(VI) by the green algae Chlorella vulgaris in dependence of pH value and cell activity. Science of the Total Environment 409, 385–396.

Wolery, T. J. (1992). EQ 3/6 A software package for geochemical modeling of aqueous system. Lawerence Livermore National Laboratory report UCRL-MA-110662 part 1, Livermore, California, USA.

Sustainability: the Balanced Approach to Modern Uranium Mining

Peter Waggitt

Abstract. In recent times the world has searched for increased sources of energy to support improving lifestyles and increased development, whilst also seeking to reduce greenhouse gas emissions and mitigate possible effects of fossil fuels on climate. Consequently, nuclear power has come back onto the agenda frequently as Governments plan future energy strategies. A further consequence has been an increased level of activity in the mining of uranium. Always a sensitive topic uranium mining projects are regarded with suspicion by stakeholders who point at the legacy sites from former operations as evidence that mining is not "environmentally sustainable". However, modern operations are now obliged to take an holistic approach to project development with "sustainability" as a key concept; although the idea is better expressed as "balanced" development since mining is, by its very nature, detrimental to the environment but this need not mean there is a permanent loss of use of the land. This modern balanced, approach requires that social, economic and environmental factors be considered in unison. This paper looks at a selection of past legacies and then provides an overview of modern "balanced" outcomes in the modern uranium mining industry.

Introduction

In recent times the world has searched for increased sources of energy to support improving lifestyles and increased development, whilst also seeking to reduce greenhouse gas emissions and mitigate possible effects of fossil fuels on climate. Consequently, nuclear power has come back onto the agenda ever more frequently

---

Peter Waggitt
Consultant, Darwin, Australia

B. Merkel, M. Schipek, *The New Uranium Mining Boom*
DOI 10.1007/978-3-642-22122-4

as Governments plan their futures. With this increasing interest in nuclear power has come an increased level of activity in the mining of the raw material for the fuel – uranium. Uranium mining has always been a sensitive topic. The modern era, which started in the late 1940s, has been no different to any other mineral expansion campaign. With some justification; all mine development projects are regarded with suspicion by stakeholders who are able to point at the legacy sites from former operations as evidence that mining is not "environmentally sustainable", and uranium carries the specter of radioactivity as an additional hazard and source of public concern.

Worldwide, uranium exploration, resource development and mining have increased significantly since about 2003 when the surge of interest in nuclear power ramped up and the shortfall of annual uranium production against annual power fuel demand became apparent. Modern operations are now obliged to take an holistic approach to project development with "sustainability" as a key concept. The notion is perhaps better expressed as "balanced" development since mining is, by its very nature, detrimental to the environment but this need not mean there is a permanent loss of use of the land. This modern balanced, approach requires that social, economic and environmental factors be considered in unison, resulting in projects whose outcomes can be regarded as balanced in these terms. This paper looks at a selection of past legacies and then provides an overview of modern "balanced" outcomes in the modern uranium mining industry.

Uranium Mining and Remediation – Past Experiences

The history of uranium mining is similar to many other minerals – a series of "boom and bust" cycles depending on the state of the specific commodity market. In the case of uranium there was a small demand prior to World War 2 which increased dramatically after 1945 as several nations became interested to develop nuclear weapons. The frenzy of exploration and mining was maintained into the late 1950s at which time the industry went into a recession. In the late 1960s there was another rush of activity as the demand grew for raw material for nuclear power plants that were planned by many governments. This rush was sustained until the early 1980s when nuclear accidents at Chernobyl in the then Soviet Union and Three Mile Island in the USA saw a rapid cut back in development plans and a consequent crash in the demand for uranium; the uranium mining industry stagnated. Sadly one feature of many of these earlier developments was lack of adequate environmental regulation and controls. As a result many uranium productions sites were simply abandoned to become liabilities and legacies of the former activities, often having adverse impacts on their surrounding environments. These were not "sustainable" activities and the outcomes could not be considered "balanced".

As the next round of interest in uranium began to grow basic environmental legislation had been introduced by some countries. By the 1980s the requirement to assure some basic level of remediation of sites was coming into force in several jurisdictions. Those uranium mining operations that closed down in France and Germany in the late 1980s and 1990s, for example, were expected to have outcomes that were at least environmentally stable and non-polluting. In the USA some legacy sites were also listed for remediation under the Federal Uranium Mine Tailings Remediation Act, and in Australia the "new" uranium mines were obliged to comply with a long list of environmental requirements which included planning and assured funding for remediation. But in all these cases there was considerable variation in the attitude of regulators towards the concept of final land use and the need for a balanced outcome in terms of sustainability. Most authorities were content to end up with a site that would be safe, stable and require little or no maintenance rather than look for some ongoing useful activity. This longer term plan acquires greater significance today when stakeholders are more regularly consulted and better informed. These stakeholders wish to see no new legacies created and they also want some replacement economic activity to maintain their communities after the uranium mines have closed.

A major concern has been the cost of remediation, especially when it is undertaken as an operation that was not fully included in the original scheme of works. Looking at the costs of the WISMUT remediation program in Germany (€ 6.4 billion) UMTRA in the USA (US$ 1.6 billion); or the € 300 million plus that Areva has spent on remediation of former uranium mines around the world; and even the Aus$ 24 million spent at Mary Kathleen in Australia, one can see how costly these activities are when undertaken as additional activities after mining. Incorporation of remediation activity throughout operations does have the potential to reduce overall remediation costs provided the final objective is clear and that it remains more or less unchanged. The idea that mining, including uranium mining should *not* be a one-time user of land is an essential part of this philosophy of the balanced approach. Thus the remediated sites should be seen as locations for future use rather than a series of liabilities for future generations to manage. It should also be pointed out that not all of the projects mentioned above had a further post-mining land use as their final objective and thus do not comply with the modern idea of a balanced and sustainable remediation outcome. Some however have become good examples of what can be achieved.

Modern Uranium Mining Remediation Examples

The primary goal of every mine site remediation should be that it maximizes the opportunities for the site to be used again in a beneficial way, preferably one that is sound both technically and economically, to the greatest extent practicable. And let us not forget that this idea of a balanced approach is not a modern one. The early writer on mining Agricola (1552), pointed out in the fifteen hundreds that in Italy it was against the law to uproot a vineyard to extract minerals located below if the value of the minerals was going to be less than the long term value of the wine and grape production! However, it is only in the last twenty years or so that ideas such as sustainability and remediation planning have taken on a more inventive air and begun to introduce such concepts in relation to productive long term post-mining land use. Now as resources are limiting in terms of finance and technical expertise due to other constraints in the market place. The following sections describe briefly some of the more innovative ideas that have been developed in relatively recent times as modern uranium miners strive to achieve balanced outcomes.

Solar Power Site

Many authorities have expressed interest in developing renewable energy resources and in particular solar power. One feature of solar power farms is that they require a considerable area for the solar panels and available infrastructure for power distribution. A remediated tailings site (not necessarily a uranium related-site admittedly) may offer these characteristics. In the case of the town of Rifle in Colorado the former uranium processing site on the banks of the Colorado River was remediated as a part of the UMTRA program between 1992 and 1996. The site was declared remediated, but a designated contaminated place, and handed to the town council in 2004. A long term monitoring and surveillance program (LTSM) was maintained at the site, which extended over about 9 ha (USEPA, 2009). There was also another 13 ha tailings pile not far away which was also remediated. The town council decided that part of the site could be used for a waste water treatment plant (WWTP). It was then decided that about 4.8 ha of solar cells could be placed on the remainder of the site and provide some of the power to operate the WWTP. The site provides 1.72 MW which represents 60% of the daytime power requirement for the WWTP at a cost less than power from the National Grid (USEPA, 2009). As a result of this success other remediated tailings pond sites (including non-uranium sites) are being examined for their suitability for similar projects One major consideration is the need for either on-site use as at Rifle or the having existing distribution infrastructure available nearby for distribution of the power (Momayez et al. 2009).

Recreation-Sport Angling

Another approach is to make the remediated site fir for use as a recreation facility. Parks and forest or open space are reasonably common choices for objectives. However, at Puy de L'Age in France a rather more innovative solution was implemented. The site had been an open cut uranium mine operated by Areva from 1977–1993 with a pit extending down to about 30 m. When mining was ended in 1993 the site was cleared, decontaminated and remediated to a new landform which fitted into the surrounding topography. The mine pit filled with water and soon attracted the attention of local anglers. After some discussions and studies the pond was stocked with sport fishing species (trout and black bass) and a fly fishing club established to manage the facility. The club grew quickly and now has many members and attracts anglers from well beyond the local district. The club operates a "catch and release" policy, primarily to protect the fish stocks from over-exploitation and reduce the need for frequent re-stocking. Water and fish quality are monitored under an LTSM and no problems have been reported to date. The site provides not only recreation for the local population but also attracts anglers and other tourists to the region thus promoting business and creating employment in associated trades and activities such as angling supply stores and tourist accommodation.

Showgrounds

A further development of the open space concept may be seen at the Ronneberg site in Germany, part of the former WISMUT uranium mining industry. Here the Lichtenberg open pit was about 2000 m long, 1000 m wide and 240 m deep at the time work ended in 1976. Until 1990 the pit was used as a dump for waste rock from other operations in the area. The objective of the remediation plan, which began in 1991, was that the site of some 175 ha should be completed in time to be used as the showground for the national horticulture show known as BUGA 2007 (BUGA 2007, 2010) which were planned for April to October 2007. The strategy called for the filling of the pit with material from above ground waste stockpiles in the vicinity and then landscaping of the final landform to include a hill; about 125 M cubic meters of material were used in the construction process (Wismut 2011). The program was not wholly completed by the time the BUGA opened in April 2007 but the main features of the showgrounds, display gardens, climbing tower, arboretum and various other amenities were all functioning for the opening. That summer many thousands of people visited the site and most would have been unaware of the final remediation earthworks being carried out at the rear of the site. Work was finally completed in 2008. The site remains as a public park and open space area with no permanent final use determined as yet, although several options have been considered it is understood.

Small Business Premises

The former uranium mine sites of Areva in France have been mentioned before. Another example of a balanced approach can be seen at some locations in the Limousin area of France. It was realized that a number of infrastructure elements could be decontaminated at low cost and be considered for use by successive industries. One example is an area of workshops and storage yards that has been handed over for use by a small scale construction enterprise and another for use by a highway maintenance unit. In both cases the sites have remained productive and offer alternative employment for some of those workers who were retrenched with the closing of the uranium mining operations.

At Grand Junction, Colorado in the USA the former ore processing suite was remediated as part of the UMTRA program for Title 1 sites. The location of the site near the centre of town and alongside the Colorado River lent itself to the development of a public open space for recreation. This site was built as a grass parkland with bicycle tails and a new bridge across the river for cycle and pedestrian traffic.

Golf Course

Another imaginative scheme may be viewed in the area of Schlema, Germany. Here a waste rock pile was reshaped and developed as a golf course. The history of the operation reveals a rather slow development process but given the innovation of the idea, the need for careful risk analysis and the costs and time required to design and construct such a facility perhaps that is not surprising. The development of a golf course on an existing agricultural site may take only a couple of years. In the case of Schlema it took rather longer. The original idea began in 1997 when a local group of golf enthusiasts decided they wanted their own course, so they started to look for a possible site. Once a potential site had been selected the golf club had been formed and the initial concept for the project was developed. In 2004 the group acquired an 86 ha site, the flattened top of one of the local waste rock piles, and decided it was suitable for the construction of the course. By 2007 the detailed planning had been completed, which included all the additional risk assessment, and construction work finally began in 2008. In 2009 the course of 9 holes was finished and officially opened. Facilities now present at the site include practice areas, caddie hut and a clubhouse all used by an enthusiastic membership. The club website describes the history and provides some excellent pictures of the development (Schlema Golf Club 2011).The whole project stands as a tribute is a testimony to the ingenuity and vision of the original group of enthusiasts and the dedication and professionalism of the regulators, the mining company and others, who worked to create this very different solution as an example of the balanced approach to land use after uranium mining.

Other Ideas and Conclusions

There have been some other ideas discussed including the use of some caverns a gas storage, and development of marinas etc on waterside sites. The water-filled former mine pit at Rum Jungle Creek South in Australia was used for many years as a water sports training facility before various public health concerns led to its closure earlier this year. The important point is to be aware that there is no reason why a uranium mining and/or processing facility should not be considered for redevelopment in the same way as any other location of former extractive industry activity. The safety standards are set and the process of risk assessment is well understood and established. Once the hazards related to a site have identified and evaluated and stakeholders consulted, there is often potential for innovative post mining land use to be considered. The decontamination and pollution issues may be too expensive in a conventional sense but perhaps from time to time financial feasibility may be modified in the light of other social considerations. The only initial barrier is the imagination of those facing the task of what to do with former mining sites.

References

Agricola, Georgius (1552) *De Re Metallica*. Translated by H.C. Hoover & L.H. Hoover, 1912). Re-published by Dover, 1982, USA

Areva (2011) http://www.areva.com/EN/operations-673/redevelopment-of-former-uranium-mines-in-france-a-public-service-provided-by-areva.html

BUGA 2007 (2007) http://www.buga-gera.info/www/buga/english/buga07/

Carson, Rachel (1962) Silent Spring. Pub: Houghton Mifflin USA

Momayez, M., Wilson, T., Cronin, A., Annavarapu, S., & Conant, B. (2009) An investigation to use tailings ponds as solar photovoltaic farms. *In proceedings* Revitalizing the Environment: Proven Solutions and Innovative Approaches, Americam Society of Mining and Reclamation, Billings Montana 30 May–5 June 2009. pub: ASMR Lexington, KY, USA

Schlema Golf Club (2011) http://www.golfclub-bad-schlema.de

UMTRA (2008) http://www.infomine.com/publications/articles/36/3563/umtra.uranium.tailings/the.uranium.mill.aspx

USEPA (2009) Success stories – Siting Renewable Energy on Contaminated Land. New Rifle Uranium Mill Tailings Radiation Control Act (UMTRCA) Title 1 Site, Colorado http://www.epa.gov/renewableenergyland/docs/success_newrifle_co.pdf

USEIA (2005) United States Energy Information Administration http://www.eia.doe.gov/cneaf/nuclear/page/umtra/grandjunction_title1.html

WISMUT (2011) http://www.wismut.de/index_english.htm

Uranium Mining Life-Cycle Energy Cost vs. Uranium Resources

W. Eberhard Falck

Abstract. The long-term viability of nuclear energy systems depends on the availability of uranium and on the question, whether the overall energy balance of the fuel cycle is positive, taking into account the full life-cycle energy costs. The fundamental question is: how much (fractional) energy units do we need to invest in order to produce one energy unit in a useable form, i.e. heat or electricity? The individual process steps that lead from the undiscovered resource to yellow cake are well established for "conventional" resources, but the energy costs and associated greenhouse gas emissions are not very well known in quantitative terms. Resources estimates are usually made on the basis of economic costs of recovery, but from a global energy supply sustainability point of view, neither commercial nor national strategic considerations are really relevant. Comparison with historic data for gold and silver or oil and gas indicate that uranium reserves would increase by orders of magnitude, if the same level of investment into exploration was made. Oil and gas prices have shown that society can and will accommodate fuel price increases by one order of magnitude over the span of half a century. A comprehensive assessment of the full-life cycle energy costs of uranium mining, milling and subsequent decommissioning and remediation is required in order to help settle the debate on the net energy balance of nuclear energy systems that dominates both, the debate on resource availability and the sustainability of its exploitation. This paper illuminates some of the crucial issues.

W. Eberhard Falck
Laboratoire REEDS, Université de Versailles St. Quentin-en-Yvelines, France

B. Merkel, M. Schipek, *The New Uranium Mining Boom*
DOI 10.1007/978-3-642-22122-4

Introduction

Energy is a fundamental need of human civilization. A simple to-the-root analysis shows that we only have four basic sources of energy: natural nuclear fission/ fusion (i.e. solar radiation and geothermal flux), artificially-induced nuclear fission ("nuclear power"), gravity, and the earth's rotational inertia. All fossil fuel-based and the so-called "renewable" energy conversion systems rely on solar radiation as the primary source of energy, merely storing energy in a convenient form, e.g. reduced carbon or water at elevated heights. Two fundamental conditions control the utility of all energy conversion systems: a) the overall energy balance of their utilization must be positive and b) they must deliver energy in a useful form at the time and location where it is needed. Generally, choices on energy systems are made by society without scientific reasoning on long-term sustainability and on the basis of expedience and short-term preference only.

Though the need to move away from carbon-oxidizing to other types of energy systems is recognized, decision-making in energy policy is still much dominated by considerations of economic competitiveness and short-term profits. Nuclear energy systems are discussed as low-carbon emitting systems, but significant environmental impacts have arisen in the past at their front-end (i.e. uranium mining and milling), and at their back-end from poorly managed radioactive waste. Severe accidents during power plant operation and their impacts remain a concern, as has been vividly demonstrated in Japan recently. They are critical issues for the development of nuclear energy systems (IAEA 2003).

Defining the System for Sustainability Analyses

Sustainability as a concept still remains somewhat vague in spite of many efforts to better define the conceptual approaches. Sets of indicators have been developed that *inter alia* cover aspects such as "exhaustion of resources", "production of non-degradable waste" and "societal impacts" (e.g. Kröger 2001; IAEA 2003).

Assessing an energy system requires first of all appropriate system boundaries to be drawn. Where the boundaries are drawn, depends on the purpose of the assessment, on the underlying conceptual system model and also often on political or ideological intentions. Boundaries can be drawn to exclude or include aspects that may make the system under investigation appear "good" or "bad", depending on the objectives. The fundamental question is: how many energy units do we need to invest in order to produce one energy unit in a useable form, i.e. heat or electricity. Or: what is the energy cost of providing 1 Joule worth of fuel for oil/gas/coal/nu-clear/biofuel-based energy conversion systems?

Energy conversion systems in industrial societies are complex, integrated and have global interrelations. All systems rely on the same industrial infrastructure and supply chains. Therefore, it is difficult to delineate an individual system for the purpose of life-cycle assessment (LCA). The basis for LCAs are materials flow assessments (MFA) of where and how much of a particular substance, or energy, enters or leaves an environmental compartment or an (industrial) process, and where it appears in products, intermediates, residues and wastes.

In recent years, the definition of the "overall" life-cycle energy balance of energy conversion systems has become the focus of scientific, political and ideological debate (IAEA 1994). There is a considerable divergence of views, whether particular systems are net energy producers or not. The question is complicated by the fact that our energy systems have developed since the beginning of man-kind and carry with them a burden of legacies, but also of endowments.

For nuclear energy systems LCAs have been carried out early on in order to understand possible impacts of radioactive materials and elaborate provisions are made for their decommissioning, waste management and remediation of affected sites. There is a considerable ethical debate over environmental remediation targets and the provisions needed for the long-term management of mining and milling residues at near-surface disposal sites (IAEA 2006), and whether, indeed, near-surface disposal is acceptable at all. The effect on the energy balance of extreme remediation targets and worst case assumptions, namely, when all residues are to be back-filled into the mine and sites are to be remediated to "greenfield" targets can be seen in the calculations by Storm van Leeuwen and Smith (2008). Dones (2007) provides more realistic estimates, but strongly emphasizes the absence of real and reliable data to support such calculations. This absence of public domain data again was noted by Sovacool (2008) in his review of nuclear system LCAs. His main interest was greenhouse gas emissions, but his reflections apply to other aspects of the life-cycle cost of nuclear energy systems as well.

Life-Cycle Energy Costs of Uranium Mining and Milling

The life-cycle of nuclear energy systems in broad terms consists of those steps that provide the fuel (uranium mining and milling), the preparation of the fuel ("yellow cake" conversion into UF_6, enrichment, fuel element fabrication), the construction of facilities, the power plant operation, the storage and/or reprocessing of spent nuclear fuel and, finally, the conditioning and disposal of wastes and spent-fuel (if not reporcessed). It is a characteristic of nuclear energy systems that significant energy and materials' costs arise before and long after the useful energy production. The major consumer is the enrichment step (Dones 2007), where the technique, either gas centrifugation or diffusion, and its primary energy supplier (hydro power, fossil fuel or nuclear) is of key importance. The "back-end" includes decommissioning of facilities and installations as well as remediation of all sites, though some investigators treat this separate, as is commonly done for other energy systems. Dones (2007) and Sovacool (2008) noted, however, that "hard" data for energy consumption of mining, milling, waste management and remediation are virtually unavailable (in the public domain).

The following provides a short overview over theses various consumers.

Exploration

Often there is no direct path from exploration to the actual mining, so it is difficult to attribute exploration costs. The construction of a mine may be delayed by many years pending favorable market conditions. A considerable amount of exploration also will not result in discovery of a resources. The oil industry uses estimates of how many meters of exploratory boreholes and how many kilometers of seismic profiling are needed to discover an oil or gas resource.

Mining

There are three principal forms of mining: underground mining, open-pit mining and in situ-leaching (ISL). Energy costs are predominantly associated with the moving of rocks (unproductive, overburden, ore), water (sumping) and air contaminated by e.g. radon. Where possible, unproductive rocks are moved within the mine and used e.g. to back-fill mined-out areas. There is obviously a strong economic incentive to move as little material as possible. A considerable amount of energy is needed to break down the rock by drilling and blasting, the actual amount depending on the rock and ore type. Blasting is a process of releasing chemical energy stored in explosives. In the future, higher lifting costs will be incurred as mines will have to reach deeper down for resources. Above-ground also a considerable amount of earthmoving is required for the construction of retaining dams and similar civil engineering structures.

The amount of energy embedded in infrastructure depends on the host rock: weak rocks require a large amount of lining etc. In the past often "renewables" were used, that is wooden props, which might actually influence the carbon balance of a mine positively. While the operational energy expenditure of a mine can be assessed easily through the consumption figures for e.g. electricity or diesel fuel, the energy imbedded in mine infrastructure is more complex to assess.

For conventional mining operations a comprehensive database of life-cycle energy requirements has been compiled (c.f. Dones 2007). Attempts to compile a database on real energy costs of uranium mining so far have failed due to the commercial sensitivity of such detailed investigations.

Milling

The ore is wet-ground to a very fine grain size. The resulting slurry is separated from the supernatant solution, subject to (acid) leaching and the uranium is recovered by solvent extraction or ion exchange. In both cases, a re-extraction of the uranium is needed. In the final step the uranium is precipitated as ammonium- or sodiumdiuranate, the "yellow cake". Energy is required to drive equipment such as crushers and pumps and for the production and shipping of the various process chemicals. The residues, the tailings, are pumped to pond-like disposal facilities (IAEA 2004b). Some conditioning to improve flow or settling behavior may be needed as well as neutralization of residual acids. Energy is required for the construction of retaining structures (dams) and for providing auxiliary materials, such as plastic liners, rock covers or fertilizer for revegetation.

Generic energy consumption data for the supporting industrial processes are available from compilations such as the ECOINVENT database (Dones 2007). Uranium-specific data, however, are virtually unobtainable due to their commercial sensitivity.

In situ Leaching (ISL) Operations

For ISL a leaching solution, usually sulfuric acid, is injected into the ore zone through a set of boreholes and the "pregnant" solution is recovered through another set of boreholes. The major energy expenditure occurs during the drilling of the boreholes and during the forced circulation of the leaching solution. A considerable amount of energy is also embedded in the leaching chemicals and neutralization agents. Very little specific information on energy requirements for the remediation of ISL exists. In the past energy needs from similar operations, e.g. copper mining, were used (e.g. Dones, 2007).

Remediation

For centuries it has been the practice to simply walk away from exhausted mines and to abandon the associated residues, but this is not acceptable anymore. Remediation energy needs are determined by local circumstances and previous disposal practices. During the operational phase future energy needs can be reduced by adequate residue management practices. The energy needs are determined *inter alia* by the remediation target, i.e. how much residual contamination may remain. The long-term management of closed tailings ponds is a well-recognized problem (IAEA 2004b, 2006) that has occasionally been used to inflate estimates of the (energy) cost of uranium production (e.g. Storm van Leeuwen and Smith 2008). Industry estimates are around 10 US$/kg uranium (Lersow and Märten 2008). In the case of open-pit mining the tailings can be put back into the mined-out pits that are closed out afterwards. The necessary long-term stewardship (IAEA 2006) entails also long-term, trans-generational energy costs for maintenance etc. It is largely an ethical question how to internalize these costs.

One must be careful to not confuse the often immense remediation (energy) costs for legacy sites with the comparatively much lower costs for the orderly close-out of ongoing operations. These are now almost always guaranteed by bonds and similar financial instruments (IAEA 2006). So far, the total energy requirements of close-out and remediation of uranium mining and milling sites has not been addressed explicitly in the literature. No hard data have been collated yet from actual cases and all calculations in LCAs are based on proxies and assumptions (e.g. Dones 2007).

Meeting the Energy Requirements of Uranium Processing

Since much of the (uranium) mining and milling today takes place in remote locations, the efficient provision of energy becomes a crucial variable in the sustainability equation. Typically, such operations cannot be connected to large-scale electricity grids. Hence, their energy needs are met from (local) fossil fuel sources or imported diesel. In order to reduce the CO_2-intensity of uranium mining and milling one could explore the use of renewables, such as photovoltaics (in arid areas), wind turbines or geothermal energy.

Uranium Resources Availability

Arguably the most authoritative compilation on uranium resources is the so-called "Red Book" (OEDC-NEA and IAEA 2010). As for all natural resources, there is a considerable difference between the global inventory in terms of substance and what is recoverable. Definitions and terminology for resources vary according to the reporting organization and context. Common to all definitions is that they are made on an economic basis, usually a current or assumed acceptable cost of recovery. This makes sense in a short- to medium-term economic context, but is not helpful in strategic long-term resources availability assessments.

The average crustal abundances of uranium is about 2.8 ppm, compared to gold and silver at 0.004 and 0.06 ppm respectively (Wedepohl 1969), but only a fraction of the total inventory is known in terms of its location and extent. Assuming that a layer of 1000 m depth is mineable in principle, results in a global inventory of 1.1×10^{18} t U, 2.4×10^{16} t Ag, and 1.6×10^{15} t Au respectively. It is interesting to compare these numbers with retrospective estimates of total mined amounts in human history: 140,000 t for Au, 1,325,630 t for Ag (Zurbuchen 2006), and 2.2×10^6 t U since 1945 (Price *et al.* 2006). The estimated recoverable resources (at 130 US$/kg U) are in the order of 5.5×10^6 t U (OEDC-NEA and IAEA 2010).

The prices per kg for U, Ag and Au were in May 2011 about US$ 21 (www.uranium.info), US$ 1110 (www.silverprice.org) and US$ 48,000 (www.goldprice.org) respectively. In other words, we are currently prepared to pay 53 times more for a commodity that is produced at 50% (20,900 t Ag in

2008; USGS 2009) and 2285 times more for one that is produced at 5.3% (2209 t 5-year average; http://www.gold.org/) the rate of uranium (42,000 t in 2007; www.world-nuclear.org/info/uprod.html). If we were prepared to spend comparable amounts of money on the exploration and recovery of uranium as on silver or gold, the resource base is likely to increase by orders of magnitude (Falck 2009). In the past a doubling of resources for every two decades due to prospecting and exploration was observed (Leersow and Mättig 2008), but very little money (compared to general energy demand) has been invested into uranium prospecting between the early 1980s and the mid-2000s.

The energy required to convert a uranium resource into yellow cake may be more useful as a resource categorization basis. While production techniques are constantly being improved, there are limitations that are dictated by ore mineralogy, geology and thermodynamics. For technical and thermodynamic reasons process losses at each step are unavoidable. Uranium recovery rates can be as high as 95%, but also as low as 60% (IAEA 2004a). The extraction efficiency drops with decreasing ore grade. Losses at consecutive processing steps are multiplicative, so that leaching, extraction and precipitation with 95% efficiency each results in a net recovery of only 85%. Mudd and Diesendorf (2008) have assessed the energy requirements for current operations, but this assessment will need to be extended to lower ore grades.

Reporting of Uranium Resources

Until the end of the Cold War uranium was not only a economic, but above all also a military strategic resource. Although in many countries a free market economy for uranium has developed since, there still remains the aspect of a national strategic resource resulting in non-market led behavior (Preston and Baruya 2006). The data in the "Red Book" (OECD-NEA and IAEA 2010) are "official" ones, reported by national representatives and there is very little scope for independent assessment. It is not known, whether there are still political constraints on reporting in some countries. There could also be an economic interest by mining companies to report low figures with a view to keep prices high. Conversely, junior mining companies might report optimistic figures in order to attract investors. When assessing the quality and reliability of the data, it is helpful to reflect on (a) who owns the data? (b) who generated the data? (c) who collated the data and for what purpose? (d) are the data updated regularly? (e) who determines which data are generated/collated? (Falck 2009). It is clear, however, that the data for a meaningful assessment can only come from the industry itself, notwithstanding commercial confidentiality issues. The Strategic Nuclear Energy-Technology Platform (SNE-TP, www.snetp.eu) initiative of the European Commission might open an opportunity to tap into data sources, as some of the major industrial players in the nuclear area are involved.

Summary and Conclusions

The viability of nuclear energy systems is hinged on the availability of uranium as fuel and on whether the overall energy balance of the respective fuel cycle is positive, taking into account the full life-cycle energy costs, meaning: what is the energy cost of providing 1 Joule worth of fuel for the nuclear energy system?

Resources estimates are commonly made on the basis of economic cost of recovery. From a global energy supply sustainability point of view neither commercial nor national strategic considerations nor time considerations are really relevant. As oil prices have shown, society can and will accommodate fuel price increases by one order of magnitude over the span of half a century. Comparisons with crustal abundances and historic mining data for gold and silver lead to the expectation that uranium reserves would increase by orders of magnitude, if comparable investments into exploration would be made. However, one also needs to keep in mind that due to process losses less than 80% of the original uranium content of the ore body can be utilized.

To date hard data on the energy expenditure in the full life-cycle of nuclear energy systems are scarce in the public domain, prohibiting an unbiased strategic assessment. The lack of such data for the “front end” and “back end” (including remediation) of the fuel cycle has given raise to unsubstantiated claims over the sustainability or otherwise of the nuclear fuel cycle. Relevant data need to be urgently collated and put into the public domain for policy making.

The overall greenhouse gas emission of nuclear energy systems are determined by the energy system(s) that provide the process energy. Possible scenarios, technical feasibilities and logistics for low-carbon energy supplies to uranium mines and mills could be explored together with the industry.

In summary, it is concluded that a comprehensive assessment of the full-life cycle energy costs of uranium mining, milling and subsequent decommissioning and remediation of the related infrastructure is required to inform an unbiased assessment of the sustainability of nuclear energy systems.

References

Dones, R (2007) Sachbilanzen von Energiesystemen: Grundlagen für den ökologischen Vergleich von Energiesystemen und den Einbezug von Energiesystemen in Ökobilanzen für die Schweiz, Final Report ecoinvent v2.0, Swiss Centre for Life Cycle Inventories, Duebendorf, CH.

Falck, WE (2006): Closing the Cycle: Life-Cycle Impact Assessment of Materials Used in Nuclear Energy Systems. In: Uranium Production and Raw Materials for the Nuclear Fuel Cycle – Supply and Demand, Economics, the Environment and Energy Security, Proc. Symp., Vienna, 20–24 June 2005: 76–83.

Falck, WE (2009) Towards a Sustainable Front-End of Nuclear Energy Systems. European Commission, Report EUR 23955 EN, Luxembourg.

IAEA (1994) Net Energy Analysis of Different Electricity Generation Systems. IAEA-TECDOC-753, Vienna.

IAEA (2003) Sustainability and Environment. Guidance for the evaluation of innovative nuclear reactors and fuel cycles, Report of Phase 1 A of the International Project on Innovative Nuclear Reactors and Fuel Cycles (INPRO). IAEA-TECDOC-1362, Vienna.

IAEA (2004a) Recent Developments in Uranium Resources and Production with Emphasis on In Situ Leach Mining. IAEA-TECDOC-1396, Vienna.

IAEA (2004b) The Long-Term Stabilization of Uranium Mill Tailings, Final Report on the Coordinated Research Project 2000–2004. – Report IAEA-TECDOC-1403, Vienna.

IAEA (2006) Management of Long-Term Radiological Liabilities: Stewardship Challenges. Report IAEA-TRS-450, Vienna.

Kröger W (2001) Measuring the Sustainability of Energy Systems. NEA News 19(1): 21–24, OECD-Nuclear Energy Agency, Paris.

Leersow M, Märten H (2008) Energiequelle Uran – Ressourcen, Gewinnung und Reichweiten im Blickwinkel der technologischen Entwicklung. Glückauf, 144(3): 116–122.

Mudd GM, Diesendorf M (2008) Sustainability of Uranium Mining and Milling: Toward Quantifying Resources and Eco-Efficiency. Environ. Sci. Technol. 42: 2624–2630.

OEDC-NEA, IAEA (2010) Uranium 2009: Resources, Production and Demand. Report OECD-NEA, Paris.

Price R, Barthel F, Blaise J-R, McMurray J (2006) Forty Years of Uranium Resources, Production and Demand in Perspective. OECD-NEA Newsletter 24(1): 4–6, Paris.

Preston F, Baruya P (2006) Paper 8: Uranium resource availability. In: Sustainable Development Commission UK [Ed.], The Role of Nuclear Power in a Low Carbon Economy, London, http://www.sd-commission.org.uk/publications/downloads/Nuclear-paper8-UraniumResourceAvailability.pdf (accessed 18.05.11).

Storm van Leeuwen JW, Smith P (2008) Nuclear Power – the Energy Balance http://www.stormsmith.nl/ (accessed 18.05.11).

Sovacool BK (2008) Valuing the Greenhouse Gas Emissions from Nuclear Power: A Critical Survey. Energy Policy 36: 2950–2963.

Zurbuchen D (2006) The World's Cumulative Gold and Silver Production. http://www.gold-eagle.com/editorials_05/zurbuchen011506.html (accessed 18.05.11).

Making Uranium-Mining More Sustainable – The FP7 Project EO-MINERS

W. Eberhard Falck, Henk Coetzee

Abstract Every mining operation impacts the environment and the adjacent communities to a lesser or greater degree and minimizing such impacts is at the core of initiatives to make mining, including uranium mining more sustainable. Over the past two decades work has focused on remediating uranium mining legacies in the wake of mine closures. However, with the demand for uranium increasing again, old mines are revived and new mines opened. In order to avoid the mistakes and poor practices of the past, we need to look into methods to make uranium mining operations more sustainable. The European Commission Framework Programme 7 project EO-MINERS aims to support stakeholder dialogue by providing independent information based mainly on earth-observation techniques with a focus on remote sensing. Typical information that can be gathered includes mine land-use, the state of remediation including recultivation success, the dispersal of acid mine drainage, surface radioactivity, risks from spoil heaps and dams, landscape fragmentation due to mine infrastructure encroachment of informal settlements on mine-affected land, etc. This paper describes the processes and procedures that are being developed in the EO-MINERS project for making such earth-observation techniques useful for deliberative stakeholder processes.

W. Eberhard Falck
Laboratoire REEDS, Université de Versailles St. Quentin-en-Yvelines, France

Henk Coetzee
Council for Geoscience, Pretoria, South Africa

B. Merkel, M. Schipek, *The New Uranium Mining Boom*
DOI 10.1007/978-3-642-22122-4

Introduction

For centuries mining has been and continues to be one of the bases of economic and social development. It is the starting point of many value chains. Due to their ever increasing energy demand, the demand for uranium is expected to shift gradually from the old "nuclear" countries towards emerging economies, such as China and India. Exploration is increasing and various countries are joining or re-joining the league of (potential) uranium producers (NEA and IAEA 2010). Most of these activities take place outside Europe as it has become increasingly difficult to obtain social license for new mining operations or even the extension of existing ones. In addition, uranium ore grades in Europe are generally low, making exploitation not economically viable under current market conditions.

The underlying challenges with which mining in general is confronted are land-use and resource use, as well as socio-political conflicts. Indeed, throughout the world, mining faces public acceptance challenges and growing social criticism in both, developed and developing countries. These can be traced back to environmental issues caused by a nonchalant attitude in the past of many mining companies towards environmental protection. Mining can have a significant impact on the surrounding environment and often devalues large areas of land. Historically, inadequate actions were undertaken to remediate mining-affected landscapes (IAEA 2005). Today, however, at least in the more developed countries, closure and remediation plans are an integral element of the operational license and financial securities are required to obtain it. Furthermore in developed countries, political and financial pressure often forces foreign operators have to apply the same stringent regulations abroad that would apply at home. However, having regulations in place is not sufficient and their enforcement remains a critical issue.

The global dimension of mining impacts is increasingly being recognized in the EU and world-wide. Various initiatives aimed at making mining more sustainable have been launched (e.g. European Commission 2011). For uranium mining, international organizations, most notably the IAEA, have supported the development and application of good practice (e.g. IAEA 2010). Otherwise industry itself is recognizing the issues and undertakes to address them on a voluntary basis through self-commitment, for instance as members of the International Council on Mining & Metals (ICMM, http://www.icmm.com/).

The Group on Earth Observations (GEO, http://www.earthobservations.org/) is concerned with the responsible management of natural resources. Sound environmental management of mining and milling can avoid environmental and social impacts and the resultant high remediation costs. Understanding and monitoring the impacts is, therefore, of concern to all stakeholders (government bodies or agencies, local authorities, industry, environmental groups, individual citizens). However, the technology platform to support such environmental monitoring is diverse, geographically inconsistent, site specific, lacks integration across technologies and is hence far from complete. A gap also exists within GEO's Global Earth Observing System of Systems (GEOSS) that currently concentrates on natural hazards and climate change (http://www.earthobservations.org/geoss.shtml/).

Uranium Production Life-Cycle Impacts

Any materials flow due to (uranium) mining is likely to cause some environmental impact. A careful engineering of the life-cycle (Fig. 1) of mineral resources-based products can help to minimize such impacts (Falck 2006, 2009). Minimizing (waste) material flows is also in the interest of mine operators, as they are a significant cost factor. There are, however, limitations to this, as high-grade resources close to the surface gradually become depleted world-wide, requiring reliance on lower-grade ores and to those at greater depth. Both factors result in more unwanted extraction, though certain techniques applicable to uranium ores, such as *in situ*-leaching (ISL) can keep this low (IAEA 2004). While a trend to lower-grade ores is unavoidable, responsible management of the resulting mining and milling residues will help to reduce the land-use and environmental impacts.

Major impacts can arise at all stages of the life-cycle:

- *Exploration* – including surveys, field studies, drilling and exploratory excavations; some impacts and waste generation already occur at this stage.
- *Project development* – includes construction of roads, access tunnels and shafts, erection of hoisting machinery, treatment plants and ancillary buildings (laboratories, administration, social), construction of service infrastructure (process and drinking water supply, sewerage, power generation and distribution), and construction of waste management facilities (mine waste dumps, tailings ponds, leach pads).
- *Mine operation* – underground or open-cast excavations, mine dewatering, *in situ* or heap-leach operation.
- *Beneficiation* – on-site processing may include comminution to reduce particle size, flotation using selected chemicals, ore leaching with a variety of chemical solutions; associated transport and storage of ore and concentrates may be a handling risk and can result in localized site contamination.
- *Mine closure* – remediation is best done progressively rather than at the end of life of the mine. While the closure and remediation is intended to mitigate environmental impacts, it is important that it does not itself create secondary effects through excessive fertilizer use, spread of weeds, silting and incompatible landscape features.
- *Long-term stewardship* – the chosen remediation strategy and associated technical solutions for mining and milling residues have to monitored for their continued effectiveness; failure of technical solutions and the stewardship program supporting them can lead to environmental impacts.

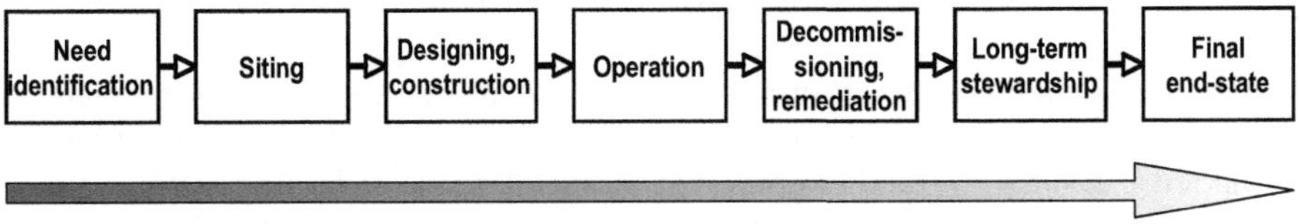

Fig. 1 The life-cycle of a uranium production facility (IAEA 2006)

The Project EO-MINERS

Though not specifically addressing uranium mining, the European Commission Framework Programme 7 project EO-MINERS (http://www.eo-miners.eu/) aims at integrating new and existing Earth Observation (EO) tools to improve best practice and to reduce mining-related environmental and social impacts by

- introducing innovative EO tools and services to the mining industry,
- providing accuracy and quality measures for EO products,
- demonstrating the application of EO in different case studies,
- fostering the dialogue between mining industry and environmental organizations based on EO-derived information, and
- generalizing the results obtained for use in operational mining applications.

Forecasting impacts, land-use and relevant remediation measures requires the development of predictive tools. GIS using EO data will enable the visualization of prospective evolutions over time. Cumulative impacts must be adequately addressed at regional scale.

The scientific and technical objectives of EO-MINERS include *inter alia*:

- To assess policies and strategies of different stakeholders (operators, regulators, public) and resulting information needs;
- To assess environmental, socio-economic and other sustainable development issues surrounding mining operations and to define indicators for their development that can be addressed using EO-techniques;
- To further demonstrate the capabilities of integrated EO-based methods and tools in monitoring, managing and thus contributing to reducing environmental and social impacts of mining during its whole life-cycle;
- To provide reliable and objective information on affected ecosystems and societies that will serve as a basis for a sound dialogue between industry, regulators and the public;
- To summarize and document procedures of both, technical and social nature as a baseline for a compendium of best practices that will assist and inform the ongoing dialogue between the public, regulators and the industry.

Three study sites in the Czech Republic (open-cast lignite mining), South Africa (open-cast coal mining) and Kyrgyzstan (gold mining) provide the focus for the development work.

Enabling Stakeholders Through Information

Different stakeholders are informed to different levels about issues relevant to mining or milling. Lack of information, especially independent and unbiased information can hinder effective processes of social decision-finding. This problem not only affects the general public, but also local or regional authorities who are not directly involved in licensing procedures, but whose area of jurisdiction is or will be affected by the operations. Providing reliable and objective information is one element in enabling stakeholders to participate in decision finding processes in a meaningful way. Information is more than data; it is the context (i.e. reference cases, benchmarks, best practice examples, etc.) that gives meaning to the data, generating understanding and knowledge for stakeholders who may not have the training to understand or interpret (raw) data (IAEA 2002).

Meaningful information on complex environmental or social issues can often be provided in the form of indicators. Indicators provide a metric of the state of (complex) systems or of issues, or for trends of their development when measurements are repeated over time. Indicators are useful tools to reduce a complex set of diverse data into a manageable set for policy making and to monitor changes during policy implementation.

Hence, indicators must be based on measurable quantities in order to be useful and as such can be intensive or extensive properties, being either independent of or dependent on the size or volume of the system under consideration. For instance, the pH-value is an intensive property of water and can indicate the general state of surface and groundwaters for instance with respect to their life-supporting functions. Conversely, the number of schools in a particular region has to be related to some meaningful extensive property, e.g. the number of people living in that region, in order to be an indicator for the provision of education. One has to carefully to distinguish between intensive and extensive properties. For instance, the amount of mine waste generated is meaningless as indicator, if not put into relation to e.g. the total of metal value recovered. If related to the ore grade, it would allow the assessment of the efficiency of the exploitation of the resource. Even then it will need to be related to the quality of mine waste management since sometimes a smaller quantity of poorly managed waste could pose a higher environmental risk than a larger quantity of well managed waste.

Thus indicators can be single parameter values, requiring a single measuring technique only, or may need to be compounded from various parameters that are measured using different techniques.

The EO-MINERS Strategy to Derive Indicators

It is widely recognized that the development of indicators that are meaningful to a wide variety of stakeholders is a social and not an engineering process. The social process defines what has to be indicated for whom and why. However, scientists and engineers, who themselves are stakeholders in the process, have to evaluate whether the candidate indicators can be related to measurable quantities, thus making the indicators operational. Hence, several work packages (WPs) in EO-MINERS inform the process of deriving and defining indicators. EO-MINERS employs a multi-pronged approach to develop candidate indicators (Falck et al. in prep), consisting of a heuristic development by expert elucidation (WP 1), examination of site-specific conceptual models for the three study sites to inform the indicators (WP3), and a

Table 1 Categories of candidate indicators for possible mining-related impacts

Category		Category	
A	Land-use	G	Geotechnical hazards and accidents
B	Mass Flows	H	Industrial and other accidents
C	Energy Flows	I	Social impacts
D	Air quality and other nuisances	J	Regional development
E	Water quality	K	Economic vulnerability/resilience
F	Transport		

semi-deliberative approach (WP1/5), elucidating input from stakeholders outside the project team. In this way not only the various technical stakeholders within EO-MINERS will contribute to a valid and useful set of indicators, but also a wide variety of stakeholders outside the project team. This process is going through several iterations in order to consolidate the set of indicators. This consolidated set of candidate indicators, after it is reviewed by EO specialists in order to assess their measurability using EO techniques, will inform the process of EO services development and will be subject to a final stakeholder evaluation towards the end of the project. Table 1 gives the an overview over the categories for candidate indicators.

Pertinence in an Uranium Production Context

EO offers a unique opportunity to collect spatial data for better assessment of mining-related environmental and social impacts during all phases of a facility's life-cycle. The tools and processes under development will help to inform deliberative decision-finding procedures as stipulated by agencies such as thc in IAEA or local regulators. The IAEA have noted that in order to arrive at sustained and accepted long-term solutions, in particular in a remediation context, all stakeholders have to be involved in the decision-finding processes (IAEA 2002, 2006).

Decision-making along the life-cycle of a uranium production facility faces a number of technical and societal challenges. There are

- uncertainties regarding the result of site assessment under normal conditions leading to the decision today (e.g. data gaps in the inventory, insufficient site characterization, integrity of engineering, …), and
- uncertainties about the future; this covers both nature and the range of natural phenomena/"events" in the future and the influence of time on the internal evolution of the designed structures/"processes".

At the same time managers are faced with the challenge to

- obtain and maintain public trust;
- achieve institutional constancy or to ensure continuity of e.g. long-term stewardship activities over many generations; and
- learn from past and ongoing experience as technological and management means for implementation are developed.

Independent access to information about the site concerned will support the development of mutual trust, allowing the public and the regulators to monitor whether the site develops as indicated and anticipated by the operator. It also facilitates mediation in the case of dispute. Taylor-made earth observation services in particular allow the monitoring of important parameters of site development at relatively low cost and often in near real-time. Such aspects can be crucial in maintaining mutual trust, as often critical changes, e.g. to the geotechnical state of tailings ponds or to the distribution of surface radiation are not readily visible from the surface or not observable without a dedicated measurement campaign.

Monitoring and surveillance of environmental impacts, including radiological impacts, tend to be more closely scrutinized in a uranium production context than in the context of other raw material production. Though this is not necessarily valid from a scientific point of view, this is certainly the case from a public perception and often from a regulatory perspective. Adequately visualized EO products allow the general public and often also the regulators to better "see" what is happening at a site. GIS-supported visualization also allows the stakeholders to better see how site developments might relate to their personal situation, e.g. distances to and possible impacts on their private home or their community.

Conclusions

The project EO-MINERS is still at an early stage. However, first experiences with stakeholder interaction and confronting stakeholders with possible EO-services have been gained. Though the situations at the study sites may not be as controversial as perhaps at uranium production sites, the situation in South Africa may be confrontational enough to draw some conclusions. The fact that during a second round of stakeholder interaction in South Africa very few suggestions for amendment to the candidate list of indicators were made, places a certain confidence into their relevance with respect to scope and coverage.

A particular conceptual difficulty arises insofar as the project team interacts with local mining-related social processes and may itself become a stakeholder in these. The project naturally has only a limited life-time and it is not clear, if and how a stakeholder dialogue will or can be sustained by the local project partners beyond the project life-time. The risk is that processes are started and expectations are raised among the various stakeholders that cannot be fulfilled by a research project. Furthermore, these project-induced processes may interfere with already existing mining-related social processes and may turn out to be counterproductive. Indeed, it was the fear of some stakeholders interviewed, in particular the mine operators, that the project would upset the local situation. Nevertheless some local interest groups expected that the project would help them to achieve their interests and goals. These expectations indeed indicate the need for shared information and in this sense validates the project's objectives. In a real situation, beyond the study sites, it would be essential that a clear mandate is given to those developing and fostering these social processes.

In a sense, projects such as EO-MINERS are "a solution in search of a problem". This is typically the starting point encountered in the interaction between the project team and outside stakeholders. The majority of stakeholders interviewed had not been aware of the possibilities of EO techniques and in particular of remote sensing techniques. This clearly indicates the need for a sustained dialogue between EO service providers and stakeholders outside the project, if the project aim of enabling these stakeholders should be achieved.

Acknowledgements The authors wish to acknowledge the various contributions and fruitful discussions during the course of the project by the whole EO-MINERS project team (http://www.eo-miners.eu/).

References

European Commission (2011) "Tackling the challenges in commodity markets and on raw materials", Communication from the Commission to the European Parliament, the Council, the European Economic and Regional Committee and the Committee of the Regions, COM(2011) 25 final.

Falck WE (2006) Closing the Cycle: Life-Cycle Impact Assessment of Materials Used in Nuclear Energy Systems. In: Uranium Production and Raw Materials for the Nuclear Fuel Cycle – Supply and Demand, Economics, the Environment and Energy Security, Proc. Symp., Vienna, 20–24 June 2005: 76–83.

Falck WE (2009) Towards a Sustainable Front-End of Nuclear Energy Systems. European Commission, Report EUR 23955 EN: 38 p. Luxembourg (Luxembourg).

Falck WE, Spangenberg J, Solar SV, Wittmer D (in prep.) Candidate Indicators for Mining-related Impacts. Earth Observation for Monitoring and Observing Environmental and Societal Impacts of Mineral Resources Exploration and Exploitation, CEC FP7 Project EO-MINERS, Deliverable D1.9.

IAEA (2010) Best Practice in Environmental Management of Uranium Mining. International Atomic Energy Agency Nuclear Energy Series No. NF-T-1.2: 47 p, Vienna.

IAEA (2006) Management of Long-Term Radiological Liabilities: Stewardship Challenges. International Atomic Energy Agency, Report IAEA-TRS-450, Vienna.

IAEA (2005) Environmental Contamination from Uranium Production Facilities and their Remediation. International Atomic Energy Agency, Proceedings of an International Workshop held in Lisbon, February 2004, Proceedings Series STI/PUB/1228, 262 p.

IAEA (2004) Recent Developments in Uranium Resources and Production with Emphasis on In Situ Leach Mining Details. International Atomic Energy Agency, Report IAEA-TECDOC-1396, Vienna.

IAEA (2002): Non-technical Factors Impacting on the Decision Making Processes in Environmental Remediation. International Atomic Energy Agency, Report IAEA-TECDOC-1279, Vienna.

OECD-NEA (2010): Uranium 2009: Resources, Production and Demand. OECD Nuclear Energy Agency and International Atomic Energy Agency, Paris.

Groundwater Monitoring Data and Screening Radionuclide Transport Modeling Analyses for the Uranium Mill Tailings at the Pridneprovsky Chemical Plant Site (Dneprodzerzhinsk, Ukraine)

Oleksandr Skalskji, Dmitri Bugai, Oleg Voitsekhovitch, Viktor Ryazantsev, Rodolfo Avila

Abstract This paper presents results of groundwater monitoring and screening modeling studies carried out in 2008–2010 for the uranium production legacy site Pridneprovsky Chemical Plant (Dneprodzerzhinsk city, Ukraine). The simulated offsite radiological risks caused by groundwater pathway are estimated to be low because of the long travel time to discharge contours (hundreds-thousands of years) and large dilution of contaminants in the Dnieper River system. The preliminary recommendation is stabilization and storage of tailings sites *in-situ*. Further groundwater monitoring and risk assessment studies are opportune.

Introduction

Former Pridneprovsky chemical plant (PChP) was one of the first Soviet uranium ore processing facilities. The plant was built in 1947 in the industrial part of

Oleksandr Skalskji
Institute of Geological Sciences, Gonchara Str. 55-b, Kiev, 01054 Ukraine

Dmitri Bugai
Institute of Geological Sciences, Gonchara Str. 55-b, Kiev, 01054 Ukraine

Oleg Voitsekhovitch
Ukrainian Institute of Hydrometeorology, Nauki Prospekt 37, 252620 Kiev, Ukraine

Viktor Ryazantsev
State Nuclear Regulatory Inspectorate of Ukraine, Arsenalna Str. 9/11, 01011 Kiev, Ukraine

Rodolfo Avila
Facilia AB, Gustavslundsvagen 151C, 16751, Bromma, Sweden

B. Merkel, M. Schipek, *The New Uranium Mining Boom*
DOI 10.1007/978-3-642-22122-4 © Springer 2011

Dneprodzerzhinsk City in Dnepropetrovsk Region, Ukraine. The PChP ceased main activities connected to extraction of uranium from ores in 1991. The resulting uranium mill tailings and chemical wastes are stored on the plant territory ("Western", "Central Yar" and "South-East" tailings) and outside plant site ("Dneprovskoe" and "Sukhachevskoe" tailings). The storage site of uranium ores "Baza S", which also belongs to the PChP infrastructure, contains the uranium ore residues and contaminated soils. The general characteristics of tailings of the PChP are listed in Table 1. The schematic map of the territory of the PChP is shown at Fig. 1.

Most of the discussed above waste sites represent sources of migration of radionuclides of the uranium decay series to groundwater. The discharge contours for the contaminated groundwater are Konoplianka River and Dnepr River (Fig. 1).

Table 1 General characteristics of main tailings of the former PChP facility

Main uranium mill tailings and waste sites	Exploitation period	Area, hectare	Amount of wastes, 10^6 t	Volume, 10^6 m^3	Total radioactivity, TBq
"Western"	1949–54	6.0	0.77	0.35	180
"Central Yar"	1951–54	2.4	0.22	0.10	104
"South-East"	1956–80	3.6	0.33	0.15	67
"Sukhachevskoe" (Sect. 1)	1968–83	90	19.0	8.6	710
"Sukhachevskoe" (Sect. 2)	1983–92	70	9.6	4.4	270
"Baza S"	1960–91	25	0.3	0.15	440
"Dneprovskoe" (D)	1954–68	73	12.0	5.9	1400

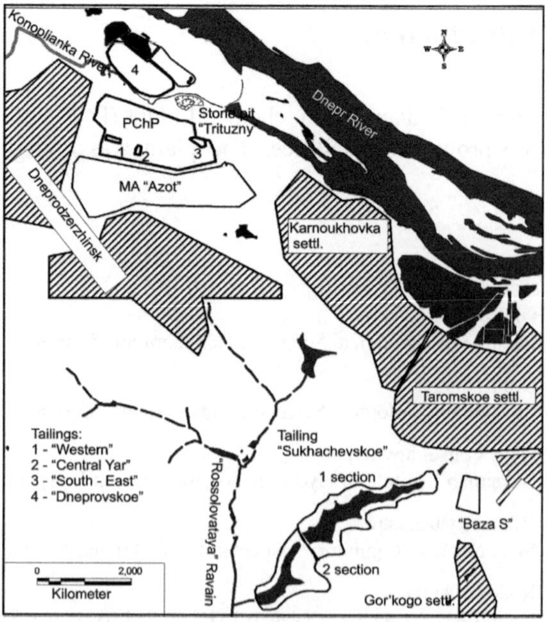

Fig. 1 Location of objects and industrial territory of PChP

Below we present results of groundwater monitoring and modeling studies carried out in 2008–2010 for the PChP site in the frame of the Ukrainian national site remediation program and Sweden-Ukrainian technical assistance project ENSURE. These works were carried out in the frame of more broad studies aimed at risk assessment and development of remediation strategies for this radioactively contaminated site.

Hydrogeological Settings and Groundwater Monitoring System

The industrial site of the PChP occupies the slope and plateau of the Dnepr River terrace. The tailing "Dneprovskoe" is situated at the flood plain of the Dnepr River. The "Sukhachevskoe" tailings are located in ravines of the plateau (Fig. 1).

The geological structure of the site includes the Precambrian crystalline rocks overlain by deposits of Paleogene, Neogene and Quaternary systems.

Hydrogeological conditions of the study area are characterized by presence of the system of interconnected aquifers in the "man-made deposits" (wastes of uranium ore processing), Quaternary deposits (alluvial sands, sandy loams), Neogene deposits (sands, sandy loams), Paleogene deposits (sands, clays) and the Precambrian fractured crystalline rocks with the zone of weathering.

The different waste sites are characterized by a variety of specific hydrogeological settings. The uranium ore processing wastes in "Western", "Dneprovskoe" and "Sukhachevskoe" tailings are situated in partly water-saturated conditions. The bodies of "Central Yar" and "South-Eastern" tailings are situated in unsaturated conditions. The contaminated soils at the "Baza S" site are located at the ground surface, and are separated from the aquifer by 10–12 m of loess soils.

At present time only the "Western" and "Central Yar" tailings have adequate groundwater monitoring systems, which allow characterizing groundwater quality upstream from the tailings, below the waste sites, and also along the groundwater flow path to the discharge contour.

The "Dneprovskoe" tailing has sufficient amount of monitoring wells in the source area (uranium processing wastes, which are partly saturated by groundwater) and in the underlying alluvial aquifer. However the observation well network outside the tailings does not provide reliable information about the groundwater transport towards the Dnepr River. The "South-Eastern", "Sukhachevskoe" tailings and "Baza S" do not possess adequate monitoring well networks.

The groundwater monitoring program at the site foresees annual determination of chemical composition (major ions), toxic metals and radionuclides of uranium decay series. Groundwater level, pH, Eh, temperature and electrical conductivity and are also measured systematically.

Groundwater on the PChP industrial site and within the flood plain of Dnepr River (where the "Dneprovskoe" tailing is located) within the zone of influence of tailings typically has very high total dissolved solids (TDS) levels. Concentration of cations and anions often exceed maximum permissible concentrations (MPC) in drinking water. The dominant anions are chlorine, sulfates and nitrates. Exceeding MPC for cations (Na^+, K^+, Ca^{2+}, Mg^{2+} and NH^{4+}) is also common. Groundwater is thus not suitable for domestic and agricultural usages.

Concentrations of toxic metals in groundwater (beryllium, boron and aluminum, cobalt, lead and cadmium, manganese, iron and nickel) also often exceed the MPC by one to two orders of magnitude.

Radionuclide contamination of the groundwater caused by migration from the tailings sites are observed at all contaminated sites except the "South-Eastern" tailings and "Baza S" uranium ore storage site (at the last site the local aquifer in loess deposits is protected by 10–12 m thick unsaturated zone). Reliable conclusions on radionuclide migration from the "Soukhachevskoe" tailings cannot be made, because of the inadequate groundwater monitoring system.

The main groundwater contamination hazard is posed by migration of uranium. Highest concentrations of uranium are observed in the vicinity of the "Western" tailings where the activity of uranium in 2009–2010 was about 600 Bq/l in well 2-ZP (tailings body) and 1100 Bq/l in (well 3-ZP, alluvial aquifer downstream from the tailings) (Fig. 2). Here due to the large gradients of hydraulic head in the alluvial aquifer (0.02–0.04 m/m) uranium from wastes has migrated along the flow path to the well 30d at the distance of about 940 m from the source (see Fig. 2).

Approximately one order of magnitude lower levels of uranium in groundwater compared to the "Western" tailings were observed in the source areas (water saturated wastes) of the "Dneprovskoe" and "Soukhachevskoe" tailings.

Migration of uranium from the "Central Yar" tailings is somewhat less pronounced compared to "Western" tailings, and is limited to the distance of approximately 200 m from tailings.

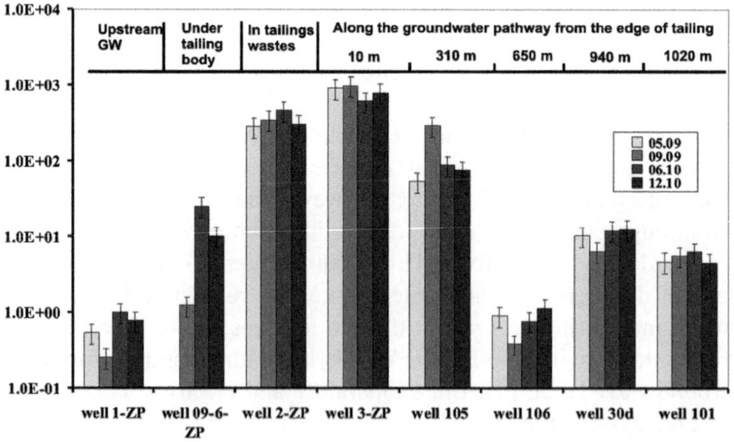

Fig. 2 Specific activity of uranium 238 + 234 in groundwater along the flow path from the "Western" tailings to Konoplianka River (Bq/l)

Groundwater Modeling Methodology

Screening-level radionuclide transport modeling analyses were carried out for all described above contamination sites, and included two modeling steps:

- Development of the three-dimensional filtration models for the PChP site and for the "Sukhachevskoe" tailings sites using Visual Modflow software, and calculation of groundwater flow lines from the waste sites;
- Modeling of migration of radionuclides of uranium decay series along of the groundwater flow lines using Ecolego-based flow tube models.

Filtration Models

For implementation of filtration models the Visual Modflow Pro 3.0 software (by Waterloo Hydrogeologic Inc.) was used (Guiger and Franz 1996). Two different filtration models were developed: (1) for the PChP industrial site (including the region of "Dneprovskoe" tailings) and (2) for the "Sukhachevskoe" tailings site (including the "Baza S" site) (Fig. 3).

The filtration model for the PChP site encompasses the territory of the Plant and the adjacent floodplain of the Dnepr River (Fig. 3a). The numerical grid size of model varies from 15 to 62 m. The geological structure of the site is represented in the model by five layers (see previous paragraph).

The filtration model of the "Sukhachevskoe" tailings site (Fig. 3b) accounts for two main aquifers: upper aquifer in Quaternary loess deposits, and the lower

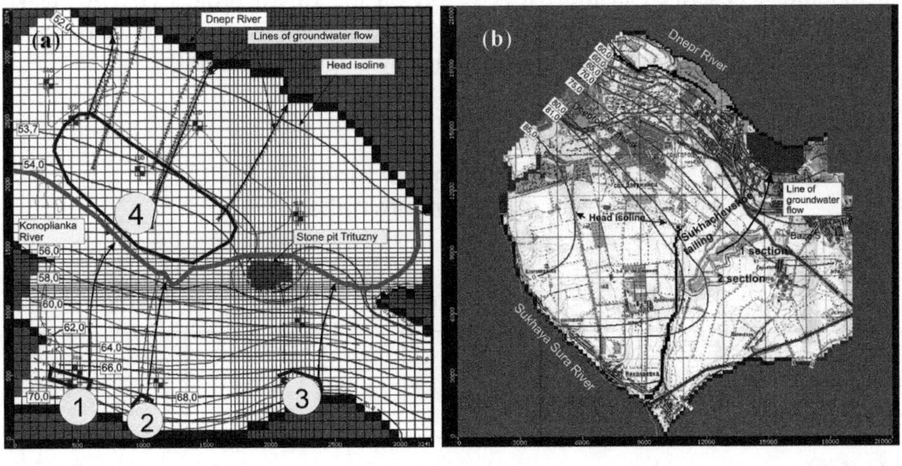

Fig. 3 Filtration models for the PChP industrial site (**a**) and "Sukhachevskoe" tailings site (**b**) showing hydraulic head isolines and groundwater flow lines from tailings sites. Tailings: 1 – "Western", 2 – "Central Yar", 3 – "South-Eastern", 4 – "Dneprovskoe"

aquifer in the Neogene deposits and in the fractured zone of Precambrian crystalline rocks. The main discharge contour for the Neogene aquifer, which represents the main interest in the context of radionuclide migration from the "Sukhachevskoe" site, is the Dnepr River. The numerical grid size of model varied from 100 to 200 m. The geological structure of the site is represented in the model by four layers.

Groundwater modeling allowed estimating groundwater travel times from the tailings sites to discharge contours. Groundwater travel time from the "Western" tailings to the Konoplianka River is estimated at 7 years, from the "Central Yar" tailing – 6 years, and from the "South-East" tailings – 4 years. The groundwater flow "Dneprovskoe" tailing needs approximately 40–50 years to reach Dnepr River. The groundwater from the first section of "Soukhachevskoe" tailings needs approximately 100 years to reach the Dnepr River, and from the second section – 340 years. The calculated flow lines allowed modeling radionuclide transport in groundwater using the methodology described in the next paragraph.

Methodology for Modeling Radionuclide Transport in Groundwater

For screening calculations of radionuclide transport in groundwater from the uranium mill tailings the Ecolego software (Facilia AB 2007) was used, which represents a software package for development of radioecological compartmental deterministic and stochastic dynamic models described by the first order ordinary differential equations.

To model radionuclide transport in groundwater using Ecolego we used the methodology described in the IAEA report on the ISAM project (IAEA 2004). This methodology has been already successfully applied for modeling of radionuclide transport from the uranium mill tailings by Bugai et al. (2008).

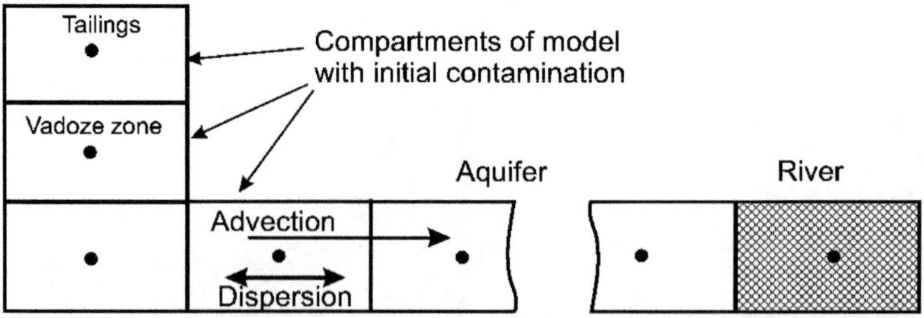

Fig. 4 A conceptual scheme of the Ecolego-based flow tube model for radionuclide migration from uranium mill tailings

Groundwater migration of radionuclides from tailings is governed by processes of advective-dispersive transport, sorption on soils and radioactive decay. The scheme of the flow tube model of radionuclide migration implemented using Ecolego is shown on Fig. 4.

The model accounted for radionuclide leaching from the solid matrix of tailings, vertical infiltration of contaminated porous solutions from the body of tailing to the vadoze zone, vertical contaminant transport in the vadoze zone to the aquifer, and lateral advective-dispersive transport of radionuclides in the aquifer.

The model assumed steady-state hydrodynamic conditions in the aquifer, which corresponded to the present day hydrogeological conditions. The model geometry was defined based on the results of groundwater flow modeling using the filtration models described in the previous paragraph. The modeling has taken into account all main radionuclides of the uranium decay series (^{238}U, ^{234}U, ^{230}Th, ^{226}Ra, ^{210}Pb, ^{210}Po), including the ingrowths of daughter radionuclides.

Radionuclide Transport Modeling

Modeling Scenarios

The most sensitive parameter of groundwater transport models is radionuclide sorption distribution coefficient (Kd) for the "soil – solution" system. The Kd values for modeling were selected as follows. The Kd values for the tailings materials were estimated on the basis of the *in-situ* measurements of the radioactivity of the solid phase and the tailings porous solutions. The uranium Kd values for the alluvial aquifer were back-calculated using the data on the extension of radionuclide plume in aquifer tin the vicinity of the "Western" tailings (as in 2009). The Kd-s of other radionuclides of uranium decay series for the alluvial aquifer were taken from literature (EPA 1999). Thus, two sets of radionuclide Kd-s were formed (Table 2): "base case" scenario (that is a most probable one) and "conservative" scenario (or "worst case" scenario).

Along with normal evolution scenario for the tailings we have modeled some accidental and remedial scenarios (waste excavation, establishment of the soil cover on the top of tailings etc.). Hydrodynamic and sorption parameters corresponding to different scenarios are summarized in Table 3.

The following parameters were estimated using modeling: – radionuclide concentrations in groundwater at different distances from the tailings; – potential exposure doses resulting from using groundwater as a drinking water source; – long-term groundwater fluxes of radionuclides to the Konoplyanka River and Dnieper River. The simulated period ranged from 1000 to 5000 years.

Modeling predictions were carried out for all uranium mill tailings at the PChP site. Below some example modeling results for the "Western" tailings are presented.

Table 2 Radionuclide Kd-s for modeling RN migration from tailings

Radionuclide	Tailings matrix	Soils of the vadoze zone	Aquifer ("base case")	Aquifer ("conservative")
U-238	5	1	10	5
U-234	5	1	10	5
Th-230	10,000	3000	3000	600
Ra-226	6300	200	200	40
Pb-210	3500	500	500	125
Po-210	2500	100	100	80

Table 3 Parameter values for different scenarios of radionuclide migration in groundwater from tailings

Scenario	Set of Kd (aquifer)	Recharge, mm/y	Other parameters
Base case	Table 2	50	Standard
Conservative	Table 2	100	Standard
Accident: degradation of cover	Base case, Table 2	100	Standard
Modeling of remedial measures			
Waste extraction	Base case, Table 2	50	Source of migration in tailing = 0
Soil cover design	Base case, Table 2	20	Standard

Modeling Results

Below some example modeling results for the "Western" tailings are presented.

The Fig. 5a shows the potential effective exposure dose resulting from using groundwater from the well located at the distance 150 m downstream from "Western" tailings ("base case" scenario). The simulated maximum dose reaches 10 mSv/y, which shows that groundwater contamination in the vicinity of tailings poses a potentially significant health hazard, and suggests the need to maintain the institutional control at the PChP site precluding the usage of local groundwaters and surface waters (such as Konoplyanka River). In fact, the alluvial aquifer at the PChP site is currently not used for domestic or agricultural purposes due to high concentrations of total dissolved solids and toxic chemicals in groundwater.

Modeling of remedial measures has shown that excavation of wastes (as well as covering of tailings by soil screen) do not significantly affect concentrations of uranium in the aquifer below the tailings in the first 300 years of the forecast (Fig. 5b), and at the distance of 1000 m (Konoplianka River) – influence is not essential during the 1000 y period (Fig. 5c). This is due to the fact that the uranium transport in groundwater during the considered period is governed first of all by the source of migration situated in the vadoze zone and alluvial aquifer below the tailings. This source is related to the radioactive contamination of soils of vadoze

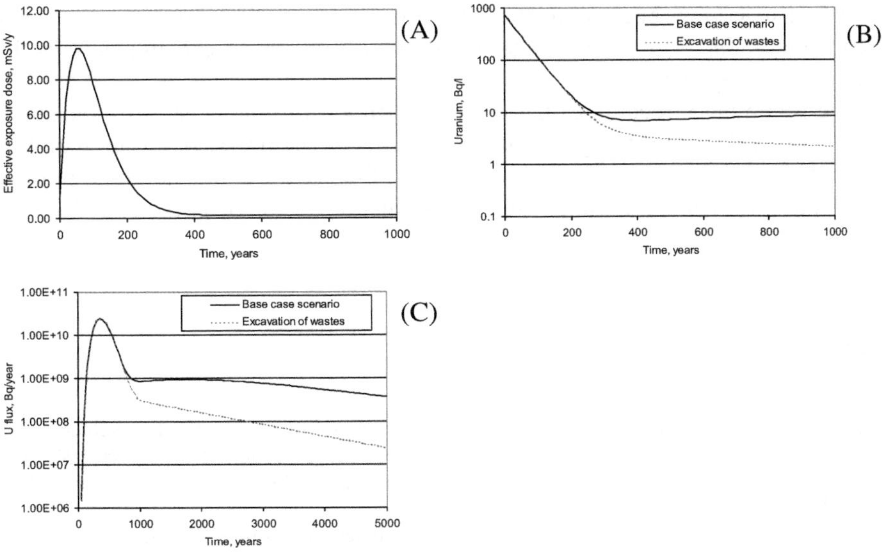

Fig. 5 Example modeling results for the "Western" tailings: (**a**) – exposure dose due to consumption of groundwater from the well at the distance of 150 m downstream from tailings; (**b**) uranium concentrations in groundwater below the tailings for the remediation scenario; (**c**) uranium flux from the tailings to Konoplyanka River

zone and alluvial aquifer, which was formed during the operational period, when the damping of considerable volumes of contaminated technological waters to tailing was carried out. Eventually, this source is not affected by the simulated remedial measures.

Similar calculations were carried out for all contamination sources at the PChP site, listed in the previous paragraphs of this article. Conclusions from modeling studies are summarized below.

Conclusions

The groundwater monitoring data and results of the screening-level radionuclide transport modeling analyses for the groundwater pathway suggest that the uranium mill tailings of the former Pridneprovsky Chemical Plant facility "Sukhachevskoe", "Baza S" and "South-East" represent relatively safe sites. On the contrary "Dneprovskoe", "Western" and "Central Yar" uranium mill tailings represent serious sources of radionuclide (in particular uranium) and toxic chemical substances migration to groundwater. It is important to maintain long-term institutional control at the site, which should preclude the potential exposure of humans to the contaminated groundwater and surface water in the immediate vicinity of tailings. However, simulated offsite radiological risks caused by groundwater pathway are estimated to be low because of the long travel time to discharge contours (hundreds-thousands of years) and due to the large dilution of contaminants in the Dnieper River system.

The preliminary recommendation is stabilization of tailings sites by means of soil covers and *in-situ* storage. Further site characterization, groundwater monitoring and risk assessment studies are opportune, first of all for the identified priority sites. The problem of groundwater contamination by toxic chemicals deserves further attention and additional assessments.

References

Bugai D.A., Skalskyy A.S., Avila R. (2008). Modeling the migration of radionuclides of the uranium decay chain from the "Dneprovskoe" tailings impoundment (Dneprodzerginsk City) into groundwater and the Dnieper River. Environmental Ecology and Safety, 2008/6, 39–45 (in Russian).

EPA (1999). Understanding variation in partition coefficient, Kd, values. Volume II. Review of geochemistry and available Kd values for cadmium, cesium, chromium, lead, plutonium, radon, strontium, thorium, tritium, and uranium. Report EPA 402-R-99-004B. U.S. Environmental Protection Agency, Washington, 1999.

Facilia AB (2007). Ecolego user guide. Stockholm: Facilia AB, 151 p. (www.facilia.se).

Guiger N., Franz T. (1996). User's manual for Visual MODFLOW. Waterloo Hydrogeologic Inc.

IAEA (2004). Safety Assessment Methodologies for Near Surface Disposal Facilities. Results of a co-ordinated research project Volume 1 Review and enhancement of safety assessment approaches and tools. International Atomic Energy Agency, Vienna.

Assessment of Failure Modes of the Ak-Tyuz Tailing Ponds in Kyrgyzstan in Preparation of Remediation Measures

Isakbek Torgoev, Alex T. Jakubick

Abstract. The results of an appraisal of the state of the Ak-Tyuz tailing ponds are summarized in the paper. The tailing site is located in the watershed of the Chu River that is prone to earthquakes, landslides and mudflows. The Chu River has the potential to carry contamination from the tailings site across the border to Kazakhstan. The aim of the paper is to propose specific remediation measures in order to stabilize the 4 tailings impoundments present at the site thus preventing a environmental impact in the watershed in case of a natural disaster.

Introduction

The Ak Tyuz mine/mill area is located at approximately 42.37° north latitude and 76.13° east longitude. The area has some of the legacy tailings sites of Kyrgyzstan that are very much in need of remediation; the general location of the Ak Tyuz sites is presented in Fig. 1a.

In the Ak Tyuz district, in the upper part of the River Kichi Kemin on the both sides of the river four tailings impoundments are located (Fig. 1b), that were generated by processing of polymetallic ores. The ores contained minerals containing thorium, cadmium, lead, molybdenum, zinc, beryllium and other elements.

The industrial area was active from 1942 until 1993. As seen in Fig. 1b there are four tailings deposits with an area of approximately 51 ha and an estimated volume of 4.17 Mio. m^3. Main properties of tailing sites are presented in Table 1.

Isakbek Torgoev
Scientific Engineering Center GEOPRIBOR, Bishkek, Kyrgyz Republic

Alex T. Jakubick
Uranium Mining and Remediation Exchange Group (UMREG), Germany

B. Merkel, M. Schipek, *The New Uranium Mining Boom*
DOI 10.1007/978-3-642-22122-4 © Springer 2011

The tailings deposits are situated in very rugged mountainous terrain in strongly sloping canyon areas often in elevated positions above the Kichi Kemin River.

During operation, the tailings impoundments were serviced with various configurations of roads, tailings delivery pipelines, return water collection, settling and conveyance systems.

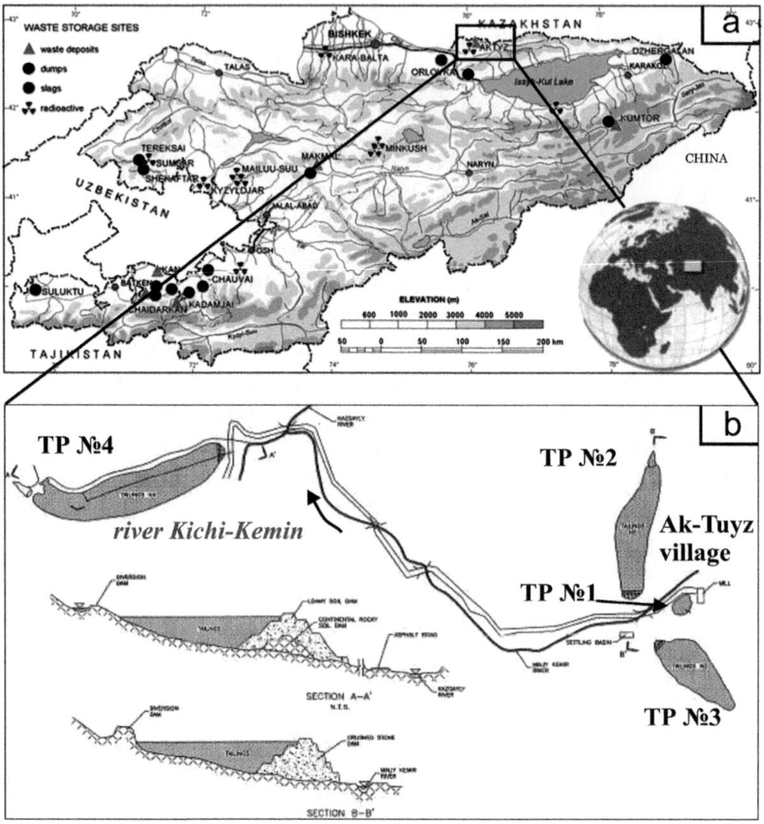

Fig. 1 Site map of U production legacy sites in Kyrgyzstan (**a**) and locations of the tailings ponds (TP) at the Ak-Tyuz site and cross sections of TP No. 2 and TP No. 4 (**b**)

Table 1 Physical characteristics of the Ak-Tyuz tailings impoundments and total alpha activity in the tailings

Tailing No.	Dam height (m)	Tailing area (ha)	Volume of material (Mio. m^3)	Total α –activity (Ci)
1	51	6	0.4	1300
2	72	10	0.5	1600
3	82	10	0.8	2100
4	59	25	2.5	6300

Tailings Impoundments and Water Diversion Structures

Accept for the tailing No. 1 all other tailings impoundments were constructed as valley-type upstream dams. The dams of tailings No. 1, 2, and 3 have been sequentially raised as centerline embankments built of sandy tailings. Generally, auto-stabilized centerline dams have an adequate performance under seismic conditions but are unsuitable for permanent water storage. Temporary flood storage is acceptable with properly functioning water diversion structures. However, they are definitely less stable than the water retention dams fully constructed prior to tailings discharge (Torgoev et al. 2005). The valley type dams are usually based directly on the alluvium to provide for dam seepage and thus controlling the phreatic water table in the tailings pond (in case of contaminated seepage provisions must be made for capture and treatment of the seepage). The failure to ensure and maintain the continuous dewatering of a valley type of tailings impoundment (caused by inadequate construction or failure of the engineered dewatering schemes) has a negative effect on the stability of the dam and can endanger the entire tailing impoundment. In addition the auto-stabilized dams have the inherent problem that they are raised while the tailings are still unconsolidated. Due to the simultaneous consolidation of the tailings and of the dam the height ratio between tailings surface and dam changes with time and the stability of the entire dam decreases.

Another construction based problem of the Ak-Tyuz tailing dams is that they were built too steep and have only a minimal protection against surface erosion (Fig. 2). Therefore, an important element of in situ remediation will be to give the dams a gentle slope and provide a protective cover to ensure a long-term stability.

There were several types of roads, transportation pipelines and hydro-technical structures (retention-ponds, dam seepage ponds and others) constructed during building and operation of the tailings ponds. This often led to undercutting of the slopes in the vicinity of the dams, thus causing small scale landslides and rock-

Fig. 2 The badly eroded tailings dam No. 2 in a side valley facing the Kichi Kemin River

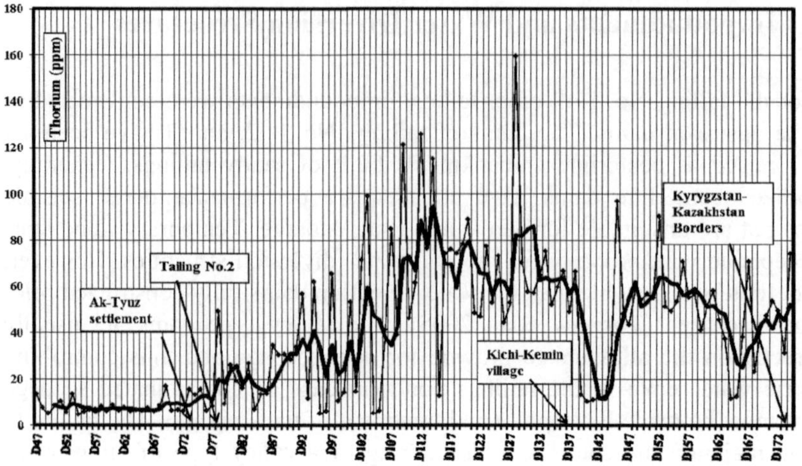

Fig. 3 Thorium concentration and distribution in bottom sediments of Kichi-Kemin river between the place of failure of the tailing dam No. 2 in 1964 and Kazakhstan border

slides threatening the stability of the tailings impoundment and/or damaging water diversion schemes (Torgoev et al. 2005).

All factors discussed, the implementation of tailings dam designs unsuitable for the site, shortcomings during construction and undercutting of slopes without geomechanical considerations make the tailing dam No. 2 vulnerable to natural disasters. In December 1964, an earthquake triggered three sequential dam failures releasing 600,000 m^3 of tailing materials (approximately 60% of the tailing volume) into Kichi-Kemin River. The tailings contained high concentrations of thorium and heavy metals. The contaminants were spread along the Kichi-Kemin River over a distance of 40 km down to the confluence of the river with the Chu River, which flows into the neighboring Kazakhstan (Fig. 3). The tailings were spread over an area of 10 ha, approximately 3600 ha of agricultural land were contaminated and the population of the settlements in the alluvial fan of the Kichi-Kemin River impacted.

Tailing Mass

The tailings were generated during processing of polymetallic and rare metal ores and successively discharged into the 4 tailings ponds. The grain-size distribution comprises 50% silt particles of 0.3–0.15 mm size; 33% particles of 0.1–0.074 mm size; 7.3% particles of 0.043–0.02 mm size; 1.4% particles of 0.01–0.005 mm size and 2.4% particles below the 0.005 mm size. The average grain size is 0.138 mm and the average density 1.65 g/cm^3.

The geochemical characteristics of the thorium and other metals in the Ak-Tyuz tailings were estimated by the Czech company GEOMIN (Abraham et al.

Table 2 Concentrations of metals, rare metals and thorium (in ppm) in the tailings impoundments and in the vicinity of the tailings areas 1, 3 and 4 (Abraham et al. 2007)

Element (ppm)	Tailing 1		Tailing 3		Tailing 4
	Concentrate	Dam	Concentrate	Surface	Surface
Ag	<2	18	<2	11	10
As	409	927	530	94	110
Be	105	70	42	13	14
Cd	7	24	16	2	3
Co	21	27	19	10	8
Cr	<2	79	20	16	6
Cu	455	2473	348	69	188
La	2254	234	1454	176	173
Mo	173	274	332	26	95
Ni	38	62	35	50	45
Pb	4637	16161	5491	846	1490
Sb	237	82	88	47	46
Sc	18	16	20	7	3
Sn	58	122	49	47	55
Te	82	<10	27	16	17
V	75	143	95	50	38
W	213	359	363	<10	<10
Y	5456	438	3826	304	257
Zn	1491	6551	3472	372	601
Zr	4787	1274	1114	1001	832
Th	70667	121	866	135	176
EDR[1]	11.40	0.65	4.10	0.80	1.02

[1] Equivalent gamma dose rate ($\mu Sv\ h^{-1}$)

2007). The results are summarized in Table 2. The spread of concentrations of the heavy metal salts is: lead (up to 5000 ppm), zinc (1500–3500 ppm), copper (350–450 ppm), molybdenum (170–330 ppm), arsenic (530–930 ppm) and beryllium (40–100 ppm). In addition, the tailing material contains high concentrations of cadmium (15–25 ppm), tungsten (210–360 ppm), yttrium (3800–5500 ppm). The concentration of rare-earth elements was also estimated. The highest concentration of rare-earth elements was found in the remnants of the tailings spilled during the dam failure of the tailing pond No. 2 in 1964 (ppm): Ce 1461, Pr 155, Nd 537, Sm 135, Eu 6, Gd 136, Tb 30, Dy 213, Ho 48, Er 144, Tm 23, Yb 132, Lu 19. The highest concentrations from all radioactive elements were associated with Thorium (800–7000 ppm) and Zircon (1100–4800 ppm).

Table 2 also presents the measurements of the equivalent gamma dose rate. The dose rate on tailings surfaces varies between 0.6 and 1.0 μSv/h, i.e. the dose rate is below the limit set for remediated tailings ponds in Kyrgyzstan. Based on the measurements a dosimetric map of the area between the processing plant and the Ak Tyuz settlement was constructed. The dosimetric map shows that high dose rates (> 2.5 μSv/h) are associated with the storage areas of rare-metal concentrates, tailings area No. 1 (11.4 μSv/h), and spots along the Kichi-Kemin River where the remnants of the tailings spill after the dam failure of 1964 can be found. Gamma-spectrometric measurements were carried out by the Czech company GEOMIN to specify the thorium (Th), uranium (U), kalium (K^{40}) concentrations and the total gamma activity in the tailings. The measurements show that the main source of ionizing radiation is thorium; the presence of uranium is rare (Abraham et al. 2007).

This indicates that the enduring thorium and heavy metal contamination of the Kichi Kemin and Chu rivers are still the aftermaths of the dam failure of the tailing area No. 2 in 1964 probably enhanced by the ongoing seepage and erosion of the tailings ponds (Fig. 3). Because the state of the tailings ponds at the Ak Tyuz remained unchanged – in fact it probably deteriorated even more – a repeated tailings containment failure at Ak-Tyuz must be assumed likely. Due to long term accumulation of the contaminants in the river sediments to possible cross boundary impact can be expected to be higher than in 1964.

Qualitative Risk Appraisal of the Ak-Tyuz Tailings Site

Because of the shortage of reliable and systematic measurements at the site, a quantitative risk assessment of the tailings management areas and a full scale Environmental/Radiological Impact Assessments (EIA) is not possible at this point. However, a preliminary expert opinion in this regard has been expressed in a pre-feasibility study of remediation carried out within the UNDP tailings safety project during 2009 and submitted to EBRD as "A Project Proposal for Remediation of the Ak Tyuz Legacy Tailings Site in the Kyrgyz Republic" (Jakubick 2009).

In addition, the health and environmental risk appraisal of the Ak Tyuz site must go beyond the radiological categories of risk assessment because factors such as cross-boundary, socio-economic impacts and possible economic benefits from re-treatment of legacy waste for recovery of valuable constituents (Au, Ag, Mo, V, rare earth element residues, etc.) must be taken into account.

Although qualitative in nature, the subsequent appraisal follows the principles of risk assessment:

$$\text{Risk} = \text{Likelihood} \times \text{Consequences}$$

where

Likelihood or Probability = [Frequency of impact] × [(Frequency of failure events) / (Lifetime of waste facility)]

Frequency of Impact = How often the release of contaminants occurs and

Consequences or Impact expressed in terms of:

- Severity (how much contamination is released);
- Extent (how many people are affected, size of territory involved etc.); and,
- Duration (of damage, exposure after release and/or cumulative effect of continuous small releases).

Based on the above risk approach, the appraisal of the Ak-Tuyz legacy tailings site is carried out by analyzing the following processes:

- Tailings dams failure and tailings release into the environment;
- Continuous release of contaminants via the water pathway;
- Dispersion of contaminated dust and radon/thoron emissions via the air pathway.

Failure of Tailings Containment and Release into the Environment

Mainly because of the questionable dam stability and neglected maintenance, the tailings dams at the Ak Tyuz site must be considered vulnerable. This is aggravated by the fact that the Ak-Tuyz site is located in a 9 balls seismic zone. Historically, the site suffered very strong earthquakes, e.g. the Kemin earthquake in 1911 had a magnitude of $M = 8.3 - 8.9$. But even the occurrence of an earthquake of $M > 6.5$ when coupled with an extreme precipitation event, such as a torrent would very likely to lead to failure of the tailings dams No. 2 and 4.

A subsequent tailings liquefaction and release into the environment in form of a tailings-mudflow is very likely to follow because the fine tailings are soaked with water; the seasonal runoff from the mountains enters the tailings impoundment easily because most of the water diversion canals are dysfunctional.

The impact of the dam failure can be assumed to have the same extent as the 1964 failure (Fig. 3). The first impact of the potential tailings-flow will consist of the contamination and physical damages to the Kichi-Kemin River valley and downstream plains. The population living along the river would be exposed to radiation and uptake of toxic elements.

The extent of the impact will reach the Chu River and stretch along the Chu River across the border to Kazakhstan. This is very likely to lead to cross border disputes between Kyrgyzstan and Kazakhstan.

With this in mind the risk of failure of the tailings dams No. 2 and 4 must be considered to be high to critical (Table 3).

Table 3 Risk appraisal of the tailings ponds located in Ak-Tyuz in their present state

RISK category	TP 1	TP 2	TP 3	TP 4
Dam failure and tailings-mudflow	High	High	Moderate	High
Tailings pond and seepage waters	High	Moderate	Moderate	High
Dust and radon/thoron emission	Low	Low	Low	High
Contamination of food chain	Moderate	Moderate	Low	Moderate
Human factor	High	Moderate	Low	Moderate
Total risk		Moderate	Low	High

Continuous Release of Contaminants via Water Pathway

Most of the tailings on the site are fine grained processing residues generated from radioactive ores containing heavy metals. The uncovered tailings ponds act as recharge areas commonly collecting so much water that the entire surface of the tailings is covered. The infiltrating water during percolation through the tailings is leaching this material and is partly released as bottom seepage. Both the tailings pond and seepage water carry the contaminants into the Kichi-Kemin River and ground water, where a large portion of the contaminants becomes bound on the sediments. The river sediments function both as accumulators of the contaminants and carriers of contamination down the water stream.

The contaminated tailings pond and seepage water serves occasionally as drinking water to the domestic animals grazing on the tailings site.

The small scale release of contaminants into the river and uptake by the sediments and soil is continuous, thus the likelihood of release is 1 (or 100%). Although the incremental effect is small, the cumulative impact is steadily growing and, without remediation, it will become more and more significant.

For this reason the risk of contaminants release via release of the tailings pond and seepage waters must be considered for tailing ponds No. 1, 4 to be high (Table 3).

Dispersion of Contaminated Dust and Radon/Thoron Emissions via the Air Pathway

The surface of tailings ponds No. 1, 2 and 4 is not covered and during the dry season and periods of strong wind the dusting of fine tailing particles has been observed. The contaminated dust particles settle in the vicinity of tailings ponds. The location of the uncovered tailings ponds No. 2 and 4 is far from inhabited areas and high concentrations of resuspended dust can only be transported by strong winds in the near-surface atmosphere. However, the annual number of days with strong winds (> 5 m/sec) at the site is limited and the local topography prevents fast air circulation in the mountain-valley. In accordance with this we assume that dusting is a low impact and low risk component of the overall risk to the general public.

Another relevant exposure pathway is the inhalation of radon and radon decay products and in the specific case of Ak Tyuz – because of the prevailing thorium content of the ore – the inhalation of thoron. The radon/thoron emission from the Ak Tyuz tailings ponds is continuous but relatively low. Considering the distance to the inhabited areas and the air circulation above the tailings – which is sufficient to disperse the radon/thoron concentration – it can be concluded that the radiological impact of dusting and radon/thoron emission can be assumed to be small.

Nonetheless, it should be mentioned that no systematic investigations of the radiological impact via the air pathway have been done at the site and this aspect may be of importance for the professional exposure of the people working on the site during remediation. A thorough monitoring of the radon/thoron emissions on the site would have to be implemented in preparation of the remediation.

For this preliminary qualitative risk appraisal we feel justified to assume that the air pathway contributes little to the overall risk of the legacy site and for now, we consider this risk to be low (Table 3).

It should be noted that the results of the present risk study compare well with the risk rating carried out by IAEA for the uranium production legacy sites in Central Asia. The IAEA identified the Ak-Tyuz tailings ponds as a high risk objects in need of remediation (IAEA 2009).

Table 3 shows that tailings pond No. 4 presents the highest risk tailings pond on the site and in greatest need of remediation. Prudent risk management requires a preferential remediation of the tailing pond (TP) No. 4 followed by TPs No. 1, 2. The TP No. 3 is not a candidate for an urgent remedial intervention at this point.

Recommended Remedial Works

The first step toward remediation of the Ak Tyuz site should be the development of an Environmental Impact Assessment (EIA) and remedial Feasibility Study (FS) in preparation of the implementation of remedial works at the Site. Preferentially, an integrated EIA/FS approach should be selected to facilitate the commencement of the implementation project. The integrated EIA/FS should dependably characterize the present status of the legacy tailings ponds at the site and quantify their health and environmental impact. Based on this, an EIA is to be prepared, which would allow the formulation of the remedial options for the legacy tailings ponds.

From a risk management point of view, an imminent objective should be to find site specific solutions for minimization of the present and potential future health and environmental impacts. This is to be achieved by developing efficient remedial concepts within a FS. From among the feasible remediation concepts the preferred option is to be selected in cooperation with the national regulatory authority, under consideration of stakeholders preferences.

This preparatory project should end with proposing the ToR for the remedial works, thus preparing the remediation works project for procurement.

The remediation of the Ak-Tyuz tailings ponds is expected to substantially contribute to improvement of the public health and environmental conditions in the Kichi Kemin valley, along the Chu River and prevent a cross-boundary impact in the Kazakhstan portion of the Chu River. By investigating the long range dispersion of the contaminants from the Ak Tyuz site the issue of the long range impact on the quality of the water, river sediments and soils in the cultivated plains in the foothills of the mountains will become quantified as well. In this respect, the EIA/FS will provide dependable answers concerning the present and future impacts for the area associated with the possible new failure of the tailings dams at the Ak Tyuz site.

Based on the EIA/FS the interim and long term measures required to decrease the risk presented by the Ak-Tyuz site will be developed. These may include in the interim a better enforcement of the site security, establishment of a regular maintenance and monitoring program, evaluation of the feasibility of re-treatment of the tailings within an overall remediation program.

The interim and remedial works may include the following types of specific measures:

A. Dewatering of tailings No. 2, 4;
B. Design and/or repair of surface water and slope runoff drainage systems;
C. Repair and/or stabilization measures for retain dams;
D. Stabilization of potential landslide areas;
E. Repair and substitution of drainage systems;
F. Erosion protection measures of slopes, embankments and waterways;
G. Restore Tailings Cover;
H. Re-vegetation of tailings slopes and surface;
I. Installation of a fence;
J. Removal of remnant of tailing material in river valley and rare metal-thorium concentrate from tailing No. 1;
K. Ecological and radiation monitoring and regular on-site inspections; and
L. Annual Maintenance and Inspection.

The optimal and preferred types of remedial works will follow from the quantitative EIA/FS study and will be defined within the Terms of Reference prepared for the remedial works.

Because of the high probability of a renewed failure of the tailings containment and expected tailings release in form of tailings mudflow into the Kichi-Kemin River valley, it is strongly recommended as an immediate measure to discontinue any further disposal of tailing into the existing Ak Tyuz tailings ponds.

References

Abraham M., Veleba B., Zacek M., Curda S., Hlupacova M., Kasparec I., *et al.* (2007) Geoecological investigation and environmental impact assessment of mining complexes in Kyrgyzstan (in Russian). Final Report of the Czech Ministry of Environmental Protection for the Republic of Kyrgyzstan, 120 p

IAEA (2009) Overview and Recommendations for Uranium Production Legacy Sites in Central Asia: An International Approach Monitoring and Regulatory Structure, Remediation, Legacy Sites Risk Assessment.2009, 134 p

Jakubick A.T. (2009) Project Proposal for Remediation of the Ak Tyuz Legacy Tailings Site in the Kyrghyz Republic. A pre-feasibility study of remediation and costing by UMREG, Uranium Mining and Remediation Exchange Group prepared for UNDP, Bishkek. Submitted to EBRD, London

Torgoev I.A., Aleshin U.G., Kovalenko D., Chervontsev P. (2005) Risk assessment of emergency situation initiation in the uranium tailings of Kyrgyzstan. Uranium in the Environment. Mining Impact and Consequences, Springer Berlin, Heidelberg, pp 563–570

The IAEA ENVIRONET Network Supporting the Remediation of Uranium Mining and Milling Sites

Horst Monken Fernandes, Philip Michael Carson

Abstract. This paper describes ENVIRONET, a new initiative of the International Atomic Energy Agency to facilitate information exchange between persons concerned with environmental remediation projects in different counties. The rationale for the development of the ENVIRONET network and the main the functions and facilities of the network are described.

Introduction

The development of mining and milling operations in the past, without proper consideration of the environmental dimension (externalities), led to the generation of legacy sites that need now to be remediated. Some countries have succeeded in implementing (large) remediation projects. But in many others the pace of environmental remediation of former mining and milling operations is very slow. This situation has been created, in part, because regulatory framework did not exist or eventually regulations were not enforced adequately.

Considering environmental remediation (ER) and decommissioning (D&D) only at the end of operations does not allow for proper development of mining and milling activities in such a way that use of natural resources is optimized, waste generation minimized and contamination of environmental media avoided. An-

---

Horst Monken Fernandes
International Atomic Energy Agency – Wagramer Strasse 5 – 1400 – Vienna – Austria

Philip Michael Carson
International Atomic Energy Agency – Wagramer Strasse 5 – 1400 – Vienna – Austria

B. Merkel, M. Schipek, *The New Uranium Mining Boom*
DOI 10.1007/978-3-642-22122-4 © Springer 2011

other problem is the lack of financial provisions to deal with both ER and D&D. By not assigning the necessary financial resources to cope with both activities and by not including ER and D&D in the overall life-cycle planning of the operation it was (and still is) quite common that the operating organization did not have the resources to implement ER and D&D activities leaving the costs of both to be faced by the government (i.e. the taxpayers).

Once a legacy site is generated the successful implementation of environmental remediation projects depends on the appropriate combination of factors from both technical and non-technical nature. In most cases, the major constraint in implementing these projects is the lack of financial resources (as mentioned above), as environmental remediation activities tend to be very expensive. However, even if financial resources are available there are a number of other factors to be considered. They include, but are not limited to: the technologies to be used, the availability of appropriate staffing and the regulatory infrastructure (including laws, regulations, and regulatory organizations). Planning is a topic of crucial importance for the successful implementation of environmental remediation projects but in many cases it is not appropriately considered by those in charge of the implementation of projects. Finally, consideration has to be given to the role to be played by different stakeholders in the decision making process. Attention must be paid to procedures for taking their views into consideration and for involving them effectively in the process. Solutions have to be found that are simultaneously technically and scientifically sound. They should also be understood and accepted by the different stakeholders.

Taking all these elements into account the International Atomic Energy Agency (IAEA) has been working to provide its Member States with relevant support in this area. The IAEA tries to fulfill this important mission by: i) publishing technical documents (IAEA 1998, 1999a, b, 2002, 2009) and safety standards (IAEA 2003), ii) organizing training courses, workshops and seminars and providing for the implementation of scientific visits and expert missions (under the activities of its Technical Cooperation Department), iii) organizing international conferences and iv) providing and supporting peer-review missions.

Despite the huge efforts implemented by the IAEA, it has been noted that there is room for improvements in the type of assistance that the IAEA provides to its Member States. Therefore, the IAEA has recently implemented different networks in the fields of Waste Management, Decommissioning and Environmental Remediation. This paper will describe the Network on Environmental Management and Remediation – ENVIRONET and explore the potential that networking has to aid in the implementation of clean-up programs.

An Insight in Central Asia Legacy Sites

A very specific example of uranium legacy sites generated in the past deals with the countries of Central Asia. They have been the main suppliers of uranium in the former USS and therefore large-scale uranium ore mining and processing in the region gave rise to the formation of a vast amount of radioactive waste (RW) in the form of mill tailings that have been deposited in piles and impoundments. These materials represent a great hazard because they are weakly protected against natural disasters; they are close to main water bodies, towns and state borders.

It should be noted that the uranium legacy sites in Central Asia are concentrated in the densely-populated Fergana Valley, particularly in the Sogdian Region of Tajikistan, in the Jalal-Abad Region of Kyrgyzstan and in the Tashkent and Navoi Regions of Uzbekistan.

These tailings are fraught with potential dangers for people and the environment. There is a threat of contamination of ground water and rivers of the water basins of some regions in Central Asia. There is a certain risk of possible accidents with dire consequences for the resident populations in these regions.

The issue of uranium tailings and toxic industrial waste in Central Asia is acute and effective solutions need to be provided. Countries in the region do not have sufficient financial and technical resources to ensure the proper maintenance and reconstruction of RW storage facilities and rehabilitate contaminated areas. Therefore the set of uranium legacy issues in the region can only be solved with significant, goal-oriented and coordinated assistance from the international community. Practical guidelines for the re-cultivation of uranium tailings, rehabilitation of contaminated areas, organization of radiological-ecological monitoring and improvements to the RW management mechanisms, are needed. Therefore, coordinated actions of potential donors and stakeholders from the private sector and the high-level processing of uranium tailings and waste dumps is needed.

The IAEA has been supporting the countries in the region by means of different projects under its Technical Cooperation Department. The ongoing project – RER 3010 “Supporting Preparation for Remediation of Uranium Production Legacy Sites” is aimed at enhancing the regulatory framework and planning process in relation to remediation of uranium production legacy sites. The IAEA has also been engaged with other UN organizations e.g. UNDP, OSCE, WB in producing a Framework Document (FD) aimed at providing an analysis of the state of RW storage facilities in the regions countries and their impact on people's daily lives and the environment.

Uranium Mining Remediation Projects in Other Countries

There are also other IAEA MS’s faced with the challenge of implementing remediation projects to deal with the uranium mining legacy sites. One of them is Brazil that is being assisted by the Agency under the project BRA 3013 “Providing Practical Guidance for the Implementation of a Decommissioning and Remediation Plan for the Minas Gerais Uranium Mining and Milling Production Centre”.

The main objective of this project is to provide Brazil with a strategy to deal with the decommissioning and environmental remediation of the uranium production center of Pocos de Caldas. The Project RLA 3006 Regional Upgrading of Uranium Exploration, Exploitation and Yellowcake Production Techniques taking Environmental Problems into Account also deals with remediation aspects as it is aimed at dealing with the Self-supplying of uranium for the nuclear plan in execution the region in an environmentally sustainable way. In Africa, the uranium mining is dealt with by means of the project RAF 3007 Strengthening Regional Capabilities for Uranium Mining, Milling and Regulation of Related Activities. The project is centered at assisting African Member States in strengthening their capabilities for an effective and efficient management of uranium resources and other radioactive ores, and to build the legislative framework to effectively regulate related activities.

The Life-Cycle Approach

Uranium mining and milling/recovery is recognized as one of the most environmental burdening processes in the nuclear fuel cycle (NFC). Therefore the IAEA is fostering the adoption of the life-cycle thinking in a way to reduce the needs of extensive remediation programs after closure at the same time that the generation of new legacy sites is avoided. The concept focuses on an up-front/early consideration of environmental impacts and associated externalities. This reduces life cycle costs of the uranium mining and milling/recovery operation thus increasing sustainability of the front-end of the NFC as well as of the NFC at large.

The three important contributors to an integrated life cycle management process, are: remediation and closure planning, remediation and closure costing and stakeholder interaction, in each of the major life cycle stages of a (uranium) mining and milling/recovery operation in a coherent way. In this approach, introducing life cycle thinking into the whole operation, remediation planning and associated costs will be addressed already pro-actively right from the start of the project, taking into account also all of the following life cycle stages of the envisaged operation including the post-closure phase. This is in contrast to the reactive planning approaches used often in the past, when remediation planning started only in the decommissioning and closure phase.

The concrete remediation and closure measures to be planned and costed and the environmental issues to be addressed are rather site specific and depend on the type of ore deposit exploited and the applied mining and milling/recovery methods. Factors to be taken into account comprise e.g. the amount of tailings and waste rock generated and their acid generation potential, hydrological, hydrogeological and climatic conditions, natural habitats, ecosystem vulnerability etc. The approach is especially recommended for the extraction of low grade uranium deposits, which may become increasingly important for the front-end of the NFC in the future after complete exploitation of known high and medium grade deposits, due to the generally increased environmental burdens associated with their exploitation (e.g. due to potentially higher specific amounts of tailings/waste rocks, energy consumption per unit product, contamination of groundwater resources in ISL operations, etc.).

Stakeholder interaction has to be understood not as simple information policy (public relation) but as an active process of interrelating with the various stakeholders involved (e.g. regulator bodies, local communities, NGOs etc.) and include them in the development and evaluation of all project stages (e.g. in community committees, public hearings, etc.). Benefits from this approach may comprise reduction of averseness to the project, e.g. averseness due to cultural reasons or differing public expectations regarding the future use and of the mining site after decommissioning, but also benefits in the planning and development stages by inclusion of indigenous traditional knowledge of the territory into the planning process. In relation to this aspect, it is also suggested to the project owner to implement and promote public engagement during the permitting process, even if not required by the respective regulator, to enhance public acceptance of the future operation. Stakeholder mapping and analyses should be reviewed and updated periodically throughout the life cycle stages to address potential changes in stakeholder composition/attitudes/expectations (e.g. due to increase in population and workforce from outside during construction/operation), taking into account that – even disregarding the post-closure phase – the life cycle stages from exploration to decommissioning of a mining operation usually encompass several decades.

The ENVIRONET

The ENVIRONET, launched in 2009, is an information network dealing with the remediation not only of legacy sites (existing contaminated sites) but also with operating sites (by taking remediation under a life-cycle perspective). The ENVIRONET is intended to aid Member States in solving the problems created in the past by improving the efficiency of information exchange and the transfer of knowledge and assistance. Topics covered by the ENVIRONET include: i) Life-cycle planning of facility operations and environmental remediation projects; ii) Project planning (quality control and assurance); iii) Data management, iv) Integration and communication; v) Site characterization; vi) Modeling (fate and transport of pollutants, engineering design and economics); vii) Risk assessment; viii) Remediation technology development and selection; ix) Monitoring; x) Stakeholder involvement and communication; xi) Regulation and policy development; xii) Risk communication; xii) Stewardship – Institutional Control and xiii) Funding.

The intended methods of work include a variety of products and services aimed at expediting the exchange of information and experiences. The ultimate goal is to build capacity in Member States and to facilitate the full implementation of remediation projects. These products and services include: i) Website (Document repository of educational materials); ii) Discussion Forum; iii) Partners Directory (online profiles); iv) Schedule of Events; v) workshops, vi) conferences, vii) training sessions, viii) long-distance training, ix) fellowships/internships and x) peer reviews. The products will be: proceedings; dedicated publications; training materials, case studies, annual reports and updates.

The vision underlying the implementation of networks is that the involvement of interested groups through the network on a specific topic will promote an effective flow of information and experience sharing. ENVIRONET is also aimed at transferring knowledge while providing for the education of those in need. In principle, information will flow from more experienced participants to those with less experience. However, in many situations the advantages of integrating a network is mutual, as a lot of positive and valuable information can flow in different directions. With the aid of the new tools provided by the Internet, virtual forums can be established, allowing for an intensive circulation of information amongst the participants of the network. Modern web-base resources will also allow the operation of educational tools, such as video materials, to be posted on the web. Finally, networking will allow the IAEA to capture the needs of a greater number of technical people in each Member State leading to an even more focused construction of its activities.

ENVIRONET Structure

To implement the mode of operation of the network a Steering Committee was created. The Steering Committee develops the "Statute" or Guidelines of the network along with the Strategic and Operation Plan. Elements like Outline Program, Success Indicators, Evaluation Process, etc. are also to be elaborated.

The ENVIRONET is composed by organizations and individuals that are involved or interested in environmental remediation activities. Other networks can also integrate the ENVIRONET community. Sponsors are taken as those multilateral organizations that can technically and/or economically support the activities of the ENVIRONET. Regional and Technical Working Groups are being formed. The first one is to allow the network to focus and capture specific needs of a geographical region in the world and consequently structure a set of activities that can be relevant to meet these specific needs. The Working Groups will allow the network to capture specific themes of general interest and elaborate activities and/or publications on selected topics. The arrangement of the ENVIRONET is shown in Fig. 1.

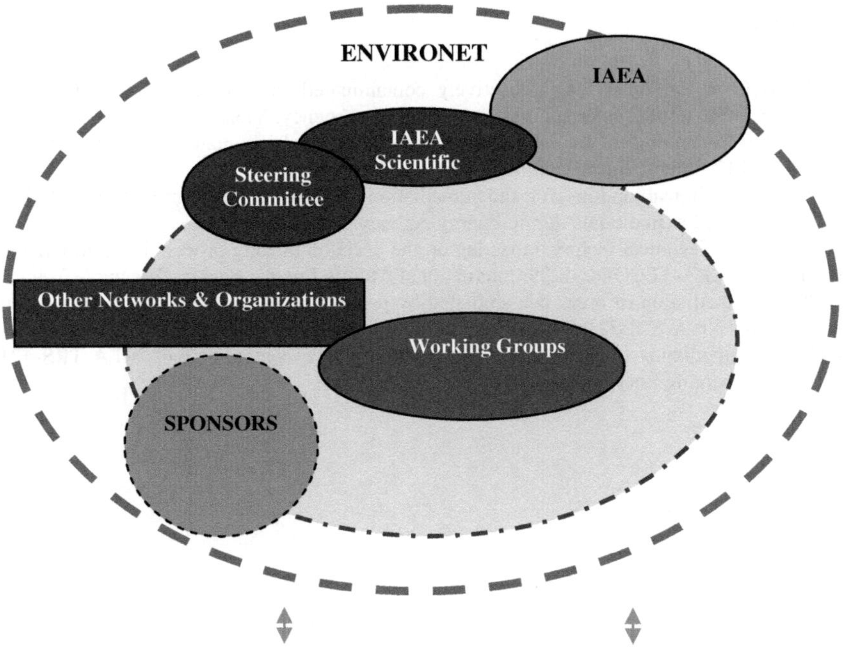

Fig. 1 Structure of the ENVIRONET

Conclusions

Some of the past uranium production operations have caused extensive environmental problems. The lack of appropriate regulatory framework in addition to the fact that environmental issues were not conveniently addressed in those operations contributed to this situation. Nowadays, this situation has changed dramatically and lessons learned from the past have led to the implementation of responsible operations from both environmental and social perspectives. Involvement of different stakeholders in the decision making process turned out to be a mandatory issue in many countries. With the so called “Renascence of Nuclear Power” new production sites will come into play. The sustainability of the uranium industry will depend on the adoption of good practices in these operations under a life-cycle perspective. The recently launched IAEA initiative – the ENVIRONET is aimed at contributing to expedite the transfer of experience amongst its members. It brings together private and state-owned companies, research institutes, and governmental organizations providing a forum for information and experience exchange. Sharing of practical experience is to be addressed by means of training courses and workshops. In addition to this long distance training and educational material will be made available.

References

IAEA (1998) Characterization of radioactively contaminated sites for remediation purposes. IAEA-TECDOC – 1086. International Atomic Energy Agency. Vienna.

IAEA (1999a) Technologies for remediation of radioactively contaminated sites, IAEA-TECDOC 1017. International Atomic Energy Agency. Vienna.

IAEA (1999b) Technical options for the remediation of contaminated groundwater. IAEA-TECDOC-1088. International Atomic Energy Agency. Vienna.

IAEA (2002) Non-Technical factors impacting on the decision making process in environmental remediation. IAEA-TECODC-1279. International Atomic Energy Agency. Vienna.

IAEA (2003) Remediation of areas contaminated by past activities and accidents. IAEA Safety Requirements No. WS-R-3. International Atomic Energy Agency. Vienna.

IAEA (2004) Remediation of sites with dispersed radioactive contamination. IAEA TRS-424. International Atomic Energy Agency. Vienna.

Environmental Management Systems in Uranium Exploration at AREVA Resources Canada Inc. (ARC)

Peter Wollenberg

Abstract. AREVA Resources Canada Inc. (ARC) has successfully implemented an Environmental Management System for all aspects of its uranium exploration in Canada. The company is certified for these activities according to the ISO 14001:2004 International Standard of Environmental Management. Years of operational experience have demonstrated the value of the process for the various aspects of the activities and the system remains dynamic adapting to changes in business, corporate and regulatory requirements.

Introduction

AREVA Resources Canada Inc. (ARC) is a Canadian company. Its head office is located in Saskatoon, Saskatchewan, Canada. ARC is a subsidiary of AREVA, a world leader in CO_2 free energy production and active in over 100 countries worldwide.

ARC is the operator and majority owner of the McClean Lake uranium operation located some 700 km north of Saskatoon. The company also owns the Cluff Lake uranium mine which has been decommissioned and operates the Midwest uranium project. It is a shareholder in the Cigar Lake, McArthur River and Key Lake uranium mines and all mine sites are located in Northern Saskatchewan. ARC is also active in uranium exploration at various other locations within Canada.

ARC has adopted an integrated approach to environmental management of its operational activities. This comprises a complete cycle which begins with uranium

Peter Wollenberg
AREVA Resources Canada Inc., Box 9204, 817 – 45th Street West, Saskatoon, SK, S7 K 3X5, Canada

B. Merkel, M. Schipek, *The New Uranium Mining Boom*
DOI 10.1007/978-3-642-22122-4 © Springer 2011

exploration projects, continues with the McClean Lake site, a uranium mining and milling facility which began production in 1999 and is completed with the now decommissioned Cluff Lake uranium mine which ceased production in 2002.

In the late 1990s ARC commenced the development of a comprehensive Quality Management System for its operational sites. This initiative was a response to the business and operational requirements for the new McClean Lake uranium mine and mill and the regulatory requirements of the Canadian Nuclear Safety Commission (CNSC). This provided a starting point to meet regulatory, corporate and public expectations. Stringent environmental management processes were a core component of the system from the beginning.

ARC is a leader in Canada in the environmental management of uranium exploration and mining sites. It has achieved multiple certifications for environmental management following the ISO 14001:2004 International Standard. This includes the operation of its McClean site (first for a uranium mine in Canada), the decommissioning of the Cluff Lake mine site (first time in Canada), the uranium exploration activities in Saskatchewan (first time for this province) and the uranium exploration and development activities in Nunavut (first time for this territory). In addition the McClean site is also certified for Occupational Health and Safety, OHSAS 18001 (first time for uranium mine). Efforts are underway to certify the exploration activities for OHSAS 18001 within the very near future.

How Was It Developed?

After considerable evaluation of the range of available management systems it became evident that the specific requirements of the various operations required an Integrated Quality Management System. Three specific objectives guided the development:

- All business activities were to be conducted in a safe and efficient manner in order to meet all regulatory and corporate requirements.
- The ISO 14001 standard for environmental management must be employed.
- Sustainable Development was to be implemented throughout the organization.

The organization of the system along with the creation of all related documentation and necessary training of personnel was tasked to all levels of management; from the President and CEO to the line supervisors.

How Does It Function?

ARC's values and those of the entire AREVA group are derived from the concept of sustainable development. To this end AREVA strives to maintain profitable growth, while simultaneously fulfilling its social responsibility and protecting the environment.

ARC has established a number of policies of which the Environmental Policy recognizes that continued economic and social development depends upon a healthy environment. Environmental considerations are incorporated into company activities to ensure sustainable development. Effects of the exploration and mining activities on the environment are mitigated by continuous efforts to apply improved technology and methods to the process.

Extensive documentation has been developed both to specify and record the various management processes. A corporate manual is the top level document and provides the overview and the description of policies, responsibilities and requirements applicable to the various operational and supporting units. It prescribes the guidelines for the activities of the corporate senior management, sets requirements and provides guidance to the rest of the organization. ARC's corporate manual is approved by the President and CEO of ARC. A manager is responsible for the corporate manual and is supported by a committee of senior managers.

Tier I documents consist of manuals which address the company's core operational and support activities. Manuals are in place in all major departments and large projects. The responsibility for these manuals rests with the heads of the individual departments and projects.

Tier II and III level documents consist of procedures, work instructions and forms. They describe how the work is carried out, what information is being recorded and where it is filed for easy access. Responsibility for these documents rests with senior line managers who have assigned responsibilities to dedicated personnel who undertake the daily and ongoing administration of the documents.

Monitoring of the management system is provided by corporate senior management that meets regularly to administer and assess the status of the system. Annual self-assessments are provided in conjunction with annual objectives and an ongoing progress review. Reviews and audits of the management system are conducted on a regular basis, both by internal and external parties. The ARC system is audited in its entirety once every three years.

Operational Experiences and Examples

The Environmental Management System has been in place for years. The initial resistance of employees to change dissipated quickly as staff recognized the benefits of taking more and more ownership of the system particularly in the operations or projects where the system was implemented from the beginning. Procedures and work instructions are written primarily by supervisors and operators who draw upon their long time personal experience and front line knowledge. Through this they gain considerable appreciation and acceptance of the process. The system is introduced to new employees through dedicated training and information sessions. Standardized procedures, work instructions and forms are used throughout the organization. This has provided considerable assistance in the integration of personnel when transferred between operations or working sites. Internal auditing and review processes are standardized throughout the organization assuring consistent assessment all the way through.

The rigorous application of the prescribed environmental management processes in the company's workings is reflected throughout a typical day in an exploration project. Such projects are located in geographically remote areas where specific heightened sensitivity and awareness is required to ensure that no permanent, irreversible negative impact occurs to the natural environment.

In Saskatchewan and in Nunavut, for example, geographical areas of interest for the exploration group are originally scanned not only for their geological merits and mineral potential but also reviewed against a database of rare and endangered species as well as archeological sensitive areas, supplied by the regulatory agencies. This information is important in the planning of the exploration work to avoid any potential interference.

Drilling activities, an essential part of exploration programs, have the potential to harm the environment if not managed properly. It is in this area particular effort is made to ensure sustainability. Extensive documentation within the Environmental Management System provides direction in setting out how to proceed. This commences with the careful selection of the actual drill collar locations, which must be situated a minimum distance from any body of water. Whenever possible drilling fluids are circulated within a closed system and only biodegradable additives are used. Berms are constructed around the drill machine to prevent accidental release of drilling fluids into the environment. Depending on local conditions silt barriers are employed in nearby water bodies to ensure that in the unlikely event of the escape of drilling muds from the drill site, potential contamination is properly contained and easily managed. Water necessary for the cooling of the drill bit is taken from nearby bodies of water using pumps with screened in take hoses 30 cm above the bottom to prevent the intake of any aquatic life. In areas where drill cuttings return to surface drilling muds and drill cuttings are managed in such a way that radioactive material is separated from non mineralized material by using a cyclone separator. This is done directly at the drill site. The radioactive material is packed in large textile bags and stored within a fenced compound and is accessible to authorized personnel only.

Particularly significant improvements in the environmental performance for drilling activities have been achieved through the application of the directional drilling technique. It allows the drilling of up to 18 different targets from one drill pad. This method is a direct result of the continuous improvement process required by the ISO 14001 environmental management standard. In this technique one pilot hole is drilled and it is from this single hole that multiple targets can be accessed by using directional motors within the drill string to direct the drill bit. This method requires the clearing of only one fairly small drill pad and one access trail is necessary for the large number of drill holes. The reduction in environmental impact such as land clearing is very significant as some 35 hectares of forest did not need to be cleared for drill pads or access roads over the last years in one project alone. Equally important are the savings in water used for the drilling, reduced from some 300,000 l/day to 40,000 l/day. Very significant savings are also recorded for the diesel fuel used in the drill rig as the pilot hole is drilled only once per set up. In one exploration project savings of over 100,000 m of drilling have been realized over the last years resulting in the saving of over 2 million liters of diesel fuel. Stringent procedures are followed at the end of the drilling campaign when the sites are rehabilitated. These efforts are made to rehabilitate the sites as closely as possible to their pre-drilling state. Reclamation may include the spreading of slash or straw and other ground cover across the drill pads to prevent erosion and to enhance the re-growth of the natural vegetation.

Proper management of fuel such as gasoline, diesel and Jet B is extremely important and has been identified as a central part of the daily field operations. Fuel is transported to the sites in bulk quantity inside double walled tanks. It is stored in specifically designed environmental tanks which are monitored on a daily basis for the volume of the fuel and the technical conditions of the containments themselves. Transfer of fuel from a larger to a smaller containments e.g. into 45 gallon drums for aircraft takes place within a bermed area and the drums are stored inside a special base designed to collect any potential leakage. Waste oil must be contained for eventual removal from site and disposed of in accordance with applicable regulations. Spill removal kits are supplied at each location where fuel is handled and all field personnel received training in the use of the kits. Heavy equipment must be parked over absorbent matting to catch any potential oil or hydraulic fluid leakages. In the main exploration camp in Nunavut the typical oil based heating system for each individual building has been replaced by electric heat derived from one central generator, fed from a double walled bermed diesel tank. This has reduced considerably the potential for fuel spills.

Waste management is another important focus in the environmental protection processes. Reusing materials is strongly encouraged and recycling items such as batteries, printing cartridges and beverage containers is mandatory. Wherever possible waste is incinerated at remote sites and eventually taken to landfill sites for proper disposal.

The protection of biodiversity in naturally sensitive areas, such as the tundra or the northern boreal forests, is achieved in many ways. During winter exploration programs creeks are crossed by using temporary mobile bridges or seasonal ice bridges. In the Nunavut project all land transportation to the exploration areas is allowed only in the winter months when the tundra and water bodies are frozen. In the summer months all transportation is by air only, usually by helicopter. Wooden walkways between the different buildings and work sites in the main exploration camp ensure that the fragile tundra is not harmed during the summer months when a thin zone of surface thaw exists.

Exploration occurs in areas where wildlife exists and stringent measures are established to minimize interaction between people and animals. To this effect measures have been taken to prevent approaching, feeding, harassment or hunting of animals. In the Nunavut tundra the company adheres to its caribou protection commitment by working closely with local caribou management boards. Traditional knowledge of the First Nations people is employed in the environmental protection and monitoring of caribou and other wildlife. Exploration activities such as drilling, geophysical surveys or flights of aircraft are reduced or temporarily halted altogether should larger groups of caribou be present within a 1 km radius of the activity. Operations will resume only once the animals have left the area.

Further Developments

The Environmental Management System is continuously evolving and adjusting to changes in regulatory, public, business and corporate requirements. Future improvements will include more refined risk assessment approaches and heightened hazard analysis. The extended promotion of the system to contractors provides additional improvement. Awareness and training sessions for staff are also ongoing using state of the art media techniques. More and more of the administration and documentation work is being transferred from the traditional paper trail to fully electronic media and this will continue.

Summary

The Environmental Management System at AREVA Resources Canada Inc., which is part of a much larger Integrated Management Process, is now firmly established and is a central part of the operations of the company. The efforts of the company and its leadership in environmental management in uranium exploration and mining and milling are widely recognized by employees, peers and regulatory bodies. Some of the processes and requirements, originally designed for internal company use have now been adopted by some provincial and territorial regulatory agencies as the base for their own applications. ARC's leadership and initiative in uranium exploration in Canada has set the standard for others to follow, enticing competitors to establish and implement their own Environmental Management Systems demonstrating clearly their awareness and responsibility to the public.

The general environmental performance of ARC has consistently improved and strengthened since the system has been in place. It has promoted industry-governmental relations and has facilitated the attainment of work permits and licenses from regulatory bodies. The company has demonstrated reasonable care and has been able to maintain and improve the trust of the general public as well as foster relations with stakeholders and communities impacted by the operations.

The Environmental Management System along with its parent Integrated Management Process adapts constantly to changes in business, corporate and regulatory requirements.

Part 2
Phosphate Mining and Uranium Recovery

Developing an "Alternative" Mining Method

Franz-Werner Gerressen, Heiko Kopfmüller

Abstract. For its Mclean Lake project, challenging due to its geotechnical nature and the applying environmental and health related restrictions, AREVA Resources Canada Ltd. addressed BAUER Maschinen GmbH for the development of an alternative mining method for small, pocket-shaped, high-grade uranium deposits, embedded in sandstone end of 2007. In close cooperation with AREVA Resources Canada Ltd., BAUER Maschinen GmbH designed and manufactured a system to mine the uranium ore, the so called High Pressure Reverse Circulation (HPRC) system.

Introduction

Even in the days of the Japanese tragedy, the increasing global demand for energy can realistically only be satisfied with nuclear energy production and the public focus will force the nuclear industry to look on more environmental friendly and economic methods for mining the uranium deposits.

Canada had a 25% share of the 2007 world uranium production and the richest deposits can be found in the northern part of the Saskatchewan province.

In 2008 the French energy company AREVA, represented by its Canadian subsidiary AREVA RESOURCES CANADA Ltd. and BAUER Maschinen GmbH joined forces for the development of an alternative mining method to exploit these high-grade, geotechnical challenging deposits.

Franz-Werner Gerressen
BAUER Maschinen GmbH

Heiko Kopfmüller
BAUER Maschinen GmbH

B. Merkel, M. Schipek, *The New Uranium Mining Boom*
DOI 10.1007/978-3-642-22122-4

Saskatchewan Uranium Deposits

The Athabasca Basin is host to unconformity-associated type Uranium deposits. Mineralization occurs at, above or below the unconformity which separates the Proterozoic Athabasca Sandstone Group from the underlying metamorphosed gneissic rock. Sandstone and basement hosted mineralization is usually associated with graphitic faults, pathfinder alteration minerals and brittle reactivation of older ductile fault systems. Deposits range in shape from massive pitchblende lens-shaped deposits with uranium contents of 20% (and even "rich ore zones" with 50%) to discrete zones of pitchblende veins. Dimensions of the typical lens-shaped ore body are 20 × 50 m and 10–15 m thickness, in a depth of 150–180 m.

AREVA Resources Canada Ltd., a subsidiary of the AREVA group, is one of the world's leading uranium exploration, mining and milling companies, partner and shareholder in several Joint Ventures in the Athabasca basin. It operates the newest, most technologically advanced uranium mill at McClean Lake, producing "Yellow Cake" from its own three local open pit operations as well as for other companies.

High Pressure Mining

With the need for an economic feasibly mining method for high-grade, small sized uranium deposits with respect to the strict legal restraints regarding environmental impact, health and safety requirements, AREVA addressed BAUER Maschinen GmbH for the development in late 2007.

Fig. 1 Inspecting a mined Kaolin cavern

Three years before, BAUER Maschinen GmbH already gained experience in "wet-mining", developing a high pressure mining method and technology for GEBRÜDER DORFNER GmbH, a German Kaolin miner (Fig. 1), to economically mine Kaolin deposits, avoiding huge open pits by modifying the existing and well developed Jet Grouting foundation technology into a mining application.

RC Sampling

For a diamond exploration project in Saskatchewan Canada, BAUER Maschinen GmbH 2005 designed and manufactured a sampling system on basis of its BG36 drill rig, capable of drilling 1.2 m diameter to a depth of 360 m, the BG36 RC (Fig. 2). With an RC – reverse circulation – ore recovery and automated screening and bagging system, two units drilled for three years in an all-year, 24/7 operation, recovering more than 75,000 t of kimberlite ore.

Fig. 2 BG36 RC drilling in Saskatchewan

New Mining System

Technical discussions between AREVA and BAUER started 2008. The input of AREVA's records of its own field trials and BAUER's engineering expertise, soon lead to a solution. With respect to the specific requirements of the deposits the method had already worked well, but the technical solution needed improvement. Combining the already known and above mentioned methods, BAUER created the HPRC system: **H**igh **P**ressure water jet and **RC** ore recovery system.

The HPRC system comprises

- HPRC power swivel w/flushing heads and crowd system;
- HPRC operation control unit in 20' container, incl. HP pump remote control and HPRC hydraulics;

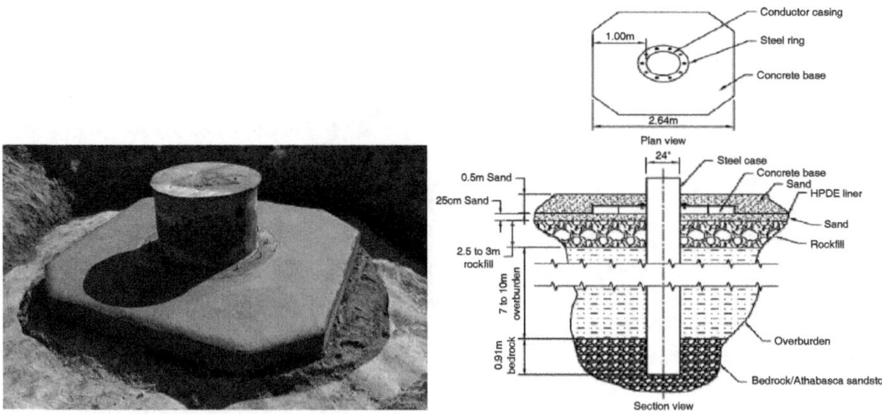

Fig. 3 Well head installation

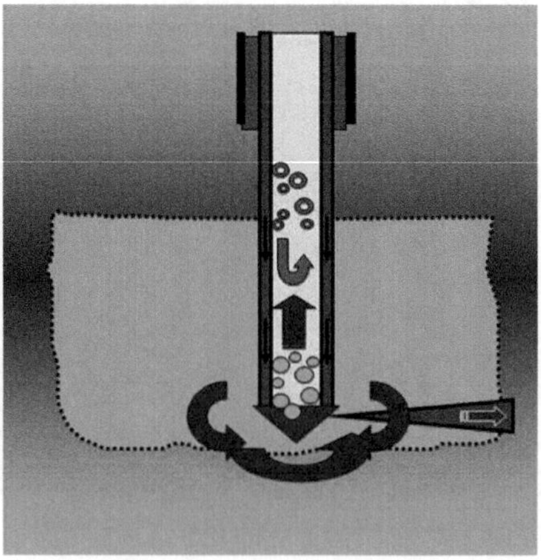

Fig. 4 HPRC principle

- 36 HPRC multi-channel mining rods;
- HPRC mining heads (4 different configurations); and
- HPRC air inlet assembly.

The mining procedure is based on a two-step sequence. For the 2009 field tests four holes have been prepared. The first step – the preparation step – was to drill and fully case from surface to ore body. Sealed with a wellhead installation on top and cementation of the casing (Fig. 3), a possible contamination of the environment and the aquifers was prevented.

For the second step – the actual mining – the BAUER HPRC system had to be adapted to a 3rd party drill rig and inserted in the predrilled boreholes. The mining process can also be divided in two, but simultaneously executed steps.

The jetting process, where the up to 600 bar HP water jet cuts the uranium ore in a 360+ oscillating motion, and the discharge process, where the produced ore is transported to surface by the RC subsystem (Fig. 4) and directly discharged into a screening and dewatering system, ready for collection and further processing.

Technical Features

The HPRC power swivel (Fig. 5) is a modified BAUER KDK 145, equipped with a clamping head and tiltable for easy and safe rod handling and incorporates also an independent feed system for high-accuracy vertical positioning. With respect to the eccentric, multi-channel configuration of the HPRC system, the power swivel was designed to oscillate with adjustable rotation speed 360°+.

The HPRC operation control unit is placed in a climate controlled 20' container (Fig. 6) that also hosts the HP pump remote control center and the HPRC hydraulic and electronic control compartment. All necessary hydraulic and electric power is routed from the base machine to the HPRC system via the OP container. Big windows give excellent view to the mining rig and the attached HPRC system.

Fig. 5 HPRC power swivel

Fig. 6 Mining Control

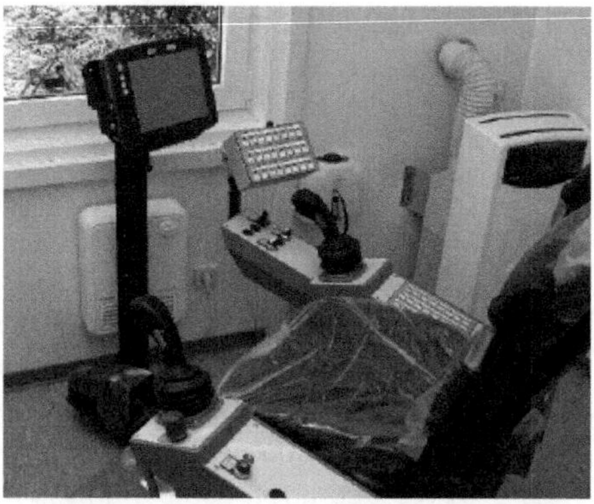

Fig. 7 Operator seat

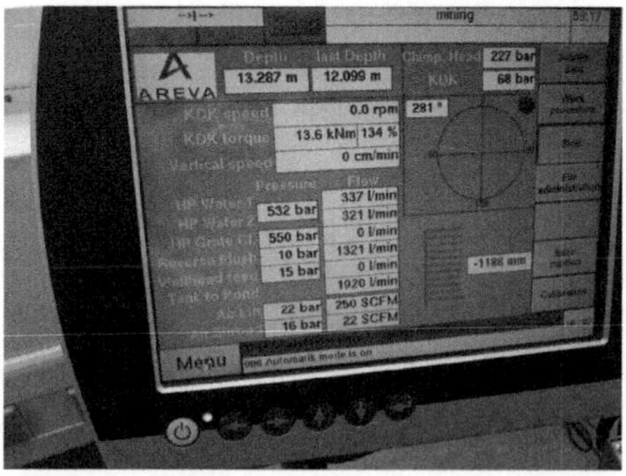

Fig. 8 BAUER B-Tronic Mining Control System

The operator controls the HPRC system incl. basic rig functions from a central operator's seat (Fig. 7), equipped with a modified for purpose BAUER B-Tronic system (Fig. 8), that displays, monitors and records all system and operational parameter and data.

The design of the mining rods and all down hole tools was quite some challenge regarding the selection of the connectors. A solution for the problem, the

Fig. 9 BAUER connector

multi channel, eccentric design of the system in combination with the safety related necessity for 100% leak proof connection of the HP lines – operated with 600 bar – soon was discovered to be already in BAUER´s equipment shelf. With some modification, the connectors (Fig. 9) provided the necessary perfect alignment for pressure-tightness of the mining rod channels. Quality control after manufacturing included pressure testing (e.g. 1000 bar for the HP lines).

Future Design

The 2009 field trials at McClean Lake showed superior performance to the former AREVA system and was followed by a second design stage for system improvements for 2010. The improvement design included, beside others, adaption of a mud motor, a modified mining head, software modification and was validated in a successive test program.

Currently AREVA and BAUER Maschinen are discussing the third design stage to take the step from a test system to a production system. New features will include GPS integration, enhanced visualization, processing of digital mining plans and more. The updated system will be tested in 2012. The final HPRC system is planned to start production 2013 with parallel mining units.

Assessment of Radiological Impact on the Environment During Recovery of Uranium from Phosphate Rocks and Phosphoric Acid

A.H. Khan

Abstract. Increasing demand for nuclear power has necessitated a fresh look at non-conventional sources of uranium such as phosphate rocks and phosphoric acid. Phosphate rocks are known to contain uranium around 50 to 200 parts per million (ppm); sometimes even up to about 800 ppm. About 150 million t of rock phosphates containing about 120 ppm of uranium amounting to a total of about 18,000 t of uranium are mined and processed around the world annually. The world potential for recovery of uranium from weak phosphoric acid is reported to be around 7000 t per annum. During production of phosphoric acid, while uranium goes with the acid, ^{226}Ra and other environmentally important radionuclides emerge with the phosphogypsum. As all the decay products of uranium are present in the phosphate rocks, processing of large quantities of phosphate rocks has a potential for contamination of the surroundings with the enhanced natural radioactivity. An assessment of the potential exposure of workers to radiation and the impact on the environment from the radionuclides present in the rock phosphates and the product materials is considered important.

Introduction

Phosphate rocks used for manufacture of phosphoric acid and certain fertilizers mined on a large scale in many parts of the world are known to contain uranium in the range of 50–200 parts per million (ppm), sometimes as high as 600–800 ppm.

A.H. Khan
Raja Ramanna Fellow, Environmental Assessment Division, Bhabha Atomic Research centre, Trombay, Mumbai – 400085, India
E-mail: ahkhan1214@gmail.com, ahkhan@barc.gov.in

B. Merkel, M. Schipek, *The New Uranium Mining Boom*
DOI 10.1007/978-3-642-22122-4 © Springer 2011

Uranium inventories in the phosphate deposits of major phosphate mining countries are estimated to be about 9 million t (wise-uranium.org 2010). Hence, rock phosphates and phosphoric acid produced from it are considered an important secondary source of uranium. Recovery of uranium from the phosphoric acid will help in reducing the environmental burden of radioactivity which would otherwise find its way to the environment along with the fertilizers. As the specific activity of the rock phosphate is generally low, its impact on the environment is also expected to be only marginal. But the large quantity of phosphogypsum produced needs to be disposed safely and may require an assessment of its potential impact on the environment. Radiological impact on the environment during recovery of uranium from phosphates and phosphoric acid can be assessed in the light of IAEA TECDOC (IAEA 1997), though this document is basically meant for the mining and processing of uranium ores. It requires the evaluation of baseline levels of uranium, radium and other long-lived radionuclides of the uranium decay chain in soil, water and air as well as their concentrations in the flora and fauna in the vicinity of the proposed facility. Since most of the radionuclides other than uranium go with the phosphogypsum, it is important to consider the end use of phosphogypsum; whether it is used for land fill, soil modification or in construction material, after meeting the requirements of exemption. During recovery of uranium from the weak phosphoric acid there is low potential for radioactivity reaching the environment. However, environmental monitoring is desirable to evaluate the realistic radiological impact, to ensure compliance with regulatory requirements and serve as a confidence building measure among the stake holders. This paper intends to summarize the environmental impact assessment and the applicable international regulatory norms associated with the recovery of uranium from phosphate rocks and phosphoric acid.

Radioactivity in the Phosphate Rocks

A survey of literature gives varying quantities of uranium and radium content of the rock phosphates produced in different countries. From a study of over 300 samples Menzel (1968) has given median contents of uranium, thorium and radium in phosphate rocks obtained from different countries as 59 ppm (1475 Bq/kg), 8 ppm (64 Bq/g) and 18 ng/kg (666 Bq/kg), respectively. Many other authors have reported similar values. Table 1 gives a summary of uranium and ^{226}Ra present in some phosphate rocks of different origins (Cioroianu et al. 1998).

As reported by Gupta and Singh (2003), phosphate rock deposits of Mussorie in northern India have 25–840 ppm U (avg: 408 ppm $\approx$ 10.2 kBq/kg) and those in Karnataka in southern India 290–840 ppm U ($\approx$ 7.2–21.0 kBq/kg). Phosphate rock production in India is around 1.2 million t per annum (wise 2009).

Table 1 Uranium and radium and thorium content of some phosphate rocks

Origin of phosphate rocks	U content ($Bq{\cdot}kg^{-1}$)	^{226}Ra ($Bq{\cdot}kg^{-1}$)
Florida	2000–3750	750–1500
Carolina	2000–3000	600–1000
Morocco	2500–4000	800–1600
Tunis	750–1250	250–350
Algeria	2500–3000	700–850
Israel	2000–3500	800–1200
Jordan	2000–2750	800–900
Togo	2500–2750	950–1000
Senegal	2500–3000	950–1100
Kola (Volcanic)	20	70
	(1 mg = 25 Bq)	

Uranium Recovery from Phosphate Mineral

The finely ground phosphate rock is treated with sulfuric acid at 70–80°C to manufacture phosphoric acid according to the following chemical reaction,

$$Ca_{10}\left(PO_4\right)_6 F_2 + 10H_2SO_4 + 20H_2O \rightarrow 6H_3PO_4 + 10CaSO_4 \cdot 2H_2O + 2HF\,.$$

Nearly 5 t of phosphogypsum is generated per ton of the phosphoric acid produced. Most of the uranium and nearly 10% of the radium present in the mineral rock goes with the phosphoric acid. Nearly 90% of the radium-226 and a small fraction (≈ 10%) of uranium originally present are retained in the phosphogypsum. The phosphoric acid so produced is mostly used for the production of fertilizers and the radioactivity may find its way to the environment. However, the uranium present in the phosphoric acid can be recovered by solvent extraction process described in detail by different authors (Gupta and Singh 2003; Cioroianu et al. 1998). Ion exchange process can also be used to recover uranium.

While serving as an additional source of uranium for the nuclear power industry, recovery of uranium from phosphates or phosphoric acid helps in reducing the environmental burden as it would otherwise go with the fertilizers and get dispersed in the environment. Large scale use of uranium containing fertilizer has a potential to contaminate the farm lands and near surface ground waters. It has also a small potential for pickup by plants. Hence, uranium recovery is receiving importance.

As most of the radium-226 and other radionuclides go with the phosphogypsum, radiation exposure of workers is expected to be only in the final stages of uranium recovery, precipitation, drying and packing operations. Uranium being predominantly an alpha emitter potential for external radiation exposure during uranium recovery is small. But contamination of floor and equipment surfaces and inhalation of radioactive aerosols or fine dust are the potential hazards. Therefore, monitoring of the workplace is necessary to evaluate the personnel exposure and efficacy of the control measures.

Presence of radionuclides in the phosphogypsum has a small potential for public exposure and environmental contamination near the stockpiles.

Essential Elements of the EIA

Environmental impact assessment is an integral part of any large industry today. The extent of the assessment however depends on the potential hazards associated with the industry. Every country has evolved regulations for the environmental impact assessment for different industries which take care of the interests of all the stake holders like the members of the public, the project proponent and the regulatory authority. Presence of radioactivity adds yet another dimension to the impact assessment. Hence a baseline survey before the operations, workplace and environmental monitoring during operations and after decommissioning are the essential elements of environmental impact assessment.

A baseline survey of the environment with regard to the topography, geology, surface hydrology, local flora and fauna, population in the immediate vicinity and the existing status is generally carried out for most of the industries. For plants dealing with radioactive material, measurement of background concentrations of radionuclides and trace metals in soil, water, sediments, air and local biota are carried out and recorded for future reference (IAEA 1997). Monitoring of gamma radiation and radon with its progeny may also be carried out to evaluate the doses to the public prior to the commissioning of the plant and a record maintained to compare with those during and post operational phase. In view of the low level of radioactivity present in the raw material and the in the wastes, the regulatory authority may decide the extent and frequency of such monitoring program. Plan for storage and transport of the product, disposal of the phosphogypsum containing low levels of radioactivity and decommissioning operations is to be prepared at the start of the operations and reviewed periodically. Disposal of phosphogypsum in public domain and for unrestricted use should comply with the exemption levels specified for different purposes by the national and international authorities (IAEA 1996; AERB 2009).

The initial process of recovery of uranium from the phosphate rock is the same as production of phosphoric acid for production of fertilizers. When uranium is recovered, the environmental impact is mostly from the disposal of large quantities of phosphogypsum. Environmental impact assessment of phosphogypsum disposal and its utilization as building construction material in India have been reported by a few investigators (Haridasan and Paul 1994; Shukla et al. 1996; Haridasan et al. 2000). The essential features of these studies are given here as a case study.

A Case Study: Environmental Impact of Phosphogypsum Disposal

Studies for the recovery of uranium from phosphate rocks are in the initial stages in India but there are many fertilizer plants in the country producing nearly 0.11 million t of phosphoric acid and about 0.50 million t of phosphogypsum annually. Environmental impact assessment of fertilizer plants processing rock phosphate is summarized here as a case study (Haridasan et al. 2000). Imported rock phosphate containing appreciable quantities of uranium is processed in the plants by wet process. Typical uranium (^{238}U) and ^{226}Ra content of the raw material and the product from these plants is summarized in Table 2.

Large quantities of phosphogypsum are disposed on land acquired by the plant. The environmental monitoring around the disposal site comprises of gamma radiation and radon progeny measurements. The gamma radiation measured over the uncovered phosphogypsum pile ranged from 0.3 to 0.5 $\mu Gy \cdot h^{-1}$ and from 0.1 to 0.15 $\mu Gy \cdot h^{-1}$ over the pile covered with 0.5 m thick soil layer. A cover of 1 m thick soil layer reduced the gamma radiation level to 0.05 to 0.1 $\mu Gy \cdot h^{-1}$, which is the same as the local background. Radon progeny concentrations measured about 300 m away from the phosphogypsum pile during the course of a day varied from 1.54 mWL (06.00 pm) to 13.05 mWL (03.00 am) with the average at 7.6 mWL.

The stockpiling of large quantities of phosphogypsum in the open is subject to influences of rain and other weather conditions. Leachability of radium-226 from phosphogypsum in distilled water and rain water was studied under different conditions such as contact time and solid to liquid ratio. In laboratory experiments with small quantity of sample with distilled water it was observed the radium-226 in leachate varied from 0.08 to 0.38 $Bq \cdot l^{-1}$ compared to a concentration of 0.09 to 0.28 $Bq \cdot l^{-1}$ (Haridasan et al. 2002). Radium-226 leached out from the phosphogypsum piles is likely to reach the nearby water body and may slightly elevate the radium-226 concentrations in the water and sediments (Haridasan and Paul 1994).

Regulatory Requirement for Use of Phosphogypsum

The phosphogypsum can be used for different purposes including that as a construction material if it meets the exemption level specified by the regulatory body.

Table.2 Typical uranium and radium concentrations in the raw materials and the products in the fertilizer plants (Haridasan et al. 2000)

Material	^{238}U ($Bq{\cdot}kg^{-1}$)	^{226}Ra ($Bq{\cdot}kg^{-1}$)
Rock phosphate	1013–1704	1285–1370
Phosphogypsum	140–205	449–939
Phosphoric acid (P_2O_5)	2280–2578	3.8–10.7
Fertilizer product (P_2O_5 – 20%)	741–1013	8–14

The Atomic Energy Regulatory Board (AERB 2009) requires that processing industries in India analyze uranium-238 and radium-226 content in each imported consignment of rock phosphate. It prescribed an exemption level of $1\,Bq \cdot g^{-1}$ for radium-226. If it is more the seller must mix with other ingredients to meet limit. It further exempts the manufacturing and use of phosphogypsum panels or blocks with radium-226 content below $40\,kBq \cdot m^{-2}$.

Conclusion

In view of the presence of significant levels of uranium, radium and other radionuclides it is desirable to carry out environmental impact assessment during processing of rock phosphates for recovery of uranium or for manufacture of fertilizers to demonstrate that the radiation exposure in the public domain is below the public dose limit of $1\,mSv \cdot y^{-1}$. Periodical workplace monitoring is also desirable to evaluate the radiation exposure of workers in the uranium precipitation, drying and packing areas, though the exposure levels are expected to be low. The phosphogypsum may be used if radium-226 content meets the exemption level.

Acknowledgement The support provided by "DAE-BRNS Raja Ramanna Fellowship" is gratefully acknowledged. Thanks are due to Dr. A. K. Ghosh, Director, HS&EG, and Mr. V. D. Puranik, Head, EAD, BARC for the encouragement received.

References

AERB (2009), Atomic Energy regulatory Board India), Directive 01/2009.

Cioroianu, T. M., Bunu, F., Filip, D. and Filip, Gh., (1998), Environmental considerations on uranium and radium from phosphate fertilizers, Proc. of a Tech. Comm. Meeting, Vienna, September 14–17, 1998, IAEA-TECDOC-1244/2001.

Gupta, C. K. and Singh, H., (2003), Uranium Resource Processing – secondary resources, Springer, Berlin/Heidelberg, ISBN 3-540-67966-9.

Haridasan, P. P. and Paul, A.C., (1994), Environmental radiological surveillance at a phosphatic fertilizer unit, Proc. 3rd National Symposium on Environment, Thiruananthapuram, March 1994, pp. 1–4.

Haridasan, P. P., Maniyan, C. G., Pillai, P. M. B. and Khan, A. H., (2000), An evaluation of the radiological impact of phosphogypsum disposal, Proc. of the National Symp. on Environment, Bangalore Univ., June 2000, pp. 211–214.

Haridasan, P. P., Maniyan, C. G., Pillai, P. M. B. and Khan A. H., (2002), Dissolution characteristics of ^{226}Ra from phosphogypsum, Jour. of Environmental Radioactivity, 62, (2002), pp. 287–294.

IAEA (1996), Basic Safety Standards, IAEA Safety Series 115, 1996.

IAEA (1997), Environmental impact assessment for uranium mine, mill and in situ leach projects, IAEA-TECDOC-979, November 1997.

Menzel, R. G., (1968), Uranium, radium and thorium content in phosphate rocks and their possible radiation hazard, J. Agr. Food Chem, Vol. 16, No.2, 1968.
Shukla, V. K., Murthy, M. V. R. and Kamath, R. R., (1996), Natural radioactivity in phosphogypsum and radiological impact assessment for its utilization in building construction materials, 5th Nat. Symp. on Environment, Calcutta, Feb. 28 – March 1, 1996.
www.wise-uranium.org 2009
www.wise-uranium.org 2010

Uranium Pollution in an Estuary Affected by Two Different Contamination Sources

Guillermo Manjón, María Villa, Rafael García-Tenorio, Juan Mantero, Santigo Hurtado, Mouloud Lehritani

Abstract. Odiel River estuary has been deeply affected, from 1960s until 1990s, by direct discharges of phosphogypsum due to operation of phosphoric acid factories. When direct discharges were halted, a self-cleaning of river has been observed. However, a second source or uranium due to former mining operations, located in the Iberian Pyrite Belt, could prevent the complete uranium cleaning due to acid mine drainage of uranium from the upstream Odiel River basin. An exhaustive study, using different analyzing techniques, allows discovering this second source of uranium contamination.

Introduction

Former phosphogypsum discharges to Odiel Estuary (SW Spain) from the phosphate industries of the area increased for more than twenty years the concentration

---

Guillermo Manjón
Applied Nuclear Physics Group, University of Seville, Seville, Spain

María Villa
Applied Nuclear Physics Group, University of Seville, Seville, Spain

Rafael García-Tenorio
Applied Nuclear Physics Group, University of Seville, Seville, Spain

Juan Mantero
Applied Nuclear Physics Group, University of Seville, Seville, Spain

Santigo Hurtado
Applied Nuclear Physics Group, University of Seville, Seville, Spain

Mouloud Lehritani
Applied Nuclear Physics Group, University of Seville, Seville, Spain

B. Merkel, M. Schipek, *The New Uranium Mining Boom*
DOI 10.1007/978-3-642-22122-4 © Springer 2011

in natural radioisotopes, such as uranium isotopes, in the Tinto and Odiel river mouths that form the Odiel Estuary (Martínez-Aguirre et al. 1994, 1995, 1999) But in the last decade the new waste management policy banned direct phosphogypsum releases to both rivers, being expected a decreasing in uranium concentration levels in the Estuary to the natural levels typical of uncontaminated areas (Absi et al. 2004; Absi 2005; Villa et al. 2005, 2009, 2010). This gives, in principle, a unique opportunity for the study of the response of a contaminated environmental compartment after the cessation of the main source of pollution.

In this work uranium concentrations, in samples collected in 2005 and 2008, in the estuary sediments are presented. These data, levels of U-isotopes and isotope activity ratios, are compared to data obtained in 2010 along the Odiel River basin, from the river sources until the estuary. Previous uranium concentration in the estuary are also reviewed to study the time evolution of uranium concentration in order to supply new data of the mechanisms of decontamination of a scenario formerly affected by Naturally Occurring Radioactive Materials (NORM) dis-

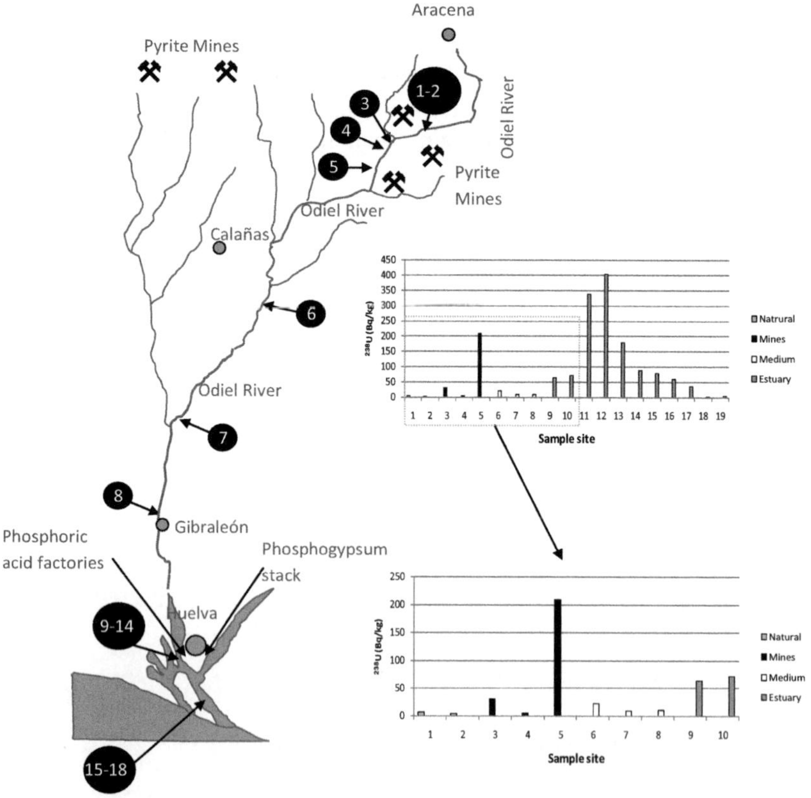

Fig. 1 Samples collected along the Odiel River basin (samples 1–8 collected in 2010 and samples 9–18 collected in 2005) and ^{238}U activity concentration in sediments collected along the Odiel River basin. The estuary of Odiel River is shown in Fig. 2

charges. On a first stage there was observed a steep decreasing of the uranium concentration in most of the Estuary. However the cleaning of the area for uranium isotopes is not as fast as expected in comparison to other radionuclides.

Experimental

The sampling stations are presented in Fig. 1 (the whole Odiel River basin) and Fig. 2 (only the Odiel River estuary).

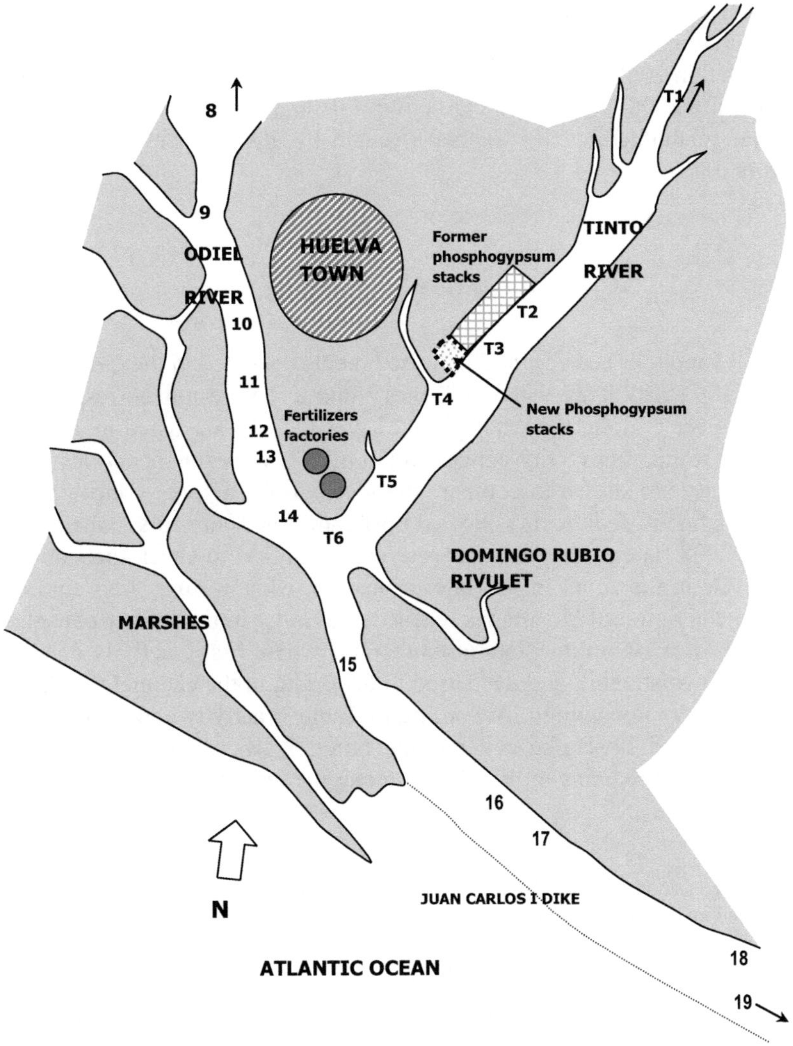

Fig. 2 Samples collected in the Odiel River estuary. Tinto River has been also sampled, but the corresponding results are not taken into account in this paper

Two different zones must be considered in Fig. 1: the estuary of Odiel River, from station number 9 to station number 18, and the rest of the Odiel River basin, from the sources of river to the estuary (stations 1–8). In Fig. 2 the distribution of sampling stations in the estuary can be observed.

The phosphoric acid wastes were directly discharged in front of sampling station number 12, additionally, wastes (without radioactive material) from a sulfuric acid factory were released in front of station number 13. The same sampling stations were considered in the whole sampling campaigns, from 1989 to 2008.

Water and sediment samples were collected in the successive sampling campaigns. Sediments were dried and homogenized before analyses. The radiochemical method applied to sediments, collected in the period 1989–2005, is based in the separation of uranium from the sample solution with TBP (Villa et al. 2010). The samples collected in 2008 were analyzed using UTEVA chromatographic columns (Lehritani Hamine 2008). ^{238}U and ^{234}U activity of samples were measured in an Alpha Analyst (manufactured by Canberra) chain by alpha-spectrometry, using PIPS detectors.

Results

U-isotopes (^{238}U and ^{234}U) corresponding to sediments collected in the Odiel River estuary from 1989 to 2008 are listed in Tables 1 and 2. The results corresponding to sediments collected along the whole Odiel River in 2010 are listed in Table 3. In the Fig. 1 the results of activity concentration of ^{238}U in sediments collected in 2005 in the estuary are shown to compare the levels to the activity concentration of same radionuclide in sediments collected upstream. The sediments, collected in the Estuary in 2008 (later estuary data), were only analyzed to check the stability of uranium levels in the zone. In this later campaign, where estuary was studied, the activity concentration of U-isotopes reached 173 Bq/kg in the former phosphogypsum releases area (sampling station number 11, where firstly activity concentration increase downstream), and decreased to 6.9 Bq/kg in the channel that Odiel and Tinto rivers have in common. According to Table 3, activity concentration of ^{238}U upstream reached 209 Bq/kg in sampling station 5, where mining wastes are mainly deposited, and remains in dissolution along the rest of the Odiel river (ac-

Table 1 Activity concentration of U-isotopes (Bq/kg) and ^{234}U/^{238}U activity ratio in samples of sediments collected in 2008 in the Odiel River estuary

Sample	Activity concentration (Bq/kg) ^{238}U	^{234}U	Activity ratio ^{234}U/^{238}U
11	160 ± 7	173 ± 7	1.08 ± 0.06
15	103 ± 3	110 ± 3	1.07 ± 0.04
19	6.9 ± 0.4	7.6 ± 0.5	1.10 ± 0.09

Table 2 Activity concentration of ^{238}U (Bq/kg) in samples of sediments collected in 1989, 1999 and 2005 in the Odiel River estuary (Villa et al. 2009)

	^{238}U Activity concentration (Bq/kg)		
Sample	2005	1999	1989
9	64 ± 1	36 ± 1	25 ± 1
10	71 ± 2	192 ± 5	548 ± 14
11	337 ± 7	475 ± 12	840 ± 25
12	403 ± 8	565 ± 10	631 ± 21
13	179 ± 4	258 ± 5	1106 ± 35
14	88 ± 2	1050 ± 20	79 ± 4
15	80 ± 2	118 ± 3	1079 ± 41
16	60 ± 2	136 ± 4	299.0 ± 13.0
17	35 ± 2	251 ± 8	299 ± 13
18	2.4 ± 0.1	4.1 ± 0.2	25 ± 1
19	5.7 ± 0.4	46 ± 1	548 ± 14

Table 3 Activity concentration of U-isotopes (Bq/kg) and $^{234}U/^{238}U$ activity ratio in samples of sediments collected upstream in the Odiel River in 2010

	Activity concentration (Bq/kg)		Activity ratio
Sample	^{238}U	^{234}U	$^{234}U/^{238}U$
1	5.9 ± 0.3	6.3 ± 0.3	1.07 ± 0.07
2	3.3 ± 0.3	4.3 ± 0.3	1.30 ± 0.11
3	31 ± 1	44 ± 2	1.45 ± 0.06
4	5.0 ± 0.4	6.4 ± 0.4	1.28 ± 0.10
5	209 ± 10	211 ± 10	1.01 ± 0.07
6	22 ± 1	33 ± 2	1.51 ± 0.08
7	9.2 ± 0.7	15.6 ± 0.9	1.70 ± 0.09
8	10.2 ± 0.6	15.9 ± 0.8	1.56 ± 0.08

tivity concentration in sediments about 10 Bq/kg in sampling stations 7 and 8). However, the ^{238}U levels obtained in samples collected before the halt of direct discharges, corresponding to the 1989 sampling campaign, were clearly higher: 1106 B/kg in the sampling station number 13.

Discussion

The most abundant natural radionuclides in phosphogypsum are ^{226}Ra and ^{210}Pb (80% of activity in the raw material), whereas only 20% of raw material activity (Rentería-Villalobos et al. 2010) of U-isotopes is associated to this byproduct. For that the first studies on the radioactive impact of phosphoric acid factories were based on the distribution and time – evolution of these radioisotopes

(Martínez Aguirre et al. 1994; Absi et al. 2004). Although the radioactive impact of uranium is lower, its study can be important if an additional source of this radionuclide in water and sediments of estuary must be traced. Indeed, uranium is also transferred into the Odiel estuary by lixiviation in the pyrite mining area located upstream the river (Sainz et al. 2003, 2004; Sánchez España et al. 2006; Santos Bermejo et al. 2003; Ferreira da Silva et al. 2009). The presence of uranium in mine lixiviates is well established (Ulrich et al. 2006) and depends of water pH, oxidation state and the presence of iron. Although results from Tinto River are not shown in this paper, a similar behavior of radionuclides is expected in this river (Villa et al. 2011).

High activity concentration levels in sample 5 (Table 3), with a $^{234}U/^{238}U$ activity ratio of the unity, is reflecting a co-precipitation of uranium with iron at pH = 3 (Ulrich et al. 2006). According to $^{234}U/^{238}U$ activity ratio, the uranium was dissolved from shaft soils at pH basic and next was transferred to the river through acid drainages (Yamamoto et al. 2010).

Uranium concentrations remain low in sediments from sample 6 until the estuary (sample 8). In this trajectory (sample 6 to sample 8), additional uranium input from other mines is dissolved (pH of water is 1–3), and the total uranium amount dissolved is then precipitated in the estuary, according to results obtained in samples 9–16, when pH increase (pH = 6–7) in front of the phosphoric acid factories. Then, a typical activity concentration of 100–400 Bq/kg in the estuarine sediments would be related to impact of acid mine drainages, whereas the maximum level observed when the direct releases were operating was about 1100 Bq/kg (Table 2).

Finally, typical natural (not contaminated) sediments could be 3–6 Bq/kg in upstream river (samples 1 and 2 in Table 3), and 2–6 Bq/kg at the end of the estuary (samples 18 and 19 in Table 2).

Conclusions

Once halted the direct discharges of phosphogypsum into the Odiel River, the uranium concentration levels have been decreased in the sediments of the Odiel River estuary. But this decrease may be, in the case of uranium, slightly lower than for other radionuclides. This different time-evolution is related, firstly, to the lixivation of uranium from old phosphogypsum stacks, close to Tinto River, and, secondly, the input of upstream waters of Odiel and Tinto River uranium, from the acid mine drainage in the Iberian Pyrite belt.

Acknowledgements This work has been partially supported by AECID (Spanish Agency for International Cooperation and Development, Ministry of Foreign Affairs of Government of Spain). The participation in sampling campaigns of Dr. I. Vioque, Dr. M. Rentería, Mrs. Díaz and Mr. Díaz is deeply acknowledged.

References

Absi A (2005) Evolución del Impacto Radiactivo Ambiental en la Ría de Huelva Tras el Cambio en la Gestión de los Residuos de las Industrias de Producción de Ácido Fosfórico. Ph.D. Thesis. Universidad de Sevilla (in Spanish).

Absi A, Villa M, Manjón G, Moreno H P, Periañez R (2004) Self-cleaning in an estuarine area formerly affected by ^{226}Ra anthropogenic enhancements. Science of the Total Environment 329:183–195.

Ferreira da Silva E, Bobos I, Matos J X, Patinha C, Reis A P, Cardoso Fonseca E (2009) Mineralogy and geochemistry of trace metals and REE in volcanic massive sulfide host rocks, stream sediments, stream waters and acid mine drainage from the Lousal mine area (Iberian Pyrite Belt, Portugal). Applied Geochemistry 24:383–401.

Ulrich K-U, Rossberg A, Foerstendorf H, Zänker H, Scheinost A C (2006) Molecular characterization of uranium(VI) sorption complexes on iron(III)-rich acid mine water colloids. Geochimica et Cosmochimica Acta 70:5469–5487.

Lehritani Hamine M (2008). Separación secuencial de uranio y torio en muestras ambientales mediante extracción cromatográfica para su determinación por espectrometría alfa. In the book: V Jornadas de Calidad en el Control de la Radiactividad Ambiental. Ed. Consejo de Seguridad Nuclear y Universidad de Zaragoza. Pag. 289–300. In Spanish.

Martínez-Aguirre A, Periañez R (1999) Distribution of natural radionuclides in sequentially extracted fractions of sediments from a marsh area in Southwest Spain: U isotopes. Journal of Environmental Radioactivity 45:67–80.

Martínez-Aguirre A, Garcia-León M, Ivanovich M (1995) U- and Th-speciation in river sediments. Science of the Total Environment 173–174:203–209.

Martínez-Aguirre A, García-León M, Ivanovich M (1994). The distribution of U, Th, ^{226}Ra derived from the phosphate fertilizer industries on an estuarine system in southwest Spain. Journal of Environmental Radioactivity 22:155–177.

Rentería-Villalobos M, Vioque I, Mantero J, Manjón G (2010) Radiological, chemical and morphological characterizations of phosphate rock and phosphogypsum from phosphoric acid factories in SW Spain. Journal of Hazardous Materials 181:193–203.

Sainz A, Grande J A, de la Torre M L (2003) Odiel River, acid mine drainage and current characterisation by means of univariate analysis. Environment International 29:51–59.

Sainz A, Grande J A, de la Torre, M L (2004) Characterisation of heavy metal discharge into the Ria of Huelva. Environment International 30:557–566.

Sánchez España J, López Pamo E, Santofimia Pastor E, Reyes Andrés J, Martín Rubí J A (2006) The impact of acid mine drainage on the water quality of the Odiel River (Huelva, Spain): evolution of precipitate mineralogy and aqueous geochemistry along the Concepción-Tintillo segment. Water, Air, and Soil Pollution 173:121–149.

Santos Bermejo J C, Beltrán R, Gómez Ariza J L (2003) Spatial variations of heavy metals contamination in sediments from Odiel river (Southwest Spain). Environment International 29:69–77.

Villa M, Absi A, Manjón G, Moreno H P, Periañez R, García-Tenorio R (2005). Natural restoration of a Spanish estuary affected by anthropogenic inputs of NORM. In the book: Natural Occurring Radioactive Materials (NORM IV). IAEA-TECDOC-1472. International Atomic Energy Agency (IAEA) Ed., Vienna. ISBN 92-0-110305-0.

Villa M, Mosqueda F, Hurtado S, Mantero J, Manjón G, Periañez R, Vaca F, García-Tenorio R (2009) Contamination and restoration of an estuary affected by phosphogypsum releases. Science of the Total Environment 408:69–77.

Villa M, Hurtado S, Mas J L, Casacuberta N, Masqué P, García-Tenorio R (2010) Sequential extraction and measurement of ^{226}Ra, ^{210}Po, U, and Th isotopes in NORM matrices. In the book: Proceedings of the Naturally Occurring Radionuclide Materials V.

Villa M, Manjón G, Hurtado S, García-Tenorio R (2011) Uranium pollution in an estuary affected by pyrite acid mine drainage and releases of naturally occurring radioactive materials. Marine Pollution Bulletin 62(7): 1521–1529.

Yamamoto M, Sakaguchi A, Kofuji H (2010) Uranium in acidic mine drainage at the former Ogoya Mine in Ishikawa Prefecture of Japan. J Radioanal Nucl Chem 283:699–705.

Distribution of Uranium Related to Particle Size of Phosphogypsum from Phosphoric Acid Production (Huelva, SW Spain)

Marusia Rentería-Villalobos, Ignacio Vioque, Juan Mantero, Guillermo Manjón

Abstract. In this work, both uranium and some trace elements distribution along with particle sizes were studied on phosphogypsum, in order to improve the knowledge on the behavior of toxic elements. ^{238}U and ^{234}U concentrations were obtained by alpha spectrometry, whereas Cr, Cu, Zn, Sr, Y, Zr, Cd, Ba, and Pb were determined using X-ray fluorescence. Determination of toxic elements in phosphogypsum was obtained in bulk and in three particle fractions: coarse (> 53 µm), medium (53–20 µm) and fine (< 20 µm). Average values obtained in the different particle fractions for ^{238}U were 100 Bq/kg in bulk, 94 Bq/kg in the coarse fraction, 115 Bq/kg in the medium one and 160 Bq/kg in the fine fraction. ^{234}U concentration average values, in the same order, 105, 96, 121 and 174 Bq/kg, respectively. Elements such as Zn, Sr, Y, Zr, Ba and Pb tend to be concentrated in the fine fraction, whereas the Cd is distributed homogeneously in the three fractions. Results obtained by these techniques have demonstrated that the most toxic elements are not distributed homogeneously into phosphogypsum and that most of

Marusia Rentería-Villalobos
Advanced Material Research Center, CIMAV, Miguel de Cervantes 120, 31109 Chihuahua, Chihuahua, Mexico

Ignacio Vioque
Applied Nuclear Physics Group, University of Seville, ETS Arquitectura, Departamento de Física Aplicada 2, Avda. Reina Mercedes s/n, 41012 Seville, Spain

Juan Mantero
Applied Nuclear Physics Group, University of Seville, ETS Arquitectura, Departamento de Física Aplicada 2, Avda. Reina Mercedes s/n, 41012 Seville, Spain

Guillermo Manjón
Applied Nuclear Physics Group, University of Seville, ETS Arquitectura, Departamento de Física Aplicada 2, Avda. Reina Mercedes s/n, 41012 Seville, Spain

B. Merkel, M. Schipek, *The New Uranium Mining Boom*
DOI 10.1007/978-3-642-22122-4 © Springer 2011

these elements are concentrated in the fine fraction (particles <20 μm), which could be easily mobilized by leaching and/or erosion.

Introduction

Some factories related to phosphoric acid and fertilizer manufacture are located in the estuary of the Tinto and Odiel rivers (Huelva, SW Spain). It is well established that these industries of phosphoric acid and fertilizers use, as raw material, phosphate rock (of marine sedimentary origin) mainly from Morocco and Senegal (Bolívar et al. 2009). The phosphate rock is, due to its origin, affected by high concentrations of naturally-occurring radionuclides and other so-called trace elements (Rutherford et al. 1994; Martín et al. 1999). The industrial processes, applied to the phosphate rock in these factories, therefore also lead to high concentrations of toxic elements in both products and wastes. Hence, the presence of trace elements and natural radionuclide concentrations, at higher than natural levels, would be considered as environmental contamination. In the town of Huelva, these industries use the "di-hydrated" process to obtain phosphoric acid, whereby the phosphate rock is treated chemically with sulfuric acid (Rutherford et al. 1994). Other products obtained from reaction are hydrogen fluoride, gypsum: calcium sulfate dihydrate ($CaSO_4 \cdot 2H_2O$) and impurities. The whole of these mentioned species is thereafter called Phosphogypsum (PG). About of 90% of PG is made up of gypsum; mainly impurities include Si, P, Fe, Al, Na, and with less content, K and Ti, as well as trace elements, rare earth elements and certain naturally-occurring radionuclides. The concentration of these elements into PG may vary mainly depending on the origin of the phosphate rock and, to a lesser extent, on factors such as plant operation and PG age (Rutherford et al. 1995). In previous studies, the fluxes and contents of U in the phosphoric acid production process have been determined. The results show that the U content is in greater proportion associated to phosphoric acid (>80%), while the U content in PG is lower (Weterings and Janssen 1985; Bolivar et al. 1995; Rutherford et al. 1996; Bolívar et al. 2009). However, if the whole U content in PG is considered, then 23% is bounded to the bioavailable fraction (Pérez-López et al. 2007).

In our study, the phosphogypsum is stored in open-air stack near to the factories. It is estimated that annually 2.5 million tons of PG are deposited in these piles. Since the 1980's, elemental, chemical and radiochemical analyses have been applied in order to monitor the possible radioactive impact to the estuary (Bolívar et al. 1995). However, further information is required on the distribution of particles in stockpiled phosphogypsum, which can help towards a better understanding of the possible environmental contamination. Improvements in the data, such as sample heterogeneity, are necessary to fully evaluate the environmental impact and risks (Salbu 2007).

Several preliminary studies have been developed to reuse the PG. The first consequence of these applications would be the decrease in the amount of stored waste and, therefore, a reduction in the impact of stacks on its close environment. Thus, as far as we know, the most common uses of phosphogypsum are: in agriculture (Papastefanou et al. 2006; Abril et al. 2009); as a substitute of gypsum in building; in the production of cement (Kuryatnyk et al. 2008); and as a conditioner of soil for road construction (Degirmencia et al. 2007). In addition, it can be used in landfills related with urban solid wastes (www1.fipr.state.fl.us/fipr/fipr1.nsf 2004). Due to these reuses, PG becomes a by-product of phosphoric acid manufacture.

However, the main restrictions for the applications of phosphogypsum are related to its high content of both toxic elements and radioactivity, and hence it is necessary to obtain a deeper understanding of physical-chemical characteristics of this by-product.

In this work, the determination of toxic elements (^{238}U, ^{234}U, Cr, Cu, Zn, Sr, Y, Zr, Cd, Ba, and Pb) was obtained in bulk and in three particle fractions of stockpiled phosphogypsum: coarse (> 53 μm), medium (53–20 μm) and fine (< 20 μm).

Materials and Methods

Phosphogypsum samples (PG) were taken from a stack belonging to a phosphoric acid industry, located next to the Tinto river (37° 15.3'N, 6° 54'W). This PG stack has been inactive since the beginning of the 1990's. Sampling was carried out in 2005, following the procedure described by Abril (2009). Elemental analysis was performed in bulk and in three particle sizes of PG. The samples were divided by dry-sieving in: > 53 μm (coarse fraction), 53–20 μm (medium fraction), and < 20 μm (fine fraction). All fractions were dried at 50°C for 48 h.

Concentrations of trace elements and isotopic uranium were obtained by X-ray fluorescence (XRF) and alpha-spectrometry, respectively.

XRF was employed to determine trace elements in dried PG samples. The analysis was carried out in a Panalytical AXIOS with an Rh tube. ^{238}U and ^{234}U were obtained by alpha spectrometry, where it was used a radioanalytical analysis procedure. PG samples were spiked with known activity of ^{232}U and co-precipitated with $Fe(OH)_3$ at pH = 9. Total sample dissolution was performed by atmospheric acid digestion using HF and HNO_3. UTEVAS resins were used to extract the uranium. Then, uranium was electrodeposited on stainless steel planchets (Hallstadius 1984). An alpha-spectrometry chain Alpha Analyst (CANBERRA) was used for alpha activity measurements. Radiochemical yield was determined taking into count the ^{232}U counting rate.

Results and Discussion

Particle Size Fractionation

Figure 1 shows the results of PG particle size fractions. The average of total mass recovery (sum of every size fractions) was between 92 and 100%. From results, in

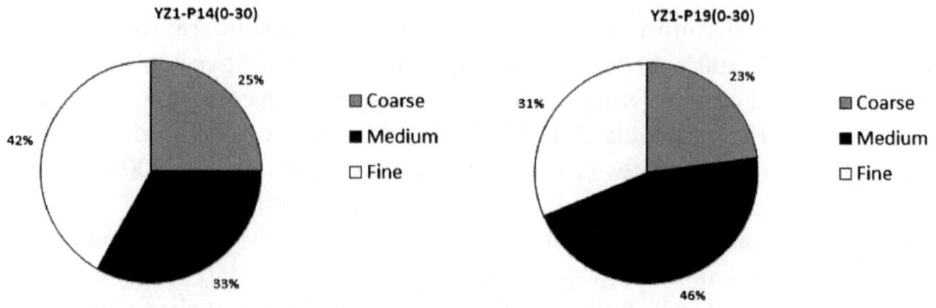

Fig. 1 Percentage of particle fractions found into PG

average, medium fraction (20–53 μm) was the dominant size fractions by up 40%, whereas coarse (> 53 μm) and fine (< 20 μm) together constituted less than 60% of the sample masses.

Taken into account the estimated amount of produced phosphogypsum per year, 2.5 million $t\,y^{-1}$, the quantities for coarse, medium, and fine fractions would be 0.8 $t\,y^{-1}$, 1.00 $t\,y^{-1}$, and 0.7 $t\,y^{-1}$, respectively.

Isotopic U by Alpha Spectrometry

Distribution of uranium activity concentrations are showed in Table 1.

From results of Table 1 a typical ^{238}U activity concentration in PG is observed, when sedimentary phosphate rock is used as raw material (Rutherford et al. 1996). Also, radioactive equilibrium was showed between ^{234}U and ^{238}U. From particle fractionation, it is important to point out that the uranium contents increase in opposite direction to the particle size. Thus, it is observed that small particles show a high concentration of uranium (Table 1). Easily, a ^{238}U fractionation can be deduced: 26% would be associated to coarse fraction, 38% to medium fraction and 36% to fine fraction. Therefore, although most of uranium is concentrated in fine size fraction 160 Bq/kg, this radionuclide also can be slightly associated to medium fraction (Table 2).

Elemental Characterization by XRF

Table 3 shows the mean content of some trace elements for bulk PG and for various particle size fractions.

According to our results, trace elements with higher concentrations (65–829 ppm) in order of abundance were Sr > Y > Ba, whereas Cr, Cu, Zn, Pb and Ba are present in lesser contents (1–35 ppm).

Table 1 Activity concentrations (Bq/kg) of isotopic U into different particle fractions of PG samples, as well as their activity ratios

Sample Code	Fraction	^{234}U	^{235}U	^{238}U	$^{234}U/^{238}U$
YZ1-P14(0-30)	Bulk	118 ± 4	2.3 ± 0.4	112 ± 4	1.06
	Coarse	117 ± 9	3.1 ± 0.9	114 ± 9	1.02
	Medium	128 ± 7	1.2 ± 0.4	116 ± 7	1.10
	Fine	226 ± 14	8.1 ± 1.1	210 ± 13	1.08
YZ1-P19(0-30)	Bulk	91 ± 5	2.3 ± 1.6	88 ± 5	1.04
	Coarse	75 ± 2	1.7 ± 0.3	73 ± 2	1.03
	Medium	114 ± 10	3.3 ± 1.1	115 ± 10	1.00
	Fine	122 ± 5	2.8 ± 0.6	110 ± 5	1.11

Table 2 Mass concentration of U into different particle fractions of PG samples

Sample Code	Fraction	$\frac{\text{mg of U}}{\text{kg of PG}}$	$\frac{\text{mg of U}}{\text{kg of PG}}$
YZ1-P14(0-30)	Coarse	2.9	
	Medium	4.2	
	Fine	5.2	
		Sum	12.5
		Bulk	9.6
YZ1-P19(0-30)	Coarse	1.9	
	Medium	3.7	
	Fine	2.8	
		Sum	8.5
		Bulk	7.5

Table 3 Concentration (mg of element per kg of fraction mass) of some trace elements presents in fractionated PG samples (bulk, coarse, medium and fine) as well as their values maximum and minimum

Element	Bulk	Range	Coarse	Range	Medium	Range	Fine	Range
Cd	2 ± 1	< 1–3	6 ± 1	5–7	5 ± 1	3–7	5 ± 1	4–8
Pb	5 ± 1	5–6	5 ± 1	4–5	5 ± 1	4–7	6 ± 1	5–9
Cu	8 ± 1	7–9	10 ± 1	7–13	8 ± 2	5–12	10 ± 1	7–16
Zn	7 ± 1	5–9	7 ± 1	6–9	7 ± 1	6–9	8 ± 1	6–11
U[a]		7–10		6.1–9.5		9.3–10.4		10–19
Zr	10 ± 1	7–16	5 ± 1	3–11	6 ± 1	3–14	7 ± 1	3–10
Cr	19 ± 2	8–26	15 ± 1	8–25	17 ± 2	11–35	24 ± 2	11–35
Ba	113 ± 27	85–142	84 ± 20	65–115	98 ± 23	67–168	138 ± 33	91–219
Y	135 ± 36	125–156	129 ± 34	114–145	136 ± 36	117–153	195 ± 52	153–236
Sr	647 ± 105	559–736	665 ± 108	621–788	687 ± 111	617–829	726 ± 118	644–829

[a] Calculated from activity concentration of ^{238}U.

The distribution of trace elements varies for each size fraction. In general, most of trace elements are enriched into fine fraction in comparison with bulk PG, and in comparison with coarse and medium size fraction (see Table 3). Concentrations of strontium, cadmium, cupper, and zinc are higher into coarse and medium size fraction than into bulk PG. On the contrary, chromium, yttrium, and barium contents in bulk PG are slightly higher than those contents in coarse and medium fractions. Content of lead is homogeneously distributed in bulk PG, as well as in coarse and medium size fractions.

Table 4 Mass concentration of trace elements related to each percent of particle fractions into PG samples, as well as the sum of fractionated PG samples (bulk, coarse, medium and fine)

Fraction	Cr	Cu	Zn	Zr	Cd	Pb	Y	Ba	Sr	U*	U**
Bulk	19.0	8.0	7.0	10.0	2.0	5.0	135.0	113.0	647.0	9.6	7.5
Coarse	4.7	3.1	2.2	1.6	1.9	1.6	40.4	26.3	208.5	3.0	1.9
Medium	6.9	3.2	2.8	2.4	2.0	2.0	54.8	39.5	276.8	4.2	3.7
Fine	6.8	2.8	2.3	2.0	1.4	1.7	55.3	39.1	205.9	5.2	2.8
Total	18.4	9.2	7.3	6.00	5.3	5.3	150.5	105.0	691.2	12.4	8.5

U*, uranium contents into the sample YZ1-P14(0-30)
U**, uranium contents into the sample YZ1-P19(0-30)

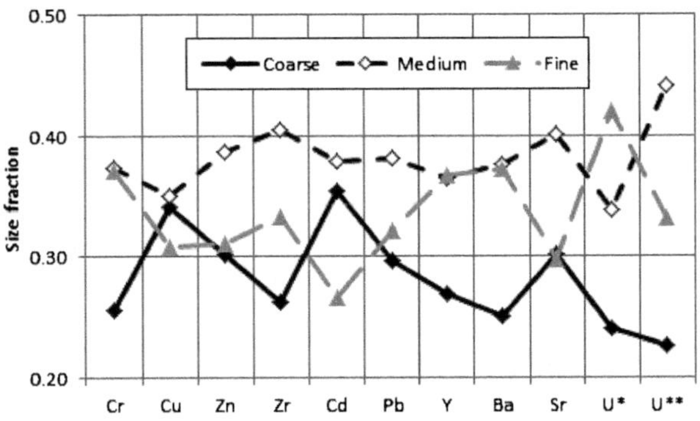

Fig. 2 Distribution of concentration (kg of element in 1 kg of PG) of trace elements

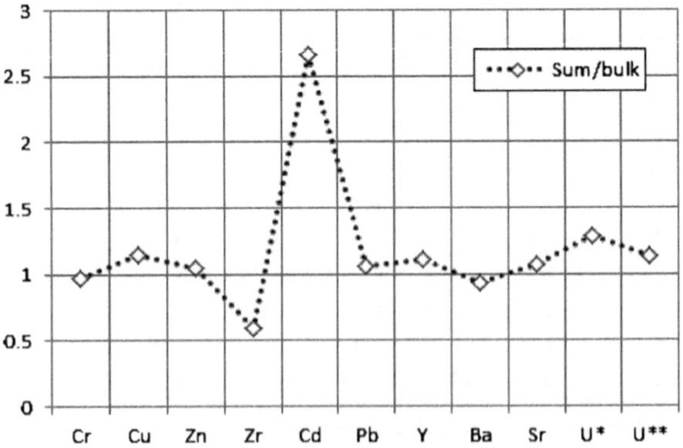

Fig. 3 Ratio between the total amount of fraction concentrations (per kg of PG) and the concentration in bulk samples (also per kg of PG)

Information about contaminants such as Cd has been reported in PG from Huelva, where it is found in high concentrations. Abril 2009 reported a Cd concentration of 2.8 ppm in stockpiled PG, which is according to our results in bulk PG that ranged from <1 to 3 ppm. In the same manner, Ba concentration of 92 ppm was determined (Pérez-López et al. 2007), which is confirmed by that found in this work.

The heavy metals such as Pb, Cd, Cu, and Cr, are considered hazardous elements and of environmental concern due to their high toxicity, since they are of great risk to human health (Wilde and Benemann 1993). Examples are shown graphically for distribution of these elements (Fig. 3).

Hence, high amounts of these toxic elements are being produced together with PG. Likewise, the distribution of those trace elements and uranium is similar, where all of them tend to be enriched into fine fraction, see Table 3.

However, taking into account the percent of each particle size fraction, the high contents of trace elements are given by medium particle fraction, Table 4 and Fig. 2.

The Fig. 3 shows the equality between the concentration experimentally determined in bulk samples and the concentration evaluated by using the experimental results obtained in the three fractions (coarse, medium and fine). The results are in a good coincidence considering the unity the best value. Only Zr, ratio lower than 1, and Cd, ratio higher than 1, are outliers. In these cases, the distribution of the elements could not be homogeneous.

By calculating the annual production of these trace elements in bulk PG, it can be concluded that these amounts would be: U, 20 t; Sr, 1618; Ba, 284; Y, 338; Cu, 8; Cr, 48; Cd, 4; Zr, 24; and Zn at around 7 t. All these amounts of toxic elements are being stockpiled within the estuary of the Tinto and Odiel Rivers, thereby representing latent risk of contamination.

Additionally, it is known that PG is used in some agricultural activities. The continued use of this by-product also provides and increases the toxic element concentration in soils, and modifies those contents in plants, and as a consequence, in the whole food chain.

Conclusions

Our results show that PG is composed by particles fraction in: coarse 30%, medium 40% and fine 28%. In PG, most of the uranium and the other toxic elements tend to appear concentrated in particles $<20\ \mu m$. Although the fine size fraction is less than 30% in mass of PG samples, its contents of toxic elements represent high levels of contamination. From an environmental point of view, a clear risk of environmental impact is present due to the enhancement and weak bond of uranium in phosphogypsum. Indeed, these elements would be concentrated within small particles with a high surface area, which implies easy mobilization by leaching and/or erosion.

References

Abril, J.M., García-Tenorio, R., Manjón, G., 2009. Extensive radioactive characterization of a phosphogypsum stack in SW Spain: ^{226}Ra, ^{238}U, ^{210}Po concentrations and ^{222}Rn exhalation rate. J. Hazard. Mater.: 164, 790–797.

Bolivar, J.P., Garcia-Tenorio, R., Garcia-Leon, M., 1995. Fluxes and distribution of natural radionuclides in the production and use of fertilizers. Appl. Radiat. Isot.: 46, 717–718.

Bolívar, J.P., Martín, J.E., García-Tenorio, R., Pérez-Moreno, J.P., Mas, J.L., 2009. Behaviour and fluxes of natural radionuclides in the production process of a phosphoric acid plant. Appl. Radiat. Isot.: 67, 345–356.

Degirmencia, N., Okucu, A., Turabi, A., 2007. Application of phosphogypsum in soil stabilization. Build. Environ: 42, 3393–3398.

Hallstadius, L., 1984. A method for the electrodeposition of actinides. Nucl. Instrum. Methods in Phys. Res.: 223, 266–267.

Kuryatnyk, T., Angulski da Luz, C., Ambroise, J., Pera, J., 2008. Valorization of phosphogypsum as hydraulic binder. J. Hazard. Mater.: 160, 681–687.

Martín, J.E., García-Tenorio, R., Respaldiza, M.A., Ontalba, M.A., Bolívar, J.P., da Silva, M.F., 1999. TTPIXE analysis of phosphate rocks and phosphogypsum. Appl. Radiat. Isot.: 50, 445–449.

Papastefanou, C., Stoulos, S., Ioannidou, A., Manolopoulou, M., 2006. The application of phosphogypsum in agriculture and the radiological impact. J. Environ. Radioact.: 89, 188–198.

Pérez-López, R., Álvarez-Valero, A.M., Miguel Nieto, J., 2007. Changes in mobility of toxic elements during the production of phosphoric acid in the fertilizer industry of Huelva (SW Spain) and environmental impact of phosphogypsum wastes. J. Hazard. Mater.: 148, 745–750.

Rutherford, P.M., Dudas, M.J., Arocena, J.M., 1996. Heterogeneous distribution of radionuclides, barium and strontium in phosphogypsum by-product. Sci. Total Environ.: 180, 201–209.

Rutherford, P.M., Dudas, M.J., Arocena, M., 1995. Radioactivity and elemental composition of phosphogypsum produced from three phosphate rock sources. Waste Manag. Res.: 13, 407–423.

Rutherford, P.M., Dudas, M.J., Samek, R.A., 1994. Environmental impacts of phosphogypsum. Sci. Total Environ.: 149, 1–38.

Salbu, B., 2007. Speciation of radionuclides e analytical challenges within environmental impact and risk assessments. J. Environ. Radioact.: 96, 47–53.

Weterings, K., Janssen, J., 1985. Recovery of Uranium, Vanadium, Yttrium and Rare Earths from phosphoric acid by a precipitation method. Hydrometallurgy: 15, 173–190.

Wilde, E.W., Benemann, J.R., 1993. Bioremoval of heavy metals by the use of microalgae. Biatech. Adv.: 11, 781–812.

www1.fipr.state.fl.us/fipr/fipr1.nsf, 2004. 93-01-109, Utilization of Phosphogypsum to Enhance Biological Decomposition and Increase Capacity Recovery at Municipal Solid Waste Landfills. Chemical Processing & Phosphogypsum.

Lithofacies Study of the Natural Phosphates: Quantification, Genetic Involvement and Distribution of Natural Radionuclides

Said Fakhi, Rabie Outayad, Elmehdi Fait, Mustapha Mouflih, Marusia Rentaria, Ignacio Vioque, Abdelghani Adib Idrissi, Moncef Benmansour, Abderrahim Bouih, Hassan Elhadi, Abdelmjid Nourreddine

Said Fakhi
Unit of Radiochemistry University Hassan II Mohammedia, Casablanca, Faculty of Science Ben Sik, Morocco
E-mail: fakhisaid@gmail.com

Rabie Outayad
Unit of Radiochemistry University Hassan II Mohammedia, Casablanca, Faculty of Science Ben Sik, Morocco

Elmehdi Fait
Unit of Radiochemistry University Hassan II Mohammedia, Casablanca, Faculty of Science Ben Sik, Morocco
Laboratory of Geochemistry, Applied Geology and Environment, University Hassan II Casablanca-Mohammedia, Faculty of Sciences Ben Sik, Moroco

Mustapha Mouflih
Laboratory of Géoressources and Sedimentary Environment, University Hassan II Casablanca-Mohammedia, Faculty of Sciences Ben Sik, Morocco

Marusia Rentaria
Départment de Física Aplicada II, Escuela Técnica Superior de Arquitectura de Sevilla Universidad de Sevilla

Ignacio Vioque
Départment de Física Aplicada II, Escuela Técnica Superior de Arquitectura de Sevilla Universidad de Sevilla

Abdelghani Adib Idrissi
Unit of Radiochemistry University Hassan II Mohammedia, Casablanca, Faculty of Science Ben Sik, Morocco

Moncef Benmansour
National Centre for Energy, Sciences and Nuclear Techniques (CNESTN-Morocco)

B. Merkel, M. Schipek, *The New Uranium Mining Boom*
DOI 10.1007/978-3-642-22122-4 © Springer 2011

Abstract. The Danian–Persian (–65 to –46 MA) Moroccan phosphate deposits of Ghar Tajer was geochemically characterized. Sediments are formed of Phosphate intercalated with limestone and marl and argillaceous rocks. Specific activity and isotopes ratios: $^{234}U/^{238}U$, ^{230}Th,$^{232}Th/^{238}U$ were determined at 20 sediments samples collected at different depth. In this presentation, we give the preliminary results of the behavior of uranium in sedimentary rocks associated with samples deposits by measuring ^{238}U, ^{234U}, ^{235}U, ^{232}Th and ^{230}Th activities and their isotopes ratios. A first attempt to correlate the behavior of the radionuclides measured by alpha spectrometry and lithology of samples of phosphate was accomplished.

Introduction

The work is part of the general investigation of the behavior of radioactive metals at the interface solid/water. It presents the results of a study on the distribution of radionuclides (^{238}U, ^{235}U, ^{232}Th, …) in a series of sediments (Danian Persian: –65 to –46 MA) collected at different depths from the Ghar Tajer phosphate mine. Specific activities determined by alpha spectrometry and isotopic ratio U-Th was combined with other data: depth, lithology and grain-size analysis. These data allow approaching the geochemical aspects of these isotopes, depending on the sedimentary paleodeposition of this compartment. For this purpose, 20 samples from different depths and different facies from the mine Ghar Tajer were analyzed for their uranium and thorium isotope content and for their isotopic ratio: $^{234}U/^{238}U$, $^{230}Th/^{238}U$ and $^{232}Th/^{238}U$.

Geographical and Geological Context of Samples

The sedimentary deposits of Ouled Abdoun is formed by several areas that differ only by the lithology of the layers and the phosphate P_2O_5 concentration. The layers were deposited at the end of the Mesozoic, (–77 MA) and Maastrichtian (–65 MA), and the beginning of the Cenozoic (–53 MA), middle Eocene Ypresian (–46 MA); (Mountasir et al. 1977; Moumni 1979; Belfkira 1980; Jourani and Al. 1987).

Abderrahim Bouih
National Centre for Energy, Sciences and Nuclear Techniques (CNESTN-Morocco)

Hassan Elhadi
Laboratory of Geochemistry, Applied Geology and Environment, University Hassan II Casablanca-Mohammedia, Faculty of Sciences Ben Sik, Morocco

Abdelmjid Nourreddine
Institut Pluridisciplinaire Hubert-Curien UMR 7178 ULP/CNRS-In2p3
23 rue de Lœss BP 28, F-67037 Strasbourg Cedex France

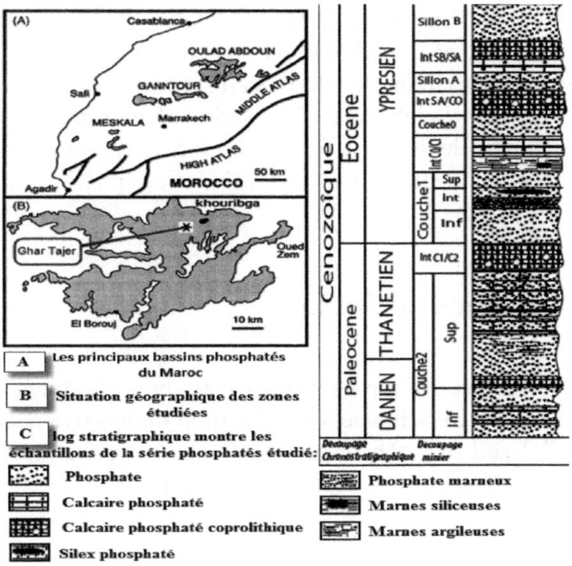

Fig. 1 Study site location and representation of stratigraphic layers of Ghar Tajer wells

Table 1 Samples analysis, geological and lithological description

Sample	Depth (m)	Lithology
GT20	8.9	Phosphate with light gray, fossiliferous coprolite
GT19	9.9	Marls compact appearance lightweight roof bioturbated
GT18	10.5	Phosphate-enriched limestone grayish marly
GT17	11	Limestone phosphatized coprolite silicified to very basic, with cherty very distorted
GT16	11.3	Marl interlayer laminating lenticular lamination discreet surmounted by roof Boujniba facies (phosphatic siliceous marls)
GT15	11.9	Phosphate very whitish gray marly, bioturbated at the base
GT14	12.4	Indurated calcareous phosphate coprolite corrugated wall, more marly towards the roof
GT13	13	Clay marl, green, gray lighter with disseminated coprolithes
GT12	13.5	Yellow coarse grained Phosphate at the base very coprolite fin and marly at the top with red oxidations and discreet laminations
GT11	14.5	Marly gray phosphate with two <silex> layers brown
GT10	14.8	Marley limestone coprolite of Ypresian age, with silex layers
GT9	15.3	phosphate more coarse grained at the base and fined layered at the top
GT8	16	Marly Limestone at the base very coprolite at top
GT7	17.2	Limestone very hard, is bioturbated at the base
GT6	18.1	Phosphate marly-enriched limestone indurated, gray yellow with red oxidations
GT5	19	Marly phosphate gray white cutting in little sequences of 20 à 30 cm
GT4	19.5	Phosphate-enriched limestone very hard
GT3	19.7	Sandy phosphate lightly marly with yellow color and red oxidations
GT2	20.1	Phosphate-enriched limestone with red oxidations
GT1	20.6	Marly phosphate facies, wet

Ghar Tajer wells is part of Oulad Abdoun Basin (Fig. 1), it is located 8 km southeast of the Khouribga city. It is limited by the old underground mining of phosphate. In the North, it is crossed by the road No. 11 connecting Casablanca to Fkih Ben Saleh town. The series of phosphates protected against erosion by a natural dale, is in the form of horizontal strata with different levels of intercalation of limestone, marl and clay with varied silicifications Ypresian (Table l).

Geochemical Studies

The subject of the geochemical study of the phosphate series is to determine the behavior of radionuclides of natural radioactive decay series, particularly series of ^{238}U and of ^{232}Th respectively. The radionuclides were identified and measured radiochemically by alpha spectrometry.

Techniques Used

Preparation of Thin Alpha Sources

To prepare the alpha sources from solid samples, the sediments are treated with specific radiochemical process. After dissolving the sample, the radionuclides are separated by ion exchange chromatography. Thin alpha sources are developed by the electrodepositing of alpha emitters by appropriate electroplating reaction (Talvitie 1972). Figure 2 illustrates the radiochemical processing steps.

Alpha Spectrometry

The identification and determination of radionuclides of both natural radioactive decay series were carried out by alpha spectrometry. The measurement is made using a detector (Ortec to eight channels), consisting of silicon detectors to surface barrier with their respective chains whose characteristics are as follows

- sensitive surface: 450 mm²;
- detection efficiency 20 to 30%; and
- 25 keV as energy of resolution.

The alpha spectra delivered by the detector are analyzed by the program MAESTRO. After calibration of the detection system and the energy efficiency, samples of the phosphate series are analyzed. Figure 3 illustrates a typical alpha spectrum.

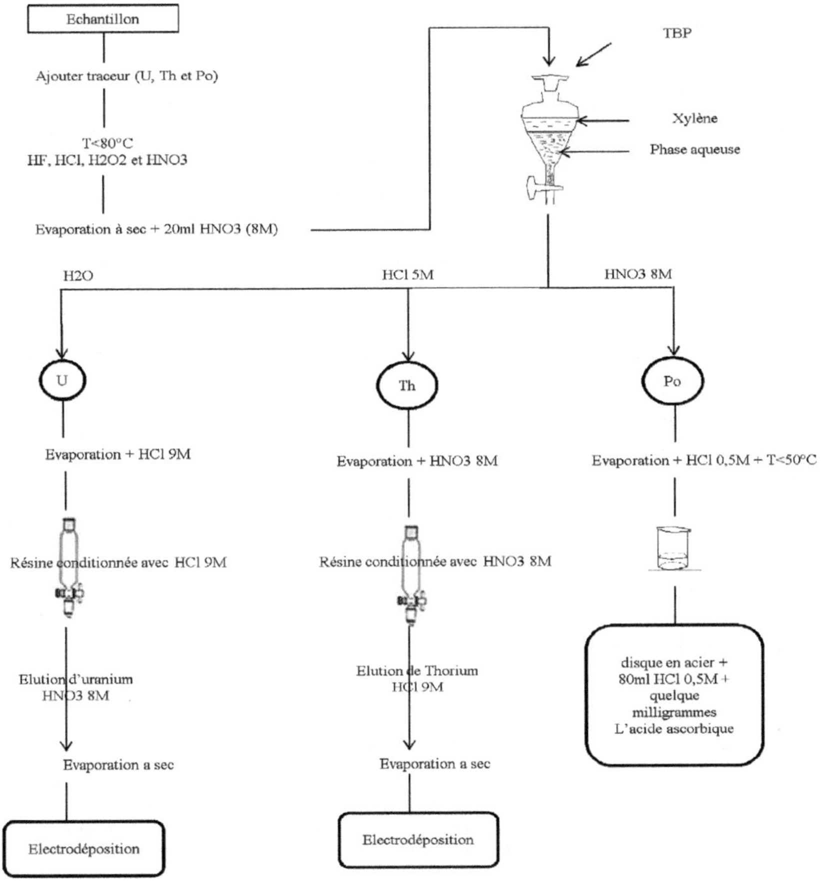

Fig. 2 Synoptic diagram of radiochemical processing steps of samples

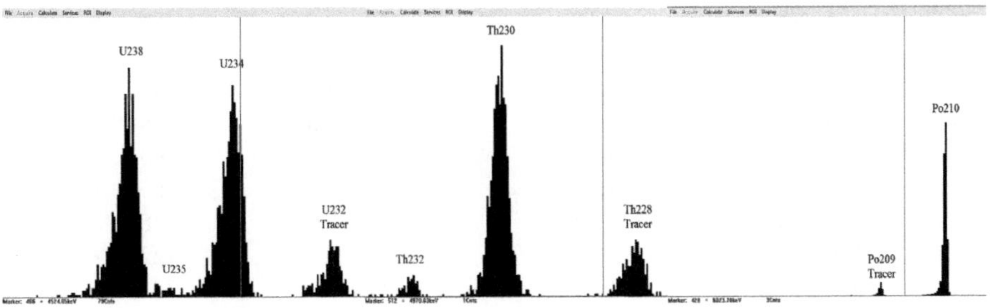

Fig. 3 Typical alpha spectrum of U and Th isotopes in the samples (^{232}U, ^{209}Po and ^{228}Th as tracers)

Results and Interpretation

To exploit the results of measurement by alpha spectrometry of the 20 samples of the phosphate series given in (Tables 2 and 3), we have shown variations in specific activities and isotopic ratios depending on the characteristics of these samples. These graphs are representative of the variation of the specific activity of each radionuclide identified "spectra Fig. 3" depending on the depth and lithology of the sample. The specific activity expressed in Bq/kg were determined with a standard deviation of 1.

Analysis of representative curves (Figs. 4 and 5) shows that the specific activities of ^{238}U, ^{234}U, ^{230}Th, ^{210}Po, ^{214}Pb and ^{226}Ra in the deposits are independent of depth. Irregular variation seems to depend on the distribution of radionuclides according to lithology.

The maximum values are generally observed in the samples, mainly composed of phosphate, and the highest value is found in sample 6 at a depth of 18.1 m. The minimum values correspond to calcareous sediments, the lowest is found in the sample 18 to 10.5 m depth and the sample of marl clay (sample 13 to 13 m).

The curve representing the variation of the activity of ^{232}Th (Fig. 5) shows that this radionuclide is also concentrated in samples of phosphate horizon: maximum values are observed in samples 20 (8.9 m), 6 (18.1 m) and 5 (19 m). As the levels

Table 2 Concentrations (Bq/kg) of U isotopes and in ppm for U in the series of natural phosphate Ghar Tajer

Ech	^{238}U (Bq/kg)	^{235}U (Bq/kg)	^{234}U (Bq/kg)	^{238}U(ppm)
GT20	1160 ± 50	53 ± 0,3	1147 ± 60	94,8
GT19	887 ± 12	42 ± 10	848 ± 12	72,4
GT18	345 ± 26	17 ± 30	304 ± 23	28,1
GT17	1450 ± 54	64 ± 0,3	1389 ± 70	118,4
GT16	905 ± 48	41 ± 4,0	874 ± 47	74
GT15	942 ± 50	45 ± 0,3	960 ± 50	77
GT14	995 ± 15	49 ± 13	946 ± 14	81
GT13	531 ± 78	24 ± 70	465 ± 69	43
GT12	1807 ± 51	85 ± 11	1746 ± 50	148
GT11	723 ± 14	30 ± 10	723 ± 14	59
GT10	985 ± 157	48 ± 13	950 ± 15	80
GT9	1564 ± 8	77 ± 0,4	1561 ± 80	128
GT8	1494 ± 19	72 ± 15	1303 ± 16	122
GT7	931 ± 10	42 ± 8	690 ± 78	76
GT6	2220 ± 14	98 ± 24	2119 ± 13	181
GT5	1335 ± 47	62 ± 15	1364 ± 48	109
GT4	1042 ± 17	49 ± 15	794 ± 13	85
GT3	1575 ± 77	74 ± 19	1730 ± 84	129
GT2	932 ± 50	43 ± 0,2	907 ± 0	76
GT1	1456 ± 70	65 ± 0,3	1433 ± 70	119

of U are generally linear with the concentration of P_2O_5. The sample 6, requires study and analysis being conducted to understand the phenomena responsible for the concentration of P_2O_5 and natural radionuclide.Thats why we thought it useful to exploit the possible variations of the isotopic ratio of the activities of radionuclides in two series of natural radioactive decay of ^{238}U and ^{232}Th.

The isotope ratios are very useful in exploiting the experimental values of specific activities:

The isotope ratios of the same series of natural radioactive decay provide information on the state of secular equilibrium between each precursor and its decay

Table 3 Concentrations (Bq/kg) of (^{230}Th, ^{232}Th) in the series of natural phosphate Ghar Tajer

Ech	^{230}Th (Bq/kg)	^{232}Th (Bq/kg)
GT20	1194 ± 114	56 ± 9
GT19	1038 ± 249	24 ± 12
GT18	399 ± 53	6.27 ± 4
GT17	1157 ± 229	27 ± 11
GT16	882 ± 105	13 ± 4
GT15	1041 ± 182	17 ± 8
GT14	982 ± 167	12 ± 6
GT13	609 ± 124	6.5 ± 4.8
GT12	1182 ± 192	19 ± 7.1
GT11	503 ± 161	9.3 ± 3.2
GT10	1143 ± 336	10.2 ± 4
GT9	1025 ± 110	8.6 ± 3.1
GT8	1364 ± 149	12.7 ± 4.70
GT7	1129 ± 152	11.5 ± 5
GT6	1822 ± 399	53.8 ± 18.8
GT5	1753 ± 225	31.3 ± 11.7
GT4	915 ± 133	12 ± 5
GT3	1787 ± 300	28 ± 9
GT2	986 ± 202	16 ± 8
GT1	1326 ± 46	17.5 ± 2.3

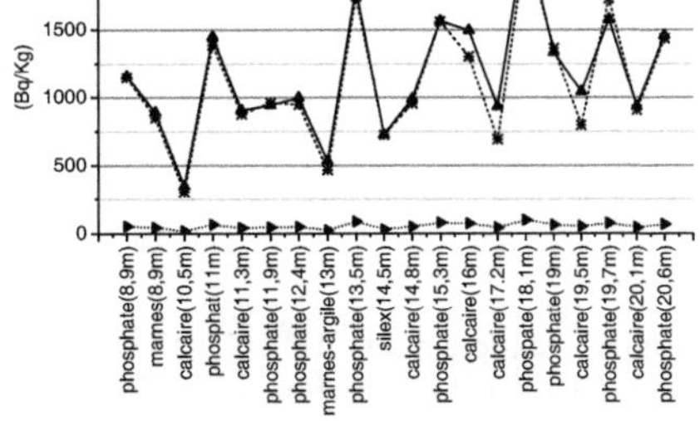

Fig. 4 Variation in ^{238}U, ^{235}U and ^{234}U as a function of depth

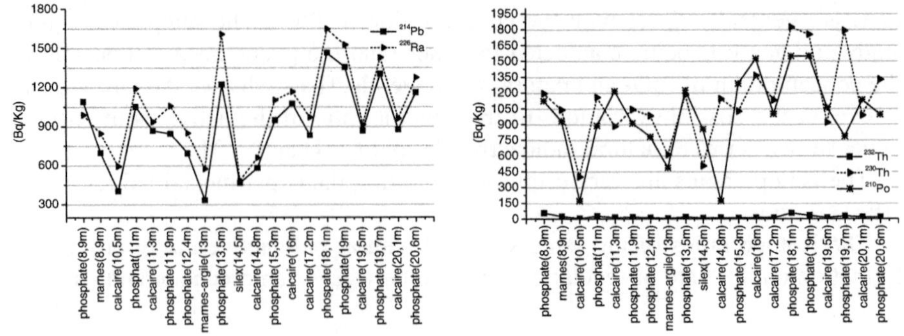

Fig. 5 Variation in ^{230}Th, ^{232}Th, ^{210}Po, ^{226}Ra and ^{214}Pb as a function of depth

products. In such cases, the activity is constant throughout the series and therefore the deposit is free from any alteration of its structures controlled by the absence of fractionation of radionuclides. The isotopic ratios of radionuclides in two separate series provide information on the origin of the deposit analyzed.

$^{232}Th/^{238}U$

In the phosphate series, the samples analyzed are of marine origin because the values of the ^{232}Th/^{238}U are very low, less than 0.05 (Fig. 6). This ratio decreases with depth with a significant increase as shown in the sample GT6, characterized by a high uranium content.

Low ratio values ^{232}Th/^{238}U, suppose that samples of drilling studied are deposited in a reducing environment (Ivanovich and Harmon 1982). The same authors have shown a correlation between the thorium content and character of the terrigenous rocks. Th is associated with argillite and carbonate fraction of sediments.

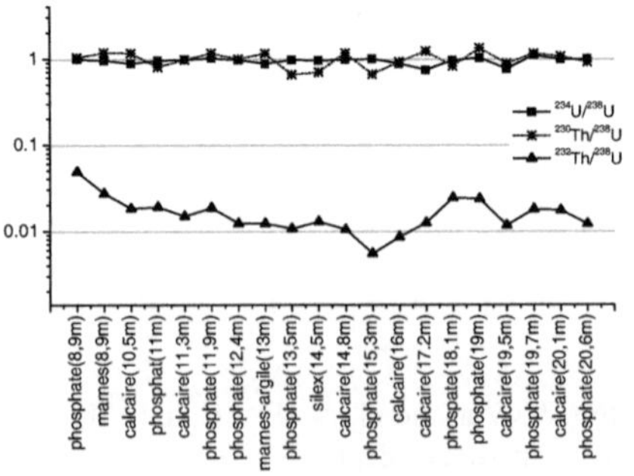

Fig. 6 Isotopic ratio ^{234}U/^{238}U, ^{232}Th/^{238}U and ^{230}Th/^{238}U as a function of depth

$^{234}U/^{238}U$

Data analysis of the variation of isotopic ratio $^{234}U/^{238}U$ as a function of depth shows that:

- On samples of predominantly phosphates are within experimental error close to unity, this implies that the sediment is not altered. The small deviation from the unity of these ratios in phosphates is attributed to alpha recoil.
- For other types of structures, in particular limestone facies have values below 1. Corresponding sediments are altered with possible leaching of uranium and its mobility to the adjacent layers below.

$^{230}Th/^{238}U$

The isotopic ratio $^{230}Th/^{238}U$ (Fig. 6) are practically equal to 1 for the phosphate horizons (samples 1, 5, 7 and 20) and limestone (samples 13 and 19). Irregular variation of this report is attributed to exchange phenomena at the interfaces of layers of different backgrounds.

- Samples located respectively at 9.9 m (marl) and 10.5 m (limestone) have a ratio higher than 1. This explains their alteration with leaching of U.
- The sample (11 m phosphate) has a ratio lower than 1. This is attributed to the displacement of uranium, possibly from the upper layers.

The same process is observed for the remaining samples of the drilling. Indeed, the horizons phosphate (13.5 m) and Flint (14.5 m) are enriched in U from the leaching of phosphate (11.9 m) and marly argillite (13 m). Leaching of U from limestone (14.8 m) enriched the sample of phosphate (15.3 m), this phenomenon is identical between adjacent layers: (sample at 17.2 m and one at 18, 1 m, as for the sample with that at 19 m to 19.5 m).

Conclusion

Before completing the research on compartment solid phase characterization of sorption of radionuclides and study their leaching from these phases, the preliminary operating results helped to establish a geochemical approach of the series of samples this system (Ghar Tajer). Radiochemical analysis by alpha spectrometry shows that:

- The specific activities of the identified radionuclides are independent depth of drilling. They seem to depend on the nature of the deposits and processes of assimilation of appropriate radionuclides – Radionuclides seem rather concentrated in the sediment of phosphates.
- The deposits are mainly of marine origin, and most of the calcareous sediments are more or less altered.

- In this work, we have shown, moreover, through the use of isotopic ratios, that the porous interface (limestone) is generally places of preferential alteration. This phenomenon is elucidated by the fractionation of U from these sites and their mobility to the lower layers.
- For the phosphate samples, the U content varies from 76 ppm to 181 ppm (GT6).
- For the intercalated layers of limestone, the U content varies from 28 ppm to 122 ppm (GT8). The higher uranium concentration in GT6 and GT8 formations relative to strata above and below results from the accumulation of phosphorites in this area due to its leaching from above and redeposited below. The process is not uniform, various sections and various horizons have accumulated more U than others.
- Marly sediments have U concentration equal to 72 ppm, whereas in the marly argillite, it is 43 ppm. Analysis of these results shows that in despite of their degree of weathering, limestones have an economic interest in the recovery process of U.
- In rocks called “DERANGEMENTS” which constitute a permanent obstacle to the exploitation of phosphate ores 1 contains 13 ppm of U.
- The study performed on the sample size of phosphate GT20, age Ypresian (−53 AM and 46 AM), shows that:

  - The fine fraction (0.05–0.1 mm), which is 2.17% of the total amount of GT20 has a specific activity of 768 Bq/kg of this fraction.
  - The average fraction (0.1–0.5 mm) which constitutes 65.78% of the total GT20 has a specific activity of 1336 Bq/kg of this fraction.
  - The fraction of diameter (0.5 to 2.25 mm) which constitutes 32% of the total amount of GT20 has a specific activity of 1220 Bq/kg of this fraction.

Acknowledgements This work is done through the collaboration programs PICS802, AECID A/026017/09, respectively, IHPC Strasbourg-France and the Laboratory of Nuclear Physics of Seville. And as part of the UMR between Radiochemistry Unit of the University Hassan II-Mohammedia and CNESTEN.

References

Benfkira O (1980) Evolution sédimentologiques et géochimiques de la série phosphatée du Maactrichtien des Ouled Abdoun (Maroc). Thèse Doct. spec. Géol. Appl, Univ. Grenoble: 164

Ivanovich M, Harmon RS (1982) Uranium series disequilibrium; application to environmental problems. Clarendon Press. Oxford: 32–39

Jourani M (1988) Les phosphates méso-cénozoïques du bassin du Mrijat et leurs paléoenvironnements. B.S.G.F., 1997, t. 168 n°5, pp: 585–600

Moumni A (1979) Etudes radiogéologique, géochimique et pétrographique des phosphates de Youssoufia (Maroc occidental). Thèse de 3ème Cycle, Nancy, p. 186. Mountain Association of Geologists 1993 Fall Field Trip: 45

Talvitie N A (1972) Electrodeposition of actinides for alpha spectrometrie determinations, Anal. Chem. 44(1972): 280

Part 3
Cleaning up Technologies for Water and Soil

Conception for Diversion of Runoff Implementing the Trünzig Uranium Tailings Pond into the Regional Catchment Area

Ulf Barnekow, Marcel Roscher, Matthias Bauroth, Gunter Merkel, Manuela Voßberg

Abstract. From 1960 to 1990 the former Soviet-German Wismut company milled uranium ore at the Seelingstädt mill in Thuringia, Germany. The mill tailings were disposed into Trünzig tailings pond from 1960 to 1967. The tailings pond is currently dry-decommissioned in situ. The paper presents the conception for diversion of runoff from the remediated Trünzig uranium tailings pond into the regional catchment area enclosing four main ditches from the pond to the receiving streams and one planned runoff retention basin to control runoff from the southern part of the tailings pond in the long term.

Ulf Barnekow
Wismut GmbH, Abt. Bergbausanierung/Altstandorte (SBE1), Jagdschänkenstraße 29, 09117 Chemnitz, Germany

Marcel Roscher
Wismut GmbH, Abt. Bergbausanierung/Altstandorte (SBE1), Jagdschänkenstraße 29, 09117 Chemnitz, Germany

Matthias Bauroth
Wismut GmbH, Abt. Bergbausanierung/Altstandorte (SBE1), Jagdschänkenstraße 29, 09117 Chemnitz, Germany

Gunter Merkel
Wismut GmbH, Niederlassung Ronneburg, Projekt Industrielle Absetzanlagen, Paitzdorfer Straße 34, 07580 Ronneburg, Germany

Manuela Voßberg
UBV – Umweltbüro GmbH Vogtland, Thossener Straße 6, 08538 Weischlitz, Germany

B. Merkel, M. Schipek, *The New Uranium Mining Boom*
DOI 10.1007/978-3-642-22122-4 © Springer 2011

Introduction

From 1947 till 1991 the Soviet-German Wismut company produced a total of 216,000 t of Uranium with about 110,000 t produced in the Seelingstädt mill from 1960 till 1990. According to the Wismut Act dated October 31, 1990 a total of € 6.6 billion were committed to the remediation of the uranium mining liabilities in south-eastern Germany. The state-owned Wismut GmbH became responsible for remediation of the uranium mining and milling sites. As part of its mine closure program Wismut GmbH is remediating the Trünzig Uranium tailings pond covering 1.2 km^2 and containing a volume of 19 million m^3 of tailings. The Trünzig tailings pond was erected in the former Uranium open pit Trünzig-Katzendorf mined out before from 1948 till 1957. From 1960 till 1967 both tailings from acid leaching and from soda-alkaline leaching of uranium ores were disposed here into separate partial ponds A and B. From 1967 till 1990 tailings disposal continued accordingly into the Culmitzsch tailings pond, located in the vicinity north of the Trünzig tailings pond. After the end of the operational phase only the partial pond B was used as a water storage pond till 2002. Air-exposed tailings beaches were covered with waste dump material from the surrounding waste dumps during the 1970s. Dry-decommissioning in situ started in the early 1990s and shall be completed by 2013. To date the Trünzig tailings pond has been interim covered, nearly completely re-contoured and partly final covered. The final cover is currently being vegetated.

The paper presents the conception for diversion of runoff from the remediated Trünzig uranium tailings pond into the regional catchment area. The directions of the main runoff ditches on the pond area were planned based on the detailed design for re-contouring and final covering of the entire tailings pond. The runoff from the pond area is diverted via constructed ditches to the two receiving streams Finkenbach creek and Culmitzsch/Pöltschbach creek. The conception encloses four main ditches from the pond to the receiving streams and one runoff retention basin to control runoff from the southern part of the tailings pond in the long term. The runoff to the East flows via ditches through six pools erected before as compensation measures for nature conservation. According to the requirements set by the permitting authorities the conception is to preserve the hydrological situation given downstream in the receiving streams before remediation (1990) with respect to flood water flow resulting from the 100 year precipitation event.

Please note that the Trünzig tailings pond and its surrounding area are located in Thuringia, except the runoff retention basin east of the tailings pond.

In 2010 Wismut filed an application for a plan approval under water law for construction of the runoff ditches from the borderline of the tailings pond to the receiving streams and for diversion of runoff. In April 2011 Wismut submitted the amendment to the above application. The paper finally presents an outlook to the future.

Site Characterization

The uranium open pit mining nearby Trünzig lasted from 1948 till 1957. One layer of uranium-rich gray mudstone or siltstone respectively was mined as uranium ore. The ore layer was part of nearly horizontally bedded sedimentary rock layers of Permian age situated in a half-graben. The Permian rocks are underlain by Ordovician shales folded in the Variscian orogenesis. The Permian sedimentary rocks at the Trünzig tailings pond consist from bottom to top of a base conglomerate, mudstone, siltstone and sandstone layers (in particular the "Culmitzsch sandstone") overlain by a thin layer of weathered dolomite residues, a weathered mudstone and a sandstone layer of lower Triassic age.

Due to open pit mining overburden materials were relocated onto the waste rock dumps called Westhalde, Osthalde, Nordhalde and on waste rock dumps located inside the open pit. Four waste rock dams were erected from 1957 till 1960 enclosing two partial ponds. The dams include the West dam, East dam, Karbonathauptdamm and the separating dam. The separating dam has a vertical sealing wall in the dam centre. Tailings disposal lasted from 1960 till 1967 including the erection of the autostable tailings dam called North dam.

The Trünzig tailings pond is located in the hydrological catchment area of the Culmitzsch/Pöltschbach creek. The Culmitzsch/Pöltschbach creek is classified as a river of second order with some feeder rivers like the Finkenbach creek. The Culmitzsch/Pöltschbach creek flows into the river Weiße Elster. The respective catchment area is 40.8 km^2 (including the area of the Trünzig tailings pond and the south dam of the Culmitzsch tailings pond). The entire length of the creek is about 11 km. The preferential land use in the catchment area is agriculture covering 75% plus 15% used as forest land and 10% residential areas.

Dry Remediation In Situ

The remediation has to ensure the safe long-term storage of the tailings reducing the additional equivalent dose to the population from all pathways to less than 1 mSv/year. Requirements for waste water discharge into the receiving streams were set by the authorities of Thuringia. Dry-remediation in situ started with first securing measures against acute risks in 1991 and will last till 2013. Measures against acute risks comprised, among others, the (re-)construction of catchment systems to collect the entire seepage and runoff from the tailings ponds and construction of the water treatment plant as well as the installation of a monitoring system to track the impacts of the tailings ponds before, during and after remediation. The preferred option for remediation was agreed by the authorities and encloses the following fundamental remedial steps:

- Expelling of pond water, collection of seepage and runoff including water treatment before being discharged into the receiving streams.
- Interim covering of remaining air-exposed (fine) tailings surfaces to create a trafficable working platform for further remedial works.

- Reshaping of dams to grant sufficient seismic dam stability to the long term and re-contouring of the pond surfaces with respect to the functionality and erosion stability of the final cover and landscaping aspects.
- Final covering including construction of water diversion systems, access roads and vegetation with respect to (restricted) re-use of the covered surface.
- Construction of a runoff diversion system including a runoff retention basin downstream of the tailings pond.

Figure 1 shows an aerial view on the Trünzig tailings pond in 2011 (view from NE to SW). One can see the final covered and vegetated areas on the tailings pond and its surroundings dams in the foreground and the still ongoing re-contouring in the background. The Culmitzsch/Pöltschbach valley is located in the foreground (north of the Trünzig tailings pond), the Finkenbach valley on the left hand side.

The final cover design depends on the underlying tailings types. On sandy tailings beach zones (in foreground in Fig. 1) the final cover consists of a ca. 2 m thick storage layer including a compacted bottom lift (0.5 m) above a compacted interim cover of 1 m thickness. The final cover material consists of waste dump material (mixed-grained soil) from the Lokhalde waste dump located in the vicinity north of the Tailings pond. On fine tailings zones the functionality of the final cover shall be granted by a minimum 2.5 m thick layer consisting of mixed-grained waste dump material (from Lokhalde waste dump and from the nearby Finkenbach waste dump). Among other requirements the final cover shall grant erosion stability of the surfaces and shall provide sufficient plant available water during growing seasons.

Fig. 1 Aerial view from NE to SW on the Trünzig tailings pond

Vegetation and Landscaping

After initial establishment of grass for granting erosion protection forestation is planned for the final cover on tailings beaches in order to assist for controlling the percolation rate. On fine tailings areas grassland will be established. Currently it is discussed with the authorities to allow for an establishment of trees on a certain part of those areas. Fig. 2 presents the generalized vegetation plan for the long term status (after establishment of vegetation).

Based on a detailed ecological survey of the entire site the landscape planning and vegetation of the Trünzig tailings pond and its surrounding dams and waste rock dumps was designed and is currently being realized with respect to balancing ecological intervention measures and compensatory measures as well as public interests.

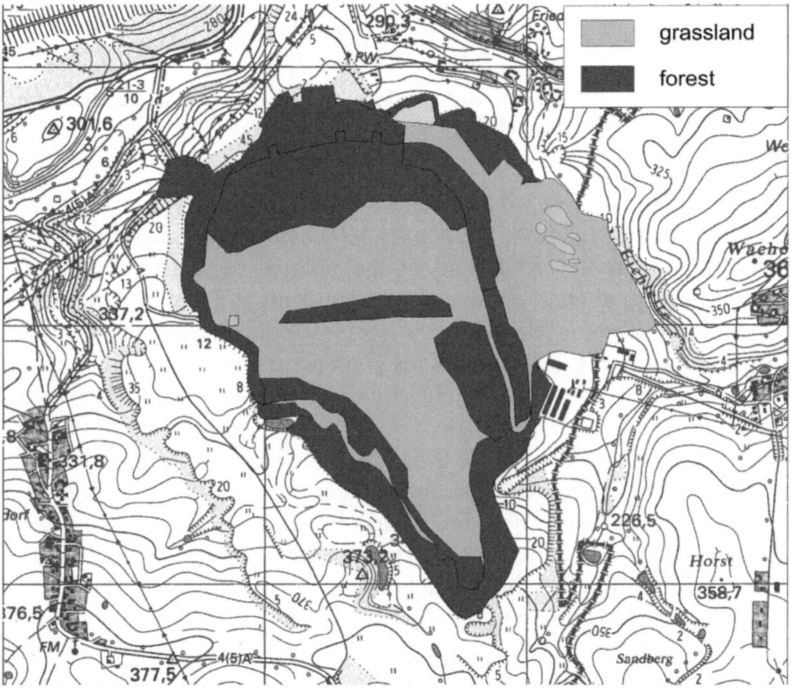

Fig. 2 Landscaping and vegetation plan for Trünzig TP and surrounding dumps

Conception for Diversion of Runoff

The directions of the ditches for diversion of runoff on the pond area were planned based on the detailed design for re-contouring and final covering of the entire tailings pond. The runoff from the borderline of tailings pond is diverted via con-

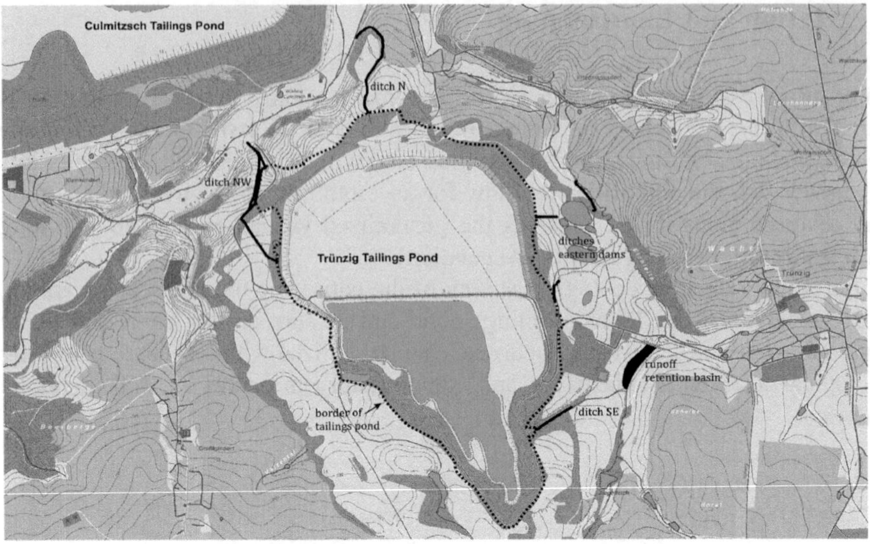

Fig. 3 Topographic map of the Trünzig tailings pond including the surrounding catchment area (The runoff ditches and the runoff retention basin are marked in black)

structed ditches to the two receiving streams Finkenbach creek (East) and Culmitzsch/Pöltschbach creek (North). The conception encloses four main ditches for diversion of runoff from the borderline of the tailings pond to the receiving streams: ditch northwest, ditch north, ditches eastern dams and ditch southeast.

In addition the conception encloses a runoff retention basin located in the Finkenbach valley east of the Trünzig tailings pond across the border on Saxonian territory. The locations of the runoff ditches and of the runoff retention basin are shown in Fig. 3.

The ditches are to divert the runoff from the remediated tailings pond to the Culmitzsch/Pöltschbach creek and Finkenbach creek safely in the long term. For hydraulic stability the surface of the ditches has been or will be reinforced with gravel or stones, locally with concrete along steep slopes. Along flat reaches grass has been or will be established at the trench bottom.

The runoff to the East is diverted via ditches through six pools. The water diversion system at the Eastern dams was erected before as a compensation measure for nature conservation. The runoff ditch to the north was erected beforehand. The respective catchment areas are connected with the receiving streams.

One fundamental requirement set by the permitting authorities of Thuringia for the conception for diversion of runoff is to preserve the hydrological situation given downstream in the receiving streams (Culmitzsch/Pöltschbach creek and Finkenbach creek) before remediation (1990) with respect to flood water flow resulting from the 100 year precipitation event. For this a hydrological modeling of the surface runoff was prepared for the catchment area of the Culmitzsch/Pöltschbach creek. The hydrological modeling was carried out for the state before remediation

Table 1 Runoff at selected points of the streams with respect to flood water flow resulting from the 100 year precipitation event (without runoff remediation basin)

Selected points of the streams	Runoff [m^3/s] state before remediation (1990)	Runoff [m^3/s] interim state after remediation	Runoff [m^3/s] long term state after remediation
Finkenbach creek before confluence with Culmitzsch/Pöltschbach creek	9.3	10.0	9.9
Culmitzsch/Pöltschbach creek before confluence with the river Weiße Elster	37	37	36

(1990) and two states after remediation – an interim state with only grass-vegetation on the remediated areas and the long term state with establishment of grown-up trees (see Fig. 2) – with respect to flood water flow resulting from the 100 year precipitation event. The results of the hydrological modeling at selected points are listed in Table 1.

The results in Table 1 show that the hydrological situation before remediation (1990) could be preserved for the Culmitzsch/Pöltschbach creek. In the Finkenbach creek the runoff will be increased after remediation. Therefore a runoff retention basin is planned to be constructed in the Finkenbach valley to control runoff from the southern part of the tailings pond into the Finkenbach creek in the long term. The planned runoff retention basin shall reduce the runoff in the Finkenbach creek to 8.9 m^3/s for the two states after remediation, thus preserving the hydrological situation before remediation (1990) for the Finkenbach creek in accordance with the requirements set by the authorities.

In 2010 Wismut filed an application for plan approval under water law for construction of the runoff ditches from the borderline of the tailings pond to the receiving streams and for diversion of runoff. In April 2011 Wismut filed an amendment to this application. The plan approval procedure is currently discussed with the authorities. Wismut currently prepares a separate application for plan approval under water law for the construction of the runoff retention basin, which is located in Saxony.

Outlook

Wismut is currently looking forward to receive a first partial permit for the NW ditch. This would connect the catchment area of the northern partial pond to the Culmitzsch/Pöltschbach creek. It is planned to construct the remaining parts of the NW ditch in 2011.

The plan approval procedure for the erection of the runoff retention basin will start in summer 2011. The situation will be complicated, because Wismut will have to handle to separate but interconnected plan approval procedures in Thuringia and Saxony for the diversion of runoff from the Trünzig tailings pond.

Acknowledgements We would like to thank our colleagues involved in the preparation of technical documents related to the applications for plan approvals under water law and for other permits, namely Mr. Mirko Köhler, Mr. Bodo Raßmann and Mr. Jürgen Priester. The conception for diversion of runoff was based on numerous technical documents prepared before by external contractors, in particular on several technical designs prepared by C & E GmbH (Chemnitz). In addition the technical design for the runoff retention basin in the Finkenbach valley was prepared by IWU GmbH (Chemnitz).

Hydrogeological Evaluation of Flooded Uranium Mine Cavities in Hungary

Gábor Földing, Gabriella Szegvári, Mihály Csővári

Abstract. In Hungary the uranium ore mining was terminated in 1997. Taking into account of the rate of water level rising, it can be assumed that deep mines will have been flooded between 2015–2020. The flooded mine water will leave the mines through common adit after the flooding. For hydrogeological and geochemical evaluation of the processes taking place in the deep mines, drillings from the surface was carried out in 2008 aiming at mine water sampling from deep mines. On the bases of the obtained data hydrogeological effects of the flooding of mine cavities as well as numerical hydrogeological modeling was performed.

Introduction

In Hungary (MECSEK-ÖKO Zrt) the uranium ore mining was terminated in 1997. From that time the deep mines, total cavity volume of which is app. 14 million m^3, are under flooding. For the time being app. 10,2 million m^3 cavity volume has been flooded. The mine water will leave the mines through common adit after the end of flooding. Over the mining period, the mine water from the area in question was contaminated with uranium (2–3 mg/l). In the case if this overflow would be also contaminated, it has to be treated. It is expected that the total volume of the water from these mines will be app. 0,8–1 million m^3 yearly. There is no extra

Gábor Földing
MECSEK-ÖKO Co., 7633 Pécs, Esztergár L. u. 19., Hungary

Gabriella Szegvári
MECSEK-ÖKO Co., 7633 Pécs, Esztergár L. u. 19., Hungary

Mihály Csővári
MECSEK-ÖKO Co., 7633 Pécs, Esztergár L. u. 19., Hungary

B. Merkel, M. Schipek, *The New Uranium Mining Boom*
DOI 10.1007/978-3-642-22122-4 © Springer 2011

capacity on the existing water treatment station for treatment of this additional mine water, therefore the station might be enlarged.

For the planning of the future mine water treatment it is important to have accurate database for the parameters of the expected mine water from this area. For this hydrogeological and geochemical evaluation of the processes, drillings from the surface (Fig. 1) was carried out in 2008 aiming at mine water sampling from these

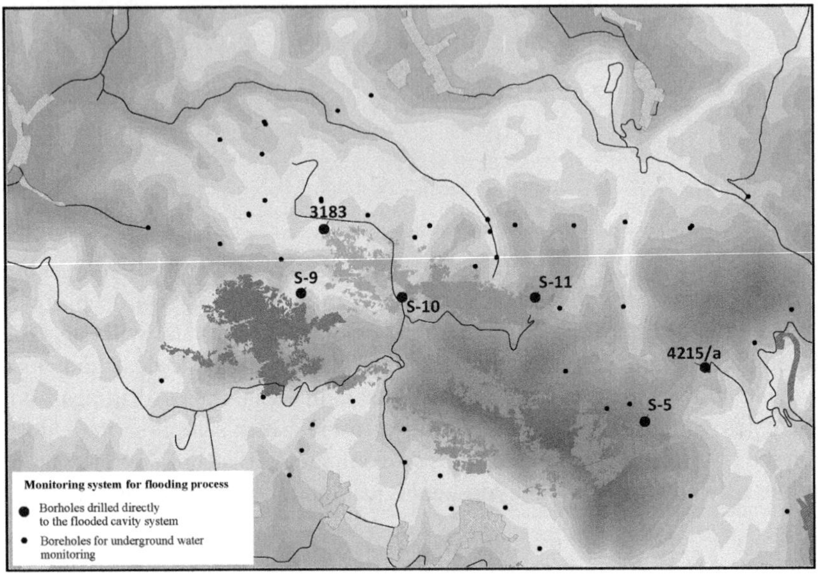

Fig. 1 Hydrogeological monitoring system of the northern mines

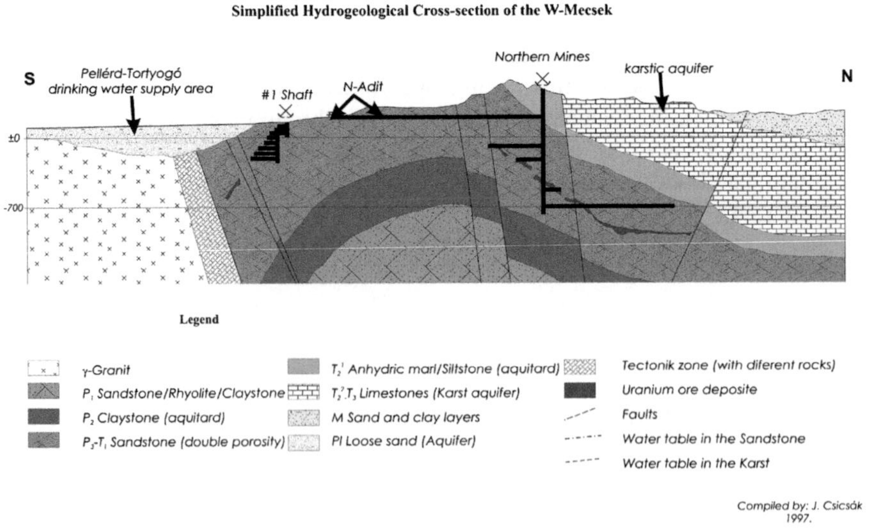

Fig. 2 Simplified hydrogeological cross-section of the Western-Mecsek

deep mines. On the bases of the obtained data hydrogeological and hydrochemical effects of the flooding of mine cavities as well as numerical hydrogeological modeling was performed.

The simplified geologic-hydrogeologic situation and the most important mine cavities and adit level is shown in Fig. 2. From Shaft No. I water-pumping is proceeding causing depression to protect Pellérd-Tortyogó drinking water resource situating southward from Shaft No. I. The deep northern mines (mine No. II-III-IV-V) can be leave to oneself, because the original hydraulic conditions in the future won't restore. With regard to the northern and eastern adit, which will keep their draining function for a long time, even for some hundred years, the mining cavities will always have a drainage in the elevation of 219–240 m according to average Baltic Sea level. This drainage will protect the northern karstic aquifers from polluted, high uranium containing water.

State of Flooding

In the case of mine No. II-IV-Vwe have data from measuring water-levels in boreholes drilled to cavity system. In these mines cavity system are interconnected, it can be supposed that the measured water-level represent more or less the whole system. Because of the complete flooding of mine No. IV-V systems the velocity of the water rise is app. 4.7 m/month, 56.4 m/year. Calculating with this value the reach of 500 m height to the adit level takes 8.9 years, so the mine water might flow out in 2019. In the case of mine No. II there is 515 m height difference and 3.65 million m^3 cavity volume is between present water level and the adit level. Calculating with the velocity of 4 m/month it takes 11 years to reach the adit level, so it might be in 2021 (Fig. 3).

As it can be seen on Fig. 3, data we gained from water level measurements and the predicted calculations based on cavity volumes about the data of the end of flooding is in good correlation. So the outflowing of mine water from northern mines can be expected around 2019–2021.

The cavity system of mining territory No. III is not in connection with the underground with those of No. II, IV, V. Even the up-watering process has started sooner in this area, in the year of 1996. The water level in borehole S-5 (drilled to mine No. III) was 142.58 mBf in the end of 2010. The velocity of flooding in 2010 was 1.8 m/month (it was 1 m/month in 2009). There is 20–40 m difference between the measured and the estimated water-level. The cavity volume below the adit level is 201,600 m^3, this means 85,4 m in depth. If the rate of flooding remains this 1.79 m/month, then it takes 4 years to reach the adit level (21.5 m/year), so water could flow out by the beginning of 2015.

Since the end of 1999 only seepage water flowing out from the northern adit, volume of this is 182,500–292,000 m^3/year and strongly affected by precipitation. Based on water-pumping data during mining activity there is an estimation about the total volume of water flowing out of the northern adit, that will be around 730,000–912,500 m^3/year after the end of flooding.

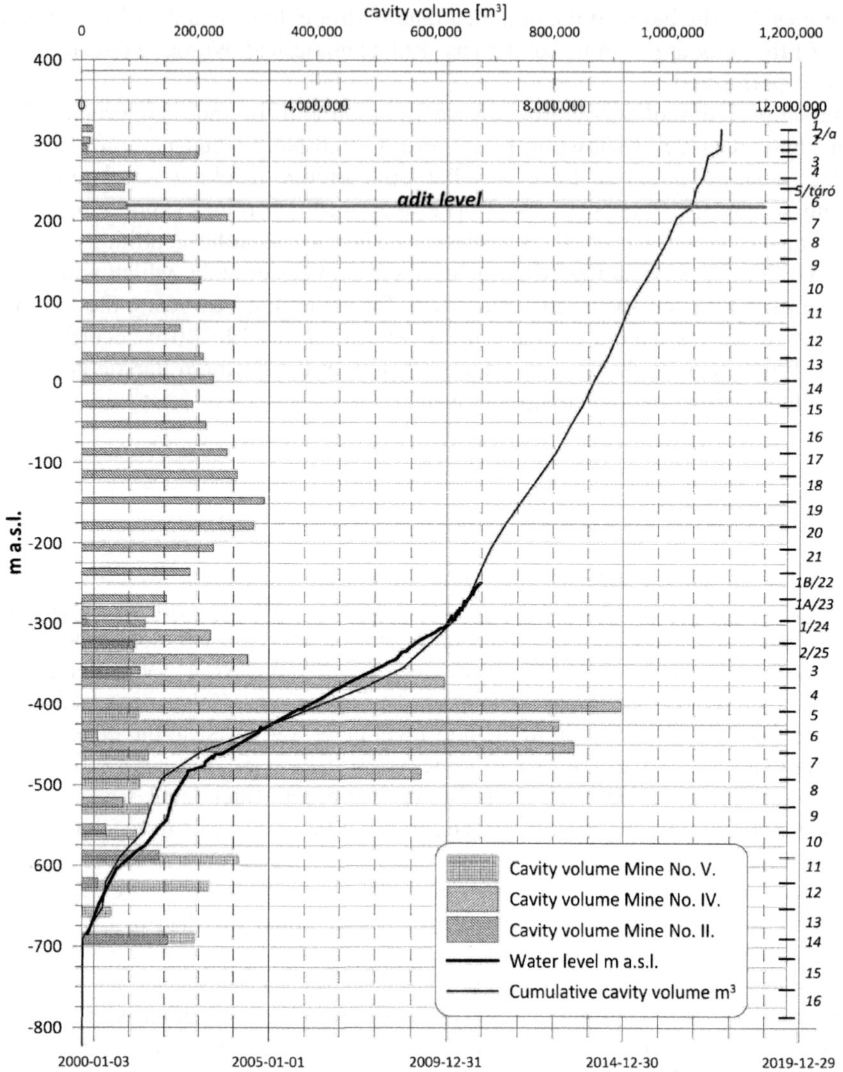

Fig. 3 Measured water level and cumulative cavity volume in the northern mines, prediction for the date of end of flooding

Mine Water Quality in the Deep Mines

The mine water composition in deep mines can be estimated on the bases of the data of drillings trial in 2008 when water samples were obtained from Mine No. III, IV. Uranium in samples from mine No. III was 10 mg/l, while that of for mine No. IV was 5 mg/l. Though the sampling procedure has to be repeated as the first approximation these values are used for prediction of the uranium concentration in the two mining areas.

The main parameters of the voids and the water inflow and the hydraulic retention time are presented in Table 1.

Accepting that after flooding concentration vs. time will be governed only by dilution alone (ideally mixed tank scenario) without any uranium inflow, the uranium concentration in mine water from that areas will be developed as shown in Fig. 4 using Eq. (1) (taking $T = 16$ years in average):

$$C_{(t)} = C_{(0)} * e^{-1/T} \quad (1)$$

(It is worth mentioning that for practical uses the $C_{(t)}$ can be calculated also by the $C_{(t)} = C_{(0)} * q^t$ equation, where the $q = 1 - t/T$, especially if $T > 10$.)

In the case of geochemical processes resulting in the precipitation of the dissolved uranium the attenuation of the uranium concentration in mine water would be faster. Though such processes cannot be excluded but taking into account the low pyrite content of the host rock their role will be likely marginal.

Using the above-mentioned conditions it is expected that from deep mining areas the uranium in mine water leaving the area will be dropped below 2 mg/l in

Table 1 Hydraulic retention time for the mining areas mine No. II, IV and V as well for No. III

Mine	Volume of mine cavities V (Mm^3)	Mean mine water inflow Q (Mm^3/a)	Hydraulic retention time T (a)
Mine No. II, IV and V	10.3	0.4	14.26
Mine No. III	3.6	0.2	18

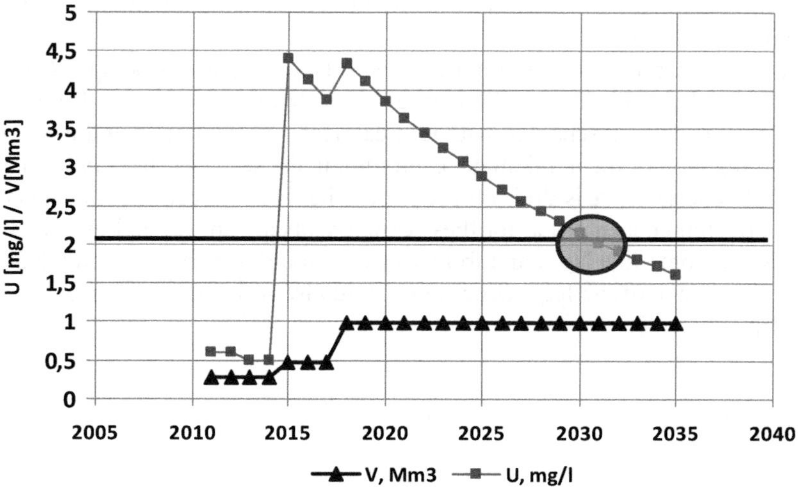

Fig. 4 Development of the uranium concentration vs. time plots for mine waters (Mines No. II-IV-V; Mine No. III) after flooding of the areas

app. 16–18 years in the case of mines No. II, IV and V while for these 22–24 years is needed in the case of the mine No. III.

Taking into account that the mines will be flooded (according to the calculations) in 2015 (Mine No. III) and in 2019 (Mines No. II, IV and V), the expected uranium concentration will be developed in the northern deep mines according to Fig. 4. It means that the today's discharge limit (2 mg/l) will be met between 2035–2040 for this water.

Of course, additional sampling and evaluation of the data has to be carried out in the near future to get data that are more reliable for the development of the composition of the mine water in northern deep mining areas. This is important from the point of view of the planning of the possibly needed water treatment capacity for treatment of the additional water from the deep mines.

Water Treatment

It can be expected that the uranium concentration in mine water from deep mines No. II, IV and V. will drop below 2 mg/l around 2035–2040. However because these waters will be mixed with the seepage from the adit (uranium concentration of which will be less than 1 mg/l) therefore the resulting uranium concentration in combined mine water from this area will be lower. The total volume of this combined water from deep mine areas will amount to 0,88 million m^3/a.

It can be seen that the uranium concentration in the outflow reach its highest value (4–5 mg/l) in between 2015–2020, and then drops under the limit value around 2034. However, this time depends on the filling up of the mine cavities: if this would happen later than the limit value will be reach also later. It is also important to underline that the calculation is based on the ideally mixed state (ideally mixed tank), Therefore, this figure shows only the very first approximation of the concentration development.

It is important that the volume of mine water to be treated in period app. 2018–2030 will be on the level of 1.4 million m^3/a. After this the mine water from the deep mine's area can be discharged without treatment. For the time being it is not known the mitigation of the uranium concentration in the seepages and mine water in shaft No. I. Therefore, it is difficult predicting the fate of those waters. Nevertheless the uranium concentration for these components are above the limit, therefore mine water from mine No. I and from seepage should be treated beyond 2030. Of course, the change of the legal conditions could also influence the longevity of the treatment.

Numerical Modeling

Numerical three-dimensional hydrogeological modeling was performed to evaluate the rate of flooding and the hydrogeological effects of polluted mine water around the cavity system. The model was performed with Feflow 5.4 and 6.0 software. The extension of the model was 30 × 30 km, and it consisted of 27 layers. Figure 5 shows the cross-section of the grid. Hydraulical and geologic parameters and boundary conditions were gained from local database.

One of the most important results of simulation was the prediction of flooding period. In the model it was examined at different storage capacities (Fig. 6) from the termination of water pumping in deep mines. The calculated longevity of flooding varies between 23–34 years.

Comparing these calculations with our measured water levels and flooding velocities, $S = 10^{-7} - 10^{-4}$ storage shows the best correlation with the measured values.

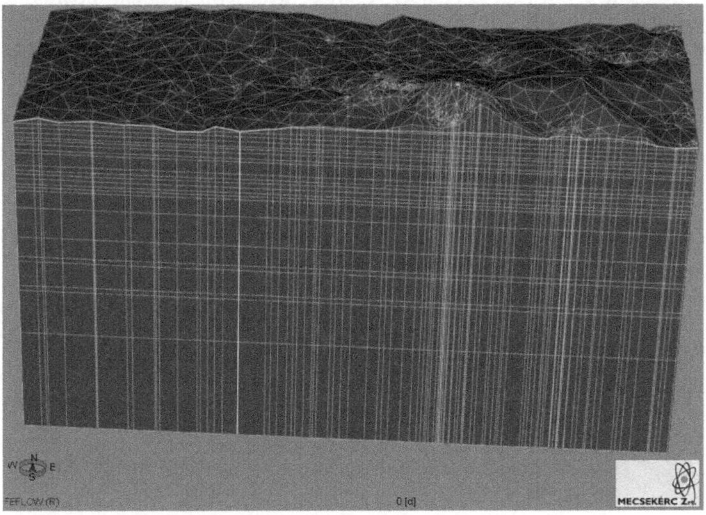

Fig. 5 3D cross-section of the modeled area

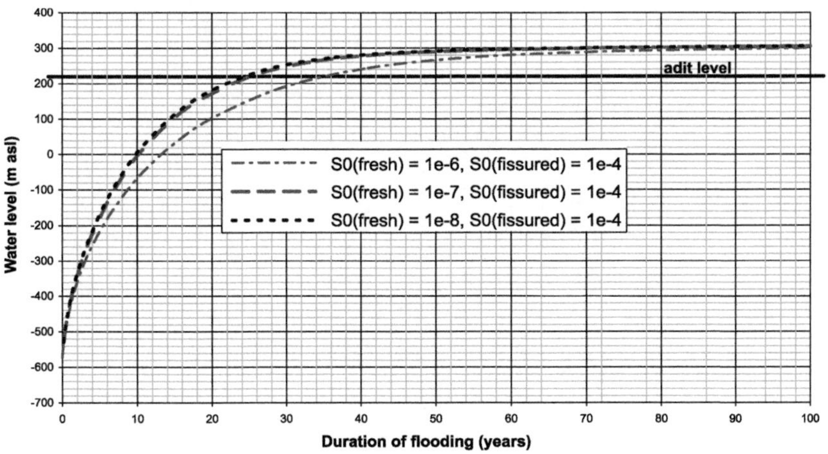

Fig. 6 Calculated mine water level in the center of depression at different storage

References

Csicsák J, Csővári M, Éberfalvi J, Lendvai Zs (2000) Mine Water Treatment with yellow cake by-production, Hungary. International symposium on the uranium production cycle and the environment. 141–142

Földing G, Csicsák J (1998) Short-term programme on the qualification of Boda Aleurolite Formation. Final research report. Volume 5. Hydrogeological, hydrogeochemical monitoring programme

Golder Associates (Hungary) Co. (2003) Hydraulical modelling of abandoned uranium mines and its surroundings

Mecsekérc Co. (2004) Final report on the recultivation and abandonment of underground facilities of uranium mining in Mecsek

Mecsek-Öko Co. and Mecsekérc Co. (2011) Annual report of long term uranium monitoring activity. Summary. 6–11

Paul M, Berta Zs (2010) Annual report on the cooperation between Wismut GmbH and Mecsek-Öko Co. 32–40

Natural Radioactive Elements in the Region of Closed Uranium Mines on Stara Planina, Eastern Serbia

Jovan Kovačević, Zoran Nikić, Petar Papić

Abstract. The region of the Mezdreja, Gabrovnica and Srneći Do uranium mines on Stara Planina Mt., eastern Serbia, was prospected for concentrations of uranium, thorium, equivalent radium, radon and potassium in subsurface and surface waters, stream deposits, soil (humus), mine spoil heaps and air. Field activities included radiometric survey, emanation and hydrochemical tests, and sampling for laboratory analyses. The extent, level and kind of water, soil and air contamination by radioactive elements are considered on the basis of field and laboratory data. The presence of radionuclides around the closed uranium mines was established through geological investigation.

Introduction

The closed uranium mines of Mezdreja, Srneći Do and Gabrovnica in the Janja ore field are centrally located on Stara Planina, eastern Serbia. Concentration levels of the natural radioactive elements in water, soil, stream deposits, waste heaps and air were measured only once near the old mines.

Jovan Kovačević
Geological Institute of Serbia, 11000 Belgrade, Rovinjska 12, Serbia
E-mail: jovan.kovacevic@sbb.rs

Zoran Nikić
University of Belgrade, Faculty of Forestry, Department of ecological engineering in soil and water resources protection, 11030 Belgrade, Kneza Višeslava 1, Serbia
E-mail: zoran.nikic@sfb.bg.ac.rs

Petar Papić
University of Belgrade, Faculty of Mining and Geology, 11000 Belgrade, Đušina 7, Serbia
E-mail: ppapic@rgf.bg.ac.rs

B. Merkel, M. Schipek, *The New Uranium Mining Boom*
DOI 10.1007/978-3-642-22122-4 © Springer 2011

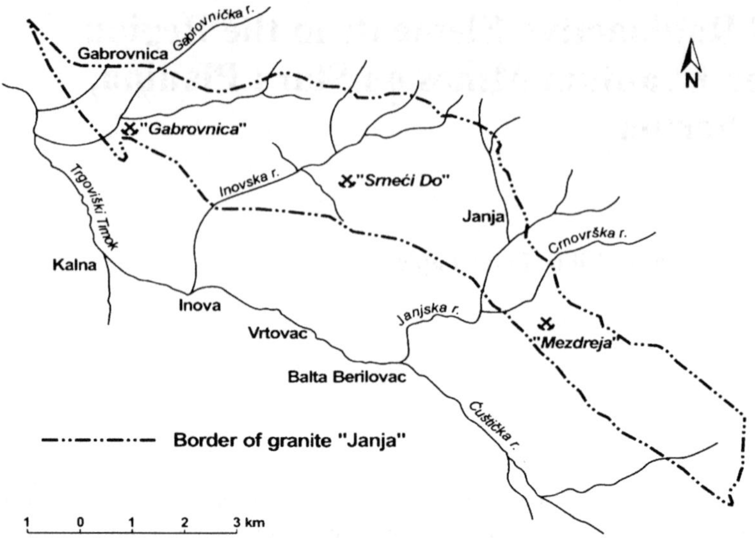

Fig. 1 Schematic locations of the closed mines

First excavations in the search of uranium were made in the Janja ore field on Stara Planina in 1949. The Mezdreja, Gabrovnica and Srneći Do uranium deposits were explored, opened and worked from 1951 to 1958. A small ore-dressing plant capacity 60 t/day and a larger plant 200 t/day were erected in Mezdreja and Gabrovnica respectfully, but in the late sixties the mine operations were suspended and the plants covered.

Mine employees, miners in particular, had their health regularly examined during and after the mine operation period. The area, however, was not geoecologically investigated for the state of the environment either before or after the operation of the mines.

The ore field Janja is located at altitudes between 500 and 950 m. It is drained by the Grabovačka, Inovska, Janjska and Crnovrška streams, the tributaries of the Trgoviški Timok. The major communities are Kalna, Grabovica, Inova, Vrtovac, Janja, Balta Berilovac and Mezdreja (Fig. 1).

Material and Method

The task of the study was to determine the level of natural radioactivity in the surroundings of the closed uranium mines of Mezdreja, Gabrovnica and Srneći Do (Fig. 1) and to assess the environmental impact. Geochemical, radiometric and hydrochemical methods were used in the prospecting. Areas designated for prospecting were locations of the old mine workings related to local hydrography and land configuration (Antonović 1990). The Mezdreja mine surroundings were investigated directly in spoil heaps, the mine, old dressing plant and the Crnovrška downriver from the mine to the Trgoviški Timok. The location Srneći Do was investigated in the mine and spoil dump, the Inovska downriver to the Trgoviški Timok.

The location Gabrovnica was prospected within the mine perimeter, in all waste heaps and the Gabrovnička downriver to the Trgoviški Timok. Alluvial plain of the Trgoviški Timok was prospected from the Crnovrška to the Gabrovnička river mouths and further 0.5 km downriver. Groundwater, surface water bodies and air were measured in all the mentioned locations.

Prospecting activities in the study area were the following:

A. Field activities: 1) sampling soil and stream deposits, 90 samples; 2) sampling underground and surface waters for full analyses, 25 samples; 3) sampling air (emanation), 25 points; 4) measuring radon concentration in water, 25 tests; and 5) radiometry in spoil heaps, soil and mine workings, 620 sites; and
B. Laboratory tests: 1) mechanical preparation of soil and stream deposit samples, 90 samples; 2) radiometry on uranium (U), thorium (Th), potassium (K), radon (Rn), 90 analyses and equivalent radium (eRa); and 3) full chemical analysis of underground and surface waters including microelements, 25 samples.

A Canadian-make scintillation counter GR 110 was used for radiometric prospecting of all waste heaps, old workings, surrounding fields, meadows and stream deposits in a regular network 20 × 20 m.

Radon from air was measured using a Scintrex emanometer ETR 1; the measurements are given in Emans.

The degree, extent and kind of environmental impact on soil, water and air were assessed based on the prospecting results.

Geology of the Janja Ore Field Introduction

The Janja ore field occupies an area of about 30 km^2 in central Stara Planina. In addition to Janja granite that builds most of the field, the terrain is made up of amphibolite and amphibolite schist, gneiss, serpentinite, grabbroid rocks, alkali metasomatic rocks, less of diabase, syenite and quartz porphyry. The ore field is bordered by metamorphic rocks (Kovačević 2005).

Amphibolites, most abundant NW in the field, consist of amphibole (hornblende), plagioclase, orthoclase, andesine, quartz, somewhat less of biotite, and are commonly dark gray in color. Gneisses also are most abundant in NW, in biotite, muscovite or two-mica varieties, all altered and commonly with large amount of limonite. Serpentinites occur in minor masses at the contact of granite and amphibolite or gneiss, most exposed NE in the field. Gabbroid rocks, bordering on the granitoid massif, are much altered, often with visible silicification, pyritization and carbonatization. Granitic rocks of Janja are the metallogenic hallmark of the Janja ore field greater. Most of the rocks are represented by medium-grained calc-alkalic biotitic granite, which consists of quartz, potassic feldspar, acid plagioclase and biotite (Krstić et al. 1974, 1976). Main accessory minerals are sphene, apatite, allanite, zircon and magnetite. The presence of aplite and pegmatite vein rocks and of quartz-feldspar and lamprophyre veins is characteristic of the granitoid massif.

The entire Janja field is intersected by faults of different trends and sizes. The faults form in places fault zones of crushed, brecciated materials or clay mass. The fault systems include numerous secondary fractures of different origin. Most fractures are filled with rock fragments to which very modest and shallow percolation of surface water is related (Nikić et al. 2008). This also explains the paucity of constant, weak springs that depend on the amount of atmospheric precipitation.

Fractures and fault zones in the Janja granite massif have metallogenic significance, because they contain uranium mineral, particularly the dislocations filled with crushed rocks. Zones of high mineralisation are near the granite contact with gabbroid rocks.

Prospecting Results

Concentrations of natural radioactive elements (U, Th, K, Ra) were measured in spoil heaps, water, humus, stream deposits and air near the closed uranium mines.

Mezdreja

The Mezdreja location includes an old mine shaft and three waste heaps (H1, H2 and H3). The mine entrance was closed by planned collapsing, with only 7–8 m left free. The old mine building and installations were removed, and the mine spoil heaps left near the shaft. Prospecting operations in Mezdreja were the following: detail radiometry (200 measurements), sampling water (10 samples) for full chemical analysis and soil and stream deposits (30 samples) for radiometry, and emanation measurement (in 9 points).

Spoil heap H1 in the immediate proximity of the old mine shaft contains about 16,500 m^3 mine waste over an area of 2800 m^2. Radioactivity in H1 was measured continuously at each 10 m, or in 45 sites, and it varied from 70 cps (counts per second) to 170 cps.

Spoil heap H2 is about 300 meters downstream of the shaft, amounting to about 1800 m^3 over some 280 m^2. Radioactivity was measured continuously in a regular 5×5 m network of 47 points. The counts varied from 270 to 600 cps, which suggested that the spoil consisted of poor ore not dressed during the exploitation.

Spoil heap H3 of 43,000 m^3 on 3200 m^2 is located about 800 m from the mine shaft downriver. Radioactivity was measured continuously in a regular 5×5 m network of 89 points. The measured values varied within the range from 130 cps to 280 cps.

Radioactivity of biotitite granite surrounding the old Mezdreja mine approximated 140 cps, which is the background count for this type of rock.

Table 1 gives maximum and minimum counts for U, Th, eRa, K and eRa/U and Th/U ratios for deposits and humus. It also gives measured radioactivity for the mine samples and a sample taken at the grinding mill.

More elevated concentrations of U, Th and eRa were detected only in samples from the mine and at the grinding plant. Fluvial deposit and humus samples had the studied elements higher than the background (1 ppm).

Figure 2 shows U, Th and Ra concentrations in the samples collected immediately around the Mezdreja mine to illustrate efficiency of the prospecting methods and of detecting weak radioactivity. Concentrations of Ra are high near the old mine grinding plant, somewhat lower of U, and far lower are Th concentrations. The decreasing trend is opposite to maximum concentrations; concentrations of Ra, followed by U and Th decline most rapidly.

Table 1 Amounts of U, Th, eRa, K and eRa/U and Th/U in different kinds of samples from Mezdreja

Kind of sample	U (ppm)	Th (ppm)	eRa (ppm)	K (%)	eRa/U	Th/U
Stream deposit	3.75–11.08	11.68–12.95	13.95–15.06	1.69–1.72	1.36–3.71	1.17–3.11
Humus	3.86–18.91	12.51–26.91	9.28–12.67	1.73–3.22	0.49–2.89	1.42–3.65
Mine excavation material	187.22	25.26	384.70	1.98	2.08	0.13
Material from near grinding mill	171.19	33.35	176.81	2.55	1.03	0.19

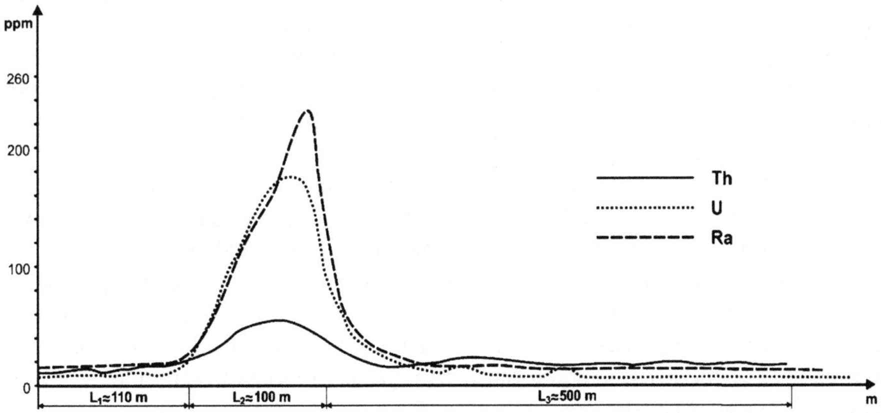

Fig. 2 Concentrations of U, Th and Ra immediately around the Mezdreja Mine.Legend: L_1 – upstream of the old mine; L_2 – old mine immediate surroundings; L_3 – downstream of the old mine

Srneći Do

This location is some 4 km from the Inovo village upstream the Inovska river. What has remained from the old mine workings is the complete collapse of the adit and a spoil heap beside it. There were no mine installations. Rocks building the area are porphyroid and fine-grained granites, often altered.

Waste material from the adit forms a heap of about 120 m^3 over an area of 90 m^2. Radioactivity measured in 100 sites ranged from 85 to 140 cps, the values of a material not significantly radioactive.

In relation to the natural background, the measured radioactivity was lower in humus and somewhat elevated in fluvial deposit (Table 2). This may have resulted from the local geology in the Inovska stream basin, made up of granite abounding in mineral veins, besides the mining activity. Eroded minerals transport U to the places chemically suitable for its precipitation.

Table 2 Amounts of U, Th, eRa, K and eRa/U and Th/U ratios in stream deposit and humus samples from Srneći Do

Kind of sample	U (ppm)	Th (ppm)	eRa (ppm)	K (%)	eRa/U	Th/U
Stream deposit	5.32–24.61	10.48–26.00	5.46–17.56	1.95–3.39	0.71–1.65	0.81–2.95
Humus	~5.11	~24.49	~10.67	~2.41	~2.09	~4.79

Gabrovnica

The area of Gabrovnica accommodates the closed Gabrovnica Mine, spoil heap and mine installations. The mine entrance was made to collapse and close it.

The location was prospected in detail by radiometry (over 320 sites) and a smaller volume of emanation measurements (11 points). Radiometric analysis was made on humus, stream deposit and mine waste samples (40 samples), and hydrochemical analysis (10 samples).

Radiometry in the closed mine buildings and installation yard varied from 120 to 170 cps. The railtrack area between the old shaft and the grinding plant, however, indicated a radioactivity within the range from 450 to 500 cps.

Two spoil heaps in a length of 120 m, near the mine, were prospected in a 10 × 10 m network; the radioactivity varied from 110 to 250 cps, with the elevated values only in places. This may indicate either the efficient ore separation from gangue or the low threshold that was regarded minimum economic concentration of the useful constituent.

Radiometry was applied to humus and fluvial deposits downstream the spoil heap on either side of the Gabrovnička to the Trgoviški Timok. The values obtained were two to three times lower (from 40 to 50 cps) than in the waste heaps or immediately around the mine installations.

Table 3 Amounts of U, Th, eRa, K and eRa/U and Th/U ratios in different samples from Gabrovnica location

Kind of sample	U (ppm)	Th (ppm)	eRa (ppm)	K (%)	eRa/U	Th/U
Humus	3.84–3.96	11.26–12.56	8.31–16.93	1.73–1.76	2.16–4.27	2.92–3.16
Stream deposit	4.16–15.63	11.84–18.57	3.41–37.12	1.67–2.32	0.38–3.50	1.18–3.97
Mine spoil material	29.85–36.25	15.91–34.90	57.74–83.61	2.68–3.64	1.59–2.80	0.53–0.96
Mine excavation material	~12.44	~31.30	~8.91	~3.81	~0.71	~2.51

Table 3 gives maximum and minimum U, Th, eRa, K radioactivity and eRa/U and Th/U for different samples from Gabrovnica.

Elevated radioactivity in Gabrovnica location was detected in samples from the mine and the spoil heap, which is normal for the materials excavated in the mine.

For humus and alluvium, radioactivity higher than the background was measured mainly in samples with an appreciable amount of organic material, which was a direct consequence of the uranium geochemistry.

The Trgoviški Timok Alluvial Plain from Balta Berilovac to Gabrovnica

Humus, alluvial deposits, subsurface and surface waters were sampled some 14 km along the Trgoviški Timok, from Balta Berilovac to the Gabrovička mouth (Fig. 1). Radiation of U, Th, eRa and K was measured on 35 samples of soil and alluvium, and the results given in Table 4.

Humus and alluvium from the Trgoviški Timok had elevated uranium in relation to the background, a likely consequence of the previous mining activities. Granites of Janja with ascertained uranium mineral, however, are the source area of streams flowing into the Trgoviški Timok. Consequently, even before the mine operation, a natural process of combined erosion, transport and accumulation could have led to the elevated uranium. Imperfect knowledge of "the zero level" before the mine operations started and of the environmental condition upon the mine closure increases the difficulty of quantifying the mining effect on uranium elevation.

Radioactivity of underground and surface waters was measured in all three locations of the closed mines and from Balta Berilovac to the Gabrovnička mouth along the Trgoviški Timok. Water samples were taken from mine-water outflows, dug wells and streams. Figure 3 shows U concentrations in waters flowing out of the old mines and in groundwater sampled from wells between the Mezdreja and the Gabrovnica Mines.

Table 4 Amounts of U, Th, eRa and eRa/U and Th/U in humus and alluvium along the Trgoviški Timok from Mezdreja to the Gabrovnička mouth

Kind of sample	U (ppm)	Th (ppm)	eRa (ppm)	K (%)	eRa/U	Th/U
Humus	1.75–9.55	7.72–17.45	3.12–23.32	1.33–2.43	0.45–11.48	0.97–7.31
Stream deposit	2.64–6.55	12.01–13.68	5.96–12.44	2.11–2.20	0.91–4.70	2.08–4.58

Table 5 Measured radioactivity in water samples from the closed Mezdreja and Gabrovnica mines

Water from closed mine	U (μg/L)	Ra (Bq/L)	Rn (Bq/L)
Mezdreja	93.8	0.07	125.8
Gabrovnica	27.2	4.42	46.44

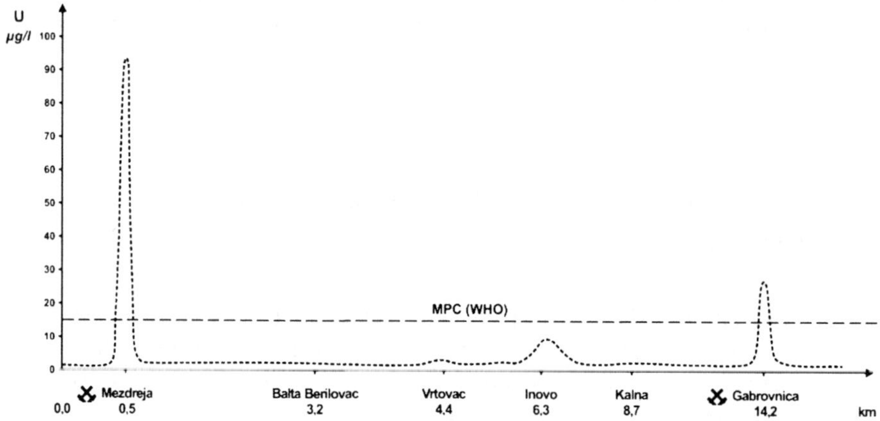

Fig. 3 Uranium concentrations in water between the closed Mezdreja and Gabrovnica Mines

The highest U concentrations measured in water from the closed Mezdreja and Gabrovnica Mines (Table 5) were six times and twice higher than allowed for drinking water (WHO 2008), respectively. Below MAC but increased U in water was detected in the wells of Inovo community, probably deriving from a trial adit of the Srneći Do by the Inovska upstream to the village. Water flowing out from the collapsed mine has eRa 45 Bq/l.

Table 6 gives measured concentrations of U, Ra and Rn in groundwater and river water in the area of the closed mines and in the Trgoviški Timok alluvium (Fig. 2). The measurements indicate a significant reduction in the radioactive elements' concentration in waters downstream the closed mines. It is probably due also to the relatively small discharge of 0.2 to 0.3 l/s from the collapsed mine shafts. Generally, these amounts of mine water are significantly affecting the quality of both underground and surface waters that are more removed from the closed mine, as corroborated by the measurements.

Table 6 Amounts of U, Ra and Rn in underground and surface waters of the closed mines

Location (River)	U (μg/L)	Ra (Bq/L)	Rn (Bq/L)
Crnovrška	0.3–2.2	0.07–0.08	14.8–29.6
Inovska	~9.7	~0.09	~14.91
Gabrovnička	~3.5	~0.63	~14.25
Trgoviški Timok from Crnovrška to Gabrovnička	0.3–2.5	<0.05–0.07	7.45–7.86
Trgoviški Timok downstream Gabrovnička mouth	~0.5	~0.08	~8.14

Emanation was measured within the zone of each closed mine. Radon from air was measured in seven sites of Mezdreja, and was highest, about 20 Emans, in waste heap H2. In Srneći Do, measured in five sites, maximum emanation was in the spoil heap and around it. Measuring sites in Gabrovnica were in front of the old mine, in the mine waste heap and downstream of it where the tailing lake had been. Maximum emanation was 18 Emans in the heap. The measured radioactivity in air was not significant.

Conclusion

Concentrations of natural radioactive elements were measured to establish the level of contamination in the closed uranium mines environment on Stara Planina. The prospecting covered areas of the Mezdreja, Srneći Do and Gabrovnica uranium mines proper and greater, where the closed mine shafts, mine spoil heaps and minor mine installations were located. Natural radioactive elements U, Th, eRa and K were prospected in humus, alluvial deposits, mine-waste heaps, underground and surface waters and air. The prospecting consisted of radiometry, emanation and hydrothermal analyses, field and laboratory assays.

Conclusions based on the prospecting data are the following:

- Radioactive elements are concentrated mainly immediately around the closed mine shafts, spoil heaps and old grinding plants.
- Uranium concentrations in water above MAC (WHO for drinking water) were detected only in discharges from the closed mine shafts, and below MAC elsewhere in underground and surface waters.
- Concentrations of radioactive elements in humus and alluvial deposits were elevated above the natural background in all prospected sites.
- No significant radioactivity was detected in air over the prospected area.

Principal conclusion is that after the closure of uranium mines the planned decontamination and remediation activities in the whole area and each location were not completed. Additional planned remediation is necessary in the mine buildings and installation yards and in the spoil heaps.

Acknowledgement This work is supported by Serbian Ministry of Science within the projects OI 176019 and 176018.

References

Antonović A (1990) Natural Radioactivity – Important for Geological Prospecting (in Serbian). Geoinstitut, Vol. 12, Belgrade.

Kovačević J (2005) Metallogeny of Stara Planina Region. Doctoral thesis, University of Belgrade Faculty of Mining and Geology. Belgrade.

Krstić B, Kalenić M, Rakić B, Rajčević D, Banković V (1974) Base geological map at 1:100,000, sheet Belogradchik and textual explanation, (in Serbian). Federal Geological Survey. Belgrade.

Krstić B, Kalenić M, Rakić B, Rajčević D, Banković V (1976) Base geological map at 1:100,000, sheet Knjaževac and textual explanation, (in Serbian). Federal Geological Survey. Belgrade.

Nikić Z, Kovačević J, Papić P (2008) Uranium in the groundwater of Permo-Triassic aquifers of the Visok region, Stara Planina, eastern Serbia. Water Air Soil Poll 192: 47–58.

WHO (2008) Drinking – water quality. 3rd edition. Recommendations. Vol. 1. Geneva.

Solubility Controls of Arsenic, Nickel, and Iron in Uranium Mine Tailings

Joseph Essilfie-Dughan, M. Jim Hendry, Jeff Warner, Tom Kotzer

Abstract. Uranium mill tailings from northern Saskatchewan, Canada contain elevated levels of As and Ni. The potential mobilization of these elements from the tailings management facilities to regional groundwater systems is an environmental concern for the uranium mining industry. In this study, electron microprobe analysis and synchrotron-based microfocussing X-ray fluorescence mapping and absorption spectroscopy (μ-XRF; μ-XAS) were used to identify and characterize the chemical composition of As-, Ni-, and Fe-bearing mineral phases at the micron scale in mine tailings samples from the Deilmann Tailings Management Facility (DTMF) located at the Key Lake mill of Cameco Corp. in northern Saskatchewan, Canada.

Joseph Essilfie-Dughan
Department of Geological Sciences, University of Saskatchewan, 114 Science Place, Saskatoon, SK, S7N 5E2,Canada
E-mail: joe377@mail.usask.ca

M. Jim Hendry
Department of Geological Sciences, University of Saskatchewan, 114 Science Place, Saskatoon, SK, S7N 5E2,Canada

Jeff Warner
Canadian Light Source Inc., University of Saskatchewan, 101 Perimeter Road, Saskatoon, SK, S7N 0X4, Canada

Tom Kotzer
Department of Geological Sciences, University of Saskatchewan, 114 Science Place, Saskatoon, SK, S7N 5E2,Canada
Cameco Corporation, 2121 – 11th Street West, Saskatoon, SK, S7M 1J3, Canada

B. Merkel, M. Schipek, *The New Uranium Mining Boom*
DOI 10.1007/978-3-642-22122-4 © Springer 2011

Introduction

Anthropogenic activities such as mining have the potential to release contaminants such as arsenic (As) and nickel (Ni) to the environment. Arsenic is a carcinogen, with exposure via drinking water shown to cause cancer of the bladder, lungs, skin, kidney, and liver in humans (USEPA 1998). Nickel is an essential element for several animal species (CCME 1999), but is toxic to both plants and animals at elevated concentrations (Barceloux 1999). The uranium (U) ore milled at Cameco's Key Lake operations in northern Saskatchewan, Canada, can contain high concentrations of As and Ni. The strong acid and oxidizing conditions used in the milling oxidize the metal sulfides and arsenides and liberate metals, such as As and Ni, from the ore. As a result, the hydrometallurgical waste solutions (raffinate) after U extraction can contain elevated concentrations of these elements of concern (EOC). Current practice is to add ferric sulfate and neutralize the pre-discharge mine tailings (raffinate/ residue) with lime, resulting in the formation of secondary minerals. These minerals serve as solubility controls to maintain the concentrations of As and Ni at very low levels in the tailings porewaters. However, the potential for the long-term mobilization of As and Ni from the tailings to regional groundwater systems remains an environmental concern for the U industry. Thus it is essential to identify and characterize the minerals phases of these elements to evaluate their solubility and long-term stability within the tailings mass.

The objectives of this study were to apply electron microprobe analysis (EMPA) and synchrotron-based spectroscopic techniques (μ-XRF, and μ-XAS) to determine the spatial distribution as well as identify and characterize the chemical composition of As-, Ni and Fe-bearing precipitates and mineral phases in mine tailings samples from the Deilmann Tailings Management Facility (DTMF) located at Key Lake. Electron microprobe analysis (EMPA) is a technique for chemically analyzing small selected areas of solid samples, in which x-rays are excited by a focused electron beam. Spatial distribution of specific elements can be recorded as two-dimensional x-ray "maps" using either energy dispersive spectroscopy (EDS) or wavelength dispersive spectroscopy (WDS). The spatial scale of analysis, combined with the ability to create detailed images of the sample, permits quantitative analyses of fine-grained geological materials such as mine tailings (Reed 1996). Similar to EMPA, synchrotron-based μ-XRF can also be used to generate elemental distribution maps EOC in heterogeneous geological materials such as mine tailings. However, on the basis of elemental distribution data generated by μ-XRF (unlike EMPA), μ-XAFS spectroscopic interrogation can be conducted on regions with elevated localized concentrations of EOCs to extract information on their speciation (redox and molecular) (Bertsch and Hunter 2001).

Accurate determination of the distribution of As-, Ni-, and Fe-bearing precipitates as well as their speciation (redox and molecular) in the tailings solids will help delineate the complexation and equilibrium geochemistry controlling their solubilities within the tailings porewaters. This will facilitate characterization and quantification of the long-term migration of these EOCs from the DTMF porewaters to adjacent groundwater systems. The information will also aid in the design of both mill processes to sequester the EOCs as well as uranium tailings storage facilities, with the aim to ultimately minimize migration of these elements into groundwater in the post-operational decommissioning phase. Additional details of the material presented in this extended abstract are presented in Essilfie-Dughan et al. 2011.

Experimental

EMPA Analyses

Five tailing samples collected during 2004/ 2005 (E2-GC74, E2-GC84, E2-GC98) and 2008 (E6-GC29, E6-GC30) DTMF drilling programs were selected for EPMA analyses. These samples were selected based on sample depth (to represent tailings solids of different ore bodies milled at Key lake) and sampling site. The samples were dried and embedded in polished epoxy cylindrical plugs. Backscattered electron (BSE) images as well as elemental maps of As, Fe, and Ni (EDS/ WDS) were collected using a JEOL 8600 Superprobe electron microprobe analyzer (Dept. of Geological Sciences, University of Saskatchewan). Quantitative WDS and qualitative EDS analyses were conducted on select features within the tailings samples to determine elemental ratios of co-localized EOCs and in mineralogical evaluation.

μXRF and μXAS Analyses on Tailings Samples

μ-XRF and μ-XAS analyses were conducted at the microprobe endstation of the hard X-ray microanalysis beamline (HXMA-06ID-1) at the Canadian Light Source (CLS; University of Saskatchewan). The incident X-ray energy was set to 17,000 eV, utilizing the Si(220) double crystal monochromator and the As, Ca, Fe, and Ni $K_{\alpha 1}$ fluorescence lines, as well as the intensity of the total scattered X-rays, were monitored using a four-element Vortex detector (SII NanoTechnology USA Inc.). A microfocussed beam ~3 μm in diameter was defined using a Kirkpatrick-Baez focussing mirror pair, with the samples oriented 45° to the incident beam. The μ-XRF data for elemental distribution were collected using a 5-μm step size on select areas containing mineralogical features previously characterized using EMPA and optical microscopy. As, Fe, and Ni K-edge micro-X-ray absorption near edge structure (μ-XANES) data were collected at various points on mapped features of interest. The beam was detuned by 50% for these measurements to reduce harmonics. The monochromator step size was 0.5 eV through the XANES region for all three edges. As for the μ-XRF data, μ-XANES data were collected at ambient temperature and pressure with the simultaneous measurement of As, Fe, and Ni reference foil spectra for energy calibration of the respective sample spectra. Reduction of the μ-XANES data (including energy calibration, averaging of multiple scans, background subtraction, and normalization) and linear combination fit (LCF) analysis were conducted with ATHENA software (Ravel and Newville 2005).

Results and Discussion

EMPA Analyses

BSE images of sample E2-GC98 (Fig. 1) indicate the presence of 10–200 μm diameter brightly rimmed nodule-like features. These features are present in all samples studied. Elemental X-ray maps (WDS) (Fig. 2) indicate that the dominant elements in the nodules are Ca and S (as gypsum), whereas the bright rims are mainly metal-bearing and contain higher and variable proportions of As, Fe, and Ni.

These nodules are proposed to have formed during the milling process, during neutralization of the hydrometallurgical acidic (H_2SO_4) mill waste streams (raffinates) via the addition of excess slaked lime ($Ca(OH)_2$). Geochemical modeling of this neutralization process at Cameco's Rabbit Lake U mill, which uses a similar tailings neutralization process as the Key Lake mill, indicates that excess Ca^{2+} and SO_4^{2-} ions combine to form gypsum ($CaSO_4 \cdot 2H_2O$) over the pH range of ~1.4 to 10 (Moldovan and Hendry 2005). Quantitative XRD analyses substantiate this observation, with DTMF mill tailings containing approximately 20% (w/w) gypsum (Shaw et al. 2011).

Empirical data and modeling results of the hydrometallurgical leaching and neutralization process of acidic raffinates within the Rabbit Lake operations (Moldovan and Hendry 2005) further show that concentrations of As and Fe in the raffinate are largely controlled by precipitation of amorphous ferric hydroxides and accompanying surface adsorption of arsenate (As^{5+}) to ferrihydrite, with the majority of the arsenate removed from solution at pH < 5. If it is similarly assumed

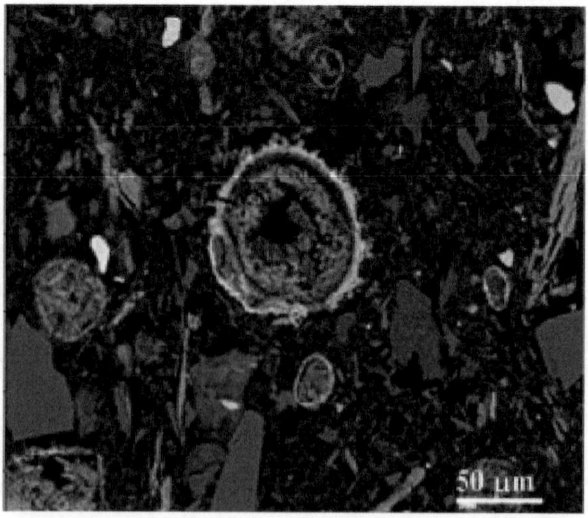

Fig. 1 BSE image of mine tailings sample E2-GC98 showing the presence of different sizes of gypsum nodules with surrounding bright metallic rims

that development of HFO-coated gypsum nodules identified in the DTMF tailings samples is concurrent with increasing neutralization and that iron is largely precipitated as ferrihydrite, progressively less As should be adsorbed to the surface of the ferrihydrite as the pH increases. Within the rims, this should equate with larger Fe/As ratios in the outer versus inner parts of the metal rims, and vice-versa. The bright rims (Fig. 3) contain greater amounts of Fe compared to As and Ni, with higher concentrations of Fe and larger Fe/As ratios on the outer part of the metallic rims. The centers of the nodules have high Ca and S concentrations, confirming the nodule composition is mainly gypsum.

Excluding the metal-rich rims surrounding the gypsum nodules, EDS spectra collected on various bright features distributed throughout the tailings matrix (Fig. 4) indicate the presence of primary As-, Cu-, Fe-, and Ni-bearing minerals (chalcopyrite, FeCuS; pyrite, FeS_2; and gersdorffite, NiAsS). The presence of these minerals, which also found within the original U ores, indicates that the primary sulfide and arsenide minerals are not completely solubilized during the milling process.

μXRF and μXAS Analyses on Tailings Samples

As shown with the combined electron microprobe BSE and X-ray analyses (Fig. 1, 2), the μXRF – generated elemental maps of As, Ca, Fe, and Ni (Fig. 5) clearly show the nodule centre is largely composed of Ca (i.e. gypsum/lime) and As, Fe, and Ni within the surrounding rims.

Although the overlapping elemental micro-distributions provide evidence of associations between Fe, As, and Ni, they cannot identify the chemical speciation (redox and molecular) of these elements. Thus, μ-XANES spectra were collected at the K-edges for Fe, As, and Ni on discrete spots (Fig. 6, A–E) on and outside of

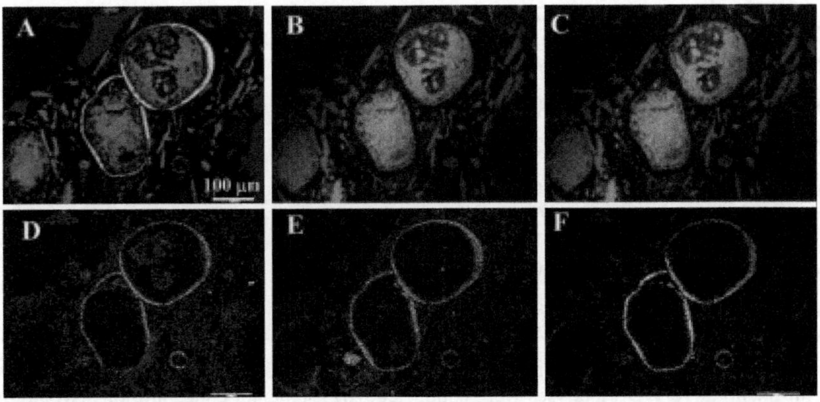

Fig. 2 EMPA analyses of tailings sample E2-GC84 showing (**A**) BSE image of nodule-like structures and elemental X-ray maps (WDS) of (**B**) Ca, (**C**) S, (**D**) As, (**E**) Fe, and (**F**) Ni

a nodule in tailings sample E2-GC84 as well as for relevant reference compounds. The As K-edge μ-XANES spectra collected on the discrete spots indicate that As^{5+} (arsenate; AsO_4^{3-}) is the dominant form of arsenic within and near the metallic rims. The spectra also indicate the presence of As^{3+} species (arsenite; AsO_3^{2-}). Fe and Ni K-edge μ-XANES spectra indicate that Fe^{3+} and Ni^{2+} are the dominant forms of iron and nickel, respectively, within and proximal to the metallic rims surrounding nodules.

LCF analysis was conducted to fit the K-edges for Fe, As, and Ni μ-XANES spectra collected on discrete spots (Fig. 6, A–E) to their respective reference compounds and thus determine the secondary mineral or complexes present. The reference compounds for the Fe K-edge represent the iron minerals (Fig. 7a) most

(a)

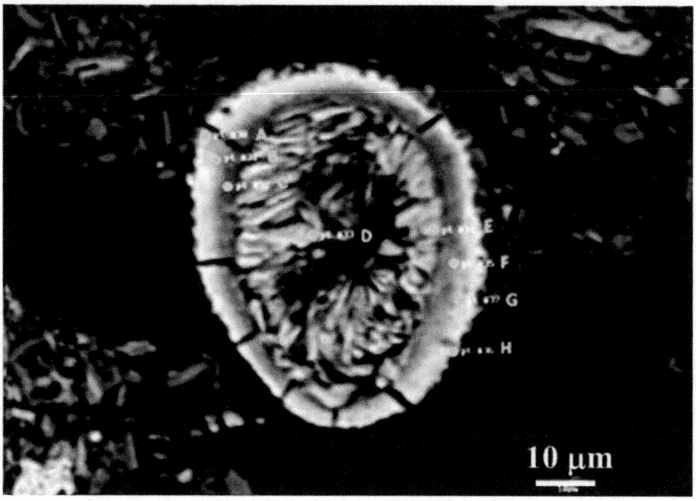

(b)

Points	Fe	As	Ni	Ca	S	Fe/As
A (outer rim)	15.91	3.75	6.42	3.19	0.48	4.24
B (middle rim)	14.03	3.94	5.03	5.48	0.40	3.56
C (inner rim)	5.93	2.10	4.15	5.50	0.53	2.82
D (gypsum)	0.09	0.12	0.04	27.04	12.47	0.75
E (inner rim)	0.90	0.86	3.54	4.13	0.59	1.05
F (inner rim)	7.67	2.75	4.49	2.50	0.97	2.79
G (middle rim)	13.19	3.60	6.14	2.21	0.45	3.66
H (outer rim)	14.91	3.29	13.10	1.76	0.70	4.53

Fig. 3 (**a**) BSE image of a gypsum nodule within tailings sample E6-GC30 showing positions where semi-quantitative WDS analyses were conducted. (**b**) Elemental concentrations (wt %) of Fe, As, Ni, Ca, and S on metal-rich rims surrounding nodules in tailings sample E6-GC30 (1–2 μm diameter analysis) from WDS analyses. The locations of points are shown in (**a**)

likely to be found in the tailings. LCF analysis of the μ-XANES spectra from all five spots (Fig. 7b) indicates that iron is present predominantly as ferrihydrite on the metallic rims and at the adjacent spot (D) in the general tailings matrix. LCF analysis conducted to fit the Ni K-edge μ-XANES spectra to Ni reference compounds (Fig. 7c) suggests that Ni adsorbed onto ferrihydrite is dominant at all five spots. LCF analysis of As K-edge μ-XANES spectra to As reference compounds (Fig. 7e) suggests arsenate (As^{5+}) adsorbed onto ferrihydrite (Fig. 7f) is the dominant As phase within and nearby the metallic rims. However, the results also suggest the presence of arsenite (As^{3+}), possibly adsorbed onto the ferrihydrite. The presence of As^{3+} is attributed to partially oxidized arsenic-bearing minerals, such

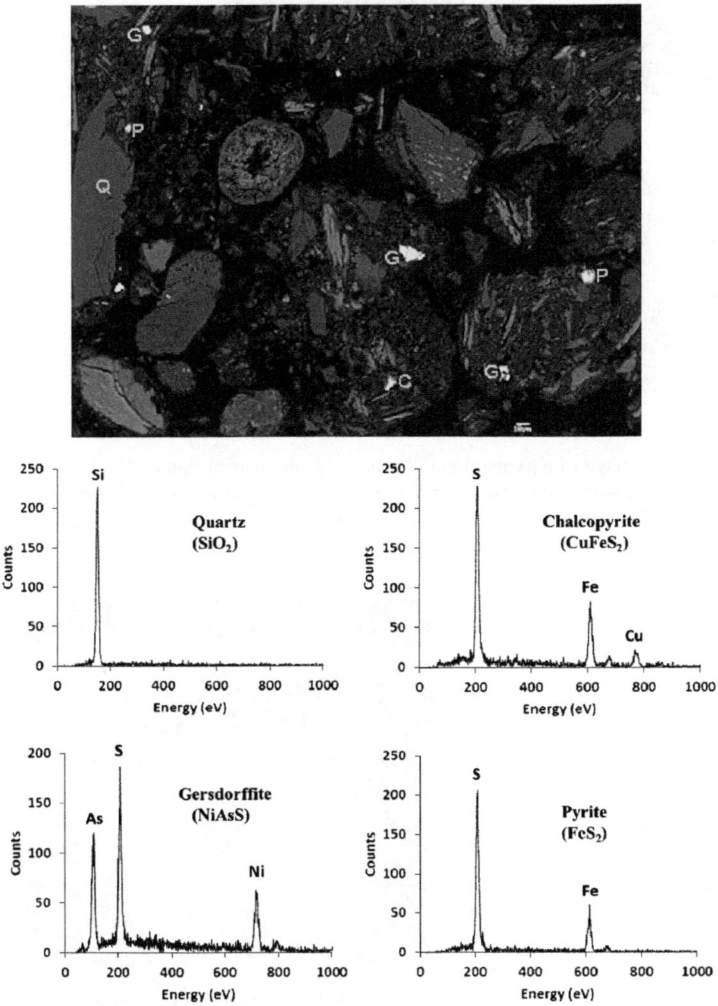

Fig. 4 BSE image of a gypsum node within tailings sample E6-GC30 showing the presence of quartz (Q), gersdorffite (G), pyrite (P), and chalcopyrite (C) (*top panel*). EDS spectra obtained from the various spots (*bottom panel*)

as gersdoffite (NiAsS; Fig. 4), derived from the original ores. Spectroscopic analyses on bulk tailings from the JEB TMF at nearby McClean Lake indicate the dominance of As^{5+} with lesser amounts of As^{3+} (Warner and Rowson 2007). The presence of As adsorbed onto ferrihydrite identified here is consistent with study results from tailings in the RLITMF (Moldovan et al. 2003).

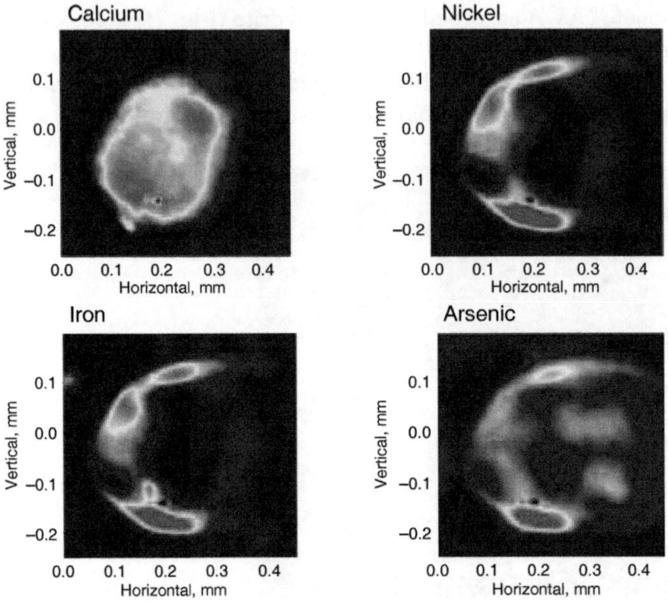

Fig. 5 X-ray fluorescence image of a gypsum nodule showing the distribution of Ca, Fe, Ni, As within the nodule centre and surrounding metal rims. Sample analyzed is E2-GC84 and represents tailings from Ni-rich Deilmann orebody. Note that images are false color are not calibrated amongst the images

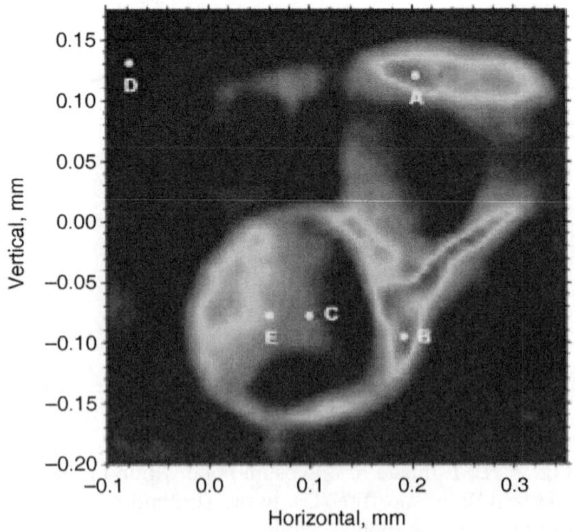

Fig. 6 Detailed X-ray fluorescence image of gypsum nodule as in Fig. 5 above, delineating locations of 2–3 micron spots where As K-edge and Fe K-edge μXANES spectroscopic data was collected

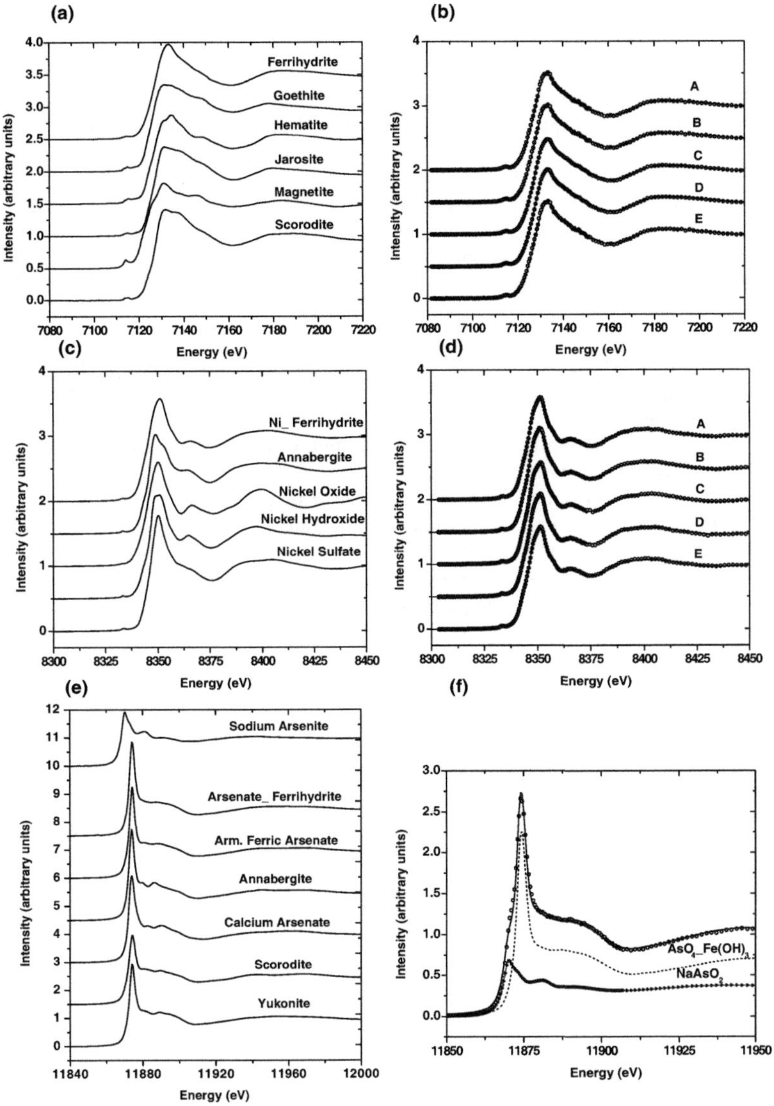

Fig. 7 X-ray absorption near-edge structure of reference compounds of Fe (**a**), Ni (**c**), and As (**e**) as well as experimental (*open circles*) and linear combination fits (*solid curves*) for Fe (**b**) and Ni (**d**) K-edge μ-XANES spectra collected on spots (A–E) in Fig. 6 and As K-edge μ-XANES spectra of spot A (**f**) together with spectra of the three standards, scaled according to their fractional contributions to the fitted spectra

References

Barceloux D G (1999) Nickel. J Toxicol Clin Toxicol 37: 239–258

Bertsch P M, Hunter D B (2001) Applications of Synchrotron-Based X-ray Microprobes. Chem Rev 101: 1809–1842

CCME (1999) Canadian Environmental Quality Guidelines. Environment Canada, Ottawa

Essilfie-Dughan J, Hendry M J, Warner J, Kotzer T (2011) Microscale mineralogical characterization of As, Fe, and Ni in uranium mine tailings using Electron Microprobe, μ-XRF and μ-XAS. Environ Sci Technol (Submitted)

Moldovan B J, Hendry M J (2005) Characterizing and quantifying controls on arsenic solubility over a pH range of 1–11 in a uranium mill-scale experiment. Environ Sci Technol 39: 4913–4920

Moldovan B J, Jiang D T, Hendry M J (2003) Mineralogical characterization of arsenic in uranium mine tailings precipitated from iron-rich hydrometallurgical solutions. Environ. Sci. Technol 37: 873–879

Reed S J B (1996) Electron microprobe analysis and scanning electron microscopy in geology. New York: Cambridge University Press

Shaw S, Hendry M J, Essilfie-Dughan J, Wallschläger D, Kotzer T (2011) Distribution and geochemical controls of As, Se, and Mo in uranium mine tailings, Key Lake, Saskatchewan, Canada. Appl Geochem (in press)

USEPA (1998) Arsenic in Drinking Water. Arsenic Research Plan. Office of Water

Warner J, Rowson J (2007) In-situ monitoring and validation of a uranium mill tailings management facility design using X-ray Absorption Near Edge Structure (XANES) Spectrsoscopy. Synchrotron Radiation News 20: 14–17

Study of Radionuclide Transport by Underground Water at the Semipalatinsk Test Site

Ella Gorbunova, Sergey Subbotin

Abstract. The results of the study of radionuclide transport by underground water at the Test Site "Degelen" and the adjacent territory of the Semipalatinsk Test Site (STS) are considered in the article. Monitoring data obtained in 1996–2008 show an uneven pollution of surface streams and underground water of the "Degelen" massif.

Introduction

The results of complex monitoring of radiation environmental conditions at the STS carried out by Institute of Radiation Safety and Ecology of National Nuclear Center of the Republic of Kazakhstan during the last decade gain a special significance and urgency in connection with the forthcoming economic usage of the site.

To define the buffer area around underground nuclear test sites it is necessary to have a full knowledge of radiative contamination of the soils and underground water sources. In particular, a high level of radionuclides (by far exceeding the safety level of drinking water) have been traced in the underground water of

---

Ella Gorbunova
Institute for Dynamics of Geospheres, Russian Academy of Sciences, 38-1 Leninsky prosp., Moscow, Russia
E-mail: emgorbunova@bk.ru

Sergey Subbotin
Institute of Radiation Safety and Ecology of National Nuclear Center of Ministry of Energy and Mineral Resources of the Republic of Kazakhstan, 2, Krasnormeiskaya str., Kurchatov, the Republic of Kazakhstan
E-mail: subbotin@nnc.kz

B. Merkel, M. Schipek, *The New Uranium Mining Boom*
DOI 10.1007/978-3-642-22122-4 © Springer 2011

"Degelen" massif (^{137}Cs up to 700 Bq/l, ^{90}Sr up to 2000 Bq/l, ^{3}H up to 1300 kBq/l, and $^{239+240}$Pu up to 110 Bq/l) (Subbotin and Lukashenko 2006). 295 underground nuclear explosions (UNE) have been carried out within the limits of the site during 1961–1989 (Shkolnik 2003).

In the framework of regeneration procedures (1996–1998) the tunnel closing and resurfacing of the landscape have been fulfilled. However, these procedures had a low efficiency for some tunnels with a permanent water inflow. The infiltration of contaminated water through dams has been registered at the tunnel portals. Special filters consisting of drain pit filled sorbent and coarse materials have been constructed in front of the tunnel portals. The radiation control stations have been equipped. Spring outlets and wells are also included in the control network. Occasionally, sampling for of superficial streams for radiochemical analysis was carried out occasionally.

Geological and Hydrogeological Conditions of site "Degelen"

The "Degelen" massif represents an erosionally partitioned cupola with a relative height of 500 m and a slope steepness varying from 45–600 up to 15–200 m. The central part of the massif consists of intrusive formation. Paleozoic effusive and effusive-sedimentary rocks spread to the periphery. In the upper part the chinked rocks are eroded to a depth of 30–50 m; in depressions the rocks are covered by Quaternary deposits. The thickness of the soft rock which underlay Neogene clays reached 30 m and more in the valleys.

The massif consisting of two main blocks (northern and southern) is broken by series of fractures mainly of northwest and sublatitudinal strike (Fig. 1) (Gorbunova et al. 2005). The most elevated part of the massif is a source of underground

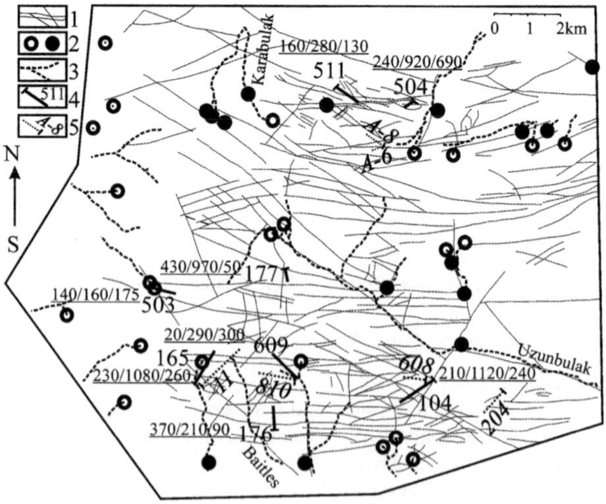

Fig. 1 Scheme of the site "Degelen" (1 – fracture; 2 – extinct and existent springs; 3 – brook; 4, 5 – tunnel and its number: 4 – with water-inflow, numerals – content of 3H (kBq/l)/90Sr (Bq/l)/137Cs (Bq/l), 5 – with worst-case situation)

water replenishment. Gentle slopes favor the underground water accumulation with a steady regime and a mature thickness of the aquifer. The massif is characterized of the radial type of underground water runoff. Underground water is partially unloaded by springs in natural conditions and through the tunnels in antropogenically disturbed areas. The surface streams are basically of a seasonal nature.

Within the massif limits, water from fractures beds (related to erosional zones) and water from fractured veins (related to intermittent discontinuities) prevail. Pore water (unconfined ground water) spreads in the soft rocks of the valleys. Underground water are connected hydraulically at the places of the wedge-out of aquiclude and form a uniform aquifer with the same level corresponding to the smoothed relief.

The “ageing” of the “Degelen” site marked by the landscape change is connected with the UNEs during 28 years. An amount of the rock crests, breakages and juts has decreased considerably. The increase of the taluses areas of at the foot of steep slopes and the formation of new taluses above the UNE epicenters have been observed. The hydrological inspection of the massif revealed a dislocation of outputs and springs’ disappearance, a reduction of a general extent of the surface streams. A partial transition of streams from the surface to the underground indirectly testify a redistribution of underground water system.

Underground Water Regime

The tunnels cut practically waterless rocks with an incline in a direction of natural drains – draws and valleys. The line of wedge-out of the aquifer is substantially below the tunnel location. The constant water-inflows in the form of jet seeps of various intensity indicate an opening of water-encroached fractures (Shpakovsky et al. 2000).

For example, the unloading of hydrogeologically active fractures of sublatitudinal strike takes place through tunnels 165, 176 and 609 instead of previously existing springs (Fig. 1). An average annual inflow from tunnels depends on an absolute mark opening of fractured vein water. A reverse linear dependence between the aquifer level and the registered water-inflow is shown in Fig. 2.

The mid-annual values of the water-inflow, received from quarter measurements during 12 years (1997–2008) allow to separate two regimes: the plain one with relatively stable inflow without significant variations, and the foothill regime with sharp changes (Fig. 3).

Meteorological factors (such as precipitation and temperature) exert a significant influence on the regime of fractured vein water. The extremes in water-inflow and in precipitation coincide for the tunnel 609 which is closed to the area of underground water replenishment. The maximal water-inflow from the tunnel 104 in 2001 has been provoked by a significant amount of precipitation in the previous year. Variations of water-inflows from the tunnels 177, 176 and 165 are identical. On the flat slopes (tunnels 503, 504 and 511) the water runoff is steady.

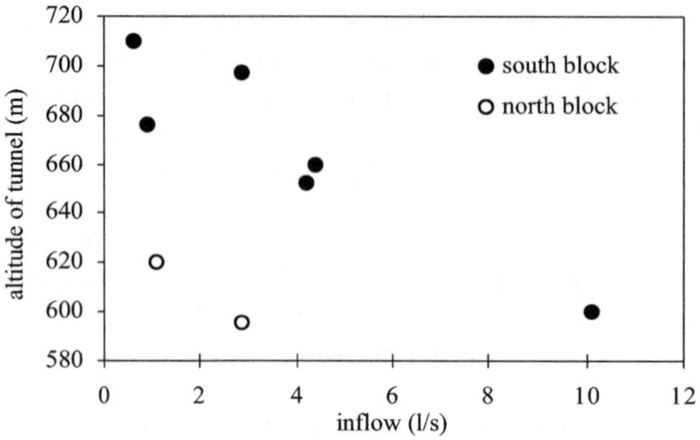

Fig. 2 Relationship between inflow and altitude of tunnel

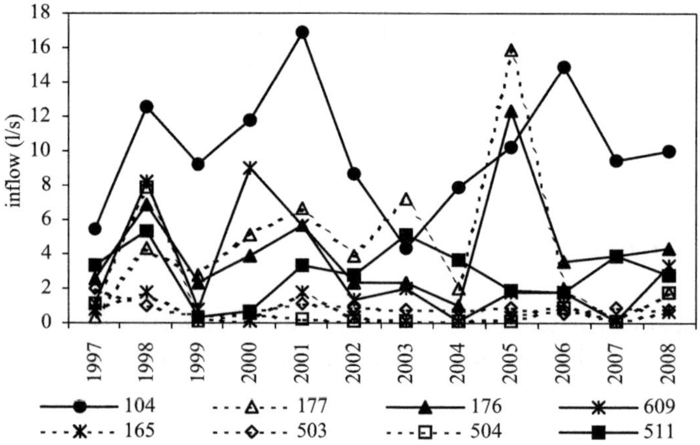

Fig. 3 Inflow from tunnels

Radioactive Pollution of Underground Water Within the Massif

Results of the long-term radiation monitoring during 12 years indicate that radionuclide ^{3}H, ^{90}Sr and ^{137}Cs continue to seep from the tunnels into surrounding area (Fig. 4).

The ^{3}H content in the water-inflow from the tunnels decreases during the period of observations. The average level of pollution compounded 400 kBq/l in 1996, and it didn't exceed 200 kBk/l in 2008. The most significant decrease of ^{3}H is tracked in the water-inflow from the tunnel 177.

The steady increase of ^{90}Sr content is registered in the water-inflow from the tunnels 503 and 177 confined to the northern slope of the southern block. In contrary, the ^{90}Sr content in the water-inflow from the tunnels 104 decreased down to

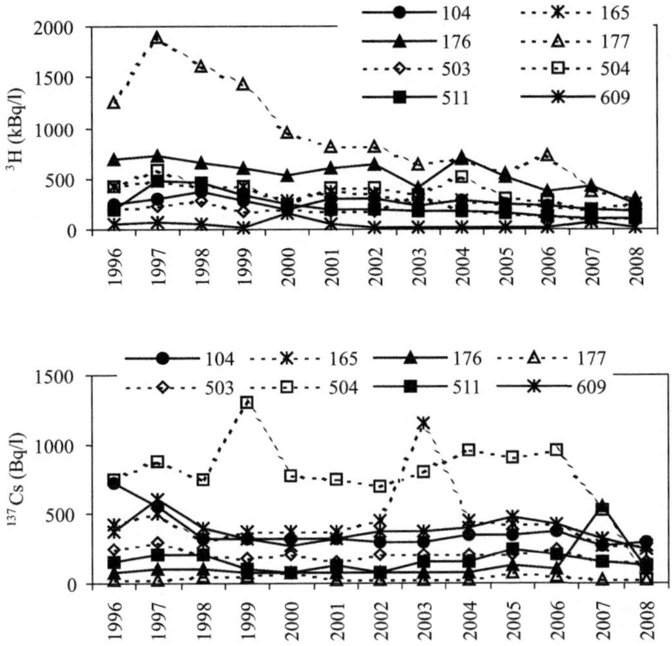

Fig. 4 Content of radionuclides in water-inflow from tunnels

600 Bq/l. Variations ^{90}Sr in water-inflow from the tunnels 165 and 609 are synchronous. Separate emissions with ^{90}Sr increase are fixed in the tunnel 165 and coincide with an increase of water-inflow (Fig. 3). The ^{137}Cs content in water-inflow from all tunnels is rather stable with the exception of the tunnels 176, 165 and 504 in which significant variations of this radionuclide are observed.

On the scheme of underground water radioactive pollution (Fig. 1) the elevated content ^3H is marked in water-inflow from the tunnels of the southern block drained by sublatitudinal fractures on the lowest hypsometric marks. The ^3H content in the fractured vein water unloading through portals of the tunnels 176 and 177 reaches 370–430 kBq/l. The elevated ^3H content is fixed in water-inflow from the tunnel 104 with raised discharge (Fig. 3).

Along the northern slope of the massif ^3H content in water-inflow from the tunnels varies from 160 up to 240 kBq/l, while in the springs at a distance of 5 km it varies from 60 up to 74 kBq/l.

The results of seasonal probes in 2000–2001 show that the ^3H content in the valleys Baitles and Usunbulak varies from 142 to 173 kBq/l; it decreases down to 86 kBq/l in the valley Karabulak although still exceeds the level of the radioactive pollution in springs.

^{90}Sr and ^{137}Cs distribution in underground water is non-uniformly. A high content of radionuclide are observed in water-inflow from the tunnels 165 and 104, cutting the zones influenced by the worst-case situations 11 and 608 (Logachev 1997) and opening the system of southern fractures of sublatitudinal strike.

The elevated ^{90}Sr and ^{137}Cs content is kept in water-inflow from the tunnels 609 and 176 along the southern slope of the massif. Beyond the intrusive massif limits, radionuclides content in underground water decreases sharply. ^{90}Sr content doesn't exceed 1.4 Bq/l, ^{137}Cs – 0.4 Bq/l in the flowing well. The radionuclides content in the southern spring is lower than the level of maximal permissible activity (MPA).

The observed data show that the process of radioactive pollution of underground water of the "Degelen" massif continues and has a rather stable character (Subbotin and Pestov 2005). The systems of sublatitudinal fractures are hydrogeologically active and potentially radioactive in connection with registered increase of radionuclide content in water-inflow from the tunnels.

Transport of Radionuclides Outside the Limits of Massif "Degelen"

Along with radiation monitoring of the surface runoff confined to the springs and the tunnels, the observations of radionuclide transportation by underground water were carried out. The well depths from 3 m up to 52 m opening the unconfined ground water and fractured sheet underground water within the limits of the basic valleys of the brooks Uzunbulak, Baitles and Karabulak. The difference between water table and potentiometric surface of the uneven age aquifers changes from –0.7 to –1.5 m up to +0.7 to +2.3 m on distance up 30 km.

The unconfined ground water is subjected to the greatest pollution within the limits of the "Degelen" site (Fig. 5). In the valley Baitles, at the distance of 4.3 km to the south from the tunnel 609 portal, the ^{3}H content at a depth of 3 m is 260 kBq/l. In the valley Uzunbulak, at the distances of 0.4–0.8 km to the northeast from the tunnel 104 portal, ^{3}H content at a depth of 1.5 m decreases from 185 down to 80 kBq/l.

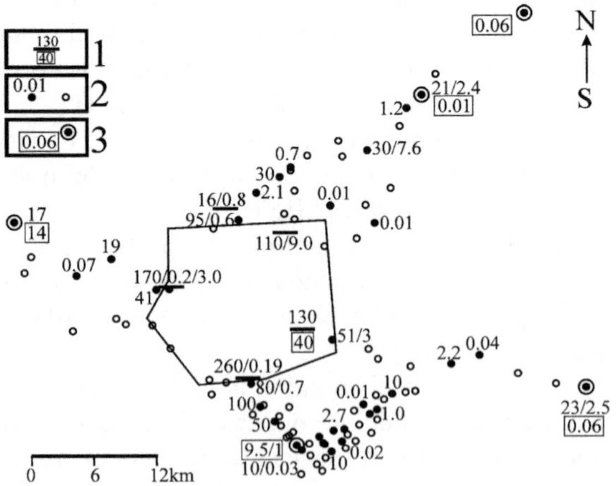

Fig. 5 Distribution of radionuclides in underground water outside the limits of the massif "Degelen" (1–3 – content of ^{3}H (kBq/l)/^{90}Sr (Bq/l)/^{137}Cs (Bq/l) in:
1 – section of wells;
2 – wells equipped unconfined ground water and waterless; 3 – wells equipped fractured sheet underground water)

To the west of the massif the ^{3}H content at a depth of 1.5 m is 170 kBq/l at a distance of 3.5 km to the northwest from the portal of the tunnel 503. In the valley Karabulak the ^{3}H content at a depth of 0.8–0.9 m changes from 110 kBq/l at a distance of 2.6 km to the northeast from the portal of the tunnel 504 down to 16 kBq/l at a distance of 4.1 km to the north from the portal of the tunnel 511.

The ^{3}H content in unconfined ground water varies from 80–95 kBq/l along the southern and northern borders of the test site down to 41–51 kBq/l along the western and eastern borders. During the last 22 years after the end of UNE, the ^{3}H content, exceeding the MPA level of 7.7 kBq/l, is observed outside the limits of the "Degelen" massif up to a distance of 7.4 km.

The high ^{3}H content in unconfined ground water, established in the wells of the southeast section, decreases from 100 down to 10 kBq/l at a distance of 7.4 km. A significantly elevated ^{3}H content (up to 19 and 30 kBq/l) are observed at distances of 5, 8 and 4.3 km to the west and to the north of the test site. The elevated ^{3}H content (from 17 and 23 kBq/lbly) in unconfined ground water at the distances of 16 km and 25 km in the northwest and southeast directions, is related to the proximity of the test sites "7" and "Telkem" where the excavation explosions took place. To the northeast, at the distances from 11.2 km up to 18.6 km, the elevated ^{3}H content (21–30 kBq/l) in unconfined ground water is, probably, connected to the surface soil pollution by a radioactive trail of the surface explosion (Shkolnik 2003).

The results of monitoring in the fractured bed of underground water show the high ^{3}H content (up to 40 kBq/l) in the well located in the valley of the Uzunbulak brook at a distance of 2.5 km to the east from the portal of the tunnel 104. These data confirm the presence of a hydraulic connection between unconfined ground water, fractured bed and fractured vein underground water.

Along the north-west direction outside the limits of the massif the ^{3}H content in the fractured bed underground water is 9.5 and 14 kBq/l at the distances of 7.4 and 16 km, respectively. In the latter case the presence of ^{3}H at a significant distance can be connected to the stream of radioactively polluted underground water from the site "7". The ^{3}H content in the fractured bed underground water in three wells located in the holes Uzunbulak and Karabulak at the distances from 18.6 up to 30.4 km from the "Degelen" massif doesn't exceed 0.06 kBq/l (Fig. 5).

The content of other radionuclides (^{90}Sr, ^{137}Cs) in the underground water outside the limits of the test site doesn't exceed the MPA level except for the wells influenced by the UNEs from the adjacent sites. Also, it is necessary to point out the wells located in hole Baitles at a distance of 7.4 km from the "Degelen" massif in which the ^{90}S content is 0.1 Bq/l in the unconfined ground water and 1 Bq/l in the fractured bed water.

The main pollution of unconfined ground water is tracked on the periphery of the massif. The elevated ^{90}Sr content from 3 up to 9 Bq/l is fixed along its south and northeast borders. Along the western and southern borders the pollution varies from 0.2 up to 0.7 Bq/l. The ^{137}Cs content in the unconfined ground water at the test site makes 3 Bq/l, and doesn't exceed the MPA level in the Uzunburack valley.

Conclusions

Currently, regular water-inflow is fixed at the portals of 8 tunnels by the results of radiation monitoring of underground water at the test site "Degelen". The unloading of underground water is carried out in the direction of natural drains – valleys of the brooks Uzunbulak, Karabulak, Baitles. The basic surface runoffs – brooks and springs have a seasonal nature.

The transport of radionuclide is curried with underground water and surface runoff. The radioactive pollution of the underflow in the valleys is established. The formation of auras of radioactive pollution in underground water is supervised by the northern and southern systems of hydrogeologically active fractures of sublatitude strike.

The basic radioactive water pollutant is ^{3}H. The ^{90}Sr and ^{137}Cs contents do not exceed the permissible values for potable water and do not pose radiative hazard. However, the migration of radionuclides, in particularly ^{90}Sr, is high enough and may lead in the future to the formation of a joint pool of underground water polluted by the sites "7", "Degelen" and "Telkem".

References

Gorbunova E M, Ivanchenko G N, Godunova L D (2005) Analysis of geological structure of site "Degelen" in respect to conditions of radionuclide transport with underground water. Proceedings ground water. Proceedings of the II International Conference 2005, Kurchatov 6–8 September 2005, the Republic of Kazahstan: 33–38

Logachev V A (1997) Nuclear tests in USSR: Semipalatinsk site. Moscow: IzdAt: 320

Schoolnik B C (2003) Semipalatinsk Test Site: creation, activities, conversion Almaty: 344

Shpakovsky B I, Shpakovskay R S, Gorbunova E M (2000) Hydrodynamics situation under the anthropogenic conditions. Proceedings of the International Conference 2000, Sankt-Peterburg: 738: 615–616

Subbotin S B, Lukashenko S N (2006) Some features of radioactive pollution of underground and surface water at the territory of former Semipalatinsk Test Site. Seminar NATO: Radioactive risks in the central Asia 2006, Almaty 20–22 June 2006. Institute on Nuclear Physics of the Republic of Kazahstan

Subbotin S B, Pestov E U (2005) The organization of monitoring of underground water at the territory of former Semipalatinsk Test Site. Kurchatov 2005, ISTC Project K-893

Rehabilitation of Uranium Mining Waste and Restoration of the Plohnbach Valley (Germany): Design Principles, Ecological Requirements and Construction in a Fauna-Flora-Habitat Area

N. Gottschalk, P. Schneider, R. Löser, J. Schreyer, S. Anders, B. Tunger

Abstract. The Plohnbach river downstream the Lengenfeld tailings pond was polluted by radioactive tailings caused from a tailings dam break in 1954. The area in its original condition was a popular excursion and recreational area. The option analysis for the rehabilitation concluded as preferable option the relocation of the polluted tailings material from the river floodplain and the Lenkteich pond to the waste heap dump Lengenfeld. Beside the fact of the realization of the remediation there is the requirement to allow for a later re-use of the area for recreational and touristic purposes. Due to the sensitive location in the FFH area "Göltzschtal" it was needed to consider special nature conservation requirements during the rehabilitation works, with emphasis on species conservation and the achievement of long–term development objectives of the FFH area.

N. Gottschalk
C&E Consulting und Engineering GmbH, Jagdschänkenstr. 52, D-09117 Chemnitz, Germany

P. Schneider
C&E Consulting und Engineering GmbH, Jagdschänkenstr. 52, D-09117 Chemnitz, Germany

R. Löser
C&E Consulting und Engineering GmbH, Jagdschänkenstr. 52, D-09117 Chemnitz, Germany

J. Schreyer
Wismut GmbH, Jagdschänkenstr. 27, D-09117 Chemnitz, Germany

S. Anders
Wismut GmbH, Jagdschänkenstr. 27, D-09117 Chemnitz, Germany

B. Tunger
Wismut GmbH, Jagdschänkenstr. 27, D-09117 Chemnitz, Germany

B. Merkel, M. Schipek, *The New Uranium Mining Boom*
DOI 10.1007/978-3-642-22122-4 © Springer 2011

Introduction

The rehabilitation area is located in the Vogtland in the south of the Free State of Saxony in the south East of Germany, about 15 km southwest of Zwickau, in the vicinity of the city Lengenfeld.

The investigation area includes the tailings pond Lengenfeld, the mine dump Lengenfeld as well as the downstream located Plohnbach river, which is partially located in the Flora-Fauna-Habitat-area “Göltzschtal”. Included is the so-called Lenkteich which was a small potable water reservoir downstream the Plohnbach

Fig. 1 Location of the Study Area in the South East of Germany

river until the year 1947. The site in its original condition was a popular excursion and recreational area.

Further, in the vicinity of the Plohnbach Valley is located the so-called North Mine Dump. It was created in the period 1953–1961 for the deposition of radioactive tailings, as well. During a flood in 1954, the dam break of the Lengenfeld tailings pond caused the transport of the tailings into the Lenkteich pond, and later the radioactive material was transported along the Plohnbach river and polluted the sediments of the floodplain. About 50,000 m^3 of tailings flowed up to 4 km in the Plohnbach valley and formed also a new wetland by damming of the river. Additionally, in 1985 a small fracture on the western embankment of the Lengenfeld tailings pond happened, with only minor effects on the environment. As a result, during a dose assessment in 2002 an effective dose of 4.04 mSv/yr in the Plohnbachaue area for members of the public was investigated, which requires a sanitation.

In an own administrative agreement the Free State of Saxony (Germany) started in 2003 the remediation measures of so called Wismut Abandoned Industrial Sites. Until 2012 78 Mio. Euro is available in total. The investigations are founded by the Free State of Saxony and by the Federal Republic of Germany one half each. The responsibility for the coordination of the rehabilitation of those sites has been taken over to the Wismut GmbH. So the residues of the former Soviet Uranium Mining in East Germany can be remediated which were shut down before 1963. According to the prevailing case law there is no responsibility for anyone with regard to these abandoned industrial sites, because they are not part of

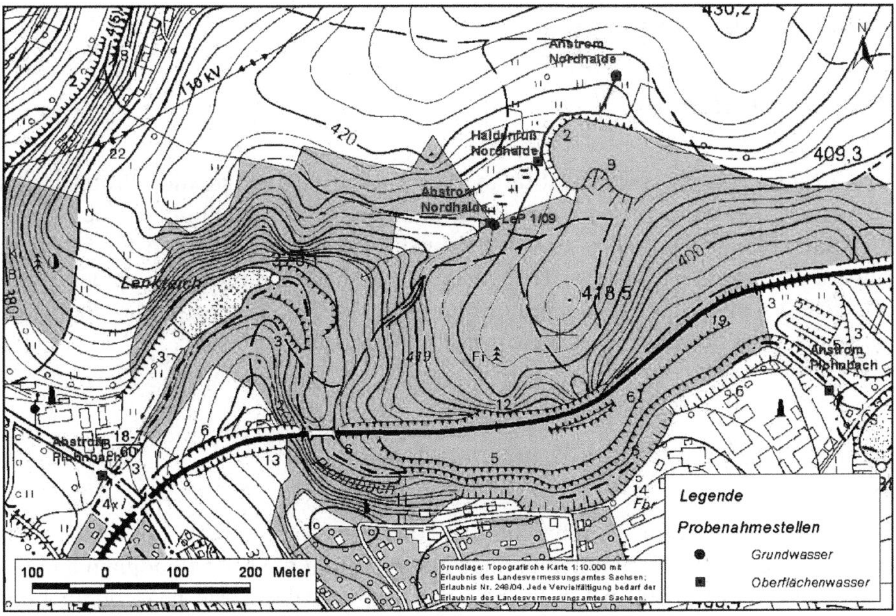

Fig. 2 Location Redevelopment Area Lenkteich

Fig. 3 Historic picture of the Lenkteich area, dated 22.11.1929 (Source: Archives of the Construction Authory of Lengenfeld town; Photo: Karl Effenberg)

the Wismut-Law from December 12th 1991. The administrative agreement predominantly has a focus on the remediation of old tailing ponds, waste dumps, mine openings and dewatering galleries.

Rehabilitation Concept

Need of Sanitation

As a result of a dose assessment by the Federal Office for Radiation Protection in 2002, the range of the effective dose for the members of the public in the area of Lenkteich/Plohnbach Valley was determined to be 4.04 mSv/yr and for Northern Mine Dump to be 1.6 mSv/yr. Those values are above the reference value for the effective dose of 1 mSv/yr and require sanitation methods. In particular, the exposure causes a high risk to the public as the tailings were accessible.

Option Analysis for the Sanitation Design

The option analysis for the rehabilitation concluded as preferable option the relocation of the 53,000 m^3 polluted tailings material from the river floodplain and the Lenkteich pond (26,000 m^3) onto the north slope of the waste heap dump Lengenfeld, which further has to be covered after the removal. Beside the fact of the realization of the remediation there is the requirement to allow a later re-use of the area for recreational and touristic purposes. Due to the sensitive location in the FFH area "Göltzschtal" there had to be considered special nature conservation requirements during the rehabilitation works, with emphasis on species conservation and the achievement of long–term development objectives of the FFH area. The construction of a fish pass in the Plohnbach river is subject of the ecological rehabilitation design.

The Northern Mine Dump is located in an open flat valley to the west above the Plohnbachaue. Downstream damp meadows are located, the dewatering takes place on a nameless creek. Before leaving the valley, this stream is lost in a swamp area that is shaded by willows. The wetland hosts a population of protected species of *Solanum dulcamara*. The marsh and the above lying wetlands are classified as a protected habitat in accordance with § 26 Saxonian Nature Conservation Law.

Detailed Sanitation Design

The remediation measures included after the clearing of the existing vegetation, the construction of the transport routes, the relocation of approximately 52,000 m^3 of tailings from the Lenkteich Pond and parts of the Plohnbach Valley onto the Northern Mine Dump Lengenfeld, and the local coverage there. The relocation area covers a surface of about 26,000 m^2. The area to be rehabilitated at the Northern Mine Dump Lengenfeld has a size of approximately 24,000 m^3. On the Northern Mine Dump Lengenfeld, after the installation of the tailings from the area of Plohnbach Valley and Lenkteich Pond, has to be installed a mineral soil cover of about 1.0 m thickness. With this design, the radiological situation is significantly improved, since the relevant exposure pathways (dust, ingestion) are interrupted. To provide data for the assessment of the sanitation results a monitoring concept with monthly survey was established.

Fig. 4 Bypass riverbed of river Plohnbach

After removal of the contaminated material and the construction of a dam Lenkteich Pond distract becomes a pond again respecting the requirements of the EC Water Framework Directive. The design includes the construction of a fish pass in the Plohnbach river and the reforestation of the upper floodplain in accordance with the development objectives of the Flora-Fauna-Habitat-Plan.

Remediation Works

Status of Sanitation

Currently, the rehabilitation project is in the implementation phase after several years of planning and approval procedures. Once in winter 2009/2010 the clearing took place, in spring 2011 the relocation of the tailings onto the Northern Mine Dump Lengenfeld was started. Currently works are the construction works for the hydraulic elements and the fish pass. By the end of 2011, the tasks will contain the

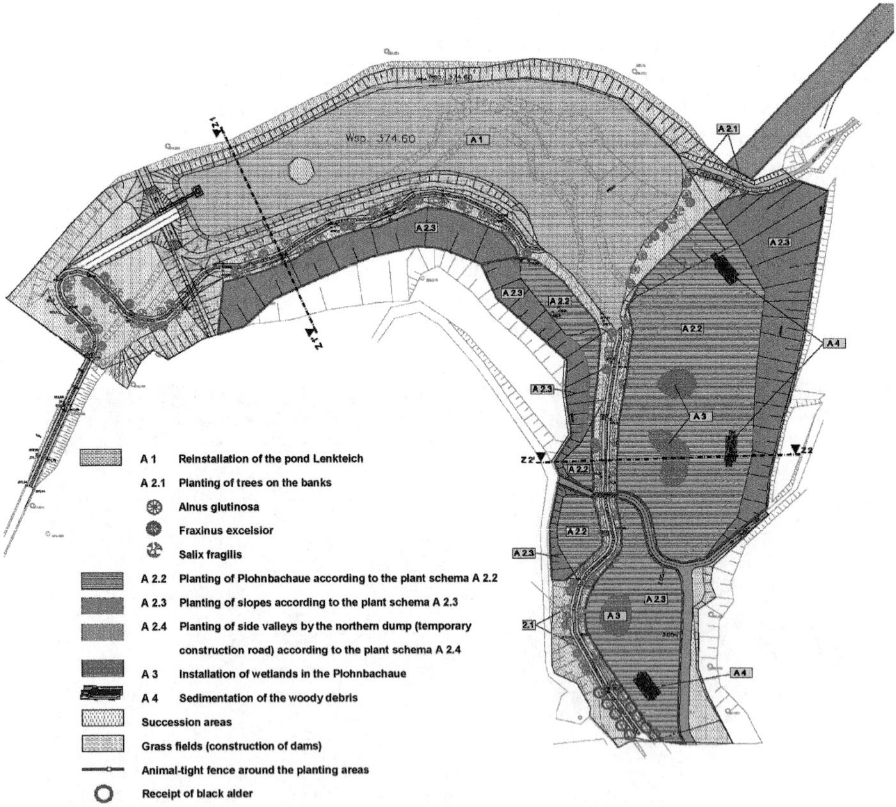

Fig. 5 Compensation measures in terms of nature protection

covering of the Northern Mine Dump Lengenfeld, to re-establish the Lenkteich Pond as well as reforestation and landscaping will be completed.

Compensation Measures for Nature Protection

To minimize the environmental impact of the remediation measure several compensation measures in terms of nature protection were realized, for instance regarding the preservation of the continuity of the stream flow in Pohnbach River. An overview on the compensation measures is given in Fig. 5. The measures contain for instance reforestation, construction of wetlands, and the agglomeration of dead wood as home for insects.

Conclusions

After finalization of the remediation works, the Plohnbachaue Valley will be a public recreation area again. The sanitation project is a pilot project for the realization of the interface of the requirements of the EC Water Framework Directive as well as the FFH Directive.

References

EUROPEAN COMMUNITY (2000): Directive 2000/60/EC of the European Parliament and Council of 23.10.2000 establishing a framework for Community action in the field of water policy, Official Journal of the European Communities, L 327, 22.12.2000, Luxembourg

EUROPEAN COMMUNITY (2005): Climate Change and the European Water Dimension, A Report to the European Water Directors, Steven J. Eisenreich, JRC, Ispra Italy, EU Report No. 21553, Printed in Italy, 2005

Technical University Bergakademie Freiberg (2002): Remediation of uranium milling seite Lengenfeld/Vogtland, needs and objectives, Merkel, B.; Planer-Friedrich, B.; Wolkersdorfer, C.: Uranium in the Aquatic Environment, Department of Geology, ISBN 3-540-43927-7, Springer Verlag Berlin Heidelberg New York, 2002

Generation and Prevention of Acid Drainage from Mining Wastes in a Uranium Deposit

Marina Nicolova, Irena Spasova, Plamen Georgiev, Stoyan Groudev

Abstract. A rich-in-pyrite dump of mining wastes located in the uranium deposit Curilo, Bulgaria, was after natural precipitations a source of heavily polluted acid drainage waters. The generation of these waters was connected with the activity of the indigenous chemolithotrophic bacteria which oxidized the pyrite, other sulfides and tetravalent uranium present in the mining wastes. Different variants of covers with an alkaline or neutral pH and containing biodegradable organic substrates were constructed on the surface and in the top layers of the dump and were tested for their ability to prevent the generation of acid drainage waters.

Introduction

The uranium deposit Curilo, Bulgaria, was for a long period of time a side of intensive mining activities including both open-pit and underground mining techniques. In 1962 the "classical mining" was stopped due to the exhaustion of the high-grade ores. However, at that time there were some low-grade ores in the deposit containing about 0.02% of uranium. For that reason, in 1985 an operation for

---

Marina Nicolova
University of Mining and geology, Sofia 1700, Bulgaria

Irena Spasova
University of Mining and geology, Sofia 1700, Bulgaria

Plamen Georgiev
University of Mining and geology, Sofia 1700, Bulgaria

Stoyan Groudev
University of Mining and geology, Sofia 1700, Bulgaria

B. Merkel, M. Schipek, *The New Uranium Mining Boom*
DOI 10.1007/978-3-642-22122-4 © Springer 2011

in situ leaching of these ores was started. In this operation the ore layers located between the bottom of the former open-pit mine and a gallery (mentioned as No. 5) located about 15 m beneath this bottom were crushed using explosives. Leach solutions consisting of diluted sulfuric acid were injected through numerous boreholes to irrigate the ore mass and dissolve uranium. The pregnant solutions were collected by a system of drainage boreholes and were treated by ion exchange for recovering uranium. In 1990 this leaching operations was ended but since that time and until recently the deposit was a permanent source of acid drainage. These waters had a low pH (usually in the range of about 2–4) and contained radionuclides, several heavy metals, arsenic and sulfates in concentrations much higher than the relevant permissible levels for waters intended for use in agriculture and/or industry. The generation of these waters was connected mainly with the bacterial oxidation of pyrite and other sulfide and uranium-bearing minerals present in the fractured ore body and the underground mining works as well as in several dumps consisting of rich-in-pyrite mining wastes and low-grade ores. Different methods to prevent the generation of such waters are known (Yanful et al. 1999; Peppas et al. 2000; Shippers et al. 2001) and some of them were applied under the real conditions in the deposit towards some of the dumps, which were densely populated by pyrite-oxidizing chemolithotrophic bacteria (Nicolova et al. 2005, 2011). These methods were connected with changes of the environmental conditions in the relevant dumps. These changes were achieved by the periodic additions of solid biodegradable organic substrates (cow manure, plant compost, straw) and crushed limestone to these dumps, which as a result of these additions were turned into ecosystems not suitable for the acidophilic pyrite-oxidizing chemolithotrophic bacteria.

This paper contains some data about field experiments carried out in the deposit and intended to prevent the generation of acid drainage from rich-in-pyrite dump of mining wastes by means of different variants of covers consisting of biodegradable organic substrates and alkalizing agents (limestone or alkaline soil) and constructed on the surface and in the top layers of this dump.

Materials and Methods

A dump consisting of about 2400 t of rich-in-pyrite mining wastes was used in the field experiments. The dump had the shape of a truncated pyramid and contained wastes with a particle size of about 90% less than 30 mm. The dump was located on an impermeable rock and was surrounded by ditches intended to collect the dump effluents. The dump was divided in four sectors with a relatively equal amount of mining wastes each. The dump material was irrigated by the natural precipitations which in the period 2000–2003 were in the range of about 680–750 mm per year.

Pyrite was the main ore mineral in the dump. The main uranium-bearing minerals in the dump were nasturane, torbernite, metatorbernite, pitch-blende, metaotunate and basetite. Chalcopyrite was the main copper-bearing mineral but some secondary copper sulfides such as covellite, chalcocite and bornite were also present, together with some copper oxide minerals. Sphalerite, galena and arsenopyrite were present as main minerals containing zinc, lead and arsenic, respectively. Quartz was the main mineral of the host rock. Clay minerals and some iron hydroxides were also present in the dump.

Data about the chemical composition of the dump material are shown in Table 1.

In one section of the dump (mentioned as No. 1) a mixture consisting of solid biodegradable organic substrates (cow manure, plant compost, straw) and crushed limestone was added to and mixed with the top layer of the mining wastes to form a relatively homogeneous organic-mineral layer to about 20 cm depth (named A-2 horizon). This A-2 horizon was covered by an approximately 15 cm layer (named A-1 horizon) consisting only of the additives, i. e. the organic substrates and the crushed limestone. The organic substrates contained rich microbial communities consisting mainly of various heterotrophic microorganisms.

The second dump section (mentioned as No. 2) was treated in the same way as the section No. 1 but each year, in the spring (at the end of March) fresh batches of the above-mentioned organic substrates and limestone were added to the horizon A-1.

The horizon A-2 of the dump section No. 3 was also prepared in the way used for this horizon of the section No. 1 and No. 2. However, the horizon A-1 of the section No. 3 was prepared in a different ways in comparison with the relevant horizon of the dump section No. 1 and No. 2. In the dump section No. 3 the horizon A-1 was 25 cm deep and consisted of a mixture of the above-mentioned solid biodegradable organic substrates with a fresh slightly alkaline (pH 7.3) cinnamic forest soil and contained 7.5% organic carbon, 0.6% total nitrogen, 0.4% phosporous, 1.4% potassium, with a C/N ratio of 12.5 and pH of 6.8. Within a month after its formation, the horizon A-1 contained a numerous and diverse microflora consisting mainly of typical soil microorganisms. Iron-oxidizing chemolithotrophic bacteria were not present in this horizon. Some herbaceous plants (related to the genera *Medicago, Trifolium* and *Festuca*) were introduced to this horizon and the classical melioration procedures such as cultivation, grassing, irrigation and addi-

Table 1 Data about the chemical composition of the dump material

Component	Content, %	Component	Content, g/t
SiO_2	67.7	Cu	945
Al_2O_3	5.45	Zn	640
CaO	0.35	Cd	28
MgO	0.21	Pb	505
S total	2.51	Ni	109
S sulfidic	1.94	Co	86
Fe	3.87	Mo	50
U	181 g/t	Mn	684
Ra	120 g/t	As	114

tion of fertilizers were applied to maintain this new ecosystem for long periods of time. The horizon A-2 was rich-in-organic carbon (within the range of 7.5–8.0%) and was a transition zone from the aerobic A-1 horizon to the deeply located dump layers which were turned into typical anaerobic zones with a slightly alkaline pH (in the range of about 7.2–7.5) and a positive net neutralization potential (30–35 kg $CaCO_3$/t). The horizon A-2 and the deeply located dump layers were inhabited by a microbial community consisting mainly of sulfate-reducing bacteria and other metabolically interdependent microorganisms.

The quality of the dump effluents was monitored at least once per month in the period April 1999–October 2004. The parameters measured in situ included pH, Eh, dissolved oxygen, total dissolved solids and temperature.

Elemental analysis was done by atomic absorption spectrometry and induced coupled plasma spectrometry in the laboratory. The radioactivity of the samples was measured, using the solid residues after their evaporation, by means of a low background gamma-spectrophotometer ORTEC (HpGe – detector with a high distinguishing ability). The specific activity of ^{226}Ra was measured using a 10 l ionization chamber.

The isolation, identification and enumeration of different microorganisms were carried out by methods described elsewhere (Karavaiko et al. 1988). Phylogenetic analyses based on 16S rRNA gene sequence were performed for speciation of the prevalent isolates from the dump (Hallberg and Johnson 2001). The bacterial activity in situ was determined by the techniques described elsewhere by Karavaiko and Moshniakova (1979) with some modifications (Groudev and Groudeva 1993).

Results and Discussion

The dump used in this study was, before the start of the experiments and after rainfall, an intensive source of acid drainage waters (the relevant columns of data in Tables 2–4). This was also valid for the control dump section No. 4 during the whole experimental period. The generation of these waters was connected with the oxidative activity of the acidophilic chemolithotrophic bacteria inhabiting the dump (Tables 5 and 6). *Acidithiobacillus ferrooxidans, Leptospirillum ferrooxidans* and *Acidithiobacillus thiooxidans* were the prevalent species in this microbial community. Moderately thermophilic chemolithotrophic bacteria related to the species *Sulfobacillus thermosulfidooxidans* and *Acidithiobacillus caldus* were not detected in the dump. However, such bacteria were found in some rich-in-pyrite sections of other dumps in this deposit. Some heterotrophic bacteria were also found but in much lower numbers. The distribution of the chemolithotrophic bacteria in the dump was not homogeneous and their density was highest in the top layers of the dump (over 10^9 cells/g dump material). The number of chemolithotrophs decreased with increasing depth but was still considerable (within the range of 10^4–10^6 cells/g) even in the deeply located layers of the dump in which the content of dissolved oxygen in the pore solutions was low (about 1–3 mg/l).

The highest bacterial number and activity were found in zones with a water content of about 65–75%. After heavy rainfall the bacterial densities in the dump effluents were decreased but the total amount of bacteria washed out from the dump was increased. The bacterial number and activity usually were increased after rest (dry) periods in the dump irrigation. This was connected with the mass transfer of oxygen and fluxes of reactants and products within the dump material.

This was due to the fact that the natural concentrations of these ions, although relatively low (less than 15 mg/l for each of the ions) were still sufficient to maintain the bacterial growth and activity in the dump at relatively high levels.

The temperature inside the dump varied during the different climatic seasons usually within the range of about 8–12°C. The variations in the dump parts located near to the external atmosphere were much higher – usually from 4–17°C. These considerable changes were connected with the relatively small sizes of the experimental dump and with the great differences of the outside temperatures during

Table 2 Composition of the drainage waters from the dump section No. 1 treated in 2000 but without further additions of solid organic substrates and limestone

Parameters	Monitoring periods			
	Before treatment	After treatment		
	2000	2001	2002	2003
pH	1.90–3.87	6.59–7.01	6.02–6.75	4.91–6.15
SO_4^{2-}, mg/l	424–1307	288–680	378–819	404–857
U, mg/l	0.37–4.46	<0.1–0.62	0.28–0.91	0.51–1.58
Fe, mg/l	170–785	9.9–44	18–68	28–95
Mn, mg/l	1.04–18.9	0.25–2.8	0.37–4.10	0.95–6.20
Cu, mg/l	0.68–8.24	0.12–0.45	0.32–0.87	0.51–2.46
Zn, mg/l	1.70–15.2	0.25–0.82	0.78–2.15	1.24–4.44
C org. mg/l	0.4–1.2	10–35	7.1–19	9.1–23

Table 3 Composition of the drainage waters from the dump section No. 2 subjected to periodic additions of solid organic substrates and limestone

Parameters	Monitoring periods			
	Before treatment	After treatment		
	2000	2001	2002	2003
pH	1.94–4.01	6.80–7.16	6.84–7.25	6.87–7.34
SO_4^{2-}, mg/l	410–1292	181–482	140–392	125–347
U, mg/l	0.35–4.10	<0.1–0.37	<0.1	<0.1
Fe, mg/l	152–875	3.7–17	5.5–21	3.5–14
Mn, mg/l	0.95–16.7	0.12–1.7	<0.1–0.4	<0.1–0.8
Cu, mg/l	0.62–9.25	<0.1–0.23	<0.1–0.18	<0.1–0.21
Zn, mg/l	1.90–17.8	<0.1–0.37	<0.1–0.28	<0.1–0.23
C org. mg/l	0.4–1.4	14–55	10–37	14–44

the different climatic seasons. The chemolithotrophic bacteria were active even during the cold winter months but at much lower levels than those observed during the warmer months.

The construction of the rich-in-organics layers with a slightly alkaline neutral pH and a positive net neutralization potential completely changed the existing conditions in the dump sections No. 1–3. The constructed horizons of the A-1 type were inhabited mainly by different heterotrophic microorganisms. Acidophilic iron- and/or sulphur-oxidizing chemolithotrophic bacteria were not detected in this horizon. Their number and activity were very low in the horizon A-2 and in the deeply located dump layers. Some sulfates were still generated as a result of the biological oxidation of the elemental sulphur formed as a result of the chemical and electrochemical oxidation of the sulfides. This biological oxidation was carried out by some neutrophilic chemolithotrophic bacteria, mainly of the species *Thiobacillus thioparus*, and by some heterotrophic bacteria, both present in the horizon A-2 in relatively small numbers. Sulfate-reducing bacteria and other metabolically interdependent microorganisms were the main inhabitants of this enlarged anaerobic zone. These microorganisms used the organic compounds, mainly monomers, dissolved in the horizon A-1 and A-2 and transported deeply in the anaerobic zone by means of the drainage waters. Some bacteria able to reduce the Fe^{3+} and Mn^{4+} ions

Table 4 Composition of the drainage waters from the dump section No. 3 (treated by formation of a rich-in-organic and mineral nutrients soil cover) and from the control non-treated dump section No. 4

Parameters	Monitoring periods		
	Before treatment 2000	After treatment (2001–2003)	
		Treated dump	Non-treated dump
pH	1.19–4.04	6.84–7.25	1.94–4.10
Eh, mV	(+373)–(+590)	(−86)–(−210)	(+361)–(+581)
DO_2, mg/l	1.4–4.4	0.2–0.5	1.2–4.1
TDS, mg/l	747–2444	299–894	765–2394
Solids, mg/l	35–140	28–88	32–122
DC org, mg/l	0.4–1.4	17–55	0.4–1.2
SO_4^{2-}, mg/l	410–1292	122–442	415–1280
U, mg/l	0.32–4.10	<0.1–0.25	0.35–4.01
Ra, Bq/l	0.10–0.45	<0.03–0.05	0.08–0.50
Cu, mg/l	0.62–9.14	<0.1–0.21	0.59–7.85
Zn, mg/l	1.85–17.0	<0.1–0.32	1.90–15.9
Cd, mg/l	0.02–0.12	<0.005–0.01	0.02–0.12
Pb, mg/l	0.10–0.51	<0.05–0.10	0.08–0.41
Ni, mg/l	0.35–1.61	<0.05–0.14	0.32–1.63
Co, mg/l	0.28–1.25	<0.05–0.14	0.28–1.11
Fe, mg/l	152–875	3.7–27	140–794
Mn, mg/l	0.91–16.3	<0.1–1.0	0.95–15.0
As, mg/l	0.07–0.51	<0.01–0.08	0.07–0.48

were also found in this zone. The reduction of iron and manganese was connected with the process of anaerobic respiration performed by these bacteria. It must be noted that, apart from the dissolved Fe^{3+} and Mn^{4+} ions, the relevant insoluble forms of these elements, especially when they were present in recently formed crystalline structures such as $Fe(OH)_3$ and MnO_2, were also used as electron acceptors during the anaerobic respiration. This resulted in solubilization of iron and manganese in the anaerobic zone of the dump. Some anaerobic heterotrophic bacteria inhabiting this zone were able to solubilize some heavy metals from the mining wastes by means of secreted organic metabolites, mainly organic acids. The concentrations of such pollutants in the dump effluents, with a few exceptions (for manganese, iron and sulfates in some cases) were negligible for a relatively long period after the treatment of the relevant dump section (Tables 2–4). This was due to the precipitation of the heavy metals as the relevant insoluble sulfides by the hydrogen sulfide produced in the process of the microbial dissimilatory sulfate reduction or as the relevant hydroxides and carbonates as a result of the alkalinization of the dump. The generation of alkalinity was connected not only with the addition of limestone to the treated dump sections but was also due to the processes of microbial dissimilatory sulfate reduction and ammonification taking place in these anaerobic systems.

The comparison of the effluents from the dump section No. 1 with the effluents from the dump section No. 2 revealed that the systems of this type intended to prevent the generation of acid drainage can be efficient for long periods of time if solid biodegradable organic substrates and limestone are added periodically to the relevant sources of sulfuric acid (mainly pyrite and other sulfide minerals). However, the best results were achieved in the dump section No. 3 in which a real soil cover consisting of a slightly alkaline rich in organic and mineral nutrients soil was established. This soil was able to support the efficient growth of some herbaceous plants which produced large amounts of biomass and in this way facilitated

Table 5 Microorganisms in the drainage waters from the dump section No. 2 subjected to periodic additions of solid organic substrates and limestone

Microorganisms	Before treatment (2000)	After treatment (2001–2003)
	Cells/ml	
Fe^{2+} – oxidizing chemolithotrophs (at pH 2)	10^4–10^8	0–10^2
S^0 – oxidizing chemolithotrophs (at pH 2)	10^4–10^8	0–10^2
Aerobic heterotrophs (at pH 2)	10^1–10^3	0–10^2
Aerobic cellulose-degrading bacteria (at pH 2)	ND	0–10^3
Aerobic heterotrophs (at pH 7)	0–10^1	10^1–10^4
Anaerobic heterotrophs (at pH 7)	0–10^2	10^4–10^7
Sulfate-reducing bacteria	0–10^2	10^3–10^7
Denitrifying bacteria	0–10^2	10^2–10^4
Fe^{3+}-reducing bacteria	0–10^2	10^2–10^4
Methanogenic bacteria	ND	0–10^2
Fungi	0–10^1	0–10^2

Table 6 Bacterial activity in situ in the dump section No. 2 subjected to periodic additions of solid organic substrates and limestone

Temperature, °C	Fe^{2+} oxidized for 5 days, g/l	$^{14}CO_2$ fixed for 5 days, counts/min.g
Before treatment – 2000		
4–8	0.37–1.14	860–2600
8–12	1.04–4.22	2400–9500
12–17	3.52–8.44	7800–18500
After treatment (2001–2003)		
4–8	0–0.14	0–440
8–12	0.10–0.21	300–640
12–17	0.14–0.31	480–880

the development of a microbial community consisting of typical, mainly heterotrophic and neutrophilic, soil microorganisms.

Conclusion

The construction of covers consisting of solid biodegradable organic substrates (cow manure, plant compost, straw) and alkalizing agent (crushed limestone or an alkaline soil) on the surface and in the top layers of a rich-in-pyrite dump of mining wastes was a very efficient way to prevent the generation of acid drainage waters in this dump. The system created in this way was characterized by environmental conditions not suitable for the growth and activity of the acidophilic chemolithotrophic bacteria able to oxidize pyrite and other sulfide minerals. Microbial communities consisting mainly of sulfate-reducing bacteria and other metabolically interdependent microorganisms were established in this new preferentially anaerobic system. This new system can exist for long periods of time if solid biodegradable organic substrates and alkalizing agent (e. g. limestone) are added periodically. A suitable way to maintain such systems is to construct a real soil cover consisting of a soil rich in organic and mineral nutrients, with a pH close to the neutral point and able to support the growth of suitable herbaceous plants (such as *Medicago sativa*) producing large amounts of biodegradable biomass. This biomass supports the growth of different heterotrophic bacteria, some of which (mainly the sulfate-reducing and ammonifying bacteria) participate actively in the removal of pollutants from the drainage waters. The classical melioration procedures can maintain such ecosystems for long periods.

References

Groudev S N, Groudeva V I (1993) Microbial communities in four industrial copper dump leaching operations in Bulgaria. FEMS Microbiology Reviews 11:261–268.

Hallberg K B, Johnson D B (2001) Novel acidophiles isolated from a constructed wetland receiving acid mine drainage, in: Biohydrometallurgy: Fundamentals, Technology and Sustainable Development (eds: V S T Ciminelli and O Gracia Jr), Part A, pp. 433–441 (Elsevier: Amsterdam).

Karavaiko G I, Moshniakova S A (1971) A study on chemosynthesis and rate of bacterial and chemical oxidative processes under conditions of copper-nickel ore deposits of Kolsky peninsula. Mikrobiologiya 40:551–557 (in Russian).

Karavaiko G I, Rossi G, Agate A D, Groudev S N, Avakyan Z A (eds), (1988) Biogeotechnology of Metals, Manual, 350 p (Centre for International Projects GKNT: Moscow).

Nicolova M V, Spasova I I, Georgiev P S, Groudev S N (2011) Prevention of the acid drainage from a heap of mining wastes in a uranium deposit by inhabiting the indigenous acidophilic chemolithotrophic bacteria, in: Proc XIV Balkan Mineral Processing Congress, Tuzla, June 14–16, 2011 (in print).

Nicolova M V, Spasova I I, Groudev S N (2005) Prevention of acid drainage in a uranium deposit, in: Safety and Security Engineering (eds: C Brebbia, T Bucciarelli, F Grazia and M Guarascio), pp. 665–673 (WITpress: Southampton).

Peppas A, Komnitsas K, Halkia I (2000) Use of organic covers for acid mine drainage controls. Minerals Engineering 13:563–574.

Shippers A, Jozsa P-G, Kovac Z-M, Jelea M, Sand W (2001) Large-scale experiments for microbial evaluation of measures for safe-guarding sulphuric mine waste. Waste Management 21:139–146.

Yanful E K, Simms P H, Payant S C (1999) Soil covers for controlling acid generation in mine tailings: a laboratory evaluation of the physics and geochemistry. Water, Air and Soil Pollution 114:347–375.

Portable XRF to Guide a Groundwater Source Removal Action at the Cañon City, CO, USA Uranium Mill

G. Greg Lord, Kenneth E. Karp, John Elmer, John Hamrick

Abstract. The Cotter Corporation began uranium ore milling in 1958 at the Cañon City Uranium Mill in Cañon City, Colorado, USA. Over the years, uranium (U) and molybdenum (Mo) leached from the tailings and impacted the shallow groundwater which eventually migrated off site. In 2008, the S.M. Stoller Corporation (Stoller) performed a source removal action by remediating 225,000 cubic yards (CY) of contaminated soil from the former settling pond that was a potential source of Mo and U groundwater contamination. Stoller used portable X-ray fluorescence (XRF) to provide real-time data that reduced the amount of soil required to be removed by 30%. This paper examines the process by which the sampling was designed, the statistical analysis, and the lessons learned.

Site Location, Cleanup History, and Regulatory Background

Site Location

Cotter Corporation (Cotter) owns and operates the Cañon City uranium mill facility (mill) located in south-central Colorado. Former milling activities at this facil-

G. Greg Lord
S.M. Stoller Corporation, 105 Technology Drive, Suite 190, Broomfield, Colorado 80021

Kenneth E. Karp
S.M. Stoller Corporation, 105 Technology Drive, Suite 190, Broomfield, Colorado 80021

John Elmer
S.M. Stoller Corporation, 105 Technology Drive, Suite 190, Broomfield, Colorado 80021

John Hamrick
John Hamrick; Cotter Corporation; PO Box 1750; Canon City, Colorado 81215

B. Merkel, M. Schipek, *The New Uranium Mining Boom*
DOI 10.1007/978-3-642-22122-4 © Springer 2011

Fig. 1 The Old Pond Area Located in Canon City, Colorado

ity resulted in soil and groundwater contamination. Remediation efforts at this site have been ongoing since the 1980s; however, groundwater in the area continues to be impacted by residual contamination in the Old Pond Area (OPA) of the site and in the down-gradient ground water. Cotter and the Colorado Department of Public Health and Environment (CDPHE) negotiated a settlement that included a source removal as the optimal approach for continued remediation efforts at the OPA, and Cotter selected Stoller to implement the selected alternative.

The OPA is situated on the 2,500-acre Cotter mill site located in Fremont County, Colorado, approximately 3 miles south of Cañon City, and approximately 1.5 miles southwest of the Lincoln Park community (Fig. 1). Populations nearby are primarily north and northeast of the site, although residential growth has increased to the west of the site in recent years.

From 1958 through 1979, mill wastes (tailings) were discharged to a series of unlined tailings ponds covering approximately 175 acres, now known as the OPA. Seepage of liquids from the unlined ponds resulted in contamination of shallow groundwater with two contaminants of concern (COCs), U and Mo. Use of the OPA for tailings disposal ceased in 1979 when a new uranium mill and new lined ponds were built on the site.

Site Cleanup History

Remedial activities were previously performed to remove large portions of the source term materials. Between 1981 and 1983, approximately 2.5 million CY of alkaline tailings were excavated from the OPA and placed into the secondary impoundment cell. In 1988, an additional 2 feet of soil was excavated from the OPA and placed in the lined Primary Impoundment (PI). In 1999, approximately 100,000 CY of soil from selected zones within the OPA having high concentrations of U and Mo in near-surface soils were removed and placed into the PI. These excavation projects have resulted in a reduction of total U and Mo in the OPA source term.

Regulatory Setting and Cleanup Goals

The Cotter uranium mill is regulated by the Mine Safety and Health Administration (MSHA) for safety and by CDPHE for radiological operations. The project involved remediation under Comprehensive Environmental Response, Compensation, and Liability Act (CERCLA). Cleanup standards in soil for U and Mo were derived during the Alternatives Assessment, based on EPA developed site-specific soil screening calculations and a hydraulic conductivity value of 3×10^{-2} cm/s. After the Alternatives Assessment and prior to remediation, the CDPHE lowered the standards slightly as discussed below.

Remediation Objectives and Approach

The objective of the source removal action was to relocate impacted soils containing elevated U and Mo from the OPA and place them in the PI. Stoller performed the following work tasks to achieve the objectives:

- Prepared the site (e.g., widened roads, bermed road shoulders, etc.) for hauling;
- Correlated field XRF instrument data with laboratory data and quantified biases;
- Conducted additional measurements of metals in soils to refine the lateral extent of COCs exceeding the project cleanup goals;
- Excavated areas with soil delineated as exceeding cleanup goals;
- Transported contaminated soil to the PI;
- Collected verification samples for laboratory confirmatory analyses;
- Re-established drainage and applied dust suppression (soil glue) on all exposed soil.

Defining Extent of Contamination

Existing Site Data

Existing data from the Engineering Management Support, Inc. report, *Alternatives Assessment Old Ponds Area, Cotter Cañon City Mill Site* (EMS 2005), were used as the starting point for refining the lateral and vertical extent of contamination boundary. The refined boundary of contamination was delineated based on the evaluation of 1,151 soil samples collected from borings located on 200 to 250 ft centers. Uranium concentrations in OPA soil and bedrock averaged 6.3 milligrams per kilogram (mg/kg) or part per million (ppm), and ranged from 0.5 to 340 ppm. Mo concentrations averaged 16.7 ppm and ranged from 1.4 to 530 ppm. The highest concentrations of both U and Mo were generally found in the 0 to 5 ft-depth interval (TQEM Report, 1995). TQEM estimated the total volume of contaminated soil remaining, exceeding concentrations of 30 ppm U and 100 ppm Mo, was approximately 344,000 CY over approximately 36 acres. Subsequently, CDPHE selected risk based soil concentrations for U and Mo of 27 ppm and 86 ppm, respectively, as the cleanup goals for the removal action at the OPA. The estimated total volume of soils to be remediated was not re-calculated, however, it would likely have increased slightly based on the lower standards.

Refinement of Soil Contamination Boundary

The lateral extent of soil contamination as defined in the Alternatives Assessment (EMS 2005) was used as the starting point for determining the areas that would require excavation. Refinement of these areas was determined to be necessary for two reasons. First, the soil samples in the Alternative Assessment study were collected from borings placed on 200 to 250 ft centers, and the lateral extent of soil contamination was inferred from these data. As a result, the estimate of lateral extent of soil contamination was only approximate. Additionally, remediation activities were conducted in the OPA since the time that these samples were taken.

The surface location of the lateral extent of soil contamination as defined in the Alternatives Assessment was staked in the field and used as starting points for refinement. Refining the location of the extent of contamination in surface soils consisted of performing field screening XRF measurements for U and Mo at selected intervals along the staked boundary. If the screening measurement concentration was less than the cleanup for Mo or U, additional screening measurements were performed every 10 ft in the direction toward the borehole exhibiting contamination until contaminant levels exceeding the cleanup goals were identified. Conversely, if the screening measurement concentration of the COCs along the staked boundary exceeded the cleanup goals, additional screening measurements were performed systematically every 10 ft in the direction away from the staked boundary until the COC concentrations were determined to be less than the applicable soil cleanup goal. This process was repeated every 40 ft along the staked extent as defined in the Alternatives Assessment, thus producing a refined extent of contamination.

Following delineation of contaminated areas, the contaminated material was excavated in 2.5 ft lifts and stockpiled in the lined PI. Upon completion of each excavation lift, the bottoms of the excavations were surveyed with the field XRF instrument to identify additional impacted materials that required removal. This continued until bedrock, groundwater, or soil below the cleanup goals was encountered. Each sample location was uniquely identified with a sample number and its exact location was determined using the site GPS.

Sampling and Analysis Operations

Sampling and analysis activities associated with the delineation, characterization and excavation of the OPA were primarily based on in-situ XRF field screening measurements. At 10% of the sample locations, a soil sample was collected 30 cm deep in as small an area as would remain open, homogenized, and placed in a quart size re-sealable plastic bag. All of the bagged soil samples were then analyzed for U and Mo with the field XRF analyzer and 10% of the bagged soil samples were sent to an analytical laboratory for analysis of U by EPA 3050B/6020 ICP-MS metals and Mo by EPA 3050B/6010B ICP metals. Results of the laboratory sample analyses were used to correlate with the field XRF results on a continuous basis.

Field Portable XRF

The field portable XRF analyzer was used as a cost effective and efficient technology to measure in-situ U and Mo concentrations in soil. This instrument was used initially to delineate and refine the extent of contamination boundaries that had been previously identified during earlier studies. The XRF analyzer was also used to direct excavation of additional material during remedial actions.

Although U is a radioactive element, the low specific activity of natural U coupled with the low cleanup goal negates the use of radiometric techniques for field detection. Therefore, U was included in the suite of metals of interest sampled with the XRF analyzer. The actual detection limits obtained with this instrument were affected by field conditions, such as soil moisture content, soil temperature and organic matter and the specific chemical makeup of the soil but in general were lower (i.e., better) than the manufacturer's published nominal limits of detection for U (10–20 ppm) and Mo (10–20 ppm).

The XRF analyzers were standardized each morning prior to use and were standardized again mid-day prior to the afternoon sampling. Innov-X-Systems, the manufacturer of the XRF analyzers, provided NIST traceable soil standards used to standardize the XRF analyzers. The standards included NIST Standard SRM 2711 and a silicon dioxide blank. The SRM 2711 standard provided a general soil analysis background standard for in-situ XRF measurements. Standardization was required every time the instrument was re-started by scanning the reference sample for one minute. If the instrument did not standardize properly, an error message would appear and the instrument would not allow scanning to begin.

Instrument Bias/Method Correlation Sampling

Due to matrix interferences, in-situ XRF data can be biased when compared to analytical laboratory results of the same sample. The magnitude of this bias is dependent on the work area and the sample conditions, as well as the initial instrument settings that affect instrument sensitivity. Possible XRF biases are discussed in EPA Method 6200. To correct for biases, soil samples were collected in areas of known metals contamination and in non-impacted areas so that these soils could be evaluated and quantified for the potential bias in the OPA. Laboratory results were compared to the field XRF results using linear regression analysis and the resulting correlation coefficients. This correlation was refined three times during the project to ensure a valid correlation factor between the field XRF results and the laboratory results.

The initial correlation study compared 42 paired sample results obtained from laboratory and field XRF determinations. Soil samples used for the comparison were collected in three types of areas: 1) areas that were known to be impacted, 2) areas that were slightly impacted (close to the action level), and 3) areas that were believed to be not impacted. Sample locations were biased to collect most of the samples close to the action level, as this was the main area of concern that

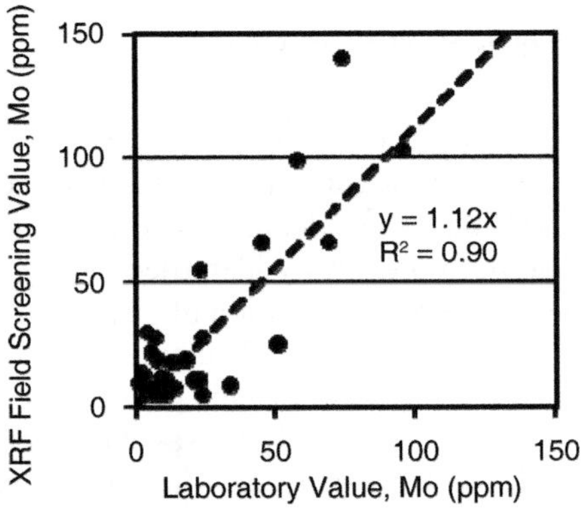

Fig. 2 Molybdenum Correlation, at concentrations less than 150 ppm

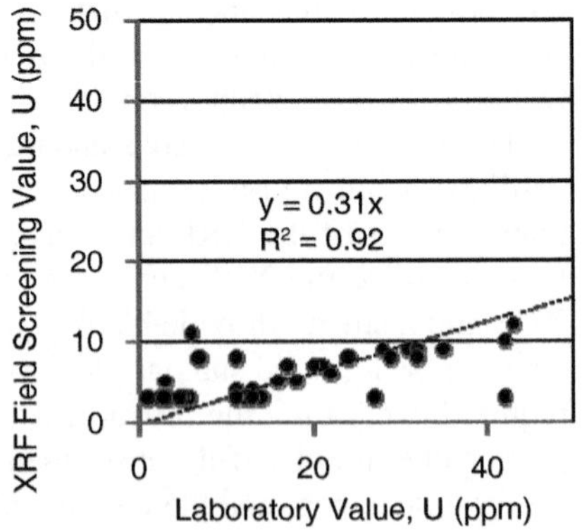

Fig. 3 Uranium Correlation, at concentrations less than 50 ppm

required the highest correlation between field and laboratory results. The study compared the laboratory results to four types of field measurements performed with the XRF analyzer on the soil samples; 1) in-situ measurement, 2) bagged soil samples, 3) bagged and dried soil samples and 4) bagged, dried and homogenized soil samples. The highest correlation coefficient with the laboratory results was obtained using either the in-situ measurements or the bagged soil samples. Typical correlations between the laboratory results and the field XRF screening results for the bagged soil samples are presented in Figs. 2 and 3. In preparing for the extensive field sampling effort, Stoller refined the study by measuring, collecting and testing 30 more soil samples in the laboratory. During the course of the project, additional laboratory data again refined the correlation.

As the project came to an end, Stoller determined using the Visual Sampling Plan (VSP) software program that only 1% of the locations where field XRF measurements were performed (10% of the XRF soil sample locations were bagged and 10% of the bagged soil samples were sent for laboratory analysis) required laboratory analysis to provide verification that the cleanup goals were achieved. Upon being analyzed in the laboratory, these samples resulted in a third, and final, refinement of the correlation. Approximately 90 samples were tested in the laboratory. Samples were collected and archived for possible future use. Results of the correlation studies indicated the screening instruments allowed reliable semi-quantitative assessment of contaminant levels; therefore, the correlations and biases for each instrument were documented and the field work completed.

Excavation Support Sampling

Excavation support sampling included *in-situ* XRF analysis, field analysis of bagged soil samples, and soil samples submitted to the analytical laboratory for analysis. Potholes were excavated in areas of known concern to verify historic borehole data and to better direct excavation efforts.

Upon completion of each excavation lift, confirmatory sampling for metals contamination was performed in the bottom of the excavation. In large excavation areas, the sampling was done on a 30 ft grid established with the GPS surveying equipment. In smaller excavation areas, the grid was adapted to the specific area, with an overall sample location density equal to or greater than a 30 ft grid. *In-situ* XRF measurements were performed at the sampling locations. The exact position of each sampling location was determined and recorded with the GPS surveying instrument.

When the in-situ measurements (corrected for identified bias) indicated that the area was contaminated above the cleanup goals for the specified metals, another lift in that area was excavated and moved to the PI. The correction for bias was accomplished by using the regression equation between the laboratory analysis and the XRF field screening analysis. A pass/fail (PF) value was calculated by multiplying the cleanup goal for the COC of interest by the slope of the regression analysis. XRF field screening results greater than the PF value indicate soil concentrations are above the cleanup goal while XRF field screening results less than the PF value indicate the cleanup criteria has been achieved. The process of sampling the bottom of the excavation, followed by excavation of the contaminated soil, was repeated until in-situ measurements indicated the U and Mo concentrations were below the cleanup goals, i.e. less than the PF value. The iterative process was also terminated if bedrock or groundwater was encountered that prevented further excavation.

Confirmatory Samples

In-situ XRF measurements were performed at the completion of each excavation lift. When the XRF measurement results were less than the cleanup goals and visual indications confirmed that no impacted areas remained, confirmatory samples were obtained from the uppermost 15-cm of surface soil at 10% of the sample locations measured with the field XRF. Soil samples collected from these locations were placed in a sample bag and analyzed with the field XRF. Approximately 10% of these bagged soil samples were submitted to the analytical laboratory for COC analyses. A duplicate (split) was collected for 10% of the samples and sent to the laboratory for internal QA/QC purposes. The field *in-situ*, bagged XRF measurements, and analytical laboratory data were used to develop the XRF correlation relationship and to increase the confidence that the remaining soils were not contaminated above the cleanup goals. A total of 74 confirmatory samples were sent to the analytical laboratory. Many areas were excavated to bedrock. Confirmatory samples for bedrock locations were limited to field measurements only, thus laboratory data are not available for these areas.

Visual Sampling Plan Statistical Analysis

Visual Sampling Plan (VSP) is a software tool that supports the development of a defensible sampling plan based on statistical sampling theory and the statistical analysis of sample results. VSP supports confident decision making, and is the EPA-preferred software for environmental characterization and remediation on radiometric sites. VSP was used in this project to verify that data quality objectives had been met. During excavation activities, VSP was used to:

- Determine and compare averages to the screening level for each analyte and sample area.
- Compare individual sample results against the cleanup goal.
- Develop geospatial contaminant concentration maps for each excavated area, based on confirmatory (e.g., *in-situ,* non-bedrock) samples.

- Evaluated the number of samples collected and the distribution of the samples to determine if the sampling was statistically significant. If not, the number of additional samples and the location of the samples to satisfy the statistical criteria was determined.
- Perform a Sign Test – MARSSIM version, to evaluate the sampling scheme in order to determine, statistically, that an adequate number of samples were collected in each area. Then perform Walsh's Test to determine if any outliers exist. Any Outliers found by Walsh's Test are then evaluated individually to establish whether the outlier impacts the results. No outliers were identified during the analysis, so based on the confirmatory sample data, the excavated areas were found to be statistically clean. Clean is determined by evaluating the median or mean value at the site and comparing it against the cleanup standards established by CDPHE.
- Establish, based on the confirmatory sampled data, that excavated areas are statistically clean. Clean is determined by evaluating the median or mean value at the site and comparing it against the cleanup standards established by CDPHE.

Used in conjunction with the field XRF measurement results, VSP reports were generated as the project progressed to verify that each area's confirmatory sample locations demonstrated that the average concentration was less than the cleanup goals established by CDPHE for U and Mo, of 27 and 86 μg/g, respectively. When the confirmatory sampling showed the area was statistically clean, a standardized VSP final report was generated documenting each excavation area, the confirmatory sample locations, diagnostic graphics, statistical assumptions, formulas used, sensitivity analyses, and the "clean" result confirmation.

VSP results demonstrated that sufficient soil was excavated from each area to ensure that the source contaminants were removed in non-bedrock areas. The VSP final reports provide documentation that the correct type, quality, and quantity of data were gathered to confidently cease soil excavation in what were previously contaminated areas.

Final Site Summary

The total volume of soil excavated and moved to the PI was approximately 30% less than anticipated based on the previously defined extent of contamination. Confirmatory field XRF measurements performed in each excavation area demonstrated that the contaminated soils in the delineated areas of concern had been removed. Laboratory data confirmed the field XRF measurements, and spatial analysis of sampling distribution, and results confirmed the excavated areas were adequately remediated. A few XRF samples still exhibited elevated U and Mo results; however, most of these results were within areas excavated to bedrock, at which point, excavation stopped according to parameters outlined in the work plan.

References

EMS (Engineering Management Support, Inc.) Alternatives Assessment Old Ponds Area, Cotter Canon City Mill Site, originally prepared September 9, 2004, and revised March 17, 2005.

EPA (U.S. Environmental Protection Agency), 1986. Test Methods for Evaluating Solid Waste, SW-846: Physical/Chemical Methods, Volume II: Field Methods, 3rd Edition, Office of Solid Waste and Emergency Response. Washington, D.C.

EPA (U.S. Environmental Protection Agency), 1999. USEPA Contract Laboratory Program National Functional Guidelines for Organic Data Review. EPA540/R-99/008.

PNL, Visual Sampling Plan (VSP), Version 2, Pacific Northwest Laboratory, Contract DE-AC06-76RL01830.

TQEM, Canon City Milling Facility Total Quality Environmental Management Project Summary Report, prepared by Cotter Corporation, Colorado Radiation Control Division, and U.S. Environmental Protection Agency, November 1995.

Characterization of Thorium Binding by Sequential Extractions in Uranium Tailings of Schneckenstein, Germany

Taoufik Naamoun

Abstract. The chemical binding forms of Thorium in uranium tailings of Schneckenstein sites were investigated regarding a remediation strategy. A modified Tessier's seven steps sequential extraction procedure was used for the fractionation of different classes of thorium species present in the sediments. The consecutive steps included leaching of the water-soluble thorium, liberation of thorium from exchange sites on inorganic soil constituents, extraction of carbonates bound species, release of specifically adsorbed thorium on manganese and iron oxides, removal of species associated with insoluble soil organic matter and sulfides and finally extraction of metal intrinsic in the soil matrix present in the structure of primary and secondary minerals. The results indicate that no thorium in the mobile phases. An appreciable amount of its total content is in the moderately reducible phase. Whereas, from 73 to 92% of the thorium is attached to clays and partly to resistant minerals.

Introduction

From 1946 to 1957, the German Democratic Republic (GDR) developed to be the third largest uranium mining province of the world after the US and Canada (Barthel 1993). In the southwest of Saxony in eastern Germany, the Mining Company WISMUT processed approximately 1.2 Mt of uranium ore between opening and final closure. During peak times, WISMUT produced 7000 t of uranium per year. And, its total production was in the order of 220,000 t. The Schneckenstein mill produced 660 thousands of tons of uranium at a capacity of 150,000 t of ore

---

Taoufik Naamoun
Faculty of Sciences B.P.1171 Sfax 3000 Tunisia

B. Merkel, M. Schipek, *The New Uranium Mining Boom*

DOI 10.1007/978-3-642-22122-4 © Springer 2011

per year. The resultant 1.96 Mt of tailings had been deposited in an elevated 108 ha two decantation basins adjacent to the site. The total volume of the residue is about 700,000 m^3 (Merkel et al. 1998).

According to Salvarredy-Aranguren et al. (2008) and Nriagu (1996), mining activities are known to be a major source of environmental metal contamination. Thus, these tailings may act as a storage pool for toxic elements, which get into the surrounding rivers by drainage and to the groundwater by infiltration.

On the other hand, the behavior of metals in soils (e.g., mobility, bioavailability) cannot be reliably predicted on the basis of their total concentrations (Tack & Verloo 1995; Van Peijnenburg et al. 1997; Krishnamurti & Naidu 2003; Meers et al. 2007). Also, the uptake and toxicity of many metals show marked dependence on speciation of the metals and these responses often correlate best with the activity of free metal ions (Laxen & Harrison 1981; Knight & McGrath 1995; Parker & Pedler 1997; Prokop et al. 2001). The process of identifying and quantifying these different species of metals in a sample is referred to as speciation. Also, according to the findings of Janssen et al. (1997) & Janssen et al. (1997), the bioavailability of metals in soils or sediments is often expressed in terms of concentration in a water phase.

The present investigation sought to outline the chemical behavior of the radioactive element thorium in tailing sediments. This paper focuses on the chemical fractionation of thorium in uranium residue by using extraction chemical procedure.

The Study Area

The area of investigation, the uranium tailing Schneckenstein, is located in the Boda valley north of the village of Tannenbergsthal/county of Vogtland, southwest of Saxony (Fig. 1).

The Boda valley is boardered by the Runder Hübel (837 m above sea level) and the Kiel (943 m a.s.l.) to the Northwest and Southeast respectively. The area is situated within the southern branch of the Boda valley at an altitude of 740 to 815 m above sea level.

Geologically, the area belongs to the transverse zone of the southern Vogtland and the western Ore Mountains; a subunit of the anticlinal zone of the Fichtelgebirge – Ore Mountains. It is located on the Southwest border of the Eibenstock granite, a biotite-syenogranite which belongs to the younger granites (J G 1). It is medium to coarse-grained, serial-porphyric and tourmaline bearing. Its intrusion is related to the Upper Carboniferous series.

Also, the investigated sites lie in the area of the watershed between the Zwickauer basin and the Eger. It belongs to the highest precipitation area of the entire Erzgebirge region. Depending on the altitude, 960 to 1160 mm precipitation fall annually and the study area receives an average annual precipitation of 1053 mm.

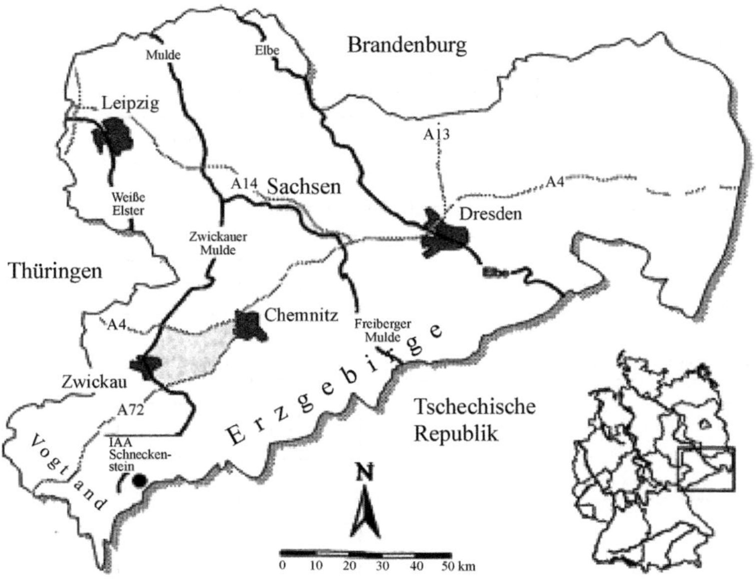

Fig. 1 The position of the Uranium tailing Schneckenstein in Saxony, Germany

Methodology

Based on the contrast of their mineral contents and metal concentrations, nine samples were selected for the procedure. They were freeze dried. To avoid any contamination, a polyethylene spoon was used. They were ground in agate mortar till a size of ≤63 μm and homogenized. A seven-step extraction procedure was used following the Salomon & Forstner (1984) findings. The leaching scheme and reagents are illustrated in Fig. 2.

Results and Discussion

The gross thorium content in the treated samples varies between 12.7 and 20.5 ppm and non Th was detected in the pore water phase. This is because Th is well known for its extremely low solubility in natural waters (Kaufman 1969). Although Th has the tendency to form complexes with sulfates chiefly at pH below 5 (Boyle 1982) and to be adsorbed on to organic matter above pH 7 (Langmuir & Herman 1980), No Th was found in the sulfide-organic phase. This may be due to the alkalinity and the low organic content in the study areas as demonstrated by the low TOC content in the most tailings environment. In addition, no Th was detected in the carbonatic phase (Fig. 3). This is attributed to the low amount of carbonates in the area after the processing of mineral and its low affinity for the above compounds (Boyle 1982).

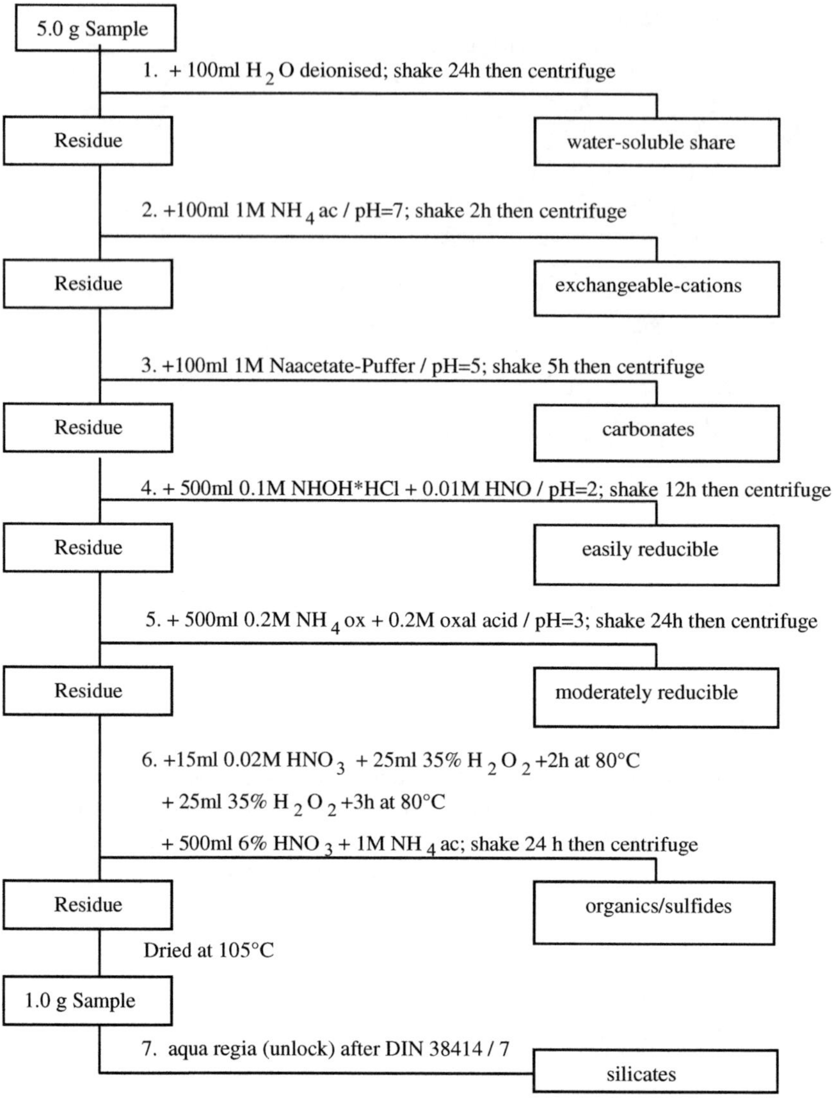

Fig. 2 Sequential extraction scheme

Also, Th has a high ability to be deposited on to clays and organics by adsorption and precipitation mechanisms resulting in the non-detection of Th in the exchangeable phase.

Moreover, since an appreciable amount of Th is frequently associated with silicates such as quartz, feldspars (Boyle 1982), and micas (mainly in inclusion) by direct cation substitution in the silicate lattice as well as by adsorption in lattice defects. Hence lithogenous fraction of Th ranges from 73 to 92%. Also since, an appreciable amount of Th was bound to the initial hematite content; thus after the

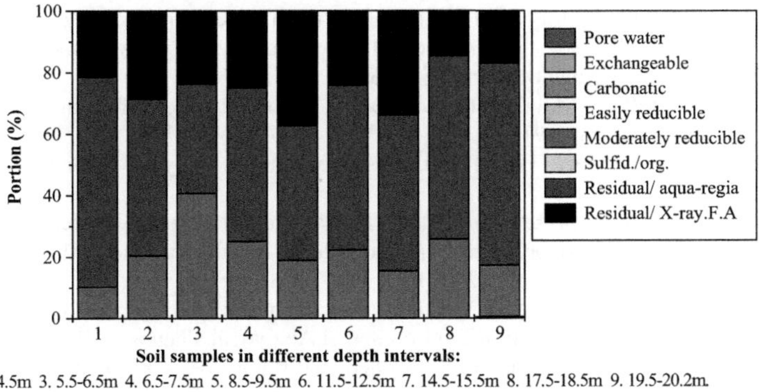

Fig. 3 Uranium Tailings Schneckenstein Selective Extraction Procedures (thorium Element)

ore processing apparently the low mobile Th was fixed to the freshly formed hydrous ferric oxides. Therefore, the remaining Th is found to be in association with the moderately reducible phase.

It is worth noting that thorium is largely attached to clays and partly to resistant minerals since the most amount of Th was dissolved by the total HF dissolution.

References

Barthel, F.H., (1993): Die Urangewinnung auf dem Gebiet der ehemaligen DDR von 1945 bis 1990, Geologisches Jahrbuch A142, Hannover.

Boyle, R. W., (1982): Geochemical prospecting for thorium and uranium deposits, Develop. Econ. Geol., 16. Elsevier-Scientific Publ Co, Amsterdam-Oxford-New York, 498 pp.

Janssen, R.P.T., et al., (1997a). Equilibrium partitioning of heavy metals in duch field soils, II: prediction of metal accumulation in earthworms. Environ. Toxicol. Chem. 16, pp. 2479–2488.

Janssen, R.P.T., et al., (1997b). Equilibrium partitioning of heavy metals in duch field soils, I: relationship between heavy metal partition coefficients and soil characteristics. Environ. Toxicol. Chem. 16, pp. 2479–2488.

Kaufman, A., (1969): Thorium –232 concentration of surface ocean water. Geochim. Cosmochim. Acta, 33, pp. 717–24.

Knight, B., McGrath, S.P (1995). A method to buffer the concentrations of free Zn and Cd ions using a cation exchange resin in bacterial toxicity studies. Environ. Toxicol. 14, pp. 2033–2039.

Krishnamurti, G.S.R. & Naidu, R. (2003). Solid-solution equilibria of cadmium in soils. Geoderma 113, 17–30.

Langmuir, D. & Herman, J. S., (1980): The mobility of thorium in natural waters at low temperatures, Geochim. Cosmochim. Acta, 44, pp. 1753–1766.

Laxen, D.P.H. & Harrison, R.M. (1981). The physicochemical speciation of Cd, Pb, Cu, Fe and Mn in the final effluent of a sewage treatment works and its impact on speciation in receiving river. Water Res. 15, pp. 1053–1065.

Meers, E. et al. (2007). Comparison of cadmium extractability from soils by commonly used single extraction protocols. Geoderma 141, pp. 247–259.

Merkel et al., (1998): Natural leaching of uranium from the Schneckenstein Uranium mine Tailing, Uranium Mining and Hydrogeology II, Proc. Of the Intern. Conference and Workshop, Freiberg, Germany, Verlag Sven von Loga, Köln.

Nriagu, J.O., (1996). A history of global metal pollution. Science pp. 272, 223.
Parker, D.R. & Pedler, J.F., (1997). Reevaluating the free-ion activity model of trace metal availability to higher plants. Plant Soil 196, pp. 223–228.
Prokop, Z. et al., (2001). Mobility, bioavailability, and toxic effects of cadmium in soil samples. Environmental Research 91, pp. 119–126.
Salomons, W. & Forstner, U., (1984). Metals in the hydrocycle, Springer Verlag, Berlin-Heidelberg, 349 pp.
Salvarredyal-Aranguren, M.A. et al., (2008). Contamination of surface waters by mining wastes in the Milluni Valley (Cordillera Real, Bolivia). Mineralogical and hydrological influences. Applied Geochemistry 23, pp. 1299–1324.
Tack, F.M.G. Verloo, M.G. 1995. Chemical speciation and fractionation in soil and sediment heavy metal analysis. Journal of Environmental Analytical Chemistry 59, pp. 225–238.
Van Pejinenbur, W.J.G.M., et al., (1997). Conceptual framework for implementation of bioavailability of metals for environmental management purposes. Ecotoxicology and Environmental Safety 37, pp. 163–172.

Safe Handling of Low pH Mill Tailings Pore Fluids Moab Uranium Mill Tailings Remedial Action Project

Donald Metzler, Joseph Ritchey

Abstract. During excavation of the uranium mill tailings pile in Moab, Utah. Free pore fluids were encountered with a pH as low as 2.5. The fluids began flowing from intercepted wick drains that were installed in 1998. Fluids were allowed to flow to temporary ponds prior to pumping to a 1.6 ha (4 ac) evaporation pond located on top of the tailings pile. Special precautions were taken to blend the low pH water to an acceptable level for spreading on top of the tailings pile to evaporate.

Introduction

The U.S. Department of Energy (DOE) Moab UMTRA Project site is a former uranium-ore processing facility located approximately 5 km (3 mi) northwest of the city of Moab in southeastern Utah. The tailings pile occupies 53 ha (130 ac) and is up to 30 m in thickness. Uranium ore was processed from 1956 to 1984. Excavation, conditioning, and relocation of the mill tailings pile began in 2009 to a disposal facility 50 km (30 mi) north near Crescent Junction, Utah. As of June 2011 over 25% of the pile has been relocated.

Donald Metzler
United States Department of Energy, Grand Junction, Colorado, 81643, USA

Joseph Ritchey
S&K Aerospace, LLC, Grand Junction, Colorado, 81643, USA

B. Merkel, M. Schipek, *The New Uranium Mining Boom*
DOI 10.1007/978-3-642-22122-4

Background

During the uranium extraction process, tailings were slurried from the mill to a retention pond where the liquids were allowed to evaporate. Due to the low porosity, fine-grained nature of the tailings, they retained large internal fluid quantities after liquids evaporated from the surface. Interim cover placement on the pile began in August 1989 (NRC, 1999) and was completed in phases as the center of the pile dried up. Placement of the cover was finished in November 1995. In July 1990, Atlas began pumping water from wells in the pile to the top of the pile for evaporation to accelerate dewatering and consolidation of the pile. They initially used 10 extractions wells, but the number of wells was subsequently reduced to seven. Estimated pumping rates were between 500,000 and 550,000 l (131,000 and 145,000 gal) per month (Canonie, 1994).

The composition of the pore water was reported to be pH values just above 2 and total dissolved solid (TDS) concentrations over 22,000 mg/L most as sulfate. Ammonia concentrations were also high with values above 2,100 mg/l. Estimates of the volume of pore water contained within the tailings ranges from 72 to 106 million l (19 to 28 million gal).

In 1998 Atlas Corporation filed for bankruptcy and in 1999 they terminated their license with the Nuclear Regulatory Commission and transferred certain assets to a Reclamation Trust. PricewaterhouseCoopers was named the trustee. From September to December 2000 the trustee installed a system of 17,000 "wicks" (vertical band drains) that penetrate center of the tailings pile to a depth of 15 m (50 ft) in a triangular pattern spaced 3 m (10 ft) apart (Figs. 1 and 2). The wicks were connected to a manifold that allowed gravity drainage to a sump. A small evaporation pond was constructed for elimination of collected water. Design of the wick system was based on 160 million l (42 million gal) of water captured by the wicks with 90% consolidation in about 18 months. The wick system captured over 5.3 million l (1.4 million gal). The amount of pore water removed through the wick system declined over several years as the pressure dissipated.

Data from the wick system was first collected starting in March 2002. Initially the pumping rate from the sump was approximately 0.03 l per second (2 gal per minute). Within nine months, the pumping rate gradually decreased, averaging approximately 0.03 l per second (0.5 gpm) by the end of 2002. By 2006 the pumping rate was only 0.006 l per second (0.1 gpm) as the wicks became more silted in. After the tailings started being removed from the site, large sections of the wick system were removed.

Ammonia concentrations in samples collected from the wick sump decreased from approximately 15,000 to 3,400 mg/l between September 2002 and September 2009, and uranium concentrations also decreased approximately 80% from 15 to 3.6 mg/l during this same time period. The ammonia and uranium concentrations decreased approximately 50% from 6,500 to 3,400 mg/l and from 8.5 and 3.6 mg/l between June 2009 and September 2009, respectively. This significant reduction suggests that surface water may be infiltrating the wick system.

Fig. 1 Installation of wicks in December 2000

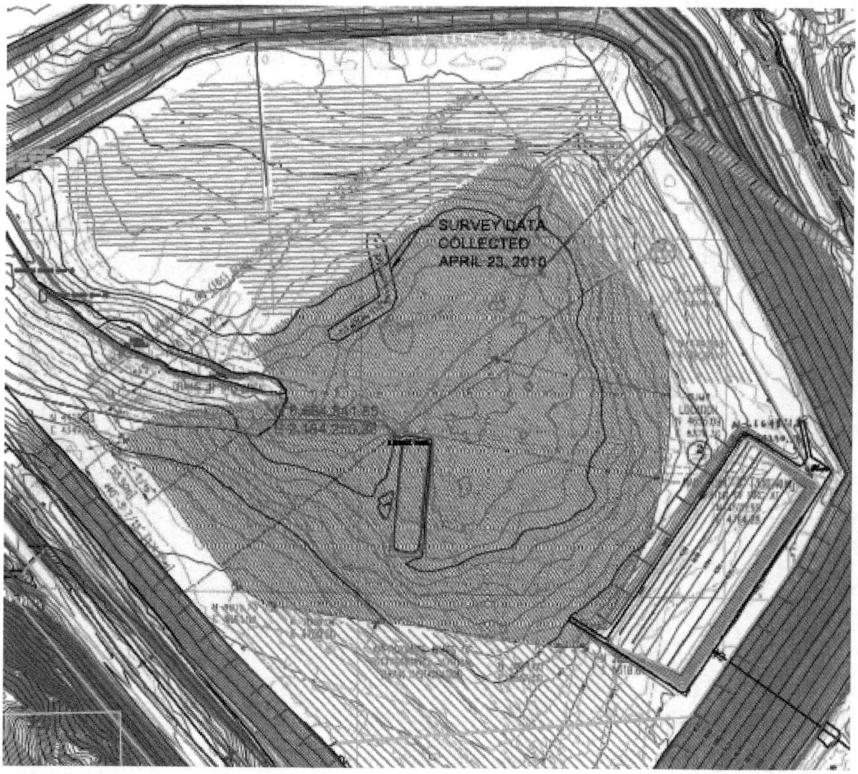

Fig. 2 Location of wicks (*gray shading*) in December 2000

Pore Water Sampling

Excavation of the pile in April 2010 cut into the wicks about 6 meters (20 ft) below the pre-excavation tailings pile surface. Several of the exposed wicks began to discharge pore water (Fig. 3) some with a flow rate estimated to be as large as 1 l per second. The pore fluids were highly acidic with a pH as low as 2.5, and also had a high concentration of ammonia, uranium, and salts.

The location of exposed wicks were surveyed and compared to existing maps of the location of the wick system. Figure 2 shows the approximate location of the flowing wicks, indicated as "survey data collected April 23, 2010". Samples were collected as shown in Fig. 4. Partial sample results are presented in Table 1 (DOE 2010a, b).

Table 1 Summary of pore water samples

Analyte	Concentration April 22, 2010 Sampling	Concentration June 17 and 24, 2010 Sampling
Ammonia (as N)	10,000 mg/l	12,000 mg/l
Uranium	8.9 mg/l	5.6 mg/l
Total Dissolved Solids	260,000 mg/l	250,000 mg/l
pH	2.71	3.24

Fig. 3 Several exposed wicks on the benched excavation. Ponded water at left is pore water from the exposed wicks at the far left

Fig. 4 Sampling the pore water in April 2010

Pore Water Management

Rather than allowing the pore water to remain uncontrolled increasing risk for personnel dermal exposure and reduced equipment operational lifespan, the project implemented a plan to safely handle the water while allowing excavation and conditioning activities to continue unhindered.

Earthen berms of fine-grained tailings were constructed within the contaminated area near the base of the excavation to contain the pore water. A total of seven ponds were constructed capable of holding over 2.4 million l (650,000 gal) while preparations were made to eliminate the water.

As part of the project's ground water remediation program, ground water contaminated with ammonia and uranium is pumped to a 1.6 ha (4 ac) lined pond located on top of the tailings pile. The pond promotes evaporation to eliminate the water. It also acts as temporary storage until site trucks withdraw the water and apply it on the tailings pile and roads in the contaminated areas of the site for dust control. Two forced air evaporators also extract water from the evaporation pond.

The project estimated that over 16 million l (4.3 million gal) of pore water was transferred to the evaporation pond in 2010. During a part of that time the volume of intercepted pore water exceeded the project's capacity to evaporate water such that ground water extraction had to be curtailed. The reduced ground water flow rate to the evaporation pond caused the pH of the pond water to drop to less than 4. This low pH was a health concern for elimination through the forced-air evaporators. To address this concern the evaporators were re-plumbed so that they functioned using water directly from the extraction wells eliminating the mixing with the pore water.

By October 2010 the amount of pore water encountered was substantially reduced so that most of the ponds were eliminated or consolidated into two ponds containing less than 380,000 l (100,000 gal) of pore water. As of June 2011 little additional pore water has been encountered even though in places the excavation has intercepted wicks at a lower elevation within the tailings pile.

References

Canonie Environmental Services Corp (1994) Atlas Corporation Ground Water Corrective Action Plan, Uranium Mill and Tailings Disposal Area. July, 1994

Department of Energy (2010a) Moab UMTRA Project April 2010 Validation Data Package for the Routine Ground Water and Surface Water Sampling Event. Office of Environmental Management, July 2010, DOE-EM/GJTAC1917

Department of Energy (2010b) Moab UMTRA Project June 2010 Validation Data Package for Performance Assessment of the Monthly Sampling for the Ground Water Interim Action and Excavation Seep Sampling Events. September 2010, DOE-EM/GJTAC1937

Nuclear Regulatory Commission (1999) Final Environmental Impact Statement Related to Reclamation of the Uranium Mill Tailings at the Atlas Site, Moab, Utah. Office of Nuclear Material Safety and Safeguards, March 1999, NUREG-1531

Development and Migration of a Critical Habitat and Its Affect on the Colorado River Protection Strategy at the Moab Uranium Mill Tailings Remedial Action Project Site

Joseph Ritchey, Donald Metzler

Abstract. Moab Wash, an ephemeral stream, transects the site of a former uranium-ore processing facility and existing mill tailings pile before entering the Colorado River near Moab, Utah, USA. Dissolved ammonia in ground water emanating from the tailings pile discharges to backwater channels of the river. The channel morphology and therefore presence of critical habitat is dependent on the flow of the wash and the river. Since 2001, the habitat area has changed such that DOE has adjusted its river protection strategy from freshwater diversion into backwater channels to include freshwater injection via wells into ground water.

Introduction

Seepage of fluids from a uranium mill tailings pile is the source of a plume of contaminated ground water that discharges to the Colorado River near Moab, Utah. The DOE Moab Uranium Mill Tailing Remedial Action Project scope is to relocate the 14.5 million tonne (16 million ton) tailings pile to an off-site disposal cell and to clean up contaminated ground water, especially ammonia and uranium, at the Moab site.

Moab Wash, which transects the site, is an intermittent stream that only flows significantly once annually; however, these flows can deposit several tons of sediment and rock in the river, and contribute to the formation of backwater channels along the river's edge at the site. Because the water in these backwater chan-

---

Joseph Ritchey
S&K Aerospace, LLC, Grand Junction, Colorado 81501

Donald Metzler
United States Department of Energy, Grand Junction, Colorado 81501

B. Merkel, M. Schipek, *The New Uranium Mining Boom*
DOI 10.1007/978-3-642-22122-4 © Springer 2011

nels is often stagnant, concentrations of ammonia in the water can exceed toxic levels for young-of-year fish, including several endangered species.

Since 2003, DOE has implemented surface water and ground water remedial actions as interim measures to reduce the ecological risk to sensitive aquatic species that may inhabit the backwater channels. The combination of significant flows in Moab Wash along with the lack of major flows in the Colorado River between 2005 and 2010 has altered the location of the critical habitat area.

Setting

The Moab uranium mill tailings site is located in southeast Utah in the Colorado Plateau geologic province. The Moab area is characterized by sparse vegetation (except along the river), steep canyon walls, and 230 mm (9 in) of annual precipitation. The 175 ha (430 ac) former mill site is bordered to the south and east by the Colorado River. The mill ceased processing ore in 1984 and an interim cover was placed on the 50 ha (130 ac) tailings pile in 1995.

Moab Wash

Moab Wash drains a 13 km^2 (5 mi^2) area of Moab Canyon. The portion that flows through the former millsite is now largely a confined channel whereas before the mill, the channel was braided. The tailings pile contributes to confinement of the channel, resulting in higher flow velocities and a greater sediment load into the Colorado River (DOE 2006).

Significant flows in Moab Wash occur about once per year usually associated with summertime thunderstorms that can be intense, over 2.5 cm (1 in) per hour, with short duration (less than 30 min). There are no gages on the wash so there are no measurements of the flow. Table 1 provides an estimate of the velocity during various flood frequencies for Moab Wash based on a regional estimate that is adjusted for data from Courthouse Wash, which enters the Colorado River about 0.9 km (0.5 mi) upstream from Moab Wash.

Table 1 Flood frequency data for Moab Wash

Flood Frequency	Estimated Flow Rate (cubic meters per second)	Estimated Velocity (meters per second)
2	11.4	2.3
5	30.0	
10	49.6	
25	84.8	2.4
100	162.3	

Colorado River

The Colorado River flow is measured at a gage several miles upstream from the former millsite. However, just a few perennial streams join the river between the gage and the Moab site such that it is reasonably used as a measure of flow for the site. The river drains a 62,400 km^2 (24,100 mi^2) area. The river has a base flow of 85 to 113 m^3/s (3000 to 4000 ft^3/s) with a mean peak flow of about 850 m^3/s (30,000 cfs). Peak flows are the result of spring melt-water from a portion of the Rocky Mountains. Table 2 provides the peak flow of the river for various flood frequencies.

During the past 15 years (1996–2010), the Colorado River peak flow has been below normal. Higher flow rates tend to scour the river channel, whereas lower-than-normal rates contribute to deposition. The last year of substantial peak flow was 1995 having a recorded rate of 1470 m^3/s (51,900 cfs). However, historical aerial photographs indicate that many of the channel features have been similar since major flooding in 1994–1995. The 1994 flooding caused major erosion and deposition in the area just downstream of the confluence with Moab Wash (Fig. 1).

Fig. 1 Aerial view of the border of the site with the Colorado River (May 2011)

Table 2 Flood frequency data for Colorado River near the Moab site

Flood Frequency	Peak Flow Rate (m^3/s) [a]
2	200
5	310
10	405
25	610
100	2200

[a] Based on data from 1914 to 2010.

Identification and Formation of Critical Habitat Areas

In 2001 surface and topographic features in the sandbar area were surveyed to identify potentially suitable fish habitat areas that may develop into calm backwaters adjacent to the site. Aerial photographs taken in September 2001 were also used to provide more detail in a sand bar area.

Spawning of the endangered fish in the Colorado River has been confirmed to occur in bouldery areas with fast moving water which occurs more than 160 km (100 mi) upriver. Therefore, the presence of spawning life stage in the Colorado River is not of primary concern in the vicinity of the site with respect to habitat protection. Similarly, the presence of larval stage is not a primary concern in the backwater areas adjacent to the site, since the distance from the spawning areas to the site is approaching the maximum potential range of larval drift. However, the segment of river adjacent to the site appears to provide suitable habitat for young-of-the-year and adults.

The Utah Division of Wildlife Resources in December 2001 evaluated the most likely areas adjacent to the site that could provide suitable fish habitat for endangered young-of-the-year and adult Colorado pikeminnow and razorback sucker (DOE 2002). Three areas were identified along the channel that have potential to develop into suitable habitat during a normal river runoff based on defined physical characteristics of: (1) backwater or slow-moving eddies, (2) sandy-silt substrate, and (3) water depths of less than 0.6 m (2 ft). In 2001, the potentially suitable areas extend for approximately 200 m (650 ft) downstream of Moab Wash.

In 2005 with the preparation of an environmental impact statement including a biologic opinion regarding the effects on endangered species, DOE agreed to divert river water to critical habitat areas. At that time, the critical areas which were identified in 2001 had not changed though the presence of critical habitat is always dependent on the river stage during the receding waters after the spring runoff. The young-of-the-year concerns ended in the end of September when fry were no longer considered to inhabit the backwater areas.

Monitoring

The objective of surface water monitoring is to observe the channel morphology and evaluate the impact of the site activities on river water quality. DOE monitors estimated Colorado River flow rates for the Cisco, Utah, gaging station (based on upstream flow rates and weather systems impacting the Colorado River basin) using data available from the National Oceanic and Atmospheric Administration (NOAA). In addition, DOE monitors the data provided by the NOAA Western Water Supply Forecast, which provides a long-term seasonal estimate of runoff volume.

From 2005 to 2009 the river water was introduced only for river flows between 280 m^3/s (10,000 cfs) and 425 m^3/s (15,000 cfs), as measured at the Cisco, Utah, gaging station (09180500), where critical habitat areas are most likely to emerge. At flows greater than 425 m^3/s (15,000 cfs), areas were completely submerged and at flows less than 280 m^3/s (10,000 cfs), area that were habitat typically are dry.

Monitoring each year showed that the location of the critical habitat area was migrating down river as sediment slowly filled in the channels. In 2007 the critical habitat was still located near the confluence with Moab Wash (DOE 2008). In 2009 the critical habitat was formed 335 m (1100 ft) downstream from the confluence with Moab Wash and extended for 175 m (DOE 2010).

A probe (Hach SensIon) is used to analyze ammonia, as N, concentrations in surface water and ground water samples. By using the probe, the real-time ammonia concentrations are obtained compared to waiting for laboratory results, which usually takes 28 days. On occasion, ammonia samples are collected for analysis by both the probe and the laboratory to assure the quality of the data.

Protection Strategy

The initial protection strategy of the endangered species was diversion of river water to dilute contaminates to acceptable levels. Initially the formation of critical habitat areas was not in the local where the majority of contaminates were discharging to the river. Extraction of ground water was implemented as a separate mass recovery remedial action.

Surface Water

The surface water remedial action strategy from 2005 to 2009 consisted of diverting river water into the backwater channels to dilute ammonia concentrations in critical habitat areas formed after the peak runoff until late September. The pumping and distribution system was often also used as a preventative measure if a backwater channel is approaching a designation as a critical habitat.

Habitat flows vary from year to year based on erosion and deposition in the backwater channels. The original area of concern was a single backwater channel directly downstream of the confluence of Moab Wash. However, deposition has over the years with sediment contributions from periodic flows of Moab Wash have caused critical habitat areas to form further downstream. Depending on river and wash flows the critical habitats may not form at all adjacent to the site or they may retreat closer to the wash confluence.

Ground Water

In 2009, seven new extraction wells were installed to augment the existing system of 41 extraction/injection wells and one infiltration trench. The new wells were located further away from the river and allowed the wells along the river to be converted to freshwater injection wells. This facilitated considering a new river protection strategy of injecting water along the river to both block ground water discharge to the river including during the presence of any critical habitat. This new strategy is being evaluated in 2011. This year the Colorado River had a peak discharge of 1320 m^3/s (46,600 cfs) on June 9th which may result in changed channel topography along the site. The presence of critical habitats will be monitored with the availability of the traditional surface water diversion if needed.

References

Department of Energy (2002) Work Plan for Implementation of the Initial Action in the Sandbar Area Adjacent to the Moab Project Site. Grand Junction Office GJO-2002-299-TAR

Department of Energy (2006) Conceptual Design for the Moab Wash Realignment, Moab Utah. Office of Environmental Management. DOE-EM/GJ1366-2006

Department of Energy (2008) 2007 Initial Action Final Report. Office of Environmental Management

Department of Energy (2010) Moab UMTRA Project 2009 Ground Water Program Report. Office of Environmental Management DOE-EM/GJTAC1941

Critical Challenges of Acid Mine Drainage in South Africa's Witwatersrand Gold Mines and Mpumalanga Coal Fields and Possible Research Areas for Collaboration Between South African and German Researchers and Expert Teams

Thibedi Ramontja, Detlef Eberle, Henk Coetzee, Rüdiger Schwarz, Axel Juch

Abstract. More than a century of mining of gold, uranium, coal and other minerals in South Africa has left significant mining legacies. One of the most serious environmental legacies is that due to acid mine drainage, which has received significant public attention in recent years. The long histories of mining in both South Africa and Germany present excellent opportunities for collaboration between researchers to allow the identification and implementation of optimal solutions to these problems. In the past two decades German mining industry has implemented a number of successful rehabilitation projects, the focus of which has been on the treatment of mine water. In doing so various solution strategies, methodical concepts and technologies have been developed, which – as part of the bilateral METSI Project – will be made available to South Africa's mining industry.

Thibedi Ramontja
Council for Geoscience, Private Bag X112, Pretoria 0001, South Africa

Detlef Eberle
Council for Geoscience, Private Bag X112, Pretoria 0001, South Africa

Henk Coetzee
Council for Geoscience, Private Bag X112, Pretoria 0001, South Africa

Rüdiger Schwarz
geotec Rohstoffe GmbH, Friedrichstraße 95, 10117 Berlin, Germany

Axel Juch
geotec Rohstoffe GmbH, Friedrichstraße 95, 10117 Berlin, Germany

B. Merkel, M. Schipek, *The New Uranium Mining Boom*
DOI 10.1007/978-3-642-22122-4 © Springer 2011

Introduction

Acid mine drainage (AMD) has been identified and has received increased public scrutiny as an environmental risk in a number of mining regions in South Africa. These regions include the Witwatersrand gold mining areas in the Gauteng and Free State provinces, the coal fields of the Mpumalanga and Kwazulu-Natal provinces and defunct copper mines in the Northern Cape. The Witwatersrand gold mines and the Mpumalanga and Kwazulu Natal coal fields are of particular concern as many of the problem areas are in close proximity to communities and/or urban settlements.

Generation of Acid Mine Drainage in the Gold and Coal Mines

Acid mine drainage in the Witwatersrand gold mines is generated both in the deep interconnected mine voids left by more than a century of mining and in the surface mine residue deposits (waste rock and tailings) created by the mining, milling and processing of the ores. These mines are situated across a Palaeoproterozoic sedimentary basin comprising nine discrete gold fields that have been described by Robb and Robb (1998). The unique features of the Witwatersrand, which lead to specific problems relating to AMD are the large size of the operations, the degree of interconnection of mines, with the largest of interconnected voids extending approximately 40 km along strike and the association of the gold ores with uranium, in some cases in economic quantities (Cole 1998; Coetzee et al. 2006).

Mining of gold in the Witwatersrand commenced in 1886 and still continues in some of the gold mining areas. In the oldest mining areas, little or no underground mining is still taking place and gold and uranium production is limited to tailings reclamation operations. As mines have stopped or scaled down their operations over time, the pumping of water from the underground workings has been stopped in some gold fields. In these cases, the large interconnected voids start to flood. In the West Rand Gold Field the water in the void day lighted in September 2002, with serious ecological impacts, while in the neighboring Central and East Rand Gold Fields, the water is currently rising, with water levels several hundred meters below the surface. As water rises in the mine voids, it reacts, together with oxygen, with sulfide-bearing minerals, exposed by mining, occurring largely in interlayered quartzites, and conglomerates, to produce acid mine drainage. In some cases AMD generated in surface mine residue deposits recharges the mine void, adding to the problem of AMD generated in the underground workings.

South Africa's coal resources are concentrated in Permian rocks of the Karoo Supergroup. These cover large portions of the country, with economically accessible coal seams being present in the northern and eastern parts of South Africa. Historically, most of the coal mining has taken place in the Mpumalanga and Kwazulu-Natal provinces, with some mining in the eastern and southern portions of Gauteng and the northern part of the Eastern Cape. The scale of the operations is large, with coal providing the basis for South Africa's energy supply via large thermal power stations and oil-from-coal plants. Coal is also used as a domestic fuel for heating and cooking as well as being an important mineral export product.

In the coal mines, acid mine drainage is produced by the oxidation of sulfides contained in both the shallow coal seams and the interlayered sedimentary rocks. The presence of sulfides has been described in the coal seam by many researchers including Zhao et al. (2010). In recent years, this has been identified as a significant source of pollution in the Vaal River catchment which provides water to the economic hub of South Africa and in the Olifants River which flows to the north and east, eventually flowing through the Kruger National Park into Mozambique.

Risks Associated with Acid Mine Drainage

A South African team of experts has recently assessed the risk associated with acid mine drainage in the Witwatersrand gold mines and developed recommendations to alleviate the situation in prioritized areas (Council for Geoscience 2010). The identified risks can be subdivided two groups, namely, risks associated with flooding and risks associated with decanting. With respect to the flooding of mines, the identified risks are:

- Contamination of shallow groundwater resources,
- Geotechnical impacts,
- Increased seismic activity.

Risks owing to the decant of acid mine drainage are:

- Serious negative ecological impacts,
- Regional impacts on major river systems,
- Local flooding of low-lying areas.

The management of risks in the prioritized areas involves the following:

- Decant prevention and management by pumping water to maintain safe water levels within the flooded mine voids,
- Water quality management,
- Water ingress control to reduce the volumes of water which need to be pumped and treated.

In the coal mines many researchers are involved in a number of research programs. Acid mine drainage from coal mining has:

- Negative ecological impacts,
- Regional impacts of major rivers.

Other risks related to mining include:

- Negative geotechnical impacts such as subsidence related to undermining,
- Spontaneous combustion.

A number of initiatives have been put in place to address the risks, including the desalination of AMD from a number of operational and abandoned mines, while programs to address many of the most acute impacts of coal mining are in preparation.

Possible Collaboration Between the South African and German Researchers

AMD is a common problem at abandoned mine sites in many parts of the world. For example, in Germany following the reunification in 1990, the legacy of uranium mining in the former GDR included 1400 km^2 of open mine workings, 311 million m^3 of waste rock, and 160 million m^3 of radioactive tailings located in densely populated areas (Hagen and Jacubick 2006; Wismut 2008). There are many other countries which have or still experiencing acid mine drainage. It is become apparent that the success in addressing acid mine drainage requires the implementation of collaborative research programs both nationally and internationally. This enables the transfer and exchange of skills. It is against this background that the Council for Geoscience, which is actively involved in acid mine drainage research, is exploring collaboration with German researchers. This is especially in the area of monitoring water, water geochemistry, soil quality, large-scale identification of contaminant soil and water, remediation measures and hydrogeological modeling for management purposes as well as assessing and managing the impact on human health in AMD affected areas.

The METSI-Project

The METSI project documents were compiled according to the CLIENT call of the German Federal Ministry of Education and Research (BMBF 2010) and submitted to the ministry in August 2010. A total of 25 German and South African companies and institutions declared their willingness to participate in the implementation of the project.

The basic objectives of the German-South African cooperation in the framework of "Dialogue for Sustainability" are described as follows (Minutes 2009):

"Germany has placed sustainability at the core of its economic and social development agenda. Specifically, the German Research for sustainability is aimed at:

1. Developing strategies for social action with the goal of meeting fundamental needs while minimizing the risks for the long-term stability of nature and society.
2. Gearing globalized value chains and production systems to sustainability and at the same time securing the long-term competitiveness of the German economy.
3. Shaping regional development with the goal of improving the quality of human life and at the same time stabilizing the natural, social and economic bases of this quality of life on a long-term basis.
4. Managing natural resources with the goal of maintaining their functions on a long-term basis and at the same time preserving and fostering their regeneration capacity as much as possible."

"METSI" is Setswana, one of the eleven official languages in South Africa, and means "water". It stands also for "Management System for the Implementation of Environmentally Sound Water Supply Technologies in South African Mining Areas with Involvement of all Relevant Stakeholders".

The project focuses on development, design and implementation of mathematical models that consolidate and evaluate existing data and generate substantiated projections of "water" amount and quality. METSI covers a wide range of scientific, technical, social as well as economic questions related to "water in mining" (Schwarz et al. 2011).

After research work done by METSI project team the use of complex models and DSS that describe flow conditions and transport of pollutants is in an early stage in South Africa and does not meet current needs and requirements. Considerable area extension of underground mines, mining activities at great depths (down to 4000 m below surface) and extremely large groundwater catchment areas present a great challenge for the quality of hydrological models.

The METSI project offers an excellent opportunity to assist South Africa in its effort to meet the challenge described above.

The METSI project – as a starting point for technology and know-how transfer of German water purification and treatment technologies – aims to address these interfaces. Moreover, the project team is well aware of the fact that existing market needs provide favorable conditions for opening up a new market for medium-sized enterprises. The South African project partners expect the cooperation with Germany to spark off an innovation boom for an optimization of water management particularly in regard to the mining company–municipality interface.

Preparatory Work

The members of METSI project team have a wealth of experience gained during many years of practical work in the field of planning and management of national and international projects:

- Integrated water management for the mining industry and in mining regions.
- Specialized knowledge about uranium and AMD issues.

- Rehabilitation technologies, equipment and management of water treatment plants.
- Financial management on behalf of public authorities.
- Authority-corporation interaction management and party to project teams.
- Knowledge of area and country, on-site experience since the early 1990ies.

The idea for the METSI project proposal was born in 2007 when three official provincial government delegations of Free State and North West visited the German Free State of Saxony. A workshop held in Johannesburg in autum of 2008 and organized by the German Embassy in Pretoria, presented results from the R&D project "Bergbau und Umwelt in Südafrika – Mining and Environment in South Africa". Also, in autumn of 2008 (14th October until 17th October 2008) the International Conference "Mine Closure 2008" was held in Johannesburg. At this conference German and South African project partners presented technical papers on results from national and international projects in a number of presentations and poster sessions (Deissmann and Kistinger 2008; Gerth and Hebner 2008; Grießl and Schwarz 2008; Kahnt and Paul 2008; Kuyumcu 2008; Schwarz et al. 2008; Stoch et al. 2008). At the International Mine Water Conference held in Pretoria, South Africa, from October 19th to 23rd 2009 three presentations of German and South African project partners have been published in the conference proceedings (Jenk et al. 2009; Schwarz et al. 2009; Winde 2009).

Objectives and Solutions

The Project's objectives are defined as follows:

1. Export of technology and experience in the field of complex water management in mining areas, know-how transfer in the field of employment of specialized sensors and data management software and DSS, know-how transfer in the field of alternative, ecologically and economically sound use of disused mines (use of the energetic potential of hydraulic conditions and methane gas for the generation of energy in an environmentally and climate-friendly manner), know-how transfer in the field of recultivation of post-mining landscapes (ecological reduction of pollutant emissions, adding to the value chain and creating additional economic benefit by implementing supplementary measures such as cultivation of energy crops on devastated land).
2. Organization of know-how transfer and capacity building on the South African side (e.g. authorities, community representatives, universities).
3. Development, design and implementation of one or more mathematical models that consolidate and evaluate existing data (e.g. data stored at authorities, mining companies, universities) and generate substantiated projections of mining water amount and quality.

4. Registering all relevant parameters on the surface (inter alia water catchment areas, environmental impacts caused by mine sites, TSF, dumps, and so on) and underground (inter alia shafts, "decant" issue, acid mine drainage, hydraulic connections between shafts), and sorting according to amount of waters and parameters (inter alia chemistry, pollution level = toxic metals, uranium).
5. Adjustment of German water treatment and purification technologies to conditions at model or pilot sites, determination of scaling factors.
6. Stakeholder involvement (know-how transfer in regard to a fair reconciliation of interests between all parties involved, authority-corporation interrelation management).

The METSI project emphasizes the following areas:

1. Water and water management,
2. Mining, use of resources (obtaining and managing resources) and land use,
3. Sustainability,
4. Social development (with a focus on mining areas).

In terms of methodology "Integrated Water Management" means the interaction of science, decision-making (politics) and practical activities (economy) with regard to various aspects of water use. The project application comprises three subject areas and/or key tasks:

1. Modeling,
2. Work program implementation at model sites, application of German environmental technologies,
3. Involvement of stakeholders (participation), capacity building, skills transfer.

Development and design of one or more complex hydrogeological/hydraulic models will be in the focus of the **first project stage**. Partner institutions from the sciences and research will make a major contribution to this working package while partners from the mining industry and consultancies will pool their hand-on experiences gained while applying models and decision support (DSS) and measurement systems in the national and international mining industry. Modeling of the hydrogeological conditions in their entirety as requested by the first package is a task German engineering can solve, due to years of practical experience. Fundamental prerequisites are access to appropriate and evaluated data and compatibility of mathematical "tools" with conditions at the South African partners. Modeling will include risk assessment and analysis of the probability of occurrence, the quantitative or qualitative value of risk, and effects on various social compartments that may react sensitively to environmental risks. Results of modeling will provide secured statements on today's situation (actual condition) and forecast periods. Forecast periods can be chosen at will. Apart from water amount information (categorized by storey and/or by water catchment areas) the models will show material conditions of water (inter alia degree of pollution/contamination). The data will illustrate any individual material e.g. uranium, metals, sulfate, nitrate, chloride and/or produce summary results. These model calculations will generate descriptions of what the material conditions in the investigated mining areas are like today or will be like in the future.

A **second work package** deals with the adaption of German innovative measurement, water management and control technologies as well as soil decontamination and water treatment technologies to the specific conditions in the RSA. Germany has excellent conditions to be successful: the engineering industry offers a broad range of expertise, management know-how and production facilities, which enables companies to produce the very machines and plants that meet any specialized technically and technologically requirements requested by customers around the world. Treatment and management technologies will undergo tests and evaluation at selected pilot and model sites. Work will include renaturalization and ecological revitalization of devastated areas. The second work package comprises investigations and tests concerning further economic use of mining sites including collieries. In this connection, investigations will be conducted in order to be able to assess the energetic potential of mining water (geothermal and flow energy) and the possibilities of converting it into electrical energy. Another field of investigation will deal with the cultivation of energy crops which, beside the forthcoming climate protection effects ("green effects"), will add to the value chain.

All in all, these tasks make the project an exchange project ("transfer project").

The **third package** will deal with what is called authority-corporation interrelation management (which includes proposals for the development of laws, statutory regulations, and recommendations for actions), and a fair system for the reconciliation of interests between parties involved ("stakeholder involvement" or "involvement of parties affected"). To this purpose, the project partners consider training of parties affected – whether working for authorities, administrative bodies or participating companies – and the development of new and application of proven methods of "stakeholder involvement" a key task. Based on experience of many years project partners and involved South African partners from provincial governments, municipalities and technical authorities will make use of internationally proven moderation and meditation procedures (e.g. "Open Space"). Important issues touch on social development, fight against poverty and the generation of jobs for the "green economy" in particular. Within the framework of the project we will draw upon a wealth of experience gained from employment schemes that have been implemented in Germany in connection with the remediation of mining areas. Moreover, the package includes basic and advanced training, skills transfer, public relations and promotion measures. Major contributions will come from participating universities and "networkers". On the German side this will mainly be done by the German Federation of International Mining and Mineral Resources (FAB), the GBMU e.V. as representative of the MineWaterTec business initiative (MineWaterTec 2011), which is supported and promoted by the Federal Ministry of Economics and Technology (BMWi), and, on the South African side, by partners such as the Southern African – German Chamber of Commerce and Industry Johannesburg (AHK).

The METSI project will make available the wealth of German experience to the Republic of South Africa, which is based on a century-long history of active mining and, in a simplified manner, may be summarized as follows:

- 800 years of metal ore mining (e.g. Ag, Au non-ferrous metals, Fe) and natural stone;
- 200 years of coal mining (hard coal and lignite). Annual mining of 330 m tons of lignite made Germany the world market leader of the 1980ies;
- 150 years of potash and rock salt mining. And for decades Germany has ranked among the top 3 salt producers in the world; and
- 50 years of uranium mining and Germany was among the world top 3 uranium producers in the years of 1950 to 1990.

It shall be underlined here that 800 years of mining mean 800 years of meeting continuously new challenges of social development, of various life cycles of mines, of fighting negative impacts of mining in densely populated areas (environmental issues, sustainability, water protection); and of gaining experience in arranging relations of coexistence among parties involved ("stakeholder involvement") aiming at a fair reconciliation of interests.

Since the reunification of Germany in 1990 complex remediation projects have been planned and industrially implemented: uranium mining area (WISMUT), German lignite mines (LMBV, VATTENFALL), rock salt industry locations (K+S). Projects implemented in Germany and serving now as reference projects comprised a huge number of tasks of national and economic importance. German experience attests that a modern mining industry may become a model of successful structural change.

Outlook

The implementation of the METSI concept in close cooperation with the Council for Geoscience in one of the major mining regions in South Africa offers an excellent basis to reduce the tension between the ecology and economy through integrated approaches and programs (and ensures environmentally friendly and sustainable eco-social development). Large-scale mine rehabilitation projects in Germany (mining of uranium, copper slate, potash, rock salt, and coal) required comprehensive planning and have produced a wealth of experience, which provides to the German raw materials industries and consultancies a fair – but temporal – advantage on international markets. Apart from challenging scientific issues – in particular in regard to model theory – the METSI project puts heavy emphasis on a fair balance of interests of all parties involved by applying proven methods such as "stakeholder involvement" and "public participation".

References

BMBF (2010) http://www.bmbf.de/en/furtherance/14892.php

Coetzee H, Winde F, Wade P (2006) An assessment of sources, pathways, mechanisms and risks of current and future pollution of water and sediments in the Wonderfonteinspruit Catchment. Pretoria, Water Research Commission. WRC Report No. 1214/1/06: 202 pp

Cole DI (1998) Uranium. The mineral resources of South Africa. MGC Wilson and CR Anhausser, Council for Geoscience. Handbook 16: 642–652

Council for Geoscience (2010) Mine water management in the Witwatersrand Gold Fields with special emphasis on acid mine drainage. Report to the inter-ministerial Committee on acid mine drainage. Pretoria: 128 pp

Deissmann G, Kistinger S (2008) Evaluation of Sustainable Remediation Measures for Mine Closure and Derelict Mines – Application of Predictive Geochemical Models and Risk-Based Cost-Benefit Analyses for Decision Making, Mine Closure 2008. Proceedings of the Third International Seminar on Mine Closure. 14–17 October 2008. Johannesburg, South Africa, 2008, 511–529

Gerth A, Hebner A (2008) The Potential of Plants in Constructed Wetlands for the Removal of Contaminants from Mine Water in Germany, Mine Closure 2008. Proceedings of the Third International Seminar on Mine Closure. 14–17 October 2008. Johannesburg, South Africa, 2008, 599–605

Grießl D; Schwarz R (2008) Creating a New Landscape Architecture – A Holistic Approach for Mine Closure Operations, Abstract Poster Presentation, Mine Closure 2008. Proceedings of the Third International Seminar on Mine Closure. 14–17 October 2008. Johannesburg, South Africa, 2008

Hagen M, Jacubick AT (2006) Returning the WISMUT legacy to productive use. In: Uranium in the Environment. BJ Merkel and A Hasche-Berger (Eds). Berlin., Heidelberg. Springer: 11–26

Jenk U, Paul M, Schoepke R (2009) Development of a Novel Approach to Source Manipulation to be Applied at the Flooded Underground Uranium Mine at Königstein, Germany, in: "International Mine Water Conference" Pretoria, South Africa 2009, Water Institute of Southern Africa's Mine Water Division & International Mine Water Association (2009): International Mine Water Conference [CD ROM] – 976 pp; Pretoria (Document Transformation Technologies), ISBN 978-0-09802623-5-3, 801–806

Kahnt R, Paul M (2008) Integrated Methodology for the Optimization of Mine Closure, Mine Closure 2008. Proceedings of the Third International Seminar on Mine Closure. 14–17 October 2008. Johannesburg, South Africa, 2008, 297–307

Kuyumcu M (2008) Remediation of Abandoned Lignite Mines in Eastern Germany, Mine Closure 2008. Proceedings of the Third International Seminar on Mine Closure. 14–17 October 2008. Johannesburg, South Africa, 2008, 419–427

MineWaterTec (2011) www.minewatertec.com

Minutes (2009) Minutes: Innovation for Sustainability in a Changing World – Second South African – German Dialogue on Science for Sustainability, 26 and 27 October 2009, Pretoria, South Africa

Robb LJ, Robb VM (1998) Gold in the Witwatersrand Basin. MGC Wilson and CR Anhausser, Council for Geoscience, Handbook 16, 294–349

Schwarz R, Gerth A, Hebner A, Mgudlwa L (2008) Integrated Water Management – Assignments of Water Management in Mining Regions, Mine Closure 2008. Proceedings of the Third International Seminar on Mine Closure. 14–17 Oktober 2008. Johannesburg, South Africa, 2008, 547–553

Schwarz R, Gerth A, Morgenstern S, Hebner A (2009) Strategies for Managing Environmental Problems and Water Treatment in Mining, in: "International Mine Water Conference" Pretoria, South Africa 2009, Water Institute of Southern Africa's Mine Water Division & International Mine Water Association (2009): International Mine Water Conference [CD ROM] – 976 pp; Pretoria (Document Transformation Technologies), ISBN 978-0-09802623-5-3, 557–566

Schwarz R, Pusch E, Juch A (2011) METSI – Innovative solutions and technologies for management of mining related water in South Africa, SDIMI 2011, Aachen 14–17 June 2011, abstractvolume

Stoch E J, Winde F, Erasmus E (2008) Karst, Mining and Conflict – A Historical Perspective of the Consequences of Mining on the Far West Rand, Mine Closure 2008. Proceedings of the Third International Seminar on Mine Closure. 14–17 Oktober 2008. Johannesburg, South Africa, 2008, 69–83

Winde F (2009) Uranium Pollution of Water Resources in Mined-Out and Active Goldfields of South Africa – A Case Study in the Wonderfonteinspruit Catchment on Extent and Sources of U-Contamination and Associated Health Risks, in: "International Mine Water Conference" Pretoria, South Africa 2009, Water Institute of Southern Africa's Mine Water Division & International Mine Water Association (2009): International Mine Water Conference [CD ROM] – 976 pp; Pretoria (Document Transformation Technologies), ISBN 978-0-09802623-5-3, 772–781

Wismut (2008) Wismut GmbH – a government-financed company in Saxony and Thuringia, www.wismut.de

Zhao Z et al. (2010) Evaluation and validation of geochemical prediction techniques for underground coal mines in the Witbank/Vryheid Regions. Pretoria, Water Research Commission. WRC Report No. 1249/1/10. 212 pp

Legacy of Uranium Extraction and Environmental Security in the Republic of Tajikistan

M.M. Yunusov

Abstract. Some results of investigating uranium tailings in the North of Tajikistan are presented. Results show that the majority of radioactivity of solid waste is associated with a certain particle size. Investigation of soil samples indicates that contamination of the area by gamma-radiation is between 95 and 98% due to Ra-226. The contents of the uranium and thorium in aerosol have been investigated as well. By hydro chemical sampling the dynamics of groundwater pollution was traced. Migration of polluting components from the tailings into groundwater was determined. Furthermore the distribution of the pollutants in the food-chain (tailing – soil – plant) and the distribution on Rn-22 in the tailings was studied.

Introduction

Studying the effect of toxic elements, particularly radionuclide in the environment and human living near radioactive tailings, is an important issue and of interest for many scientists, NGO's and international experts.

The Republic of Tajikistan was during the Soviet era, as well as some other republics, a major supplier of strategic raw uranium for the nuclear industry in the USSR. After the collapse of the USSR the legacy of uranium production and radioactive waste remained. Currently in northern Tajikistan in the cities of Taboshar, Gafurov Chkalovsk and in the village of Adrasman 54.8 million t of low activity uranium waste with an activity of 6240 Ku is stockpiled. The total occupied area is 182 ha. From 10 existing tailings eight are covered but two are uncovered and thus subject to erosion due to climatic conditions.

M.M. Yunusov
SE "Vostokredmet", Republic of Tajikistan

B. Merkel, M. Schipek, *The New Uranium Mining Boom*
DOI 10.1007/978-3-642-22122-4 © Springer 2011

Uncovered radioactive tailings “Digmay”, located near the city of Khujand, and a waste plant with low-grade ores near the town of Taboshar are considered to be the most dangerous of man-made radioactive legacies. It should be noted that the tailings “Digmay” refers to the category “action needed” and 36 million t are stockpiled on 90 ha and another 2.2 million t of waste “factories low-grade ores” on 3.7 ha respectively.

These two objects have a negative impact on the environment in particular during floods, prolonged rains, and strong winds. Due to erosion and wind there is a probability carrying over of radionuclides and heavy elements far from the sanitary protection zone (SPZ) around the above objects. In this case an impact on the health of populations living near the objects and of course on the biosphere can be stated.

All of these sites are on the agenda of SE “Vostokredmet”, which since its development more than 60 years ago, is engaged in radiation safety of workers, the population adjacent to industrial facilities settlements, and natural environment.

Results

Waste tailings are a complex in chemical composition. Apart from natural radionuclide they contain high concentrations of accompanying toxic elements. The results of the sieve and radiometric analysis show that most of the radioactivity of solid waste (from 18 to 30,000 Bq/kg) is associated with fine fractions with particle sizes less than 1 mm (see Table 1), which are most exposed to haze removal from the surface of the tailings.

In this regard, we have investigated the quantitative and qualitative compositions of dust containing radionuclides around waste tailing paradigm and the city of Taboshar. It should be noted that nowadays there is no data on radionuclide concentrations in aerosols around the radioactive tailings of northern Tajikistan in the literature.

Selective determination previously made by IAEA experts related to suspended air particles and radionuclide's “paradigm” of Digmay and Taboshar were general in nature, meaning they depended on the mass of Th and U in the air, not dividing them into fractions. Therefore, the previously obtained experimental data do not provide accurate and complete picture of the content of Th and U in the air above the objects.

Table 1 Fractional composition and activity of Digmay waste tailings

Fractions	Class	Gamma-activity, mkR/h	Total-activity, Bq/kg
Rough gross	> +30	150–400	1200–3000
Gravel	+1.0–30	300–600	3000–6000
Sandy	+0.025–1.0		
Sandy-silt	+0.075–0.025	450–2800	18,000–30,000
Clayed-silt	< 0.075		

Effect of inhalation of radioactive aerosols in the organs of the respiratory system depends on the size, shape and density of the particles. Additional risks may represent deposition of these aerosols on the skin and sucking its radioactive components. To determine the qualitative and quantitative composition of Th-232 and U-238, we carried out an experiment to capture particulate filters of varying subtlety. It is known that aerosols smaller than 1 micron when ingested through the lungs have a severe impact. At our study sites we found dust particles smaller than 1 micron contain between 35 and 50% thorium and uranium.

In our case, the investigated objects on top of the slope of the tailing "Digmay" the total concentration of U-238 in aerosols was 3.624 ng/m^3 and in aerosols less than 1 micron 1.556 ng/m^3, respectively 0.343 and 0.184 ng.m^3 in the "farm"; 0.483 and 0.269 ng/m^3 near the village Gozien, and 0.609 and 0.353 ng/m^3 in Taboshar (see Table 2).

In order to determine the boundaries of the contaminated region in 1991, 1993, and 2010 within the sanitary protection zone an areal gamma-ray survey was performed. Investigation of soil samples indicates that contamination of the area by gamma-radiation at 95–98% was due to Ra-226. The background value was determined by taking samples outside of industrial pollution.

The fact that the pollution is caused by leaching of radioactive material from the surface of the tailings dam is shown by the distribution of values of exposure dose (VED) and the activity of natural radionuclide's (ANR) at a depth of pits traversed in the SPZ. The maximum activity of soils and subsoil in the range of 0–0.1 m at a depth of 0.2–0.3 m VED values are close to background (Fig. 1).

Consequently, the activity of soil in the range of 0–0.1 m is due to deposition of dust particles on Earth's surface, and their destruction in the physical and chemical processes that occur as a result of interaction with the sediments and oxygen, and subsequent migration of ERN in the topsoil.

Table 2 The volume concentration of radionuclides, ng/m^3

Filters	Filter pore size filters, micron	"Digmay"		Village of Goziyon		Farm in "Digmay" paradigm		Taboshar	
		Th-232	U-238	Th-232	U-238	Th-232	U-238	Th-232	U-238
	10.2–∞	0.096	0.404	0.049	0.043	0.041	0.049	0.090	0.059
	4.2–10.2	0.153	0.660	0.098	0.058	0.051	0.036	0.122	0.076
	2.1–4.2	0.136	0.572	0.102	0.060	0.037	0.036	0.089	0.063
	1.3–2.1	0.088	0.431	0.103	0.054	0.031	0.038	0.072	0.059
	0.69–1.3	0.066	0.335	0.072	0.044	0.028	0.039	0.058	0.056
	0.39–0.69	–	–	0.033	0.038	–	–	–	–
	0.0–0.39	0.228	1.221	0.377	0.186	0.113	0.145	0.215	0.297
	Inhalation ways:	0.293	1.556	0.483	0.269	0.141	0.184	0.273	0.353
	Scores:	0.765	3.624	0.836	0.483	0.300	0.343	0.646	0.609

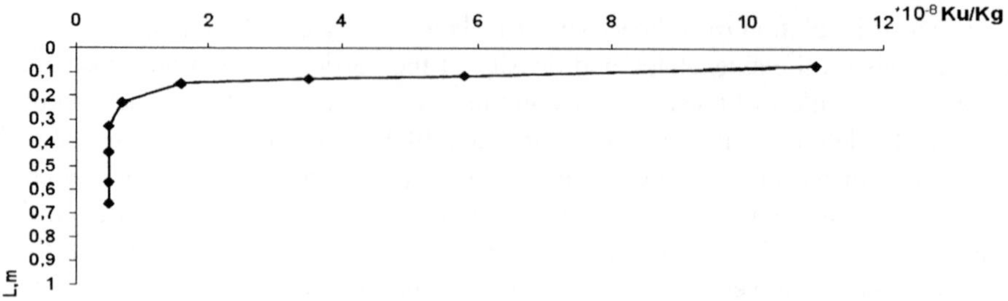

Fig. 1 Distribution of the total activity of A_{ff} over the depth of pit tailings of Digmay, Hole 1

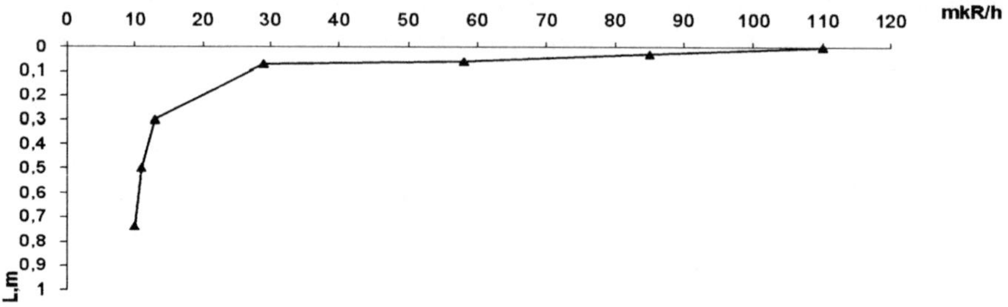

Fig. 2 Distribution of gamma radiation over the depth of pit tailings of Digmay, Hole 1

Results of the determination of the waste isotopes Th-230, Po-210 and Pb-210 suggest that the waste has seen a significant accumulation of decay products, in particular Po-210 and Pb-210, to a lesser extent Th-230. This is associated with a shift of the radioactive equilibrium in the Ra-226 decay products which are Rn-222 and further Pb-210 and Po-210. Th-230 is formed by the decay of U-238, a lack of which is noted in the waste. Due to the fact that the leading role in the formation of contamination by gamma radiation adjacent to the tailing area belongs to Ra-226, Pb-210 and Po-210, which are beta and alpha emitters, their contribution to the radioactivity area is not so significant. Therefore, in excess of background values the contribution of these radionuclides should be considered when calculating the feed and food chains load with internal radiation.

With reference to uranium as health indicator elements found in the samples were split into the following groups and presented in Table 3.

Table 3 Group of the elements pollutants

Group of elements					
Sin genetically	Entering to the structure of crystal lattice	Strictly regulated metals and toxic elements	Regulated metals	Radio-active	Anomalous for the region's water
Pb, Mo, Zn	Zn, Be	V, Se, Cd, Be, Mo, Hg, As, Sb	Co, Ni, Cr, Cu, Pb, Zn, Mn	U, Ra, Th	Al, Ag, Bi, Mn, Sn, Sr, Ba, F.

Of the elements listed in Table 3, we highlighted the association of elements directly linked to tailings and waste have a significant impact on the distribution of elements in the "tailing-soil-plant". This association includes Mn, Ni, V, Cr, Mo, Pb, Zn, U.

Biogeochemical studies on the tailings "Digmay" made in the period from 1991 to 2010, included the study of vegetation in the vicinity of the tailings and features of the distribution of elements-pollutants in the "tailing-soil-plant".

At geobotanical map in 1991 in most characteristic of low-ephemeral wormwood (*Artemisia sogdiana* + *Carex pachystilis* + *Poa bulbosa*) plant communities were predominant. Projective cover of the soil by plants mainly is 40–60%. In the south-western part of the study area the projective cover reached 8% mainly in depressions and channels of ephemeral streams.

Significant changes in plant communities, both in the SPZ, and beyond have been taken place in the area of tailings during the period from 1991 to 2010. In the north-westerly direction at the site, located under the tailings dams, previously occupied by Parnolistniks, according to a study in 2010 is dominated by ephemeral-sagebrush community. The south of the survey area is currently dominated by ephemeral with sagebrush community. Vegetation changes are the consequence of drying ponds as a source of moisture and increasing environmental pressures on the environment from the storage of waste.

Currently a biogeochemical map of the Digmay tailing area is compiled based on an analysis of ten most prevalent plant species in the area of the tailings regarding the four pollutants U, Cd, Pb and Zn, taken into account the morphology, the mechanical composition of surface sediments, and wind directions.

This map helps getting an idea of the extent of the dust pollution caused by the wind blow out of radioactive material from a surface of the tailings. At the same time a significant role in the formation of the contamination are the physical and chemical processes causing the surface, groundwater and subsurface migration of pollutants and their plants accumulation.

The negative impact of tailings on the hydrosphere is due to the leaching of accumulated radioactive waste producing zonal distribution of contaminations of individual components. This applies especially to contaminants in solution that are controlled by pH. By major macro components zoning has the form Fe^{3+}, Fe^{2+}, Ca^{2+}, Cl^{-}, NO_3^{-}, and SO_4^{2-}. The remaining macro-and micro components, including radionuclide's are related to sulfate.

Results showed high efficiency of applying various integrated physical and chemical methods in the study of pollution of the hydrosphere, in particular hydro chemical sampling and vertical electric sounding (VES). Detailed sectors of the surface of mineralized layer had a sufficiently high contrast and showed good agreement with the results of hydro chemical sampling.

Analysis of the received information indicates the presence of infiltration of saline waters in the north-west and north of the tailings and the presence of tailings in the area of the two sources of groundwater pollution; the first of which is associated with the tailings, and the second related to artificial irrigation.

According to the results of hydro chemical sampling at observation wells dynamics of groundwater pollution for the period from 2005 to 2010 were traced. To this end, we chose the cross-section in the north-western and northern directions, observing most intensive migration of polluting components of the tailings. For the analysis we chose SO_4^{2-}, Cl^-, and NO_3^-, both directly involved in the process of entering the tailings with liquid phase slurry and as dry residue reflecting the net impact of other components. From the normalized values of pollutant concentrations the total function of groundwater pollution macro components was calculated.

Furthermore, we have studied the physical and chemical processes that reflect the dynamics of mass transfer of radon and its decay products from the tailings into the environment. Under calm and dry weather conditions measurements reflect the concentration of radon in the surface layer of the tailing and its values are 300–350 Bq/m^3 at a height of 2.0 m. Simultaneously, with the normative model UPM-86 and the software package "Era", we calculated the ground-level concentrations of radon above the tailing and beyond. The average value of annual release of radon on the results of simulation and practical measurements is 8.14×10^{12} Bq. The relative discrepancy between measurement and simulation is estimated at 7–8%.

Along with experimental investigations mathematical modeling of stratified distribution of radon concentration using the program "Ecolego" was carried out. Satisfactory convergence of experimental and calculated data was observed. It should be noted that the experimental data are largely dependent on weather conditions prevailing during the measurements and terrain.

In order to assess the diffusion properties of clays composing the neighborhood Digmay's hill and using them for the preservation of the tailings, as well as to assess the real rate of diffusion of radon in a neutral loam research on a tank model was carried out. To determine the effect of granulometric composition of the layer of neutral soil on the migration of radon, we have performed investigations on three models. In all models an emanating neutral layer was covered with soil containing fractions of various size classes a) model 1 (class-50 25 mm); b) model 2 (class-25 + 10 mm), c) model 3 (class-10 mm).

Table 4 Calculated values of the diffusion coefficient D*

No п/п	The layer	Spacing, см	$D^* 10^{-3}$, см$^2 \times$ s^{-1}
1	1	120–100	22.670
2	2	100–80	16.776
3	3	80–60	8.647
4	4	60–40	12.709
5	5	40–0,0	12.335
			14.627

* average D

After establishment of diffusion equilibrium across the model tube air samples were collected, concentrations of radon determined, and apparent diffusion coefficients calculated. The average diffusion coefficient for model 3 is equal to $14.7 \times 10^{-3}\,cm^2 \times s^{-1}$. This means that for at least 100-fold attenuation the emission of radon required backfilling of waste neutral loam layer thickness of about 3,7–3,8 m.

As can be seen from Table 4 the most packed bed is the third. It corresponds to the smallest value of D^*. The first layer is sealed to the highest diffusion coefficient.

Studies and mathematical modeling indicate that the choice of material for preservative capping is the most important parameter. Performance of the neutral layer is largely determined by packing density.

Preparation of a Safety and Environmental Impact Assessment and Design of Remediation Works for Uranium Tailings at AMCO Site, Zambia

Uwe Walter, Rolf Zurl, Michael Paul

Abstract. The Government of the Republic of Zambia has obtained the support of the World Bank and the Nordic Development Fund for the Copperbelt Environment Project (CEP), to address the environmental liabilities and obligations which have remained with ZCCM Investments Holdings Plc. (ZCCM-IH) and the Government of the Republic of Zambia following the sale of mining assets. The Consortium WISUTEC – C&E with assistance of local consultants has been contracted by ZCCM-IH to implement this project. The scope of this assignment is to assist ZCCM-IH Environmental Coordination Unit (ZECU) to develop strategies and protocols for mitigating environmental impacts emanating from AMCO's uranium tailings for the sitting, design, operations and closure of the proposed AMCO uranium tailings waste management facility.

Introduction

The scope of this assignment is to assist ZCCM-IH Environmental Coordination Unit (ZECU) to develop strategies and protocols for mitigating environmental impacts emanating from AMCO's uranium tailings for the sitting, design, opera-

Uwe Walter
WISUTEC GmbH, Jagdschänkenstr. 33, 09117 Chemnitz, Germany

Rolf Zurl
C&E Consulting und Engineering GmbH, Jagdschänkenstr. 52, 09117 Chemnitz, Germany

Michael Paul
WISMUT GmbH, Jagdschänkenstr. 29, 09117 Chemnitz, Germany

B. Merkel, M. Schipek, *The New Uranium Mining Boom*
DOI 10.1007/978-3-642-22122-4

tions and closure of the proposed AMCO uranium tailings waste management facility. These strategies and protocols are necessary to protect workers, the public and the environment, presently and in the future, from adverse radiation hazards in accordance with the requirements and recommendations of the IAEA and the International Commission on Radiological Protection (ICRP), as well as from other adverse human health and environmental impacts associated with the chemical and physical properties of the uranium tailings and their spill-over area.

The project was structured into four tasks which partly overlap. These tasks were:

1. To conduct an environmental impact assessment (EIA) of the AMCO uranium tailings (TD 11 and TD 13) and the layout of feasible remediation options.
2. To conduct a safety assessment of the proposed preferred remediation option chosen for the AMCO uranium tailings waste management facilities.
3. To prepare design and bidding documents for this preferred remediation option.
4. Supervision of remediation works.

The presentation will highlight the experiences of the project team according to the specific conditions in Zambia and for the location inside the Copperbelt region.

The Environmental Impact Assessment (EIA)

The scope of the EIA was to identify the type (radiological and non-radiological), level and significance of potential environmental impacts to be assessed, the physical extent of the impacted area, the main concerns of key stakeholders, and also information gaps that require additional data collection and further study.

Scoping is designed to answer three main questions:

- What effects could this project have on the environment?
- Which effects are likely to be most important and therefore need most attention in the environmental impact study?
- Which alternatives ought to be considered in developing the remediation proposals for the project?

Internationally accepted standard procedures in the scoping stage have been followed which take radiological and non-radiological aspects into account.

One of the most common ways of identifying the likely impacts from a project is to consider:

- all the activities that will be involved in implementing the project,
- all the components of its receiving environment that could be affected, and
- all the potential interactions between activities and components.

For this purpose a Scoping Checklist (complemented by an auxiliary list) was developed.

Project Area

The project area is located in the close vicinity of the city of Kitwe in the central part of the Copperbelt province of Zambia. The district covers an area of 777 km^2 and has borders with the Democratic Republic of Congo.

The AMCO sites are located adjacent to a residential area, at the northwestern periphery of Kitwe some 500 m north of the Mindolo Stream (see Fig. 1). Both AMCO sites are situated approximately 5 km from the town center of Kitwe.

TD (Tailings Dams) and the increased levels of activity in its close vicinity were discovered only two years ago, as part of the preparation of a consolidated environmental management plan for the ZCCM mines.

The former AMCO village, which has been demolished in the meanwhile, consisted of 89 households with a population of 583 of all age groups. The settlement was located between the tailings deposit of TD 13 and the Mindolo Stream. Children had free access to the deposit and used it as a playground. Chickens and ducks roamed the area. The first houses were less than 50 m down-gradient from the tailings facility, while the Mindolo Stream is located about 500 m down-gradient from the former village. The river, being fed by the seepage of the Mindola Lake, is perennial, but a water point in the former village existed, and therefore the villagers did not depend on river water for drinking or washing purposes. However, they used the river for irrigation (gardening) and swimming. Fishing in the river is quite common.

Table 1 summarizes the baseline information available concerning TDs 11 and 13.

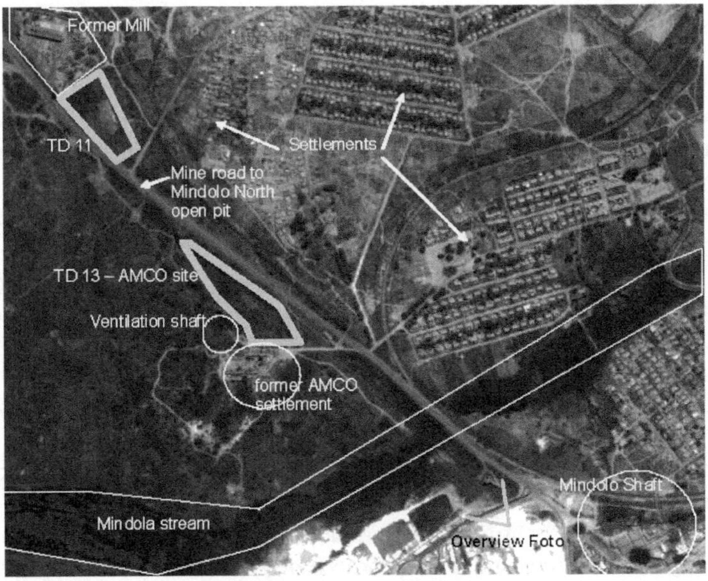

Fig.1 Project area (Based on Google Earth Image)

Table 1 Baseline information of TD 11 and TD 13

	TD 11	TD 13
Origin	Residue from acid leaching of Uranium ore	Residue from acid leaching of Uranium ore
Construction	"Paddock-style" TMF, tailings dam formed from tailings	"Paddock-style" TMF, tailings dam formed from tailings
Active	1956–1960	1956–1960
Area	7100 m^2 * 7630 m^2 ***	9100 m^2 * 8500 m^2 ** 11,140 m^2 ***
Perimeter	520 m ***	570 m ***
Average Thickness (estimated)	3.0 m	3.0 m
Volume	20,600 m^3 * 22,000 m^3 ***	19,900 m^3 * 30,000 m^3 ** 33,000 m^3 ***
Current state	Uncovered, freely accessible, vegetated, partially used for agriculture	Uncovered, freely accessible, vegetated, erosion patterns

* Data Sheet from May 1, 1972 received from Mopani Copper Mine.
** Consolidated Environmental Management Plan Phase 2 (CEMP II), 2005.
*** Calculated using GPS-measurements.

Tailings Material

The tailings material has a fine sandy texture. As shown in Fig. 2, the tailings facility is generally well covered with vegetation. Disturbances of the vegetation were observed, they are mainly due to human activity, which may result in re-suspension and deposition of airborne contaminated dust, especially during the dry windy months. In the IAEA Mission Report it is also mentioned that an excavation of the tailings (presumably to be used as building material) has been observed.

According to the IAEA Mission Report 0 the radioactivity of the tailings is in the range between 4 and 10 μSv/h (microsievert per hour), or 35–88 Bq/g (Becquerel per gram) respectively. Over the years, tailings from this tailings pond have been spilled or washed out downslope and now they occur as a thin veneer over an area of more than 1000 m^2. The edge of the former AMCO village is located only a few meters from the tailings' edge and within this "spill" area. The radioactivity within the village (outside the cottages, as no measurements were carried inside) varies between 0.4 to 1.0 μSv/h.

Despite the small size of the tailings heaps, the situation is relatively complex due to the high specific activity of the tailings (35–80 Bq/g). This requires special attention to the radiological risks in the existing situation and induces long-term risks associated with the leaching and water-borne transport of the radionuclides by groundwater. Moreover, any remediation activity which leads to the handling

Fig. 2 Tailings material at TD 13

Fig. 3 Mindola stream

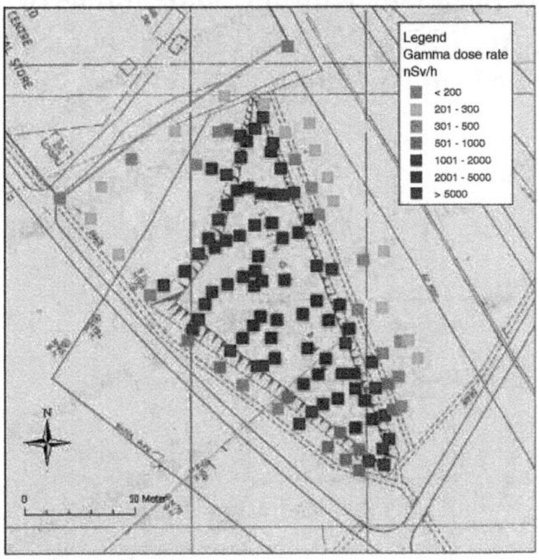

Fig. 4 Gamma dose Rates at TD 11

of the tailings material (reshaping, covering, relocation) deserve special attention with respect to risks to workers and the public. In particular, the design of the remediation project must foresee a comprehensive set of measures to minimize exposure during the remediation works.

The following Radiological measurements were done:

- Gamma dose rate (Fig. 4),
- Radon concentration,
- Radon flux (Rn exhalation),

- Nuclide contents of tailings samples,
- Nuclide contents of bio samples, and
- U concentration in waters.

General Approach and Assessment Criteria

The general approach to assess radiological risks at sites contaminated by uranium mining and milling activities is an exposure analysis. It starts with the identification of the radiological key parameters that characterize the contaminated material and its dispersion at the site.

The effective doses are compared with the "1 mSv criterion". This follows international practice for a given, i.e., already existing situation, when the radiation protection principles of intervention are valid. According to the ALARA principle in radiation protection (ALARA – as low as reasonably achievable), in existing situations of contamination the remedial measures must do more good than harm.

In Zambia, the latest Radiation Protection Act 0 does not provide any guidance on how to deal with justification and optimization of remedial measures at sites contaminated with uranium mining and milling residues. Guidelines addressing the issue are still under development. Thus, during a joint meeting of the Consultant with the Directorate of the Zambian Radiation Protection Agency on April 27, 2007, in Lusaka, agreement was reached to follow international practice and to apply the 1 mSv/a criterion for decision about the justification of

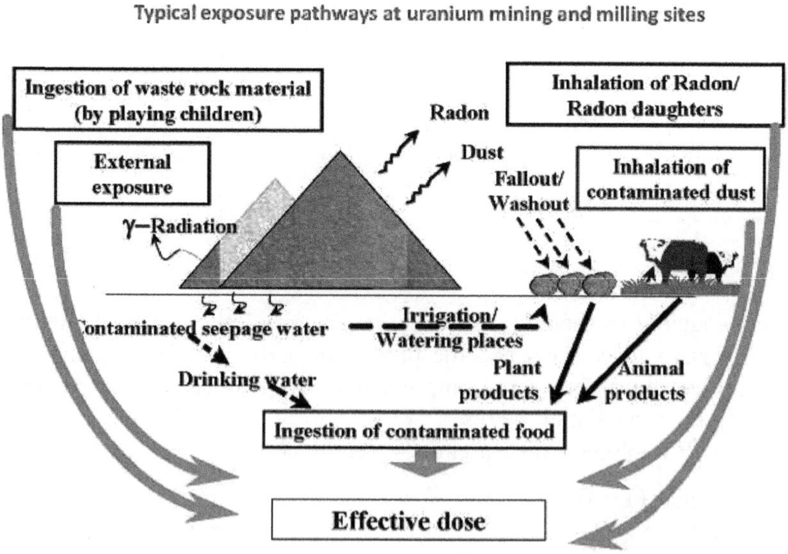

Fig. 5 Typical Dose Model for uranium mining and milling sites

remedial measures at TD 13/TD 11, as well as to apply the ALARA principle in intervention situation to optimize the measures from the point of view of radiation protection.

Figure 5 illustrates the typical pathways for an uranium mining and milling site. Within the exposure analysis exclusively those pathways shall be taken into account which are under consideration of the site-specific circumstances for propagation of radioactivity as well as under observance of the living conditions of the local people really of relevance.

The identification of the relevant exposure pathway was based on the situation as found by the Consultant during the missions in April and June 2007.

a) The inspection of the effective doses shows that the radiological impact of TD 13 and TD 11 results for all reference persons in effective doses which are clearly higher than 1 mSv per annum. Even the temporary stay at the site alone (primary exposure pathways) may cause doses greater than 1 mSv per year. This result is already sufficient to justify remedial measures from a radiation protection point of view.
b) The highest effective doses were estimated for babies, and the lowest for adults, respectively. This is the consequence of a higher radiation sensitivity of the baby's body, with corresponding higher dose conversion coefficients for the intake of radioactivity, expressed in the unit sievert per bequerel. As a consequence, the effective doses for the scenarios "ingestion of contaminated food" and "direct ingestion" are dominating the exposure for babies, but also for children of the age group "2–7 years". On the other hand, for babies the special scenario "sleeping or crawling on the contaminated tailings facility ground" (with inhalation of elevated radon concentrations), contributes in an amount to the effective dose, so that the babies dose becomes higher than those for infants and adults.

The Proposed Preferred Remediation Option

The development of a suitable technology for the remediation of the tailings dams was based on state of the art technologies. The goals of these technologies are:

- to minimize the impact of the tailings dams to the environment,
- to break the pathway for the transfer of nuclides to any human being, and
- to create a long term stable deposit of the tailings material.

Given the local environmental, radiological and socio-economic conditions at the AMCO sites as well as current legislation, the following remediation options could be considered.

During project development the following rehabilitation options were discussed and evaluated:

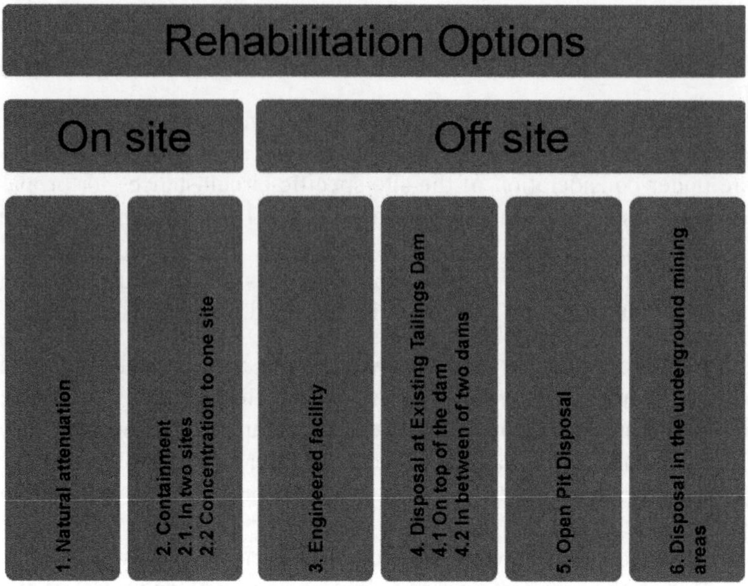

In this project various options for remediation of the AMCO tailings are being investigated. In order to account for various influencing factors of different remediation options, it is considered appropriate to use a Multi-Attribute Utility Analysis (MAUA) approach to pre-select the preferred re-mediation option.

The following attributes and ratings are used to determine a preferable remediation option (Table 2):

Table 2 Ranking Criteria

Attribute	Weighting low 2, medium 4, high 8	Rating very bad (1), bad (2), medium (3), good (4), very good (5) most expensive (1), cheapest (13)
Compliance with the legal system in Zambia	Exclusion factor	0/1
Compliance with international legal systems (e.g. EU)	Exclusion factor	0/1
Environmental impact on groundwater	8	1–5
Environmental impact on air	8	1–5
Environmental impact on soil	4	1–5
Working conditions and health risk to workers	4	1–5
Comparative risk of handling- and transport accidents	4	1–5
Sustainability	8	1–5
Realization time	4	1–5
Costs (investment and operational)	8	1–13

With reference to the estimated costs the ranking of the various options the "cheapest" option received the highest ranking in the table above, and the most expensive one the lowest. This ranking was then used in the overall application of the MAUA. In the bottom line of the conclusion table the results of the evaluation are summarized.

Including costs, the two most feasible options are the Disposal at Existing Tailings Dam and the Open Pit Disposal.

These options are:

- comparably cheap,
- environmentally friendly,
- safe to build and operate, and
- sustainable.

After discussion with stakeholders and beneficiaries the Disposal at Existing Tailings Dam was finally evaluated.

The Design

The preferred option involves the complete removal of all tailings from TD 11 and 13 including any other material that may result from the clean-up of both sites and its disposal to the nearby (distance ~7 km) active tailings dam complex TD 15A operated by MCM (Mopani Copper Mine). The uranium tailings will be disposed in a sealed containment within the active tailings management facility.

Both existing tailings sites (TD 11 and TD 13) have to be cleaned up down to clearance levels. The clearance levels are set to guarantee that no radioactive material remains at the sites.

After its placement within the tailings management facility the uranium tailings will be covered with tailings from copper-cobalt-processing with a minimum thickness of 2 m. Final closure works (final cover construction, re-vegetation etc.) will be established later under the closure plan for the active tailings dam.

The removed tailings have to be placed in a non-operating part of the tailings site TD 15A above the actual water level, otherwise measures would be required to prevent adverse effects from the active copper tailings disposal during the construction period.

The following conditions related to the deposit area (Fig. 6) can be specified:

- Footprint area: 30,000 m^2,
- area of tailings deposit: 20,700 m^2,
- dam slope: 1:3,
- dam height: 3.5 m, and
- entire storage volume: approx. 60,000 m^3.

The disposal cell requires the construction of a sealing layer (thickness 0.5 m) at its bottom, the slopes and the top. Also a protection layer at the top is required.

Different soil materials were sampled to determine relevant parameters for the design of the disposal cell (Fig. 7) and the proposed cover. The results of the geotechnical and the geochemical analysis are given in App. 6.

The following list summarizes the necessary closure works:

- Clearing of vegetation at the tailings sites TD 11, TD 13 and TD 15A.
- Construction/reconstruction of transport routes between the tailings sites TD 11/ TD 13 and TD 15A.
- Construction of the basal sealing layer of the disposal cell using selected clayey laterite.
- Excavation and transport of tailings and contaminated soil from TD 11 and TD 13 sites to TD 15A.
- Compacted placement of the tailings within the disposal cell in thin lifts.
- Covering of the stored tailings with a compacted sealing layer on top using clayey lateritic soil.
- Construction of a protection layer on top of the sealing layer consisting of sandy tailings from copper-cobalt processing.

The construction works must be accompanied by:

- Radiological clearance measurements at the decontaminated areas of TD 11 and TD 13 to demonstrate the complete removal of the tailings from the sites.
- Geotechnical investigations during the construction works of the disposal cell.
- Measurements to avoid/minimize spreading of contamination into the surroundings (dust, contaminated soil, surface contamination on machinery and equipment).
- Working place measurements to assess the radiation exposure of workers engaged with the remediation activities.

The radiological and geotechnical responsibilities for these measures have to be assigned by the client to a qualified and experienced contractor.

The key parameters of the remediation works related to Option 7 are:

• Volume of tailings and additional contaminated material to be transported	60,000 m^3
• Basal area of the tailings body	16,250 m^2
• Final volume of the tailings body	60,000 m^3
• Surface area of the basal sealing layer	20,000 m^2
• Thickness of the basal sealing layer	0.5 m
• Surface area of the top sealing layer	26,400 m^2
• Thickness of the top sealing layer	0.5 m
• Surface area of the protection layer	34,400 m^2
• Thickness of the protection layer	2 m

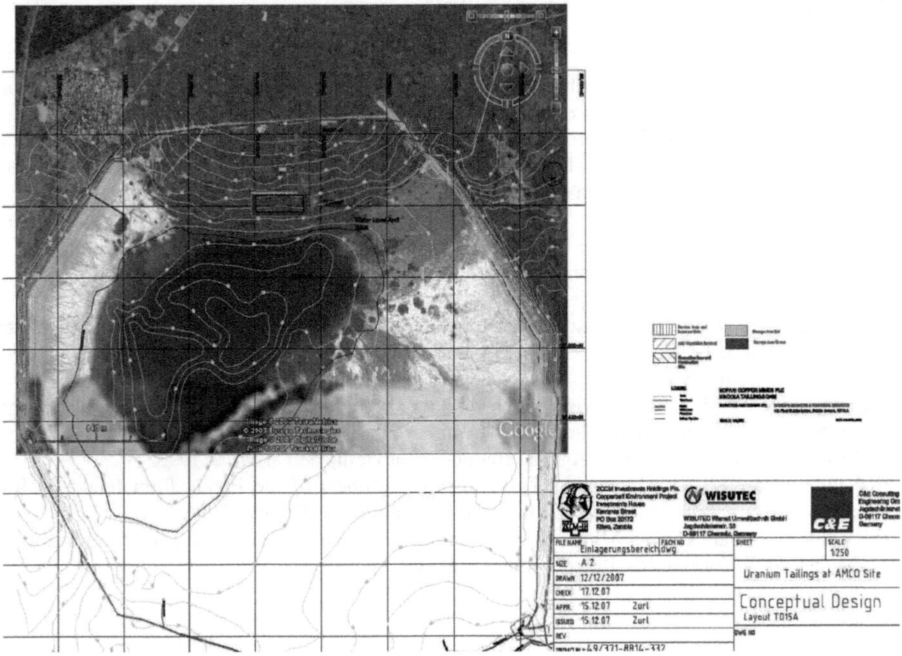

Fig. 6 Design of Disposal Area

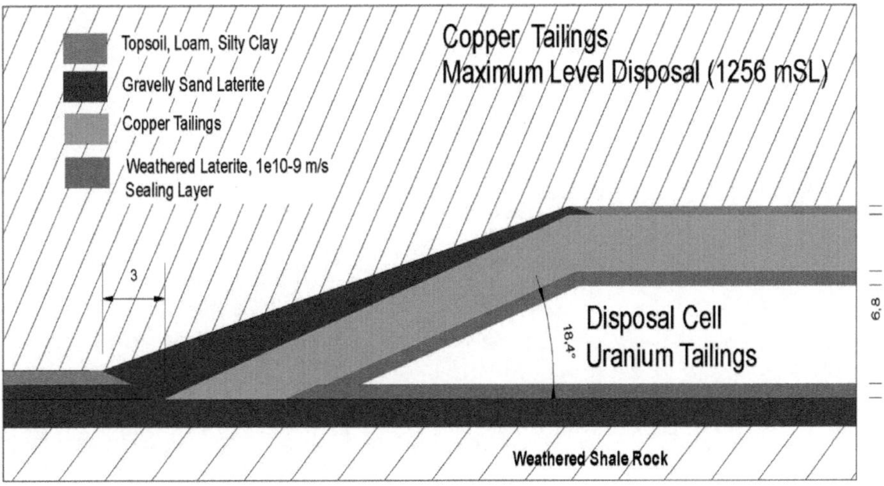

Fig. 7 Design of Disposal Cell

References

Preparation of Phase 2 of a Consolidated Environmental Management Plan, Project Report, Zambian Consolidated Copper Mines Investment Holdings ZCCM-IH, Kitwe, December 2005.

Project RAF/3/005 03 01, Mission Report; J. van Blerk, D. Muleya: Rev 2, Nov. 2005: Remediation Options for the AMCO (Kitwe) Uranium Mining Residues and Environmental Impact Assessments for the Lumwana Copper Project.

Geology of Copperbelt.

ICRP 60: Recommendation of the International Commission on Radiological Protection, 1990.

The Ionising Radiation Protection Act, Government of Zambia, Act No. 16 of 2005, Lusaka, 17th October 2005.

German Federal Ministry for Nature, Reactor Safety and Environmental Protection: Guidelines on How to Assess Radiation Exposure Caused by Natural Radioactivity at Mining Sites (in German), Berlin 1999.

M. Lambarki, Mission Report, visit of the AMCO site before the preparation of the Technical Proposal of the Consultant, Chemnitz, November 2006.

ICRP Publication 65: Protection Against Radon-222 at Home and at Work,1993.

IRP Publication 72: Age-Dependent Doses to the Members of the Public from Intake of Radionuclides – Compilation of Ingestion and Inhalation Coefficients, 1996.

[SRK 2003] Mopani Copper Mines Plc Environmental Impact Statement, Impact Assessment for Nkana, SRK Consultants, Illovo/South Africa. July 2003.

Phytoextraction of Heavy Metals by Dominating Perennial Herbs

Gerhard Gramss, Klaus-Dieter Voigt, Dirk Merten

Abstract. The cosmopolitan perennials, giant goldenrod and stinging nettle reach with 6.2–11.3 t ha^{-1} (DW) the biomass production of those annual crops which are recommended for heavy metal extractions from soils but share with those the intolerable cleanup times. Due to their economic and environmental advantages, cropping rod and nettle, e.g., as renewable energy sources, is an interesting variant in the (remediative) use of heavy-metal contaminated soils.

Introduction

Cultivated soils contaminated with AsCdCrCuNiPbZn by human activities recommend themselves for phytoextraction treatments with high-biomass plants. This also applies to areas damaged with the long-lived radionuclides ^{137}Cs (half life 30 yr) and ^{90}Sr (28.6 yr) from radioactive fallout (Dushenkov 2003; Zhu and Shaw 2000). In 7 examined soils, 4–20% of the total content in CdCuNiPbZn but not Cr appeared in the soluble/exchangeable fraction (Malandrino et al. 2011), a partial, but in the case of many elements statistically non-significant, indicator of plant availability (Meers et al. 2005). Soil-bound arsenic reaches solubility rates of < 1% in 10 mM $CaCl_2$ (Nam et al. 2010). Co-precipitation with (hydr)oxides and cation

Gerhard Gramss
Institute of Geological Sciences, Friedrich-Schiller-University, Burgweg 11, D-07749 Jena, Germany

Klaus-Dieter Voigt
Food GmbH Jena, Orlaweg 2, D-07743 Jena, Germany

Dirk Merten
Institute of Geological Sciences, Friedrich-Schiller-University, Burgweg 11, D-07749 Jena, Germany

B. Merkel, M. Schipek, *The New Uranium Mining Boom*
DOI 10.1007/978-3-642-22122-4 © Springer 2011

exchange with clay minerals reduces the solubility of most transition metal and actinide radioisotopes to negligible amounts (Dushenkov 2003).

Supplements of synthetic and phytochelants which make soil heavy metals (HM) soluble, mobile in the mass flow, and thus more plant available are no common practice. In two sparing field applications of EDTA at 0.7–9 mmol kg^{-1} soil, plant uptake by oilseed rape was 1–2 orders of magnitude lower than the concentration of EDTA-solubilized HM. The chelant-induced labile metal fractions retained the intolerable leaching rates for >2 yr (Wenzel et al. 2003). Field-grown maize lost by 78% in biomass (at 9 mmol EDTA), reducing Cd uptake from 3.7 in the untreated control to 1.4 g ha^{-1} (Neugschwandtner et al. 2008). Critical soil concentrations in the micronutrients, (Cd)CoCuMnNiZn can nevertheless be controlled as they are part of plant metalloproteids. Chinese cabbage enriched in N_{org} up to the toxic threshold of 4% (w/w) by ammonium fertilization showed proportional increases in shoot concentrations of the respective elements, somewhat in CsSr, but virtually not in non-essential traces. Concomitant increases in dry wt × N_{org} raised shoot uptake of the essential transition metals to the 7.65-fold. Leaching diminished to 60% (Gramss and Bergmann 2008).

The annual plant uptake of a trace element does not exceed a few ‰ of its total concentration in soil, prolonging the cleanup time to hundreds to thousands of years (Duquene et al. 2009). Cultivated soils had thus to be treated with innumerable rotations of high-biomass crops or perennials (e.g., Herzig et al. 2005) rather than with a single (engineered) hyperaccumulator in monoculture. In a field test, willow and tobacco were superior to poplar, sunflower, rape, and maize in shoot Cd content and biomass production (Herzig et al. 2005). *Salix* and *Populus* clones with their deep root systems are therefore increasingly applied in the restoration of ecosystems, in biofuel production, and CdZn phytoextraction (Klang-Westin and Eriksson 2003; Meers et al. 2005). Swedish biofuel cultures of *Salix viminalis* L. on arable are estimated to allow for 6–7 partial harvests over 25 yr with the harvest time in winter. This comprises losses of the Cd (21–48%) bound in the cast-off foliage (Klang-Westin and Eriksson 2003).

No attempts have been made to assess the sanitary effects of the cosmopolitan high-biomass perennials, stinging nettle (*Urtica dioica* L.) and goldenrods (*Solidago* sp.). Their ancient monocultures occupy undisturbed sites of river banks, roadsides, and rangeland (Weber 2001) for decades without signs of self-toxification or exhaustion and control competing plants by height and density of their colonies. Damages by herbivores and microbial pests are negligible (Jakobs et al. 2004).

In this study, nettle and giant goldenrod (rod; *S. gigantea* Ait.) were potted on a metalliferous soil to be compared with field-grown plants from non-contaminated river sediments. It was the goal to determine shoot biomass production and trace metal content under field conditions in regard to harvest time and soil HM content, effects of ammonium application on shoot N_{org} and the resulting trace metal content, soil cleanup times also in relation to the half life of radionuclides, and the return rates of shoot macronutrients from ageing stalks to the ground as a mechanism of the colonies' self-preservation. Intolerable cleanup times may force to the long-term (remediative) use of contaminated soils for industrial crops.

Materials and Methods

In the outskirts of Jena (Germany), triplicate 1-m^2 plots were marked in undisturbed colonies of rod (located in Burgau) and nettle (Göschwitz) where plants reached 2.10 m in length on the non-contaminated soils of alluvial river sediments. The biomass of untreated controls harvested at different maturity stages was compared with the yield of NH_4Cl-treated plots over 2 growth periods. Washed rootstocks of both plants were also inserted into triplicate 4-l pots with the metalliferous Settendorf soil from uranium mining. Their shoots were collected twice within one year (Table 1). The Settendorf/Burgau/Göschwitz soils of pH_{aqu} 7.08/7.60/7.40 and C_{org} (%) 3.7/2.8/4.0 had the following content in macronutrients (mg kg^{-1} DW): Ca, 10,300/28,300/20,900; K, 4790/6000/3370; Mg, 4680/6460/3320; P, 840/730/620; NO_3-N, 140/14/18 (compare Section Results for heavy metal contents). Metal concentrations were determined (by ICP-MS) for duplicate *aqua-regia* extracted 5 g samples of soils and 300-mg samples of microwave digested meals prepared from at least 10 whole shoots of the plants. Soil NO_3^- concentrations (spectrophotometrically; Gramss and Bergmann 2008) and N_{org} in soil and shoot meals (Kjeldahl) were also recorded. Duplicate 5-g samples of shoot meals were ashed at 550°C.

Table 1 Production of rod and nettle crops in pot and field cultures during 2009

Soil	Mode of cultivation	Fertilization		Annual shoot dry wt g kg^{-1} soil ±SD	t ha^{-1}
	Rod				
Settendorf	Triplicate 4-l pots, crops from Jun 5 and Sep 20 combined, plants per pot separately analyzed	Control (non-amended)		8.10 ± 1.38	–
		NH_4-N treated, 40 mg N kg^{-1} soil, applied 3 times		11.04 ± 0.49	–
Burgau	Undisturbed rod colony 7 × 360 m formed within > 20 yr. Triplicate 1-m^2 plots randomly selected	Control, date of harvest	June 5	4.59 ± 0.02[a]	6.9
			August 15	8.36 ± 0	12.5
			November 1	7.50 ± 0.65	11.3
			Jun 5/Sep20	5.95 ± 0.05	8.9
		3 × 4 g N m^{-2}	Jun 5/Sep 20	7.73 ± 0.02	11.6
	Nettle				
Settendorf	Triplicate 4-l pots, crops from July 1 and Oct 1 combined as above	Control (non-amended)		4.99 ± 0.12	–
		NH_4-N treated as above		7.19 ± 0.10	–
Göschwitz	Ancient undisturbed river bank colony 7 m wide. Triplicate 1-m^2 plots randomly selected	Control, date of harvest	July 1	3.22 ± 0.47	4.8
			October 1	4.12 ± 0.47	6.2
			July 1/Oct 1	4.35 ± 0.64	6.5
		3 × 4 g N m^{-2}	October 1	6.50 ± 0.67	9.8

[a] Shoot DW of field cultures (g kg^{-1} soil) calculated with 150 kg soil m^{-2}.

Results

The biomass of field-grown rods reached 12.5 t ha^{-1} (DW) at Aug 15 and diminished non-significantly by Nov 1 (Table 1). June 5 clippings led to losses in total biomass due to the slow regeneration of secondary shoots. Applications of NH_4Cl-N at 120 kg ha^{-1} resulted in 30-% increases of herbage. Congruent values were obtained from rod cultures potted on Settendorf soil (Table 1). Nettle yields of 6.2–6.5 t ha^{-1} from field plots were not impaired by July 1 clipping (Table 1). Ammonium applications as above improved the biomass production to the 1.4/1.6-fold both in pot/field cultures by promoting stand density and shoot length.

Table 2 Organic N and HM concentrations (mg kg^{-1} DW) in Settendorf and Burgau soil and in N- and non-treated shoots of the respective annual rod plants harvested at different maturity stages. Impact of treatments on whole-shoot ash content (% by DW)

Element	Settendorf soil	Pot plants of		Burgau soil	Field-grown plants of				
		June 5/Sept 20	June 5/Sept 20, N		June 5	Aug 5	Nov 1	June 5/Sept 20	June 5/Sept 20, N
	Column 1	2	3	4	5	6	7	8	9
N_{org}	1900	13,700	16,800[a]	ND	12,800	7500[b]	5900[b]	13,240	18,100[c]
Micronutrients and Cd									
Cd	36.5 (3)	4.29	3.11[a]	0.306	0.047	0.034	0.082	0.059	0.058
Co	28.8	0.248	0.156[a]	8.81	0.069	0.077	0.079	0.077	0.081
Cu	278 (100)	9.57	8.67	25.8	8.19	5.67[b]	7.61	8.85	9.31
Mn	1874	61.5	78.3[a]	754	20.5	13.5	24.9	26.7	32.9
Ni	49.6	2.03	2.92[a]	20.9	1.79	0.910[b]	2.69	2.02	2.22
Zn	2829 (300)	314	253[a]	72.2	22.7	16.5[b]	28.3[b]	28.1[b]	21.7[c]
Sum	5096	391.6	346.2	882	53.3	36.7	63.7	65.8	66.3
± SD	± 20	± 2.6	± 3.3	± 3	± 3.6	± 1.6	± 3.2	± 2.4	± 2.8
Non-essential traces and ash									
As	171 (20)	1.12	1.23	9.42	0.30	0.30	0.063	0.042	0.064
Ba	42.1	5.65	4.84	158	11.4	9.90[b]	12.7	12.3	16.5[c]
Cr	18.3	0.816	0.753[a]	29.9	0.245	0.175	0.633[b]	0.365[b]	0.417
Cs	35.7	4.36	6.10[a]	2.40	0.024	0.004	0.009	0.031	0.014
Pb	151 (100)	0.627	0.532[a]	41.5	0.280	0.147	0.333	0.293	0.418
Sr	57.3	39.4	33.5[a]	78.6	42.0	38.5[b]	47.9[b]	43.1	48.5[c]
Th	10.8	0.032	0.024	6.50	0.004	0.003	0.008	0.006	0.008
U	54.0 (?)	0.094	0.071	1.30	0.003	0.003	0.003	0.003	0.002
Sum	540	52.1	47.1	328	54.0	48.8	61.6	56.1	65.9
± SD	± 2	± 0.5	± 0.7	± 1.1	± 1.0	± 0.5	± 2.2	± 1.2	± 2.0
Ash %	–	8.71	8.90	–	8.34	6.10[b]	3.92[b]	9.34	10.98[c]

At $p \leq 0.05$, significant differences denote [a] values of table column 3 to column 2; [b] values of columns 6, 7, and 8 to column 5; [c] values of column 9 to column 8; () permissible soil HM concentrations (Kloke 1979); ND, not determined. Comparable limits for HM toxicants in grain (mg kg^{-1}): As, 0.5; Cd, 0.1; Pb, 0.3 (Schachtschabel et al. 1998, p. 318).

Trace metal concentrations denote Burgau and Göschwitz soils as non-contaminated. Settendorf soil contained the 5-fold amount in total HM. Concentrations of AsCdCuPbUZn exceed permissible limits (Kloke 1979) but they were not taken up in phytotoxic levels (Schachtschabel et al. 1998, p. 336). Concentrations in AsCdPb in rod and nettle from Settendorf soil would be non-permissible for food grains (Tables 2 and 3). Identical ammonium applications raised the N_{org} content of potted/field grown herbage of rod (to the 1.2/1.4-fold) and nettle (1.2/1.9-fold) apparently not enough to induce significant concentration increases in all of the micronutrients (Cd)CoCuMnNiZn and the K/Ca analogous Cs/Sr. The ash content in both plant species correlated with shoot CaKMgP concentrations (data not shown), was promoted by ammonium treatment, and declined in the autumnal plants of rod more dramatically than in nettle (Tables 2 and 3).

Referred to the total (*aqua regia*) content of the soils, the annual shoot biomass of rod and nettle from Settendorf soil contained not more than 2–4‰ of the trace metal groups determined (Fig. 1a, b).

Table 3 Organic N and HM concentrations (mg kg^{-1} DW) in Settendorf and Göschwitz soil and in N- and non-treated shoots of the respective annual nettle plants harvested at different maturity stages. Impact of treatments on whole-shoot ash content (% by DW)

Element	Settendorf soil	Pot plants of		Göschwitz soil	Field-grown plants of			
		July 1/ Oct 1	July 1/ Oct1,N		July 1	Oct 1	Oct 1, N	July 1/ Oct 1
	Column 1	2	3	4	5	6	7	8
N_{org}	1900	17,300	20,750^{a}	ND	15,740	14,200^{b}	26,800^{c}	17,400^{b}
Micronutrients and Cd								
Cd	36,5	0.786	1.65a	0.391	0.011	0.089	0.038	0.033
Co	28.8	0.639	0.757	6.39	0.088	0.610^{b}	1.10^{c}	0.358^{b}
Cu	278	10.8	10.0	21.5	6.38	4.51	7.03^{c}	6.30
Mn	1874	160	218^{a}	511	32.5	62.5^{b}	78.5^{c}	32.1
Ni	49.6	3.24	4.36^{a}	14.4	1.01	1.45^{b}	4.15^{c}	1.21
Zn	2829	226	314^{a}	152	16.2	18.7^{b}	25.3^{c}	16.7
Sum	5096	401.5	548.8	706	56.2	87.9	116.1	56.7
± SD	± 20	± 10.6	± 20.5	± 9	± 3.3	± 0.7	± 3	± 3.8
Non-essential traces and ash								
As	171	1.20	1.32	8.14	0.145	0.250^{b}	0.234	0.168
Ba	42.1	15.2	16.1	167	43.4	64.2^{b}	67.4	44.9
Cr	18.3	0.770	0.550^{a}	19.2	0.885	0.740	0.954	0.873
Cs	35.7	0.913	1.26^{a}	1.10	0.015	0.012	0.020	0.016
Pb	151	0.728	0.623	37.9	0.340	0.823^{b}	0.756	0.466^{b}
Sr	57.3	127	114^{a}	95.0	142	126^{b}	154^{c}	154^{b}
Th	10.8	0.050	0.026	3.50	0.019	0.016	0.035	0.020
U	54.0	0.197	0.157	0.800	0.009	0.007	0.009	0.009
Sum ± SD	540 ± 2	146 ± 1.1	134 ± 0.3	333 ± 1.2	187 ± 1.7	192 ± 3.2	223 ± 3.0	201 ± 1.9
Ash %	–	16.1	16.9	–	12.7	11.5	18.7^{c}	13.8

At $p \leq 0.05$, significant differences denote [a]values of table column 3 to column 2; [b]values of columns 6 and 8 to column 5; [c]values of column 7 to column 6; ND, not determined

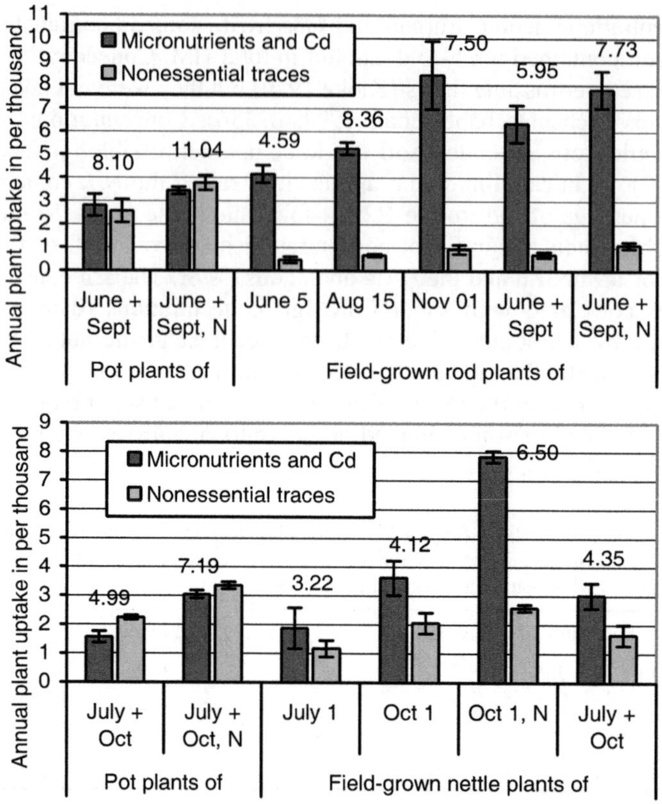

Fig. 1 **(a)** (rod, *above*) and **(b)** (nettle) Annual uptake of heavy metals in ‰ to the soils' *aqua-regia* metal concentrations (compare Tables 2 and 3). Contributing shoot biomass (DW) given in g kg^{-1} soil at column heads (compare Table 1). Error bars represent confidence intervals of 95%. Totals of "Nonessential traces" do not comprise values of Sr

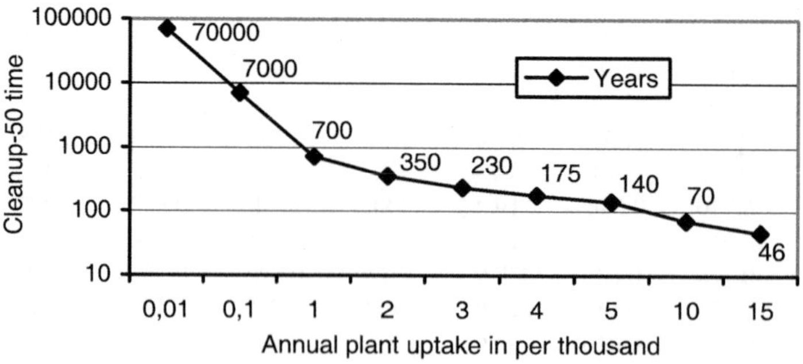

Fig. 2 Cleanup-50 time (years) necessary to reduce the soils' *aqua-regia* metal concentrations to 50% with constant relative uptake rates (‰) of the shoot biomass, rates that decline nevertheless absolutely with the soils' residual metal content (compound interest calculation)

Table 4 Annual uptake of individual HM in ‰ to the soils' total metal concentrations by the shoot biomass of N-treated rod and nettle crops (compare Fig. 1a, 1b).

Element	Rod from		Nettle from	
	Pot culture	Field crop	Pot culture	Field crop
	June 5/Sept 20, N-treated	June 5/Sept 20, N-treated	July 1/Oct 1, N-treated	Oct 1, N-treated
	Column 1	2	3	4
Micronutrients and Cd				
Cd	0.940	1.47[a]	0.325	0.632[b]
Co	0.060	0.071	0.189	1.12[b]
Cu	0.344	2.79[a]	0.259	2.13[b]
Mn	0.461	0.337	0.836	1.00[b]
Ni	0.651	0.823	0.632	1.88[b]
Zn	0.988	2.33[a]	0.799	1.08[b]
Sum ± SD	3.44 ± 0.08	7.82 ± 0.40	3.03 ± 0.07	7.84 ± 0.10
Non-essential traces				
As	0.080	0.053[a]	0.055	0.187[b]
Ba	1.27	0.806[a]	2.75	2.63
Cr	0.455	0.108[a]	0.216	0.322[b]
Cs	1.89	0.045[a]	0.253	0.119[b]
Pb	0.039	0.078[a]	0.030	0.130[b]
Sr	6.45	4.77[a]	14.3	10.5[b]
Th	0.025	0.010	0.017	0.067[b]
U	0.015	0.012	0.021	0.074[b]
Sum ± SD[c]	3.77 ± 0.15	1.11 ± 0.08	3.34 ± 0.05	3.53 ± 0.08

At $p \leq 0.05$, significant differences denote [a]values of table column 2 to column 1; [b]values of columns 4 to column 3; [c]Sum without the values of Sr.

Field-grown rods from the whole growing season came to 6.4–8.4‰ of the micronutrients, and to 1‰ in non-essential traces. Comparable field-grown nettle plants were less efficient but expressed stimulation by ammonium treatment. Elements such as AsCoPbThU were predominantly extracted with uptake rates < 0.1‰ (Table 4), i.e., with values that correspond to cleanup-50 times » 7000 yr (Fig. 2). Annual uptake rates of 0.2–2.8‰ (3500–245 yr) denote the control of most other HM. Cleanup-50 times of Sr by rod (5‰; 140 yr) and nettle (14‰; 50 yr) approach the half life of its ^{90}Sr isotope (28.6 yr).

Discussion

Rod and nettle plants with annual yields of 6.6–12.5 t ha^{-1} (DW) on non-fertilized and non-contaminated rangeland are primary candidates for the treatment of HM contaminated arable rather than of mineral mine deposits (Tables 1–4). The plants

Table 5 Annual extraction of 5 elements from field soils with different HM concentrations which had been planted to rod, nettle (compare Tables 2 and 3), rape (Wenzel et al. 2003), corn (maize) (Neugschwandtner et al. 2008), or willow (*Salix viminalis*; without cast-off leaves; Klang-Westin and Eriksson 2003). Values refer to DW. ND, not determined

Element		Rod		Nettle		Rape	Corn	Willow
	Crop (t ha^{-1})	11.3		9.8		8	9.4	2.6–8.7
	Soil origin	Settendorf	Burgau	Settendorf	Göschwitz			
Cd	Shoot (mg kg^{-1})	4.29	0.082	1.65	0.089	ND	0.390	0.60–4.10
	Soil (mg kg^{-1})	36.5	0.306	36.5	0.391		5.00	0.17–0.45
	Extract. (g ha^{-1})	48.5	0.900	16.2	0.900		3.70	4.20–16.8
Cs	Shoot	6.10	0.031	1.26	0.020	ND	ND	ND
	Soil	35.7	2.40	35.7	1.10			
	Extraction	69.0	0.400	12.3	0.200			
Pb	Shoot	0.627	0.418	0.728	0.823	2.00	6.30	ND
	Soil	151	41.5	151	37.9	77.1	544	
	Extraction	7.10	4.70	7.10	8.10	16.0	59.2	
Sr	Shoot	39.4	48.5	127	154	ND	ND	ND
	Soil	57.3	78.6	57.3	95.0			
	Extraction	445	548	1245	1509			
Zn	Shoot	314	28.3	314	25.3	120	ND	ND
	Soil	2829	72.2	2829	152	344		
	Extraction	3548	320	3077	248	960		
N application (kg ha^{-1})		0 and 120		0 and 120		21	128	60–120

tolerated, too, the metalliferous Settendorf soil with the 5-fold content in total HM whereby phytotoxic concentrations in the shoots but also signs of hyperaccumulation were not observed (Tables 2 and 3).

Acquisition rates of HM could even be somewhat lower than in oilseed rape, maize, and willow with a comparable biomass production on fertilized arable (Table 5), but the diversity of the underlying soil-shoot transfer factors vary too much with pH, the concentration of HM, and their sorption to the soil matrix (Schachtschabel et al. 1998, p. 324) to make their uptake rates from different soils comparable.

In Chinese cabbage, ammonium fertilization was the major tool to increase shoot concentrations of the metalloproteid-associated micronutrients, (Cd)CoCuMnNiZn near-proportionally with N_{org} and linearly with the higher biomass (Gramss and Bergmann 2008). Ammonium treated rods from Settendorf/Burgau soils showed the 1.23/1.37-fold N_{org} concentrations to the controls with no consistent effects on uptake rates of all micronutrients (means = 0.96 ± 0.29/1.03 ± 0.14-fold; Table 2). Net uptake reached nevertheless 130% due to the higher biomass. Nettles from Settendorf/Göschwitz soils responded to increases in N_{org} (1.20/1.89-fold) with elevated micronutrient uptake rates (1.38 ± 0.36/1.77 ± 0.58-fold without Cd) to correspond with a net extraction of 166/279% due to the higher biomass (Table 3). The N_{org} values of both plants remained below the phytotoxic limit of 4% by DW

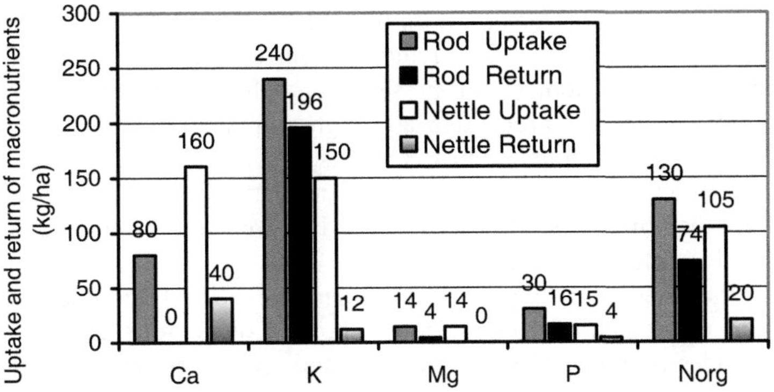

Fig. 4 Uptake of macronutrients (kg ha^{-1}) from undisturbed Burgau and Göschwitz soils by non-fertilized annual rod (10 t ha^{-1}) and nettle (6 t ha^{-1}) crops clipped twice in summer and fall, and return rates (kg ha^{-1}) of macronutrients by ageing shoots of rod (harvested at Nov 1) and nettle (from Oct 1) to root/soil systems

(Gramss and Bergmann, 2008) to give room for higher HM extraction rates by intensified N applications.

Plants harvested at the end of the growing season combined highest shoot HM concentrations with near-optimum biomass values (Tables 1–3) to make a preceding harvest in summer unnecessary. Contrary to the autumnally senescing nettle plants, rods translocated up to 59% of its macronutrients (Fig. 4) but apparently no trace metals back to the rhizome network reducing the expense for mineral fertilization by continuous clipping. Diluting HM concentrations in Settendorf soil to 60/20% with builder's sand caused adequate reductions in shoot HM concentrations of Chinese cabbage (Gramss and Bergmann 2010). Calculating cleanup-50 times under consideration of the gradually declining total HM concentration of the soil but with constant bioaccumulation rates (Fig. 2) may therefore lead to huge but realistic time spans. Reducing soil Cd from 36.5 to 3 mg kg^{-1} by rod (Table 2) with annual uptake rates of 0.940‰ would take up 2650 yr. In the case of the plant-available Ca-analogous Sr, cleanup-50 times of 50 yr for its beta radiation emitting isotope are low but do not justify the exposure of personnel to seriously contaminated soils.

Summary and Conclusions

Goldenrod and stinging nettle are perennial herbs which form persistent colonies on the undisturbed rangeland around Jena (Germany). Triplicate 1-m^2 plots were marked within the colonies to estimate biomass production, optimum harvest time, and uptake rates of heavy metals (HM) from the non-contaminated soils. The plants were also potted on Settendorf soil with intolerable AsCdCuPbUZn concentrations. Field-grown rods/nettles harvested on Nov 1/Oct 1 with signs of senescence yielded 11.3/6.6 t ha^{-1} in a biomass of optimum HM content. Contemporarily, 59/17% in the sum of CaKMgNP had returned from the herbage to the rhizomes, a mechanism of the colonies' self-preservation. More than rods, nettles responded with gains in biomass production and higher uptake rates of proteid-associated transition metals to ammonium treatments. Referred to the total HM content of the soil, the annual uptake rates of the plant matter corresponded with those reported for other high-biomass crops but led to intolerable remediation times for the seriously contaminated soil.

It is concluded that the use of the dominant perennials in the HM control of farmland rather than of organic-poor mine deposits holds promise for stable herbage yields over decades by preventing leaching and soil erosion around the year. The treatment saves expenses for successive establishments of annual crops, pesticides, and mineral fertilizers. Conventional selection may lead to nettle cv. which accumulate CdCuPbZn at 30/250/2000/3000 mg kg^{-1} (Laskowski and Hopkin 1996). Moreover, rod hybrids reached 3 m in height and a latex content enhanced from 5 to 12% (Van Beilen and Poirier 2007). Nevertheless, the non-realistic remediation times leave it to the calculating economist to use the given soil for the (remediative) cultivation of oilseed or high-biomass crops such as maize and herbaceous or woody perennials for industrial and energy projects.

References

Duquène L, Vandenhove H, Tack F, Meers E, Baeten J, Wannijn J (2009) Enhanced phytoextraction of uranium and selected heavy metals by Indian mustard and ryegrass using biodegradable soil amendments. Sci Total Environ 407: 1496–1505

Dushenkov S (2003) Trends in phytoremediation of radionuclides. Plant Soil 249: 167–175

Gramss G, Bergmann H (2008) Applications of NH_4Cl and citrate: Keys to acceptable phytoextraction techniques? pp. 321–331. In Merkel BJ, Hasche-Berger A (eds): Uranium, Mining and Hydrogeology. Springer, Berlin

Gramss G, Bergmann H (2010) Optimizing the content of nutritionally essential trace metals in Chinese cabbage (*Brassica chinensis* L.) with soil applications of compost or nitrogen, pp. 361–378. In Bundgaard K, Isaksen L (eds): Agriculture Research and Technology. Nova Science Publishers, New York

Herzig R, Nehnevajova E, Vangronsveld J, Ruttens A, Mastretta C (2005) pp. 160–190. In PHYTAC-development of systems to improve phytoremediation of metal contaminated soils through improved phytoremediation. Final report of the 5th Framework Programme, Projects Nr QLRT-2001-00429 and QLRT-2001-02778 (NAS), December 2005

Jakobs G, Weber E, Edwards PJ (2004) Introduced plants of the invasive *Solidago gigantea* (Asteraceae) are larger and grow denser than conspecifics in the native range. Divers Distribut 10: 11–19

Klang-Westin E, Eriksson J (2003) Potential of *Salix* as phytoextractor for Cd on moderately contaminated soils. Plant Soil 249: 127–137

Kloke A (1979) Contents of arsenic, cadmium, chromium, fluorine, lead, mercury and nickel in plants grown on contaminated soil. Paper presented at United Nations – ECE Symp. on Effects of Air-borne Pollution on Vegetation, Warsaw

Laskowski R, Hopkin SP (1996) Effect of Zn, Cu, Pb, and Cd on fitness in snails (*Helix aspersa*). Ecotoxicol Environ Safety 34: 59–69

Malandrino M, Abollino O, Buoso S, Giacomino A, La Gioia C, Mentasti E (2011) Accumulation of heavy metals from contaminated soil to plants and evaluation of soil remediation by vermiculite. Chemosphere 82: 169–178

Meers E, Lamsal S, Vervaeke P, Hopgood M, Lust N, Tack FMG (2005) Availability of heavy metals for uptake by *Salix viminalis* on a moderately contaminated dredged sediment disposal site. Environ Pollut 137: 354–364

Nam SM, Kim M, Hyun S, Lee S-H (2010) Chemical attenuation of arsenic by soils across two abandoned mine sites in Korea. Chemosphere 81: 1124–1130

Neugschwandtner RW, Tlustoš P, Komárek M, Száková J (2008) Phytoextraction of Pb and Cd from a contaminated agricultural soil using different EDTA application regimes: Laboratory versus field scale measures of efficiency. Geoderma 144: 446–454

Schachtschabel P, Blume H-P, Brümmer G, Hartge KH, Schwertmann U (1998) Lehrbuch der Bodenkunde, 14th ed. Enke, Stuttgart, Germany

Van Beilen JB, Poirier Y (2007) Prospects for biopolymer production in plants. Adv Biochem Engin/Biotechnol 107: 133–151

Weber E (2001) Current and potential ranges of three exotic goldenrods (*Solidago*) in Europe. Conserv Biol 15: 122–128

Wenzel WW, Unterbrunner R, Sommer P, Sacco P (2003) Chelate-assisted phytoextraction using canola (*Brassica napus* L.) in outdoors pot and lysimeter experiments. Plant Soil 249: 83–96

Zhu YG, Shaw G (2000) Soil contamination with radionuclides and potential remediation. Chemosphere 41: 121–128

Field Scale Phytoremediation of Soils Contaminated with Heavy Metals and Radionuclides and Further Utilization of the Plant Residues

Daniel Mirgorodsky, Lukasz Jablonski, Delphine Ollivier, Juliane Wittig, Sabine Willscher, Dirk Merten, Georg Büchel, Peter Werner

Abstract. Phytoremediation is applied to a site in the former uranium mining area of Ronneburg in Eastern Thuringia, Germany, slightly contaminated with heavy metals and radionuclides (HM/R). In a joint research project, remediation of

Daniel Mirgorodsky
Friedrich Schiller University Jena, Institute of Geosciences,
Burgweg 11, D-07749 Jena, Germany

Lukasz Jablonski
Technical University Dresden, Institute of Waste Management
and Contaminated Site Treatment, Pratzwitzer Straße 15, D-01796 Pirna, Germany

Delphine Ollivier
Friedrich Schiller University Jena, Institute of Geosciences,
Burgweg 11, D-07749 Jena, Germany

Juliane Wittig
Technical University Dresden, Institute of Waste Management
and Contaminated Site Treatment, Pratzwitzer Straße 15, D-01796 Pirna, Germany

Sabine Willscher
Technical University Dresden, Institute of Waste Management
and Contaminated Site Treatment, Pratzwitzer Straße 15, D-01796 Pirna, Germany

Dirk Merten
Friedrich Schiller University Jena, Institute of Geosciences,
Burgweg 11, D-07749 Jena, Germany

Georg Büchel
Friedrich Schiller University Jena, Institute of Geosciences,
Burgweg 11, D-07749 Jena, Germany

Peter Werner
Technical University Dresden, Institute of Waste Management
and Contaminated Site Treatment, Pratzwitzer Straße 15, D-01796 Pirna, Germany

B. Merkel, M. Schipek, *The New Uranium Mining Boom*
DOI 10.1007/978-3-642-22122-4 © Springer 2011

HM/R-contaminated sites is investigated and concepts for the subsequent utilization of the contaminated plant residues are developed. To minimize HM/R-accumulation in soil and to reduce groundwater contamination a combination of phytostabilization and phytoextraction methods is applied and lysimeter experiments are performed to demonstrate the reduction of seepage water rate and load. The final utilization of HM/R loaded plant residues after harvests was studied by biogas and ethanolic fermentations and by combustion of the plant material. The fate of HM/R in the different by-products was investigated.

Introduction

The remediation of large HM/R contaminated areas by conventional ex situ techniques (excavation and chemical treatment) is expensive; usual strategies are the set-aside of agricultural land which is not such a satisfying solution or a large-scale covering of the zone. An emerging, cost-effective and sustainable technology is the use of plants (phytoremediation) to stabilize or extract HM/R. Phytoremediation contributes to the improvement of the physical (erosion protection), chemical (quality improvement of compartments) and ecological (biodiversity) quality of the soil (Maegher 2000; Miller 1996). Moreover, the biomass produced can be utilized for production of energy.

In the area of Ronneburg, in Eastern Thuringia, Germany, over 113,000 t of uranium were produced between 1946 and 1990 (Jakubick et al. 1997). Although major efforts were conducted to remediate the area, the soil and especially ground water are still contaminated with HM/R. In a joint research project phytoremediation techniques of HM/R-contaminated sites and concepts for the subsequent utilization of the HM/R-loaded plant residues are investigated. Main topics of this research project are to study the application of phytoremediation techniques for HM/R-contaminated sites and develop concepts for the subsequent utilization of the HM/R-loaded plant residues. Some of the objectives concern the minimization of the mobile part of the HM/R, the improvement of their uptake during phytoremediation as well as to increase biomass production. A combination of phytostabilization and phytoextraction methods with aspects of microbial, soil and plant techniques are accomplished. Cycling of U and heavy metals including rare earth elements as chemical analogues of trivalent actinides in the system soil – water – plant are investigated.

The utilization of the HM/R-loaded plant residues after the harvest can contribute to the minimization of wastes and to the winning of alternative energy (Meers 2005; Van Ginneken et al. 2007), and therefore likewise to a reduction of the remediation costs. Therefore, utilization of the harvested plants from the test field site was investigated, and the fate of contaminants was measured.

Experimental Setup

In order to investigate phytostabilization and phytoextraction strategies, a test site near Ronneburg, Eastern Thuringia, Germany, was excavated, the soil was mechanically homogenized and filled in ten plots with 2 × 2 × 1 m, which are bordered at the sides to prevent lateral flow between the plots. The plots are amended with three different soil additives (three replications).

1. Test field soil (homogenized substrate), (TF).
2. Test field soil with mycorrhiza (*Glomus intraradices*) and a bacteria culture containing the strains *Streptomyces tendae* F4 and *S. acidiscabies* E13 (TF + MS).
3. Test field soil with 10 kg/m^2 calcareous topsoil (MIX).
4. Test field soil without plants (C).

and were planted with an identical plant mixture of sunflower (Helianthus annuus), Triticale (x Triticosecale) and Indian mustard (Brassica juncea), based on pot experiments. In addition all plots were fertilized with 100 kg/nitrogen-phosphor-potassium (NPK)/ha. A more detailed description of composition and homogenization of the test field substrate, on the measurement system and of selection of the genotype plants in germination and pot experiment are given in Mirgorodsky et al. (2010a, b).

Three weighable lysimeters with a surface of 0.5 m^2 and 100 cm in depth each were installed. While one lysimeter contains an undisturbed soil monolithe with accumulation of high contents of HM/R (L1), two others were filled with TF in addition of calcareous topsoil (MIX, L2) and microorganisms (TF + MS, L3). The amount of seepage water was recorded with a tipping bucket and collected in a small tank for chemical analysis. While lysimeter measurements started in October 2008, the first planting season began in April 2009. In 2010, the plots were not fertilized, and planted with an identical crop mixture of Indian mustard and Triticale (the same variety and density as the year before) without sunflower.

Samples of the first 30 cm of soil were extracted by the method of Zeien and Brümmer (1989) before seeding and after harvesting in every vegetation period to get information about the mobile and specifically adsorbed fractions.

Results of Phytoremediation Experiments

Changing HM/R Mobility in Soil by Different Amendments

After application of the different variations of soil amendments and biological additives (MIX, MS) in comparison to the homogenized substrate (TF), the HM/R mobility in soil changed characteristically. The addition of calcareous topsoil as amendment to TF decreased significantly the heavy metal mobility by increasing pH (pH 4.4 to 6.7), organic matter (3 to 10%) and CEC (21 mmol/kg to 64 mmol/kg). As a result, the concentration of available metals, especially Al, Co, Ni, Zn, and U has been reduced.

Time Depending Changes of HM/R Mobility in the Test Field Soil

Concentration of available metals, especially Co, Ni and Zn after harvesting has been reduced in a range of 91% for Co, 48% for Ni and 97% for Zn for the unamended soil (TF) and 80% for Co and 50% for Zn with the microbial strategy (TF+MS) from 2009 to 2010. Furthermore, stabilization of bioavailable metal concentrations could be achieved with microbial strategy for Cu, Ni and REE. The addition of calcareous topsoil (MIX) also avoided an increase of the mobile HM/R fraction in the topsoil. Here, concentrations of available Co (0.1 to 0.08 mg/kg), Ni (1.4 to 1.5 mg/kg), Zn (1.5 to 1.4 mg/kg), and REE (0.7 to 0.8 mg/kg) could be stabilized or variations are within standard deviation. The changes in the bioavailable fraction of U from May 2009 to October 2010 in untreated test field soil are presented in Fig. 1. In the untreated test field soil (control, no plants, no soil additives), concentration of available U increased from 0.8 to 3.5 mg/kg from 2009 to 2010. In contrast, no significant reduction or increase of bioavailable U could be quantified for the soil strategies TF and TF+MS. The addition of calcareous topsoil (MIX) showed a slight increase in uranium bioavailability, possibly due to formation of soluble uranyl carbonate complexes (Koch-Steindl and Pröhl 2001). Therefore, planting and addition of different soil amendments are suited for uranium remediation to avoid an increase of bioavailable contaminants in the soil system.

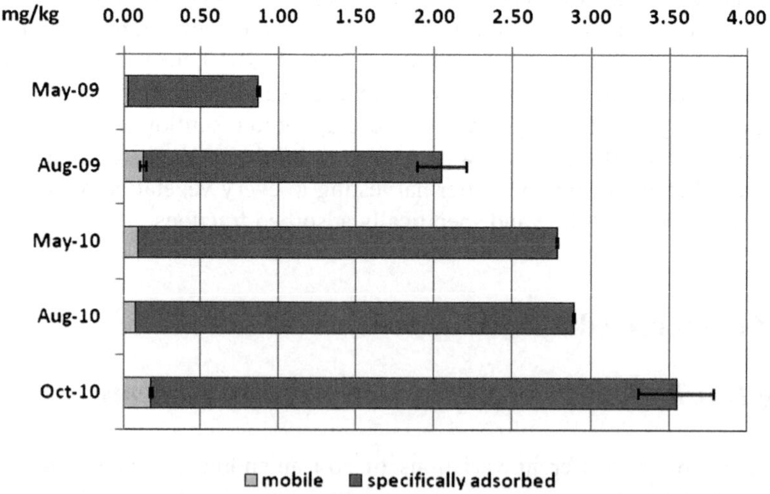

Fig. 1 Bioavailable fraction (mobile + specifically adsorbed) of uranium in the untreated soil (control) at the testsite Gessenwiese from May 2009 to October 2010 (error bars = standard deviation)

Effects of Soil Amendments on Biomass and Phytoextraction Potential

Based on germination tests and pot experiments a combination of three plant species (triticale, sunflower (data not shown) and Indian mustard) was chosen for the phytoremdiation experiments. Besides the extraction capacity for HM/R, biomass production was the most important objective.

According to germination tests with all plants, triticale is known to prefer acidic soil, and it is a fast growing plant. Indian mustard seems to be more sensitive to competition than triticale. For all these reasons, triticale took the advantage in biomass productivity over the other plants (Mirgorodsky et al. 2010). For all plant species, fertilization with NPK seems to have the major effect on biomass productivity in 2009. The other soil amendments and plant species investigated here had no significant effect on biomass production.

In contrast to 2009, the influence of soil amendments without NPK fertilizer on biomass productivity was investigated in 2010. Here, triticale produced the highest biomass yield (4.3 t DW/ha/harvest) with microbial additives (TF + MS). Second highest yield was achieved with the MIX substrate (3.1 t DW/ha/harvest). Also Indian mustard grew better after mentioned soil strategies with 1.3 and 1.6 t DW/ha/harvest, respectively (Table 1).

Shoots and roots from representative plant samples have been analyzed for HM/R content for each amendment. In order to compare results with other phytoremediation projects and to quantify the remediation potential of each plant and each amendment, the phytoextraction potential (plant concentration × biomass per area per year) was calculated (Lombi et al. 2002). To define which plants can be used for phytoextraction processes, the translocation factor ("TransF", shoot to root metal concentration ratio) was proposed (Zacarias et al. 2011). Plants with

Table 1 Biomass productivity, phytoextraction potential and translocation factors for Al, Ni and U per plant and soil amendments in 2010

Plant species/ Soil Amendment	Biomass Productivity [kg/ha]	Phytoextraction potential [g/ha/harvest]			Translocation factor "TransF"*		
	2010	2010			2010		
Triticale		Al	Ni	U [mg/ha]	Al	Ni	U
TF	1979 ± 938	580 ± 210	24 ± 1	113 ± 10	0.6	0.3	0.4
TF+MS	4346 ± 958	1007 ± 214	73 ± 4	290 ± 60	0.4	0.3	0.3
MIX	3106 ± 437	1882 ± 265	51 ± 5	370 ± 32	0.8	0.7	0.4
Indian Mustard							
TF	824 ± 263	570 ± 58	52 ± 17	102 ± 15	0.9	2.3	0.5
TF+MS	1319 ± 336	197 ± 3	68 ± 17	51 ± 10	0.2	1.8	0.1
MIX	1643 ± 127	818 ± 161	41 ± 5	110 ± 22	0.3	2.1	0.2

* TransF, translocation factor, mg metal in aerial part/mg metal in roots.

high translocation factors should be suitable for phytoextraction (Vamerali et al. 2010).

The phytoextraction potential of Triticale for U could be increased with the addition of calcareous topsoil (MIX) in 2009 and 2010. Triticale could extract 320 mg/ha/harvest of U in 2009 and 370 mg/ha/harvest in 2010. Moreover, in 2009 the extraction of Al, Ni and U by Indian mustard was quite independent toward soil amendments. Microbial strategy with mycorrhiza and streptomyces (TF+MS) seemed to have no significant effect on biomass and phytoextraction potential with all studied plants in 2009 (Mirgorodsky et al. 2010).

The translocation factors for different metals (Al, Ni, U) and plant species were calculated from the experimental results. Triticale has low TransF values for Al, Ni and U in all soil variances for 2009, and moderate TransF for Al (0.4–0.8) and U (~0.4) in 2010 after all soil amendments without fertilization (Table 1). In the experiments of 2009 and 2010, triticale showed higher TransF values for Al and U, whereas Indian mustard had higher values for Ni in all cases. According with this index Indian mustard is suitable for phytoextraction of Ni and Zn (data not shown); while triticale is more suitable for phytoextraction of Al and U, respectively.

Effects of Soil Amendments on HM/R Loads in Seepage Water

Besides the concentration of HM/R in seepage water, also the amount of seepage water formation plays a key role for groundwater contamination risks. Here, seepage water amounts of 120 l/m^2/a for L1, 64 l/m^2/a for L2 and 42 l/m^2/a for L3 were recorded in 2009. Percolation rates in 2010 of 158 l/m^2/a for L1, 72 l/m^2/a for L2 and 47 l/m^2/a for L3 with quite unchanged seepage water concentrations, except a decrease of U concentration for L2 and L3, verified a trend to an increasing HM/R output in the untreated monolithic lysimeter in comparison to the soil treated ones L2 and L3. The risk of groundwater contamination due to the inflow of seepage water decreased clearly by using soil amendments of calcareous soil and microbes. The seepage water loads for Al (36,000 g/ha/a), Ni (11,000 g/ha/a),

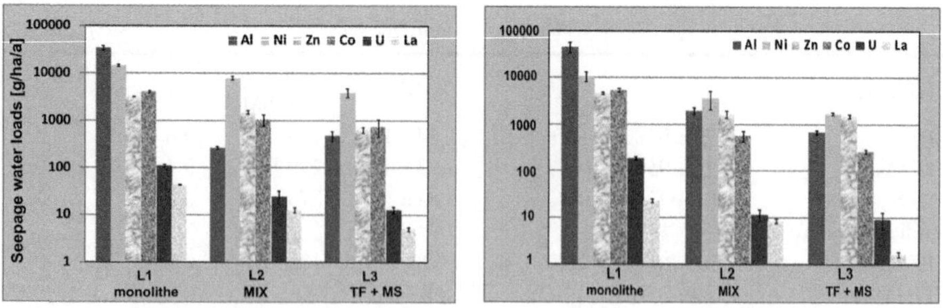

Fig. 2 **(a, b)** Seepage water loads of Al, Ni, Zn, Co, U (g/ha/a) and La (mg/ha/a) for 2009 (Fig. 2a) and 2010 (Fig. 2b)

Zn (3000 g/ha/a), Co (4000 g/ha/a), U (85 g/ha/a) and La (43 g/ha/a) are orders of magnitude higher for the monolith containing the geochemical barrier compared with soil amendments MIX and TF+MS in 2009 (see Fig. 2a).

In 2010 seepage water loads are still an order of magnitude higher for the monolithic lysimeter L1. Here, 44,000 g/ha/a Al, 10,600 g/ha/a Ni, 4600 g/ha/a Zn, 5300 g/ha/a Co, 180 g/ha/a U and 61 g/ha/a La were detected in comparison to the soil added lysimeter MIX and the lysimeter with microbial additives (TF+MS) (Fig. 2b).

Hence, the initial excavation and homogenization of the soil had increased the HM-R availability, but after addition of calcareous soil and MS the leaching of HM-R in groundwater has been reduced (Mirgorodsky et al. 2010b).

Utilization Experiments with Metal Loaded Plant Biomass from the Test Field Site "Gessenwiese"

Investigation of Ethanol Fermentation with Triticale from Experiments on the Test Field Site

In fermentation experiments with two *Triticale* fractions from the test field site "Gessenwiese" (HM/R contents of the plants 165 and 318 mg/kg Al, 5 and 6 mg/kg Ni, 30 and 28 mg/kg Zn, 0.04 and 0.06 mg/kg U, respectively), and with a *Triticale* sample from a field with no HM/R contamination for comparison, the feasibility of the fermentation of a metal loaded phytoremediation biomass was tested, and the possible inhibition of the product yield of the process was investigated.

The highest product yield was achieved with the uncontaminated material (Baltico, see Fig. 3) with 45.3 L/dt (+/– 0.4 L/dt) DW of *Triticale*. The fermentation of *Triticale* samples from the test field site "Gessenwiese" resulted in 41.4 (+/– 0.2) and 40.5 L/dt (+/– 0,2 L/dt) DW of *Triticale*, these are 91.5 and 89.5% of the yield of the uncontaminated material, respectively (see Fig. 3). The industrial standard yield of the fermentation of wheat, *Triticale* and rye is about 40 L/dt DW (see Fig. 3) (Die Österreichische Saatgutwirtschaft 2007); hence the fermentation yields are in the optimal range, and they would be economical in practice.

In further experiments, also a possible inhibition of the fermentation process by HM/R contained in the plants was investigated. For this purpose, the utilization of the carbohydrates in the fermentation suspensions of the heavy metal loaded *Triticale* from the test field site was measured. As a result, 96.8% of the usable sugars were consumed in the process, which is a nearly completely depletion of the utilizable carbohydrates in the process. Hence, no inhibition could be observed in the fermentation process of *Triticale* with such HM/R concentrations in the plant biomass. Thus, a utilization of the phytoremediation harvest from such a site for the winning of biofuels is feasible.

Fig. 3 Comparison of the ethanolic fermentation yields of Triticale harvests from the test field site "Gessenwiese" (TF III, Monokultur) with uncontaminated material ("Baltico") and the standard for uncontaminated material

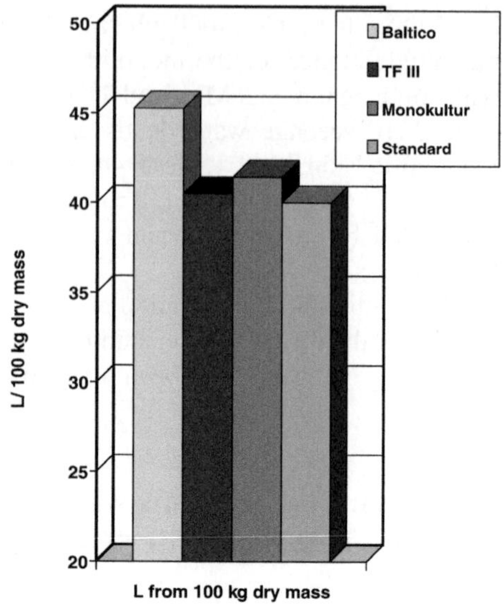

Experiments on Ethanolic Fermentation of HM/R-Amended Solanum tuberosum

To find out the limitations of the process, the general tolerance in the fermentation of heavy metal loaded *Solanum tuberosum* was investigated. A loading with the metals Cd, Cu, Fe, Mn, Ni, Zn, and Pb was examined; the metals were added in defined amounts to a freshly produced *Solanum tuberosum* pulp. As a result, at low heavy metal concentrations as they also occur in the heavy metal loaded plant biomass, no inhibition of the fermentation yield was observed (100% yield compared to uncontaminated material). It is anticipated, that low concentrations of trace metals are utilized in the process as micronutrients, and toxic metals are tolerated in low concentrations. Only at elevated concentrations of heavy metals, a significant inhibition of the fermentation process can be observed (inhibition by 33% at HM++ and 78% at HM+++, respectively). As a result, the actual heavy metal concentrations in the harvested plants do not inhibit a further fermentation process for utilization of the plant biomass from phytoremediation.

Further experiments were also carried out on the field of biogas production from plant remainders of the phytoremediation harvest, as well as experiments for the combustion of the plant remainders. An inhibition of the biogas process by the HM/R contained in the plants was not observed, which was investigated in parallel experiments with an uncontaminated sludge. The digested sludge originating from the biogas process finally represents a sink for the HM/R from the plant substrate.

The distribution of the HM/R in the different mass fluxes (flue gas, waste waters from flue gas treatment, filter dusts, ashes, slags) after the combustion of

HM/R loaded plant residuals from the test field site "Gessenwiese" was also investigated. Here, an accumulation of the U in the ashes by 70–99% was found.

Summary

In phytoremediation experiments on a HM/R contaminated test field site, different soil amendments, additions of calcareous topsoil (MIX) and fungi and microbes (TF+MS) in combination with genotype mixtures of triticale, sunflower and Indian mustard were investigated. Triticale is well adapted to the acidic conditions of the field, produced high biomass and extracted sufficient amounts of HM/R. The addition of calcareous top soil (MIX), mycorrhiza and streptomyces have reduced the leaching of HM/R, decreased the risk of groundwater contamination on this site and could be suitable for stabilizing methods.

Concerning the further utilization of HM/R loaded *Triticale* from the test field site, fermentation experiments were carried out and in comparison with *Triticale* from an uncontaminated field, 89 and 91.5% yield of bioethanol was achieved. These yields are still in the optimal range for such a process, and they would be economical in practice. An inhibition of the fermentation process by HM/R contained in slight concentrations in the plant material was not observed; 96.8% of the usable sugars were depleted in the fermentation process.

The HM/R loaded plant biomass from phytoremediation experiments on the test field site can subsequently be utilized in a biogas process. The digested sludge of the biogas process represents a sink for HM/R from loaded plant substrate. An accumulation of U in the ashes by 70–99% was observed after the combustion of HM/R loaded plant residuals from the test field site "Gessenwiese".

The combination of microbial treatment, addition of soil amendments like calcareous top soil and a crop mixture will be further investigated and could result together with a subsequent utilization of the biomass in an innovative and sustainable remediation technique.

Acknowledgements The authors are grateful for a funding of the project by the Federal Ministry of Education and Research BMBF (PTKA, FKZ 02S8538).

References

Die Österreichische Saatgutwirtschaft (2007) http://www.landnet.at/article/articleview/18798/1/5195

Koch-Steindl H and Pröhl G (2001) Considerations on the behaviour of long-lived radionuclides in soil. Rad. Environ. Biophys. 40, 93–104

Lombi E, Zhao FJ, Dunham SJ and McGrath SP (2001) Phytoremediation of heavy metal-contaminated soils: Natural hyperaccumulation versus enhanced Phytoextraction, J. Environ. Qual. 30, 1919–1926

Maegher R (2000) Phytoremediation of toxic elemental and organic pollutants. Curr. Opin. Plant Biol 3, 153/62

Meers E (2005) Phytoextraction of heavy metals from contaminated dredged sediments. PhD Thesis in: Applied Biological Sciences, Ghent University, Belgium

Miller RR (1996) Phytoremediation, technology overview report. Ground-Water Remediation Technologies, Analysis Center, Pittsburgh, PA, USA, pp 80

Mirgorodsky D, Ollivier D, Merten D, Bergmann H, Büchel G, Willscher S, Wittig J, Jablonski L and Werner P (2010a) Maßnahmen zur Strahlenschutzvorsorge radioaktiv belasteter Großflächen durch Sanierung mittels Phytoremediation und anschließende Verwertung der belasteten Pflanzenreststoffe (PHYTOREST). atw-Int. J. Nucl. Power 12 (2010), 774/78

Mirgorodsky D, Ollivier D, Merten D, Büchel G (2010b) Phytoremediation for soils contaminated with radionuclides and heavy metals. Proceedings of the 11th International Conference on Management of Soil, Groundwater and Sediment (ConSoil), Salzburg/Austria

Vamerali T, Bandiera M and Mosca G (2010) Field crops for phytoremediation of metal-contaminated land – a review. Environ. Chem. Lett. 8, 1–17

Van Ginneken L, Meers E, Guisson R, Ruttens A, Elst K, Tack FMG, Vangronsveld J, Diels L and Dejonghe W (2007) Phytoremediation for heavy metal contaminated soils combined with bioenergy production. J. Environ. Engin. Landscape Managem. Vol. XV, 227/36

Zacarias M, Beltran M, Torres LG and Gonzalez A (2011) A feasibility study of perennial/annual plant species to restore soils contaminated with heavy metals. J. Phys. Chem. Earth doi:10.1016/j.pce.2010.12.008.

Zeien H and Brümmer GW (1989) Chemische Extraktion zur Bestimmung von Schwermetallbindungsformen in Böden. Mitt. Dtsch. Bodenkdl. Ges. 59, 505/15

Free and Immobilized Microbial Systems – Potential Effective Radioactive Decontaminators

Ioana-Carmen Popescu, Georgiana Milu, Mihaela Stoica, Gheorghe Crutu, Ecaterina Militaru

Abstract. Romanian uranium extractive industry represents the major cause for the environmental radioactive pollution. Radionuclides' migration into the water and soil generates nutritional imbalances, which cause plants and animals ailment and threats people's health condition. The biotechnologies based on radioactive ions bio-accumulation, bio-fixation and bio-precipitation processes seem to be a better remediation alternative to the conventional ones. *Thricoderma harzianum*'s uranium recovery efficiency was more than 85% when it was immobilized on a polyester surface respectively of 90% when it was free.

Ioana-Carmen Popescu
R&D National Institute for Metals and Radioactive Resources – ICPMRR,
48 Şoseaua de Centură, Măgurele, Ilfov county, Romania,
E-mail: JanePopescu@gmail.com; ioana.popescu@icpmrr.ro Phone: 0720916109

Georgiana Milu
R&D National Institute for Metals and Radioactive Resources – ICPMRR,
48 Şoseaua de Centură, Măgurele, Ilfov county, Romania

Mihaela Stoica
R&D National Institute for Metals and Radioactive Resources – ICPMRR,
48 Şoseaua de Centură, Măgurele, Ilfov county, Romania

Gheorghe Crutu
R&D National Institute for Metals and Radioactive Resources – ICPMRR,
48 Şoseaua de Centură, Măgurele, Ilfov county, Romania

Ecaterina Militar
R&D National Institute for Metals and Radioactive Resources – ICPMRR,
48 Şoseaua de Centură, Măgurele, Ilfov county, Romania

B. Merkel, M. Schipek, *The New Uranium Mining Boom*
DOI 10.1007/978-3-642-22122-4 © Springer 2011

Introduction

Recently, some biotechnologies based on radioactive ions bio-accumulation, bio-fixation and bio-precipitation process recently has come into the spotlight as environmentally friendly and economically affordable remediation alternatives to the conventional radioactive decontamination methods (Brierly 1982, 1986; Crutu et al. 2008; Manea et al. 2008). Such biotechnologies use the activity of different physiological microorganisms groups in order to remove radioactive ionic species from the contaminated environment (Le Roux 1970; Ivanus 2005; Ivanus and Bica 2006). The capacity of free microbial system to concentrate the radioactive metallic ions from their diluted solutions by adsorption or direct absorption and biosorption from water and soil is a consequence of their property to fix and to precipitate either on their surface or intracellular the metallic ions in solution (Lovey et al. 1991; Thacker et al. 1998).

The first stage involves the stoichiometrical reaction between the metallic ionic species in solution and the reactive chemical groups on peptidoclycane surface.

Subsequently to the complexation stage the active sites become nucleation cores on which the ionic metallic species precipitates.

The biosorption process follows the next stages (Manea et al. 2008):

- Biomass cells growing,
- Cells cropping,
- Direct contact between the biomass and the metallic ionic species,
- Biomass regeneration.

Radioactivity in Banat Region

Romanian uranium mining has an over fifty year old tradition. Uranium mining was originally developed under the direct leadership of the former Soviet Union.

Since 2001, mining activity has almost ceased but a dangerous legacy has been left behind in the form of huge amounts of radioactive wastes stored in heaps and tailing ponds. Existing treatment plants and technologies are either too old or of insufficient capacity to function well.

The chemical analysis of mine water sampled in different hot spots of the mining site has pointed out uranium concentration ranging 0.1 to about 13 mg/l. Table 1 shows the chemical composition of the mine water samples.

Uranium and heavy metals presence into the mine waters may be explained by their complex geochemistry highly influenced by the weathering factors beside the chemical composition of the soil that hosts them. Uranium ore sequences are directly overlain by Jurrasic carbonate rocks (as showed in Fig. 1). Meteoric waters filter through the carbonates and then interact with the uranium bearing sequences and the mine wastes. Resultantly ground and stream waters contain significant levels of dissolved uranium beside carbonates and other ionic species.

Table 1 Chemical composition of mine water samples collected on the mining site in Banat

Sampling point	U	Mo	Pb	Na	Fe	Mn	Zn	K	Cd	Cu
Jitin creak 100 m downstream	1.52	0.131	0.01	170.6	0.15	0.03	–	–	–	0.01
Creak underneath waste heap	1.45	0.12	0.01	65.52	0.04	0.04	–	–	–	0.01
Mine water from 4th gallery	0.85	0.112	0.01	65.43	0.08	0.04	–	–	–	–
L1 entrance	12.8	0.586	0.04	860.8	3.8	0.13	0.33	5.27	0.01	0.89
L2 exit	2.7	0.341	0.06	736	3.52	0.07	0.27	1.75	0.01	0.78
L3 Natra creak before the confluence with Lisava creak	1.5	0.144	0.01	158.4	0.79	0.04	0.03	–	–	0.14
L4 Lisava creak after the waste dump	0.037	0.006	–	6.55	0.04	0.04	–	–	–	–
L5 Natra creak after the confluence with Lisava creak	0.544	0.047	0.04	72.01	0.25	0.04	–	–	–	0.05
Cdownstream	0.46	0.065	0.02	92.12	0.07	0.04	–	–	–	0.02
Camonte	0.017	0.007	–	3.04	0.01	0.02	–	–	–	–

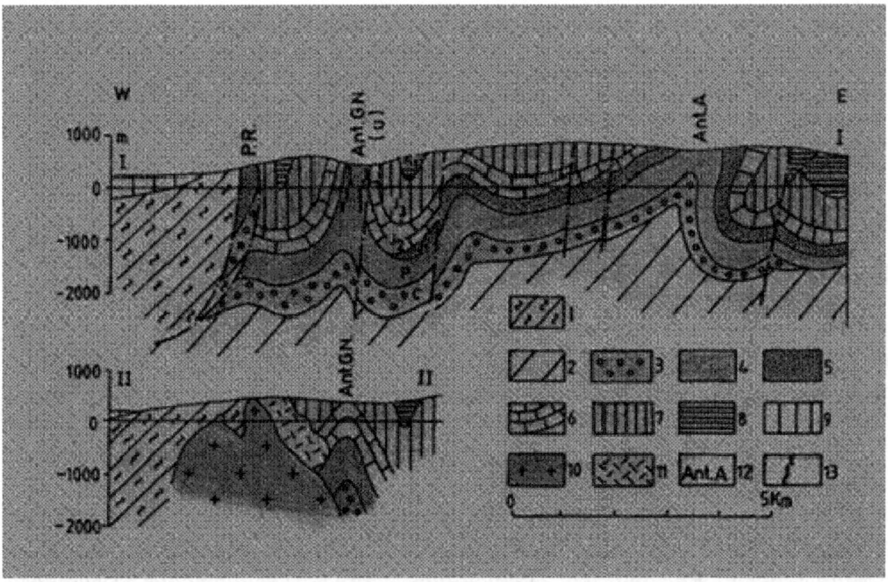

Fig. 1 Geological sections through the anticline Garliste – Natra: 1&2 – crystalline schist of different degrees with metamorphism; 3 – Superior Carboniferous deposits (C); 4 – Permian deposits (P); 5 – Inferior Jurassic deposits (J1); 6 – Middle Jurassic deposits (J2): marls, sandstones, chalkstones; 7 – Superior Jurassic deposits (J3): chalkstones; 8 – inferior Cretaceous deposits (K1): marls and chalkstones; 9 – Tertiary deposits; 10 – magmatic intrusions (granodiorites, Superior Cretaceous); 11 – skarns; 12 – Anticline Garliste-Natra with uranium; 13 – uranium mineralization. (Information provided by the Handbook titled "Studii si cercetari de geologie" = Studies and researches in geology, Tome 45, Romanian Academy Publishing House, Edited in 2000)

Experimental Activity

The biomass was obtained using a number of microbial systems insulated out of the existent micro-flora originated from the studied mining sites. The biomass was used to remove uranyl ions from mine water samples.

In order to increase biomass remediation efficiency it is necessary to eliminate the alkaline ions accumulated in its cells. Their presence, either in the cytoplasm or on the plasmatic membrane surface causes an undesirable competition with the other ions in solution.

In order to eliminate sodium and potassium ions the biomass was thermal treated. It was boiled at 100°C in distilled water for 20–30 min. The ration between the initial amount of wet biomass and the water volume was 100 g/l. After cooling the treated cells suspension was centrifuged at 3000 rot/min for 10 min and the sediment was washed twice with distilled water. The supernatant resulted after the thermal treatment and centrifugation exhibited a very high content of alkaline ions eliminated from cells body, which was reflected by the electrical conductivity increase.

The immobilization method consisted in biomasses' suspension on deionized water in a solid to liquid ratio of 15%. Then they were homogenized with 33% poly-ethylene-amine and 25% glutar-aldehyde. Then a high viscous paste was obtained. It was further extruded and dried for 24 h at 60–65°C followed by grinding and meshing, when immobilized biomass pellets resulted.

Results and Discussions

The experiments pointed out that *Thricoderma harzianum* may be considered the most efficient biomass for uranium biosorption at a pH of 4.5, as confirmed by the literature (Brierly 1982, 1986). The experiments carried out using *Thricoderma harzianum* immobilized on polyesters in percolation columns exhibited a uranium recovery yield higher than 85%.

Other biomasses that exhibited satisfactory yields for uranium biosorption from different effluents used either free or immobilized were: *Aspergillus aurata, Aspergillus niger, Rhizopus* sp., *Penicillium hergnei, Pseudomonas aeruginosa, Chaetomium distortium* and *Saccharomyces cerevisiae.*

In case when free cells suspensions used the recovery yields were of 90% as confirmed (Lovey et al. 1993; Rugge et al. 1995; Raboca 1997; Robinson et al. 1998).

The most significant parameter, which influenced uranium bio-sorption efficiency, was pH with an optimum value of 4.5–5.0.

Table 2 shows uranium amounts accumulated by the studied biomasses types at the radioactive wastewater treatment.

Table 2 Uranium accumulation by different types of biomasses

Biomass type	Microbiological component	mg U/g cells
Filamentous fungi	*Rizopus arrhizus*	180
	Penicillum digitatum	57
Bacteria	*Arthobacter RAG*	800
	Pseudomonas aeruginosa	150
	Zoologea ramigera	500
	Streptomyces chromogenes	312
	Citrobacter sp.	9,000
	Streptomyces longwoodenisis	440
Marine algae	*Chlorella regularis*	159
Yeasts	*Saccharomyces cerevisiae*	150

Fungi are able to accumulate metals and radionuclides based on physiochemical and biological mechanisms, including their binding by metabolites, polymers or specific polypeptides (Thovinen and Kelly 1974; Schippers et al. 1995).

It is thought that the fungal cell wall is composed of two major components: a skeletal network microfibrilar intracellular consisting mainly of chitin encrusted in an amorphous layer of proteins and various polysaccharides. *Sacharomyces cerevisae* proved that is best suited for bioaccumulation/biosorbtion of radioactive and heavy metals from contaminated soils and waters.

Conclusions

Romania is affected by the radioactive pollution due to the uranium mining activities carried out for nuclear fuel fabrication. Radionuclides' migration into the water and soil generates nutritional imbalances, which cause plants and animals ailment and threats people's health condition.

In order to assure safe condition for people living it is necessary to clean up the radioactive contaminated sites. The microbial systems using as active biosorbents bacteria (*Thricoderma harzianum, Arthobacter RAG, Pseudomonas aeruginosa, Zoologea ramigera, Streptomyces chromogenes, Citrobacter* sp., *Streptomyces longwoodenisis*), fungi (*Rizopus arrhizus, Penicillum digitatum*) and yeasts (*Saccharomyces cerevisiae*) proved nature's capability to "cure" itself. Those "healing" mechanisms may be capitalized by an integrated biotechnological flowsheet.

The experiments carried out with a view to reducing radioactive metals contents in industrial effluents and soils resulted from uranium mining industry pointed out that best alternative to apply the biotechnological processes is to use as biosorbent the immobilized biomasses.

Unfortunately the economic crisis has affected our research plans, therefore our work results are preliminary ones.

For a correct assessment of microbial systems' remediation capability additional funds are necessary because further studies are required to size up a remediation technology.

Acknowledgement We would like to express our gratitude to Adriana Botez, Ecaterina Militaru, Dana Aura Rodica Radu, Oana Dobre and Carmen-Mariana Nicolae, who helped us with their expertise to characterize the water samples from chemical viewpoint.

References

Brierly C. (1982). "Microbiological Mining", Scientific American, August, pp. 44–54.

Brierly C., Brierly A. (1986). "Microbian Mining in General Molecular and Applied Microbiology", John Willey.

Crutu Gh., Manea C., Groposila G., Curelea D. (2008). "Bioremediation methods", (GEOENV 2008) Greece, Heliotopus Publication, ISBN 978-960-6745-01-7, pp. 117–120.

Ivanus R. C.(2005). "Solubilizarea bacteriană" = Bacterial lixiviation (in Romanian), Editura Universitaria, Craiova.

Ivanus R. C., Bica Gh. (2006). "Ecomanagementul recuperarii şi refolosirii materialelor" = Ecomanagement of materials recovery and re-utilization (in Romanian), Editura Universitaria, Craiova.

Le Roux N. (1970). "Microbian Aspect of Metallurgy", American Elsevier Publishing Co., New York.

Lovley, D. R., Phillips, E. I. P., Gorby, Y. A. & Landa, E. R. (1991). "Microbial reduction of uranium". Nature, 350, pp. 413–416.

Lovley, D. R., Roden, E. E., Phillips, E. J. P. & Woodward, J. C. (1993). "Enzymatic iron and uranium reduction by sulfate reducing bacteria". Marine Geology, 113, pp. 41–53.

Lovley, D. R., Widman, P. K., Woodward, J. C. & Phillips, E. I. P. (1993). "Reduction of uranium by cytochrome C3 of Desulfovibrio vulgaris". Applied and Environmental Microbiology, 59, pp. 3572–3576.

Manea C., Crutu Gh. & Gropoşilă G. (2008). "Ecological Remediation Methods", International Workshop in Geoenvironmental and Geotechnics, (GEOENV 2008), Heliotopus Publication,Greece, ISBN 978-960-6745-01-7, pp. 121–124.

Rugge, K., Bjerg, P. L. & Alristensen, T. H. (1995). "Microbial redox interactions with uranium: an environmental perspective. Distribution of organic compounds from municipal solid waste in the groundwater downgradient of a landfill" (Grindsted, Denmark). Environmental Science and Technology, 29, pp. 1395–1400.

Raboca N. (1997). Europa. "Resurse metalife" = Europe "Mineral Resources"(in Romanian), Casa de editura "Sarmis", Cluj-Napoca.

Robinson, K. G., Ganesh, R. & Reed, G. D. (1998). "Impact of organic ligands on uranium removal during anaerobic biological treatment". Water Science and Technology, 37, pp. 73–80.

Schippers, A., Hallmann, R., Wentzien, S. & Sand, W. (1995). "Microbial diversity in uranium mine waste heaps". Applied and Environmental Microbiology, 61, pp. 2930–2935.

Thacker, M. D., Barton, L. L. & Thomson, B. M. (1998). "Removal of U and Mo from water by immobilized Desulfovibrio desulfuricans in column reactors". Biotechnology and Bioengineering, 60, pp. 88–96.

Thovinen, O. H., Kelly, D. P. (1974). "Studies on the growth of Thiobacillus ferrooxidans II. Toxicity of uranium to growing cultures and tolerance conferred by mutation, other metal cations and EDTA". Archives of Microbiology, 95, pp. 153–164.

Bioleaching of Shale – Impact of Carbon Source

Viktor Sjöberg, Anna Grandin, Lovisa Karlsson, Stefan Karlsson

Abstract. Bioleaching is often used for processing low-grade shale feedstock and the microbial community used for that purpose is supplied with nutrients such as sugar and/or Fe^{2+}. In the present study, the leaching efficiency was tested when crushed weathered shale was mixed with aspen wood shavings and kept moist, at the mixtures field capacity. The purpose was to investigate whether a more complex carbon source and a lower content of water may be a feasible way of lowering the cost for bioleaching. After 56 days of incubation the amount of uranium mobilized from the shale reached some 1.7% with a minimum of effort and cost.

Introduction

Bioleaching is known as a cheap method that is particularly useful for processing of low-grade feedstock's and successfully used for shale's. In general the feedstock is piled and sprayed with solutions containing critical nutrients to enhance the microbial activity. Depending on whether autotrophic or heterotrophic microorganisms are utilized the composition of the solution differs. In laboratory evalu-

Viktor Sjöberg
Man-Technology-Environment Research Centre, Örebro University, SE-701 82 Örebro, Sweden

Anna Grandin
Man-Technology-Environment Research Centre, Örebro University, SE-701 82 Örebro, Sweden

Lovisa Karlsson
Man-Technology-Environment Research Centre, Örebro University, SE-701 82 Örebro, Sweden

Stefan Karlsson
Man-Technology-Environment Research Centre, Örebro University, SE-701 82 Örebro, Sweden

B. Merkel, M. Schipek, *The New Uranium Mining Boom*
DOI 10.1007/978-3-642-22122-4

ations of different microorganisms ability to mobilize elements such as uranium from low-grade feedstock's known genera of microorganisms are inoculated in a cell culture medium containing a few percent of the feedstock corresponding to a liquid to solid ratio (L/S) of 50 to 100 (Burgstaller et al. 1992; Faramarzi et al. 2004; Wang et al. 2009; Anjum et al. 2010). As in full scale bioleaching the microbial activity has been enhanced by addition of critical nutrients, for heterotrophic microorganisms 1 to 10% organic carbon, typically as glucose or sucrose, or for autotrophic microorganisms extra Fe^{2+} (Burgstaller et al. 1992; Bosshard et al. 1996; Anjum et al. 2009, 2010; Mishra et al. 2009). To reduce the cost for large scale operations, some studies have been conducted with less refined carbon sources such as molasses, mango peels and seed cake. In general has the addition of these less refined carbon sources resulted in decreased mobilization of elements (Burgstaller et al. 1992; Mulligan et al. 2004; Anjum et al. 2009, 2010).

In fact, addition of organic carbon in the form of pine saw dust to uranium containing solutions has been found to decreases the dissolved fraction of uranium due to adsorption of the element to the saw dust itself (Bagherifam et al. 2010).

Controlled studies of naturally occurring microorganisms such as *Cladosporium oxysporium, Aspergillus flavus* and *Curvularia clavata* supplied with 3% sucrose indicate a mobilization of 71, 59 and 50% uranium, respectively, from a uranium-bearing ore mixed into the cell culture medium at L/S 10 after 10 days (Mishra et al. 2009). Isolated autotrophic *Acidithiobacillus ferrooxidans* mobilized 80% uranium from black shale after 2.5 day at L/S 10 (Choi et al. 2005). However, the leaching was strongly dependent on the addition of Fe^{2+} to the medium and a concentration of 5 g/l was necessary to obtain high mobilization (Choi et al. 2005).

However, all of the above cited works used aqueous systems with finely grained shale at high L/S and addition of refined additives such as sugar and Fe^{2+} to obtain high mobilization of elements from the feedstock. These conditions require rather expensive treatment in an industrial application why there is a need to reduce the cost for the process.

In the present work the aim was to examine the mobilization of uranium from weathered low-grade shale with aspen wood shavings as carbon source for naturally occurring microbes when the system was kept at water field capacity.

Materials & Methods

Weathered shale, containing some 30 ppm uranium, was collected from a local open pit in the county of Kumla in mid-Sweden. Before use the shale was carefully crushed and sieved to obtain a 0.25–0.56 mm fraction.

To supply the naturally occurring heterotrophic microorganisms with an unrefined easily degradable carbon source, aspen (*Populus tremula*) wood shavings was used. They were approximately 5 × 5 mm and used as received from the planer.

To obtain good contact between the crushed shale and the wood shavings approximately 5 g of each was stratified in 50 ml test tubes (Sarstedt®) with holes drilled in the bottom for drainage. After stratifying the materials, 50 ml de-ionized water was added to reach field capacity. All excess water was allowed to drain and collected for uranium, organic acid and pH analysis. To avoid dehydration of the samples, in the dry laboratory environment, the samples were kept in a humidor at approximately 20°C.

Once a week the test tubes were flushed with 50 ml de-ionized water to maintain the water content and to generate aqueous samples.

As a control treatment, shale samples were put on an end-over-end shaker with de-ionized water at L/S 10. Every third or fourth day the aqueous phase was removed, analyzed, and replaced with fresh de-ionized water.

All tests were conducted in at least duplicates and all samples for uranium analysis were preserved within minutes after collection to a final concentration of 1% nitric acid (sub-boiled distilled). Uranium was analyzed as ^{238}U with ICP-MS (Agilent 7500 cx) with plasma power set to 1500 W and with 10 ppb ^{103}Rh as internal standard. Analysis of organic acids was conducted with CE (Hewlett Packard 3D CE) using the method described by Dahlén et al. (2000).

Results & Discussion

After 56 days of incubation 1.7% of the uranium content had been mobilized from the shale into the aqueous phase and the pH had increased from 3.31 to 4.04 (Fig. 1). From the control sample 2.9% uranium was mobilized after 29 days (Fig. 1). One reason for the higher mobilization in the controls can be that abrasion between shale particles probably increased the amount and surface area of uranium-containing particles. The assumption is supported by the declining increase in mobilization rate already after eight days (Fig. 1). It has also been found that the mobilization of uranium from a similar shale in cell culture medium reaches near 20% after 2.5 days, when the samples are leached during orbital shaking (Choi et al. 2005).

This indicates that some of the uranium bearing minerals in the shale are fairly soluble. When it comes to the mobilization of uranium in the bioleaching, more focus should be put on the stabile and continuous increase in mobilization and not to the amounts. Since the mobilization in the bioleaching showed a stabile increase over time (Fig. 1) it was concluded that microbial processes were responsible for the mobilization since no mechanical abrasion occurred in those samples. The lower concentrations may in fact be caused by adsorption of mobilized uranium to the wood shavings which would lead to an underestimation of the release from the shale. It has been found that saw dust can absorb approximately 60% of uranium in solution at pH 4 when the ratio between saw dust and uranium was 150 (Bagherifam et al. 2010). In this study the corresponding ratio based on dissolved uranium reached 2.5×10^6 why the wood shavings may well act as a sorbent for uranium.

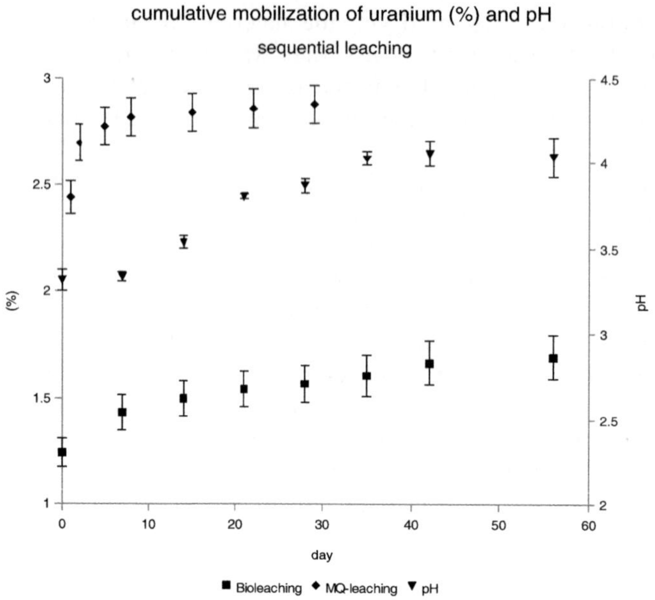

Fig. 1 Mobilization of uranium from weathered shale by bioleaching, end-over-end shaking with de-ionized water and change in pH during microbial activity

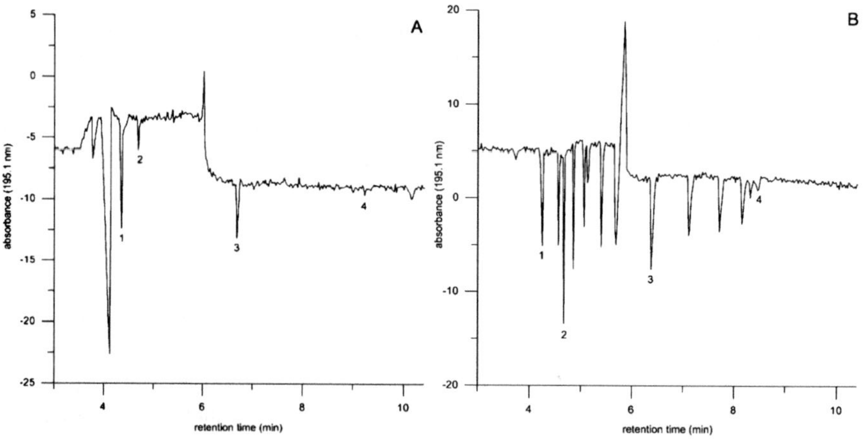

Fig. 2 (**A**) Organic acids present after 2 weeks of incubation. (**B**) Standard solution containing a mixture of organic acids, each at 50 ppb. (1) molybdate (internal standard); (2) oxalic acid; (3) acetic acid and (4) citric acid

Since many microorganisms reported from similar systems e.g. *Aspergillus niger* and *Pseudomonas putida* produce organic acids such as oxalic and citric acid (Krebs et al. 1997) the presence of organic acids in the aqueous samples was used as an indicator of microbial activity. During the second week of incubation the

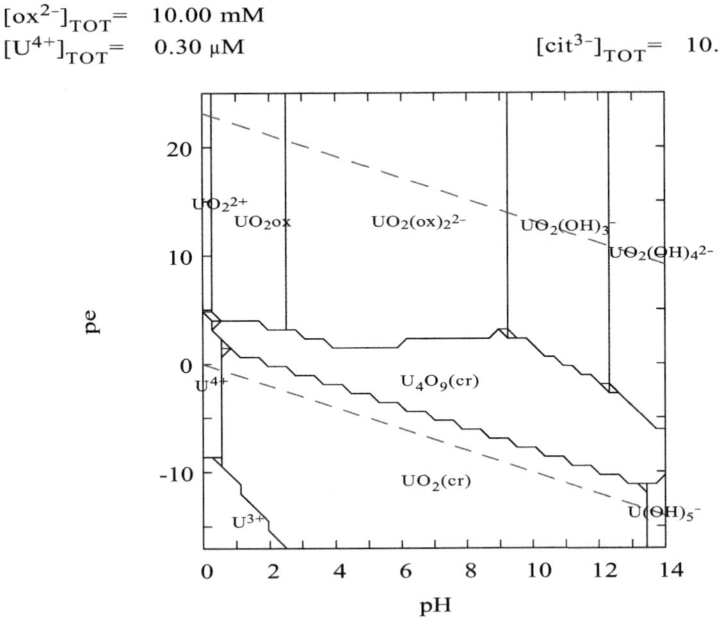

Fig. 3 Possible species of uranium in the presence of oxalic and citric acid

presence of organic acids was confirmed by CE (Fig. 2A). From the external standard solution (Fig. 2B) the acids were tentatively identified as oxalic, acetic and citric acid.

To elucidate if solid uranium species were likely to precipitate during the conditions in the systems a pe-pH diagram was constructed with the software MEDUSA (Fig. 3). Input concentration of uranium was set slightly lower than those in the control sample (0.32 μM) and the concentrations of oxalic and citric acids were set to about one tenth of the concentrations measured by Anjum et al. (2010). The resulting pe-pH diagram clearly indicates that if sufficient aeration was maintained all dissolved uranium would enter into the solution as uranium-oxalate at pH between approximately 2.5 and 9, which covered the range in the present study. However, the model did not account for the possible adsorption of uranium to the wood shavings since stability constants are unavailable.

Conclusions

Addition of organic carbon in the form of aspen wood shavings to crushed weathered shale and keeping water at field capacity result in increased mobilization of uranium. Representative distribution coefficients for the uranium/wood shavings partitioning should be produced in order to account for adsorption to the organic material on general system properties.

Since well known organic acids produced by microorganisms were found in the solutions it is concluded that the mobilization of uranium is related to microbial activity. It is not clear if the interaction is direct between the shale and the organism or indirect by the release of hydrogen ions and/or complexing agents to the water phase during wood decomposition.

As an alternative to use expensive additives such as refined sugars and maintaining aqueous systems, addition of wood shavings to the feedstock seems to be a feasible alternative for inexpensive leaching of uranium from low-grade materials. However, further optimization of critical parameters (e.g. water content, amount of wood shavings and aeration) must be done in order to establish optimum leaching conditions.

References

Anjum F, Bhatti HN, Ghauri MA, Bhatti I, Asgher M and Asi MR (2009) Bioleaching of copper, cobalt and zinc from black shale by Penicillium notatum. Afr J Biotechnol 19: 5038–5045

Anjum F, Bhatti HN, Asgher M and Shahid M (2010) Leaching of metal ions from black shale by organic acids produced by Aspergillus niger. Appl Clay Sci 47: 356–361

Bagherifam S, Lakzian A, Ahmadi SJ, Rahimi NF and Halajnia A (2010) Uranium removal from aqueous solutions by wood powder and weat straw. J Radioanal Nucl Chem 283: 289–296

Bosshard PP, Bachofen R and Brandl H (1996) Metal leaching of fly ash from municipal waste incineration by *Aspergillus niger*. Environ Sci Technol 30: 3066–3070

Burgstaller W, Strasser H, Wöbking H and Schinner F (1992) Solubilization of zinc oxide from filter dust with Penicillium simplicissimum: bioreactor leaching and stoichiometry. Environ Sci Technol 26: 340–346

Choi MS, Cho KS, Kim DS and Ryu HW (2005) Bioleaching of uranium from low grade black schist by Acidithiobacillus ferrooxidans. Worl J Microbiol and Biotechnol 21: 377–380

Dahlén J, Hagberg J and Karlsson S (2000) analysis of low molecular weight organic acids in water with capillary zone electrophoresis employing indirect photometric detection. Fresenius J Anal Chem 366: 488–493

Faramarzi MA, Stagars M, Pensini E, Krebs W and Brandl H (2004) Metal solubilization from metal-containing solid materials by cyanogenic Chromobacterium violaceum. J Biotechnol 113: 321–326

Krebs W, Brombacher C, Bosshard PP, Bachofen R and Brandl H (1997) Microbial recovery of metals from solids. FEMS Microbiol Rev 20: 605–617

Mishra A, Pradhan N, Kar RN, Sukla LB and Mishra BK (2009) Microbial recovery of uranium using native fungal strains. Hydrometallurgy 95: 175–177

Mulligan CN, Kamali M and Gibbs BF (2004) Bioleaching of heavy metals from a low-grade mining ore using Aspergillus niger. J Hazard Mat 110: 77–84

Wang J, Bai J, Xu J and Liang B (2009) Bioleaching of metals from printed wire boards by Acidithiobacillus ferrooxidans and Acidithiobacillus thiooxidans and their mixture. J Hazard Mat 172: 1100–1105

Interactions of Ionic Liquids with Uranium and Its Bioreduction

Chengdong Zhang, Arokiasamy J. Francis

Abstract. We investigated the influence of ionic liquids (ILs) 1-butyl-3-methylimidazolium hexafluorophosphate $[BMIM]^{+}[PF_6]^{-}$, N-ethylpyridinium trifluoroacetate $[EtPy]^{+}[CF_3COO]^{-}$ and N-ethylpyridinium tetrafluoroborate $[EtPy]^{+}[BF_4]^{-}$ on uranium reduction by *Clostridium* sp. under anaerobic conditions. Potentiometric titration, UV-vis spectrophotometry, LC-MS and EXAFS analyses showed monodentate complexation between uranyl and BF_4^- PF_6^-; and bidentate complexation with CF_3COO^-. Ionic liquids affected the growth of *Clostridium* sp. as evidenced by decrease in optical density, changes in pH, gas production, and the extent of U(VI) reduction and precipitation of U(IV) from solution. Reduction of U(VI) to U(IV) was observed in the presence of $[EtPy][BF_4]$ and $[BMIM][PF_6]$ but not with $[EtPy][CF_3COO]$.

Introduction

Ionic liquids (ILS) are ion-pair compounds containing an organic cation and an organic or inorganic anion. They are versatile since their chemical properties such as hydrophobicity, melting point and viscosity can be adjusted to obtain the desired reaction. They are used as green solvents in organic synthesis and in materi-

---

Chengdong Zhang
Environmental Sciences Department, Brookhaven National Laboratory, Upton, NY, USA
College of Environmental Science and Engineering, Nankai University, Tianjin, China

Arokiasamy J Francis
Environmental Sciences Department, Brookhaven National Laboratory, Upton, NY, USA
Division of Advanced Nuclear Engineering, Pohang University of Science and Technology, Pohang, Korea

B. Merkel, M. Schipek, *The New Uranium Mining Boom*

DOI 10.1007/978-3-642-22122-4

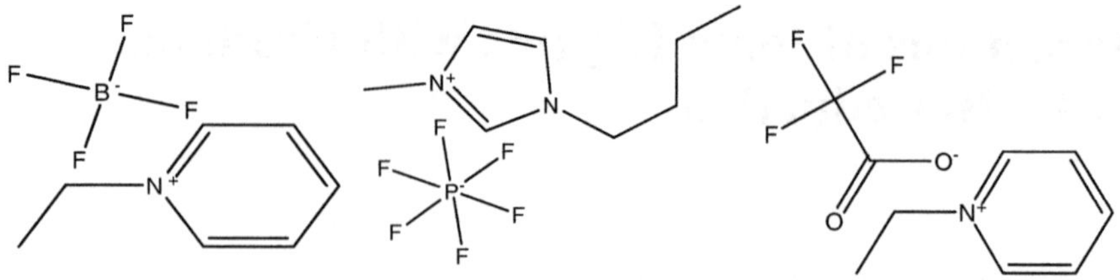

Fig. 1 Molecular structures of ILs N-ethylpyridinium tetrafluoroborate [EtPy][BF4], 1-butyl-3-methylimidazolium hexafluorophosphate [BMIM][PF6] and N-ethylpyridinium trifluoroacetate [EtPy][CF33COO]

als chemistry (Anastas and Warner 1998). Visser et al. (2003) determined the molecular association of UO_2^{2+} with hydrophobic ILs [C_4mim][PF_6] and [C_8mim] [$N(SO_2CF_3)_2$] in the presence of CMPO and TBP using EXAFS. The uranium formed a $UO_2(NO_3^-)$ (CMPO) complex in the IL phase while IL-associated anion complexation with uranium was not observed. Gaillard et al. (2005) used time-resolved emission spectroscopy and EXAFS analysis to determine the complexation ability of fluorinated anions with uranyl ion in aqueous solution at room temperature. Both $[BF_4]^-$ and $[PF_6]^-$ formed inner-sphere monodentate complexes with uranyl ion while $[Tf_2N]^-$ did not complex with uranium.

Microbial reduction of uranium is of considerable interest because of its potential application in the bioremediation of contaminated sites and in the pretreatment of uranium containing wastes. In this study, we investigated the interaction of uranium (VI) with N-ethylpyridinium tetrafluoroborate [EtPy][BF_4], 1-butyl-3-methylimidazolium hexafluorophosphate [BMIM][PF_6] and N-ethylpyridinium trifluoroacetate [EtPy][CF_3COO], (Fig. 1) in aqueous solution and examined their effects on uranium reduction by anaerobic bacteria.

Characterization of Uranium Associated with ILs

Potentiometric Titration

Potentiometric titration of uranyl nitrate showed the inflection point at 2 mM OH^-/mM U at pH 6.4 (Fig. 2). The release of 2 protons is due to the association of uranium ion with 2 OH^- in water. The titration curves for the U:[BMIM][PF_6] and the U:[EtPy][BF_4] with uranyl nitrate alone indicated no complex formation between the metal and the ILs. Addition of one-fold excess IL had no effect on the complexation. However, titration of the U in the presence of [EtPy][CF_3COO] showed a release of one more proton into the medium as indicated by the inflection point at 3 mM OH^-/mM U. This phenomenon shows that there is an interaction of the uranium with the [EtPy][CF_3COO]. The interaction increases with addition of one-fold excess [EtPy][CF_3COO] (more acid released).

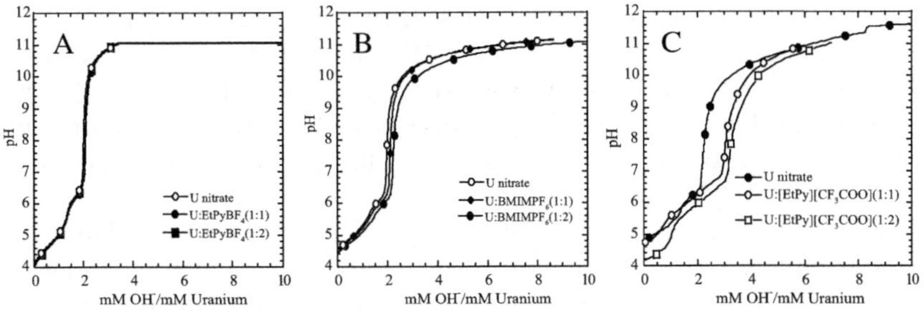

Fig. 2 Potentiometric titration curve of U and ILs mixture: (**A**) [BMIM][PF_6]; (**B**) [EtPy][BF_4]; (**C**) [EtPy][CF_3COO]

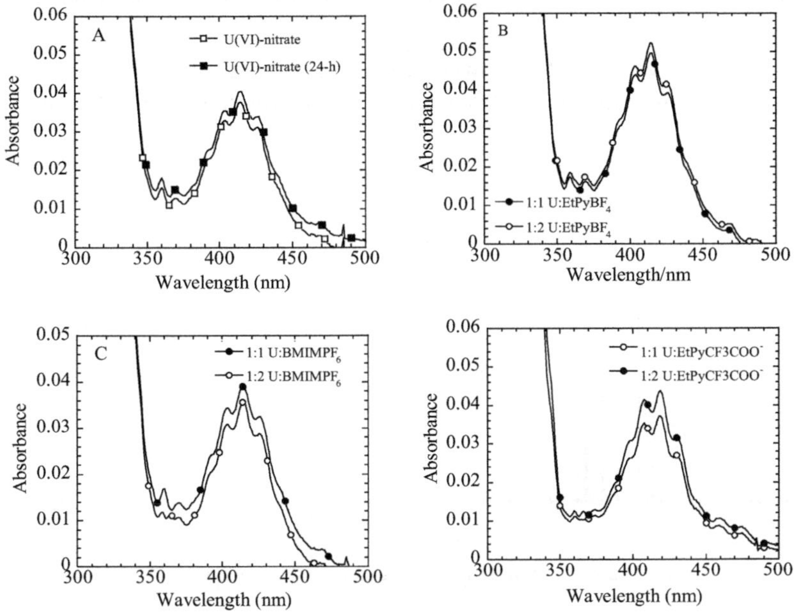

Fig. 3 UV-vis spectrophotometry of uranium nitrate and U-IL mixtures: uranium-nitrate (**A**); [EtPy][BF_4] (**B**); [BMIM][PF_6] (**C**); [EtPy][CF_3COO] (**D**)

UV-vis Absorption Spectroscopy

The UV-vis absorption spectrum for the pH unadjusted uranyl nitrate in the range of 300 to 500 nm showed fine structure in the 380 to 440 nm range characteristic of the presence of uranyl ion species. The maximum absorbance was at 413 nm and two shoulder peaks were noted at 402 and 425 nm. The UV-vis absorbance spectra for the 1 : 1 U : [EtPy][BF_4] and 1 : 2 U : [EtPy][BF_4] mixtures showed the absorption maxima for both spectra are higher by 25 to 30% compared to uranyl nitrate alone and this enhanced absorption indicates there is interaction of the IL with uranium. However, the 1 : 1 and 1 : 2 of U : [BMIM][PF_6] mixtures showed absorbance maxima similar to the uranyl nitrate (see Fig. 3). The [BMIM][PF_6] compound is the least soluble of the three ILs in this study and therefore it was difficult to determine the presence of a U-IL complex using this method. In the presence of uranyl ion and [EtPy][CF_3COO] there was a shift in absorbance maximum to 408 nm from 402 and to 419 from 425 nm compared to uranyl nitrate alone. The observed shift to lower energy is most probably due to the electron withdrawing ability of the fluoride ions, confirming the complexation of uranyl ion with the IL.

Electrospray Ionization Mass Spectrometry (ESI-MS) Analysis

In Fig. 4 uranium nitrate in water showed peaks at 270 (m/z), 287 (m/z) and 305 (m/z) due to the uranium ion and its association with hydroxyl ions in water. In the

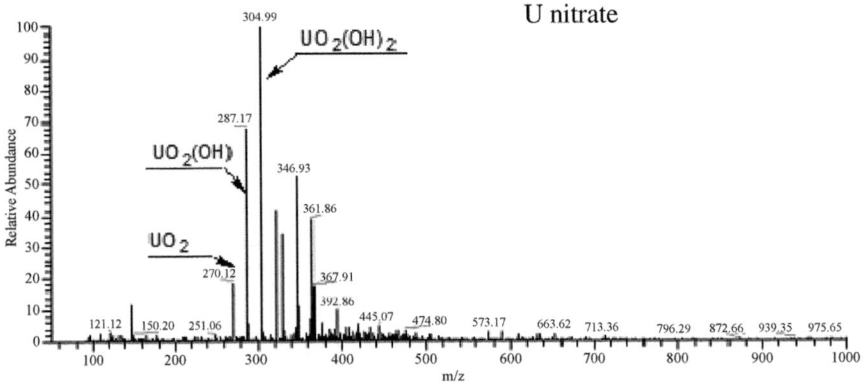

Fig. 4 Mass spectroscopy of uranium nitrate in water under positive mode

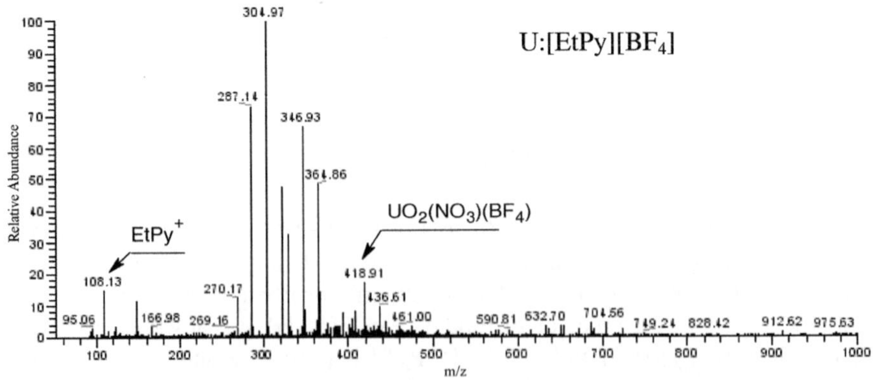

Fig. 5 Mass spectroscopy of 1 : 1 U : [EtPy][BF_4] mixture under positive mode

MS of 1 : 1 U : [EtPy][BF_4] (Fig. 5), 108 (m/z) is the molecular ion peak of N-ethyl pyridinium cation. The small peak at 418 (m/z) may come from the anion exchange between uranium nitrate and [EtPy][BF_4].

In Fig. 6, the ESI-MS spectrum under negative mode for uranyl nitrate shows peaks at 365 (m/z) $[UO_2(NO_3)(OH)_2]^-$ and 455 (m/z) $[UO_2(NO_3)_3]^-$. They are due to uranyl ion complexation with nitrate and hydroxyl ions. Figure 7 shows the ESI-MS spectrum under negative mode for U : [EtPy][CF_3COO]. In addition to the peaks observed in Fig. 6 for uranyl nitrate, there is a molecular ion peak at 413 (m/z). We believe the appearance of this peak is due to formation of $[UO_2(NO_3)(OH)(H_2O)(COO)]^-$ species with loss of the CF_3 moiety.

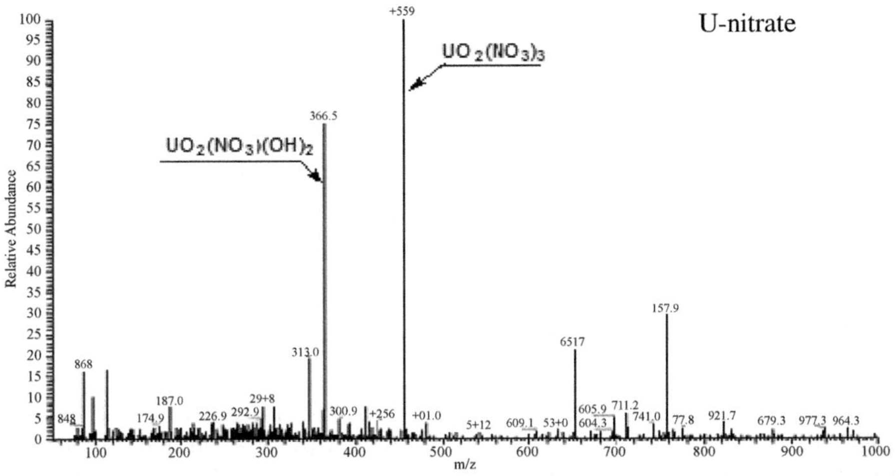

Fig. 6 Electrospray ionization (ESI-MS) spectra under negative mode showing the molecular ions produced by uranyl nitrate solution

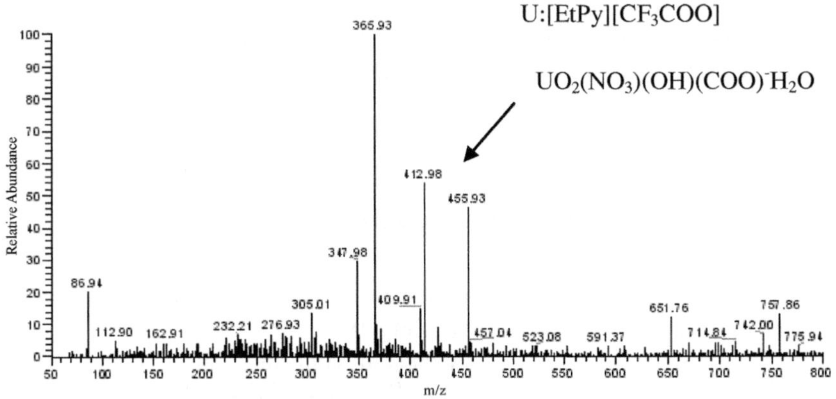

Fig. 7 Electrospray ionization (ESI-MS) spectra undernegative mode showing the molecular ions produced by U : [EtPy][CF_3COO] solution

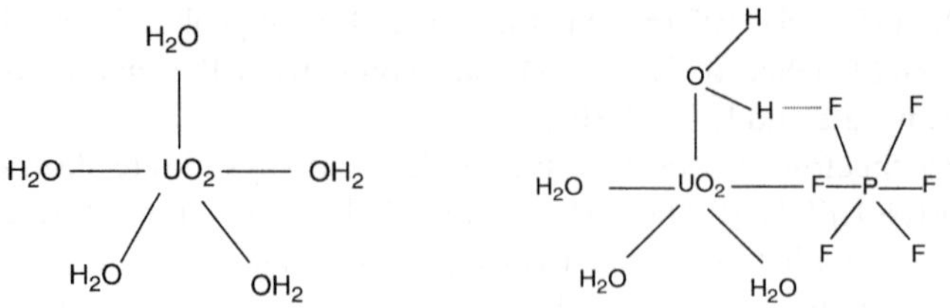

a) $\mathbf{UO_2^{2+}}$ b) $\mathbf{UO_2PF_6^{+}}$

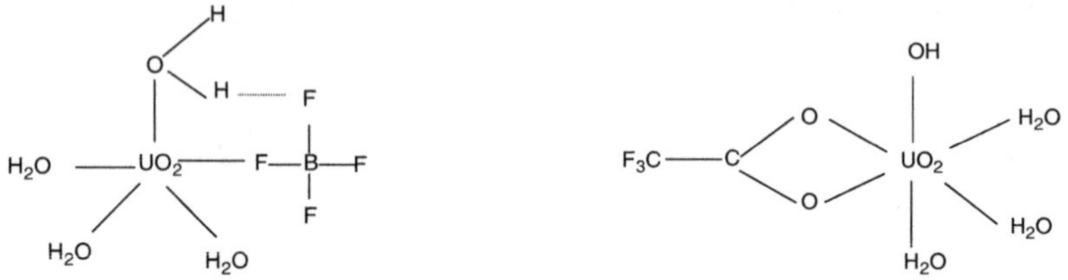

c) $\mathbf{UO_2BF_4^{+}}$ d) $\mathbf{UO_2(OH)(CF_3COO)}$

Fig. 8 Proposed structures of U-Ionic Liquid complexes

X-ray Absorption Spectroscopy

X-ray Absorption Near-edge Spectroscopy (XANES) analysis of the absorption edge energies for U(VI)-nitrate and mixtures of U(VI)-nitrate and various ILs revealed that they have the samc absorption edge energy (17,175 eV). This confirms that the U(VI) added to the IL's solution was present as U(VI) species and not reduced to U(IV). Extended X-ray absorption fine structure (EXAFS) spectroscopy analysis of the U ILs samples showed the following complexes formed between uranyl and [EtPy][BF_4], [EtPy][CF_3COO], and [BMIM][PF_6]: monodentate complex between uranyl and BF_4^-, and PF_6^-, bidentate complex between uranyl and CF_3COO. Based on the ESI-MS and EXAFS results we propose the following structure (Fig. 8).

Effect of ILs on Growth of *Clostridium* sp.

Addition of ILs to bacterial growth medium affected the growth of *Clostridium* sp. as evidenced by decrease in optical density (Fig. 9).

The optical density decreased with increase in concentration of [BMIM][PF_6]. Similar effects were observed with [EtPy][BF_4] and [EtPy][CF_3COO] (Fig. 9B and C). In comparison with [EtPy][BF_4], the two ILs, [BMIM][PF_6] and [EtPy] [CF_3COO], showed higher inhibition on the growth of *Clostridium sp.* Addition

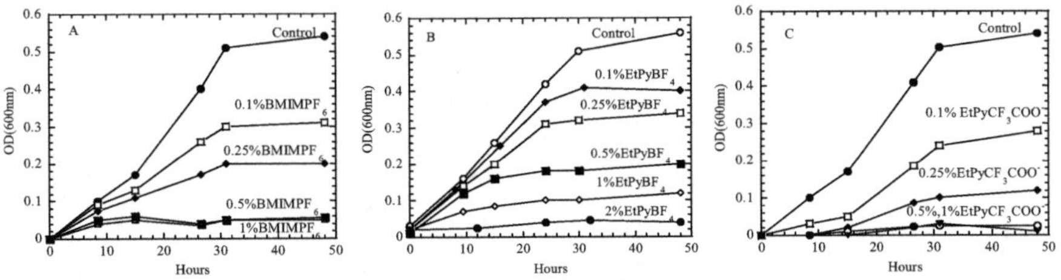

Fig. 9 Effects of IL on the optical density of *Clostridium* sp.

of 0.1% of [BMIM][PF_6] or [EtPy][CF_3COO], the optical density decreased to 0.3 and 0.25 respectively. As the concentration of these two ionic liquids increased >0.5%(V/V), no growth of bacteria was observed.

After 48 h incubation the pH of the medium changed from 5.5 to 3 due to organic acids produced from glucose metabolism. However, in presence of ILs, the growth was inhibited, and less glucose was consumed, thus less acids produced and the pH changed slowly. The drop in pH with the high concentrations of ILs may be due to the hydrolysis of ILs. Commensuration with growth, the amount of total gas, CO_2, H_2 production was less in ILs containing samples. We found that the growth of *Clostridium* sp. and *Pseudomonas fluorescens* was affected by one more of these ILs (Wang et al. 2011; Zhang et al. 2011; Nancharaih and Francis 2011).

Effects of Ionic Liquids on Reduction of U(VI) by *Clostridium* sp.

Addition of uranyl nitrate to growing culture of *Clostridium* sp. resulted in the rapid reduction of U(VI) to U(IV) which precipitated from solution (Fig. 10A). In the presence of [EtPy][BF_4] reduction of U(VI) proceeded similar to uranyl nitrate but the concentration of reduced(IV) in solution was 3-fold higher and it precipitated much more slowly (Fig. 10B). In presence of [BMIM][PF_6] U(VI) reduction followed the same pattern as uranyl nitrate (Fig. 10C). Uranium which formed a bidentate complex with [EtPy][CF_3COO] was not reduced (Fig. 10D).

Analysis of Uranium (VI) and U(IV) by UV-vis Spectroscopy

Uranium in the culture sample was extracted with citiric acid under anaerobic conditions and the absorption spectra of U(VI) and U(IV) were determined using a Hewlett Packard 8453 diode array scanning UV-vis spectrophotometer. Before bacterial reduction, the maximum absorbance was at 435 nm which is the characteristic absorbance of U(VI). After bioreduction, the absorbance shifted to 555 and 665 nm, which indicated the presence of U(IV) (Fig. 11). XANES analysis showed energy shift from 17,175 to 17,169 eV indicating the reduction of U(VI) to U(IV).

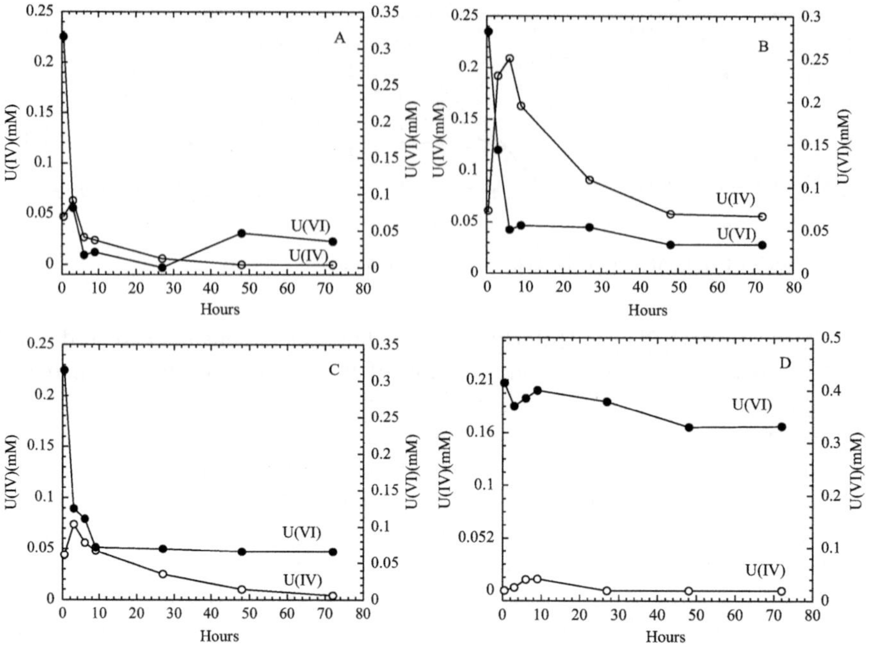

Fig. 10 Reduction of U(VI) by *Clostridium* sp. in presence of various ILs: (**A**) uranyl nitrate; (**B**) [EtPy][BF_4]; (**C**) [BMIMPF_6]; (**D**) [EtPy][CF_3COO]

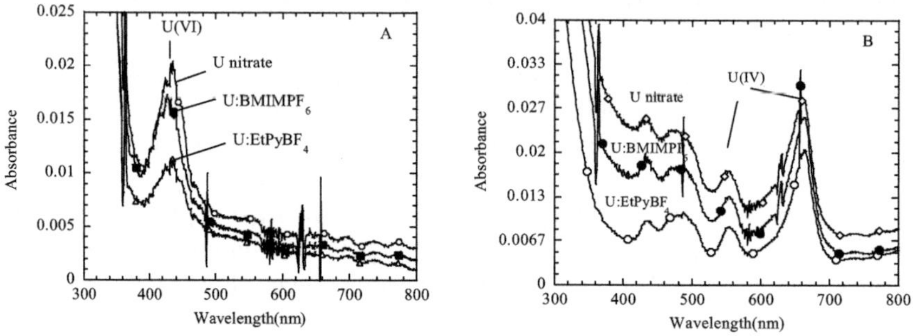

Fig. 11 UV-vis spectroscopy of precipitates extracted by citric acid (**A**) before bioreduction; (**B**) after bioreduction

These results show that the ILs affected the growth of *Clostridium* sp. as evidenced by decrease in optical density, changes in the pH of the medium, total gas production and rate and extent of reductive precipitation of uranium.

Acknowledgement We thank C.J. Dodge for assistance in XAS analysis and S.V. Malhotra for advice. This research was supported by Laboratory Directed Research and Development (LDRD) program at Brookhaven National Laboratory by the U.S. Department of Energy under contract No. DE-AC02-98CH10886 and in part by the World Class University (WCU) program through the National Research Foundation of Korea funded by the Ministry of Education, Science and Technology (R31 – 30005).

References

Anastas, P.T.; Warner, J.C. *Green-Chemistry-Theory and Practice;* Oxford University Press: New York, 1998.

Gaillard, C.; Azzi, A.E.; Billard, I.; Bolvin, H.; & Hennig, C. 2005, Inorg. Chem. 44, 852.

Nancharaiah, Y.V. and Francis, A.J. 2011, Bioresour. Technol. 102, 6573.

Visser, A.E.; Jensen, M.P.; Laszak, I.; Nash, K.L.; Choppin, G.R.; Rogers, R.D. 2003, Inorg. Chem., 42, 2197.

Wang, H.; Malhotra, S.V.; Francis, A.J. 2011, Chemosphere 82, 1597.

Zhang, C.; Malhotra, S.V.; Francis, A.J. 2011, Chemosphere 82, 1690.

Part 4
Analytics and Sensors for Uranium and Radon

Environmental Analysis of Uranium with Recombinant Antibodies

Diane A. Blake, Xiaoxia Zhu, Bhupal Ban

Abstract. The techniques of antibody phage display were used to generate antibody fragments that specifically recognized U(VI) in complex with the chelator, 2,9-dicarboxyl-1,10-phenanthroline (DCP). The single-chain fragment antibodies (scFvs) obtained by these procedures were able to bind to uranium-loaded DCP in the presence of >1000-fold molar excess of metal-free DCP. The ability of these antibodies to tolerate high levels of chelator will enhance their ability to accurately measure uranium in groundwater that has been amended to enhance U(VI) extraction. Further engineering of these antibodies will utilize yeast display techniques.

Introduction

Most long-term plans for reducing worldwide carbon emissions rely on nuclear power to supplement more intermittent but "greener" sources of electrical power,

Diane A. Blake
Department Biochemistry, Tulane University School of Medicine, 1430 Tulane Avenue, New Orleans, Louisiana 70112 USA

Xiaoxia Zhu
Department Biochemistry, Tulane University School of Medicine, 1430 Tulane Avenue, New Orleans, Louisiana 70112 USA
School of Public Health, Nantong University, Nantong 226007, China

Bhupal Ban
Department Biochemistry, Tulane University School of Medicine, 1430 Tulane Avenue, New Orleans, Louisiana 70112 USA

B. Merkel, M. Schipek, *The New Uranium Mining Boom*
DOI 10.1007/978-3-642-22122-4 © Springer 2011

which include solar cells, wind turbines, and tidal/wave action. The primary fuel for nuclear power is uranium, and demand continues to outstrip supplies. Uranium is usually found in relatively low concentrations in natural ore deposits, and it must be harvested either by conventional mining and above-ground processing or by in situ leaching activities. The ability to detect and quantify uranium is important both for monitoring remediation at previously contaminated sites and for protecting personnel and the environment during current and future mining operations.

Immunoassays have numerous advantages for quantifying the levels of environmental contaminants. Immunoassays are rapid, inexpensive and simple to perform. Rugged, field-portable immunosensors have been developed by many research groups, including our own, to provide near real-time data about contaminant levels in the field (Yu et al. 2005; Melton et al. 2009). For uranium analysis, our laboratory has generated monoclonal antibodies that bind tightly and specifically to chelated U(VI) (Blake et al. 2004). These antibodies, when incorporated into immunosensors, accurately measured the U(VI) concentrations in groundwater at a uranium mine tailings site (Melton et al. 2009).

The success of these sensors was dependent in large part upon the binding properties of the monoclonal antibody used in the environmental immunoassay. This antibody, called 12F6, bound to UO_2^{2+} in complex with the chelator 2,9-dicarboxyl-1,10-phenanthroline (DCP). The equilibrium association constant (K_a) for the interaction between 12F6 and the UO_2^{2+}-DCP complex was $1.1 \times 10^9\ M^{-1}$. The 12F6 antibody also bound relatively tightly to metal-free DCP with a K_a of $1.3 \times 10^6\ M^{-1}$ (Blake et al. 2004). The relatively high affinity that 12F6 exhibited for the metal-free chelator meant that the DCP concentrations in our field assays had to be carefully controlled at a low concentration (200 nM). Decreasing the DCP concentration in the assay provided low numbers because there was not enough chelator to complex all of the uranium in the sample. Increasing the DCP concentration yielded false positives, because the DCP itself interacted with the 12F6 monoclonal antibody. The development of new antibodies less sensitive to the concentration of DCP would enhance our assay performance. These new antibodies would increase the precision of our environmental assays and make them easier for less experienced analysts to use in the field. The goal of these studies was to develop new antibodies with the properties needed to improve the analytical method. The new antibodies should retain their tight binding to the UO_2^{2+}-DCP complex but show much weaker binding to metal-free DCP.

Recombinant antibody technology has provided an alternative method to engineer low-cost antibodies with specific binding properties (Hoogenboom 2005). A growing number of these recombinant antibodies have been produced for the analysis of food and environmental samples (Chiu et al. 2000; Tout et al. 2001; Li et al. 2006). In this chapter, we describe the generation of a phage displayed immune library and the selection of single chain variable fragment (scFv) antibodies with binding properties tailored for use in our immunosensors. One of these antibodies was subsequently used to develop a sensor-based assay for uranium in groundwater samples.

Materials and Methods

Materials

2,9-Dicarboxyl-1,10-phenanthroline (DCP) was purchased from Alpha Aesar, (Heysham, UK). ACS grade uranyl acetate was a product of Mallinckrodt Chemical Works, St. Louis, MO, USA). UO_2^{2+}-DCP-BSA conjugates were available from previous studies (Blake et al. 2004; Melton et al. 2009).

The pSD3 plasmid used to construct the phage display library was a kind gift of Prof. Garry Whitelam, University of Leicester, Leicester, UK. This vector contains sites for the insertion of antibody heavy- and light-chain variable regions at the N-terminal of coat protein III (ΔgIII) of bacteriophage M13. Between the cloning sites and the gene III coat protein are inserted 3 additional features: a c-myc tag for detection, a $(His)_6$ peptide for purification, and an amber codon to facilitate soluble scFv production (Li et al. 2000). The yeast display vector used in these studies, pDNL6-CBH1, was a kind gift of Dr. Andrew Bradbury, Los Alamos National Laboratory, Los Alamos, NM, USA. This vector contains insertion sites for the scFv coding region as a fusion protein with Agap2 yeast surface protein and tags for detection and purification.

Preparation of Antibody Library and Selection of scFv Antibodies to Chelated Uranium

The generation of a phage-displayed scFv library from the spleen tissue of rabbits immunized with UO_2^{2+}-DCP-protein conjugates was recently reported by our laboratory (Zhu et al. 2011). The method is briefly outlined in Fig. 1.

Selection for scFv antibodies that bound tightly to the UO_2^{2+}-loaded chelator but not to the metal-free chelator was accomplished by first loose and then increasingly stringent selection procedures. Maxisorp immunotubes (75 × 12 mm, Nunc, Denmark) were coated with DCP-BSA in HBS (137 mM NaCl, 3 mM KCl, 10 mM HEPES, pH 7.4) at concentrations that decreased sequentially from 100 to 1 μg/ml and the tubes were incubated at 4°C overnight. The next morning tubes were washed three times with HBS, and then blocked with HBS containing 3% non-fat dry milk w/v for 1.5 h at 37°. Uranyl acetate (1 μM) was added to the blocked tubes to load the immobilized DCP with uranium. Before binding to antigen-coated tubes, 1.5 ml of phage library (10^{13} cfu/ml) was pre-incubated in HBS buffer containing sequentially increasing concentrations of DCP (0.05 to 0.5 μM in HBS containing 3% bovine serum albumin). The preincubated phage were then added to the immunotubes and incubated at 37°C for 2 h with rotation. Phage not binding the antigen-coated tubes were removed by washing with PBST (PBS containing 0.05% Tween 20), followed by washing with HBS. The bound phage were eluted with 1 ml of 100 mM triethylamine at room temperature with rotation for

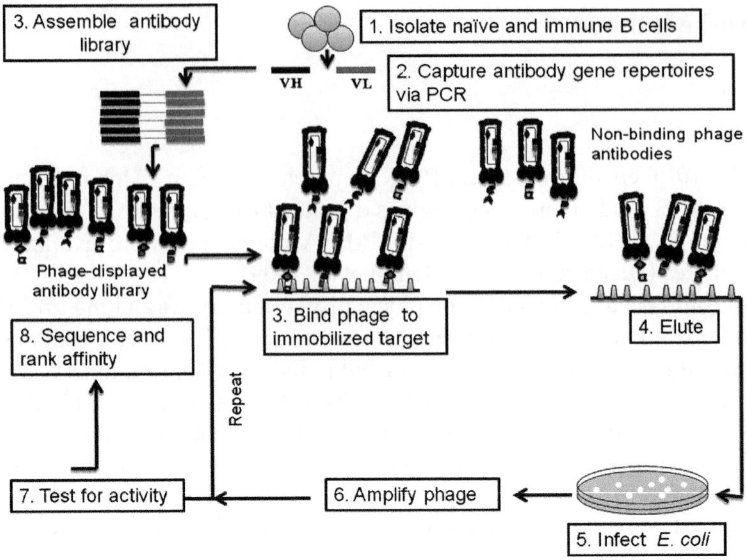

Fig. 1 Generation of an immune scFv phage display library and selection procedures. The immune repertoire of the immunized animal is captured by PCR of antibody-coding genes (Steps 1–3). scFv antibodies with the desired binding properties are then enriched by sequential rounds of selection (Steps 3–6). After the selection process yields no further enrichment, individual phage are isolated and soluble scFv antibodies are characterized by sequencing and affinity analysis (Steps 7–8)

10 min and then the solution was immediately neutralized by adding 0.5 ml of Tris-HCl (1 M, pH 7.4).

An aliquot of 0.75 ml of the neutralized phage were used to infect cultures of log-phase *E. coli* TG1 cells. An aliquot of the infected bacterial was used in serial dilutions to quantify the number of phage eluted. This value was designated "phage output". The remaining infected TG1 cells were collected by centrifugation and plated onto a 150 mm culture dish. Phage were prepared from the recovered colonies and titered by infection. This value was used in determining the "phage input" for later rounds of selection. The resultant phage pool at each round was tested by phage ELISA to evaluate the success of the panning experiment (Zhu et al. 2011).

Expression of Soluble scFv and Affinity Ranking by ELISA

Unique bacterial clones were inoculated into culture medium and grown to early exponential phase. They were subsequently induced with 1 mM isopropylthio-β-D-galactoside at 30°C for 16 h. The following day, the culture was centrifuged at 10,000 g for 10 min at 4°C. The cell pellet was resuspended in 1/50 volume of lysis buffer (50 mM sodium phosphate, pH 8.0, 0.3 M NaCl, 1 mg/ml lysozyme and protease inhibitors) and incubated on ice for 30 min. After the addition of

$MgCl_2$ (10 mM) and DNase I (20 µg/m), the solution was incubated for 20 min at room temperature and then centrifuged at 12,000 g for 30 min. The supernatant fraction was filtered and purified using His-Select nickel affinity Gel (Sigma, Saint Louis, MO), according to the manufacturer's instructions. The purified proteins were dialyzed into HBS and used immediately or stored at –70°C in 50% glycerol/50% HBS.

ELISA assays were performed with microtiter plates coated with UO_2^{2+}-DCP-BSA (2 µg/ml), blocked with 3% BSA and washed. The ability of the purified proteins to bind to the coated plates was assessed using indirect ELISA as described in (Blake et al. 2005). Those scFvs that showed a signal at least 5-fold above background were also tested by competitive ELISA. For this assay, the scFvs were incubated at a fixed concentration in the wells of the plate that contained 1 µM DCP and varying concentrations of UO_2^{2+}. After incubation for 1 h at room temperature, plates were washed and the bound antibodies were detected by horseradish peroxidase (HRP) labeled anti-myc antibody (9E10) (Bethyl Laboratory, Montgomery, TX USA), followed by incubation with SuperBlue TMB microwell peroxidase substrate (KPL, Gaitherburg, MD USA).

Incorporation of the scFvs into Immunosensors

The antibodies that showed the lowest IC_{50} values in competitive ELISA were subsequently incorporated into immunosensors based on kinetic exclusion analysis, already available in our laboratory. Principles of this method and details of the sensors have been published elsewhere (Blake et al. 1999; Yu et al. 2005; Melton et al. 2009). Acidified environmental ground water samples were available from a previously published study and standard curves for UO_2^{2+} in HBS and in diluted Rifle Artificial Ground Water (AGW) were prepared as made as previously described (Melton et al. 2009). The final concentrations in the calibration assay mixtures applied to the flow cell were as follows: scFv antibody, 0.833 nM; Cy5-labled anti-myc antibody (9E10), 10 nM; BSA, 50 µg/ml; DCP, 1 µM; UO_2^{2+} standard solutions, 0–100 nM. The signals (delta) at varying $UO_2^{2}+$ concentrations were fit using SlideWrite software (Advanced Graphics Software, Carlsbad, CA) and the following binding equation: $y = a_0 - (a_1 \times x)/(a_2 + x)$, where y is the delta at any given ligand concentration, x is the ligand concentration, a_0 is the delta when no ligand was present (y intercept) and all of the active antibody is available to bind to the beads, a_2 is the IC_{50}, and a_1 is the total change in the value of delta as the x goes from zero to infinity.

Results

An outline of the method for library production is shown in Fig. 1. Rabbits were used as a source of genetic material because: 1) a relatively small number of

V-region-specific primers are required for the generation of rabbit antibody fragment libraries; 2) hapten and peptide epitopes that do not elicit a good immune response in mice often show enhanced immunogenicity in rabbits; and 3) rabbit antibodies often have higher specificity and affinity than do mouse antibodies. Two rabbits were therefore immunized until their sera showed strong reactivity with the UO_2^{2+}-DCP complex in competitive ELISA (data not shown). Total RNA was isolated from pooled spleen tissue of two immunized rabbits for the synthesis of cDNA. A set of degenerate primers was used for isolation of rabbit immunoglobulin variable genes (Li et al. 2000) and subsequent library production. Multiple aliquots of the assembled pSD3 plasmid were transformed into *E. coli* TG1, and the transformants were pooled. Library diversity was estimated by determining the percentage of unique *Bst1N*1 fingerprints in a selected number of clones. Based on the total amount of DNA used for transformation, the calculated transformation efficiency and estimates of library diversity, the library size was calculated to be $\sim 1 \times 10^7$.

This library was subjected to 5 rounds of selection, as shown schematically in Fig. 1. The goal of this process was to select antibodies that bound with high affinity to UO_2^{2+}-DCP but with low affinity to metal-free DCP. The selection strategy combined decreasing the amount of immobilized UO_2^{2+}-DCP-BSA on the solid substratum while simultaneously increasing the concentration of metal-free DCP in the incubation solution. The ability of the phage particles to bind to an immobilized UO_2^{2+}-DCP-BSA conjugate at after the 2nd–5th round of panning is shown in Fig 2.

After the 5th round of panning, ~ 150 clones were randomly selected for further characterization. These characterizations included: 1) Phage ELISA of monoclonal

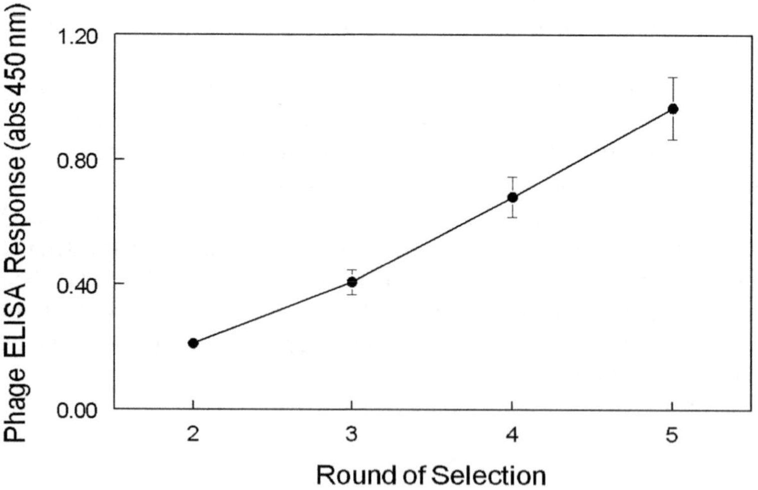

Fig. 2 The ability of phage particles to bind to UO_2^{2+}-DCP-BSA increased as the stringency of the selection increased. A fixed number of phage particles were tested for their ability to bind to microwell plates coated with 2 μg/mL UO_2^{2+}-DCP-BSA. Color was developed via an enzyme-labeled anti-c-myc antibody

phage particles; 2) Restriction enzyme analysis of plasmid DNA to identify unique clones; 3) Sequence analysis of antibody variable light- and heavy-chain genes from isolated plasmids. These further selection criteria narrowed the clones of interest to 11, and soluble scFvs were prepared from these constructs.

These soluble antibody fragments were analyzed using both indirect ELISA, which measures the ability of the antibody to bind to the UO_2^{2+}-DCP-BSA conjugate, and competitive ELISA, which measures the ability of the antibody to recognize the environmental ligand to be analyzed in the field. Representative competitive ELISAs for a subset of the clones under study are shown in Fig. 3, below. The two clones with the lowest IC_{50} values, 3A and 5B, were chosen for testing with the immunosensor.

Sensor studies showed that an scFv prepared from clone 3A had the highest affinity for the UO_2^{2+}-DCP complex, with an IC_{50} of 19.6 nM. Because of the depletion experiments performed on the library during the panning process, this scFv also tolerated very high concentrations of metal-free DCP, with an IC_{50} of 23,500 nM.

Uranium spiked into artificial groundwater and environmental water samples available from a previous study (Melton et al. 2009) were analyzed in our immunosensor using the 3A scFv. The spikes into groundwater were done multiple times, with an n of 3 for each experimental point. The error for any single spike was always below 20%; when the CV's at each point on a standard curve were averaged, the averages ranged from 4.7–6.2% in replicate experiments. The reproducibility in assays with the recombinant antibody was similar to that obtained

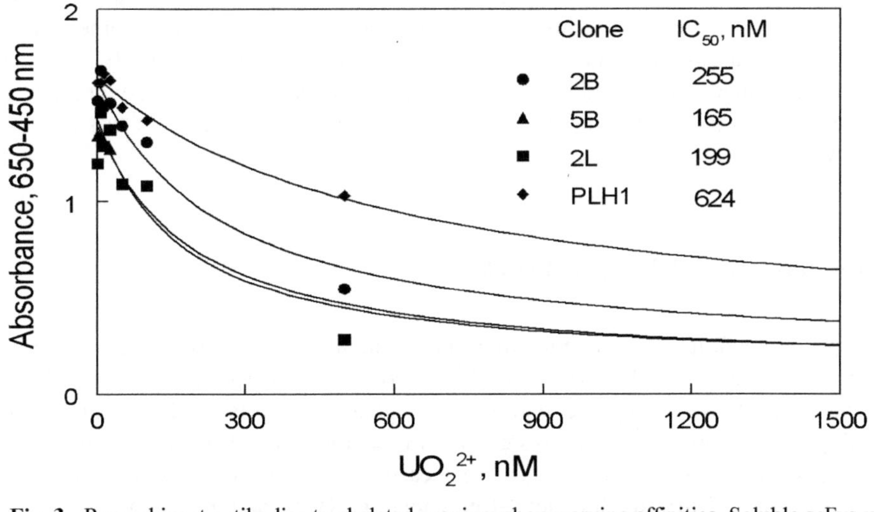

Fig. 3 Recombinant antibodies to chelated uranium show varying affinities. Soluble scFvs were mixed with 2 μM DCP and varying concentrations of UO_2^{2+} in microwells coated with UO_2^{2+}-DCP-BSA. After a wash step, the amount of recombinant antibody remaining in each well was quantified using an anti-c-myc-HRP conjugate. The concentration of UO_2^{2+} that inhibited color formation by 50% (IC50) is shown for each antibody

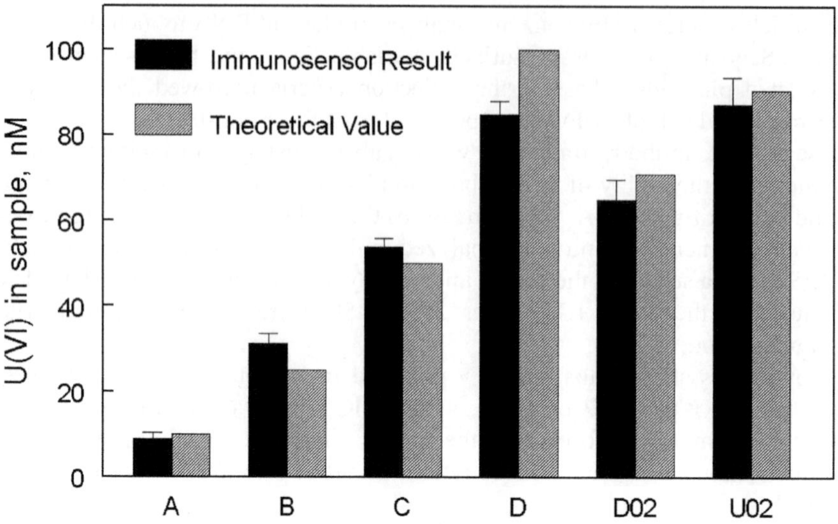

Fig. 4 Analytical recovery of uranium in spiked samples and environmental groundwater samples

previously using monoclonal antibodies (Melton et al. 2009). The minimal level of detection, defined as 3SDs below the zero value (Glass et al. 2007) was 2.2 nM, well below the US EPA action limit of 126 nM (US EPA 2000). The theoretical and actual recovery of UO_2^{2+} from spiked samples and groundwater collected at a uranium mine tailings site is shown in Fig. 4. The concentrations of uranium measured by the immunosensor were well-correlated to the spiked uranium samples (A–D) in the Fig. 4. The immunosensor data was also in good agreement with the values of uranium in the two environmental samples (D02 and U02), which were determined by an independent contractor using kinetic phosphorescence analysis (Brina and Miller 1993). The average analytical recovery in this experiment was 98.9%.

Conclusions and Future Work

The present studies demonstrate that recombinant scFvs with properties engineered for specific applications (i.e., biosensor-based measurement of metals in groundwater) can be prepared if the correct genetic material and techniques are employed. The phage display system permitted the generation of proteins with very specific binding properties (in this case, high affinity for a metal-chelate complex and low affinity for metal-free chelator). Because the recombinant antibodies described herein were expressed and purified from bacterial cultures, they will be much less expensive to mass-produce than monoclonal antibodies. The highest affinity antibody produced in this study, 3A, has a 31-fold higher tolerance

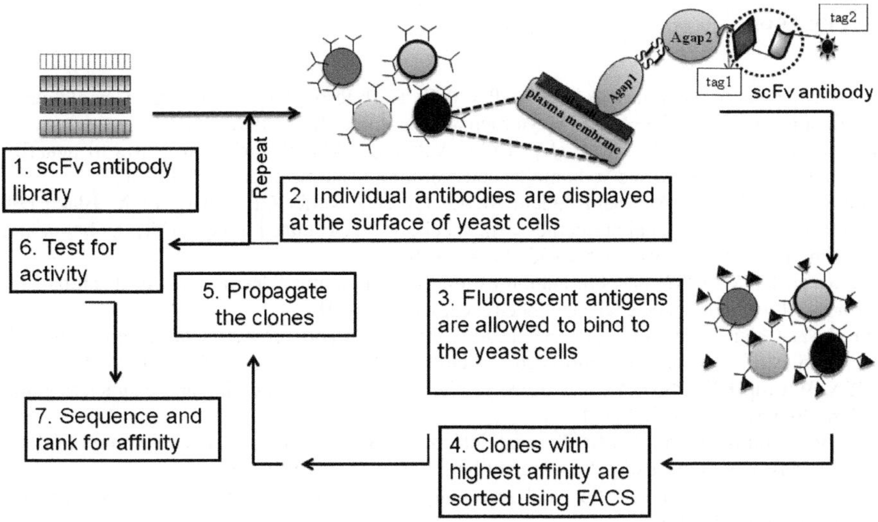

Fig. 5 Yeast display as a method to further enhance the binding properties of antibodies used for environmental analysis. The scFv is displayed as a fusion to the Agap2 protein on the surface of the yeast cell. The Agap2 protein is used by yeast to mediate cell-cell contacts and the Agap2 protein projected the scFv away from the cell, thus minimizing interactions with the presenting cell. Affinity selection is accomplished using fluorescent activated cell sorting (FACS)

for metal-free DCP than the monoclonal antibody used in our previous field studies (Melton et al. 2009). The use of higher chelator concentrations will facilitate the extraction of U(VI) from its natural complexants in groundwater. The high DCP concentrations tolerated by this antibody will also be invaluable in the development of assays for uranium during in situ leaching operations, where uranium complexants are added to groundwater in very high concentrations. The recombinant scFvs isolated in these studies will be thus the basis for improved rapid and affordable assays for the detection of residual uranium in a wide variety of environmental water samples. These antibodies could also be useful in the development of rapid tests for uranium in serum, hair and urine that could be used to monitor worker safety.

Future experiments in the development of recombinant antibodies will take advantage of a yeast display system first described by Wittrup and coworkers (Gai and Wittrup 2007) that will facilitate the identification of scFvs with the desired binding properties. The yeast display system, shown graphically in Fig. 5, presents the scFv antibodies at the surface of yeast cells. Thousands to millions of different antibodies displayed on individual cells may then be subjected to high throughput screening using a fluorescence activated cell sorter, in methods such as those described by as described by (Ayriss et al. 2007).

Acknowledgments This work was supported by a grant to D.A. Blake from the Office of Science, Department of Energy (DE-SC0004959). The authors gratefully acknowledge the gift of the pSD3 vector from Dr. Garry Whitelam, University of Leicester, Leicester, UK.

References

Ayriss J, Woods T, Bradbury A, Pavlik P (2007) High-throughput screening of single-chain antibodies using multiplexed flow cytometry. J Proteome Res 6: 1072–1082

Blake RC 2nd, Pavlov AR, Blake DA (1999) Automated kinetic exclusion assays to quantify protein binding interactions in homogeneous solution. Anal Biochem 272: 123–134

Blake RC 2nd, Pavlov AR, Khosraviani M, Ensley HA, Keifer GE, Yu H, Li X, Blake DA (2004) Novel monoclonal antibodies with specificity for chelated uranium(VI): Isolation and binding properties. Bioconj Chem 15: 1125–1136

Brina R, Miller AG (1993) Determination of uranium and lanthanides in real-world samples by kinetic phosphorescence analysis. Spectroscopy 8: 258–301

Chiu Y-W, Chen R, Li QX, Karu AE (2000) Derivation and properties of recombinant Fab antibodies to coplanar polychlorinated biphenyls. J Agric Food Chem 48: 2614–2624

Gai SA, Wittrup KD (2007) Yeast surface display for protein engineering and characterization. Curr Opin Struct Biol 4: 467–473

Glass TR, Ohmura N, Saik H (2007) Least detectable concentration and dynamic range of three immunoassay systems using the same antibody. Anal Chem 79: 1954–1960

Hoogenboom HR (2005) Selecting and screening recombinant antibody libraries. Nature Biotechnol 23: 1105–1116

Li T, Zhang Q, Liu Y, Chen D, Hu B, Blake DA, Liu F (2006) Production of recombinant scFv antibodies against methamidophos from a phage-display library of a hyperimmunized mouse. J Agric Food Chem 54: 9085–9091

Li Y, Cockburn W, Kilpatrick J, Whitelam G (2000) High affinity scFvs from a single rabbit immunized with multiple haptens. Biochem Biophys Res Comm 268: 398–404

Melton SJ, Yu H, Williams KH, Morris SA, Long PE, Blake DA (2009) Field-based detection and monitoring of uranium in contaminated groundwater using two immunosensors. Environ Sci Technol 43: 6703–6709

Tout NL, Yau KTF, Trevors JT, Lee H, Hall JC (2001) Synthesis of ligand-specific phage-display scFv against the herbicide picloram by direct cloning from hyperimmunized mouse. J Agric Food Chem 49: 3628–3637

Yu H, Jones RM, Blake DA (2005) An immunosensor for autonomous in-line detection of heavy metals: Validation for hexavalent uranium. Int J Environ Anal Chem 85: 817–830

Zhu X, Kriegel AM, Boustany CA, Blake DA (2011) Single-chain variable fragment (scFv) antibodies optimized for environmental analysis of uranium. Anal Chem 83: 3717–3724

The Routine Determination of Uranium Activity in Natural Rocks and Minerals by Gamma Spectrometry

Fatima Zahra Boujrhal, Rajaâ Cherkaoui El Moursli

Abstract. Uranium activity in various natural materials is determined by gamma spectrometry. The content of uranium in these materials, can reach some tens ppm, and which is more or less in radioactive equilibrium with its daughters, principally radium and radon (immediate daughter of radium). In this work, we present the method used to determine the content uranium and its daughters in natural rocks and minerals. Gamma spectrometry with NaI (Tl) scintillation detector 3×3", with well of 1×2", and method of regions are used. Results obtained by this technique are in good agreement with those given by Inductively Coupled Plasma Mass Spectrometry (ICP-MS).

Introduction

Uranium is a naturally occurring element found in low level within all rocks, soil, and water. Significant concentration of it occurs in some substance such as phosphate rock and mineral.

Uranium is weakly radioactive and remains so because of its long physical half-live equal to age of the earth (4.45 billion years for uranium-238) and is attempt to become stable element. It will undergo a series of 14 radioactive decays until it eventually becomes lead-206, a stable non-radioactive element.

Fatima Zahra Boujrhal
Université Sultan Moulay Slimane, Faculté des Sciences et Techniques, Laboratoire de Gestion et de Valorisation des ressources naturelle (LGVRN), B.P. 523, Béni Mellal, Morocco

Rajaâ Cherkaoui El Moursli
Université Mohammed V, Faculté des Sciences, Laboratoire de Physique Nucléaire, Av. Ibn Battouta (LPNR), B.P. 1014, Rabat, Morocco

B. Merkel, M. Schipek, *The New Uranium Mining Boom*
DOI 10.1007/978-3-642-22122-4 © Springer 2011

Uranium 238 is more or less in radioactive equilibrium with its radioactive daughters, principally radium-226 and radon-222. This uranium isotope is used to estimate the age of the earliest igneous and sedimentary rocks.

Phosphate mining and production of phosphate fertilizers are the most important items of the Moroccan industry. Characterization and controlling the quality of these products by various physical-chemical analysis is necessary. One of these is the gamma spectrometry for determination of uranium content. This technique, described in: Berrada et al. (1992, 1995) and Boujrhal et al. (1999) is developed in our laboratory as routine determination of Uranium and its daughters is natural sample specially the phosphates.

Material and Method

Material

Various natural materials are selected from different environment sites; gels from water sources, and phosphates and fossilized teeth from sedimentary deposits, in order to determine their uranium activity. Only one artificial material type, phosphogypsum (residue of the sedimentary phosphate industry), is analyzed here.

Techniques

Gamma Spectrometry

A gamma spectrometry is first developed in our laboratory in order to determine the content of uranium and some of its daughters, principally radium and radon, in Moroccan sedimentary phosphate and their by-products.

The detector used for this technique is NaI (Tl) scintillation detector because of its bad resolution. This detector is crystal scintillation $3\times3"$, with the well of $1\times2"$. The samples to be analyzed in Plexiglas's containers (2.5×5 cm) are putted in this well. The obtained spectra are divided in three regions, but only for the two first regions, a system of two equations to two unknowns is elaborated.

$$S_1/tc = m(\alpha_1 U_1 + \beta_1 Ra_1)$$

$$S_2/tc = m(\alpha_2 U_2 + \beta_2 Ra_2)$$

where,

S_1 and S_2: Areas of 1st and 2nd regions of sample analyzed respectively,
tc: Time of counting,
m: Mass of sample analyzed,

U: Uranium content in the sample in ppm,
Ra: Radium content in the sample in ppm,
α_1 and α_2: contribution coefficients of uranium respectively in regions 1 and 2,
β_1 and β_2: contribution coefficients of radium respectively in regions 1 and 2.

The resolution of this system follows to the determination of Uranium and Radium sample analyzed.

Note that the four means contribution coefficients are determined by the resolution of this system using eight standards with a known uranium and radium contents.

On the other hand, the radon emanation rate is determined by

$$T(\%) = 100(Ra - Rn)/Ra.$$

Ra: Activity of radium –226 in the sample,
Rn: Activity of radon –222 in the sample.

Inductively Coupled Plasma Mass Spectrometry (ICP-MS)

Trace element concentrations of samples can be determined ICP-MS after an acid digestion by a closed microwave oven.

For this technique, about 10 to 150 mg of sample needs to be digested before its introduction as aerosol via the nebuliser to the torch operating with argon gas as source of ionization for the majority of elements. The efficiency of this ionization depends on the charge of the solution introduced, the droplet size produced by the nebuliser and the plasma temperature. The plasma is generated using the energy of an induction bobbin to ionize the argon gas.

The ions produced by the ICP are collected by mass spectrometer giving, knowing the mass/ion charge ratio, the masse of the element of interest.

Results

Uranium and radium content as like radon emanation rate determined by gamma spectrometry are represented in Table 1, whereas the uranium content analyzed by ICP-MS (Table 2) are represented in Table 2.

Uranium content in all samples studied is less than 130 ppm, whereas, the radium content can reach 1920 ppm in the mineral 1; a dried hydrothermal gel. The radon emanation of this mineral is the higher, about 93%. That means, the mineral retain only 7% of radon produced inside its matrix.

Some results are also mentioned in: (Boujrhal et al. 2004, 2005).

The uranium content obtained by ICP-MS is in good agreement with that determined by gamma spectrometry according to the incertitude.

Table 1 Uranium – radium contents and radon emanation power determined by gamma spectrometry

Sample	Uranium content (ppm)	Radium content (ppm eU)	Radon emanation (%)
Phosphate 1	124 ± 9	137 ± 5	8.4 ± 0.7
Phosphate 2	70 ± 16	90 ± 10	5.25 ± 1.2
Fossilized teeth	122 ± 22	200 ± 7	20.5 ± 1.3
Phosphogypsum 1	<LD	70 ± 12	29.5 ± 4.3
Phosphogypsum 2	<LD	90± 8.8	22.5 ± 5.5
Mineral 1	<LD	1920 ± 60	93 ± 1.2
Mineral 2	<LD	360 ± 12	30 ± 2.5

Table 2 Uranium content determined by ICP-MS

Sample	Uranium content (ppm)	3σ%
Phosphate 1	117	2.4
Phosphate 2	74	2.5
Fossilized teeth dried at 200°C	182	2.5
Phosphogypsum 1	1.93	2.2
Phosphogypsum 2	0.84	2.4
Mineral 1	2.255	8
Mineral 2	1.32	8

Conclusion

Method of regions for Gamma spectrometry using NaI (Tl) is developed in our laboratory to determine the presence of natural radionuclides, especially Uranium-238, Radium-226 and Radon-222. It is an adequate method for this scintillation detector according to its bad efficiency. It is simple and rapid, that's why it is adopted in our laboratory as a routine method.

Results obtained by gamma spectrometry for all the samples studied are in good agreement with those given by ICP.

Uranium content is lower than detection limit in the minerals and phosphogypsum studied here, whereas, it is lower than 130 ppm in phosphate.

References

Berrada M, Boujrhal FZ, Choukri A, El Khoukhi T, Iraqi MR (1992) Radon emanation from sedimentary phosphate and corresponding phosphogypsum, in: Radon and inert Gaz in the Earth Sciences and Environment, Toelicht. Verhand. Geologische en mijnkaarten van belgie (Mém. Expl. Cartes Géologique et Minière de la Belgique), N°32, 322p/b, pp. 253–258.

Berrada M, Boujrhal FZ, Couchot P, Chambaudet A, and Mercier R (1995) Effet de la température de cuisson sur le potentiel d'émanation et le taux de dégazage en radon de phosphates sédimentaires, Dans: Gaz Geochimestry (C. Dubois, D. Clein, A. Chambaudet, M. Rebetez). Science Reviews, 335–358.

Boujrhal FZ, Berrada M, Cherkaoui El Moursli R (1999) Retention of radon by apatite structure: the case for sedimentary phosphate. Phosphorus Bulletin 10: 274–282.

Boujrhal FZ, Hlil EK, Cherkaoui El Moursli R (2004) Study of apatite behaviour in the presence of radionuclides U and Rn and local modification of their crystalline and electronic structure. Radiation Physics and Chemistry, 69: 1–6.

Boujrhal FZ, Hlil EK, Cherkaoui El Moursli R, El Khoukhi T, Sghir B (2005) A comparative study of radon retention ability of crystalline apatite and amorphous oxide materials. Materials Science Forum 480–481: 169–174.

Airborne Radiometric Surveying for the Management of Health, Safety and the Environment in the Uranium Mining Industry: Potential Applications and Limitations

Henk Coetzee, James Larkin

Abstract. Some requirements for radiological site assessment may be at least partially fulfilled using airborne radiometric surveys. These include the identification of contaminated areas and the quantification and characterization of radionuclide contamination, as well as the identification of pollution resulting from human activities. Surveys flown over areas contaminated by mining and other activities demonstrate the strengths and weaknesses of airborne surveying. Applications include the identification and mapping of contamination throughout the life-cycle of a mine.

Introduction

Airborne radiometric surveying – the measurement of gamma-rays using aircraft mounted detectors – is a well-established geophysical technique for the detection of radioactive ores and the mapping of geological units with contrasting radionuclide contents. With growing environmental awareness in recent years, as well as improvements in instrumentation and data processing techniques, the airborne radiometric method has also found application in environmental surveys, looking at NORM (Coetzee and Szczesniak 1993) as well as anthropogenic radionuclides. Smith and Tauchid (1989) describe the use of the airborne survey method for the identification of reactor fragments from the crashed Cosmos 954 satellite (Hov-

Henk Coetzee
Council for Geoscience, Pretoria, South Africa
School of Geosciences, University of the Witwatersrand, Johannesburg, South Africa

James Larkin
Radiation and Health Physics Unit, University of the Witwatersrand, Johannesburg, South Africa

B. Merkel, M. Schipek, *The New Uranium Mining Boom*
DOI 10.1007/978-3-642-22122-4 © Springer 2011

gaard 1997) and the mapping of fallout from the Chernobyl disaster in Sweden. More recently, Rawlins et al. (2009) have described the use of airborne radiometric surveying to map the distribution of fallout from Chernobyl in Northern Ireland 20 years after the accident.

This paper looks at applications of the airborne radiometric method in radiation protection and identifies some of the strengths and weaknesses of the method.

Airborne Radiometric Surveying

Instrumentation

The airborne radiometric method is a geophysical survey technique in which aircraft-mounted gamma-ray detectors are used to map the distribution of gamma emitters on the earth's surface. Initially these surveys were flown to locate anomalous areas in uranium exploration programs. In addition to the exploration application, surveys are now flown as an aid to geological mapping (Duval 1983) as well as for the monitoring of pollution related to uranium mining (Coetzee and Szczesniak 1993).

Early surveys (Cook 1952) used Geiger–Müller detectors. Since the early 1960s (Moxham 1960), NaI(Tl) detectors have been used almost exclusively for airborne surveying, while experimental use has been made of solid organic (Duval 1972), bismuth germinate (BGO) (Stettler et al. 1997; Coetzee et al. 1998; Coetzee 2008) and CsI (Koomans et al. 2008) detectors. Since the 1970s, multi-channel spectrometers have been utilized, allowing the recording of total radioactivity as well as radioactive decays from ^{40}K, ^{214}Bi (in the ^{238}U series) and ^{208}Tl (in the ^{232}Th series) (Grasty 1975), while modern instruments typically record a gamma ray spectrum in 256 or more channels with gamma energies from 0 to 3 MeV.

Data Collection

In a typical radiometric survey using modern instrumentation, one or more NaI(Tl) scintillation detectors, connected via multi-channel-analyzers to a recording system is mounted in an aircraft (fixed wing or helicopter). Typical modern systems use either 4 or 8 individual 4.2 l detectors, sometimes with an additional "upward-looking" detector used to estimate the contribution of atmospheric radon to the measured signal. Smaller detectors have also been used in ultralight aircraft installations (Coetzee 2008; Koomans et al. 2008). The data collection systems are generally integrated with navigational and other geophysical instruments.

Data are collected by flying a series of closely-spaced survey lines at low (< 120 m) altitude, with the instrument repeatedly collecting spectral data with a sampling period typically of 1 s. In a typical survey, with flight lines spaced 100–200 m apart (this spacing may be decreased for high spatial resolution surveying of smaller areas, such as those likely to be encountered in radiation protection applications), a nominal flight altitude of 80 m and an airspeed of between 150 and 250 km/h, the survey area will be fully covered, owing to the relatively large field of view, often referred to as the "circle of investigation" (Pitkin and Duval 1980) of the detector. The flight path is generally followed by the pilot of the aircraft using a real-time differential GPS system, which is also used to recover the positions of each measurement and allow the precise and accurate positioning of the survey data and integration with other spatial data.

Calibration and Data Processing

Over time, a calibration and processing methodology has developed, taking account of Compton interactions in the source, detector and intervening air layer, system sensitivity, airborne radon, aircraft altitude and other factors affecting the relationship between ground radionuclide activities and what can be measured from a moving aircraft (IAEA 2003). Recent developments in data processing also include multi-channel data processing techniques where spectral data are used to identify characteristic features of a spectrum such as radioactivity due to anthropogenic radionuclides (Hovgaard 1997) and to improve the signal-to-noise ratio in measured spectra (Hovgaard and Grasty 1997; Dickson and Taylor 1998).

In most geophysical applications, an assumption of secular equilibrium is made for the ^{238}U and ^{232}Th decay series and data are expressed as the elemental concentration of the parent radionuclide, based on this assumption. In some cases, the concentrations are expressed as eU_3O_8 and $eThO_2$, the prefix "*e*" denoting the equivalent concentration, based on the measured activity of the relevant daughter radionuclide. The data presented for the ^{238}U-series on Fig. 2 uses this convention. The calibration for these two decay series is, however, based on measurements of the activities of ^{214}Bi and ^{208}Tl respectively and in a radiation protection context it would be more valid to express data in terms of the activities of these two radionuclides. Conversion factors are also available (IAEA 2003) to express data in terms of ground-level exposure or dose rates.

Requirements for Radiation Protection Applications

The requirements of the radiation protection community differ significantly from those in the geophysical community, where anomaly identification and, in some cases characterization in terms of NORM nuclides is the key aim of radiometric surveying. A summary of some of the aims which can be partially or fully fulfilled using airborne gamma detectors is presented in Table 1.

Table 1 Requirements, applicability and applications of airborne radiometric surveying in radiation protection

Requirement	Applicability of airborne surveying	Typical applications
Identification of contaminated areas	Airborne surveying is the optimal method for the rapid coverage of large areas. The large footprint covered by an airborne detector assures complete coverage of the survey area, provided the correct survey parameters are selected	Contaminated site investigations Site monitoring Site screening surveys prior to ground follow-up
Identification of specific radionuclides	Airborne surveying is limited to the detection of gamma emitters. Multi-channel spectrometric measurements allow the identification of specific gamma-ray emitters	Identification of radionuclides Characterization of contaminants and pollution streams Apportionment of pollution to specific sources
Quantification of activity concentrations on surface	Quantitative calibration of airborne spectrometers requires the use of large ($\geq 0.5\,m^3$) calibration sources as well as a dynamic calibration range several km long with well quantified activities. This makes the quantitative calibration for NORM with the assumption of secular equilibrium possible, but complicates the determination of activities of other radionuclides where such facilities are not available	The limitations of the technique need to be acknowledged in survey design. It may not be possible to quantify the activity of small sources or localized contamination
Quantification of radiologically relevant doses	The calibration of airborne systems assumes that the emitting surface is flat, homogenous and infinite (effectively this implies a radius of several hundred meters depending on the flight altitude). Within these limitations, it is possible to quantify the activities of NORM nuclides and to use these to estimate ground-level doses	Quantification of ground-level exposure rates for larger contaminated surfaces
Detection and identification of small radioactive sources	An airborne gamma-spectrometer measures radiation emanating from a relatively large area on the ground. Where a small source emits a statistically significant signal, it will be detected, but the position within the measurement area remains uncertain. Nevertheless, the full coverage of the surface afforded by the method ensures that all significant sources within an area will be detected	Rapid full coverage of large areas assists in the location and identification of small sources

Airborne surveying has the following benefits in radiation protection applications:

1. Airborne radiometric surveying provides full coverage of the ground surface, allowing the identification of all significant anomalous sources. No ground-based method can do this, unless an entire site is surveyed with a station spacing of typically 1–2 m along lines spaced 1–2 m apart. Such a survey would be impracticable on all but the smallest contaminated sites.

2. Contaminated sites in mining environments tend to contain a suite of inorganic contaminants, which often occur in close spatial proximity. The delineation of contaminated areas within a mining environment allows focused investigation of mine-related contamination.
3. The ability to produce images allows the easy integration of radiometric data with other spatial data (Chevrel and Coetzee 1997; Coetzee et al. 2002) and the integration of these data using GIS methods.
4. Airborne data collection is relatively cheap, compared to ground surveying, where total coverage of larger areas is required. The availability of systems of different sizes, ranging from small detectors mounted in slow-low-flying ultralights to larger detector systems suitable for larger aircraft installations allows cost-effective options for a range of sites and sensitivity requirements.

A number of limitations also exist. These are primarily related to:

1. The method's inability to quantify the dose impact due to small sources on the ground.
2. The lack of calibration facilities for non-NORM radionuclides.
3. The limitation to measurement of gamma-emitting radionuclides which prevents a full radiological impact assessment from being undertaken, particularly where decay series are in secular disequilibrium.

These limitations define a niche for the method in the screening of large areas for gamma-emitting radionuclides. In most cases, ground follow-up will be required to fully quantify impacts in anomalous areas, while uncontaminated areas can be screened out of a study in a rapid, cost-effective fashion.

Examples of Application

The collection of gamma ray spectra allows the determination of radioactivity due to both NORM and artificial radionuclides. The following examples show the application of these techniques to a South African gold/uranium mining environment and a nuclear research site.

Environments Affected by the Mining of Uranium-Bearing Ores

The gold-bearing conglomerates of the Witwatersrand contain significant concentrations of uranium, occurring as uraninite. This has, at times during the more than 100 year history of mining, been exploited economically at some of the mines. Uranium has been identified as a significant contaminant in the waste streams of these mines (Coetzee et al. 2006).

Since the early 1990s, airborne radiometric surveys of the gold mining areas have identified areas contaminated with mining waste materials. Although the method does not identify uranium contamination, it is sensitive to the radium

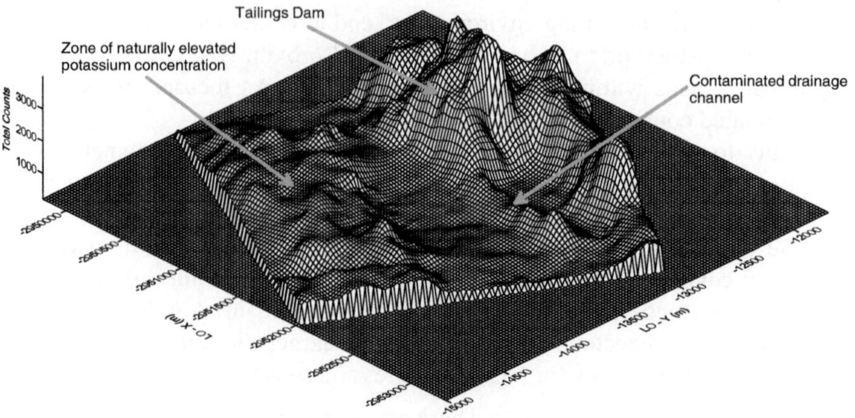

Fig. 1 3-D visualization of total count radiometric data from the abandoned New Machavie Gold Mine, located west of Johannesburg, South Africa. Note the area of greated elevation of radioactivity over the mine's tailings deposit and the contaminated drainage channel downstream. Some elevated counts are due to local elevation in natural background due to ^{40}K

daughter ^{214}Bi, which forms part of the radioactive waste stream of these mines and using these types of surveys it is possible to quantify the activity concentration of the ^{214}Bi by comparing the activity concentration measured in the uranium window of an airborne radiometric survey with activity concentrations as measured in the 351 keV photopeak of ^{214}Pb and the 1765 keV photopeak of ^{214}Bi (Larkin 2011). However because of the chemical extraction processes associated with gold production and, in some cases, leaching by acid drainage generated by the mining wastes resulting in extreme disequilibrium in the uranium decay chain, there can be no extrapolation of the ^{214}Bi activity concentrations to the other radionuclides in the uranium decay series. It is however possible to identify a number of features, including mining residues, contaminated stream sediments, footprints left after the removal and reprocessing of mine residues and urban areas where mine residue material has been used as a construction material (van Tonder et al. 2008).

Figure 1 shows a 3-D visualization of the total gamma-ray counts measured over an abandoned mine situated to the west of Johannesburg. The data were collected along lines located 50 m apart at a nominal flight altitude of 60 m. A 16 l NaI(Tl) detector was used for the survey. A zone of elevated counts is visible over the mine's tailings dam and a drainage channel running to the south. In the west of the survey area an area of slightly elevated count rates occurs due to local natural enrichment in ^{40}K. Figure 2 shows the separation between the ^{40}K and ^{214}Bi anomalies achieved using spectrometric measurements. On this type of site, dose rates could be determined over the large contaminated surfaces which can then be used in the development of a radiological protection program and would be a key element in any decision making process regarding site remediation. The source-detector geometry would most likely result in an underestimate of dose over the contaminated drainage channel. Ground follow-up surveys would be required to quantify exposure over this anomaly.

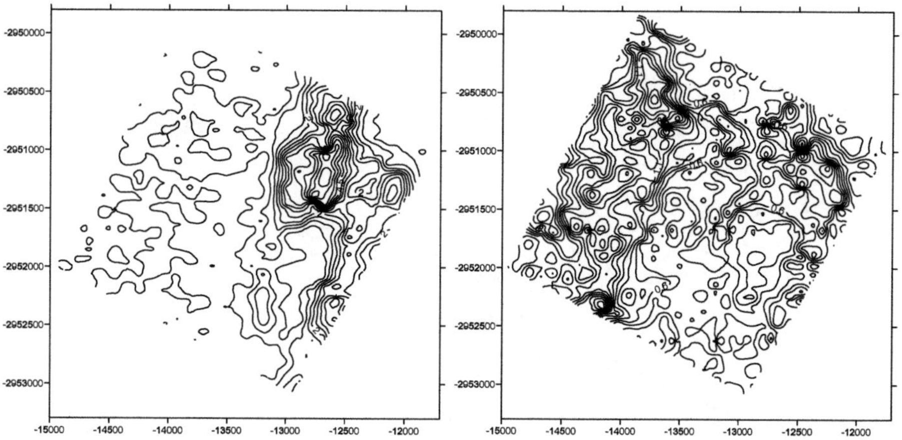

Fig. 2 Contour maps of radiometrically determined eU_3O_8 concentration (*left*) and K_2O concentration (*right*) showing different spatial distributions of the two radionuclides

Environments Contaminated with Anthropogenic Radionuclides

Figure 3 shows counts attributable to ^{137}Cs collected over a nuclear research site. The data were collected using an ultralight aircraft equipped with a 4 l NaI(Tl) detector along flight lines spaced 50 m apart at a nominal flight altitude of 50 m. Two ^{137}Cs anomalies are detected in the survey data, but no leakage out of the contaminated areas can be identified.

Unfortunately no calibration sources or ranges exist for anthropogenic radionuclides. This precludes not only the quantification of activities from the airborne data but also the separation of the signals from these nuclides from those due to NORM, using conventional calibration techniques. Coetzee (2008) proposes a method whereby the continuum of back-scattered counts can be removed from a gamma-ray spectrum, leaving only counts in targeted photopeaks. This method is applied to spectra which have been processed using Noise Adjusted Singular Value Decomposition (NASVD), as proposed by Hovgaard and Grasty (1997) to improve the signal to noise ratio, and uses the approach of Mariscotti (1967) to separate the photopeak counts from the continuum.

NASVD processing may also be used to identify and map unique spectral components (Hovgaard 1997), some of which identify the presence and/or absence of certain radionuclides in different parts of a study area. In the area shown in Fig. 3 a weak signal with the spectral signature of $^{234}Pa^{m}$ was identified over an area where yellowcake was stored. This anomaly could not be identified by examining spectral data or conventional mapping methods.

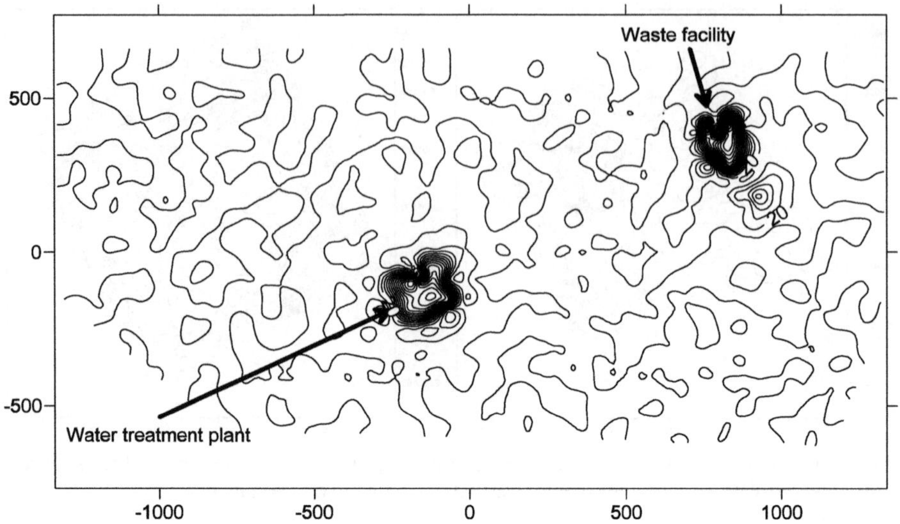

Fig. 3 Contour map of counts due to ^{137}Cs, calculated using a continuum removal technique, due covering a nuclear research centre

Conclusions

The airborne radiometric method is able to rapidly map large surface areas where radiologically significant contamination is suspected and identify significant accumulations of gamma-ray emitters. Using gamma-ray spectrometry it is possible to identify and, in some cases quantify, gamma emitters including NORM and anthropogenic radionuclides.

Spatial locations of measurements are recorded and allow the compilation of maps showing the distribution of radionuclides. Data can also be combined with other spatial data using GIS and spatial analysis techniques.

Advanced spectral processing techniques can be used to enhance the signal-to-noise ratio of radiometric survey data as well as identifying often subtle spectral characteristics due to the presence of specific radionuclides.

Owing to the source-detector geometry of an airborne detector flying above small sources, it is not always possible to accurately quantify ground-level exposure rates. However no other technique exists which is able to provide rapid and cost-effective full coverage of large areas of land, such as those often encountered in the mining of radioactive ores. This defines an important niche for the airborne radiometric method in the screening of large areas for radioactive contamination and, in some cases, the quantification of ground-level gamma-ray exposure.

References

Chevrel, S. and H. Coetzee (1997) An example of GIS use in surface and ground-water pollution sensitivity analysis: Regional impact of mining activities in South Africa. CMWR 97 (Fourth international conference on Computer Methods and Water Resources), Byblos, Lebanon.

Coetzee, H. and H. Szczesniak (1993) Detection and monitoring of pollution from mine tailings dams along rivers in the Witwatersrand gold fields using the airborne radiometric method. 16th Int. Colloquium on African Geology. Ezulwinini Valley, Swaziland.

Coetzee, H., E.H. Stettler, T.A. Grace, P.J. Kulper and M.E. Hauger (1998) Development of a lightweight high-resolution airborne geophysical system. 4th Technical Meeting of the Environmental and Engineering Geophysics Society (European Section), Barcelona, Spain, September 14–17, 1998.

Coetzee, H., S. Chevrel and F. Cottard (2002) Inter-disciplinary studies of the impact of gold and uranium mining in the Witwatersrand Goldfield. Uranium in the aquatic environment, proceedings of Uranium mining and hydrogeology III, Freiberg, Springer Verlag.

Coetzee, H., F. Winde and P. Wade (2006) An assessment of sources, pathways, mechanisms and risks of current and future pollution of water and sediments in the Wonderfonteinspruit Catchment. Pretoria, Water Research Commission: 202 pp.

Coetzee, H. (2008) Acquisition, processing and enhancement of multi-channel radiometric data collected with ultralight aircraft mounted detectors. School of Geosciences. University of the Witwatersrand, Johannesburg. PhD: 357 pp.

Cook, J.C. (1952) An analysis of airborne surveying for surface radioactivity. Geophysics 17(4): 687–706.

Dickson, B. and G. Taylor (1998) Noise Reduction of aerial gamma-ray surveys. Exploration Geophysics 29: 324–329.

Duval, J.S.J., J.M. Worden, R.B. Clark and J.A.S. Adams (1972) Experimental Comparison of NaI(Tl) and Solid Organic Scintillation Detectors for Use in Remote Sensing of Terrestrial Gamma-Rays. Geophysics 37(5): 879–888.

Duval, J.S. (1983) Composite Color Images of Aerial Gamma-Ray Spectrometric Data. Geophysics 48(6): 722–735.

Grasty, R.L. (1975) Uranium Measurement by Airborne Gamma-Ray Spectrometry. Geophysics 40(3): 503–519.

Hovgaard, J. (1997) A new processing technique for airborne gamma-ray spectrometer data (Noise adjusted singular value decomposition). Am Nucl. Soc. Sixth topical meeting on Emergency Preparedness and Response, San Francisco.

Hovgaard, J. and R.L. Grasty (1997) Reducing statistical noise in airborne gamma-ray data through spectral component analysis. Proceedings of Exploration 97: Fourth Decennial Conference on Mineral Exploration, Toronto.

International Atomic Energy Agency (2003) Guidelines for radioelement mapping using gamma ray spectrometry. Vienna, International Atomic Energy Agency: 173 pp.

Koomans, R.L., J. Limburg, and E.R.v.d. Graaf. Towards lightweight airborne gamma spectrometry: A case study. 2008 (cited 2008 30 May 2008); Available from: http://www.medusa-online.com/files/systemen/05_TowardsLightweightAirborneGamma.pdf.

Larkin, J. (2011) Comparative Investigation of Airborne and Ground-based Radiometric Survey Techniques. University of the Witwatersrand, Johannesburg. MSc: 132 pp.

Mariscotti, M.A. (1967) A method for automatic identification of peaks in the presence of background and its application to spectrum analysis. Nuclear Instruments and Methods 50: 309–320.

Moxham, R.M. (1960) Airborne radioactivity surveys in geologic exploration. Geophysics 25(2): 408–432.

Pitkin, J.A. and J.S. Duval (1980) Design Parameters for Aerial Gamma-Ray Surveys Geophysics 45(9): 1427–1439.

Rawlins, B.G., C. Scheib, A.N. Tyler, D. Jones, R. Webster and M.E. Young (2009) The spatial distribution of caesium-137 over Northern Ireland from fallout from the Chernobyl nuclear accident. Geophysical Research Abstracts Vol 11, EGU2009-2471.

Smith, A.Y. and M. Tauchid (1989) Uranium exploration and technology: Preserving the "know-how". IAEA Bulletin 1989(1): 16–19.

Stettler, E.H., M.E. Hauger, H. Coetzee, T. Grace, R.J.D. Meijer, J. Limburg, P.J. Kulper and P. Cole (1997) Recent magnetometer and spectrometer experiments using an airborne microlight platform. 4th Technical Meeting, South African Geophysical Association, Swakopmund, Namibia, 29–30 September, 1997.

van Tonder, D.M., H. Coetzee, L.K.C. Strachan, T. Mafanya, P.W. Wade, N. Msezane, M.G. Kwata, M. Roos, D. Sebake, H. Mengistu, B. Yibas, and S. Esterhuyse (2008) Regional gold mining closure strategy for the Witwatersrand Basin: Department of Minerals and Energy, Pretoria, Council for Geoscience Report No. 2008-0179, pp. 140.

Coca-Cola® for Determining Bioaccessible Uranium in Contaminated Mine Soils

Bernd Lottermoser, Ewald Schnug, Silvia Haneklaus

Abstract. There is an increasing need for low-cost soil testing procedures that accurately determine the bioaccessibility of metals in contaminated land. This study demonstrates that the use of Coca-Cola Classic® in single extraction tests provides an excellent indication of bioaccessible U in contaminated mine soils. Moreover, soft drinks are superior in terms of availability, costs, preparation and disposal compared to traditional chemicals (e.g. DTPA). Coca-Cola® extraction tests may be used for environmental impact assessments of U mine sites, nuclear fuel processing plants and waste storage and disposal facilities.

Introduction

Since the 1950s, numerous soil chemical tests have been developed for diagnostic and prognostic applications in the soil and environmental sciences. These tests continue to rely on high-quality chemical reagents (e.g. EDTA, DTPA, CH_3COOH, NH_4F, $CaCl_2$), meticulous experimental protocols, sophisticated laboratory infrastructure and trained technical staff. Such requirements are not necessarily readily

Bernd Lottermoser
School of Earth Sciences, University of Tasmania, Private Bag 79, Hobart, Tasmania 7001, Australia

Ewald Schnug
Institute for Crop and Soil Science, Federal Institute for Cultivated Plants, Julius Kühn-Institute (JKI), Bundesallee 50, D-38116 Braunschweig, Germany

Silvia Haneklaus
Institute for Crop and Soil Science, Federal Institute for Cultivated Plants, Julius Kühn-Institute (JKI), Bundesallee 50, D-38116 Braunschweig, Germany

B. Merkel, M. Schipek, *The New Uranium Mining Boom*
DOI 10.1007/978-3-642-22122-4

available throughout the world and particularly in developing nations, where ever increasing land contamination issues need to be understood.

Chemical testing of soils is to identify and assess the limitations to the use of soil and hence serves a number of purposes: a) it provides insights on constraints to plant growth; b) it allows monitoring of temporal and spatial trends in soil chemical properties; c) it permits informed choices on land capability options; d) it allows the prediction of corrective action and the need for and application rate of fertilizers and soil amendments; and e) it quantifies ecological, plant, animal and human health risks associated with resource degradation or land contamination.

Most chemical soil test procedures rely on the extraction of nutrients or contaminants from the solid soil phases by using a liquid reagent, followed by analysis of the element in the extractant. In particular, single reagent extractions using ethylenediamine tetraacetic acid (EDTA) and diethylenetriamine pentaacetic acid (DTPA) were originally designed to extract metals from agricultural soils to evaluate the soils' micronutrient status (e.g. Trierweiler and Lindsay 1968; Brown et al. 1970; Haq and Miller 1972). Today, such extraction techniques are also used in the evaluation of contaminated sites to establish the bioaccessibility of metals (i.e. the fraction that is soluble in the soil and available for plant uptake) in contaminated substrates (e.g. Rauret 1998).

The purpose of this study was to explore the suitability of low-cost, readily available soft drinks (Diet Coke®, Coke Zero®, Coca-Cola Classic®) as indicators of bioaccessible U and other trace elements (As, Ce, Cu, La, Mn, Ni, Pb, Th, Y,

Fig. 1 Partly soil covered waste rock dump at the base of local hills, Mary Kathleen uranium mine site (field of view ~ 200 m). Erosion from the waste rock piles leads to elemental loadings of surrounding soils

Zn) in contaminated soils of the Mary Kathleen U mine site, Australia (Lottermoser et al. 2005; Fig. 1). Data acquired using these unconventional reagents were compared with those obtained using single DTPA and $CaCl_2$ extraction procedures. Hence, this research adds to our understanding of trace element bioaccessibility on contaminated sites and provides an evaluation of novel and unconventional extraction tests as possible tools for evaluating contaminated lands.

Materials and Methods

The Mary Kathleen mine site is situated in the dry tropics region of western Queensland, Australia, centered on 20° 45.1' S, 140° 0.6' E. The mine ceased operations in 1982 and underwent rehabilitation between 1982 and 1985. The mine area now consists of an abandoned open pit, partly filled by a pit lake, numerous waste rock dumps, a capped tailings storage facility, a former mill site, and a closed evaporation pond area. The open pit is largely self-contained, with internal drainage, but waste dumps and evaporation pond areas are prone to physical erosion and chemical reactions involving the atmosphere and seasonal rain. Physical dispersion of mineralized particulates from, as well as chemical reactions in waste dumps and naturally mineralized areas, have led to local enrichment of U, Cu, light rare earth elements (LREE), As, Mn and S in adjacent soils (Lottermoser et al. 2005; Fig. 1).

In all, twenty-six soil sampling sites were chosen from the Mary Kathleen mine site. Soil samples were taken from three environments: 1) waste rock dumps, 2) evaporation ponds, and 3) background areas of the former township. Representative aliquots of soil samples were chemically extracted using traditional DTPA and $CaCl_2$ soil tests as well as novel Diet Coke®, Coke Zero® and Coca-Cola Classic® extraction procedures as documented by Rayment and Higginson (1992), Houba et al. (2000) and Schnug et al. (1996, 1998), respectively. Extractants were analyzed for selected elements (As, Ce, Cu, Fe, La, Mn, Ni, Pb, Th, U, Y, Zn) using ICPMS and ICPAES techniques.

Results

Collected topsoils are sandy loams and possess slightly alkaline pH values due to the presence of calcrete. Mine soils from waste rock dumps and evaporation ponds contain median trace element values that exceed those of background samples (As: 4-fold; Ce: 4-fold; Cu: 2-fold; La: 5-fold; Mn: 4-fold; U: 25-fold; Zn: 1.5-fold). Significant correlation of elements (As, Ce, Cu, Fe, Pb, Th, Y) similar to those detected in wallrocks and waste rock dumps indicates that much of the trace element load of mine soils is due to the physical erosion of waste rock particles into the surrounding soils. By contrast, the lack of correlation of U with any other elements suggests that U has been leached from ore aggregates and is now held in adsorptive sites of soil particles.

Soil Extractions

Chemical extraction (DTPA, $CaCl_2$, Coca-Cola Classic®, Diet Coke®, Coke Zero®) leached variable trace element concentrations from the soil samples and reflect instantly ($CaCl_2$) and easily mobilized plant available element pools (DTPA; Coca-Cola® products). The extraction strength generally increased: a) for As, Ce, Cu, La, Ni, Y and Zn in the order of DTPA > Coca-Cola products > $CaCl_2$; and b) for U in the order of Diet Coke® > Coke Zero® > $CaCl_2$ > DTPA = Coca-Cola Classic® (Fig. 2). Data of single extraction tests using Coca-Cola Classic®, Diet Coke® and Coke Zero® demonstrate that extractable Cu, Mn and Zn concentrations correlate significantly with DTPA-extractable metals. Moreover, the correlation between DTPA-extractable U and that extracted using Coca-Cola Classic® is close to unity (+0.98), with reduced correlations for Diet Coke® (+0.66) and Coke Zero® (+0.55). Also, the correlation between $CaCl_2$-extractable U and that extracted using Coca-Cola Classic®, Diet Coke® and Coke Zero® is near unity (+0.94, +0.86 and +0.88, respectively). Coca-Cola Classic® extracts U concentrations near identical to DTPA, whereas distinctly higher U fractions were extracted using Diet Coke® and Coke Zero®.

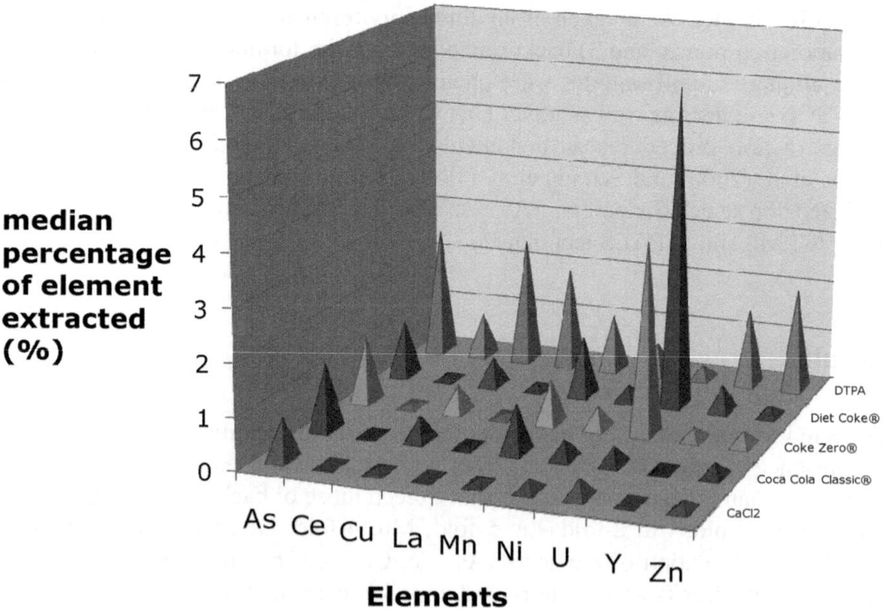

Fig. 2 Median percentages of elements extracted from Mary Kathleen mine soils (*n*: 26), using DTPA, $CaCl_2$, Coca-Cola Classic®, Diet Coke® and Coke Zero®

Discussion

The bioaccessibility of metals in soils is generally established using chemical extraction tests that apply the complexing agents EDTA and DTPA. Complexation of metals in the rhizosphere is a critical mechanism by which plant roots access metals. Chemical extraction tests aim to mimic the extraction and complexing of metals by plant roots. In this study, complexing of U by phosphate in Cola soft drinks may have led to the detected U concentrations. Increased extraction of U during Diet Coke® and Coke Zero® extractions compared to Coca-Cola Classic® was possibly due to the presence of artificial sweeteners (acesulfame potassium, aspartame), benzoate and citrate. An addition of citric acid is already known to solubilize more U from contaminated soils (e.g. Ebbs et al. 1998; Huang et al. 1998; Shahandeh and Hossner 2002).

Validation of Cola Extraction Tests

New chemical soil tests need to be calibrated against data generated by established techniques. Results of this study demonstrate significant correlation between DTPA-extractable U and that extracted using Coca-Cola Classic®, with reduced correlations for Diet Coke® and Coke Zero®. Coca-Cola Classic® extracts U concentrations near identical to DTPA, whereas distinctly higher U fractions were extracted using Diet Coke® and Coke Zero® (possibly due to the presence of citric acid in these soft drinks). Thus, extraction of U from contaminated mine soils using Coca-Cola Classic® mimics that of a well-established chemical extraction method. An indication of success or failure of a new chemical soil test should also be established by comparing its convenience to which it is superior to established chemical extraction methods. Previous research has already documented that Coca-Cola® is a suitable extractant to determine the bioaccessibility of soil micronutrients (Cu, Mn, Zn) to cereal crops (Schnug et al. 1996, 1998). The unconventional reagent is superior in terms of ease of use, expenditure, preparation and disposal compared to traditional chemicals (e.g. DTPA). Results of this study further strengthen the fact that the use of Coca-Cola Classic® in single extraction tests provides an excellent indication of bioaccessible U in contaminated mine soils. The use of soft drinks in bioaccessibility studies is especially attractive to those regions and developing nations where access to distilled water and high-quality chemical reagents is difficult to accomplish.

Conclusions

Increasing land contamination issues, particularly in developing nations, require low-cost, readily available extractants to quantify health risks. Moreover, environmental impact assessments and rehabilitation of U mine sites require a solid understanding of chemical speciation of U for assessing plant uptake and mobility in soils. Results of this study demonstrate that U concentrations extracted from contaminated mine soils using Coca-Cola Classic® correlate with those of the established DTPA extraction method. Therefore, Coca-Cola Classic® has potential applications in single extraction tests of contaminated soils and may be used for environmental impact assessments of U mine sites, nuclear fuel processing plants and waste storage and disposal facilities.

Acknowledgements Support for this project was given by CRC CARE Project 2-03-06-08/09 and the Alexander von Humboldt Foundation. Dr. N.C. Munksgaard conducted the soil extractions.

References

Brown AL, Quick J, Eddings JL (1970) A comparison of analytical methods for soil zinc. Soil Sci Soc Am J 35: 105–107

Ebbs SD, Norvell WA, Kochian LV (1998) The effect of acidification and chelating agents on the solubilisation of uranium from contaminated soil. J Environ Qual 27: 1486–1494

Haq AU, Miller MH (1972) Prediction of available soil Zn, Cu, and Mn using chemical extractants. Agron J 64: 779–782

Houba VJG, Temminghoff EJM, Gaikhorst GA, van Vark W (2000) Soil analysis procedures using 0.01 M calcium chloride as extraction reagent. Commun Soil Sci Plant Anal 31: 1299–1396

Huang FYC, Brady PV, Lindgren ER, Guerra P (1998) Biodegradation of uranium-citrate complexes: implications for extraction of uranium from soils. Environ Sci Technol 32: 379–382

Lottermoser BG, Ashley PM, Costelloe MT (2005) Contaminant dispersion at the rehabilitated Mary Kathleen uranium mine, Australia. Environ Geol 48: 748–761

Rauret G (1998) Extraction procedures for the determination of heavy metals in contaminated soil and sediment. Talanta 46: 449–455

Rayment GE, Higginson FR (1992) Australian laboratory handbook of soil and water chemical methods. Port Melbourne: Inkata Press

Schnug E, Fleckenstein J, Haneklaus S (1996) Coca Cola® is it! The ubiquitous extractant for micronutrients in soil. Commun Soil Sci Plant Anal 27: 1721–1730

Schnug E, Fleckenstein J, Haneklaus S (1998) Factors affecting available micronutrient concentrations in soils using Coca-Cola® as extractant. Commun Soil Sci Plant Anal 29: 1891–1896

Shahandeh H, Hossner LR (2002) Enhancement of uranium phytoaccumulation from contaminated soils. Soil Sci 167: 269–280

Trierweiler JF, Lindsay WL (1968) EDTA-ammonium carbonate soil test for zinc. Soil Sci Soc Am J 33: 49–54

Development of an Immunochromatographic Strip for Rapid, Instrument-Free, On-Site Detection of Uranium

Xiaoxia Zhu, Bhupal Ban, Xiaoxi Yang, Sergey S. Shevkoplas, Diane A. Blake

Abstract. A one-step assay has been developed for on-site detection of environmental uranium. A monoclonal antibody with specificity for U(VI) complexed to 2,9-dicarboxyl-1,10-phenanthroline (DCP) was conjugated to colloidal gold and subsequently mixed with a sample containing U(VI) and DCP. The mixture was allowed to migrate up a nitrocellulose strip and intensity of the antibody captured at a band of immobilized U(IV)-DCP was inversely proportional to the amount of U(VI) (50–200 nM) in the sample. The strips could be read visually or the images could be captured via a cell phone camera for more quantitative analysis.

Xiaoxia Zhu
Department Biochemistry, Tulane University School of Medicine, 1430 Tulane Avenue, New Orleans, Louisiana 70112 USA
School of Public Health, Nantong University, Nantong 226007, China

Bhupal Ban
Department Biochemistry, Tulane University School of Medicine, 1430 Tulane Avenue, New Orleans, Louisiana 70112 USA

Xiaoxi Yang
Department of Biomedical Engineering, Tulane University, New Orleans, Louisiana 70118 USA

Sergey S. Shevkoplas
Department of Biomedical Engineering, Tulane University, New Orleans, Louisiana 70118 USA

Diane A. Blake
Department Biochemistry, Tulane University School of Medicine, 1430 Tulane Avenue, New Orleans, Louisiana 70112 USA

B. Merkel, M. Schipek, *The New Uranium Mining Boom*
DOI 10.1007/978-3-642-22122-4 © Springer 2011

Introduction

The existing technologies to measure uranium require extensive sample pretreatment and complex instrumentation (atomic absorption spectroscopy or inductively coupled plasma emission spectroscopy) in a centralized facility. U(VI) can be specifically identified by ICP/MS (Rozmaric et al. 2009), but this method requires expensive equipment and trained operators. A simpler instrument, the Kinetic Phosphorescence Analyzer (KPA) is also used primarily in a laboratory setting; however, the reagents required by the KPA are expensive and subject to interference from sample matrixes (Brina et al. 1993). Because samples must be transported off-site for analysis, the use of these methods often leads to long turnaround times.

Our laboratory has previously described an immunosensor-based assay for uranium that has been used in the field to provide near real-time data about changes in U(VI) levels in groundwater during in situ bioremediation experiments (Melton et al. 2009). In this study, we describe an even simpler field test for U(VI) that takes advantage of existing lateral flow technology. The lateral flow assay is a proven system that has been used for the detection of a wide variety of analytes (Posthuma-Trumple et al. 2009). While lateral flow assays for U(VI) have not been reported, DNAzymes that recognize Cu^{2+}, Pb^{2+} and Hg^{2+} have been incorporated into lateral flow devices (He et al. 2011; Fang et al. 2010; Mazumdar et al. 2010). In this study, we incorporated a monoclonal antibody that recognizes U(VI) complexed to the chelator, 2,9-dicarboxyl-1, 10-phenanthroline (DCP) into a lateral flow device.

Materials and Methods

Materials

2,9-Dicarboxyl-1,10-phenanthroline (DCP) was purchased from Alpha Aesar (Heysham, UK). ACS grade uranyl acetate was a product of Mallinckrodt Chemical Works, St. Louis, MO USA). The 12F6 monoclonal antibody and U(VI)-DCP-BSA conjugate were available from previous studies (Blake et al. 2004; Melton et al. 2009). Goat anti-mouse IgG, bovine serum albumin (BSA) and gold chloride were purchased from Sigma (St. Louis, MO USA). Nitrocellulose membrane (pore size 8 μm) was a product of Sartorius Stedim (Bohemia NY USA). Conjugate and absorbent pads were provided by Millipore (Billerica, NY USA).

Preparation of Colloidal Gold and Gold-Labeled Antibody

Colloidal gold was prepared as described by (Frens 1973). Briefly, 100 ml of 0.01% (w/v) chlorauric acid was diluted in purified water and heated to boiling, and an aliquot (1.6 ml) of 1% trisodium citrate was added with constant stirring. The solution was maintained at 100°C for 15 min and then cooled. Purified water was added to bring the solution back to the original volume and sodium azide was added 0.05%. The colloidal gold suspension could be stored for months at 4°C.

Preliminary experiments were performed to determine the optimum pH and antibody concentration for the conjugation (data not shown). An aliquot of the colloidal gold suspension (10 ml) was adjusted to pH 8.0 and 400 μg of purified 12F6 was added dropwise. After incubation at room temperature for 30 min, 1 ml of 5% bovine serum albumin (BSA, w/v in H_2O) was added and stirring was continued for an additional 15 min. The mixture was centrifuged at 20,000 × g for 30–60 min, and the precipitate was resuspended in 1 ml of storage buffer (20 mM sodium borate, pH 8.0 containing 2% BSA, 0.01% sodium azide and 0.5% sucrose) and stored at 4°C.

Assembly of Immunochromatography Strip

The format of the lateral flow assay is diagramed in Fig. 1.

The sample disposition pad was prepared by dispensing the gold-labeled 12F6 antibody (0.7 μL/mm^2) onto a sheet of glass fiber (CFCP103000, Millipore). The

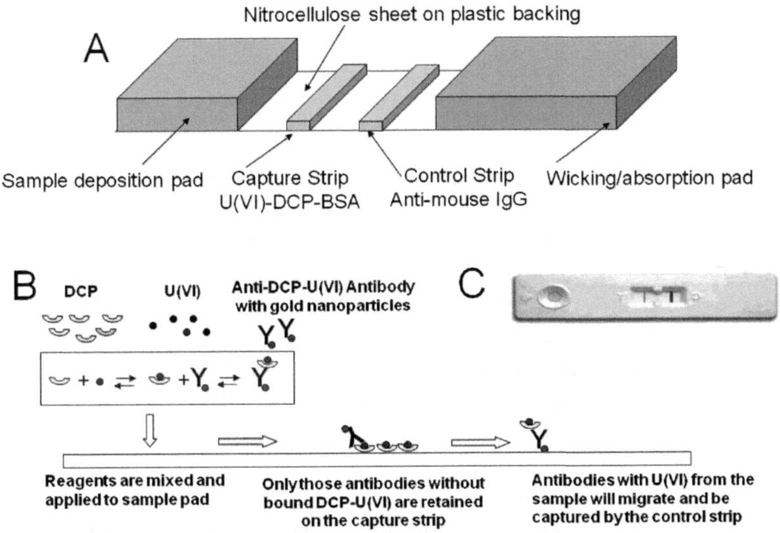

Fig. 1 (**A**) The sample deposition pad contained the colloidal gold labeled antibody. The nitrocellulose sheet had a capture strip of immobilized U(VI)-DCP-BSA and a control strip of anti-mouse IgG. A wicking/absorption pad transported the assay components from the point of application to the capture and control strips. (**B**) Binding reactions taking place during the assay. The arrows show the flow of components on the test strip. (**C**) Test strip in plastic case, showing signals at the capture and control strips

sample deposition pad was dried for 2 h at 37°C and stored at 4°C. U(VI)-DCP-BSA (1 μl per mm line, 0.25 mg/ml) and the anti-mouse IgG (1 μl per mm line, 0.5 mg/ml) were applied to the nitrocellulose membrane as the capture and control strips, respectively. The nitrocellulose membrane was dried for 30 min at 37°C and stored under dry conditions until use. The nitrocellulose membrane with capture and control strips was pasted onto the center of a self-adhesive plastic backing. The sample disposition pad with the gold-labeled antibody and a wicking/absorption pad were attached to the opposite ends of the strip, as shown in Fig. 1. The assembled strip was cut to a final size of 5 × 27 mm.

Sample Analysis

The end of the strip with the sample deposition pad was dipped into buffer containing varying concentrations of U(VI) and 1 μM DCP. The solution migrated toward the wicking/absorption pad and after 5 min, the result could be read visually. For a more quantitative analysis, the developed strips were photographed with a cell phone camera and the images were converted to RGB color values. The mean of the color intensity within each test zone was used to quantify the colorimetric response. Because data from camera phones are dependent on lighting conditions, the value of intensities of the colors were calibrated using the intensities from a test uranium strip run in the absence of uranium.

Results

Two different colloidal gold particles preparations were tested for these studies. Particles made by adding 1.6 ml of 1% trisodium citrate added to 100 ml 0.01% gold chloride solution were salmon pink, with an estimated diameter of 25 nm. When the amount of 1% trisodium citrate solution was reduced to 1 ml, the colloidal gold particles were dark red, with an estimated diameter of 40 nm. The smaller, salmon pink particles were found to be more stable after mixing with the 12F6 monoclonal antibody, and these particles were selected for further studies. Spectral analysis showed that the peak absorbance of these particles was 535 nm (data not shown).

The concentration of U(VI)-DCP-BSA dispensed onto the nitrocellulose membrane was optimized by testing dilutions of the conjugate between 0.05 and 1 mg/ml. Higher concentrations of the conjugate provided darker lines during the lateral flow assay, but also reduced the sensitivity of the final assay. A concentration of 0.25 mg/ml was chosen for the final strip. This concentration provided a line dark enough to be observed visually and sensitivity sufficient to detect uranium at the US EPA action limit (30 ppb or 126 nM). A black-and-white photograph of the completed immunochromatographic assay is shown in Fig. 2.

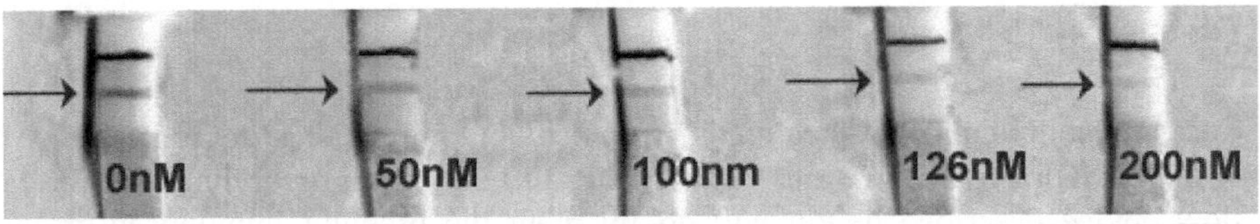

Fig. 2 U(VI) was diluted into 1 μM DCP at the indicated concentrations and applied to the test strips. The accumulation of the gold-labeled 12F6 antibody at the site of the immobilized U(VI)-DCP-BSA conjugate is shown by the arrows

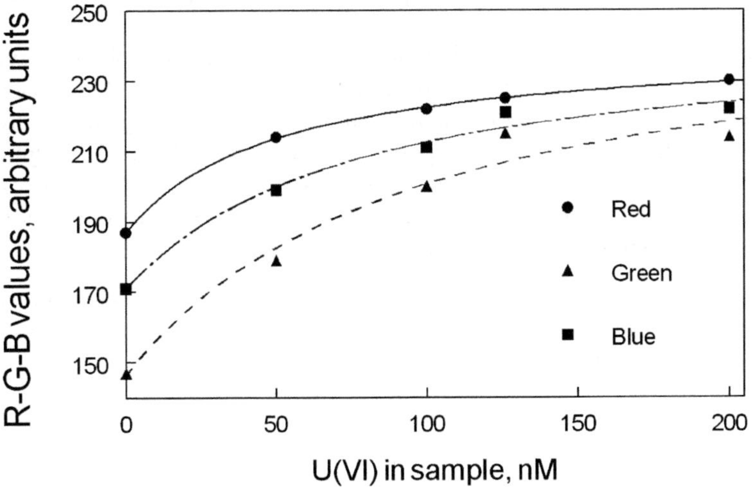

Fig. 3 Background-corrected red (●), green (▲) and blue (■) color intensities determined from the test strips

The top line on each strip is due to the capture of the gold-labeled 12F6 antibody by an anti-mouse antibody. Color development at this control line indicates that the assay is performing correctly and that there are no components present to non-specifically inhibit antibody binding. The test would be considered as invalid if there was no color line present in the control region. As the uranium concentration in the sample increased, the color in the test region (indicated by the arrows) decreased. The limit of detection in a visual assay was estimated to be ~ 100 nM.

In an effort to further quantify uranium detection, pictures of the strips were digitized and converted to RGB color values. The mean of color intensity within each test zone were used to quantify the colorimetric response, as shown in Fig. 3. Because data from cell phone cameras are dependent on lighting conditions, the value of the intensity of a test strip run in the absence of uranium (0 nm) was used to calculate a background-corrected response.

The concentrations of U(VI) that provided half-maximal color change for the red, green and blue intensities were 50, 95 and 75 nM, respectively. These data agree well with the visually read strips, where the sensitivity was estimated to be between 50 and 100 nM.

Discussion

Current technologies for detecting environmental uranium either are too expensive or not designed to analyze and quantify the concentration in real time. The lateral flow device described herein will complement the uranium immunosensors already described by our laboratory (Melton et al. 2009). The new method achieves: i) color change in uranium strip, ii) sample color data collection, iii) quantification of concentration of environmental uranium in real time. Studies are underway to validate these lateral flow devices with samples from several U(VI)-contaminated sites in the US.

Cell phones have become necessity in worldwide use and camera phones have been demonstrated to be useful tools in collecting and processing data. The system we describe is designed to address these analytical needs by combining an easy-to-use strip that provides a colorimetric response with a camera phones for digital quantification of uranium strip images. It offers new opportunities for low-cost, rapid monitoring of environmental uranium, especially in remote locations where trained personnel or complicated instruments are not available.

Acknowledgements This work was supported by the Office of Naval Research (#N00014-10-1-0270, D.A. Blake) and by a subcontract from the President and Fellows of Harvard College to S.S. Shevkoplas (under an award from the Bill & Melinda Gates Foundation).

References

Blake RC 2nd, Pavlov AR, Khosraviani M, Ensley HA, Keifer GE, Yu H, Li X, Blake DA (2004) Novel monoclonal antibodies with specificity for chelated uranium(VI): Isolation and binding properties. Bioconj Chem 15: 1125–1136

Brina R, Miller AG (1993) Determination of uranium and lanthanides in real-world samples by kinetic phosphorescence analysis. Spectroscopy 8: 258–301

Fang Z, Huang J, Lie P, Xiao Z, Ouyang C, Wu Q, Wu Y, Liu G, Zeng L (2010) Lateral flow nucleic acid biosensor for Cu^{2+} detection in aqueous solution with high sensitivity and selectivity. Chem Comm 46: 9043–9045

Frens G (1973) Controlled nucleation for the regulation of particle size in monodisperse gold suspensions. Nature Phy Sci 241: 20–22

He Y, Zhang X, Zhang S, Baloda M, Gurung AS, Zeng K, Lui G (2011) Visual detection of Hg^{2+} in aqueous solution using gold nanoparticles and thymine-rich hairpin DNA probes. Biosensor and Bioelectronics doi:10.1016/j.bios.2011.05.003

Mazumdar D, Liu J, Lu G, Zhou J, Lu Y (2010) Easy-to-use dipstick tests for detection of lead in paints using non-cross-linked gold nanoparticle-DNAzyme conjugates. Chem Comm 46: 1416–1418

Melton SJ, Yu H, Williams KH, Morris SA, Long PE, Blake DA (2009) Field-based detection and monitoring of uranium in contaminated groundwater using two immunosensors. Environ Sci Technol 43: 6703–6709

Posthuma-Trumple GA, Korf J, van Amerongen A (2009) Lateral flow (immune)assay: its strengths, weaknesses, opportunities and threats. A literature survey. Anal Bioanal Chem 393: 569–582

Rozmaric M, Ivsic AG, Grahek Z (2009) Determination of uranium and thorium in complex samples using chromatographic separation. Talanta 80: 325–362

Part 5
Modelling

Ecotoxicity of Uranium in Freshwaters: Influence of the Physico-Chemical Status of the Rivers

Karine Beaugelin-Seiller, Laureline Février, Rodolphe Gilbin, Jacqueline Garnier-Laplace

Abstract. As uranium speciation, and then its bioavailability, is highly variable in the range of the physico-chemistry of freshwaters, determining environmental protection criteria for generic ecosystems, i.e. without considering these processes, may lead to very low values. Based on the most recent knowledge on the variation of uranium bioavailability vs. its speciation, "conditional" protection criteria have been determined for 50 main physico-chemical domains, still including some identified conservatism, due to complexation properties of uranium.

Introduction

The implementation of the Water Framework Directive (EC 2000) asks for the derivation by member states of environmental protection criteria such as Environmental Quality Standards in order to allow assessments of the chemical and ecological status of water bodies that should reach a good status from 2015. The transposition of the WFD into the French legislation led to a regulatory text (MEDAD 2007) defining EQS for the 41 substances considered for the chemical

Karine Beaugelin-Seiller
Institut de Radioprotection et de Sûreté Nucléaire, DEI, SECRE, Cadarache France

Laureline Février
Institut de Radioprotection et de Sûreté Nucléaire, DEI, SECRE, Cadarache France

Rodolphe Gilbin
Institut de Radioprotection et de Sûreté Nucléaire, DEI, SECRE, Cadarache France

Jacqueline Garnier-Laplace
Institut de Radioprotection et de Sûreté Nucléaire, DEI, SECRE, Cadarache France

B. Merkel, M. Schipek, *The New Uranium Mining Boom*
DOI 10.1007/978-3-642-22122-4 © Springer 2011

status assessment of water bodies to which were added 86 other substances relevant to the national reduction program of hazardous substances in water.

Uranium, as one of those 86 substances, is regulated provisionally by an EQS of $0.3\ \mu g \cdot l^{-1}$ to use as an increment of the natural background, on the basis of freshwater ecotoxicity data. Due to the wide improvement of knowledge on uranium fate and effects in freshwaters associated to a number of recent R&D projects worldwide, the recently acquired comprehension of the ecodynamic and ecotoxicity of this element was used to update and review the uranium EQS, beginning with its PNEC for water. Based on a SSD approach, a first attempt led to a generic chronic PNEC value of $5\ \mu g\ U \cdot l^{-1}$, to be used also as an increment of the geochemical background level (Beaugelin-Seiller et al. 2009). This value, close to, if not lower than, the background in some areas in France, appeared then difficult to operate. However, regarding draft EQS derivation methodology recently released, there was a discrepancy between its determination method and those applied classically to other metals in Europe for regulatory purposes.

To converge in methods with other environmental European protection criteria for metals, the basic ecotoxicity data previously collected were reanalyzed, applying classification and selection criteria usually used in this framework to determine metal freshwater PNECs. The previous study, focused on chronic ecotoxicity, was also extended to acute aspects. Finally, acute and chronic PNECS were determined integrating the knowledge related to the relationship between physicochemical speciation of uranium, its bioavailability and in fine its ecotoxicity.

From a rigorous data selection, the first step of the study aimed to assess generic PNEC values from which the conditional PNECs were then derived. Their application to five of the main French water courses is presented.

Data and Derivation Method Selection

Two processes contribute to the final EQS value: the application of consensual criteria for the primary data selection and the way in which data could be used to determine the PNEC values.

Selection Criteria

Aiming to protect the structure and the functioning of the compartment of interest of the exposed ecosystem (biodiversity associated with the water column in our case), we considered the four classical endpoints (mortality, growth, reproduction, morbidity) generally analyzed according to their ecological relevance in terms of population dynamics (Forbes and Calow 2002). Concerning the chronic ecotoxicity, the mortality data are used only in the absence of sub-lethal effects.

Were considered acceptable data from publications giving enough information on the tested species and the experimental conditions (duration, exposure medium, endpoint, effect criterion), to judge their ecological relevance. This approach implies consequently a part of expert's judgment.

The distinction between acute and chronic exposure were based on OECD guidelines (e.g., OECD 2004) or by default on the relative information contained in the Technical Guidance Document (TGD; EC 2003). Isolated mortality data were always considered as acute, when results of 72-h tests on microalgae were assimilated to acute data for EC_{50} and chronic data for EC_{10}. More generally, chronic data collected were limited to NOEC or EC_{10} when acute data were limited to EC_{50} and LC_{50}.

Methods Applicable to Determine PNECs

Regarding acute ecotoxicity, the TGD (EC 2003) recommends only the safety factor (SF) approach to asses the acute PNEC that consists in applying a SF of 100 to the lowest EC_{50}/LC_{50} among those obtained on algae, invertebrates and fish. The literature review led to identify 177 such data, on the following taxa: algae, cnidarians, crustaceans, insects, mollusks, plants, fish. Only data on the three required taxa were then considered to apply the method.

Regarding chronic ecotoxicity, two methods are recognized to derive chronic PNECs (EC 2003), depending on the number and quality of data. Having more than 10 data on at least 8 identified taxa (fish, a second chordata family, crustaceans, insects, an animal family of an other phylum than arthropods or chordata, a family of an order of insect or other not yet represented phylum, algae, higher plants), the species sensitivity distribution approach is preferred (Postuma et al. 2002). By default, the SF approach is applied, the value of the SF depending on the nature and quantity of available data (EC 2003). As 51 chronic data were collected on only 6 of the 8 required taxa, the SF approach was imperative.

Generic PNEC Values

The European method applied for metals is declined here for the data sets of chronic and acute ecotoxicity previously collected to propose generic values of PNEC, applicable in a protective way whatever the physico-chemistry of the hydrosystem. Because of the relation between uranium speciation and its ecotoxicity, these values are overprotective for physico-chemical domains where uranium bioavailability is low vs. they are reasonably protective for domains favorable to uranium bioavailability.

Generic Acute PNEC

An homogenous corpus of acute data was collected, including an algae EC_{50} of $0.037\,mg \cdot l^{-1}$ (Pradines et al. 2005), a crustacean EC_{50} of $0.039\,mg \cdot l^{-1}$ (Markich and Camilleri 1997), a fish EC_{50} of $0.08\,mg \cdot l^{-1}$ (Bywater et al. 1991) and a cnidarian EC_{50} of $0.095\,mg \cdot l^{-1}$ (Markich and Camilleri 1997).

Applying the SF of 100 to the lowest acute ecotoxicity data led then to a generic PNEC value of $0.4\,\mu g \cdot l^{-1}$ (dissolved U).

Generic Chronic PNEC

In the absence of recent results lower than results previously consulted to derive the current uranium NQE, the lowest data of chronic ecotoxicity in freshwater remains the 7days NOEC on crustaceans of $2.7\,\mu g \cdot l^{-1}$ presented by Pickett et al. (1993). It was thus suggested to maintain the generic value of the chronic uranium PNEC in freshwaters at $0.3\,\mu g \cdot l^{-1}$, to apply as increment of the geochemical background. This value is lower than the acute one, as it is generally observed.

Conditional PNEC Values

To reduce the conservatism of the previous generic PNEC values and obtain more realistic protection criteria, the uranium speciation in natural conditions and its influence on the element bioavailability, i.e. in fine its ecotoxicity, have been taken into consideration.

Uranium Speciation in Natural Waters

Metal speciation in water depends on physico-chemistry of the medium, temperature, pH, concentrations in major ions and other water quality parameters determining the chemical species of the element. Depending on the metal, the role of each of these parameters varies. For uranium, we explored these aspects through a thermodynamic modeling using CHESS (Van der Lee and De Windt 2002) with a dedicated database (Denison 2004). Due to the lack of information on quality and quantity of organic matters, the study was realized in inorganic environment using the characteristics of French surface waters (Fig. 1), specified from internal knowledge (Beaugelin-Seiller 2004) and the part of European geochemical atlas (Salminen 2005) related to first order European rivers.

The sensibility analysis confirmed the major influence of three parameters on uranium bioavailability in French rivers, being by decreasing order pH, HCO_3^-

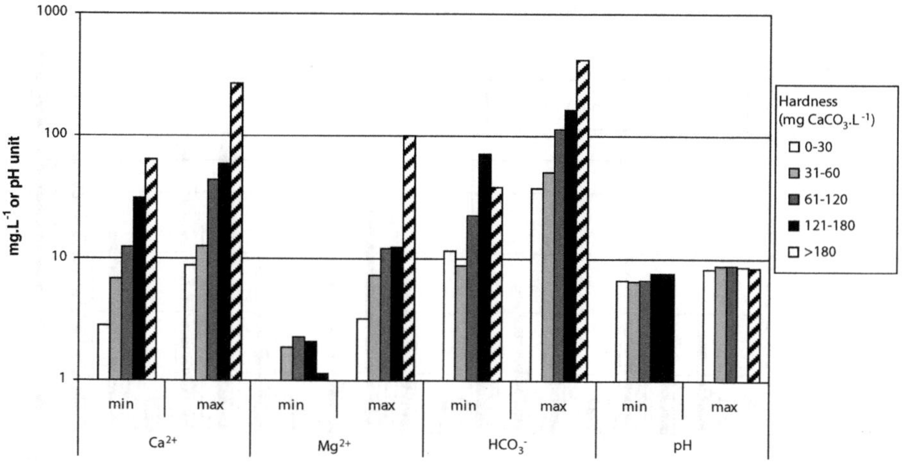

Fig. 1 Variation of the main physico-chemical characteristics impacting uranium speciation in function of hardness in French rivers

concentration and hardness (e.g., Franklin et al. 2000; Charles et al. 2002). The parameter HCO_3^-, as used throughout this paper, is defined as all the carbonated dissolved species (HCO_3^-, CO_3^{2-} and others).

Bioavailable Forms of Uranium

Because of observed inconsistencies with the free ion model (Morel and Hering 1993), some authors (Markich et al. 2000; Denison 2004; Fortin et al. 2007) as well as observations in our laboratory (Zeman et al. 2008) support the contribution of other uranium species than the free ion (uranyle ion $UO2^{2+}$) in uranium bioavailability. This hypothesis led to consider all or any combinations of the forms identified during the speciation modeling according to their relevance in terms of bioavailability.

On the basis of the current knowledge on the potential ecotoxic effects associated to uranium in connection with its bioavailability, four chemical species were considered as potentially bioavailable and thus ecotoxic: UO_2^{2+}, $UO_2(OH)^+$, $UO_2(OH)_2$ and UO_2CO_3. The fraction of the total dissolved uranium represented by these 4 species was then retained for the present study.

PNEC Determination

Considering the domains of "bioavailability" identified according to hardness, the conditional PNECs were obtained from the generic PNECs by applying the char-

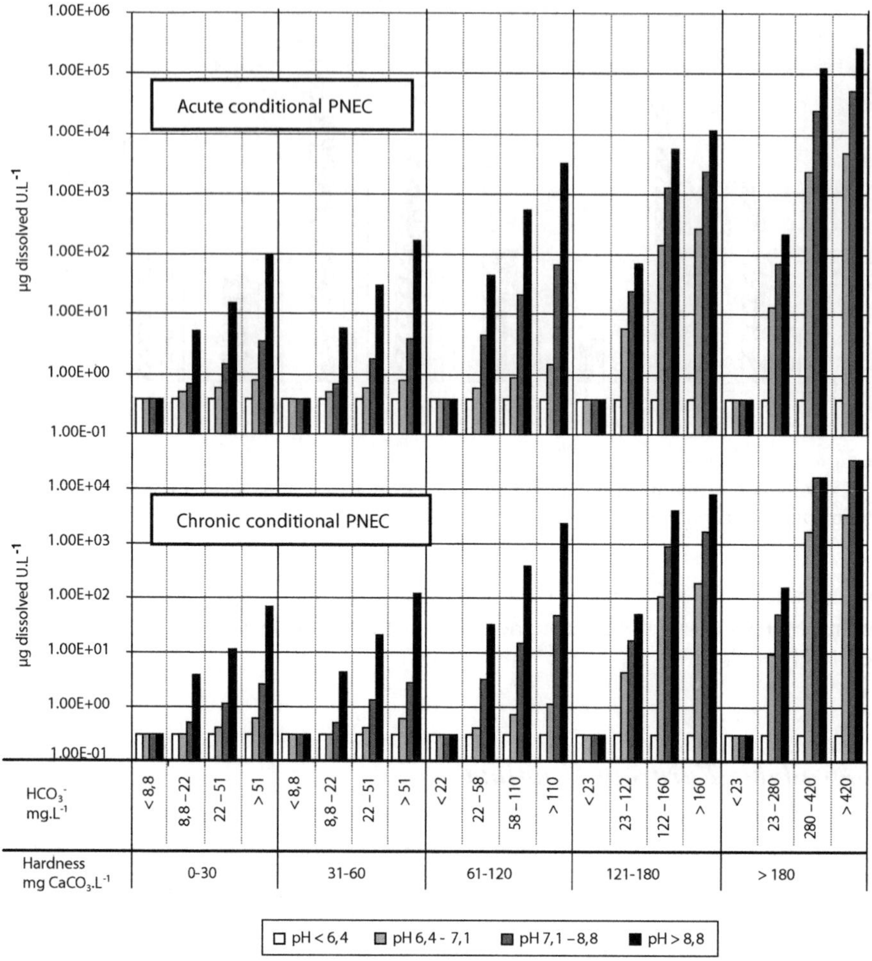

Fig. 2 Range of uranium conditional PNECs depending on the physico-chemical domain characterized by pH, hardness and carbonates concentration (*upper panel:* acute PNEC; *lower panel:* chronic PNEC)

acteristic "bioavailable" fraction of each domain, varying from less than 0.01 to about 90% of the total dissolved uranium. The values so obtained are to be used as increment of the geochemical background.

For acute exposure, conditional PNECs range for French surface waters from 0.4 to 270 000 $\mu g \cdot l^{-1}$ (dissolved uranium, Fig. 2). For chronic exposure, conditional PNECs vary from 0.3 to 8200 $\mu g \cdot l^{-1}$ (dissolved uranium, Fig. 2), as the 4 highest values were rejected, being higher than the highest tested NOEC. The meaning of the conditional PNEC values was verified towards the limit of uranium solubility in the same physico-chemical conditions. They are always lower.

Application to Some French Rivers

For illustration, the conditional chronic uranium PNECs were derived for the physico-chemical domain representative of the seven main watercourses in France (Garonne, Loire, Meuse, Moselle, Rhin, Rhône and Seine). They are summarized in Table 1.

Table 1 Chronic conditional uranium PNECs for the seven main French water courses

pH	Hardness (mg $CaCO_3 \cdot l^{-1}$)	HCO_3^- (mg $\cdot l^{-1}$)	Chronic PNEC (μg dissolved U $\cdot l^{-1}$)
	61–120	58–110	15
		> 110	49
	121–180	23–122	17
7.1 to 8.8		122–160	910
		> 160	1700
	> 180	23–280	51
	61–120	58–110	15

Dealing with the Natural Background

Because of the natural occurrence of certain substances among which metals in surface waters, the final value of their PNEC must be defined after consideration of the geochemical background. Two approaches are then recommended at the European level: if the PNEC is lower than the background or of the same order of magnitude, the way to proceed is to apply the "added risk approach" (ARA), which consists in correcting the value of PNEC by adding to it the background. Otherwise it is recommended to work with the "total risk approach" (TRA), i.e. not to correct the PNEC by the background. Whatever the scenario, it is thus necessary to always know the uranium background concentration in freshwaters to fix the water PNEC of this metal.

A practical approach consists in taking as background the 10th percentile of all the monitoring data distribution within a river or region, not influenced by known anthropological releases. Applied to the Rhône River, this led to estimate a background of about 0.2 μg $\cdot l^{-1}$ (dissolved uranium). Lower than the lowest conditional PNEC from Table 1, it constitutes an indication to apply preferably the TRA to this river.

Conclusions

From generic values of PNECs, reconducting the 0.3 μg $\cdot l^{-1}$ value for the chronic exposure and assessing a new value of 0.4 μg $\cdot l^{-1}$ for the acute one, conditional

PNECs to physico-chemical domains representative of surface freshwaters were derived, taking into consideration uranium speciation influenced by pH, hardness and carbonates.

Applying this approach to the seven main French rivers, it appears that the total risk approach will be more relevant for such water bodies, when the added risk approach could be preferable for some others with different physico-chemical characteristics. Whatever the case, it would be useful to acquire a better knowledge of these background concentrations, today not so well established, at least at a regional level.

As the geochemical modeling was realized without any organic matter, the PNEC values proposed still include some conservatism, due to the high capacity of uranium to form organic complexes (Crançon and Van der Lee 2003; Le Goff and Bonnomet 2004; Jackson et al. 2005; Ranville et al. 2007). Deeper investigations are nevertheless required in terms of R&D to solve such problems at the local level, the one of interest regarding the variability of quality and quantity of organic matters in rivers.

To conclude, the water PNEC is the first step on the long way of the determination of an integrated EQS, that should be set considering also sediment ecotoxicity, secondary poisoning and sanitary aspects in order to achieve its goal of protecting both the human health and the environment.

References

Beaugelin-Seiller K (2004) Catalogue des données "Hydrosystèmes" acquises pour l'outil CASTEAUR. report SECRE/04-04, 34 p. IRSN, Fontenay-aux-Roses, France

Beaugelin-Seiller K, Garnier-Laplace J, Gilbin R (2009) Vers la proposition d'une norme de qualité environnementale pour l'uranium en eau douce. Report DEI/SECRE/2009-015, 44 p. IRSN, Fontenay-aux-Roses, France

Bywater JF, Banaczkowski R, Bailey M (1991) Sensitivity to uranium of six species of tropical freshwater fishes and four species of cladocerans from Northern Australia. Environ Toxicol Chem 10: 1449–1458

Charles AL, Markich SJ, Stauber JL, De Filippis LF (2002) The effect of water hardness on the toxicity of uranium to a tropical freshwater algae Chlorella sp. Aquat Toxicol 60: 61–73

Crançon P, Van der Lee J (2003). Speciation and mobility of uranium(VI) in humic-containing soils. Radiochim Acta 91: 673–679

Denison F (2004) Uranium (VI) speciation: modelling, uncertainty and relevance to bioavailability models. Application to uranium uptake by the gills of a freshwater bivalve. PhD thesis, Environmental Biosciences, Chemistry and Health, 347 p. Aix-Marseille I university

EC (2000) Directive 2000/60/EC of the European Parliament and of the council of 23 October 2000 establishing a framework for Community action in the field of water policy. Official Journal of the European Communities L 327 online access via http://eur-lex.europa.eu/

EC (2003) Technical guidance document in support of Commission Directive 93/67/EEC on risk assessment for new notified substances and Commission Regulation (EC) No 1488/94 on risk assessment for existing substances, Directive 98/8/EC of the European Parliament and of the Council concerning the placing of biocidal products on the market. Part II. Luxembourg, Office for Official Publication of the European Communities

Forbes VE, Calow P (2002) Extrapolation in ecological risk assessment: balancing pragmatism and precaution in chemical controls legislation. Biosciences 52: 249–257

Fortin C, Denison FH, Garnier-Laplace J (2007) Metal – phytoplankton interactions: modelling the effect of competing ions (H+, Ca2+ and Mg2+) on uranium uptake. Environ Toxicol Chem 26: 242–248

Franklin NM, Stauber JL, Markich SJ, Lim RP (2000) pH-dependent toxicity of copper and uranium to a tropical freshwater alga (*Chlorella* sp.). Aquat Toxicol 48: 275–289

Jackson BP, Ranville JF, Bertsch PM, Sowder AG (2005). Characterization of colloidal and humic-bound Ni and U in the "dissolved" fraction of contaminated sediment extracts. Environ Sci Technol 39: 2478–2485

Le Goff F, Bonnomet V (2004) Devenir et comportement des métaux dans l'eau: biodisponibilité et modèles BLM. Rapport technique, Direction des Risques Chroniques, Unité « Évaluation des Risques Écotoxicologiques », mars 2004, 85 p. INERIS, Verneuil en Halatte, France

Markich SJ, Camilleri C (1997) Investigation of metal toxicity to tropical biota: Recommendations for revision of the Australian water quality guidelines. Supervising Scientist Report 127, Supervising Scientist, Canberra, Australia

Markich SJ, Brown PL, Jeffree RA, Lim RP (2000) Valve movement responses of Velesunio angasi (Bivalvia: Hyriidae) to manganese and uranium: An exception to the free ion activity model. Aquat Toxicol 51: 155–175

MEDAD (2007) Circulaire du 7 mai 2007 définissant les "normes de qualité environnementale provisoires (NQEp)" des 41 substances impliquées dans l'évaluation de l'état chimique des masses d'eau ainsi que des substances pertinentes du programme national de réduction des substances dangereuses dans l'eau

Morel FMM, Hering JG (1993) Principles and Applications of Aquatic Chemistry. John Wiley & Sons, New York, NY, USA

OECD (2004) Essai n° 202: Daphnia sp., essai d'immobilisation immédiate Lignes directrices pour les essais de produits chimiques/Section 2: Effets sur les systèmes biologiques. 12 p. (available on line at the following address http://www.oecdbookshop.org/oecd/display.asp?lang=fr&sf1=DI&st1=5LMQCR2K7S0W)

Postuma L, Suter II GW, Traas TP (2002) Species Sensitivity Distributions in Ecotoxicology. Lewis Publishers, Boca Raton London New York Washington DC

Pickett JB, Specht WL, Keyes JL (1993) Acute and chronic toxicity of uranium compounds to Ceriodaphnia dubia. Westinghouse Savannah River Co, report WSRC-RP-92-995, prepared for the US Departement of Energy (US-DoE), contract DE-AC09-89SR 18035

Pradines C, Wiktor V, Camilleri V, Gilbin R (2005) Development of biochemical methods to estimate the subcellular impact of uranium exposure on *Chlamydomonas reinhardtii*. Radioprotection 40: S163–S168

Ranville JF, Hendry MJ, Reszat TN, Xie Q, Honeyman BD (2007) Quantifying uranium complexation by groundwater dissolved organic carbon using asymmetrical flow field-flow fractionation. J Cont Hydrol 91: 233–246

Salminen R (2005) Geochemical Atlas of Europe. Part 1: Background Information, Methodology and Maps. Espoo: Geological Survey of Finland

Van der Lee J, de Windt L (2002) CHESS Tutorial and Cookbook/Version 3.0, users manual LHM/RD/02/13, Ecole des mines de Paris, Fontainebleau, France

Zeman F, Gilbin R, Alonzo F, Lecomte-Pradines C, Garnier-Laplace J, Aliaume C (2008) Effects of waterborne uranium on survival, growth, reproduction and physiological processes of the freshwater cladoceran Daphnia magna. Aquat Tox 86: 370–378

Radiological Impact Assessment of Mining Activities in the Wonderfonteinspruit Catchment Area, South Africa

Rainer Barthel

Abstract. An assessment of radiological impacts of gold mining activities in the Wonderfonteinspruit Catchment Area (WCA) was carried out on behalf of the National Nuclear Regulator (NNR) of South Africa. The radiological risks in the WCA comprise both the effects of current mine water discharges and diffuse emissions of seepage and runoff from slimes dams, as well as legacies of past radioactive discharges, now present as radionuclides stored in sediments and soils.

Introduction

The Wonderfonteinspruit catchment area (WCA) is a complex territory situated in the south-west of Johannesburg, with the upper section of the catchment in the Gauteng Province and the lower part in the North West Province (Fig. 1). The WCA constitutes the eastern catchment of the Mooi River. Most of the catchment area is underlain by dolomitic rocks with three of the dolomite compartments dewatered by gold mines. The catchment of the Wonderfonteinspruit covers parts of two goldfields in the Witwatersrand basin, the western part of the "West Rand" goldfield in the upper region of the catchment between Krugersdorp, Randfontein and Mohlakeng, and the "Far West Rand" goldfield, comprising gold mines near Bekkersdal, Westonaria and Carletonville.

Rainer Barthel
Brenk Systemplanung GmbH, Heider-Hof-Weg 23, D-52080 Aachen

B. Merkel, M. Schipek, *The New Uranium Mining Boom*
DOI 10.1007/978-3-642-22122-4 © Springer 2011

For more than a century, major gold mining activities have been carried out in the WCA, with the potential for pollution of surface and ground waters. Various contamination sources such as

- large active gold mines, which discharge fissure and process water into the aquatic environment, and
- numerous diffuse sources like old/abandoned mine workings and deposits of mining/milling residues (so-called slimes dams)

have led to the pollution of ground and surface waters and of sediments/soils in this area with radionuclides of the natural decay chains.

The Mooi River system and the WCA in particular have already been under investigation concerning the effects of radioactive contamination of the aquatic environment. Based on the assumption that mainly the use of contaminated water for drinking purposes gives rise to significant radiation exposures to the public, in IWQS (1999) the resulting radiation dose levels were assessed as insignificant for the majority of sites monitored. Wade et al. (2002) addressed the questions whether radionuclides from natural decay chains have been accumulated in river sediments and to which extend and under which conditions they may be remobilized and can pose a threat to human health. Coetzee et al. (2006) comprises a compilation of studies regarding inter alia the analysis of the inventory of uranium sources and pathways in the catchment area, the extent of sediment contamination, the speciation of uranium in contaminated sediments, and risk assessments for water users

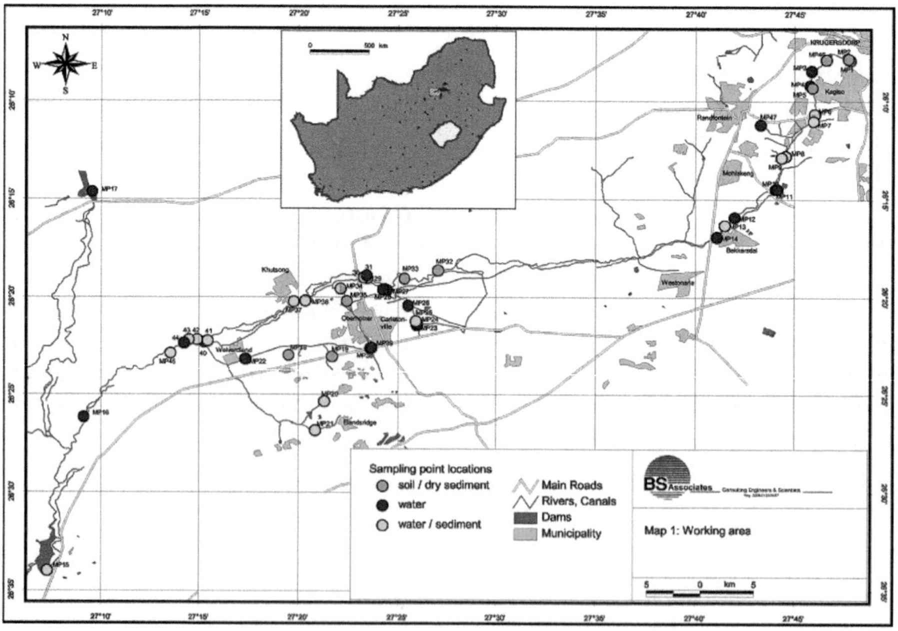

Fig. 1 Wonderfonteinspruit Catchment Area; sampling locations defined for investigating potential radiological impacts of former and present mine water discharges

with respect to the chemical toxicity of uranium. It was concluded that a significant amount of uranium (several tons per year) is entering the Wonderfonteinspruit via point discharges and large-scale diffuse discharges, that uranium is concentrated in the fluvial sediments of the river system, and that the immobilization of uranium in the sediments is not necessarily irreversible. The results of this study indicated that uranium poses a hazard to water users in the catchment area due to its chemical toxicity. It was concluded that a full radiological risk assessment, looking at both, dissolved radionuclides in water and radionuclides bound to sediments, would be required to determine current and future risks due to radiological impacts of the mine water discharges.

Objectives

People residing in the WCA make use of the contaminated water, e.g. for usage in households, agriculture, recreation, and fishing. As a consequence of concerns regarding the radiological health hazards associated with the real or potential use of contaminated water, the National Nuclear Regulator (NNR) found it necessary to conduct an independent survey and assessment of radiological impacts of the mine water discharges on the local population. In November 2006, the NNR commissioned BS Associates Ltd (subsidiary of Brenk Systemplanung GmbH) with a respective project to

- measure the levels of long-lived radionuclides of the natural decay chains of U-238, U-235 and Th-232 in water and sediment samples taken from selected areas of the WCA, and to determine typical activity ratios between relevant radionuclides, which can be applied in continuative investigations and for the design of efficient monitoring programs,
- determine likely and possible radiation exposures for members of the public arising from former and actual mining activities, especially by the discharge of contaminated mine water, but also by diffuse emissions of contaminated waters from slimes dams (seepage, runoff during events of heavy rain), and estimating effective doses for realistic and potential scenarios of usage of polluted environmental media assessing the respective exposure pathways,
- compare the effective doses calculated for members of the public with dose limits set by the regulatory bodies, and
- elaborate recommendations for mitigating adverse radiological impacts.

Sampling

Based on former investigations concerning the contamination of surface water and sediments in the gold mining area of the WCA (IWQS 1999; Wade et al. 2002; Coetzee et al. 2006) and on geographical maps, a sampling program was drafted by selecting dams along the Wonderfonteinspruit and in its catchment area, as well as canals/furrows conducting water discharged from the gold mines to the Wonderfonteinspruit, where elevated activity concentrations could be expected. From a site visit performed with a representative of the NNR, and from meetings with experts, who participated in former investigations of water and sediment contamination in the WCA, valuable knowledge was obtained about some recent changes in the surface water system compared to the former situation. Additional sampling points were defined to cover locations with a potential of radiologically relevant use of contaminated water bodies or of contaminated soil. The sampling program was established to provide a (minimal) data base necessary for

- the radiological characterization of important mine water discharges as well as of runoff/seepage water from slimes dams that influence the radioactive contamination of the Wonderfonteinspruit,
- understanding roughly the radionuclide transport in the Wonderfonteinspruit including the effects of radionuclide immobilization in sediments (especially their retention in wetlands), and of dilution with uncontaminated water entries (e.g. discharges from sewage works),
- the estimation of realistic and potential radiation exposures to the public living near to accessible surface waters like streams, canals, furrows, dams and ponds as well as in areas, where soil may be radioactively contaminated from former operations (abandoned dams, old riverbeds, floodplains, etc.),
- the specification of the natural background activity concentrations and the estimation of resulting natural radiation exposures, and
- the comparison with former activity measurements and dose estimations.

Sampling of water and/or sediment/soil was performed at 47 measurement points (denoted MP1 to MP47). The locations of the sampling sites are shown in Fig. 1, which also indicate the types of samples taken. For water samples, some general characteristics (pH, EC, temperature) were measured onsite. With respect to sediment/soil samples, we distinguished between wet samples taken from the bottom of water bodies (dams, furrows, ponds, wetlands) and dry samples, which were taken from contaminated ground, desiccated dams, and the old riverbed.

In total, at the 47 measurement points, 38 water samples and 38 sediment/soil samples were taken. Water samples comprised 10 samples taken from canals used for the transport of discharged mine water and 28 samples taken from dams, ponds, furrows and wetlands. The soil/sediment samples comprised 23 wet samples taken from dams, ponds and furrows, and 15 dry samples. At some measurement points, dry samples were taken from different depths or locations to check spatial variations of radioactive contaminations.

Laboratory Analyses

The water and sediment/soil samples were analyzed with respect to the activity concentrations (water) or specific activities (sediment/soil, related to dry mass) of the long-lived radionuclides of the natural decay chains of U-238 (U-238, U-234, Th-230, Ra-226, Pb-210, Po-210), U-235 (U-235, Pa-231, Ac-227), and Th-232 (Th-232, Ra-228, Th-228). The laboratory analyses were performed by IAF – Radioökologie GmbH. The water samples were not filtered, because both the dissolved and the suspended radioactive matter are relevant for exposure pathway analyses. High resolution γ-spectrometry was applied for all samples, allowing the measurement of activities for most of the radionuclides of interest with low uncertainties. Standard uncertainties of measurements depend on activity levels and were typically in the range of 5 to 10%. For radionuclides with low emission yields (like U-235, Th-230, Pa-231) uncertainties were in the range of 10 to 20%. For Po-210 and also for U and Th isotopes, α-spectrometric measurements were performed. For Po-210 and Th-232, α-spectrometry has to be applied, because these radionuclides are pure α-emitters. U-234 activities were also measured exclusively by α-spectrometry, due to high decision threshold of the γ-spectrometric analysis of U-234. As the α-spectrometry requires relatively expensive radiochemical sample preparations, only few samples were selected for the application of this method to measure specific activities of U isotopes (U-238, U-234 and U-235), Th isotopes (Th-232, Th-230, Th-228, and Th-227, the short-lived daughter of Ac-227) and of Po-210. Based on results of the γ-spectrometric measurements, only 12 water samples and 8 soil/sediment samples were selected for α-spectrometric measurements of the U and Th isotopes. Po-210 was measured in 5 of these 12 water samples.

Data Quality

The high quality of the performed γ- and α-spectrometric activity measurements was demonstrated by correlation analyses. This is exemplarily shown in Fig. 2 by

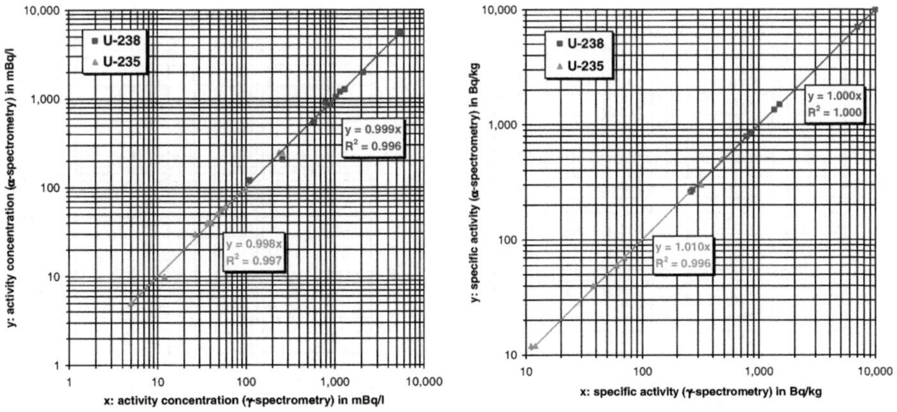

Fig. 2 Comparison of activity concentrations (12 water samples, *left*) and specific activities (8 sediment samples, *right*) measured for U-238 and U-235 by means of γ-spectrometry and α-spectrometry, respectively

comparison of α- and γ-spectrometric analyses for U-238 and U-235 in water and sediment samples. Furthermore, comparisons of specific activities or activity concentrations, which were independently measured and should exhibit physically defined ratios, were used to prove the quality of measurement methods applied.

Natural Background Values

For assessing radiological impacts resulting from gold mining, the activity levels observed at sampling sites have to be compared with natural background values.

For the relevant radionuclides, background values for activity concentrations were estimated from values measured for the Klerkskraal Dam, which is located far outside of the mining region (MP17, see Fig. 1). The background values for specific activities of sediment/soil were estimated by comparison of various sites with low values of the specific activities measured. Based on this approach estimated background values used for dose calculations and radiological assessments are compiled in Table 1.

For gamma dose rate (GDR), a background value of 80 nSv/h was estimated by comparison of GDR-measurements performed at various sampling sites with

Table 1 Background values for specific activities of sediment/soil ($Cbg_{S,r}$ in Bq/kg) and activity concentrations of water ($Cbg_{W,r}$ in mBq/l) for long-lived radionuclides (r)

	U-238, U-234	Ra-226	Pb-210	Th-230, Po-210	U-235	Pa-231, Ac-227	Ra-228	Th-232, Th-228
$Cbg_{S,r}$	40	40	40	40	2	2	20	20
$Cbg_{W,r}$	40	5	10	–	2	0	5	–

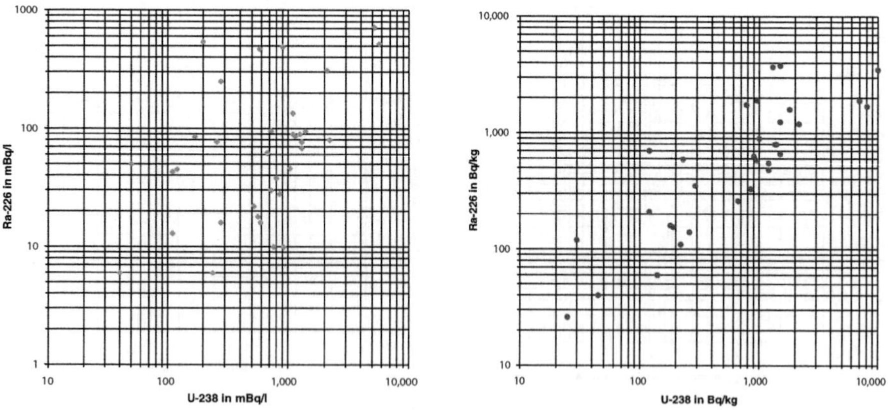

Fig. 3 Comparison of activity concentrations in water samples (*left*) and specific activities in soil/sediment samples (*right*) for U-238 and Ra-226

low/negligible external gamma radiation 1 m above the ground. This background value coincides, e.g., with the GDR measured at the Klerkskraal Dam.

U-238 and Ra-226 Activity in Water and Soil/Sediment Samples

Activity concentrations and specific activities of U-238 and Ra-226 measured in water and sediment/soil samples, respectively, are compared in Fig. 3, revealing weak correlations between these radionuclides. Their activities in the samples are determined by various site specific contamination sources and transport processes as briefly discussed in the following.

Radionuclide Migration in the WCA

The long-lasting mining related discharges of naturally occurring radionuclides from point and diffuse sources into the WCA resulted in a complex pattern of radioactive contamination of water bodies, sediments and soils throughout the catchment area with high temporal and spatial variability of contamination levels and activity ratios of long-lived radionuclides. The mining related sources of radioactive discharges into the WCA comprise

- discharges of radionuclides with mine waters and process waters,
- seepage from slimes dams bearing dissolved radionuclides, and
- erosion of contaminated particles from slimes dams and their dispersion with the run-off water after heavy rainfalls.

Radionuclides can enter the aquatic environment in various forms such as

- dissolved radionuclides, often in form of aqueous complexes,
- in colloidal form or adsorbed to other colloids (e.g. Fe-hydroxides, humic substances), or
- as particulates (e.g. eroded particles from slimes dams).

The source terms (types and activity levels of radionuclides) are very different. Radioactivity of mine waters and seepage from slimes dams is often dominated by uranium nuclides, which are dissolvable under oxidizing and acidic conditions, whereas particulate emissions from slimes dams reveal their inherited nuclide vector, which often show a residual enrichment of radium over uranium.

The further fate of the radionuclides in the environment and their transport along the aqueous pathway is influenced by various hydro-geochemical and hydrodynamic processes, which themselves are subject to diurnal and seasonal fluctuations controlled by various environmental parameters (e.g. temperature, pH, flow rates of streams, presence of anions and/or organic ligands).

The sediments in ponds and water storage dams can accumulate and act as a sink for the radionuclides discharged into the catchment area. Processes which can result in an accumulation of the waterborne contaminants in the sediment phase comprise the sedimentation of contaminated particles and the uptake of dissolved or colloidal radionuclides, e.g. by precipitation/co-precipitation, flocculation of colloids, and/or sorption especially onto Fe(III) and Mn(IV) oxi-hydroxides or the sedimentary organic matter. The extent to which the various radionuclides can be removed from the aqueous phase by sorption processes on sediments/soils is rather element specific, and the tendency of a radioelement to sorb to some type of surface can vary over several orders of magnitude.

Although the sediments can provide a (temporary) sink for the radionuclides transported with the stream water and reduce the contaminant concentration in the flowing water and prevent thus further downstream transport of radionuclides, they can provide themselves a source for environmental risks and hazards, e.g. due to the dispersion of contaminated sediments on floodplains after flooding events or uptake of suspended sediment particles by watering cattle. Furthermore it has to be taken into account that the radionuclides accumulated in the sediments can be remobilized into the aquatic phase by various physical and chemical perturbations of the systems, which can lead to desorption, dissolution or dispersion of sediment bound radionuclides. Thus the transport and immobilization of radionuclides in the aquatic environment represents a rather complex system with a high dynamic and variability, superimposed by the variability of the radionuclide source term, which can lead to substantial changes in the observed radionuclide concentrations in water bodies and sediment/soil systems in time.

Dose Assessments

The analysis and assessment of radiological impacts requires specifications of the

- radioactively contaminated environmental media (water bodies, soil),
- scenarios for human activities leading to radiation exposures, and
- exposure pathways, which specify the routes and mechanisms by which radioactive material can reach or irradiate human receptors directly from the polluted media or via the food chain (fish, crops, livestock products), leading to relevant contributions to effective doses.

The hierarchical model of the considered environmental media, scenarios and exposure pathways is schematically depicted in Fig. 4.

In Fig. 4, abbreviations for referencing to the 15 exposure pathways considered are prefixed as headline of the exposure pathway boxes. They correspond to the first two letters of the pathway steps listed. The last step (=> person) is omitted in the abbreviations. For the sake of brevity, the livestock products (milk and meat from cattle, egg and meat from poultry) are also omitted in the abbreviations.

The dose model used for estimating contributions of the particular exposure pathways to effective doses of persons is based on the Licensing Guide LG-1032 (CNS 1997), and on recommendations of the International Atomic Energy Agency (IAEA 2001), and of the German Radiation Protection Commission (RPC 2006).

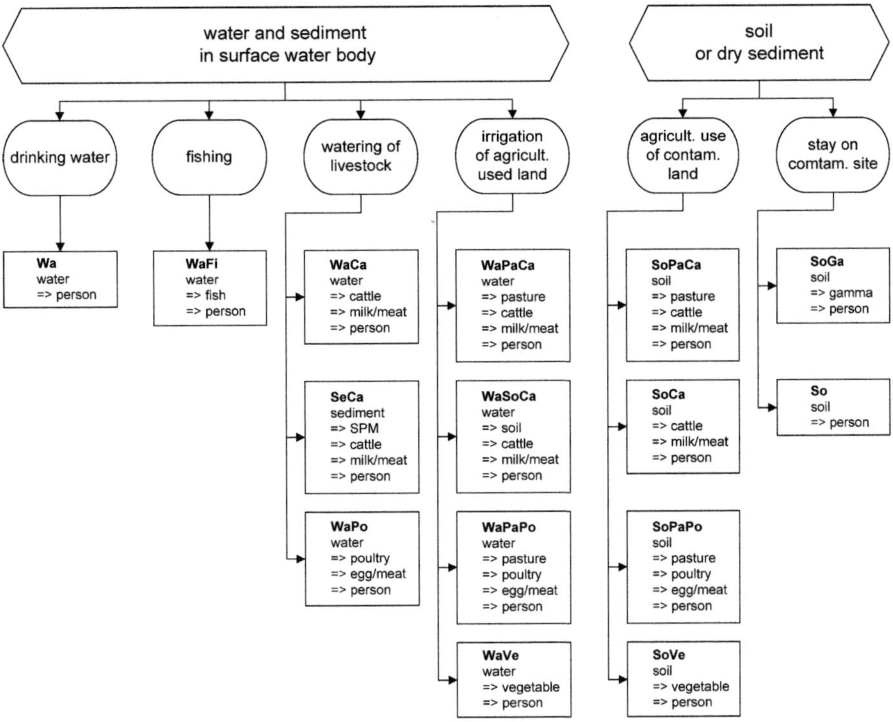

Fig. 4 Environmental media, scenarios and exposure pathways taken into account

During the sampling campaign in December 2006, the measurement points and their surroundings were assessed with respect to the realities of exposure scenarios and pathways. At many sites, realities regarding certain scenarios were observed directly and confirmed in conversations with local residents. If a high likelihood was assessed for scenarios due to the geographical conditions and habits of nearby living people, the respective exposure pathways were qualified as "realistic". For distinct sites, certain exposure pathways, albeit not assessed as realistic, were assessed as "potential". For dose calculations, each of the measurement points was considered as representing a site, and doses were calculated on a "per sample" basis, i.e., using the activity measurements for respective water and sediment/soil samples. Figure 5 shows the distribution of incremental doses (above natural background) calculated for critical age groups of the public at the sites considered.

The representativeness of the incremental effective doses estimated as "per sample" doses is limited due to spatial variations of contamination levels at the sites considered and due to seasonal and diurnal variation of mining discharges, diffuse emissions and conditions for the transport of pollutants in the environment as well. The "per sample" doses underestimate the potential radiation exposures of the public at sites, where not all relevant polluted media were sampled. The "per sample" doses also do not take into account realistic or potential exposures

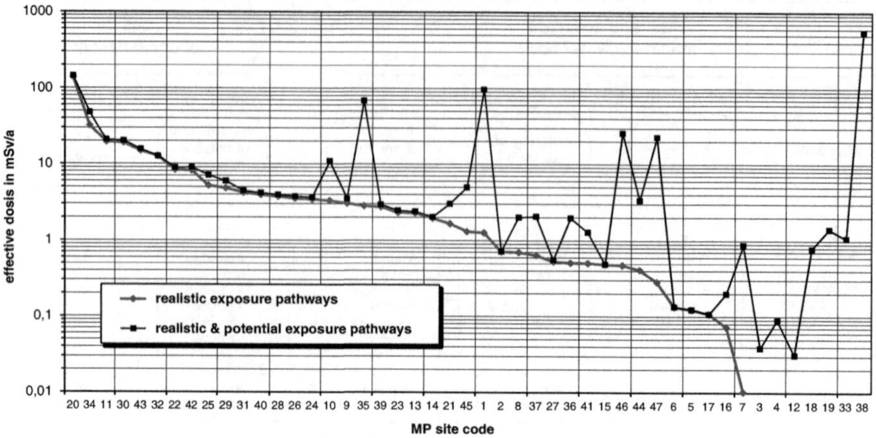

Fig. 5 Incremental effective doses of the critical age group for "realistic" and "realistic & potential" exposure pathways (using transfer factor sets from CNS (1997) and IWQS (1999))

from nearby sampling points, which may cause effective doses of the public above those estimated for a particular sampling point.

Conclusions

The incremental effective doses estimated on a "per sample" basis disagree significantly with those reported in the past. It was established that the (potential) usage of water for drinking does not dominate the radiation exposure. Three significant exposure pathways that were disregarded in former dose assessments may cause elevated contamination of food and dominate radiation exposures: 1) interception of radioactive substances by the leafs of pasture and crops irrigated with mine water, 2) uptake of radioactive suspended particulate matter by cattle during watering at the banks of surface water bodies, and 3) agricultural use of land contaminated with slimes transported by storm water runoff from slimes dams. At about 50% of the investigated sites, these three exposure pathways can lead to radiation exposures of the public above the effective dose limit of 1 mSv/a, with maximum estimated doses up to some tens of mSv/a. To prevent or diminish such radiological impacts of mine water discharges and old mining legacies as far as reasonably achievable, various protection measures were proposed.

References

CNS (1997) Licensing Guide LG-1032, Guideline on the assessment of radiological hazards to members of the public from mining and minerals processing facilities (Rev. 0), Council for Nuclear Safety of the RSA

Coetzee H, Winde F, Wade PW (2006) An assessment of sources, pathways, mechanisms and risks of current and potential future pollution of water and sediments in gold-mining areas of the Wonderfonteinspruit catchment, Water Research Commission Report No. 1214/1/06

IAEA (2001) Generic models for use in assessing the impact of discharges of radioactive substances to the environment, Safety Reports Series No. 19, Vienna, 2001

IWQS (1999) Report on the radioactivity monitoring programme in the Mooi River (Wonderfonteinspruit) catchment; Institute for Water Quality Studies, Department of Water Affairs and Forestry, July 1999

RPC (2006) Guideline for the evaluation of radiation exposures from environmental pollution caused by mining activities (Evaluation Guideline – Mining), Radiation Protection Commission (Germany), Rev. 06(1846)SSK/A3-183/U2 per 2006-11-13

Wade PW, Woodborne S, Morris WM, Vos P, Jarvis NV (2002) Tier 1 risk assessment of selected radionuclides in sediments of the Mooi River catchment; Water Research Commission Report No. 1095/1/02

Challenges in Assessing Uranium-Related Health Risks: Two Case Studies for the Aquatic Exposure Pathway from South Africa – Part I: Guideline and Toxicity Issues and the Pofadder Case Study

Frank Winde

Abstract. Comparing guidelines for U in drinking water from across the world reveals that recommended limits differ significantly. Furthermore, most current limits are based on nephrotoxic effects of U as found in animal experiments after short-term, high-dose U exposure and take not into account latest finding that U displays a much wider spectrum of toxic effects after long-term, low dose exposure. Epidemiological evidence from the Pofadder case study in South Africa suggests a link between elevated U levels in borehole water and increased incidences of leukemia in local residents.

Introduction

Recently in South Africa, a range of reports in news media including TV, radio and print media about potential impacts of mining-related uranium pollution on the environment, somewhat sensationalized at times, has unsettled the general public and raised serious questions about possibly associated health risks.

The immanent threat of highly polluted mine water decanting from several large flooded mine voids into densely populated areas in and around Johannesburg has further heightened the general sense of urgency to address these questions from a scientific point of view.

Using 2 cases studies from South Africa this 2-part paper attempts to outline some of the difficulties researcher face in reliably assessing human health risks resulting from exposure to uranium (U) polluted water.

Frank Winde
Mine Water Research Group, School of Environmental Sciences and Development, North-West University, Potchefstroom Campus, Private Bag X6001, Potchefstroom, 2520, Republic of South Africa

B. Merkel, M. Schipek, *The New Uranium Mining Boom*
DOI 10.1007/978-3-642-22122-4 © Springer 2011

In the first part guideline values for acceptable U concentrations in drinking water from various national and international sources are compared and critically analyzed in order to arrive at a benchmark for assessing locally found U levels. In a second step a brief overview on recent findings regarding the toxicity of uranium is presented.

Against this background findings from two South African case studies are reported which one is covered in this paper (part 1 of 2) while the second case study is explored in part 2 of the series. The case studies reported in this paper is largely based on the 1996-project of the Water Research Commission of South Africa (WRC, project no. K5/839) that investigated possible impacts of naturally elevated U levels in borehole wells on the health of a rural farming community consuming the well water.

Guideline Issues

Comparing recommended limits for the concentration of U in best quality drinking water reveals that recommended values vary considerably between the different sources by up to a factor of 500 (Table 1).

The U limits of WHO as well as of the U.S. EPA, Health Canada and the Umweltbundesamt (Federal Environmental Bureau) in Germany are all based on the nephrotoxicity of U observed in two 1998-studies exposing rats and rabbits for 30 to 91 days to drinking water spiked with uranyl nitrate (Gilman et al. 1998a,b; von Soosten 2008; WISE 2001). Based on this a tolerable daily intake (TDI) of

Table 1 Recommended limits for the concentration of U-238 in best quality drinking water according to different sources

Organization/Country	Year	U-238 conc. limit [μg/l]
World Health Organization (WHO) – drinking water	1998	2
	1/2003	9
	9/2004	15
Environmental Protection Agency of the USA (U.S. EPA) – drinking water	2006	30[a]
Health Canada – drinking water	2001	20
Germany; Federal Institute for Risk Evaluation – mineral water (Tafelwasserverordnung)	2008	
Germany; Federal Environmental Bureau (UBA) – drinking water limit proposed by the to EU	2008	
South Africa (Department for Water Affairs) – domestic water	1985	1000
– domestic water	1996	70
Range		2–1000

[a] In the case of the US-EPA the originally determined limit of 20 μg/l was finally set at 30 μg/l to cater for cost considerations in water treatment (WISE 2008).

6 µg/kg bodyweight per day was defined by the WHO and subsequently reduced to 0.6 µg/kg × d to allow for variations in vulnerability of exposed populations i.e. between adults and children. Being based on a short-term, high-dose animal experiments this approach to determine U limits for water consumed by humans leaves uncertainty not only regarding the transferability of results from rats/rabbits to humans and the effects of uranium speciation other than uranyl nitrate but also regarding the impacts of exposure times exceeding a three months period.

The original WHO limit for U in drinking water (2 µg/l) is now proposed by the Bundesanstalt für Risikobewertung (Federal Institute for Risk Evaluation) in Germany for mineral waters used to prepare formula-based baby food, while for drinking water the Umweltbundesamt (German Federal Environmental Bureau) proposes a limit of 10 µg/l for life-long exposure to be legislated in the European Union (EU) (von Soosten 2008; UBA 2008; BfR/BfS 2006, 2007; Konietzka et al. 2005; Dieter 2000). Both values are significantly lower than the one of the US-Environmental Protection Agency (EPA: 30 µg/l) or South Africa (70 µg/l) as two major U producers. Preceding a predicted globally renewed interest in U mining the WHO, in January 2003, increased the U limit from 2 to 9 µg/l and again in September 2004 to 15 µg/l (WHO 1998/2005/2006; UBA 2005). Coincidentally, from 2003 onwards, the spot price for U rose by more than 1000% from 10 US$/lb U_3O_8 to a peak of 136 US$/lb in June 2007 sparking questions on a possible link between the two developments (Fig. 1).

Such link may relate to a 1959-agreement between the International Atomic Energy Agency (IAEA) and the WHO, which according to Bertell (1999) constitutes a serious conflict of interest. It may constrain the work of the WHO by stipulating the following: "*... whenever either organisation proposes to initiate a programme or activity on a subject in which the organisation has or may have a substantial interest, the first party shall consult the other with a view to adjusting the matter by mutual consent*" (WHO 1999). Established by the United Na-

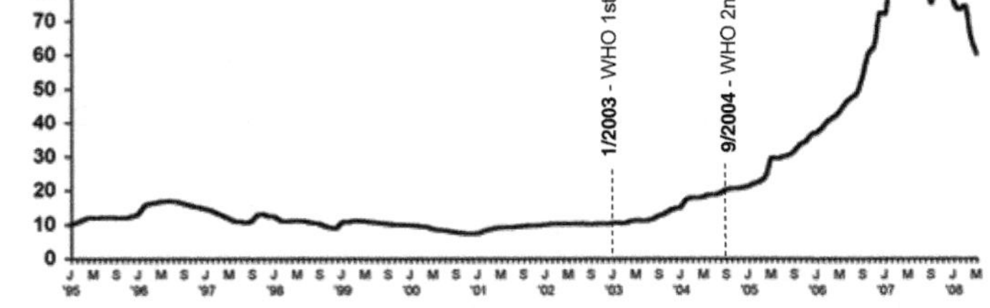

Fig. 1 The timing of the two successive increases of the WHO limit for U in drinking water in relations to the spot price for U_3O_8 (source U-price chart: WNN, 2009)

tions in 1956 the IAEA was originally tasked to prevent the proliferation of nuclear arms while promoting the peaceful use of nuclear energy. However, according to Bertell (1999), since established the promotional goals of the IAEA frequently took priority over regulatory aspects. This may explain why the U-limits set by the WHO were significantly lowered twice during a comparably short period when more U-mining was needed to satisfy a growing global demand.

According to the WHO the raise of the recommended U limit for drinking water was based on increasing evidence that drinking water is the most important source of U reducing the need to cater for possible U intake from other sources such as food. While maintaining the originally devised TDI value for U of 0.6 μg/kg per day the assumed contribution of drinking water to the total U intake was thus increased from 10% (i.e. assuming that 90% of the U intake come from other sources) initially to 50% (January 2003) and 18 months later to 80% (September 2004) (von Soosten 2008).

However, applying the revised drinking water limits of the WHO to different age groups assuming age-specific body weights and water consumption indicates that the TDI-value the WHO claims to have maintained is exceeded significantly for small children (by 25%) and even more so for babies (by up to 150%) (Table 2).

Table 2 illustrates that at the original U-limit of 2 μg/l the daily intake of U via drinking water even for formula-fed babies remained below the TDI of 0.6 μg/kg bw x d. However, this is no longer the case after the WHO first increased the limit to 9 μg/l resulting in an average daily U intake of 0.9 μg/kg bodyweight, which exceeds the TDI set by the WHO itself by 50%. This worsened when the limit, in 2004, was raised from 9 to 15 μg/l with the latter representing 750% of the original U limit. At this U level the average consumption of drinking water leads to daily U intakes of small children and babies that exceed the TDI of the WHO by 25 and 150% respectively. This questions how far the successive increases of the U limits for drinking water were indeed justified.

Table 2 Daily intake of U through drinking water complying with the increased U-limits of the WHO for different (hypothetical) age groups in relation to the TDI for U set by the WHO (0.6 μg/kg bodyweight [bw] and day). Bold: Daily U intake via drinking water that exceed the TDI value) (WHO 1998, 2006)

TDI: 0.6 μg U/kg bw × d *(based on nephrotoxicity)*	U limit drinking water [μg/l]	Corresponding U intake via drinking water [% of total U intake]	Resulting U intake [μg/kg bw × d]		
			Adult (60 kg; 2 l/d)	Child (20 kg; 1 l/d)	Baby (5 kg/0.5 l/d – formula)
1998	2	10% (food/inhal.: 90%)	0.07	0.01	0.02
2003	9	50% (food/inhal.: 50%)	0.3	0.45	*0.9* (150% TDI)
2004	15	80% (food/inhal.: 20%)	0.5	*0.75* (125% TDI)	*1.5* (250% TDI)

Uranium Toxicity: Recent Findings

Like most non-essential heavy metals U is chemo-toxic to humans and has been reported to cause irreversible damage to kidneys (nephrotoxic) if consumed above certain concentrations (UBA 2005). Schnug and Lindemann (2006) found that between 1986 and 2004 the consumption of (partly U containing) mineral water and the number of kidney replacement therapy increased at the same rate in Germany, Austria and the USA.

In addition to this U is also radioactive and as such potentially damaging to biological tissue. Generally the biochemical health effects of U are considered to overweigh the radiation-based impacts.

In order to assess radiation-based health impacts the absorbed-dose model of the International Commission on Radiological Protection (ICRP) is commonly used proposing that biological damage increases proportionally to the received and absorbed dose. Most data underlying this model are derived from life-span studies of surveyors of the atomic bomb explosion in Hiroshima and Nagasaki. However, as the studied population has been exposed to a short-term very high dose of external gamma radiation it is uncertain to what extent that can be used to assess effects of long-term exposure to low dose internal alpha radiation associated with incorporated U. Specifically addressing this gap, the French research initiative "Environhom" for the first time in radioprotection demonstrated that "*biokinetics and toxicity of radionuclides after chronic exposure may not be simply extrapolated from data acquired after acute exposure*" and "*… (Results)… showed that many deterministic effects may be induced after ingestion of small amounts of radionuclides …*" (IRSN 2005).

The limited reliability of existing absorbed dose based models is increasingly illustrated by deviations found between predictions of the model and epidemiological data. The latter increased significantly after 1990, when the collapse of the socialist block resulted in formerly classified data on the health of large numbers of eastern European uranium miners became available to researchers as well as a result of the first Gulf War 1990 where depleted uranium (DU) was first used in ammunition. This triggered a number of studies investigating a possible link between the use of DU and a combination of various cancers and health problems collectively known as "Gulf War Syndrome". Based on the world's largest cohort comprising 59,000 U miners that worked at the East German Wismut SDAG between 1945 and 1990 Jacobi et al. (1997) found a 20–70 times higher risk of contracting liver cancer from occupational U exposure than indicated in their previous study based on extrapolated data from the atomic bomb survivor study. Investigating a group of 29 Gulf War Veterans who retained fragments of DU-shrapnels in their bodies indicated continued elevation of U levels in urine for several years after first exposure and statistically correlated with lowered performance in neurocognitive examinations (McDiarmid et al. 2000). Exposing rats to U, IRSN (2005) confirmed that apart from kidneys the brain is also targeted by U-toxicity, being possibly as sensitive as the kidneys. In this context it should be noted that long-standing rumors in the gold mining town of Carletonville (South Africa) link U-polluted drinking water to an abnormally high number of children with learning problems in a mining community that relied on pumped groundwater from the mine (Tempelhoff 2008; Stoch 2001).

IRSN (2005) and Henner (2008) demonstrated that U is also genotoxic causing damage to the DNA of exposed fish. Investigating a population living in a uranium mining area William et al. (1995) found such effects also in humans. Recently Raymond-Wish et al. (2007) added U to the long list of known endocrine disruptive compounds (EDC) which rapidly emerge as a major threat to water quality worldwide. Mimicking the effects of estrogen in the body, U could possibly increase the risk of fertility problems and reproductive cancers at levels so far regarded as safe in drinking water. Since the uranyl ion produced in the gastrointestinal tract mimics the calcium ion U is regarded as "bone-seeker" with 60–80% of all incorporated U estimated to accumulate in the skeleton. Until recently it was believed that U mostly is found in the outer mineral part of the bone from where it cannot affect the marrow inside the bone due to its low penetration depth not exceeding 80 μm. However, Arruda-Neto et al. (2004) found U also in the much more sensitive bone marrow where it can affect the blood system. This is of importance for the Pofadder study discussed in this paper which established a link between elevated U levels in drinking water and increased incidences of leukemia in a rural community in an arid region of South Africa.

Case Study 1: Nonmining Related U Pollution of Borehole Water (Pofadder Study)

The study was triggered by observations of a medical doctor in Stellenbosch who, in 1995, noted that a majority of his leukemia patients came from an area around Pofadder, a rural settlement in the arid North Cape Province of South Africa close to Namibia. Approaching the Department of Water Affairs and Forestry (DWAF) and later the Water Research Commission of South Africa (WRC) to investigate a possible link between leukemia and the locally used water as common denominator, preliminary investigations funded by the DWAF started in 1996 followed by WRC project launched in January 1997 with the title: *The correlation of high uranium, arsenic and other chemical element values in groundwater with abnormal haematological values related to the occurrence of leukaemia in the North Western Cape* (Toens et al. 1998).

The study is essentially based on two sets of data relating a) to groundwater samples taken from 126 x boreholes during the early 1980s by the Atomic Energy Corporation of South Africa as part of an U exploration program and b) to blood samples from a 1993 population survey of 630 x individuals from 120 x localities in the Kenhardt magisterial district which also includes the Pofadder area. After screening all data and selecting the relevant sets for the study area the blood parameters of a total of 418 people from 120 localities (> 16 years of age in 1993)

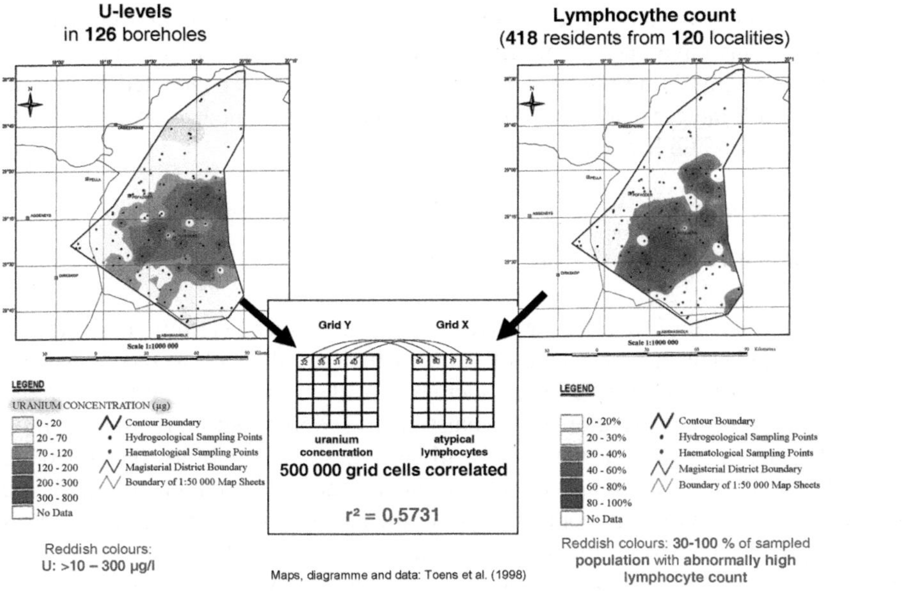

Fig. 2 GIS-based spatial correlation of the percentage of residents with atypical lymphocyte counts and the U concentration in borehole water according to Toens et al. (1998)

were correlated with various water quality parameters from 62 boreholes. Follow-up studies conducted by the Department of Water Affairs and Forestry (Sekoko et al. 2005) and the Council for Geoscience (van Wyk and Coetzee 2008) confirmed the U levels. While inter-farm migration occurs it is limited. The majority of the surveyed residents use the sampled boreholes of their farms as main and commonly only source of drinking water (Toens et al. 1998).Using a Geographical Information System (GIS) the 2 x data sets were geo-referenced and correlated employing 500,000 grid cells superimposed on the study area (Fig. 2).

Correlating all determined water quality parameters were with the proportion of residents suffering from abnormally high numbers of lymphocytes only for U a statistically significant relationship was found displaying a correlation coefficient of 0.5731 (Fig. 2). A somewhat higher correlation coefficient for the combined U-As concentration must now be discarded after van Wyk and Coetzee (2008) found that all As results were analytically flawed. Given the uncertainties a study of such nature will necessarily involve (data sets generated by different institutions for different purposes at different times are compared using their spatial distribution which is likely to introduces additional inaccuracies, migration of residents across farm boundaries etc.) the relatively low correlation coefficient does not surprise. Since the correlation is based on a large number of epidemiological data (from over 400 residents) it is suggested that the established relationship between U levels in borehole water and the risk of water users to develop leukemia is indeed significant even if the correlation is comparably weak. Apart from possible further investigations precautionary measures should therefore be taken to prevent the continued exposure of residents to this health risk. To the author's knowledge no interventions have been implemented since the conclusion of the study some 13 years ago. This is despite that fact that an immediately conducted follow-up study by Wullschleger et al. (1998) confirmed that U contaminated groundwater is widely used also in other settlements of this region.

Summary and Conclusions

Comparing limits for U in drinking water as recommended by various organizations worldwide indicates a wide range of values which are difficult to reconcile with each other. Furthermore, two recent increases of the limit by the WHO in rapid succession preceding the global U renaissance as well as an extraordinary high U-limit in SA as a major former supplier of U suggest the possibility that economic considerations may, at times, unduly influence the determination of guideline values. This is of particular concern as, over the past decade, scientific evidence is mounting that U displays a much wider spectrum of toxicity effects than was used in determining the currently applicable guideline values. While the latter was done exclusively based on nephrotoxicity observed in animal experiments following short-term and high-dose U exposure uranium is now known to damage the brain, the immune system, the DNA as well as the hormonal balance sometimes at U concentrations well below those permitted by currently existing guidelines.

Epidemiological data from an arid region of South Africa suggest that elevated U levels in drinking water is linked to increased incidences of leukemia in the resident population. While further investigations are perhaps needed to verify this finding it is suggested that precautionary measures are to be implemented urgently in order to prevent the continued exposure of a large number of people to U polluted drinking water in this area.

References

Arruda-Neto et al. (2004): The accumulation and microdistribution of uranium in the bone and marrow of beagle dog. Int J Radiat Biol, 80 (8) 567–575

BEIR V (Biological Effects of Ionising Radiation, report no. V) (1990): Health effects of exposure to low levels of ionizing radiation. Board on Radiation Effects Research, Washington: National Academic Press

Bertell (1999): Conflict of interest between IAEA and WHO. WISE News Communique on November 19, www.antenna-wise.nl, accessed 9 Feb. 2009

BfR (Bundesanstalt für Risikobewertung) und Bundesanstalt für Strahlenschutz (BfS) (2006): Gemeinsame Stellungnahme Nr. 014/2006 des BfS und des BfR vom 16. Januar 2006. BfR korrigiert Höchstmengenempfehlung für Uran in Wässern zur Zubereitung von Säuglingsnahrung

BfR (Bundesanstalt für Risikobewertung) und Bundesanstalt für Strahlenschutz (BfS) (2007): Gemeinsame Stellungnahme Nr. 020/2007 des BfS und des BfR vom 5. April 2007. BfR empfiehlt die Ableitung eines europäischen Höchstwertes für Uran in Trink- und Mineralwasser

Dieter H (2000): Toxikologische Bewertung von Radium bzw. Uran in Trink- und Mineralwasser. Umweltmedizinischer Informationsdienst 2/2000, p. 19, Umweltbundesamt, Berlin

Gilman AP, Villeneuve DC, Secours VE, Yagminas AP, Tracy BL, Quinn JM, Valli VE, Willes RJ, Moss MA (1998): Uranyl Nitrate: 28 day and 91 day toxicity studies in the Sprague Dawley Rat. Toxicol Sci, 41, 117–128

Gilman AP, Moss MA, Villeneuve DC, Secours VE, Yagminas AP, Tracy BL, Quinn JM, Long J, Valli VE, Willes RJ (1998): Uranyl Nitrate: 91 day exposure and recovery studies in the male New Zealand white rabit. Toxicol Sci, 41, 138–151

Henner P (2008): Bioaccumulation of radionuclides and induced biological effects in situations of chronuic exposure of ecosystems – a uranium case study. In: De Kok LJ, Schnug E (eds): Loads and fate of fertilizer derived uranium. Backhuys Publishers Leiden, The Netherlands

IRSN (Institute de Radioprotection et de surete nucleaire) (2005): Envirhom: Bioaccumulation of radionuclids in situations of chronic exposure of ecosystems and members of the public. Progress report 2 covering the period June 2003 – September 2005. Report DRPH 2005-07 & DEI 2005-05. p. 64

Konietzka R, Dieter HH, Voss J-U (2005): Vorschlag für einen gesundheitlichen Leitwert für Uran im Trinkwasser. Umweltmed Forsch Prax, 10 (2) 133–143

McDiarmid MA, Keogh JP, Hooper FJ, McPhaul K, Squibb K, Kane R, DiPino R, Kabat M, Kaup B, Anderson L, Hoover D, Brown L, Hamilton M, Jacobson-Kram D, Burrows B, Walsh M (2002): Health Effects of Depleted Uranium on Exposed Gulf War Veterans. Environm Research, 82, (2) 168–180

Raymond-Wish S, Mayer LP, O'Neal T, Martinez A, Sellers MA, Christian PJ, Marion SL, Begay C, Propper CR, Hoyer PB, Dyer AC (2007): Drinking water with uranium below the U.S: EPA water standard causes estrogen receptor-dependent responses in femal mice. Environ Health Persp, 115, (12) 1711–1716

Schnug E, Lindemann I (2006): Verringerung der Strahlenbelastung durch bewusstes Konsumverhalten bei Trinkwässern. Strahlentelex Nr. 476–477/2006, 4–5

Sekoko I, Mafejane A, Potgieter D, Conradie B, Kempster P, Kühn A, Diefenbach A (2005): A survey on the radiological and chemical quality of water resources in selcted sites of the Northern Cape Province. Resource Quality Service Report No. N/0000/GEQ0603, final report May 2005, DWAF, Pretoria, pp. 61, unpublished

Stoch L (2001): long-term resident in the Far West Rand mining area, April 2000, personal communication

Tempelhoff E (2007): Mynmonopolie se gru-moeras. Beeld, 2 February 2007, p. 11 (newspaper article)

Toens PD, Stadler W, Wullschleger NJ (1998): The association of groundwater chemistry and geology with atypical lymphocytes (as biological indicator) in the Poffadder area, North Western Cape, South Africa. WRC report 839/1/98

UBA (Umweltbundesamt) (2008): Uran im Trinkwasser. Kurzbegründung der gesundheitlichen Leit- und Grenzwerte, http://www.umweltdaten.de/wasser/themen/trinwasserkommission/kurzbegründung_uran_leitwert.pdf 19. August 2008

UBA (Umweltbundesamt) (2005): Uran und Human-Biomonitoring: Stellungnahme der Kommission Human-Biomonitoring des Umweltbundesamtes. Bundesgesundheitsbl – Gesundheitsforsch – Gesundheitsschutz 7-2005, 48, 822–827, DOI 10.1007/s00103-005-1069-7, Springer Medizin Verlag

Van Wyk N, Coetzee H (2008): The distribution of uranium in groundwater in the Bushmanland and Namaqualand areas, Northern Cape Province, South Africa. In: Merkel BJ, Hasche-Berger A (2008): Uranium, mining and hydrogeology. ISBN 978-3-540-87745-5, Springer, Berlin – Heidelberg, 639–644

Von Soosten C (2008): Stellungnahme der Beratungskommission der Gesellschaft für Toxikologie e.V. (GT) zur aktuellen Diskussion über eine mögliche gesundheitliche Gefährdung durch Überschreitungen des Trinkwasser-Leitwertes für Uran. http://idw-online.de/pages/de/news279231, 22.9.2008

WHO (World Health Organization) (1998): Guidelines for drinking water quality. Health criteria and other supporting information, Addendum to Vol. 2 WHO/EOS/98.1 Geneva: World Health Organization
WHO (1999): Agreement between the International Atomic Energy Agency (IAEA) and the WHO, ratified by the 12th WHO-General Assembly, 28th May 1959, Resolution WHA 1240, Organisation mondiale de la santé, "Documents fondamentaux", 42nd edition, WHO, Geneva
WHO (World Health Organization) (2006): Guidelines for drinking water quality (electronic resource): incorporating first addendum. Vol. 1, Recommendations, 3rd ed. Electronic version for the web
William WA, Lane RG, Legator MS, Whorton EB, WilkinsonGS, Gabehart GJ (1995): Biomarker monitoring of a population residing near uranium mining activities. Environ Health Perspect, 103 (5) 466–470
WISE (2008): EPA promulgates final rule for Radionuclides in Drinking Water, including uranium standard of 30 micrograms per litre http://www.wise-uranium.org/uregus03.html#EPADWRADFI, October 2009
World Nuclear News (2009): Uranium spot price 1995 to 2009, http://www.world-nuclear-news.org/nwsltrdisplay.aspx?id=24952, 12 October 2010
Wullschleger NJ, Visser D, Stadler W (1998): Groundwater health hazards of selected settlements in the province of the Northern Cape. Report to the DWAF. Unpublished

Challenges in Assessing Uranium-Related Health risks: Two Case Studies for the Aquatic Exposure Pathway from South Africa – Part II: Case Study Potchefstroom

Frank Winde

Abstract. Following on Part I this paper explores possible health risks associated with mining-related uranium contamination of water using the case study of Potchefstroom, a municipality affected by deep level gold mining in the catchment of the Wonderfonteinspruit. Owing to the dolomitic nature of the water U peaks from upstream gold mining areas are likely to slip through the water treatment plant that relies on the standard potabilization process. The installation of an early warning monitoring system is proposed to stop abstraction if and when U peaks arrive.

Introduction

Following on Part I which explores effects of naturally elevated U levels on residents in an arid farming area of South Africa this, second, part of the paper focuses on effects of mining-related water pollution using the catchment of the Wonderfonteinspruit that is, amongst others, impacted by the Far West Rand (FWR) gold mining area as an example. In contrast to the East-, Central- and West-Rand goldfields where active deep-level mining stopped and associated voids are now being flooded most gold mines in the FWR are still active.

After a brief characterization of mining-related U pollution of surface water and groundwater in the catchment upstream of Potchefstroom the paper explores the extent to which, if at all, local water users in Potchefstroom as well as the mining area are at risk of being adversely affected by U contaminated drinking water.

Frank Winde
Mine Water Research Group, School of Environmental Sciences and Development, North-West University, Potchefstroom Campus, Private Bag X6001, Potchefstroom, 2520, Republic of South Africa

B. Merkel, M. Schipek, *The New Uranium Mining Boom*
DOI 10.1007/978-3-642-22122-4 © Springer 2011

Case study 2: Gold-Mining Related U Pollution of Raw and Drinking Water of Potchefstroom (South Africa)

Uranium Levels in the Raw Water Sources of Potchefstroom

As one of only a few municipalities in South Africa Potchefstroom is also a water service provider responsible for sourcing, treating and distributing drinking water to its estimated 250,000 residents. The water is drawn exclusively from two surface water reservoirs located upstream of Potchefstroom, the Boskop Dam and the much smaller Potchefstroom Dam which both are fed by the Upper Mooi River. As a large proportion of its catchment is underlain by weathered dolomite much of the flow in the Mooi River originates from karst springs issuing dolomitic groundwater from so-called compartments (sub-units of the larger karst aquifer defined by water-tight dykes). This also applies to the most important tributary of the upper Mooi River, the Wonderfonteinspruit (WFS) which – over a distance of some 100 km – drains two major gold mining areas namely the now mined out West Rand (WR) located in its non-dolomitic headwaters and the active Far West Rand goldfield mostly located on outcropping dolomite that occurs further downstream. Exposed to the discharge of large volumes of water pumped from the underground mine void, seepage from adjacent tailings deposits in-stream U concentration range from an average of 300–400 μg/l in the head water to around 60–80 μg/l in the lower WFS just downstream of the FWR. Average U levels in most mine effluents are commonly between 100 and 400 μg/l. The significant drop in the in-stream U concentration from the upper to the lower WFS does not necessarily reflect reduced inputs of U but rather increased dilution capacity. This applies particularly to deep mines in the FWR which receive large volumes of clean dolomitic groundwater that dilutes their U-input to comparably low concentrations. This dilution capacity is lacking in the head water region of the WFS where large parts of the catchment are covered with mining residues whose (highly polluted) seepage forms a significant part of baseflow of the stream. In order to compensate for the dilution effect of the large groundwater volumes available to the FWR mines and to make mining-related U pollution between the two goldfields comparable Winde (2010a) employed the load concept rather than comparing concentrations only. Based on this the WR, in 1997, was found to have contributed some 430 kg U/a compared to 2567 kg U/a coming from mines in the FWR accounting for 83% of the total load of (dissolved) U received by the WFS. During the period 1998 to 2008 the total U load discharged into the WFS increased to 3466 kg/a (representing an increase of 13%) of which nearly half (46%; 1607 kg/a) came from the WR. The reason for the relative increase of the WR contribution is not entirely clear but may relate to the decant of mine water from the flooded void that was temporarily routed into the WFS, to mining a tailings filled dam right in the stream bed (Lancaster dam) as well as the reduction of U loads from all FWR mines.

During the past decade the WFS carried an average of over 3 tons of U per annum towards the upper Mooi River. Since this refers exclusively to dissolved U and excludes particulate U associated with tailings being washed into the stream from adjacent deposits or from pipeline spillages the actual U load is likely to be higher (Winde 2010b).

However, since April 2003 the polluted part of the WFS does no longer reach the Mooi River on surface as much of the water enters the underlying karst aquifer through caves and sinkholes following the silting of much of its canalized river bed (Winde and Erasmus 2011). While this reduced the direct U input into the Mooi River it simultaneously increased the contamination of dolomitic groundwater in the Boskop-Turrfontein compartment (BTC). Since the BTC feeds the Mooi River via the Gerhard Minnebron (GMB) Eye and the lower WFS via the 2 x Turffontein springs, U from the mining area may still reach the Mooi River indirectly. This may explain the significant increase in U levels of these springs. Compared to 1997 the average U concentration at the upper Turffontein Eye increased from 0.9 to 14 μg/l (by nearly 16 times) and at GMB from 0.5 to 2.3 μg/l (by nearly 5 times) (Winde 2010b). As the GMB spring and its associated wetland is the single largest water source for Potchefstroom providing an estimated 100–120 Ml/d this trend is reason for concern even though the absolute U levels are still comparatively low. With estimated turnover-times between 17 and 22 years for water in reservoirs feeding the 3 x dolomitic springs the situation may still worsen (Winde and Erasmus 2011).

In fact, for most of the time it appears that U levels at the GMB spring are close to natural background values (around 0.2 μg/l for dolomitic water of this region compared to 0.4 μg/l as global freshwater average) interrupted only by sporadic U peaks of which the highest reached 25 μg/l (1998–2008, Winde 2010b). It is mainly the higher magnitude of the pollution peaks rather than an increased frequency that appears to have caused the higher average U concentration. Real-time continuous data logger measurements of water quality parameters directly at the eye suggest that U peaks are triggered by rain events in the upstream catchment arriving at the eye approximately 6–7 days later. Possible pathways include the spillage of highly polluted stormwater run-off from tailing deposits that enters the karst aquifer via sinkholes formed where mine water canals frequently overflow. Stormwater runoff from tailings is a major source of U pollutions as many of the tailings dumps are covered by salt crusts with very high U concentrations from 600 to over 2000 ppm, which readily dissolve on contact with water. This explains why, despite the large influx of clean rainwater, U levels rise after rain events with maximum U levels reached in first flush water that dissolved most of the crusts. This is confirmed by high U levels found in a dam of the lower WFS that only fills partially (i.e. does not overflow) after heavy rainfall thus representing the first flush water. In late December 2010 exceptionally heavy rainfall resulted in the WFS for the first time in years reaching the Mooi River on surface again displaying an U concentration of 16 μg/l compared to maximal 4 μg/l in the 5 years before (Nel 2010).

Dynamics of In-Stream U Levels

In addition to rain events which trigger additional U inputs also other factors influence the in-stream U level. This includes pumping regimes by mines whose discharge of mine effluents show distinct diurnal and weekly patterns as much more water is pumped during night time and on weekends when cheaper electricity tariffs apply. This results in higher U inputs at times when no regular sampling takes place. Moreover, photosynthesis-triggered diurnal and seasonal fluctuations of the pH and Eh as master variables controlling the solubility and thus the concentration of dissolved U in stream water – further impact on U levels in the WFS. As a result U levels in the WFS are highly dynamic varying at a single sample site by up to 2 orders of magnitude even if samples are taken at fixed intervals at the same day of the week and the same time of the day. This renders it difficult for the current weekly sampling program employed by the Potchefstroom municipality to capture all possible U peaks moving from the mining area downstream into its water supply system.

Uranium Levels in Potchefstroom's Drinking Water

Historical data from several monitoring programs suggest that spikes of high U at times arrive at the municipal water treatment plant. A study by the Institute for Water Quality studies (IWQS) in 1997 found a maximum U concentration of 48 µg/l at the Potch dam inlet canal of the water works compared to an average of 1.9 µg/l at the Boskop Dam inlet canal (IWQS 1997). This suggests that peak concentrations occur and reach the water works. This is of concern as Baeza et al. (2008) found that ions typically occurring in dolomitic water such as Ca, Mg and bicarbonate each significantly reduce the efficiency with which standard potabilization processes (such as filtration, flocculation and chlorination) which are also used at the Potchefstroom water works remove dissolved U. Since all water used by Potchefstroom water works is of dolomitic origin it is likely that sporadically arriving U peaks may not be reduced by the treatment procedures and thus be able to enter the reticulation system. Comparing the average U concentration between raw water and treated water of the Potchefstroom water works supplied by DWAF (2004) showed for the period Jan–March 2003 ($n = 8$) that treated water showed in fact a higher average U level (2.6 µg/l) than the raw water (1.6 µg/l). A similar trend was confirmed using data provided by the Potchefstroom municipality (Nel 2001) showing that U levels in the raw water at Boskop Dam was lower than in the tap water in Potchefstroom albeit both on a very low level (below 1 µg/l). Comparing U levels in tap water of Potchefstroom for the period 1999 to 2010 from different sources (weekly monitoring program of the Potchefstroom Municipality (Nel 2011) and mainly grab sample data from other parties including the DWA) indicates a somewhat higher level for the period pre-2010 suggesting a recent improvement (Fig. 1).

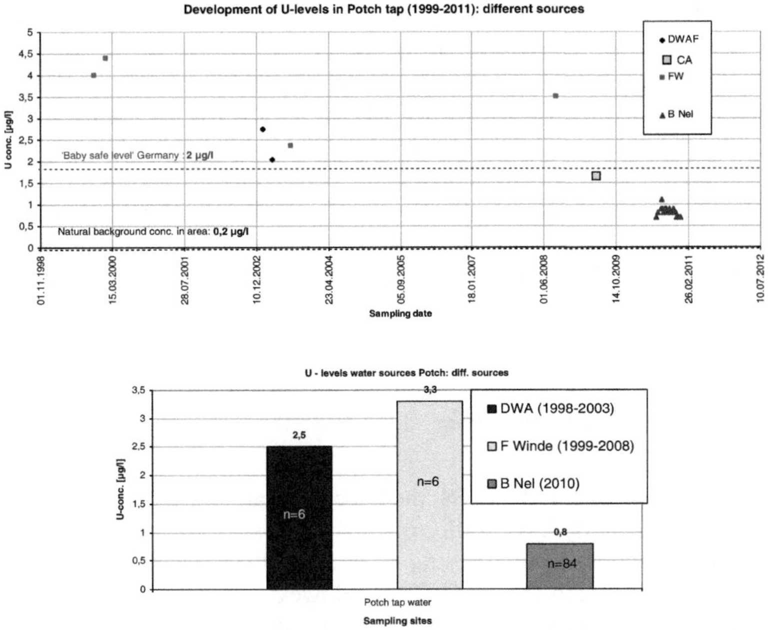

Fig. 1 Uranium concentrations in tap water of Potchefstroom (1999–2010) according to different sources (DWA – DWAF (2004); CA – Albrecht (2009), FW – Winde (own data), B Nel (Nel 2010))

While the weekly monitoring program of the municipality by far generated the highest number of data it covers only a limited period. Since data from this sources for a monitoring point for which data from other sources for the same period were available (the WFS at Muiskraal bridge, C2H161) were found to underreport it is unclear whether the observed reduction in U levels in the tap water represent a trend or perhaps are caused by analytical differences. A possible improvement could possibly relate to the fact that, since April 2003, the polluted WFS does no longer reach the Mooi River, reducing the probability of U-peaks arriving at the Potchefstroom water works. However, samples taken by Winde in 2008 and Albrecht in 2009 indicate still higher U levels than found by Nel (2010).

Since weekly sampling is prone to miss sporadic U peaks possibly arriving at the municipal water treatment plant, in 2002, the scales of water kettles were sampled and analyzed for U. The rational behind this approach is that dissolved U is commonly co-precipitated along with calcium carbonate when the dolomitic water is heated. Since this happens every time the kettle is used the U concentration in the kettle scales is an indirect indication for the average U level in the tap water over the period the kettle was used. As most kettles are used 1–2 times daily over several years U levels in kettle scales are an indicator for the long term tap water quality. Comparing scales from 10 x kettles from Potchefstroom to scales from 10 x kettles sampled in Ventersdorp, a neighboring community where dolomitic water is used that is not impacted on by mining, indicated that U levels in Potchefstroom kettle scales are 20 times higher (20 to 1 ppm). This suggests that, over longer periods (some kettles have been 10 years old), U peaks are able to slip through the treatment plant and arrive at the tap water.

Based on distribution coefficients (Kd) found for calcite-dolomitic groundwater systems of the GMB wetland (Kd = 3273, *n* = 4) the U level in the kettle scales would correspond to a long-term average of 6.1 μg/l in Potchefstroom's tap water for the 1990s to early 2000s. As this is double the concentration even of the highest average for grab sample analyses of tap water (3.3 μg/l; Fig. 1) thereafter, it is likely that U-peaks arriving at the water treatment plant during the 1990s, when the WFS still reached the Mooi River on surface, were higher and occurred perhaps more frequently.

Assessing Associated Health Risks

Compared to levels measured in the WFS, the U concentration in Potchefstroom's tap water are still significantly lower reflecting the dilution by unpolluted groundwater from two major dolomitic aquifers represented by the Bovenste Oog (feeding the upper Mooi River and the GMB Eye). However, compared against the natural background of 0.2 μg/l in pristine dolomitic groundwater even the very low levels reported by Nel (2010) for 2009–2010 (around 0.8 μg/l) represent a man-made elevation by a factor of 4. Compared to the most stringent guideline of 2 μg/l (applied in Germany for mineral water used for babies and infants under 7 years) this would not pose a health risk.

However, sporadic U peaks are likely to exceed this level by an unknown margin. Depending on the frequency and duration at which especially vulnerable user groups (such as babies and children) are exposed to such U peaks in consumed water, adverse health impacts may occur. A 2008-study of 8200 municipalities in Germany found that drinking water in 150 communities exceeded the level of 2 μg/l (with a maximum of 8.5 μg/l U) indicating the need for corrective action (Foodwatch e.V. 2008). Since the (tentative) U guideline for South Africa stipulates 70 μg/l as upper U-limit for best quality water no such (legal) need for interventions exists in Potchefstroom. Some problematic aspects associated with setting guidelines values are discussed in Part 1 of the paper series. In terms of risks one also needs to consider that U concentrations in some mineral waters, often used as substitute for tap water in areas perceived as unsafe, frequently exceed existing limits by significant margins reaching maxima of 45 μg/l in South Africa mainly due to geological conditions.

While associated health's risks in Potchefstroom at the moment may be marginal, if existing at all, this may be different in the upstream mining areas where especially informal settlers from very poor communities frequently use untreated stream and mine water with U levels well over 100 times higher than measured in Potchefstroom tap water (50–400 μg/l). However, 2 studies employing the "absorbed-dose-model" concluded in 1999 and 2007 that no health risk for the drinking water pathway (i.e. direct consumption of the water for a life-time) exists at these U levels (IWQS 1997, Barthel 2007). This is in contrast to findings in the Pofadder study of 1998 (discussed in Part 1: Toens et al. 1998), which – based on epidemiological data – established a link between U levels at the same range as found in the WFS and elevated incidences of leukemia in the residential population. This may point to shortcomings in the absorbed-dose-model for comprehensively assessing health risks associated with long-term low dose exposure to U as suggested, amongst others, by Busby et al. (2010).

Conclusions and Recommendations

While exposure to elevated U levels of residents of Potchefstroom is thought to be rather limited the same is not true for several hundreds if not thousands of poor people living in informal settlements in the mining area. Using untreated water with U levels well above guideline values this part of the population is particular at risk resulting from a combination of high exposure and increased vulnerability stemming from malnutrition, lack of medical care and high prevalence of HIV infections amongst other factors. For these communities urgent intervention is suggested aimed at taking receptors out of the exposure pathway as ongoing pollution renders it difficult to improve the water quality in the short term. Possibilities include the provision of alternative water supply systems and access control of particularly polluted water sources such as mine water canals which have been used for sourcing domestic water.

With regard to Potchefstroom it is suggested to comprehensively investigate the sources and pathways responsible for the sporadic U peaks in order to identify cost-effective ways of reducing pollution and protect the large dolomitic water resources from further contamination. This is in contrast to strategies currently explored by the Municipality to abandon the rich groundwater resources of the upstream karst aquifers and source its water from the (already overstretched and also polluted) Vaal River system using pipelines. For the interim period, until pollution prevention measures are taken, a cost-effective technological solution is proposed aimed at reducing the risk of U peaks moving through the treatment plant into the tap water system. This refers to an early warning system that prevents water abstraction during periods of elevated U levels. Since currently such peaks occur only sporadically while good quality water prevails for most of the time, a real-time monitoring system could continuously measure either U level directly (should a suitable sensor be available) or using proxy parameters such as the electrical conductivity (which frequently increases along with U levels) and stop water abstraction for periods when trigger levels are exceeded. Depending on the duration of pollution peaks that may necessitate the use of a reservoir to bridge the interruption of water treatment. Alternatively, the polluted water could be diverted through a more sophisticated treatment system able to remove U better than the standard processes. Such obviously more expensive water treatment system would only have to be used for the limited period when pollution plumes move through keeping the overall running costs at economically sustainable levels.

Lastly, data generated by the early warning system could be directly loaded onto a publicly accessible website for the information of interested or concerned residents.

References

Albrecht C (2009): U concentration for tap water from Potchefstroom sampled April/May 2009. Cancer Research Association of South Africa (CANSA), unpublished.

Baeza, A, Salas, A, Legarda F (2008): Determining factors in the elimination of uranium and radium from groundwaters during standard potabilization process. Sci Tot Env, 406, 24–34.

Barthel R (2007): Assessment of the radiological impact of the mine water discharges to members of the public living around Wonderfonteinspruit catchment area. BSA-project-no. 0607-03, BS Associates, Consulting Engineers and Scientists, Report to the National Nuclear Regulator (NNR), Centurion, South Africa.

Busby C, Bertell R, Schmitz-Feuerhake I, Scott Cato M, Yablokov A (ed.) (2010): 2010 Recommendations of the European Committee on Radiation Risk. The health effects of exposure to low doses of ionizing radiation. Regulators' edition. ISBN: 978-1-897761-16-8, Green Audit, Aberystwyth, United Kingdom, pp. 248.

DWAF (Department of Water affairs and Forestry) (2004): Mooi River surveillance. Uranium analyses for selected sites in the Mooi River catchment 1998–2004. unpublished.

Foodwatch e.V. (2008): Uranbelastung von Trinkwasser in Deutschland (Messung FAL-PB August bis November 2006, Stand 27.3.2008). Analytical results, 10 pp.

IWQS (Institute for Water Quality Studies) (1999): Report on the radioactivity monitoring programme in the Mooi River (Wonderfonteinspruit) catchment. Report No: N/C200/00/RPQ/2399. DWAF Pretoria, pp. 26, unpublished.

Nel B (2010): U concentration in weekly tap water samples of the Potchefstroom municipality, 2009–2010, unpublished.

Rosborg I, Nihlgard B, Gerhardson L, Gernersson M-L, Ohlin R, OlssonT (2005): Concentrations of inorganic elements in bottled waters on the Swedish market. Environ Geochem Health, 27, 217–227.

Toens PD, Stadler W, Wullschlege NJ (1998): The association of groundwater chemistry and geology with atypical lymphocytes (as biological indicator) in the Poffadder area, North Western Cape, South Africa. WRC report 839/1/98.

Winde F (2010a): Uranium pollution of the Wonderfonteinspruit: 1997–2008. Part I: U-toxicity, regional background and mining-related sources of U-pollution. Water SA, 36 (3), 239–256.

Winde F (2010b): Uranium pollution of the Wonderfonteinspruit: 1997 – 2008. Part II: U in water – concentrations, loads and associated risks. Water SA, 36 (3), 257–278.

Winde F, Erasmus E (2011): Peatlands as filters for polluted mine water? – A case study from an uranium-contaminated karst system in South Africa. Part I: Hydrogeological setting and U-fluxes. Water, 3, 291–322, doi:10.3390/w3010291.

Characterization of Uranium Behavior in the Ruprechtov Site (CZ)

Barbora Drtinová, Karel Štamberg, Dušan Vopálka, Alena Zavadilová

Abstract. The Ruprechtov natural analogue site is being studied because its geological and geochemical conditions resemble sedimentary sequences which can cover potential host rocks for underground waste repositories. The uranium content, sorption and isotopic exchange were studied on five selected rock samples that represented i) layers with higher content of uranium and ii) neighboring layers with practically no uranium content with the aim to discuss processes influencing transport of uranium in the body. In addition, the titration curves of rock samples were determined and modeled by SCM and so called GC and CA approaches.

Introduction

The Ruprechtov site is situated in NW of the Czech Republic in Hroznětín, part of Sokolov basin. Ruprechtov was chosen for the study of uranium behavior because

---

Barbora Drtinová
Department of Nuclear Chemistry, Faculty of Nuclear Sciences and Physical Engineering, Czech Technical University in Prague, Břehová 7, CZ-115 19 Prague 1, Czech Republic

Karel Štamberg
Department of Nuclear Chemistry, Faculty of Nuclear Sciences and Physical Engineering, Czech Technical University in Prague, Břehová 7, CZ-115 19 Prague 1, Czech Republic

Dušan Vopálka
Department of Nuclear Chemistry, Faculty of Nuclear Sciences and Physical Engineering, Czech Technical University in Prague, Břehová 7, CZ-115 19 Prague 1, Czech Republic

Alena Zavadilová
Department of Nuclear Chemistry, Faculty of Nuclear Sciences and Physical Engineering, Czech Technical University in Prague, Břehová 7, CZ-115 19 Prague 1, Czech Republic

B. Merkel, M. Schipek, *The New Uranium Mining Boom*
DOI 10.1007/978-3-642-22122-4

of its resemblance with natural analogue systems in Gorleben (Germany) and Mol (Belgium) (Noseck and Brasser 2006). The clay formation serving as a natural barrier in the vicinity of the deep nuclear waste repository provides retardation of PA (performance assessment)-relevant elements including uranium.

Uranium accumulation is distributed heterogeneously within volcanodetritic layers in the depth of 10–40 m. Uranium was found predominantly in U(IV) form. For the application of a migration model it is necessary to know the formal description of interaction of studied contaminant with the solid phase. The characterization of sorption on the surface of minerals needs evaluation of several types of experiments. In our previous study (Vopálka et al. 2008) we presented results of sorption end exchange experiments on samples of different mineralogic composition that could not be described by any formal equilibrium isotherm, although the exchange experiments with isotope ^{233}U were used for the elimination of the so-called "exchangeable uranium". The dissolution of sulfides was assumed as the main reason of the observed behavior.

In this study, the surface qualities as layer and edge sites of minerals present in samples was identified by means of evaluation of titration curves. The aim was to prepare a formal sorption model, which will be able, to explain the results of interaction experiments with uranium. From the evaluation of sorption and exchange kinetic experiments we expected the formulation of interaction isotherms for different samples. Comparison of determined values of exchangeable uranium using ^{233}U on selected samples with the results of sequential leaching could help in discussion of uranium sorption and/or mineralization in studied rock materials.

Experiments and Results

The kinetics of uranium sorption and isotopic exchange, interaction isotherms and evaluation of titration curves (that enabled to determine characteristics of the "edge" and "layer" sites) differentiate the qualities of studied types of sedimentary rocks. The sediment samples were held under Ar atmosphere, then lyofilized and crushed, powdered and sieved (< 1 mm) before experiments.

Mineralogical Characterizations

The studied sediments (Table 1) were analyzed for the elemental and mineralogical composition and the specific surface area; results are presented in Table 2.

Samples were selected with respect to the position of uranium mineralization in the borehole profile. Sample D2 represents position with the highest accumulation of uranium in the borehole NAR2, sample D5 in NAR3. Samples D1, D3 and D4 represent the layer below or above the maximum of the uranium mineralization.

Table 1 List of the studied samples

Sample	Horizon (m)	Borehole	Gama spectroscopy (CPS)	Core description
D1	25.7	NAR 2	310	Coal clay, gray to black, roughly sandy with coal seams
D2	29.86	NAR 2	920	Tuffaceous clay, sand
D3	32.4	NAR 2	680	Kaoline clay
D4	20.85	NAR 3	280	Carboniferous clay
D5	23.4	NAR 3	340	Tuffaceous clay, gray, sand

Table 2 Rock sample characterization: mineralogical composition (weight %) and the specific surface area ($m^2 \cdot g^{-1}$, BET method)

Sample	Specific Surface area	Montmorillonite	Kaolinite	Anatase	Illite	Quartz	Pyrite
D1	18.9	44.9	30.6	2.7	5.6	5.7	4.5
D2	12.5	26.4	36.3	4.1	7.6	15.0	2.8
D3	16.6	17.2	71.0	1.8	3.7	2.9	0.0
D4	16.2	48.1	21.2	7.4	0.0	0.3	18.7
D5	20.7	40.6	28.8	4.4	7.2	7.7	0.8

Titration Curves

Procedure and Conditions of Experiments

The aim of determination of titration curves and their subsequent description via the surface-complexation model was to obtain information and conception of the surface and potentially sorption properties of the samples respectively to their mineralogical composition. Due to the presence of mainly kaolinite and montmorillonite (Table 2) the mineral surfaces can be characterized by functional groups such as “layer site” and “edge site”. These are capable, depending on the pH, of ion exchange mechanism (especially “layer sites”) and surface complexation (“edge sites”) of ionic and molecular forms present in the liquid phase.

The sediments were pre-treated in order to remove carbonates and other impurities using the procedure described in (Wanner et al.1994). In the titration flask was mixed 0.5 g of the pre-treated and dried (under ambient temperature) sample with 50 ml 0.1 $mol \cdot l^{-1}$ $NaNO_3$ (applied to maintain ionic strength at almost constant value). The potentiometric titration using 0.1 $mol \cdot l^{-1}$ HNO_3 and 0.1 $mol \cdot l^{-1}$ NaOH was then performed under N_2 atmosphere on automatic titrator 845 TIM with a combined electrode pHC2001-8 (Radiometer Analytical).

Evaluation of Titration Curves

It is supposed that the protonation reactions (1), (2) and the ion-exchange reaction (3) occur on the edge and layer sites, respectively. Chemical Equilibrium Model

(CEM, non-electrostatic) was used to the modeling of reactions (1) and (2), and by means of Ion Exchange Model (IExM) the reaction (3) was described. The experimental data were fitted using the Newton-Raphson multidimensional non-linear regression procedure in the course of which the characteristic parameters of surface sites were determined. The goodness-of-fit was evaluated by quantity WSOS/DF (see Eqs. (6) and (7)) – if WSOS/DF is approx. between 0.1 – 20, then the fit can be regarded as good. The resulting values of equilibrium constants (K_1, K_2 and K_{ex}) and the total concentrations of edge (SOH_{tot}) and layer (X_{tot}) sites, being valid for selected samples, are summarized in Table 3 and the resulting dependences are demonstrated in Fig. 1. (sample D1).

$$SO^- + H^+ \leftrightarrow SOH^0, \qquad K_1 = [SOH^0]/([SO^-]\cdot[H^+]) \tag{1}$$

$$SOH^0 + H^+ \leftrightarrow SOH_2^+, \qquad K_2 = ([SOH_2^+]/[SOH^0]\cdot[H^+]) \tag{2}$$

$$XNa + H^+ \leftrightarrow XH + Na^+, \quad K_{ex} = ([XH]\cdot[Na^+])/([XNa]\cdot[H^+]) \tag{3}$$

$$SOH_{tot} = [SO^-] + [SOH^0] + [SOH_2^+] \tag{4}$$

$$X_{tot} = [XNa] + [XH] \approx [X^-] + [XH] \tag{5}$$

$$WSOS/DF = \chi^2/n_i\,,\; n_i = N - n \tag{6}$$

$$\chi^2 = \sum_i^N \frac{(SSx)_i}{(s_q)_i^2} \tag{7}$$

Here N – is the number of experimental points, n – the number of parameters sought, n_i – the number of degrees of freedom, $(SSx)_i$ – the square of the deviation of the experimental from the calculated value, $(s_q)_i$ – the relative standard deviation.

Table 3 Titration curves parameters of samples D1, D2, D3 and D5 obtained by GC Approach

Sample	Parameters of surface sites – edge sites (CEM)		
	K_1 [mol^{-1}]	K_2 [mol^{-1}]	SOH_{tot} [mol.kg^{-1}]
D1	$(1.29 \pm 0.05)\cdot 10^{9}$	$(9.43 \pm 0.16)\cdot 10^{5}$	$(1.55 \pm 0.01)\cdot 10^{-1}$
D2	$(1.98 \pm 0.02)\cdot 10^{10}$	$(6.49 \pm 0.03)\cdot 10^{2}$	$(5.20 \pm 0.02)\cdot 10^{-2}$
D3	$(8.82 \pm 0.03)\cdot 10^{10}$	$(1.20 \pm 0.01)\cdot 10^{3}$	$(7.03 \pm 1.42)\cdot 10^{-2}$
D5	$(1.03 \pm 0.03)\cdot 10^{11}$	$(5.15 \pm 0.09)\cdot 10^{3}$	$(1.25 \pm 0.01)\cdot 10^{-1}$
	Parameters of surface sites – layer sites (IExM)		WSOS/DF
	K_{ex} [–]	X_{tot} [mol/kg]	
D1	$(7.78 \pm 0.05)\cdot 10^{1}$	$(1.77 \pm 0.01)\cdot 10^{-1}$	3.39
D2	$(7.21 \pm 0.04)\cdot 10^{4}$	$(3.13 \pm 0.01)\cdot 10^{-2}$	5.16
D3	$(4.77 \pm 0.03)\cdot 10^{6}$	$(1.09 \pm 0.01)\cdot 10^{-2}$	3.74
D5	$(7.90 \pm 0.14)\cdot 10^{6}$	$(4.25 \pm 0.03)\cdot 10^{-2}$	7.01

The values of WSOS/DF indicate the relatively good agreement between experimental and calculated data or between surface properties of samples studied and surface complexation models used. Of course, if the mineralogical analysis of given samples is taken into account (see Table 2), than it is evident that it deals with modeling approach called Generalized Composite approach (GC), where the given mixture of solid phase components, e.g. montmorillonite + kaolinite etc., is considered as a compact sorbent characterized by a single set of titration and sorption parameters, which are sought by the direct fitting of the experimental data. In such a case, the second modeling procedure can be used, namely the so called Component Additivity approach (CA), composed of a weighted combination of models describing the protonation (on edge sites) + ion exchange (on layer sites) on individual solid phase components, e.g. on montmorillonite + kaolinite + etc. The surface sites parameters corresponding to individual phase components have to be obtained by fitting the appropriate experimental data or can be found in literature, e.g. in database RES^3T (RES^3T-Rossendorf 2006).

According to the mineralogical analysis of samples D1–D5 (see Table 2), at least four minerals in these samples can be designated as basic phase components on the surface of which the reactions (1)–(3) take place. From the point of view of their protolytic and ion exchange constants and concentration of sites (SOH_{tot} and X_{tot}) it can be expected that only the first four minerals out of the six components (montmorillonite, kaolinite, anatase, illite, quartz and pyrite) mentioned, play the important role; especially it holds for montmorillonite and kaolinite the mass fractions of which are relatively high.

The first results of application of CA approach to the modeling of titration curve of sample D2 are demonstrated in Fig. 2. The surface charges of individual four basic phase components as a function of experimental pH (see Fig. 2 – right) were calculated using the procedure described in detail in (Filipská and Štamberg 2005). The required parameters characterizing the each individual phase compo-

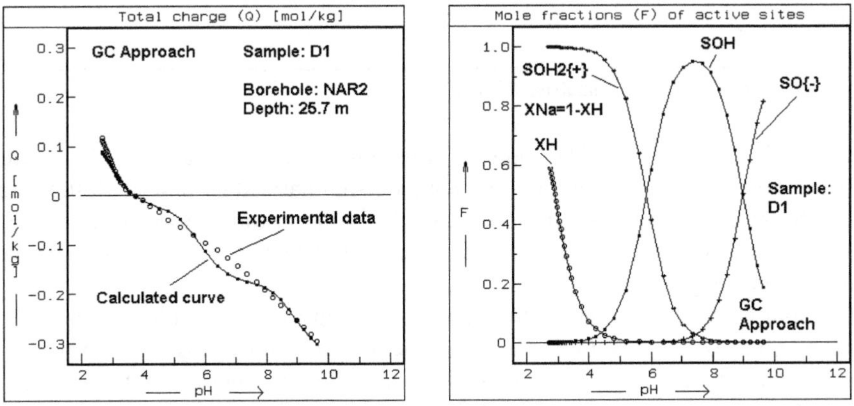

Fig. 1 Evaluation of experimental titration data (*left* – total surface charge = f(pH)) and calculation of surface sites speciation (*left* – molar fraction = f(pH)). GC approach and CEM and IExM models were used

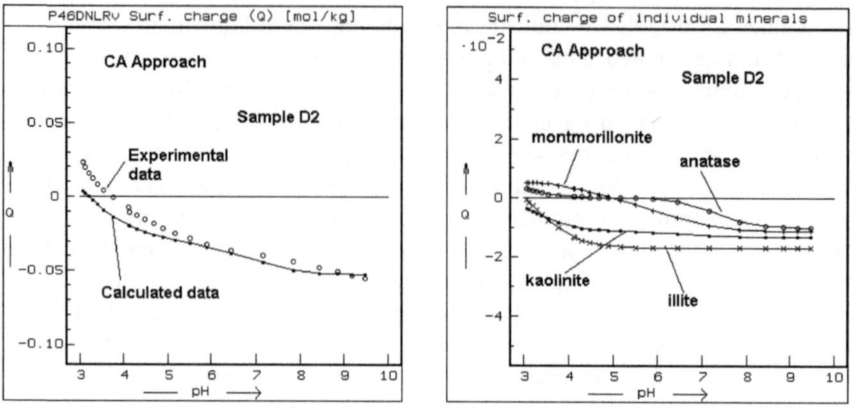

Fig. 2 Calculation of titration curve (*left* – total surface charge = f(pH)) and of pH dependences of surface charges of individual minerals (*right*). CA approach including CEM and IExM models and data from Table 4 were used

Table 4 Parameters of surface sites of selected minerals (average values ± standard deviations) taken from RES^{3}T database – input data for CA Approach calculations

Mineral	Parameters of surface sites – edge sites (CEM)		
	log K_1 [mol^{-1}]	log K_2 [mol^{-1}]	SOH_{tot} [mol · kg^{-1}]
Montmorillonite	8.29 ± 1.49	5.70 ± 1.34	(4.09 ± 2.05)·10^{-2}
Kaolinite	9.13 ± 2.79	4.94 ± 2.32	(4.71 ± 4.23)·10^{-2}
Anatase	8.16 ± 0.90	3.98 ± 1.23	(2.53 ± 0.51)·10^{-1}
Illite	7.62 ± 4.07	5.66 ± 3.22	(2.98 ± 0.74)·10^{-1}
	Parameters of surface sites – layer sites (IExM)		
	log K_{ex} [–]	X_{tot} [mol · kg^{-1}]	
Montmorillonite	4.60*	2.20·10^{-2} *	
Kaolinite	2.89 ± 0.01	(3.23 ± 0.12)·10^{-2}	
Anatase	–	–	
Illite	–	–	

* Only one value was found in database RES3T.

nent, namely K_1, K_2, K_{ex}, SOH_{tot} and X_{tot}, were taken from database RES^{3}T. In this database, almost all available literature data are summarized so first of all the data should be selected and than and then calculates the arithmetic average of each parameter, including the corresponding standard deviation. The results can be found in Table 4.

As input data into corresponding equations (Filipská and Štamberg 2005), the average values or the "average – standard deviation" values (Fig. 2), and the experimental pH values were used. In such a way, the dependences of total surface charge, Q [mol/kg], on pH (see Fig. 2 – left, where Q is sum of surface charges of four individual mineral components) and the same dependences for four mineral basic phase components (see Fig. 2 – right) were calculated. It is evident, that the relatively good conformity of experimental with the CA approach calculated data was obtained, especially if pH >5.

Interaction Kinetics and Sequential Extraction

The determination of exchangeable uranium by means of sequential leaching tests was compared with determination of that using the isotopic exchange tests in the presence of ^{233}U. An own modification of the KPA method (kinetic phosphorescence analysis) enables the determination of total uranium concentration in leachates. ^{233}U activity was determined by liquid scintillation counting (Triathler, HIDEX, coctail Rotiszint eko plus).

Uranium Determination Using Uraplex TM Agent and TRLFS

In our TRLFS laboratory (Time-Resolved Laser-Induced Fluorescence Spectrometer), we modified method of KPA often used for determination of uranium in aqueous solutions at low concentrations, based on measurement of induced radiation intensity of the sample. Solid state tunable laser based on flash lamp, pumped Q-switched Nd:YAG laser 1064 nm (with conversion possible by modules SHG (532 nm), THG (355 nm) OPO (410–2400 nm), UV (355–410 nm) and FHG (256 nm), pulse length 2–12 ns and repetition rate 1–10 Hz), was used. To reduce the influence of nonradiant processes such as quenching and thus to extend the lifetime of luminescence, a complexation agent Uraplex (pH 3) was added in a 1 : 1.5 ratio to the measured sample (Croato et al. 1997). We exited the sample on 417 nm where obtained fluorescence signal was strongest.

Conditions of Kinetic and Leaching Experiments

To determine exchangeable uranium and the relevant kinetic dependencies of sorption/isotopic exchange, synthetically prepared solutions spiked with isotope ^{233}U ($^{233}UO_2(NO_3)_2$) were used: solution A with $2\cdot10^{-7}$ mol · l^{-1} ^{233}U (pH = 8.39); solution B containing $2\cdot10^{-7}$ mol · l^{-1} ^{233}U, $1.8\cdot10^{-6}$ mol · l^{-1}U (pH = 8.41); solution C with $2\cdot10^{-7}$ mol · l^{-1} ^{233}U, $1.98\cdot10^{-5}$ mol · l^{-1}U (pH = 8.32). Basic composition of working solutions resulting from the analysis of groundwater collected in the studied area was – $1.67\cdot10^{-3}$ mol · l^{-1} $NaHCO_3$, $0.32\cdot10^{-3}$ mol · l^{-1} $KHCO_3$, $1.26\cdot10^{-3}$ mol · l^{-1} $CaCl_2$, $0.25\cdot10^{-3}$ mol · l^{-1} $MgCl_2\cdot6H_2O$ and $0.53\cdot10^{-3}$ mol · l^{-1} $MgSO_4\cdot7H_2O$.

Each solution was characterized by a speciation calculation. Batch sorption experiments in PE vials under oxic conditions at room temperature were carried out with duplicate samples.

The sequential leaching was done as a reduced BCR procedure (Ure et al. 1993). Surface uranium (“exchangeable”) and uranium bound in carbonates was determined in the first step of the leaching where 1 g of the sediment sample was contacted with 40 ml of 0.11 mol · l^{-1} CH_3COOH for 16 h (pH 3.3). In the next step, 40 ml of 0.1 mol · l^{-1} $NH_2OH \cdot HCl$ (contact time 16 h, pH 2) was used for the determination of uranium fraction bound in Fe/Mn – oxides and hydroxides. The result of the last used step of leaching was uranium fraction accessible from organic matter and sulfides (namely, by adding of 10 ml H_2O_2, pH 2, contact time 1 h, ambient temperature; afterwards 10 ml H_2O_2, 1 h, 85°C; cooling; and in the end 50 ml CH_3COONH_4, 16 h, pH 6).

Discussion of Leaching and Kinetic Experiments

A broad set of kinetic/exchange experiments were performed for three selected samples (D1, D2 and D3) with three working solutions (A, B and C) and for the phase ratio V/m from 5 to 100 ml/g. The changes of pH, exchangeable uranium q_0, distribution coefficient K_d (detailed calculation and evaluation of K_d and q_0 can be found in Vopálka et al. (2008)) and total uranium sorbed were analyzed during 28 days. This time seemed to be insufficient in some cases (see e.g. pH change on Fig. 3.) for the description of the system equilibrium. In our opinion, the great changes of pH in some experiments indicated the presence of sulfides oxidation which caused the decrease of pH values (D1, D2). As for sample D4, it was thus not studied because of its high content of pyrite – 18.7 weight % (Table 2).

On the other hand, sample D3 was free from pyrites and thus the changes in pH were negligible in comparison with sample D2 (Fig. 4.) and as a result, it was possible using the equilibrium points of kinetics to construct an equilibrium isotherm having the Langmuir shape. (The exception is the behavior of the samples D1 and D2 at a V/m = 5, namely due to small amount of sample and thus the pyrites.)

The equilibrium values of q_0 were compared with values of exchangeable uranium derived from sequential leaching – for the three sequential leaching steps

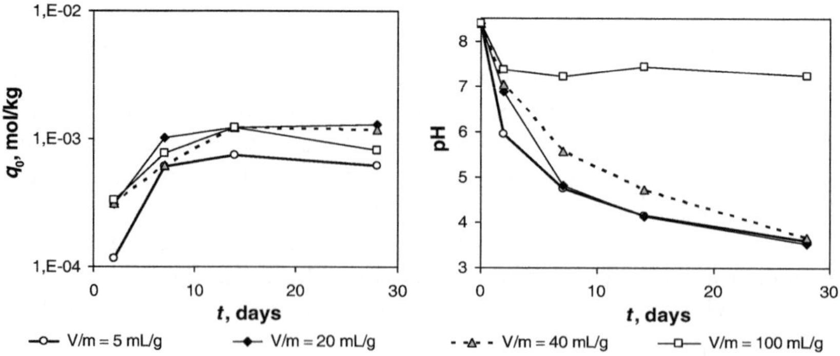

Fig. 3 Results of kinetic sorption/exchange experiments with the sample D2 and solution A for four values of phase ratio V/m

there was a very good match (Table 5). In the profile of the borehole NAR 2 was found relatively high amount of exchangeable U(VI) in the middle of the uranium mineralization represented by the sample D2.

Amount of uranium leached during BCR procedure for samples D1, D2 and D4 increased linearly from step 1 to step 3, where was the highest U concentration. On the contrary, in the samples D3 and D5 occurred the minimum of uranium concentration in the step 2 (Fig. 5).

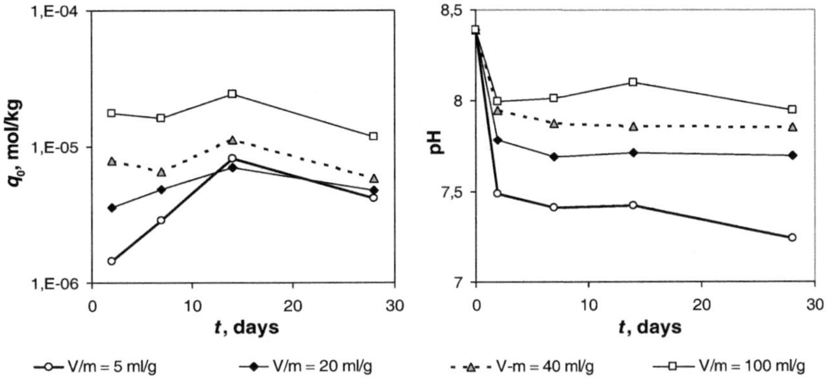

Fig. 4 Results of kinetic sorption/exchange experiments with the sample D3 and solution A for four values of phase ratio V/m

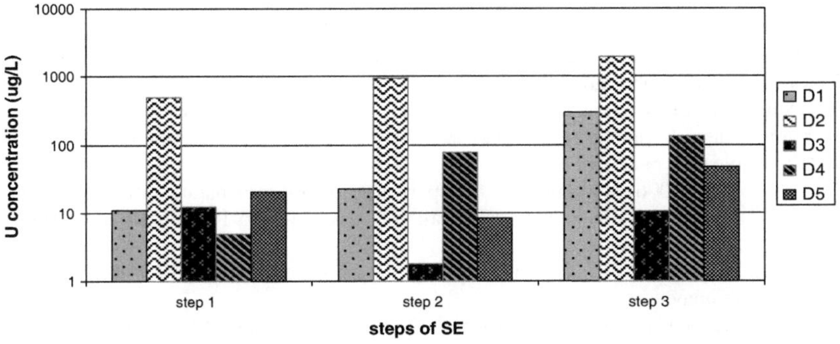

Fig. 5 Uranium fractions released during the BCR extraction

Table 5 Comparison of exchangeable uranium determinated by two methods

Sample (NAR 2)	q_0 [mol·kg^{-1}][a]	q_0 [mol·kg^{-1}] – leaching[b]
D1	$5.09 \cdot 10^{-5}$	$1.8 \cdot 10^{-6}/9.2 \cdot 10^{-5}$
D2	$7.36 \cdot 10^{-4}$	$8.4 \cdot 10^{-5}/8.3 \cdot 10^{-4}$
D3	$4.81 \cdot 10^{-6}$	$2.0 \cdot 10^{-6}/5.0 \cdot 10^{-6}$

[a] Solution A, contact time 28 days.
[b] BCR extraction: 1st step/all 3 steps.

Conclusions

i) Titration curves of selected rock samples from the Ruprechtov site were described by SCM (Surface Complexation Model) using the so called Generalized Composite approach, and the surface sites parameters were determined. In addition, the mineralogical composition of samples was taken into account and the Component Additivity approach based on literature data characterizing the surface sites of component minerals was successfully applied. The knowledge of surface sites parameters makes possible the surface complexation modeling of sorption or desorption of radionuclides to be used. ii) The kinetics of uranium interaction with selected samples was studied under different conditions. The influence of exchangeable uranium present in samples was determined by isotope exchange using ^{233}U. iii) In some samples the self-leaching of uranium, which would be caused by the oxidation of pyrite, was observed. Unfortunately, it did not enable to formulate the shape of interaction isotherm. iv) The comparison of the value of exchangeable uranium determined with the use of ^{233}U and by sequential leaching was good.

Acknowledgement The research was supported by the Ministry of Education of the Czech Republic under contract MSM 6840770020 and RAWRA, Czech Republic.

References

Croato P V, Frank I W, Johnson K D, Mason P B, Smith M M (1997) Evaluation of kinetic phosphorescence analysis of the determination of uranium. Research and Development Report, New Brunswick Laboratory, 48 pp.

Filipská H, Štamberg K (2005) Mathematical Modeling of a Cs(I) – Sr(II) – Bentonite – Magnetite Sorption System, Simulating the Processes Taking Place in a Deep Geological Repository. Acta Polytechnica 45, No. 5: 11–18.

Noseck U, Brasser T (2006) Radionuclide transport and retention in natural rock formations. Ruprechtov site. Report GRS-218. Gesellschaft für Anlagen- und Reaktorsicherheit (GRS) mbH, Braunschweig, 157 pp.

RES3T – Rossendorf Expert System for Surface and Sorption Thermodynamics (2006). Januar 24, Rossendorf, Germany.

Ure A M, Quevauviller P H, Muntau H, Griepink B (1993) Speciation of heavy metals in soils and sediments. An account of the improvement and harmonization of extraction techniques undertaken under the auspices of the BCR of the commission of the European communities. International Journal of Environmental Analytical Chemistry 51: 135–151.

Vopálka D, Havlová V, Andrlík M (2008) Characterization of U(VI) behaviour in the Ruprechtov site (CZ). In: Uranium, Mining and Hydrogeology (Merkel BJ, Hasche-Berger A, eds), Springer: 583–590.

Wanner H, Albinsson Y, Karnland O et al. (1994) The acid-base chemistry of montmorillonite. Radiochimica Acta 66/67: 157–162.

Estimation of Distribution Coefficient of Uranium and Its Correlation with Soil Parameters Around Uranium Mining Site

G.G. Pandit, S. Mishra, S. Maity and V.D. Puranik

Abstract. Distribution coefficient of uranium and its daughter products are very important for migration study around uranium mining sites. Since the distribution coefficient depends very much on the soil chemistry, generation of site specific K_d is very important. The present study emphasizes on the estimation of distribution coefficient for uranium and its correlation with various soil parameters. The distribution coefficient of uranium in top and one meter depth soil samples from above locations were estimated using laboratory batch method. The distribution coefficient of uranium varies from 69 to 5524 l/kg. No significant difference in uranium K_d values was observed for top and one meter depth soil samples. A good correlation was observed between distribution coefficient of uranium and soil parameters like pH and concentration of $CaCO_3$.

G.G. Pandit
Environmental Assessment Division, Bhabha Atomic Research Centre, Trombay, Mumbai-400085, India, Email: ggp@barc.gov.in

S. Mishra
Environmental Assessment Division, Bhabha Atomic Research Centre, Trombay, Mumbai-400085, India, Email: ggp@barc.gov.in

S. Maity
Environmental Assessment Division, Bhabha Atomic Research Centre, Trombay, Mumbai-400085, India, Email: ggp@barc.gov.in

V.D. Puranik
Environmental Assessment Division, Bhabha Atomic Research Centre, Trombay, Mumbai-400085, India, Email: ggp@barc.gov.in

B. Merkel, M. Schipek, *The New Uranium Mining Boom*
DOI 10.1007/978-3-642-22122-4 © Springer 2011

Introduction

Naturally occurring radionuclides are present in many natural resources. Activities involving the extraction, exploitation and processing of materials containing NORM include the mining and processing of uranium and metal ores, the combustion of fossil fuels, production of natural gas and oil, and the phosphate industry. From uranium mining there is a chance of contamination by trace metals and naturally occurring radionuclides (IAEA 2003). Any metals or other contaminants that are present in the uranium tailings may be leached into the soil and enter into the groundwater. The possible pathway of the metals to reach the environment is via the leaching of the metals from tailings to the aquifer below and subsequent transport in the aquifer. Migration of metals and other tailings constituents occurs through leaching and erosion. Predicting the transfer of radionuclides in the environment for normal release, accidental, disposal or remediation scenarios in order to assess exposure requires the availability of an important number of generic parameter values. One of the key parameters in environmental assessment is the solid liquid distribution coefficient, K_d, which is used to predict radionuclide–soil interaction and subsequent radionuclide transport in the soil column (Vandenhove et al. 2009). The radionuclides present at these sites can enter the food chain directly via the soil–plant–animal pathway, or indirectly by the use of contaminated groundwater or surface water for irrigation purposes or drinking water. To assess the uptake in the food chain and by biota in general and to predict human exposure, knowledge on the environmental parameters governing radionuclide mobility and uptake is indispensable. Distribution coefficient values of uranium are very much dependent on various physico-chemical parameters of soil and ground water, that's why site specific distribution coefficient is very much important for the prediction of the contaminant transport in geological matrices. Uranium sorption is affected by soil properties other than soil texture such as pH, $CaCO_3$, content of amorphous iron oxides, soil organic matter content, cation exchange capacity, and phosphate status (EPA 1999).

In the present study the soil samples and ground water samples collected around uranium mining site were characterized and distribution coefficient for uranium was estimated in soil samples. Since distribution coefficient is a strong function of soil parameters, its correlation with various soil parameters was also observed.

Materials and Methods

Sample Collection and Processing

Nine villages around uranium mining site, Jharkhand state of India were selected for the present study. From different villages surface and one meter depth soil samples were collected in duplicate. Ground water samples (well/bore well water) were also collected from the same locations. All the soil samples were collected with a stainless steel spatula and kept in polyethylene containers previously treated with a molar solution of hydrochloric acid and rinsed with distilled water (Hanns 1984). Soil samples were air dried and passed through 2 mm mesh sieve, and then sieved samples were homogenized and used for chemical characterization as well as estimation of distribution coefficient of uranium. Soil samples (<2 mm) were used for estimation of pH, cation exchange capacity, total carbon and $CaCO_3$. Filtered (Whatman 542) ground water samples were directly used for the estimation of anions (SO_4^{2-}, NO_3^-, F^-, Cl^- and HCO_3^-), inorganic carbon and organic carbon. Ground water samples were digested by using electronic grade nitric acid and metals were analyzed in the digested water samples. Soil samples (<2 mm) were microwave digested and metals were analyzed in the digested soil samples. All chemicals used were Merck, Suprapur, Analar or Electronic grade.

Methodology

Soil samples were air dried and sieved using different mesh sieves (2 mm, 300 and 45 μm) in a electromagnetic sieve shaker (EMS 08) to find out the particle size distribution in order to know the soil type. The pH was determined in water with a 1 : 2.5 soil to solution ratio as per IS 2720, Part No. 26 using pH meter (DPH-500). The calcium carbonate was estimated by titrimetric method as per IS 2720, Part No. 23. The concentration of calcium carbonate in soil was determined by taking a 1 : 20 soil to 0.1 (N) hydrochloric acid ratio, after shaking for one hour the excess hydrochloric acid was titrated by using 0.1 (N) sodium hydroxide solution. The values of CEC (cation exchange capacity) were obtained by calculation from exchangeable cations content determined by measuring the solution content after treatment of soil fraction with ammonium acetate solution. The soil samples were digested using Microwave digester (Ethos 1). All the trace and toxic metals in soil were analyzed using Atomic Absorption Spectrometer AAS (GBC-Avanta), Inductively coupled Plasma Atomic Emission Spectrometer ICP-AES (Jovin Yvon-Ultima 2) and Differential Pulse Adsorptive Stripping Voltametry DPASV (Methrom-663 VA stand).

In water samples anions (Cl^-, F^-, NO_3^-, SO_4^{2-}) were analyzed using ion chromatograph (IC Metrohm 733) and CO_3^{2-}, HCO_3^- were analyzed by titration. Organic and inorganic carbon were analyzed using carbon analyzer (Shimadzu TOC).

The sorption experiment has been carried out by laboratory batch method (EPA 1999). Natural uranium was used as a tracer. In 50 ml polypropylene centrifuge tubes soil (1 g) and natural uranium (400 μg) in 30 ml ground water (1 : 30 S/L ratio) were shaken at room temperature. The experiment was carried out in different time interval (0–100 h) to find out the equilibration time. The laboratory batch method was carried out in a dual rotating shaker bath Model SK-300 with a shaking rate of 0.8–1.2 oscillations per second. Solution mixture was centrifuged and filtered through 0.45 μ filter paper. Uranium was analyzed in the filtrate using DPASV (Methrom-663 VA stand).

Distribution coefficients are defined as the ratio of the concentration of the uranium in the sorbent and in the solution at equilibrium.

The distribution coefficients were calculated as per the following equation:

$$K_{\mathrm{d}} = \left[\left(C_0 - C_{\mathrm{e}}\right)/C_{\mathrm{e}}\right]\cdot V/m \text{ ml/g or L/kg} \tag{1}$$

where C_0 is the concentration of the initial solution (µg/ml), C_e is the concentration of the solution in equilibrium (µg/ml), V is the volume of the solution (ml), m is the amount of the adsorbent (g).

Results and Discussions

Chemical Characterization of Ground Water and Soil Samples

pH, concentration of inorganic carbon, organic carbon, bicarbonate, fluoride, chloride, nitrate and sulfate of ground water samples collected around uranium mining area are shown in the Table 1. The particle size distribution of soil samples shows that the soil samples are mostly sandy type.

Table 1 Mean and ranges for the descriptive parameters of the ground water samples around uranium mining site

Parameters	Mean	Range
pH	7.02	6.82–7.27
Inorganic carbon (mg/l)	37.51	18.36–57.2
Organic carbon (mg/l)	11.81	3.2–22.1
Bicarbonate (mg/l)	122.5	80.7–182.6
Fluoride (mg/l)	0.813	0.26–1.73
Chloride (mg/l)	88.4	3.47–442.3
Nitrate (mg/l)	32.6	0.29–197.6
Sulfate (mg/l)	55.5	10.4–232.7

Table 2 Mean and ranges for the descriptive parameters of the 18 soil samples from surface and one meter depth collected from nine different locations around Turamdihi Tailings Pond

Parameters	Surface		One meter depth	
	Mean	Range	Mean	Range
pH	6.69	5.00–8.70	6.82	5.05–8.90
$CaCO_{3\%}$	0.49	0.20–1.48	0.56	0.34–1.05
CEC meq/100 g	7.93	2.86–20.15	10.08	4.84–16.97
Total carbon mg/g	7.01	1.6–15.59	5.64	1.09–14.26

Fig. 1 Variation of Uranium sorption with respect to contact time

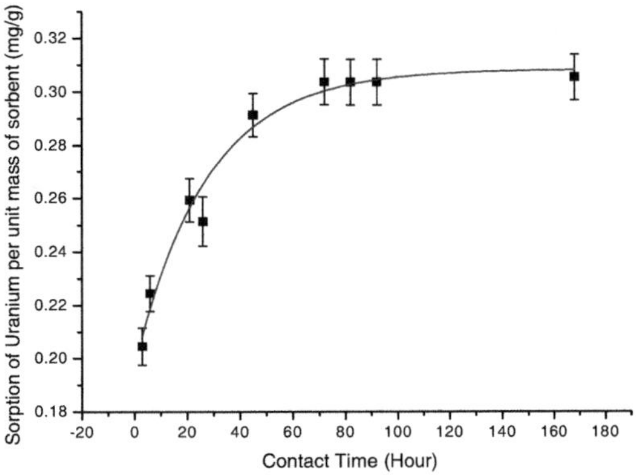

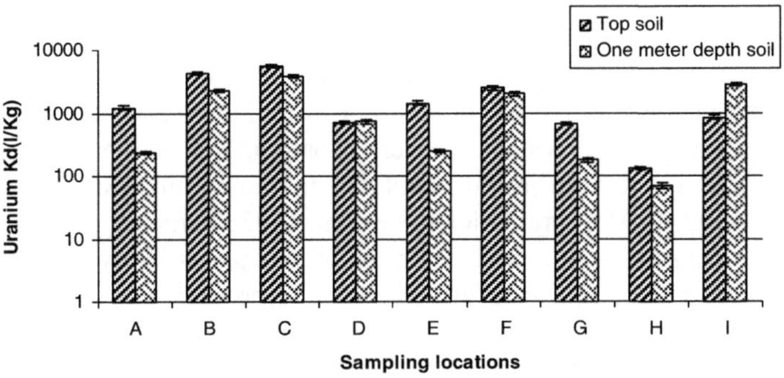

Fig. 2 Distribution coefficients of Uranium in top and one meter depth soil samples around uranium mining site

pH, concentration of calcium carbonate, CEC and total carbon for both top and one meter depth soil samples are shown in Table 2. In case of pH and concentration of $CaCO_3$ there is no significant variation observed in top and one meter depth soil samples. The concentration of total carbon of the soil samples vary from 1.09 to 15.59 mg/g.

From sorption kinetics of uranium it is observed that within 72 h uranium is coming in equilibrium with soil and water shown in Fig. 1. Distribution coefficient of uranium varies from 69 to 5524 l/kg in different sampling locations around uranium mining site shown in Fig. 2. In top and one meter depth soil samples there was no significant variation observed for distribution coefficient of uranium.

Effect of pH with Distribution Coefficient of Uranium

In the present experiment it is observed that the $K_d(U)$ values are increasing in the pH range of 5 to 6 and decreasing in the pH range of 6 to 9. In the pH range of 5 to 6 the regression equation obtained is $\log K_d(U) = 1.67(\pm 0.169) X\, pH - 6.38(\pm 0.94)$

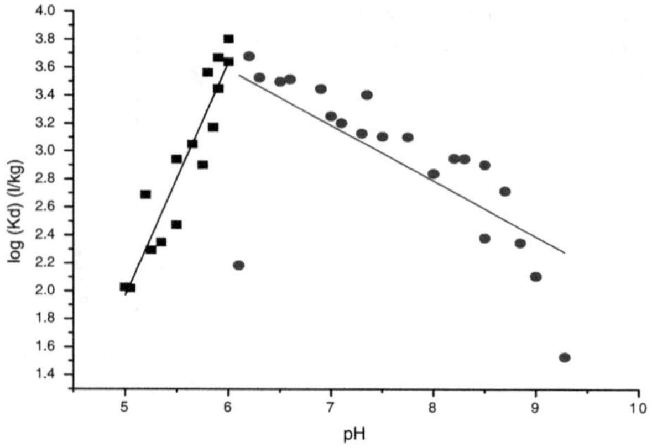

Fig. 3 Variation of distribution coefficient of Uranium with pH in soil samples around uranium mining site

with $R^2 = 0.87$ and in the pH range of 6 to 9 the regression equation is log $K_d(U) = -0.397(\pm 0.09)X\,pH + 5.969(\pm 0.7)$ with $R^2 = 0.469$ for soil samples shown in Fig. 3. This trend is similar to the in situ K_d values reported by Serkiz and Johnson (1994), and percent adsorption values measured for uranium on single mineral phases such as those reported for iron oxides (Hsi and Langmuir 1985; Tripathi 1984; Waite et al. 1992, 1994), clays (McKinley et al. 1995; Turner et al. 1996; Waite et al. 1992), and quartz (Waite et al. 1992). Aqueous pH is likely to have a profound effect on U(VI) sorption to solids. There are 2 processes by which it influences sorption. First, it has a great impact on uranium speciation such that poorer-adsorbing uranium species will likely exist at pH values between 6.5 and 10. Secondly, decreases in pH reduce the number of exchange sites on variable charged surfaces, such as iron-oxides, aluminum-oxides, and natural organic matter. (Echevarria et al. 2001) they found a significant relation between soil K_d and pH. This pH-dependent behavior is related to the pH-dependent surface charge properties of the soil minerals and complex aqueous speciation of dissolved U(VI), especially near and above neutral pH conditions where dissolved U(VI) forms strong anionic uranyl-carbonato complexes with dissolved carbonate.

Effect of Concentration of $CaCO_3$ with Distribution Coefficient of Uranium

In the soil samples it is observed that with increase in the concentration of $CaCO_3$ the distribution coefficient value of uranium decreases. In our experiment, distribution coefficient of uranium follows a good relationship with the concentration of $CaCO_3$. The variation of distribution coefficient values of uranium with respect to concentration of $CaCO_3$ for the soil sample is shown by the following equation $\log K_d(U) = -1.97(\pm 0.22)X\,CaCO_3 + 3.97(\pm 0.127)$ with $R^2 = 0.69$, which is shown in Fig. 4. This type of variation of distribution coefficient of uranium with respect to calcium carbonate in soil is observed may be due to with increasing calcium

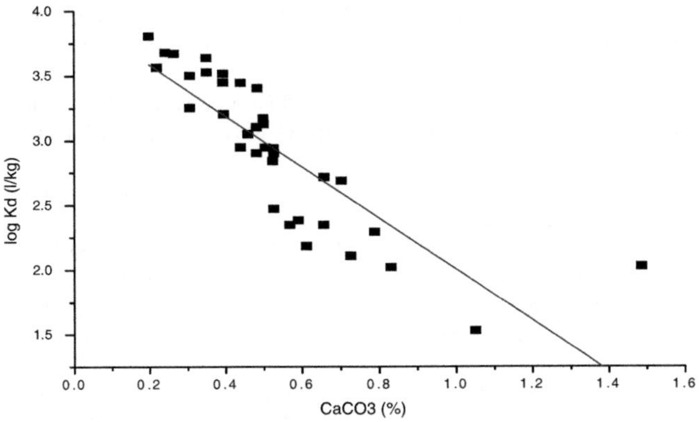

Fig. 4 Variation of distribution coefficient of Uranium with $CaCO_3$ in soil samples around uranium mining site

carbonate concentration in soil pH of soil decreases due to formation of bi carbonic acid and $K_d(U)$ decreases. Further study has to be carried out for better explanation.

Conclusions

The distribution coefficient of uranium varies from 69 to 5524 l/kg in soil samples collected from uranium mining site. No significant difference in uranium K_d values was observed for top and one meter depth soil samples. A good correlation was observed between distribution coefficient of uranium and soil parameters like pH, and concentration of $CaCO_3$. Using these correlation equations distribution coefficient values for other sites of same type of soil can be estimated. This site specific K_d values will help in the migration study of uranium from uranium tailings.

References

Echevarria G, Sheppard M, Morel JL (2001). Effect of pH on the sorption of uranium in soils. J. Environ. Radioactiv. 53:257–264.

EPA (Environmental Protection Agency) (1999). Understanding Variation in Partitioning Coefficients, K_d, Values: Volume II: Review of Geochemistry and Available K_d Values for Cadmium, Caesium, Chromium, Lead, Plutonium, Radon, Strontium, Thorium, Tritium and Uranium. US-EPA, Office of Air and Radiation, Washington, USA. EPA 402-R-99-004B.

Hsi C-KD, Langmuir D (1985). Adsorption of Uranyl onto Ferric Oxyhydroxides: Application of the Surface Complexation Site-binding Model. Geochim. Cosmochim. Acta 49:1931–1941.

IAEA (International Atomic Energy Agency) (2003). Extent of Environmental Contamination by Naturally Occurring Radioactive Material (NORM) and Technological Options for Remediation. Technical report series 419. STI/DOC/010/419, ISBN 92-0-112503-8.

McKinley JP, Zachara JM, Smith SC, Turner GD (1995). The Influence of Uranyl Hydrolysis and Multiple Site-Binding Reactions on Adsorption of U(VI) to Montmorillonite. Clays Clay Min., 43:586–598.

Serkiz SM, Johnson WH (1994). Uranium Geochemistry in Soil and Groundwater at the F and H Seepage Basins (U). EPD-SGS-94-307, Westinghouse Savannah River Company, Savannah River Site, Aiken, South Carolina.

Tripathi VS (1984). Uranium(VI) Transport Modeling: Geochemical Data and Submodels. Ph.D. Dissertation, Stanford University, Stanford, California.

Turner GD, Zachara JM, McKinley JP, Smith SC (1996). Surface-Charge Properties and UO_2^{2+} Adsorption of a Subsurface Smectite. Geochim. Cosmochim. Acta 60:3399–3414.

Vandenhove H, Gil-Garcı C, Rigol A, Vidal M (2009). New best estimates for radionuclide solid–liquid distribution coefficients in soils. Part 2. Naturally occurring radionuclides. J. Environ. Radioactivity.

Waite TD, Payne TE, Davis JA, Sekine K (1992). Alligators Rivers Analogue Project. Final Report Volume 13. Uranium Sorption. ISBN 0-642-599394 (DOE/HMIP/RR/92/0823, SKI TR 92:20–13.

Uranium (VI) Binding to Humic Substances: Speciation, Estimation of Competition, and Application to Independent Data

Pascal E. Reiller, Laura Marang, Delphine Jouvin, Marc F. Benedetti

Abstract. In groundwaters containing natural organic matter (NOM), mostly humic substances (HS), it is expected that it plays a role on the behavior of uranium in the environment. Another point is the actual effect of the competition with major cations and carbonate ions for U(VI)-NOM complexation. Our aim is to acquire experimental data for the U(VI)-HS complexation with two new different methodologies. The NICA-Donnan model is then used to predict uranium speciation in different groundwaters.

Pascal E. Reiller
Commissariat à l'Energie Atomique et aux Energies Alternatives, CE Saclay, DEN/DANS/DPC/SECR, Laboratoire de Spéciation des Radionucléides et des Molécules, BP 11, 91191 Gif sur Yvette, France

Laura Marang
Commissariat à l'Energie Atomique et aux Energies Alternatives, CE Saclay, DEN/DANS/DPC/SECR, Laboratoire de Spéciation des Radionucléides et des Molécules, BP 11, 91191 Gif sur Yvette, France
Laboratoire de Géochimie des Eaux, Université Paris Diderot, IPGP and UMR CNRS 71574, Case Postale 7052, 75205 Paris Cedex 13, France

Delphine Jouvin
Laboratoire de Géochimie des Eaux, Université Paris Diderot, IPGP and UMR CNRS 71574, Case Postale 7052, 75205 Paris Cedex 13, France

Marc F. Benedetti
Laboratoire de Géochimie des Eaux, Université Paris Diderot, IPGP and UMR CNRS 71574, Case Postale 7052, 75205 Paris Cedex 13, France

B. Merkel, M. Schipek, *The New Uranium Mining Boom*
DOI 10.1007/978-3-642-22122-4

Introduction

The human contributions to the uranium geochemical cycle can lead to an increase of uranium concentration in some environments. The information of its speciation in the field and in trace concentration is not an easy task, and the use of model can help in providing reasonable estimates. Under oxidizing conditions, uranium is essentially under the form of U(VI) complexes; under reducing conditions $U(OH)_4(aq)$ is the dominant form in solution (Guillaumont et al. 2003). U(VI) sorption onto minerals (Sylwester et al. 2000) can be well described by surface complexation and/or ion exchange model. In contrast, sorption of U(VI) to soils and sediments is more complicated because of the difficulty to fully characterize the species present and/or the properties and contribution of the solid-phase sorbents, including natural organic matter (NOM). NOM exits in natural waters, soils and sediments – mostly humic substances (HS) (MacCarthy 2001) –, and can play an important role on uranium speciation. A robust experimental data basis is needed to derive parameters to predict the behavior of trace elements in field systems. The NICA-Donnan model describes metal-NOM binding (Kinniburgh et al. 1999): it accounts for the NOM chemical heterogeneity, competition between metals, and ionic strength effects. It is based on the electrostatic interaction of metal within a Donnan gel,

$$C_{i,\mathrm{D}} = C_i \exp\left(\frac{z_i F}{RT}\Psi_\mathrm{D}\right),$$

where C_i and $C_{i,\mathrm{D}}$ are the concentration in bulk and Donnan phase, respectively. The specific bonding is described by the Non-Ideal Competition Adsorption model,

$$Q_{i,T}\sum_i Q_{\max,j} \times \frac{\left(\tilde{K}_{i,j}C_{\mathrm{D},i}\right)^{n_{i,j}}}{\sum_i\left(\tilde{K}_{i,j}C_{\mathrm{D},i}\right)^{n_{i,j}}} \times \frac{\left[\sum_i\left(\tilde{K}_{i,j}C_{\mathrm{D},i}\right)^{n_{i,j}}\right]^{p_j}}{1+\left[\sum_i\left(\tilde{K}_{i,j}C_{\mathrm{D},i}\right)^{n_{i,j}}\right]^{p_j}},$$

where log and $n_{i,j}$ are the median affinity constant and heterogeneity parameter of metal i for the jth distribution of sites, and p_j the heterogeneity parameter of the jth distribution of sites. Only a few datasets cover a sufficiently wide range of $[U]_{free}$, ionic strength, and pH to provide NICA-Donnan generic parameters (Milne et al. 2003). Our aim is to implement this experimental data basis using two different methodologies i) the Flux Donnan Membrane technique (FDM), and ii) an insolubilized humic acid (IHA). Independent data sets from the literature can be analyzed and compared to blind predictions.

Materials and Methods

$Mg(NO_3)_2$, KNO_3 (Panreac), EDTA (Sigma-Alrich), H_3PO_4 (Normapur, Prolabo), $UO_2(NO_3)_2$ (SPEX certiprep 5% HNO_3) and water (Milli Q, Millipore) were used.

The Gorleben humic acid (GHA) was extracted from one of the deep groundwaters in the Gorleben area (Wolf et al. 2004).

FDM is a dynamic modification of the standard Donnan Membrane Technique (DMT, Temminghoff et al. 2000) with both improved detection limits and faster response. A strong ligand is added on the acceptor side to increase the amount of free metal ion transported to the acceptor, and on the analysis of the metal flux. The theoretical basis for the FDM has been described in details elsewhere (Marang et al. 2006), and will not be recalled here. The cation exchange membrane (BDH, Laboratory Supplies) consists of a matrix of polystyrene and divinyl-benzene: ion exchange capacity and thickness are given by the supplier as 0.8 mM/g and 0.16 ± 0.1 mm, respectively. The area of the membrane is 7 cm^2. The membranes were preconditioned by first soaking in 0.1 M HNO_3 to remove trace metal impurities. Then, they were washed with a 1 M $Mg(NO_3)_2$ solution. Finally, the membranes were equilibrated in the final background electrolyte: 2 mM $Mg(NO_3)_2$. Both donor and acceptor solutions were circulated at a constant rate of 2.5 ml/min. The acceptor and donor volumes were 25 ml and 200 ml, respectively. The FDM was calibrated vs. equilibration time when transport on the donor side was dominated either by uranium diffusion in solution (case 1) or by uranium diffusion in the membrane (case 2). Calibration conditions are summarized in Table 1. In case 1, the calibration was performed with a high concentration of phosphoric acid (H_3PO_4) on the acceptor side. In case 2, EDTA (ethylenediamine tetraacetic acid) and H_3PO_4 were present in the donor and acceptor side, respectively. The concentration of EDTA and H_3PO_4 needed for the two calibration conditions were estimated with the help of ECOSAT code (Keizer and van Riemsdijk 1994), which includes NICA-Donnan, using known thermodynamic binding constants (Guillaumont et al. 2003; Hummel et al. 2005), and Mg-NOM complexation (Marang 2007). For the U(VI)-HS study, the donor solution contains 10 mg/l of GHA with $8\ 10^{-8} \leq [U(VI)]_{total}\,(M) \leq 10^{-5}$. The acceptor solution consisted of 10^{-2} M H_3PO_4. The *p*H was fixed at 4. The acceptor and donor sides were sampled after 24 h, and calibration 1 was used for high metal to ligand ratio or they were sampled after 48 h and calibration 2 used for low metal to ligand ratio.

Table 1 Experimental conditions of the FDM for the U(VI) calibrations for solution controlled diffusion (case 1) and for membrane-controlled diffusion (case 2), and comparison between experimental and theoretical calibrations curves slopes for uranium when the transport is solution-controlled (case 1) and membrane-controlled (case 2)

	Case 1	Case 2
[EDTA] in Donor	0	0.5 mM
[H_3PO_4] in Acceptor	100 mM	100 mM
Background Electrolyte	2 mM	2 mM
$[U]_{Total}$ in donor	5 μM	20 μM
pH	4	4
	$A_1 = (s^{-1})$	$A_2 = (s^{-1})$
	$2.3\ 10^{-6}$	$1.47\ 10^{-5}$

The synthesis and characterization of IHA are detailed elsewhere (Weber et al. 2006). A mixture of 50 mg/l of IHA was reacted with solutions of known $[U(VI)]_{total}$ in 2 mM $Mg(NO_3)_2$; *p*H ranged from 4 to 6.8. The solutions were then centrifuged during 20 min at 4000 rpm. The supernatant was removed to perform subsequent chemical analysis. Uranium concentrations were determined by time resolved laser induced spectrofluorimetry using the standard addition method in 5% H_3PO_4 (0.1 M, detection limit of 10^{-10} M) (Marang 2007). HA concentrations were analyzed by TOC (Schimadzu TOC-Vcsh).

Results and Discussion

Experimental slopes of the calibrations curves are given in Table 1. The U(VI) binding isotherm to GHA is given in Fig. 1a: data points corresponding to $[UO_2^{2+}]_{free} > 10^{-7}$, were obtained using the case 1 calibration; data point with $[UO_2^{2+}]_{free} < 10^{-7}$, were obtained with the case 2 calibration curve. The discrepancy between data points corresponding to $10^{-8} < [U]_{free} < 10^{-7}$, could be due to a mixed contribution of membrane and solution diffusion controlled flux (Marang et al. 2006). The data points corresponding to the experiments made with the IHA show a much smaller U(VI) binding under similar conditions (Fig. 1). Moreover, the slope of the binding is steeper (i.e. ≈ 0.7).

The data in Fig. 1 show that the binding increases as pH increases from pH 4 to 5 but remains constant thereafter. This difference is due to the difference of reactivity among both extracts; indeed HS are heterogeneous (Milne et al. 2001). Weber et al. (2006) have shown that the abundance of the high proton affinity sites (S2) of IHA is halved compared to the pristine HA due to condensation reactions during insolubilization. Using *p*H, $[U(VI)]_{total}$, $[Mg^{2+}]$, and C(GHA), [U(VI)-GHA]

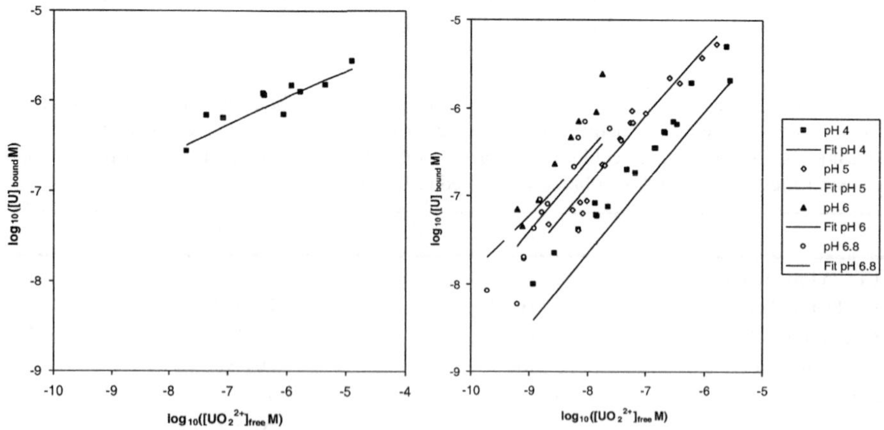

Fig. 1 Uranium(VI) complexation isotherms (log $[U]_{bound}$ vs. log $[UO_2^{2+}]_{free}$) at *p*H 4 with Gorleben HA (10 mg/l). The background electrolyte was 2 mM $Mg(NO_3)_2$

was simulated using the GHA parameters (Table 2) and the U(VI) parameters from the study of Saito et al. (2004). There is a good agreement between prediction and experimental data. Further fitting of the experimental data is only possible for the low-affinity sites (S1), keeping the S2 parameters values from Saito et al. (2004) constant. Median affinity constant, $\log \tilde{K}_1$, and heterogeneity parameter, n_1, are reported in Table 2. Fitting of $\log \tilde{K}_2$ does not improve the results due to its low importance under these $[U(VI)]_{total}$ and pH conditions.

The $\log \tilde{K}_1$ and n_1 were adjusted to the IHA data. The parameters are based on experimental data acquired for acid conditions. For this reason, the model could not describe the IHA experimental data which are acquired at higher pH. Site density (Q_{max}) and proton parameters were obtained from Weber et al. (2006), Mg^{2+} parameters are from Marang (2007). The four series were fitted altogether resulting in deviation for the pH 4 and 6.8 series (Fig. 1b). In the calculation of U(VI)-IHA binding, only free UO_2^{2+} was considered to bind specifically, and $[UO_2^{2+}]_{free}$ is calculated from $[U]_{free}$ (Guillaumont et al. 2003). For the two fittings, the resulting $\log_1$ are in accordance and are reported together with n_1 in Table 2. For IHA, the lack of increase after *p*H 6 results from the formation of $UO_2(OH)_n^{2-n}$ and $UO_2CO_3(aq)$, complexes for $10^{-8} \leq [U]_{total}$ M $\leq 10^{-5}$ (Zeh et al. 1997). Only the S1 sites are adjusted; the S2 sites are fixed after Saito et al. (2004). These new parameters can now be used to predict uranium speciation in natural samples.

In mildly acid oxic media the data from Savannah River Site (SC, USA, Jackson et al. 2005), can be used. The concentrations of anions are not reported, and ions from the carbonate system are supposed to be in equilibrium with ambient $CO_2(g)$ (0–5 and 5–10 cm depth); Eh is assumed to be oxic even if slightly reducing condition could occur. DOC values were considered as HA (50% C). UO_2-HA complexes account for more that 99.99% for the six cases in Jackson et al. (2005) (Fig. 2), which is in agreement with field data.

In more basic media, Ranville et al. (2007) quantified the association of U(VI) with NOM collected from a vertical porewater profile through a clay-rich aquitard (Hendry and Wassenaar 2000). The low measured binding of uranium by NOM implies that the competition of U(VI) carbonate complexes, which is also in

Table 2 Nica-Donnan model parameters used in this study

	S1 Low proton affinity groups	S2 High proton affinity sites (Saito et al. 2004)
Gorleben		
$\log \tilde{K}_{i,UO_2^{2+}}$	4.59 ± 0.05	6.92[a]
$^{n}UO_2^{2+}$	0.33 ± 0.13	0.39a
IHA		
$\log \tilde{K}_{i,UO_2^{2+}}$	4.72 ± 0.08	6.92[a]
$^{n}UO_2^{2+}$	0.76 ± 0.08	0.39[a]

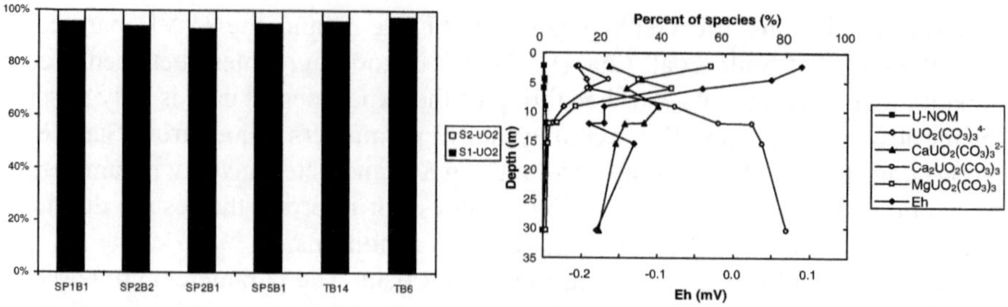

Fig. 2 Calculated speciation of uranium using ECOSAT and NICA-Donnan data from Saito et al. (2004) in the groundwaters from Jackson et al. (2005) (*left*), and from Ranville et al. (2007) (*right*)

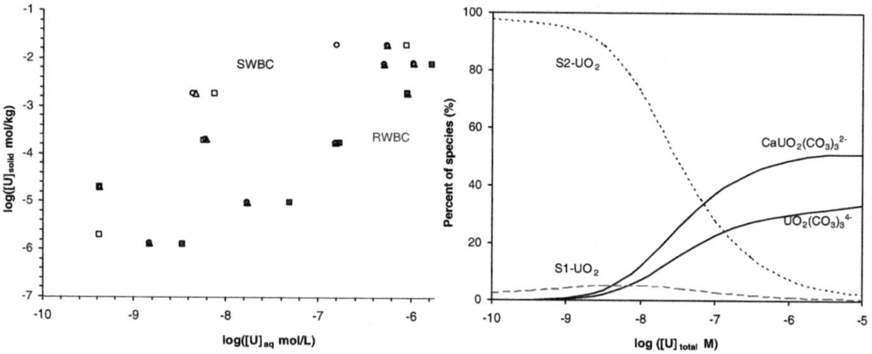

Fig. 3 Solid phase uranium concentration vs. solution phase uranium concentration obtained by Bruggeman and Maes (2010) on pyrite (5 g/l) in Synthetic (SBCW) or Real (RBCW) Boom Clay water (*left*), compared to uranium-NOM speciation estimated in the same RBCW conditions (*right*)

agreement with the modeling. Major part of uranium should be present as ion-paired complexes $(Ca,Mg)_nUO_2(CO_3)_3^{2n-4}$ (Dong and Brooks 2006), depending on depth (Fig. 2).

In a more reductive case of an Underground Research Laboratory (Mol, Belgium), Bruggeman and Maes (2010) reported an influence of NOM (mostly HA) on the U(VI) fixation to pyrite. The authors compared a synthetic Boom Clay water (SBCW), with no NOM added, to a real Boom Clay water (RBCW) (de Craen et al. 2004). Excluding NOM, U(VI) in solution should be present under the form of $CaUO_2(CO_3)_3^{2-}$ and $UO_2(CO_3)_3^{4-}$. In this laboratory experiment, according to Bruggeman and Maes (2010), the solubility seems to be controlled by a UO_{2+x} phase, mostly U_3O_8. Figure 3 presents the digitized data compared to the calculated speciation. NOM influence was anticipated for $[U]_{total} < 1$ μM. The calculated speciation is in fair agreement with the observation as less than 9% of uranium is awaited to be fixed to NOM at 1 μM. The proportion of U-NOM complexes, on S2 sites, is awaited to increase with decreasing $[U]_{total}$, and the proportion of fixed uranium to pyrite should decrease. It is difficult to propose any further interpretation as the FeS_2/HA system is complicated, but the general trend is rather satisfying.

Conclusions

This paper presents the results of the study of U(VI)-HA complexation. U(VI) is strongly bound to HA for pH < 6. For higher pH, elevated carbonate concentrations significantly limit the formation of U(VI)-HA complexes. The ECOSAT code was then used to predict the uranium speciation in different groundwaters. When pH is lower than 7, U(VI) is majorly controlled by NOM and will follow its fate in the different horizons of the soils and groundwaters. For higher pH values the situation is more complicated; for high total concentration of uranium, there is a minor amount of uranium associated to NOM. Conversely, when the total concentration of uranium is less than 10^{-7} M, the interaction of U(VI) with NOM becomes important even in reductive environments.

References

Bruggeman C, Maes N (2010) Uptake of uranium(VI) by pyrite under Boom clay conditions: Influence of dissolved organic carbon. Environ Sci Technol 44: 4210–4216.

de Craen M, Wang L, van Geet M, Moors H (2004) Geochemistry of Boom Clay pore water at the Mol site, SCK•CEN-BLG-990, SCK•CEN, Mol, Belgium.

Dong WM, Brooks SC (2006) Determination of the formation constants of ternary complexes of uranyl and carbonate with alkaline earth metals (Mg2+, Ca2+, Sr2+, and Ba2+) using anion exchange method. Environ Sci Technol 40: 4689–4695.

Guillaumont R, Fanghänel T, Fuger J, Grenthe I, Neck V, Palmer DA, Rand M (2003) Update on the chemical thermodynamics of uranium, neptunium, plutonium, americium and technetium, North-Holland, Amsterdam, The Netherlands.

Hendry MJ, Wassenaar LI (2000) Controls on the distribution of major ions in pore waters of a thick surficial aquitard. Water Resour Res 36: 503–513.

Hummel W, Anderegg G, Rao LF, Puigdomènech I, Tochiyama O (2005) Chemical thermodynamics of compounds and complexes of U, Np, Pu, Am, Tc, Se, Ni and Zr with selected organic ligands, Elsevier, Amsterdam, The Netherlands.

Jackson BP, Ranville JF, Bertsch PM, Sowder AG (2005) Characterization of colloidal and humic-bound Ni and U in the "dissolved" fraction of contaminated sediment extracts. Environ Sci Technol 39: 2478–2485.

Keizer MG, van Riemsdijk WH (1994) A computer program for the calculation of chemical speciation and transport in soil-water systems (ECOSAT 4.7). Agricultural University of Wageningen, Wageningen, The Netherlands.

Kinniburgh DG, van Riemsdijk WH, Koopal LK, Borkovec M, Benedetti MF, Avena MJ (1999) Ion binding to natural organic matter: competition, heterogeneity, stoichiometry and thermodynamic consistency. Colloid Surf A 151: 147–166.

MacCarthy P (2001) The principles of humic substances. Soil Sci 166: 738–751.

Marang L (2007) Influence de la matière organique naturelle sur la spéciation des radionucléides en contexte géochimique, PhD Thesis, Université Denis Diderot (Paris VII), and CEA-R-6187 Report. http://tel.archives-ouvertes.fr/docs/00/41/87/23/PDF/These_laura_marang_final.pdf., Paris, France.
Marang L, Reiller P, Pepe M, Benedetti MF (2006) Donnan membrane approach: From equilibrium to dynamic speciation. Environ Sci Technol 40: 5496–5501.
Milne CJ, Kinniburgh DG, Tipping E (2001) Generic NICA-Donnan model parameters for proton binding by humic substances. Environ Sci Technol 35: 2049–2059.
Milne CJ, Kinniburgh DG, van Riemsdijk WH, Tipping E (2003) Generic NICA-Donnan model parameters for metal-ion binding by humic substances. Environ Sci Technol 37: 958–971.
Ranville JF, Hendry MJ, Reszat TN, Xie QL, Honeyman BD (2007) Quantifying uranium complexation by groundwater dissolved organic carbon using asymmetrical flow field-flow fractionation. J Contam Hydrol 91: 233–246.
Saito T, Nagasaki S, Tanaka S, Koopal LK (2004) Application of the NICA-Donnan model for proton, copper and uranyl binding to humic acid. Radiochim Acta 92: 567–574.
Sylwester ER, Hudson EA, Allen PG (2000) The structure of uranium (VI) sorption complexes on silica, alumina and montmorillonite. Geochim Cosmochim Acta 64: 2431–2438.
Temminghoff EJM, Plette ACC, van Eck R, van Riemsdijk WH (2000) Determination of the chemical speciation of trace metals in aqueous systems by the Wageningen Donnan Membrane Technique. Anal Chim Acta 417: 149–157.
Weber T, Allard T, Tipping E, Benedetti MF (2006) Modeling iron binding to organic matter. Environ Sci Technol 40: 7488–7493.
Wolf M, Buckau G, Geyer S (2004) Isolation and characterization of new batches of Gohy-573 humic and fulvic acids., Buckau G ed., Humic Substances in Performance Assessment of Nuclear Waste Disposal: Actinide and Iodine Migration in the Far-Field Second Technical Progress Report, Forschungszentrum Karlsruhe, FZKA 6969, http://bibliothekfzkde/zb/berichte/FZKA6969pdf, 111–124.
Zeh P, Czerwinski KR, Kim JI (1997) Speciation of uranium in Gorleben groundwaters. Radiochim Acta 76: 37–44.

Sorption of Uranium on Iron Coated Sand in the Presence of Arsenate, Selenate, and Phosphate

Romy Schulze, Broder Merkel

Abstract. Batch and column experiments were performed with solutions containing 0.5 µmol/l uranium in the absence and presence of selenate, arsenate and phosphate with equimolar concentrations at a pH of about 6.5. Iron(hydr)oxide surfaces are known for their strong interaction with arsenate and phosphate in solution. However, uranium binds much stronger than arsenate while selenate binds only to a minor extend. Sorption of uranium differs when arsenate, selenate, or phosphate is present in solution. With arsenate present a slight decrease of uranium sorption was found, while phosphate and selenate enhanced sorption.

Introduction

Uranium may be an important contaminant concerning groundwater but in particular for uranium mining and milling sites and at nuclear waste disposal. Retardation and transport of U(VI) are affected by their sorption/desorption reactions at solid/solution interfaces. But, sorption efficiency does not only depend on pH and concentration of uranium but also on interfering and competitive ions. Mining activities are often associated with significantly increased arsenate and selenate concentrations in mine water. Because these ligands are naturally recurrent with uranium deposits, it is necessary to understand their interactions. Furthermore

---

Romy Schulze
Department of Hydrogeology, Technische Universität Bergakademie Freiberg, Gustav-Zeuner-Str. 12, 09599 Freiberg, Germany

Broder Merkel
Department of Hydrogeology, Technische Universität Bergakademie Freiberg, Gustav-Zeuner-Str. 12, 09599 Freiberg, Germany

B. Merkel, M. Schipek, *The New Uranium Mining Boom*
DOI 10.1007/978-3-642-22122-4

phosphate is a common component in aquatic environments and plays an important role in governing the mobility of U(VI) (Cheng et al. 2004).

Iron oxides, available as fine powders, have a relatively high surface area and have been used to eliminate several metals from aqueous solutions. However, adsorbent properties of these powders were not fully exploited in treatment operations because of difficulties associated with their separation from the aqueous phase and their low hydraulic conductivity. To overcome this problems coating of iron oxide onto the surface of quartz sand is an effective method.

Thus the objective of this work was to investigate sorption reactions of uranium occurring at the water/iron oxide coated sand (IOCS) interface in the absence and presence of selenium, arsenic, and phosphorus.

Materials and Methods

Materials

Sand for the iron coating procedure was a natural quartz sand termed F32 purchased from (Quarzwerke Frechen, Germany) with an average grain size of 0.24 mm and a specific theoretical surface area of 102 cm^2/g. It contains 99.7 m.% SiO_2, 0.2 m.% Al_2O_3 and 0.03 m.% Fe_2O_3. The iron coating process of this sorbent was carried out as described by Benjamin (1996), with some minor modifications.

A solution containing 120 ml of 2 M $Fe(NO_3)_3$ 9 H_2O and 3 ml of 5 M NaOH was poured over 300 g sand. After stirring for 15 min the mixture was placed in a heat resistant dish and heated for 24 h at 110°C and then at 550°C for 3 h. Upon cooling the coated sand was washed with deionized water until the water was clear and then dried again for 24 h at 110°C.

U(VI) stock solutions of $1 \cdot 10^{-3}$ mol/l were prepared by dissolving $UO_2(NO_3)_2 \cdot 6H_2O$ (Chemische Fabrik Erich Nickel, Germany) in distilled water. The solutions with equimolar concentrations of arsenate, selenate, and phosphate were prepared by dissolving $Na_2HAsO_4 \cdot 7\,H_2O$ (Fluka, Switzerland), $Na_2SeO_4 \cdot 10\,H_2O$ (Aldrich, Germany) and KH_2PO_4 (Applichem, Germany) in the appropriate volume of ultra pure water. Experimental solutions with a concentration of $5 \cdot 10^{-7}$ mol/l were prepared freshly from these stock solutions. For constant ionic strength a 0.01 molar sodium chloride (Merck, Germany) solution was added to all input solutions. The solution pH was adjusted to 6.5 with 0.1 M NaOH.

Experimental Setup

Some batch experiments were conducted to dimension the column experiments and to investigate the sorption kinetics. Sorption experiments for the kinetic study were conducted by stirring 10 g of IOCS suspended in 1 l solution containing 5 μM U and 10 mM NaCl at pH 7. Other experiments were conducted with uncoated quartz sand and without stirring, to estimate the influence of mechanical forces. Samples for the ICP-MS analysis (Thermo Scientific XSERIES 2) were taken after various time steps. Furthermore batch experiments were done with different solid to solution ratios and with solutions containing both U and As or U and Se. All experiments were conducted at room temperature (20 ± 2°C) and the samples were analyzed for dissolved uranium, arsenic, selenium, and other elements like Mg, Al, Fe, K, and Ca.

PE-columns (modified syringes) with an inner diameter of 27 mm were used in the column study as fixed-bed-flow reactors, which yielded a working volume of about 27.3 ml. Every column contained 40 g of coated or uncoated sand with an effective column length of 50 mm. Pore volume of each column was 12.3 ml with sand glass wool (Filterwatte No. 1408, Assistent, Germany) on both sides to avoid loss of substrate. All columns were rinsed with an upward directed flow rate of 0.17 ml/min by means of a peristaltic pump (IPC 24, High Precision Multichannel Dispenser, Ismatec) maintaining a total contact time of about 70 min.

Four columns were packed with IOCS and one with uncoated sand. Two columns were rinsed with 0.5 μM uranium solution (one with coated and the other with uncoated sand). The remaining three columns filled with IOCS were used for the experiments with solutions containing 0.5 μM of uranium and 0.5 μM arsenate, selenate, and phosphate respectively.

Results and Discussion

The coating of quartz sand lead to a loading of 21.4 mg_{Fe}/g_{Sand}. The coatings are assumed to be a mixture of hematite and goethite. Thus one column contained approximately 0.88 g iron. The sorption of uranyl on IOCS has a rather low kinetic. Batch reactions showed that 50% of sorption happens in the first 10 min, 70% with in 30 min, but all in all it took over 170 min to reach equilibrium. Furthermore, it could be shown that the minimization of mechanical forces (stirring) lead to a significant increase of sorption.

The major U(VI) species present at a neutral pH range in pure water are UO_2OH^+, $(UO_2)_5(OH)_5^+$, and $(UO_2)_4(OH)_7^+$ (database: NEA 2007). In the presence of carbonate, aqueous uranyl carbonate complexes predominate at pH > 6.0. PHREEQC (Parkhurst and Appelo 1999) and the NEA 2007 database (Grenthe et al. 1992) were used to plot the speciation diagram of Fig. 1. Surface reactions with distinct log-k given in Table 1 were applied for modeling the sorption at the ironhydroxide surface with PHREEQC based on surface complexation assumimg a double layer model. The adsorption of all possible uranium(VI) hydroxide complexes will result in the formation of anionic or neutral surface species. The only possible cationic surface species is monodentate uranyl, S-O-UO_2^+, where S-O denotes a ironhydroxide surface site.

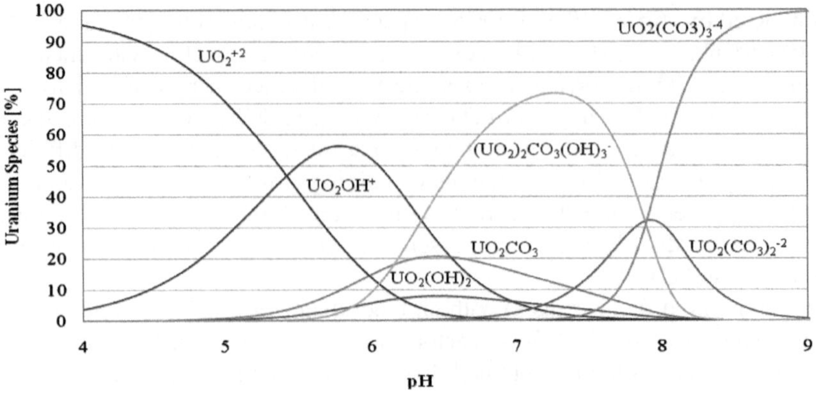

Fig. 1 U(VI) species distribution at varying pH modeled with PHREEQC (database: NEA 2007, $\sum U = 5\cdot10^{-7}$ M, 0,01 M NaCl, $p_{CO2} = 10^{-3.5}$ bar)

Table 1 Reactions and parameters used in the model calculations

SURFACE_SPECIES	log k	
strong and weak binding sites		
Hfo_sOH = Hfo_sOH	0.0	(1)
Hfo_sOH + H+ = Hfo_sOH$_2^+$	6.51[a]	(2)
Hfo_sOH = Hfo_sO$^-$ + H$^+$	–9.13[a]	(3)
Hfo_sOH + UO$_2^{2+}$ = Hfo_sOUO$_2^+$ + H$^+$	2.0[b]	(4)
Hfo_sOH + UO$_2^{2+}$ + CO$_3^{2-}$ = Hfo_sOUO$_2$CO$_3^-$ + H$^+$	7.55[b]	(5)
Hfo_sOH + UO$_2^{2+}$ + 2 CO$_3^{2-}$ = Hfo_sOUO$_2$(CO$_3$)$_2^{3-}$ + H$^+$	12.95[b]	(6)
Hfo_sOH + CO$_3^{2-}$ + H$^+$ = Hfo_sCO$_3^-$ + H$_2$O	10.94[a]	(7)
Hfo_sOH + CO$_3^{2-}$ + 2 H$^+$ = Hfo_sCO$_3$H + H$_2$O	18.93[a]	(8)

[a]From Waite et al. (1994), [b]from Wazne et al. (2003).

Cheng et al. (2004) investigated adsorption of uranium on goethite coated sand and found that uranium sorption increases with growing pH. At a pH higher than 6 over 99% of uranium ($5\cdot10^{-6}$ M) was sorbed in his experiment.

Even before the breakthrough was reached it could be seen that the sorption of uranium on IOCS is much greater than on the uncoated sand. After 160 days, 3800 µg uranium were sorbed on the IOCS, which equals around 95 µg/g_{IOCS}. On uncoated sand merely a loading of 14.7 µg/g_{sand} was achieved. In batch experiments with a solid/liquid ratio of 40 g_{IOCS}/l only 2.5 µg/g and with a ratio of 1 g/l about 45.5 µg/g uranium were sorbed. Because these results by far underestimated the sorption capacity during column experiments, the duration of the column experiments could not be calculated realistically.

The sorption of uranium differs for every competing ligand. In the presence of uranium, selenate sorbs only to a minor extend of about 10% (Fig. 2). Although arsenate is known to be a good sorbing element arsenic sorption was less than uranium sorption. Furthermore, in the presence of arsenate, uranium does not sorb as good as without arsenic (Fig. 2). First arsenic occurrence at the outlet could be de-

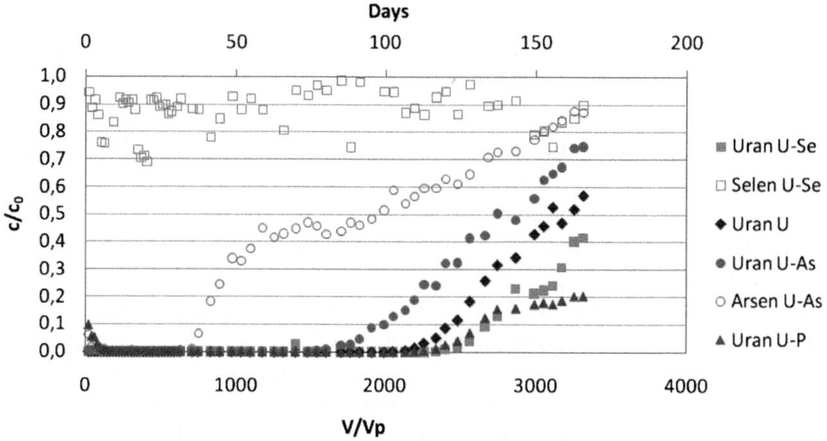

Fig. 2 Results and comparison of all column experiments

tected after ca. 35 days and then an increase during the next 25 days to half of the inlet concentration forming a plateau which lasted for around 30 days. Only after beginning of breakthrough of uranium, a further increase of arsenic was detected. Because at neutral pH uranium is present partly as negatively charged uranyl hydroxo-carbonate complex ($(UO_2)_2CO_3(OH)_3^-$), the competitive behavior with arsenic oxyanions ($HAsO_4^{2-}$, $H_2AsO_4^-$) is assumed to be one reason for less sorption of both arsenic and uranium. The shape of the curve could be explained by the assumption that arsenic is present to about 50% as free arsenate ion ($HAsO_4^-$) and 50% as charged uranyl-arsenate complex. The first increase could be explained by less sorbing of arsenate ions and the plateau due to a negative charged complex ($UO_2AsO_4^-$ analogous to the uranyl-phosphate complex $UO_2PO_4^-$) which binds better on the positively charged iron hydroxide surface. This $UO_2AsO_4^-$ species was for the first time reported by Gezahegne et al. (2009). The subsequent increase of arsenic and uranium is related to breakthrough of $UO_2AsO_4^-$ and $(UO_2)_2CO_3(OH)_3$.

In the presence of selenate, a surprising enhancement of uranium sorption could be observed, although selenate sorption was only about 10% from the very beginning. Species calculation using PHREEQC and the database NEA revealed SeO_4^{2-} as being the only relevant species (~99.9%) which obviously does not sorb and $UO_2SeO_4^0$ as the only U-Se species making less than 1 per mille. The results of the column experiments were supported by batch experiments, which have shown that uranium sorbs nearly to 100% in both U-As and U-Se systems, if 0.4 g IOCS was added to 10 ml solution similar to the column ones. However, only 71% of arsenate and less than 6% of selenate were sorbed on IOCS in this experimental setup. Our results are a proof that complex formation happens and it can be speculated in analogue to SO_4^{2-} that a considerable amount (about 10%) of selenium is forming a possible $UO_2(SeO_4)_2^{2-}$ complex that binds more effectively on iron hydroxide then $(UO_2)_2CO_3(OH)_3^-$. The existence of an anionic uranyl-selenate species is not reported in the literature so far. Thus this experiments can be seen as the first reported indication for the existence of the $UO_2(SeO_4)_2^{2-}$ species.

On contrary, in the presence of phosphate sorption of uranium is enhanced. Obviously at least two species are controlling the behavior, because a significant plateau is formed at c/c_0 of 0.2 for about 30 days before the break through continues. Relevant databases like NEA, LLNL, Hatches and others calculate $UO_2HPO_4^0$, $UO_2PO_4^-$ and $UO_2(HPO_4)_2^{2-}$ as U-P species in extremely different ratios and concentrations. Not one result out of 6 databases used can explain the chart of Fig. 2 for uranium sorption in the presence of phosphate.

Modeling the sorption graphs of uranium for the columns with U-Se and U gives a relatively good fit to experimental data. However, by modeling the sorption of selenium using constants of the database WATEQ4f (Ball and Nordstrom 1991), the sorbed amount is overestimated – maybe because the log-k's were prepared for ferrihydrate. Furthermore it was not possible to model the sorption curves for the U-As system, because all databases overrated the sorption of arsenate and the conventional equations are incapable to model plateaus like measured here.

Conclusions

This study showed that there are clues for a negatively charged uranyl-arsenate complex and a negatively charged uranyl-selenate complex at circum neutral pH. Conventional databases come to their limits by modeling uranyl phosphate systems, too. Altogether, iron coated sand is a good sorbent for uranium, but the uranium sorption could be both significantly enhanced or worsened depending on the ions in solution. Further investigations and especially column experiments at trace level are necessary to overcome the lack of data for the interactions between arsenate and uranium in solution.

References

Ball JW, Nordstrom DK (1991) WATEQ4F-User's manual. U.S. Geological Survey Open-File Report: 80–96

Benjamin MM (1996) Sorption and filtration of metals using iron-oxide-coated sand. Water Research 30 (11): 2609–2620

Cheng T, Barnett MO, Roden EE and Zhuang J (2004) Effects of phosphate on uranium (VI) adsorption to goethite-coated sand. Environ Sci Technol 38: 6059–6065

Gezahegne W, Bachmaf S, Geipel G, Planer-Friedrich B and Merkel B(2009): Detection of uranylarsenates in acidic and alkaline solutions with time resolved laser-induced fluorescence spectroscopy (TRLFS). Geochimica et Cosmochimica Acta: 73 (13)

Grenthe I, Fuger J, Konings RJM and Lemire RJ (1992) Chemical thermodynamics of uranium. North-Holland. Amsterdam

Parkhurst DL and Appelo CAJ (1999) User's guide to PHREEQC (version 2) – A computer program for speciation, batch-reaction, one-dimensional transport, and inverse geochemical calculations. u.S. Geological Survey Water-Resources Investigations Report: 99–4259

Waite TD, Davis JA, Payne TE, Waychunas GA and Xu N (1994) Geochim Cosmochim Acta 58 (24): 5465–5478

Sorption Behavior of Uranium in Agricultural Soils

Sascha Setzer, Dorit Julich, Stefan Gäth

Abstract. Because mineral phosphorous fertilizers are often loaded with uranium (U), their application could cause a significant U contamination of agricultural soils. In order to predict the mobility and plant availability of U, it is necessary to describe the liquid/solid distribution of U in soils. First results indicate a high adsorption of U onto the soil matrix. Freundlich adsorption coefficients ($\log K_f$) range between 1.3–6.7 and illustrate the different sorption capacity of the investigated soils.

Introduction

Depending on the origin and processing method of mineral phosphorous fertilizers, U concentrations could range up to 173 $mg \cdot kg^{-1}$ (Dittrich and Klose 2008). Considering an extensive P-fertilization in conventional agriculture, their use could cause a significant U input into agricultural soils. Regarding the high human- and phytotoxic potential of U, it is necessary to characterize the fate of U in the pedosphere and to identify and quantify processes like enrichment in soils, plant uptake or leaching into groundwater aquifers.

Sascha Setzer
Institute of Landscape Ecology and Resources Management, Justus-Liebig-University, Heinrich-Buff-Ring 26–32, Giessen 35392, Germany

Dorit Julich
Institute of Landscape Ecology and Resources Management, Justus-Liebig-University, Heinrich-Buff-Ring 26–32, Giessen 35392, Germany

Stefan Gäth
Institute of Landscape Ecology and Resources Management, Justus-Liebig-University, Heinrich-Buff-Ring 26–32, Giessen 35392, Germany

B. Merkel, M. Schipek, *The New Uranium Mining Boom*
DOI 10.1007/978-3-642-22122-4

In surface and soil waters the complex-ion UO_2^{2+} is the predominant form of U. Since the high tendency to hydrolysis and the formation of complexes with different ligands, the chemistry of U in soils is very complex.

The physico-chemical properties of soils may vary in a broad range and could cause very different liquid/solid distributions of U, which mainly determine its mobility and plant availability. The liquid/solid distribution coefficients K_d for U described in literature range between 0.03 and 400,000 (Echevarria et al. 2001; Thibault et al. 1990) and demonstrates the high variability of sorption capacities of different soil-types. Several studies have reported an increased U solubility with increasing carbonate content, due to the formation of negatively charged U-carbonate-complexes with a lower affinity to sorption active soil components (Echevarria et al. 2001; Yamaguchi et al. 2009). Echevarria et al. (2001) and Vandenhove et al. (2007) characterized the pH value as the predominant influencing factor, determining U K_d values in soils. Both reported of decreasing K_d values with increasing pH values. Echevarria et al. (2001) observed a strong dependence to the presence or absence of carbonates, meanwhile Vandenhove et al. (2007) found further influences of organic carbon and amorphous iron oxide contents.

In order to predict the behavior of U in agro-ecosystems, it is necessary to identify and quantify the factors mainly determining the sorption behavior of U in soils. Although several studies have investigated liquid/solid distributions of U in soils, available information are often based on small sample sizes. Furthermore, distribution coefficients are either generated by only one U concentration stage (Echevarria et al. 2001) or influencing factors (e.g. pH-value, fulvic acids) are investigated separately without considering possible co-effects (Echevarria et al. 2001; Kornilivich et al. 2001).

Therefore the current study aims to quantify sorption partition coefficients for agricultural soil by batch experiment with 5 different concentrations of U with different soils-types, varying in the physico-chemical properties, which are hypothesized to influence the sorption behavior of U.

Material and Methods

Soil Properties

The investigated soils are originated from agricultural sites in Germany. Sampling area, soil sampling and analytical procedure are described in detail by Reiher (2008) and Zörner (2010). 98 soils have been investigated; 92 surface soils with depths of 0–15 cm and 6 subsoils with depth of 15–45 cm.

Soil characteristics vary in the physico-chemical properties which are hypothesized to affect the sorption behavior of U. Figure 1 shows the most important soil properties pH value, carbonate and organic carbon content as box plot diagram. Soil pH values vary from 4.1 to 7.9, with an average of 6.5. 36 of the studied top-

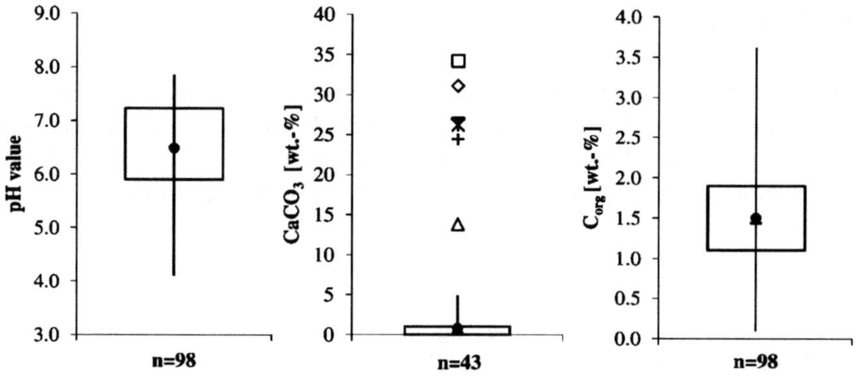

Fig. 1 Value distribution of the physico-chemical properties pH value, carbonate ($CaCO_3$) and organic carbon (C_{org}) content of the investigated soils and number n of considered soils (*point*: mean; *line*: median; *box*: 75% confidence interval; *symbols*: outliers)

soils had calcium carbonate contents with an average of 1.2% and a maximum of 5.1%. To extend the range of carbonate contents 6 subsoils were included in the study. Selected subsoils had very high carbonate contents with values higher than 10% (outliers in the boxplot). Organic carbon contents range between 0.1 and 3.6%, with an average of 1.5%.

Batch Sorption Experiment

Sorption characteristics of U were determined by batch experiments described in detail by Schug et al. (2000). A suspension of 30 ml 0.01 M $Ca(CO_3)_2$ solution and 10 g soil was contaminated with 5 different concentrations of U ($UO_2(NO_3)_2$; Co. Ultra Scientific). The concentration levels are given in Table 1. The suspension was shaken horizontally for 16 h at 175 rpm and centrifuged at 3000 rpm for 20 min. The extract was filtered through a syringe filter (0.45 μm pore size) and acidified with 100 μl HNO_3. All experiments were conducted in duplicate.

Table 1 Applied U concentrations of each levels (I–VI) in batch sorption experiment

Concentration level	U concentration [mg·kg−1]
Control	–
I	0.005
II	0.05
III	0.5
IV	5
V	25
VI	50

Data Evaluation

The U concentration of the liquid phase was determined by inductive coupled mass spectroscopy (ICP-MS Agilent 7500). The adsorbed U concentration was calculated by the sum of applied U and the natural occurring $Ca(NO_3)_2$-soluble U fraction subtracted by the determined U concentration in the liquid phase:

$$C_{\text{sorb}} = C_{\text{a}} + C_{\text{c}} - C_{\text{liq}} \tag{1}$$

with
C_{sorb} – U concentration in solid phase $[\text{mg} \cdot \text{kg}^{-1}]$,
C_{a} – Applied U concentration $[\text{mg} \cdot \text{kg}^{-1}]$,
C_{c} – $Ca(NO_3)_2$-soluble U content $[\text{mg} \cdot \text{kg}^{-1}]$,
C_{liq} – U in liquid phase related to the soil weight $[\text{mg} \cdot \text{kg}^{-1}]$.

To describe the liquid/solid distribution of U the model of the Freundlich sorption isotherm was applied to the experimental data:

$$\log C_{\text{sorb}} = \log K + 1/n \cdot \log C_{\text{liq}} \tag{2}$$

with
C_{sorb} – Amount of U adsorbed $[\text{mg} \cdot \text{kg}^{-1}]$,
K – Equilibrium distribution coefficient $[\text{L}^{n}\ \text{mg}^{1-n}\,\text{kg}^{-1}]$,
C_{liq} – U concentration in liquid phase $[\text{mg} \cdot \text{L}^{-1}]$,
$1/n$ – Sorption intensity.

Results

A class based frequency distribution of the determined logarithmic Freundlich adsorption coefficients ($\log K_f$ values) of all investigated soils is given in Table 2. K_f values range between 1.3–6.7 and indicate a high affinity of U onto the soil matrix. The equilibrium of the liquid/solid distribution is clearly located on the solid phase. 82 of the tested soil show liquid/solid distribution ratios $>$1:100. Only 16 soils have distribution ratios between 1 : 10 and 1 : 100, but still with a high sorption affinity to U.

The spectrum of determined distribution coefficient verifies the soil partition coefficient given in literature (Echevarria et al. 2001; Thibault et al. 1990) and demonstrates the high variability of sorption capacities of different soil types.

To illustrate U sorption kinetics, equilibrium Freundlich adsorption isotherms of 6 selected soils are displayed in Fig. 2. In the studied concentration range from 0.005–50 $\text{mg} \cdot \text{kg}^{-1}$ U sorption proceeds linear, indicating that the sorption maximum for U has not been reached. The Freundlich model could describe the sorption behavior of U appropriate as indicated by the high values for the coefficients of determination as given in Table 3. 98 and 99% of the variation of C_{sorb} could be explained by the models in the worst and in the best case, respectively.

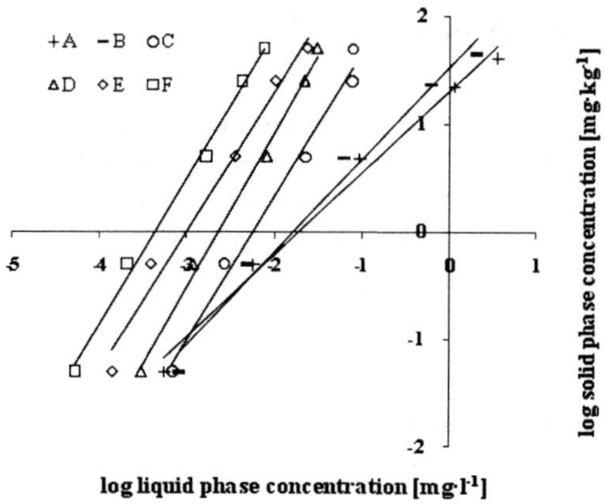

Fig. 2 Freundlich adsorption isotherms of 6 selected soils (A–F). Symbols represent means of the measured values at each concentration level

Table 2 Class based frequency distribution of Freundlich adsorption coefficient (log K_f values) of all investigated soils

log K_f value	Frequency
1–2	16
2–3	13
3–4	13
4–5	20
5–6	30
6–7	6

Table 3 Physico-chemical properties (pH, $CaCO_3$, C_{org}), K_f values and coefficients of determination (R^2) of selected soils (A–F) displayed in Fig. 1

ID	log K_f	R^2	pH	$CaCO_3$ (wt. %)	C_{org} (wt. %)
A	1.3	0.99	7.85	13.81	0.43
B	1.5	0.98	7.50	2.80	2.90
C	3.0	0.99	7.19	0.13	1.27
D	3.8	0.98	7.02	0.05	1.00
E	3.9	0.98	6.86	0.05	1.83
F	4.6	0.98	6.38	–	1.29

The dependency of U soil partition coefficients from soil pH and carbonate contents as described by Echevarria et al. (2001), Vandenhove et al. (2007) and Yamaguchi et al. (2009) is indicated by the current study. Calculated K_f values increase with decreasing pH value and carbonate content as given in Table 3. The described correlation between soil pH and/or carbonate contents and the K_f values could not be verified for all studied soils until now, but further investigations aim to find extended Freundlich isotherms including all influencing soil properties. Yet, no others soil properties could significantly be referred to correlate with soil partition coefficients.

Conclusion

U shows a high tendency to adsorb onto the soil matrix. The liquid/solid distribution of U could be described appropriate by Freundlich adsorption model. K_f values vary in a broad range depending on different soil characteristics. Soil pH and carbonate content was figured out as most important influencing factor, not related to all investigated soils.

References

Dittrich B, Klose R (2008) Schwermetalle in Düngemitteln. Schriftreihe der sächsischen Landesanstalt für Landwirtschaft 3
Echevarria G, Sheppard MI, Morel JL (2001) Effect of pH on the sorption of uranium in soils. Journal of Environmental Radioactivity 53: 257–264
Kornilivich B, Pshinko G, Koval'schuk I (2001) Effect of Fulvic Acids on Sorption of U(VI) on Clay Minerals of Soils. Radiochemistry 43: 464–467
Reiher W (2008) Entwicklung eines regionalisierten Modells zur Projektion des landnutzungsabhängigen Schwermetallstatus von Oberböden. Boden und Landschaft 52
Schug B (2000) Entwicklung von Pedotransferfunktionen zur Regionalisierung des Retentionspotenzials von Böden für Cadmium, Blei und Zink. Boden und Landschaft 30
Thibault DH, Sheppard, MI, Smith, PA (1990) Critical compilation and review of default soil solid/liquid partition coefficients, Kd for use in environmental assessments. Atomic Energy Canada, Manitoba/Canada. AECL-10125, Whiteshell Nuclear Research Establishment
Vandenhove H, Van Hees M, Wouters K, Wannijn J (2007) Can we predict uranium bioavailability based on soil parameters? Part 1: Effects of soil parmeters on soil solution uranium concentration. Environmental Pollution 145: 587–595
Yamaguchi N, Kawasaki A, Iiyama I (2009) Distribution of uranium in soil components of agricultural fields after longterm application of phosphate fertilizers. Science of the Total Environment 407: 1383–1390
Zörner, D (2010) Prognose des Schwermetallhaushaltes von Landschaften mit dem Modell ATOMIS. Boden und Landschaft 53

Kinetics of Two-Line Ferrihydrite Phase Transformation Under Alkaline Conditions: Effect of Temperature and Adsorbed Arsenate

Soumya Das, M. Jim Hendry, Joseph Essilfie-Dughan

Abstract. The rate of two-line ferrihydrite transformation with respect to temperature and adsorbed arsenate under alkaline conditions was determined via batch experiments. The rate of transformation to either goethite or hematite increases with temperature and follows first order reaction kinetics. However, as the As/Fe ratio is increased from 0.010 to 0.018, the rate of transformation decreases by two orders of magnitude with no crystallization observed at As/Fe ratios >0.018. These results are relevant for determining the long-term stability of two-line ferrihydrite in tailings from U mill operations.

Introduction

Two-line ferrihydrite, also known as amorphous ferric oxide or hydrous ferric oxide (nominal formula: $5Fe_2O_3{\cdot}9H_2O$), is a poorly crystalline Fe(III) oxy-hydroxide that occurs as very small (2–4 nm) spherical particles in a highly ag-

Soumya Das
Department of Geological Sciences, University of Saskatchewan,
114 Science Place, Saskatoon, SK S7N 5E2, CANADA

M. Jim Hendry
Department of Geological Sciences, University of Saskatchewan,
114 Science Place, Saskatoon, SK S7N 5E2, CANADA

Joseph Essilfie-Dughan
Department of Geological Sciences, University of Saskatchewan,
114 Science Place, Saskatoon, SK S7N 5E2, CANADA

B. Merkel, M. Schipek, *The New Uranium Mining Boom*
DOI 10.1007/978-3-642-22122-4 © Springer 2011

gregated form (Eggleton and Fitzpatrick 1988; Dzombak and Morel 1990; Gruenhagen 1997; Jambor and Dutrizac 1998; Janney et al. 2000). Two-line ferrihydrite (hereafter called ferrihydrite) lacks an ordered internal structural arrangement (Erbs et al. 2010), has a very high specific surface area (Davis and Leckie 1978; Dzombak and Morel 1990), and is ubiquitous in low temperature surface environments and hydrometallurgical processes (Jambor and Dutrizac 1998; Cornell and Schwertmann 2000; Michel et al. 2007). It is commonly observed in areas contaminated by acid mine drainage (Cornell and Schwertmann 2000) and mine tailings environments with pH values greater than 5 (Carlson and Schwertmann 1981).

Ferrihydrite is metastable under oxic conditions and dissolves in either acidic (pH 2–5) or alkaline (pH 10–14) conditions (either below or above its pH_{zpc}), re-precipitating as goethite at room temperature (Schwertmann and Murad 1983; Cudennec and Lecerf 2006). In contrast, the transformation at pH 7–8 favors hematite via internal rearrangement along with complete dehydration (Schwertmann and Murad 1983; Cudennec and Lecerf 2006). This phase transformation to either goethite or hematite results in a decrease in reactive surface area and an associated reduction in the ability to adsorb contaminants (Dzombak and Morel 1990; Gruenhagen 1997; Jambor and Dutrizac 1998).

Ferrihydrite plays an important role in the removal of arsenic from industrial effluents via adsorption (Jia and Demopoulos 2005). Usually, arsenic is released via leaching from arsenic-bearing mineral phases (oxides, sulfides, arsenates, and arsenides) in hydrometallurgical processes such as oxidation and acidic dissolution (Riveros et al. 2001). At Cameco Corporation's Deilmann tailings management facility (DTMF) in northern Saskatchewan, Canada, arsenic in the uranium mine tailings is elevated and has the potential to be a major contaminant of local ground waters. These tailings remain under oxic (mean Eh of +200 mV) and alkaline (pH ~9.8) conditions at an ambient temperature of 1–2°C; mean As/Fe molar ratios of solid samples from the two tailing bodies at the DTMF are ~0.019 and ~0.100 (Shaw et al. 2011).

The arsenic concentration in the tailings porewaters is maintained at an acceptable value via adsorption onto the surface of ferrihydrite, which is generated during mill raffinate neutralization processes (Moldovan et al. 2003). As such, determining the long-term stability of ferrihydrite has become increasingly important. A comprehensive assessment of the rate of transformation of ferrihydrite with respect to temperature (25, 50, 75, and 100°C) at pH 10 was performed via batch experiments. The impact of adsorbed arsenate (at As/Fe ratios of 0.500 to 0.010) on the rate of this transformation was also assessed and the results used to simulate tailings deposition conditions (i.e., pH, temperature, As/Fe ratios) at the DTMF. Our findings are relevant and useful for understanding the long-term stability of ferrihydrite in tailings from U mill operations in Saskatchewan and elsewhere.

This paper is a synthesis of research conducted by Das et al. (2011a,b) and interested readers are referred to those papers.

Materials and Methods

Synthesis and Characterization of Iron-Oxy(Hydroxides)

Ferrihydrite, goethite, and hematite were synthesized in the laboratory using the methods of Cornell and Schwertmann (2000) and characterized by comparing X-ray diffraction (XRD) patterns to standards in the Joint Committee on Powder Diffraction Standards (JCPDS) database. All precipitates were subsequently centrifuged and washed 5–6 times with double-distilled deionized (DDI) water until the pH of the mineral precipitates reached their respective pH_{zpc} values (8–8.5 for ferrihydrite, ~ 7.8 for goethite, and ~ 8.5 for hematite; Stumm 1992).

Time-Series Transformation of Ferrihydrite

Ferrihydrite-Water System

Time-series transformation of ferrihydrite with respect to temperature (25, 50, 75, and 100°C) at pH 10 was conducted in batches to establish kinetics and determine the activation energy of the phase transformation. Four batches of ferrihydrite were prepared, washed, and re-dispersed in 200 ml of DDI water in polyethylene containers and later transferred to water baths at room temperature (21–25°C) or pre-heated to 50, 75, and 100°C, respectively. The pH of the all four slurries were set and maintained at pH ~ 10 (± 0.05) by addition of trace metal grade HNO_3 or NaOH (0.1 M) in small increments (10 μl) using a pipette after 1 h (for 100°C experiments), 24 h (for 50 and 75°C experiments), or 7 d (for 25°C experiments). Sampling intervals were based on the temperature of the systems as follows: every 7 d at 25°C, 60 min at 50°C, 30 min at 75°C, and 5 min at 100°C. Aliquots (~ 15 ml) were pipetted from each reaction vessel and centrifuged at 5000 rpm for 10 min; supernatants were removed and the precipitates air-dried (24 h) for XRD analyses.

Ferrihydrite-Arsenate System

Six batches of ferrihydrite were prepared, washed, and resuspended in 200 ml DDI water in a similar fashion as above to estimate the kinetics of ferrihydrite phase transformation under varying As/Fe ratios at pH 10. The six ferrihydrite precipitates were homogenized by stirring on a stir plate at room temperature, then hydrated sodium arsenate ($Na_2HAsO_4 \cdot 7H_2O$) added to each under continued stirring to generate arsenate adsorbed to ferrihydrite at As/Fe molar ratios of 0.500, 0.100, 0.050, 0.018, 0.013, and 0.010. The pH of all slurries was raised to and maintained at pH ~ 10 (±0.05) by addition of trace metal grade NaOH (0.1 M) in small increments (10 μl). Once complete homogenization was achieved, all six polyethylene bottles were transferred to a water bath preheated to 75(±2)°C. A sampling interval of 24 h was used for As/Fe molar ratios of 0.500, 0.100, 0.050, 0.018, and 0.013 and a sampling interval of 6 h was used for the As/Fe ratio of 0.010. As above, ~15 ml aliquots were pipetted from each container until the end of the experiment and centrifuged at 5000 rpm for 10 min; the precipitates were then air-dried (24 h) for prior to XRD analyses.

X-ray Diffraction (XRD)

Air-dried samples were gently grounded in a ceramic mortar and pestle to break up larger aggregates prior to analysis. A few milligrams of ground sample were placed on the wetted slide and evenly distributed using small amount of methanol. All XRD analyses were performed on a Rigaku Rotoflex 200 XRD equipped with a Cu target and graphite monochromator. The XRD was operated at 3.2 kW (40 kV and 80 mA). Data were collected between 5 and 80 degrees 2θ at 2 degrees/min. All raw data files were converted to an Excel file from which the relative intensity was plotted versus 2θ. Ferrihydrite, goethite, and hematite were quantified in the transformation phases using the methods applied by Schwertmann and Murad (1983) and Johnston and Lewis (1983). Briefly, XRD analyses of pre-determined mixtures of pure ferrihydrite and goethite (1–90 wt%) and pure ferrihydrite and hematite (1–90 wt%) were conducted. The intensities of the individual XRD scans from the transformation experiments were calibrated against the XRD scans from the pre-determined mixtures. The lower limit of detection of both goethite and hematite was 1 wt% with an accuracy of ±5 wt%.

Results and Discussion

Time-Series Transformation of Ferrihydrite: Effect of Temperature

The observations in this study agree with limited data from studies of Schwertmann and Murad (1983) and Schwertmann et al. (2004) that indicate the degree of ferrihydrite transformation increases with temperature. Hematite is the dominant phase at higher temperatures (e.g., 75 and 100°C; Fig. 1), in keeping with the results of Manceau and Drits (1993) who also reported that hematite appears to be the stable phase when ferrihydrite is heated at temperatures <100°C. Extended X-ray absorption fine structure (EXAFS) data presented by Combes et al. (1990) show that with continued heating at 92°C, Fe-O bond distances continue to increase and after 25 h match those of hematite. Our data indicate that both hematite

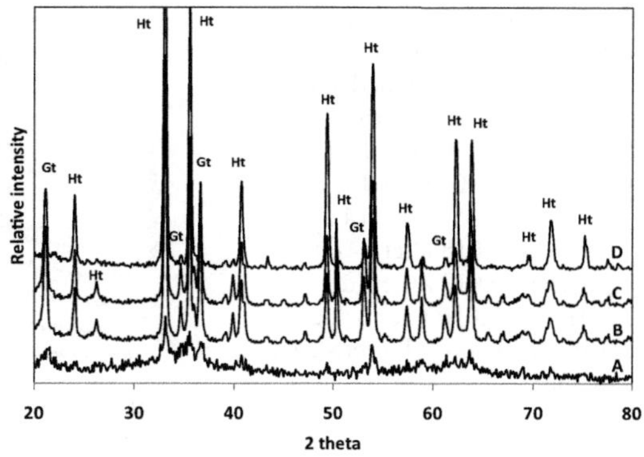

Fig. 1 Time series transformation of ferrihydrite at pH 10 at (*A*) 25°C after 21 d, (*B*) 50°C after 7 d, (*C*) 75°C after 1 d, and (*D*) 100°C after 3 h. Ht and Gt represent hematite and goethite, respectively

and goethite form at 50°C (data not presented), however hematite remains the dominant phase at the end of the experiment. Jambor and Dutrizac (1998) support the above observation and explain that ferrihydrite begins to lose hydroxyl ions from its structure with continued heating and, as a result, becomes highly destabilized and favors the formation of hematite instead of goethite or maghemite.

Time-Series Transformation of Ferrihydrite: Effect of Adsorbed Arsenate

Time series XRD analyses (Fig. 2) show that the phase transformation of ferrihydrite at pH 10 and 75°C is dependent on the concentration of adsorbed arsenate. As the As/Fe ratio increases from 0.010 to 0.018, the rate of ferrihydrite transformation decreases; for example, ~50% transformation of ferrihydrite occurs in 24 h at an As/Fe ratio of 0.010 but requires 48 h at an As/Fe ratio of 0.013. Ferrihydrite transformation is even slower at an As/Fe ratio of 0.018, where the first sign of transformation is observed after 8 d. No measurable phase transformation is identified at As/Fe ratios greater than 0.018 (i.e., 0.500, 0.100, and 0.050) even after aging at 75°C for 14 d (Fig. 2).

Interestingly, hematite appears to be the stable end product of ferrihydrite phase transformation in the presence of adsorbed arsenate (Fig. 2). The notable absence of goethite in the transformed phases is in contrast to findings under pristine condition, where goethite forms ~10% of the transformation product. The reduction in the rate of ferrihydrite transformation in the presence of adsorbed arsenate can be attributed to the inhibition effects of arsenate via adsorption and surface complexation onto ferrihydrite surfaces (Paige et al. 1996; Waychunas et al. 1996; Fendorf et al. 1997). Arsenate, similar to other solutes such as phosphate and silicate, strongly adsorbs onto ferrihydrite and produces an immobile network that inhibits ferrihydrite dissolution (a precursor for goethite formation) and promotes hematite nucleation (Cornell et al. 1987; Galvez et al. 1999).

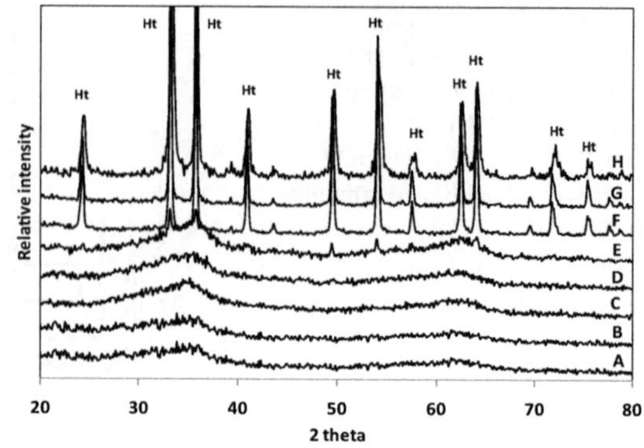

Fig. 2 Phase transformation of ferrihydrite with varying As/Fe molar ratios at pH 10 and aged at 75°C. (*A*) pure ferrihydrite, (*B*) As/Fe = 0.500 after 14 d, (*C*) As/Fe = 0.100 after 14 d, (*D*) As/Fe = 0.050 after 14 d, (*E*) As/Fe = 0.018 after 14 d, (*F*) As/Fe = 0.013 after 5 d, (*G*) As/Fe = 0.010 after 3 d, and (*H*) pure hematite. Ht represents hematite

Kinetics of Ferrihydrite Phase Transformation

Percent ferrihydrite remaining and hematite formed with respect to time at pH 10 and four different temperatures (Fig. 3) and three As/Fe ratios (Fig. 4) were calculated using Sigmaplot® with a user defined equation:

$$[A]_t = [A]_0\, e^{-kt} \tag{1}$$

where $[A]_t$ is the amount of ferrihydrite remaining at time t, $[A]_0$ is the initial amount of ferrihydrite, k is the rate constant, and t is time.

In general, the good agreement between the experimental data and the model fit indicate that both ferrihydrite transformation and hematite formation follow first order rate kinetics under pristine conditions (Fischer and Schwertmann 1975; Schwertmann and Murad 1983). The data indicate a statistically better fit (R^2 values; Table 1) for first order kinetics vs. pseudo first order (Vu et al. 2008). The rate of transformation increases with temperature at pH 10, occurring fastest at 100°C and slowest at 25°C (Table 1; Fig. 3).

The rate of ferrihydrite transformation is considerably slowed by even low concentrations of adsorbed arsenate (i.e., As/Fe ratios of 0.010 vs. 0.000; Table 1). The reduction in the rate of transformation is three orders of magnitude (from 1.18×10^{-1} to 3.00×10^{-4}) as As/Fe ratios increase from 0.000 to 0.018 (Table 1; Fig. 4). These transformations also follow first order rate kinetics. The formation of hematite can also be estimated using first order kinetics (comparable R^2 values; data not presented), with the rate of hematite formation decreasing as the As/Fe ratio increases.

The retardation of ferrihydrite transformation by adsorbed arsenate has been previously demonstrated by Paige et al. (1996) and Jia et al. (2006), among others. Paige et al. (1996) show that ferrihydrite transformation is considerably reduced with the addition of arsenate (As/total solid ratios of 3, 5, 15, and 50 mol%) under acidic (pH 1.3) conditions at room temperature, with no transformation recorded when As/(As + Fe) ratios were >1. Jia et al. (2006) also report similar inhibition

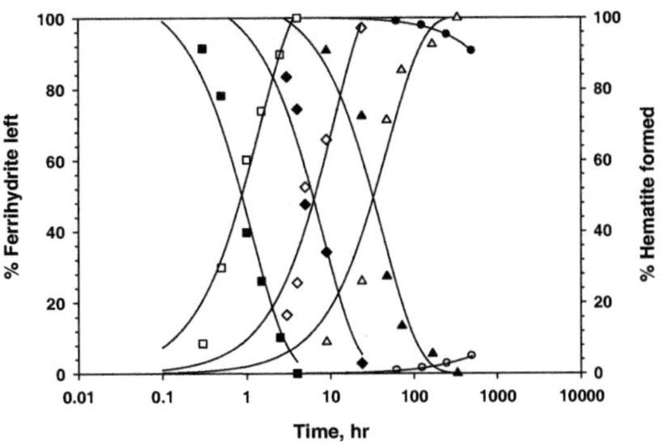

Fig. 3 Ferrihydrite transformation and hematite formation with time at pH 10. Circles, triangles, diamonds, and squares represent XRD data on samples from 25, 50, 75, and 100°C, where the solid and open symbols represent 2-line ferrihydrite transformation and hematite formation, respectively

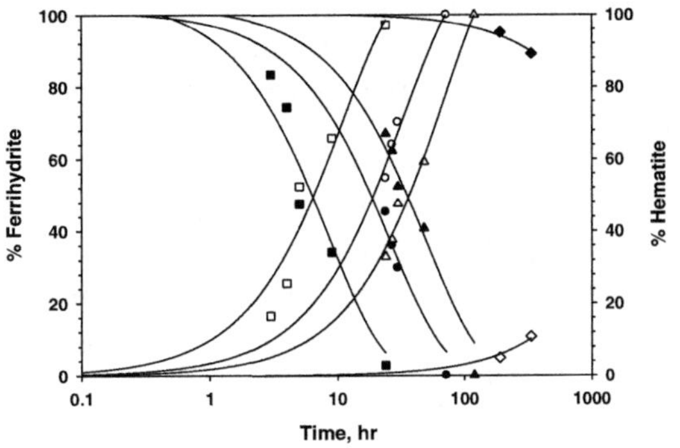

Fig. 4 The reduction in 2-line ferrihydrite and formation of hematite with time at pH 10 and 75°C for As/Fe molar ratios of 0.018, 0.013, and 0.010. Closed and open diamonds, triangles, and circles represent calculated the percent of 2-line ferrihydrite transformed and percent hematite formed from XRD analyses, respectively. Data for percent ferrihydrite transformed and hematite formed at an As/Fe molar ratio = 0.000 are represented by the solid and open squares, respectively

Table 1 First-order reaction rate constants for ferrihydrite transformation at different temperatures (25, 50, 75, and 100°C) and varied As/Fe molar ratios (0.000, 0.010, 0.013, and 0.018) at pH 10

T (°C)	As/Fe	k (h^{-1})	R^2
100	0.000	1.09×10^{1}	0.99
75	0.000	1.18×10^{-1}	0.97
50	0.000	2.13×10^{-2}	0.95
25	0.000	2.00×10^{-4}	0.96
75	0.010	3.83×10^{-2}	0.99
75	0.013	2.03×10^{-2}	0.99
75	0.018	3.00×10^{-4}	0.98

effects of arsenate and did not observe any ferrihydrite transformation even after aging ferrihydrite for two weeks (pH 8) at As/Fe molar ratios of 0.500, 0.250, and 0.125 at room temperature. These results and those from the current study illustrate that arsenate significantly affects the rate and product of ferrihydrite phase transformation.

Implications at DTMF

As the ambient temperature at the DTMF (~ 1–2°C) is considerably lower than the temperature considered in our arsenate-ferrihydrite experiment (75°C), a rearranged expression of the Arrhenius equation (Atkins and de Paula 2005) was introduced to calculate the rate constant at 1°C:

$$\ln\frac{k_x}{k_{75}} = \frac{E_a}{R}\left(\frac{1}{T_{75}} - \frac{1}{T_x}\right) \tag{2}$$

where k_x is the rate constant at temperature x, k_{75} is the rate constant at 75°C, E_a is the activation energy of ferrihydrite transformation at pH 10, R is the molar gas constant, T_x is the absolute temperature at temperature x, and T_{75} is the absolute temperature for 75°C. Application of this equation assumes that the activation energy when adsorbed arsenate is present (obtained from the slope of the Arrhenius plot of four different temperatures used here) is similar to that in pristine conditions (As/Fe = 0.000) at pH 10 (Erbs et al. 2010).

Based on the calculation of rate constants at 1°C from Eq. (2), Eq. (1) was used to calculate the time required to transform 10, 50, and 99.9% of the ferrihydrite to hematite (Table 2). The reduction in the rate of ferrihydrite transformation at 1°C is similar to that calculated for 75°C, with the rate decreasing by two orders of magnitude (from 1.38×10^{-6} to 2.04×10^{-8}) as the As/Fe ratio increases from 0.013 to 0.018 (Table 2).

The mean As/Fe ratios of the two ore bodies in the DTMF are ~0.019 and ~0.100 and, as such, no transformation of ferrihydrite is predicted under the physical (~ 1°C) and chemical (~pH 10) conditions that have been maintained in this tailings facility for the past 20 years. Thus, ferrihydrite within these tailings (including some samples with As/Fe ratios greater than 0.250) should remain stable for more than 10,000 years, representing a long-term sink for arsenate.

Table 2 First-order reaction rate constants for ferrihydrite transformation at pH 10 at 1°C under varying As/Fe ratios, and predicted time (years) needed to transform 10, 50, and 99.9% ferrihydrite under in situ tailing conditions

As/Fe	K_1 (h^{-1})	10%	50%	99.9%
0.000	8.04×10^{-6}	2	10	130
0.010	2.61×10^{-6}	5	30	400
0.013	1.38×10^{-6}	40	260	3400
0.018	2.04×10^{-8}	600	3900	52,000

References

Atkins P, de Paula J (2005) The Elements of Physical Chemistry. W. H. Freeman: New York

Carlson L, Schwertmann U (1981) Natural ferrihydrites in surface deposits from Finland and their association with silica. Geochim Cosmochim Acta 45: 421–429

Combes JM, Manceau A, Calas G (1990) Formation of ferric oxides from aqueous solutions: A polyhedral approach by X-ray Absorption Spectroscopy. II. Hematite formation form ferric gels. Geochim Cosmochim Acta 54: 1083–1091
Cornell RM, Schwertmann U (2000) Iron Oxides in the Laboratory, 2nd Ed: Wiley-VCH
Cornell RM, Giovanoli R, Schindler PW (1987) Effect of silicate species on the transformation of ferrihydrite into goethite and hematite in alkaline media. Clays Clay Miner 35: 21–28
Cudennec Y, Lecerf A (2006) The transformation of ferrihydrite into goethite or hematite, revisited. J Solid State Chem 179: 716–722
Das S, Hendry MJ, Essilfie-Dughan J (2011a) Transformation of two-line ferrihydrite to goethite and hematite as a function of pH and temperature. Environ Sci Technol 45: 268–275
Das S, Hendry MJ, Essilfie-Dughan J (2011b) Effects of adsorbed arsenate on the rate of transformation of 2-line ferrihydrite at pH 10. Environ Sci Technol 45: 5557–5563
Davis JA, Leckie JO (1978) Surface ionization and complexation at the oxide/water interface. II. Surface properties of amorphous iron oxy hydroxide and adsorption of metal ions. J Colloid Interface Sci 67: 90–107
Dzombak DA, Morel FMM (1990) Surface Complexation Modeling: Hydrous Ferric Oxide, Wiley-Interscience: New York
Eggleton RA, Fitzpatrick RW (1988) New data and a revised structural model for ferrihydrite. Clays Clay Miner 36: 111–124
Erbs JJ, Berquó TS, Reinsch BC, Lowry GV, Banerjee SK, Penn RL (2010) Reductive dissolution of arsenic-bearing ferrihydrite. Geochim Cosmochim Acta 74: 3382–3395
Fendorf S, Eick MJ, Grossl P, Sparks PDL (1997) Arsenate and chromate retention mechanisms on goethite. 1. Surface structure. Environ Sci Technol 31: 315–326
Fischer WR, Schwertmann U (1975) The formation of hematite from amorphous iron(III) hydroxide. Clays Clay Miner 23: 33–37
Galvez N, Barron V, Torrent J (1999) Effect of phosphate on the crystallization of hematite, goethite, and lepidocrocite from ferrihydrite. Clays Clay Miner 47: 304–311
Gruenhagen SE (1997) Phosphate sorption properties of ferrihydrite precipitated in the presence of various anions: PhD dissertation, Purdue University
Jambor JL, Dutrizac JE (1998) Occurrence and constitution of natural and synthetic ferrihydrite, a widespread iron oxyhydroxide. Chem Rev 98: 2549–2585
Janney DE, Cowley JM, Buseck PR (2000) TEM study of synthetic 2- and 6-line ferrihydrite. Clays Clay Miner 48: 111–119
Jia Y, Demopoulos GP (2005) Adsorption of arsenate onto ferrihydrite from aqueous solution: Influence of media (sulphate vs. nitrate), added gypsum, and pH alteration. Environ Sci Technol 39: 9523–9527
Jia Y, Xu L, Fang Z, Demopoulos GP (2006) Observation of surface precipitation of arsenate on ferrihydrite. Environ Sci Technol 40: 3248–3253
Johnston JH, Lewis DG (1983) A detailed study of the transformation of ferrihydrite to hematite in an aqueous medium at 92°C. Geochim Cosmochim Acta 47: 1823–1831
Manceau A, Drits VA (1993) Local structure of ferrihydrite and feroxyhite by EXAFS spectroscopy. Clay Miner 28: 165–184
Michel FM, Ehm L, Antao SM, Lee PL, Chupas PJ, Liu G, Strongin DR, Schoonen MAA, Phillips BL, Parise JB (2007) The structure of ferrihydrite, a nanocrystalline material. Science 316: 1726–1729
Moldovan BJ, Jiang DT, Hendry MJ (2003) Mineralogical characterization of arsenic in uranium mine tailings precipitated from iron-rich hydrometallurgical solutions. Environ Sci Technol 37: 873–879
Paige CR, Snodgrass, WJ, Nicholson RV, Scharer JM (1996) The crystallization of arsenate-contaminated iron hydroxide solids at high pH. Wat Environ Res 68: 981–987
Riveros PA, Dutrizac JE, Spencer P (2001) Arsenic disposal practices in metallurgical industry. Can Metall Q 40: 395–420
Schwertmann U, Murad E (1983) Effect of pH on the formation of goethite and hematite from ferrihydrite. Clays Clay Miner 31: 277–284

Schwertmann U, Stanjek H, Becher HH (2004) Long-term in vitro transformation of 2-line ferrihydrite to goethite/hematite at 4, 10, 15 and 25°C. Clay Miner 39: 433–438

Shaw SA, Hendry MJ, Wallschläger D, Essilfie-Dughan J, Kotzer T (2011) Distribution, characterization, and geochemical controls of As, Se, and Mo in uranium mine tailings, Key Lake, Saskatchewan, Canada. Manuscript in preparation

Stumm W (1992) Chemistry of the solid-water interface, Wiley: New York

Vu HP, Shaw S, Benning LG (2008) Transformation of ferrihydrite to hematite: an in situ investigation on the kinetics and mechanisms. Mineral Mag 72: 217–220

Waychunas GA, Fuller CC, Rea BA, Davis JA (1996) Wide angle x-ray scattering (WAXS) of "two-line" ferrihydrite structure: Effect of arsenate sorption and counterion variation and comparison with EXAFS results. Geochim Cosmochim Acta 60: 1765–1781

The Interaction of U(VI) with Some Bioligands or the Influence of Different Functional Groups on Complex Formation

Laura Frost, Alfatih A.A. Osman, Gerhard Geipel, Katrin Viehweger, Henry Moll, Gert Bernhard

Abstract. The complex formation of U(VI) with glutathione, uric acid and benzoic acid was studied using spectroscopic techniques. Additionally glutathione was derivatized to obtain different glutathione-S-conjugates in order to assess the influence of the thiol group on complex stability. In the U(VI)-glutathione system a 1 : 1 complex with a large stability constant of 39.07 ± 0.15 could be identified. Derivatization did not lead to a decrease in complex stability. Thus a significant influence of the thiol group on U(VI) coordination can be excluded. Investigating the U(VI) complexation by uric acid and benzoic acid, for both ligands a comparably weak complexation was found. Structure-related reasons for the difference in complex stability are discussed.

Laura Frost
Helmholtz-Zentrum Dresden-Rossendorf, Institute of Radiochemistry,
P.O. Box 510119, D-01314 Dresden, Germany

Alfatih A.A. Osman
Helmholtz-Zentrum Dresden-Rossendorf, Institute of Radiochemistry,
P.O. Box 510119, D-01314 Dresden, Germany

Gerhard Geipel
Helmholtz-Zentrum Dresden-Rossendorf, Institute of Radiochemistry,
P.O. Box 510119, D-01314 Dresden, Germany

Katrin Viehweger
Helmholtz-Zentrum Dresden-Rossendorf, Institute of Radiochemistry,
P.O. Box 510119, D-01314 Dresden, Germany

Henry Moll
Helmholtz-Zentrum Dresden-Rossendorf, Institute of Radiochemistry,
P.O. Box 510119, D-01314 Dresden, Germany

Gert Bernhard
Helmholtz-Zentrum Dresden-Rossendorf, Institute of Radiochemistry,
P.O. Box 510119, D-01314 Dresden, Germany

B. Merkel, M. Schipek, *The New Uranium Mining Boom*
DOI 10.1007/978-3-642-22122-4

Introduction

U(VI) released anthropogenically, e.g. through mining activities, can be accumulated for instance in plants and consequently can enter further parts of the food chain. To receive information about the effect and distribution of uranium on a cellular level, knowledge has to be gained about the uranium complexes formed.

Glutathione (GSH), an essential cellular antioxidant occurring ubiquitously in plants, mammals and microorganisms (Kidd 1997), has a great binding potential towards numerous heavy metal ions (Singh 2005). Hence the process of interaction between glutathione and uranium might influence the distribution of this metal on a cellular level. Furthermore it is a precursor for phytochelatin synthesis and thus also a model substance for the study on U(VI) phytochelatin interaction. Glutathione is a polydentate ligand offering two carboxylate, an amino, a thiol and two amide groups for coordination. Because of the high electro negativity of reduced sulfhydryl groups a coordination of metals by the thiol group is found frequently and leads to the formation of metal-glutathione complexes of high thermodynamic stability (Ballatori and Clarkson 1985). Nevertheless, uranium(VI) as a hard cation is expected to be coordinated by hard Lewis bases like carboxylate oxygens. To assess the influence of the thiol group on U(VI) coordination, it was derivatized using the reagents *β*-fluoropyruvate and monobromobimane. Glutathione, synthesized S-conjugates as well as further investigated ligands used for U(VI) complexation experiments are shown below (Fig. 1).

Uric acid (UA) was investigated as a further ligand being potentially capable of chelate formation with U(VI). UA is a heterocyclic organic molecule that is created as final and major oxidation (breakdown) product of catabolism of the purine nucleosides, adenosine and guanosine (Huang et al. 2002). It is present in bio-

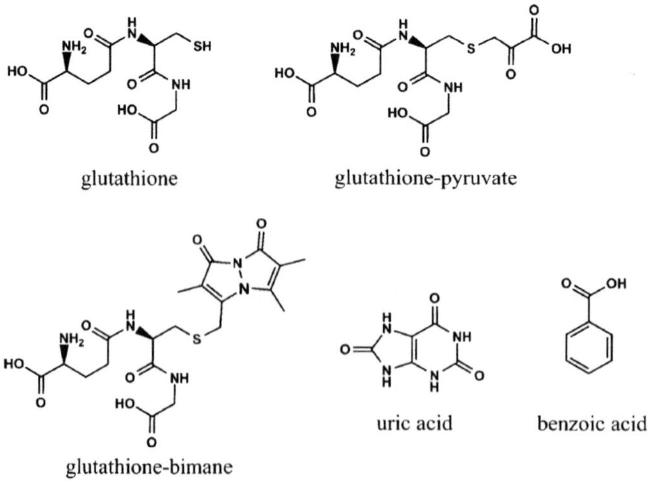

Fig. 1 Structures of glutathione, synthesized derivatives, uric acid and benzoic acid

fluids such as urine and sweat. As a bioligand, uric acid interacts with biometal ions such as Mg, Ca, Fe, Co, Cu, and Zn to form soluble and/or solid complexes (Moawad 2002; Davies et al. 1986). UA offers carbonyl and secondary amine functionalities for U(VI) coordination.

To compare these results to those gained with a ligand lacking the ability to form chelates, furthermore the interaction of U(VI) with benzoic acid was studied.

Benzoic acid (BA) occurs free and bound as benzoic acid esters in numerous plant and animal species. Furthermore it is a model ligand to study the interaction with bacterial siderophores and humic acids (Frost et al. 2011b). Benzoic acid offers one carboxyl functionality for U(VI) complexation.

In general, the study of the interaction of uranium with biologically relevant ligands is elementary for understanding the intricate interactions of uranium in biological systems on a molecular level.

Experimental

Solutions and Reagents

A stock solution of 0.1 M $UO_2(ClO_4)_2$ was used, whose exact concentration was determined by ICP-MS (Elan 6000, Perkin Elmer). Reduced glutathione was purchased from Duchefa, Netherlands. Uric acid (99.5%) and benzoic acid (p. a.) were purchased from Merck, Germany. The glutathione-S-conjugates were synthesized using the reagents β-fluoropyruvate and monobromobimane (both from Sigma-Aldrich, Germany) according to (Frost et al. 2011a). All solutions were prepared with Milli-Q-water. For all complexation experiments ionic strength was adjusted to 0.1 M with $NaClO_4$ (Sigma-Aldrich, Germany). The pH was kept constant by adding $HClO_4$ or NaOH. In case of investigations with glutathione partly the pH was adjusted to 7.4 to imitate plant cell cytoplasm pH (Roberts et al. 1980). For further preparation details and parameters see (Frost et al. 2011a,b).

UV-vis Spectroscopic Measurements

With UV-vis absorption spectroscopy the complexation of U(VI) with glutathione, glutathione-pyruvate and benzoic acid were investigated. [U(VI)] was kept constant, while varying [ligand]. Absorption spectra were recorded a) in case of complexation by glutathione or benzoic acid from 350 to 500 nm, b) in terms of complexation by glutathione-pyruvate from 290 to 330 nm. Association and stability constants of the respective U(VI) complexes and, for verification purposes, their single component spectra were determined with the factor analysis program SPECFIT (Binstead et al. 2007).

Time-Resolved Laser-Induced Fluorescence Spectroscopy (TRLFS)

In addition to room temperature nanosecond pulse TRLFS, which was very successfully applied to investigate U(VI) complexation e.g. by uric acid, two additional applications of this direct speciation technique were used for optimal complex characterization: cryo-and femtosecond pulse TRLFS. Cryo TRLFS was the preferred method in characterizing U(VI) complexation by benzoic acid, since this technique enables detecting U(VI) complexes that do not exhibit luminescence at room temperature, but do at cryogenic conditions. TRLFS with fs pulses to excite the organic ligand was very appropriate for characterizing the interaction of U(VI) with glutathione derivatized with the fluorescent label monobromobimane. The fs system used is described in detail in (Geipel et al. 2004). In contrast to all other TRLFS investigations, where the [U(VI)] was kept constant and the ligand concentration was varied, using the fs pulse technique the [U(VI)] was varied, while [ligand] was kept constant. For any further details concerning sample preparation and measurement setups and parameters see (Frost et al. 2011a,b).

Results and Discussion

UV-vis Spectroscopy

Investigations with Glutathione and -Pyruvate-S-Conjugate

In terms of the complexation of U(VI) by glutathione investigated at pH 5.0, with increasing [GSH] absorption increases and from a [GSH]:[U(VI)] of 1 : 1 on, a change of the absorption maxima is recognizable (Fig. 2), indicating complex formation. For the calculation of the association constant log K_{Ass} of a uranyl glutathione complex of a 1 : 1 constitution eight individual spectra were used. The protonation state of glutathione within the complex was deduced from the pK_A values of glutathione (Rey et al. 2004) and under the assumption, that the thiol group as a weak Lewis donor does not coordinate to UO_2^{2+} and thus remains protonated.

Hence glutathione is twofold protonated at pH 5.0 as well as at pH 7.4, since the carboxylic acid groups are deprotonated (pK_1 2.10, pK_2 3.53) and the thiol (pK_3 8.65) and amino group (pK_4 9.52) remain protonated. Regardless, whether the thiol group is derivatized, the complex association reaction can be generally given by:

$$(UO_2)_3(OH)_5^+ + 3(HGS-X)^{2-} \rightarrow 3(UO_2HGS-X) + 5OH^- \qquad (1)$$

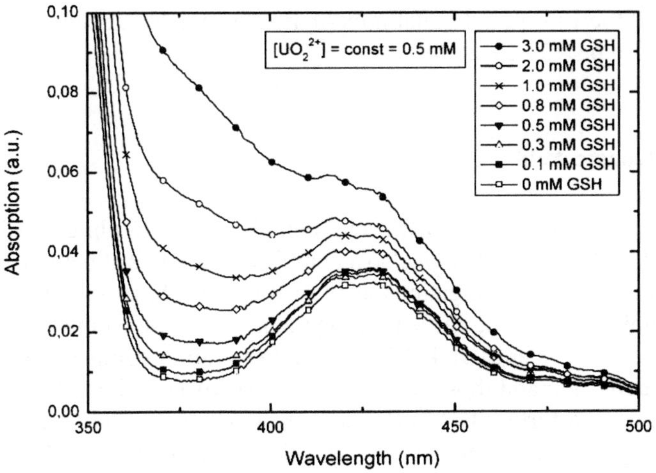

Fig. 2 UV-vis absorption spectra of 0.5 mM UO_2^{2+} as a function of glutathione concentration at pH 5.0 in 0.1 M $NaClO_4$

In case of complexation by glutathione, X is the proton of the thiol group leading to $UO_2H_2GS^+$ as the complex formed. Using SPECFIT an average log K_{Ass} of 3.82 ± 0.15 was calculated for the complexation of UO_2^{2+} by glutathione at pH 5.0. The deviation given represents the standard error of the mean, SEM. The presence of this complex was also verified through gaining the single component spectra.

Investigating the coordination of UO_2^{2+} by the derivatization product glutathione-pyruvate at pH 7.4, the spectra of the glutathione-pyruvate concentration series at constant [UO_2^{2+}] reveal an isosbestic point (not shown here), indicating the formation of a 1 : 1 complex. These spectra were evaluated with SPECFIT to gain the association constant of the UO_2HGS-pyruvate complex. The formation of the 1 : 1 complex again can be described by (1), where X now represents the pyruvate residue. With SPECFIT an association constant log K_{Ass} of 4.27 ± 0.08 was calculated. Comparing this constant to the one of the $UO_2H_2GS^+$ complex it can be concluded, that the thiol group is not involved in the coordination of the uranyl ion.

If there were a significant contribution of the thiol group to coordination, a considerable decrease of the association constant would have resulted from derivatization. This result is consistent with literature (Marzotto 1977; Bismondo and Rizzo 1992), which suggests a coordination of the uranyl ion by the carboxylato groups. Thus uranyl behaves unlike many other heavy metals as for instance Cd (Singh 2005), which are preferably coordinated by the thiol group.

All hitherto calculated constants just represent association constants. For a conversion of those into complex stability constants, the formation constant of $(UO_2)_3(OH)_5^+$ (Guillaumont and Mompean 2003) as well as the pK_A values of the thiol and amino group of glutathione have to be considered. For a detailed description of the conversion see (Frost et al. 2011a). Then an extrapolation to infinite dilution was done according to (Grenthe and Wanner 2000). Thus for the 1 : 1 uranyl glutathione complex the stability constant log β^0_{112} amounts to 39.09 ± 0.15.

The stability constant of the 1 : 1 uranyl glutathione-pyruvate complex at zero ionic strength was calculated to be > 39.24 ± 0.08.

Investigations with Benzoic Acid

Literature has stated that a 1 : 1 U(VI) benzoic acid complex exists (Vulpius et al. 2006; Glorius et al. 2007). In this work now also a complex with a 1 : 2 stoichiometry could be identified. UV-vis investigations at pH 3.0 and 4.0 have shown the presence of a 1 : 1 complex. Extending these investigations to pH 5.0, by the emergence of distinctive red-shifted bands at 432.5, 447, and 461 nm in comparison to the bands of the free uranyl ion and those of the 1 : 1 U(VI) BA complex (Glorius et al. 2007) the existence of a further U(VI)-BA species becomes evident (Fig. 3). With increasing [ligand] absorption decreases. In the absence of ligand the dominant aqueous uranyl species at pH 5.0 is $(UO_2)_3(OH)_5^+$. The spectrum of this species changes due to increasing complexation with benzoic acid.

Based on Eqs. (2) and (3):

$$UO_2^{2+} + C_6H_4COO^- \leftrightarrow UO_2\left(C_6H_4COO\right)^+ \quad \log \beta_{110} \tag{2}$$

$$UO_2^{2+} + 2C_6H_4COO^- \leftrightarrow UO_2\left(C_6H_4COO\right)_2 \quad \log \beta_{120} \tag{3}$$

the stability constants of the respective U(VI) BA complexes were calculated using SPECFIT. For further calculation input parameters see (Frost et al. 2011b). The stability constants were calculated to amount $\log \beta_{110} = 2.45 \pm 0.15$ and $\log \beta_{120} = 4.48 \pm 0.24$ (0.1 M ionic strength). After extrapolation to zero ionic strength the stability constants amount to 2.90 ± 0.15 and 5.02 ± 0.24, respectively.

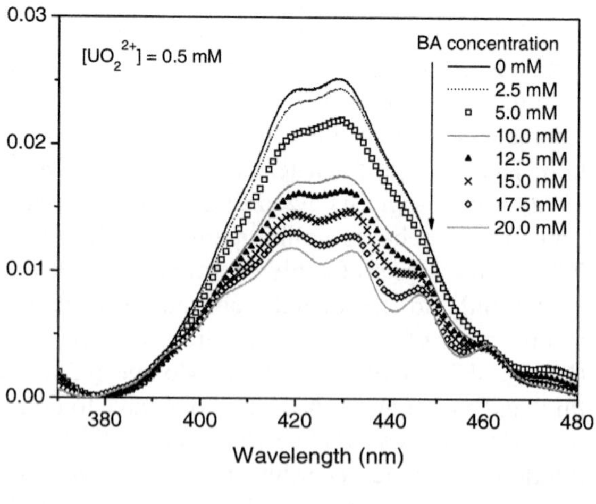

Fig. 3 BA concentration series at constant [U(VI)] at pH 5.0

TRLFS Measurements

Investigations with Glutathione and Bimane-S-Conjugate

In case of experiments on the complexation of U(VI) by glutathione at pH 7.4, where [U(VI)] was kept constant and [glutathione] was varied, the luminescence intensities of the present UO_2^{2+}-hydroxide species at each measured [glutathione] were determined, as stated in detail in (Frost et al. 2011a). For the complexation of U(VI) by glutathione-bimane-S-conjugate at pH 7.4 the fluorescence intensities of the ligand in dependence on [UO_2^{2+}] were determined. These luminescence intensities were evaluated further using the Stern–Volmer equation (Stern and Volmer 1919):

$$\frac{I_0}{I} = 1 + \mathrm{K}_D\,[Q] \tag{4}$$

where I_0 is the intensity without quencher, K_D represents the Stern–Volmer constant and $[Q]$ is the quencher/ligand concentration. If the plot of $(I_0/I{-}1)$ versus [ligand] yields a straight line, the slope K_D then directly represents the association constant of the respective complex. The appearance of a straight line furthermore indicates the sole presence of a complex with a 1 : 1 stoichiometry, and thus is a proof of such. In the following the Stern–Volmer plots based on the data gained for the U(VI)-glutathione system and the -glutathione-bimane system are given (Figs. 4 and 5).

Comparing the gained two K_D values it can be concluded, that again derivatization does not lead to a decreased complexation. Since no significant decrease of the association constant associated with derivatization of glutathione was observed, again an involvement of the thiol group in coordination of UO_2^{2+} can be excluded. Thus it could be confirmed that the SH group of glutathione does not contribute to coordination of the uranyl ion significantly. As stated before, this result is consistent with literature (Marzotto 1977; Bismondo and Rizzo 1992), which suggests a coordination of the uranyl ion by the carboxylate groups.

These association constants again were transformed to complex stability constants (for details see Frost et al. 2011b) and extrapolated to zero ionic strength

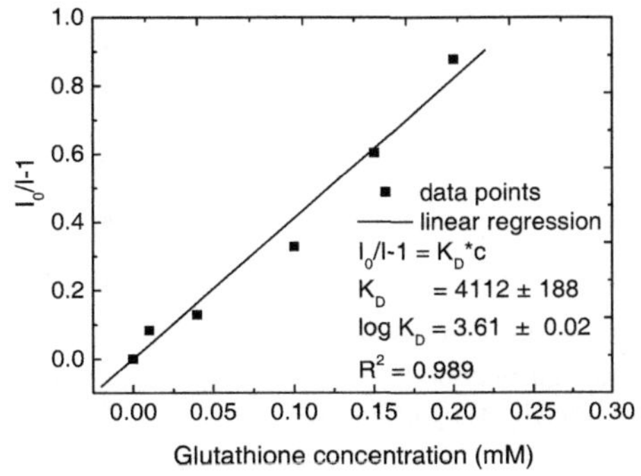

Fig. 4 Stern–Volmer plot based on $(UO_2)_3(OH)_5^+$ luminescence intensities in dependence on [glutathione] at pH 7.4

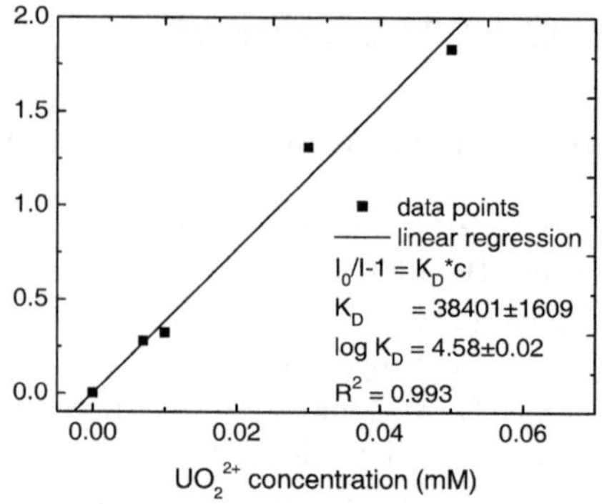

Fig. 5 Stern–Volmer plot based on glutathione-bimane fluorescence intensities in dependence on $[UO_2^{2+}]$ at pH 7.4

according to the procedure stated in (Grenthe and Wanner 2000). Finally the complex stability constant of the 1 : 1 U(VI) glutathione complex amounts to 39.04 ± 0.02, whereas the stability constant for the 1 : 1 U(VI) glutathione-bimane complex can be defined as $> 39.35 \pm 0.02$.

Investigations with Uric Acid

Investigating the complexation of U(VI) at constant [U(VI)] of $5 \cdot 10^{-5}$ M while varying the uric acid concentration at pH 3.0, binding to urate caused a decrease in the uranyl luminescence intensity. This indicates a static luminescence quenching and formation of a non-luminescent complex as well. The monoexponential decay indicated the presence of the free uranyl ion with a lifetime of 1.36 ± 0.01 μs.

As described in the previous section, the Stern–Volmer equation was used to calculate the stability constant of the present U(VI)-urate complex log β to be 4.36 ± 0.02 (Fig. 6). Furthermore, the straight line in Fig. 6 indicates a 1 : 1 complex stoichiometry.

Applying slope analysis method using mass action law, the stoichiometry of the reaction can also be estimated and the stability constant can be validated (Fig. 7).

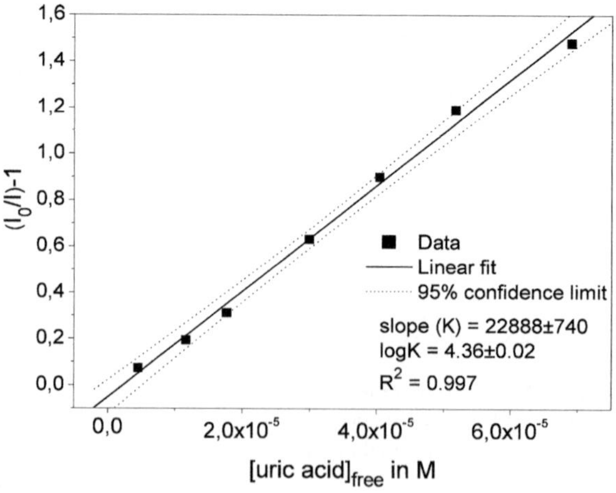

Fig. 6 Stern–Volmer plot, static quenching of uranyl(VI) by uric acid at 25°C

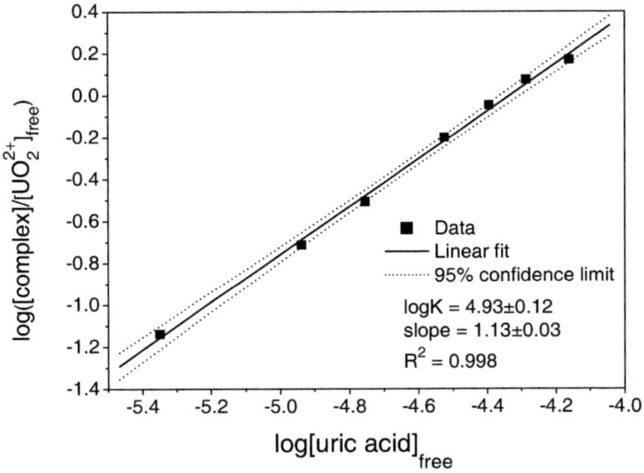

Fig. 7 Slope analysis of the complex reaction of uranium (VI) – uric acid system

The concentrations of the free uranyl ion were determined from the measured fluorescence intensities. These concentrations and their corresponding fluorescence intensities were used to calculate the corresponding concentrations of the uranyl-urate complex and the non-complexed ligand.

The slope analysis of uranyl(VI)-urate system gave a slope of 1.13 ± 0.03 (Fig. 7) which indicates a 1 : 1 complexation. The corresponding formation constant was obtained from the intersect: $\log \beta = 4.93 \pm 0.12$. For such a simple system at such weak concentrations of both the ligand and the metal ion, uranium is expected to bind N atom at position 7 and the carbonyl group >C=O at position 6.

Investigations with Benzoic Acid

In analogy to the TRLFS experiments with glutathione and uric acid, investigations on benzoic acid complexation were done keeping [U(VI)] constant, while [benzoic acid] was varied. These experiments were done at pH 3.0 and 4.0 at RT. Calculated UO_2^{2+} luminescence intensities were evaluated using the Stern–Volmer equation, as described previously. At both pH conditions (I_0/I–1) plotted versus [ligand] yields a straight line, indicating that only a 1 : 1 complex is present. Calculating the respective complex stability constant at zero ionic strength a value of $\log \beta_{110} = .27 \pm 0.20$ results. Since the maximum measurable [BA] : [U(VI)] ratio is limited by the increasing quenching of U(VI) luminescence at increasing [BA], consequently no higher U(VI) BA species can be detected with RT TRLFS.

Cryo TRLFS yielded remedy to this issue and was successfully applied for validating the existence of a 1 : 2 uranyl benzoic acid complex. Figure 8 shows the luminescence spectra in the U(VI)-BA system measured at cryogenic conditions (−18°C). Concretely, spectra of solutions with molar ratios of benzoic acid to uranyl of 20 : 1 and 220 : 1 compared to the one of the uranyl ion are shown.

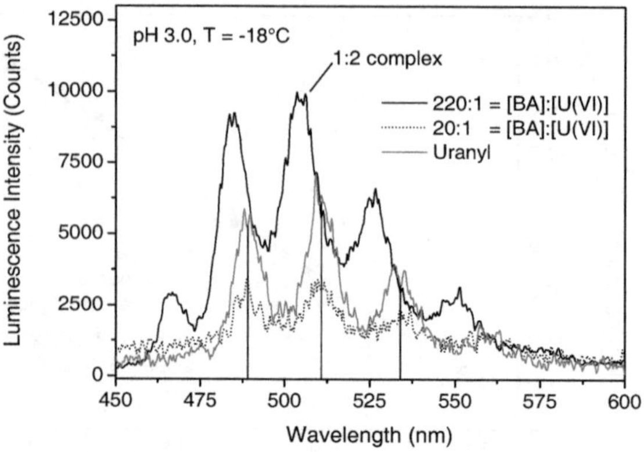

Fig. 8 Characteristic luminescence spectra of uranyl (1·10-4 M) as a function of [BA] at 0.1 M ionic strength, pH 3.0 and −18°C

Since towards a ratio of 20 : 1 of benzoic acid to uranyl only a decrease in luminescence in comparison to the spectrum of the uranyl ion is observable, it can be concluded that the 1 : 1 uranyl benzoic has no luminescence properties. If a U(VI) speciation calculation with the program MEDUSA (Puigdomènech 2004) is done for these conditions it can be found that only the 1 : 1 complex is present. There is also no shift of the spectrum recognizable. If the molar ratio of benzoic acid to uranyl is increased further towards 220 : 1 a strong increase and blue shift of the spectrum in comparison to the spectrum of the uranyl ion can be observed. The time-resolved spectrum of this solution indicates monoexponential decay meaning there is only one luminescent species present. Consequently this luminescence spectrum can clearly be assigned to the 1 : 2 uranyl benzoic acid complex.

Conclusion

UV-vis spectroscopy in combination with the factor analysis program SPECFIT has proven to be a sensitive method to explore the complexation of uranyl with various organic ligands. Besides, TRLFS is a powerful technique to characterize complexes in the aqueous phase at environmentally relevant conditions, with potential to develop further (e.g. cryo applications). Using TRLFS for the first time a stability constant for the 1 : 1 uranyl-uric acid complex could be calculated.

By conducting cryo TRLFS, the new results on the complexation of uranyl by benzoic acid gained by UV-vis spectroscopy could be validated. Referring to this, RT TRLFS did not suffice because of interfering quenching effects often observed for organic ligands at RT. This again depicts the potential of cryo TRLFS.

The major aim of this study was to assess, relating to the examined molecular ligand structures, why some ligands form complex of high thermodynamic stability with uranyl, while U(VI) interaction with other ligands is fairly weak. Representatives for these two ligand categories were investigated: on the one hand glutathione and on the other hand uric acid and benzoic acid. The study has shown that obviously glutathione forms eight to tenfold stronger complexes with the uranyl ion than uric acid and benzoic acid do. What are the structure-related reasons for such a considerable discrepancy in complex stability? Investigating the interaction of U(VI) with glutathione-S-conjugates, no decrease in complex stability in comparison to U(VI) complexation by glutathione was found. Thus here a significant involvement of the thiol group in coordination can be excluded. Besides the SH group, glutathione offers 2 carboxylic groups as well as one amine group as major potential binding sites for U(VI) complexation. Since benzoic acid offers one carboxyl functionality but exhibits a very weak U(VI) coordination, for U(VI) complexation by glutathione a chelate-like coordination with an involvement of at least one carboxyl functionality can be assumed to explain the large complex stability constant. Since uric acid coordinates as a bidentate ligand and is suggested to undergo a chelation with U(VI) as well, specifically by its 7-amine and 6-carbonyl group, it follows that chelate formation must not necessarily result in strong complexation.

In general, these results contribute to a better understanding of the intricate U(VI) interactions in biological systems on a molecular level.

Acknowledgement This work was partly funded by the BMWi under contract no. 02E10618.

References

Ballatori N, Clarkson TW (1985) Biliary secretion of glutathione and of glutathione metal-complexes. Fundam Appl Toxicol 5: 816

Binstead RA, Zuberbühler AD, Jung B (2007) SPECFIT/32, Version 3.0.40, Spectrum Software Associates, Marlborough

Bismondo A, Rizzo L (1992) Thermodynamics of the complex formation between uranyl(VI) and some polypeptides in aqueous solution. Thermochim Acta 196: 131

Davies KJ, Sevanian A, Muakkassah-Kelly SF, Hochstein P (1986) Uric acid – iron complexes. A new aspect of the antioxidant functions of uric acid. Biochem J 235: 747

Frost L, Geipel G, Viehweger K, Bernhard G (2011a) Interaction of uranium(VI) towards glutathione – an example to study different functional groups in one molecule. Radiochim Acta, accepted 2010

Frost L, Moll H, Bernhard G (2011b) A new uranyl benzoate species characterized by different spectroscopic techniques. Radiochim Acta, accepted 2011

Geipel G, Acker M, Vulpius D, Bernhard G, Nitsche H, Fanghänel T (2004) An ultrafast time-resolved fluorescence spectroscopy system for metal ion complexation studies with organic ligands. Spectrochim Acta 60: 417

Glorius M, Moll H, Bernhard G (2007) Complexation of uranium(VI) with aromatic acids in aqueous solution – a comparison of hydroxamic acids and benzoic acid. Radiochim Acta 95: 151

Grenthe, I, Wanner H (2000) Guidelines for the extrapolation to zero ionic strength. TDB-2, NEA web page document, http://www.nea.fr/html/dbtdb/guidelines/tdb2.pdf

Guillaumont R, Mompean FJ (2003) Update on the chemical thermodynamics of uranium, neptunium, plutonium, americium and technetium. Vol. 5, Elsevier, Amsterdam

Huang C, Chen M, Huang L, Mao I (2002) Uric acid and urea in human sweat. Chinese J Physiol 45: 109

Kidd PM (1997) Glutathione: systemic protectant against oxidative and free radical damage. Altern Med Rev 1: 155

Marzotto A (1977) Uranyl complexes of glutathione. J Inorg Nucl Chem 39: 2193
Puigdomènech I (2004) MEDUSA Software, 16 bit version, http://www.kemi.kth.se/medusa/
Moawad MM (2002) Complexation and thermal studies of uric acid with some divalent and trivalent metal ions of biological interest in the solid state. J Coord Chem 55: 61
Singh BK (2005) Complexation behaviour of glutathione with metal ions. Asian J Chem 17: 1
Rey NA, Howarth OW, Pereira-Maia EC (2004) Equilibrium characterization of the As(III)-cysteine and the As(III)-glutathione systems in aqueous solution. J Inorg Biochem 98: 1151
Roberts JKM, Ray PM, Wade-Jardetzky N, Jardetzky O (1980) Estimation of cytoplasmic and vacuolar pH in higher plant cells by 31P-NMR. Nature 283: 870
Stern O, Volmer M (1919) Über die Abklingzeit der Fluoreszenz. Physikal Z 20: 183
Vulpius D, Geipel G, Baraniak L, Rossberg A, Bernhard G (2006) Complex formation of uranium(VI) with 4-hydroxy-3-methoxybenzoic acid and related compounds. J Radioanal Nucl Chem 270: 661

Challenges in Detection, Structural Characterization and Determination of Complex Formation Constants of Uranyl-Arsenate Complexes in Aqueous Solutions

Wondemagegnehu A. Gezahegne, Christoph Hennig, Gerhard Geipel, Britta Planer-Friedrich, Broder J. Merkel

Abstract. Uranium forms analogous minerals with phosphate and arsenate. In aqueous solutions an analogy is expected to govern the complexes that uranium builds with these ligands. Three uranyl arsenate complexes $UO_2H_2AsO_4^+$, $UO_2HAsO_4^0$ and $UO_2(H_2AsO_4)_2^0$ were identified and reported previously with TRLFS in the pH range 1 to 3. Using a similar detection system and elevating the pH range a negatively charged fourth uranyl-arsenate complex, $UO_2AsO_4^-$ was found under circum neutral pH. Determining the complex formation constant for this complex was not possible due to the susceptibility of the fluorescence intensity to external influences and the difficulty of resolving the measured spectra into individual fluorescence contributions. By immediate shock-freezing to 15 K we succeeded to measure a reproducible EXAFS spectrum of a uranyl-arsenate species in an aqueous solution at pH 2.

Wondemagegnehu A. Gezahegne
Department of Geology, Chair of Hydrogeology, Technische Universität Bergakademie Freiberg, 09596 Freiberg, Germany

Christoph Hennig
Institute of Radiochemistry, Helmholtz-Zentrum Dresden-Rossendorf, P.O. Box 510119, D-01314 Dresden, Germany

Gerhard Geipel
Institute of Radiochemistry, Helmholtz-Zentrum Dresden-Rossendorf, P.O. Box 510119, D-01314 Dresden, Germany

Britta Planer-Friedrich
Environmental Geochemistry, Bayreuth University, 95440 Bayreuth, Germany

Broder J. Merkel
Department of Geology, Chair of Hydrogeology, Technische Universität Bergakademie Freiberg, 09596 Freiberg, Germany

B. Merkel, M. Schipek, *The New Uranium Mining Boom* 607
DOI 10.1007/978-3-642-22122-4 © Springer 2011

Introduction

The existence of minerals such as Trögerite, $H_2(UO_2AsO_4)_2 \cdot 8H_2O$ and $UO_2(HAsO_4) \cdot 4H_2O$ both in nature and as products of laboratory syntheses (Barten and Cordfunke 1980) is a clear indication of the high affinity between arsenic and uranium to form solid phases. The analogy observed in the minerals that both arsenate and phosphate form with uranium leads to the assumption that these two ligands also form similar complexes with uranium in aqueous solutions. Yet, compared to uranyl phosphate complexes, less information exists about complexation of uranium(VI) and arsenate(V) at environmentally relevant concentrations and pH ranges. One suitable non intrusive technique is time-resolved laser-induced fluorescence spectroscopy (TRLFS). It has been used for speciation studies of such actinides as uranium (Denning 1992; Meinrath 1997) and to confirm the formation of aqueous uranyl complexes with simple ligands such as sulfate (Geipel et al. 1996), silicate (Moll et al. 1998), carbonate (Kato et al. 1994; Wang et al. 2004), phosphate (Brendler et al. 1996) and hydroxide (Moulin et al. 1995).

Towards the end of the 1990s, Rutsch et al. used TRLFS to study uranium speciation in acidic solutions in the presence of excess arsenic (Rutsch et al. 1999). They were able to identify three uranyl-arsenate species with previously unknown fluorescence spectra and lifetimes. Because species identification can only be done by model analogy in the absence of commercial reference material or known thermodynamic constants, the authors assigned the identity of the new complexes by analogy to uranyl phosphates (Brendler et al. 1996; Grenthe et al. 2003). In the investigated pH range of 1.5 to 3, the authors assigned the positively charged $UO_2H_2AsO_4^+$ as the predominant complex over UO_2^{2+}. Extrapolating their measured data to pH 5, they predicted predominance of $UO_2(HAsO_4)^0$ at pH > 3.2. The $UO_2(H_2AsO_4)_2$ complex was also identified but seemed to play only a minor role at pH 2 to 3.5. As expected, no evidence for a negatively charged $UO_2AsO_4^-$ could be found in the acidic solutions investigated. They also provided complexation constants for the three uranyl arsenate species.

Since this initial study by Rutsch et al. (1999) the identification and characterization of uranyl arsenates received little further attention. Neither were the postulated complexes structurally characterized to confirm their identity nor were the studies extended to higher pH ranges. Besides, the complex formation constants have not been confirmed by any other well established method.

A comprehensive knowledge of the uranyl arsenate system is especially critical for the assessment and planning of remediation strategies for uranium mine waters, which often contain arsenic as an accompanying element. The uranium speciation in such waters is pH-dependent and the arsenic present might compete to an extent with the other ligands to form a single charged $UO_2H_2AsO_4^+$ complex and the non-charged $UO_2(HAsO_4)^0$ and $UO_2(H_2AsO_4)_2$ complexes in the acidic milieu which are expected to alter uranium mobility, surface complexation, and sorption capacity substantially. Therefore, measurements were carried out with TRLFS on synthetic solutions which were prepared in both acidic and alkaline conditions. The goal was to identify dissolved uranyl-arsenate complexes in a wider pH range and examine the reproducibility of their complexation constants using a similar measuring system. Despite its speciation capabilities TRLFS provides no direct structural information. Therefore, it was also aimed to determine the structures of the uranyl arsenates with Extended X-Ray Absorption Fine Structure spectroscopy.

Experimental Section

Samples and Sample Preparation

Solutions with constant concentrations of 5×10^{-6} M uranium (VI) and arsenate concentrations ranging from 5×10^{-6} M to 1×10^{-4} M were prepared by diluting stock solutions of 5×10^{-3} M $UO_2(ClO_4)_2$ in 0.1 M $HClO_4$ and 1×10^{-3} M As_2O_5 (Alfa Aesar), respectively. Constant ionic strength was maintained for all solutions by using 0.1 M $NaClO_4$. The pH range of the solutions was adjusted in steps of 0.5 from 1 up to 9 using NaOH and $HClO_4$. The solutions were prepared under ambient atmospheric as well as in CO_2-free conditions. All TRLFS measurements were carried out immediately after sample preparation in closed quartz curvettes. For cryogenic measurements at 153 K (–120°C) with TRLFS, samples were flash-frozen in polyethylene cuvettes. For the EXAFS measurements the U and As concentration had to be elevated, but experiments showed that. at uranium concentrations higher than 5 μM precipitate formation occurred especially at pH above 2. Therefore, samples had to be immediately flash-frozen with liquid nitrogen after filtration with 200 nm filter and filling into sample holders to hinder or slow down the kinetics of precipitation at concentrations acceptable to EXAFS measurements.

TRLFS Setup

The TRLFS system consists of a pulsed Nd:YAG diode laser system with subsequent fourth harmonic generation that produced a beam with wavelength of 266 nm which was used for the excitation of samples in quartz cuvettes. The time-resolved fluorescence radiation spectra were recorded after application of the laser pulse in the range between 420 and 600 nm with a CCD camera. 100 laser shots were averaged for each spectrum and at each delay step three spectra were recorded. The pulse energy was about 1.8 mJ. For further detail concerning the TRLFS setup, see (Baumann et al. 2008). Measurements on flash-frozen samples at 153 K were carried out with a system that used a pulsed Nd:YAG-MOPO laser system as beam source. Origin 8 (Origin Laboratory Inc.) and TRLFS processing Software developed at HZ Dresden-Rossendorf (Steinborn et al. 2008) were used for the spectral deconvolution and determination of fluorescence decay times.

EXAFS

Extended X-ray absorption fine-structure (EXAFS) spectra were recorded at the Rossendorf Beamline (ROBL) at the European Synchrotron Radiation Facility (ESRF). Uranium L_{III}-edge (17,185 eV) EXAFS spectra were collected in fluorescence mode using a 13-element Ge detector (Canberra) with a sample orientation of 45° and a detector orientation of 90° to the incident beam.

All samples were measured flash-frozen at 15 K in a closed-cycle He cryostat. Prior to averaging, the fluorescence spectra were corrected for the detector dead time. The EXAFS spectra were processed according to standard procedures (Teo 1986) using the suit of programs EXAFSPAK (George and Pickering 1995). Theoretical scattering phases and amplitudes are calculated using scattering code FEFF 8.2 (Ankudinov et al. 2002).

Results and Discussion

TRLFS Analyses for Uranium Speciation at Neutral to Alkaline pH

In general, high power, pulsed laser and time-resolved detection systems have proven to be one of the most sensitive current techniques for the analysis of fluorescing actinides such as uranium in dilute solutions (Beitz et al. 1988). They detect changes in the chemical environment of U(VI) through shifts of fluorescence spectra and variations of the lifetime of the excited state. The fluorescence intensity, bandwidth, and location of the emission bands of the uranyl fluorescence spectra and its fluorescence lifetime are highly sensitive to the bonding, symmetry, and local chemical environment of uranyl ion. Consequently, shifts in the position of fluorescence emission bands and variations in the fluorescence lifetimes are considered as manifestations of different uranyl arsenate complexes. Because the fluorescence band positions are the main attributes of the TRLFS spectrum, they are used as primary tools for identification of complexation of an element with a ligand whereas the fluorescence lifetime plays a complementary role due to its dependence on experimental temperature (Krepelova et al. 2007). These properties have been used in our experiments to differentiate between the types of uranyl arsenate complexes as well as uranyl carbonates and hydroxides.

Because of the similarities that arsenate and phosphate have in their chemical behaviors such as in protonation constants, they are conceivably expected to exhibit comparable complexation tendencies. Consequently, the following four uranyl-arsenate complexes are expected to form:

$$UO_2^{2+} + H^+ + AsO_4^{3-} \rightleftharpoons UO_2(HAsO_4)_{(aq)}$$

$$UO_2^{2+} + 2H^+ + AsO_4^{3-} \rightleftharpoons UO_2H_2AsO_4^+$$

$$UO_2^{2+} + 4H^+ + 2AsO_4^{3-} \rightleftharpoons UO_2(H_2AsO_4)_{2(aq)}$$

$$UO_2^{2+} + AsO_4^{3-} \rightleftharpoons UO_2AsO_4$$

$UO_2(HAsO_4)_{(aq)}$, $UO_2H_2AsO_4^+$, and $UO_2(H_2AsO_4)_{2(aq)}$ have already been detected previously in the pH range 1.5 to 3 (Rutsch et al. 1999). In our ambient experiments with U(VI) and arsenate concentrations of 5×10^{-6} M and 1×10^{-4} M respectively, we were able to identify these species by comparing the positions of their spectral emission bands. Increasing the pH up to 6.5 showed that the species $UO_2(H_2AsO_4)_{2(aq)}$ and $UO_2(HAsO_4)_{(aq)}$ continued to exist in dissolved state with $UO_2(HAsO_4)_{(aq)}$ being dominant between pH 4.5 and 6.5 (Table 1). As shown in Fig. 1 and Table 2, there were discernible spectral differences among the uranyl arsenate species at room temperature, making it possible to differentiate between them via their spectroscopic fluorescence signatures. Under atmospheric experimental conditions, U(VI) speciation in the neutral and alkaline milieu is known to be dominated by uranyl carbonate complexes. In comparison to the free uranyl ion which shows good fluorescence spectra with five characteristic fluorescence emission bands as shown in Fig. 1 and a characteristic life time of about 1.9 ± 0.4 µs, the carbonate-bearing uranyl complexes showed no measureable fluorescence at room temperature, which is a phenomenon known from other studies on uranyl carbonates (Bernhard et al. 1996). Contrarily, in the presence of arsenate well-resolved fluorescence spectra with well defined emission bands and enhanced intensity were measured in the pH range between 7 and 8.5 (Fig. 2). This is the primary evidence that complexation between U(VI) and arsenate does indeed occur at neutral and alkaline aqueous media.

Table 1 Solution composition of aqueous uranyl-arsenate complexes of selected samples, I = 0.1 M NaClO4

pH	Condition	Environment	U (VI)	Arsenate	Dominant species
1	298 K	Atmospheric	5×10^{-6} M		UO_2^{2+}
2	298 K	Atmospheric	5×10^{-6} M	1×10^{-4} M	UO_2^{2+}, $UO_2H_2AsO_4^+$
3.5	298 K	Atmospheric	5×10^{-6} M	1×10^{-4} M	UO_2^{2+}, $UO_2(H_2AsO_4)_2$
5.5	298 K	Atmospheric	5×10^{-6} M	1×10^{-4} M	UO_2HAsO_4
8	298 K	Atmospheric	5×10^{-6} M		No fluorescence
8	153 K	Atmospheric	5×10^{-6} M		Uranyl carbonate
8	298 K	CO_2-free	5×10^{-6} M		Uranyl hydroxide
8	153 K	CO_2-free	5×10^{-6} M		Uranyl hydroxide
8	298 K	Atmospheric	5×10^{-6} M	1×10^{-4} M	$UO_2AsO_4^-$
8	298 K	CO_2-free	5×10^{-6} M	1×10^{-4} M	$UO_2AsO_4^-$
8	153 K	Atmospheric	5×10^{-6} M	1×10^{-4} M	$UO_2AsO_4^-$
8	153 K	CO_2-free	5×10^{-6} M	1×10^{-4} M	$UO_2AsO_4^-$

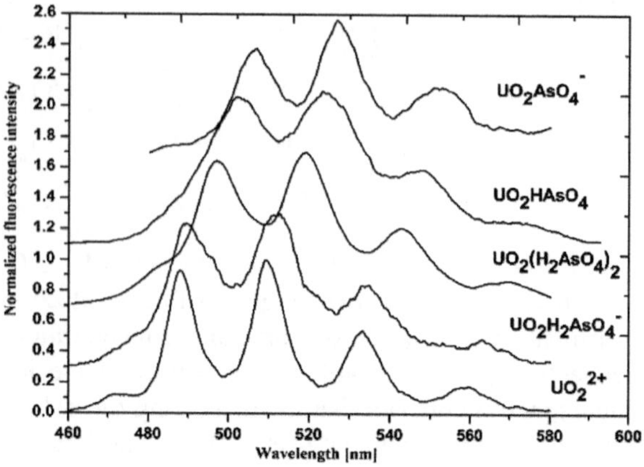

Fig. 1 Deconvoluted fluorescence spectra with characteristic positions of fluorescence emission bands of uranyl arsenate species with U(VI) and arsenate concentration of 5×10−6 M and 1×10−4 M, respectively. For pH values refer to Table 1

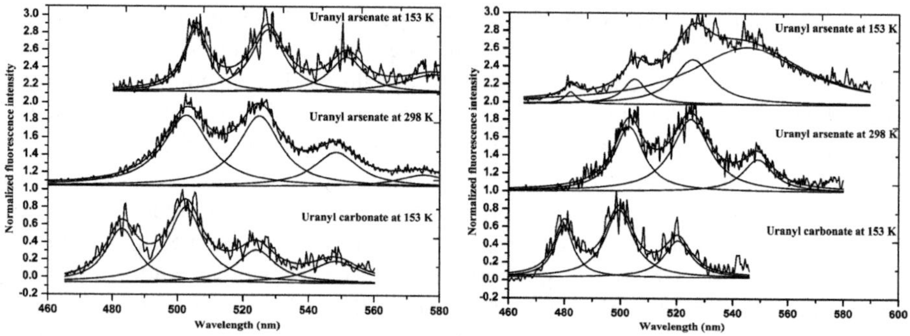

Fig. 2 Normalized fluorescence spectra of uranyl carbonate, and uranyl arsenate complexes at pH 7.5 (*left*) and 8 (*right*). The spectra are offset along the vertical axis for clarity

As shown in Fig. 2, lowering the temperature of U(VI) solutions to 153 K led to enhancement of the fluorescence intensity and well-resolved blue-shifted fluorescence bands of the aqueous uranyl carbonate complex. This observed enhancement of fluorescence intensity and spectral resolution with lowering of temperature was consistent with the observations of Wang et al. (2004), Dieke et al. (1949) and other authors. In comparison with the spectra of the uranyl carbonate complex at 153 K, the spectra of the uranyl-arsenate solution in the pH range from 7 up to 8.5 were largely red-shifted (Table 2). Therefore, there were appreciable differences in the spectral positioning of the emission bands of the U(VI) solutions measured with and without arsenate at cryogenic state. Additionally, there were significant differences in the fluorescence lifetimes between the two types of complexes formed (Table 2). The identity of the uranyl-arsenate complex that formed is assumed to be $UO_2AsO_4^-$ which is derived from analogies to the similar uranyl-phosphate system in which $UO_2PO_4^-$ is dominant at neutral and alkaline conditions in aqueous solutions.

In U(VI) solutions prepared in a CO_2-free glove-box (Fig. 3), the dominant species in the pH range between 7 and 9 was the uranyl hydroxide complex with characteristic broad fluorescence spectrum as described in literature (Moulin et al. 1995).

Table 2 Fluorescence emission band positions of uranyl arsenate complexes, I = 0.1 M

Species	pH	Fluorescence bands (nm)	Lifetime (μs)
UO_2^{2+}	1	471,488,510,534,560	1.9 ± 0.4
$UO_2H_2AsO_4^+$	2	478,493,515,538,565	8 ± 1.2
$UO_2(H_2AsO_4)_2$	3.5	485,497,518,543,567	26.3 ± 7.3
UO_2HAsO_4	6	501,523,548,573	1.6 ± 0.4
$UO_2AsO_4^-$	7.5	506,527,551,576	14.7

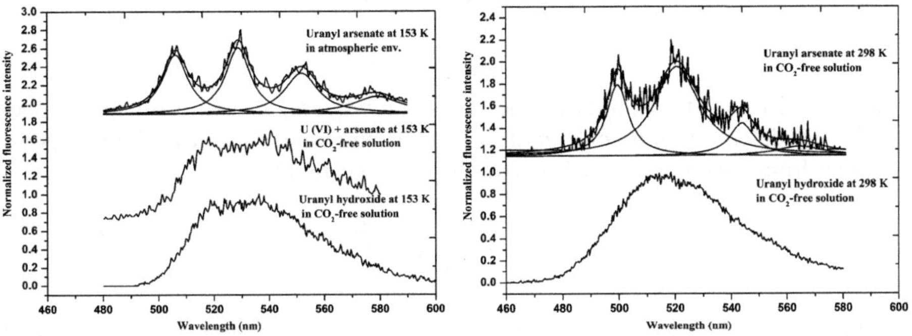

Fig. 3 Normalized fluorescence spectra of uranyl hydroxide, and uranyl arsenate complexes at pH 7.5 (*right*) and 8 (*left*). The spectra are offset along the vertical axis for clarity

Fig. 4 Fluorescence spectra of $UO_2AsO_4^-$ as function of arsenate concentration at pH 8. The spectra are shifted in ordinates for illustration. Samples were prepared in ambient atmospheric condition. The measurements were performed at 153 K

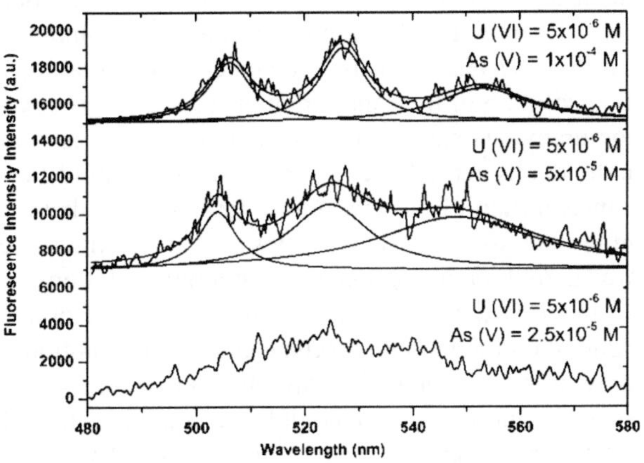

A further evidence for the formation of $UO_2AsO_4^-$ is the development of the fluorescence bands of U(VI) solution with increasing arsenate concentrations. Figure 4 shows a set of time-resolved fluorescence spectra of 5×10^{-6} M uranium (VI) solution with varying arsenate concentrations at pH 8. The solutions were prepared in an atmospheric condition with arsenate concentrations up to 1×10^{-4} M. As can be seen from Fig. 4, the development of spectral shape and resolution of the peaks is directly related to the arsenic concentration evidencing chemical bonding of U(VI) with a ligand in its vicinity. The fluorescence decay of the $UO_2AsO_4^-$ is found to be mono-exponential with a fluorescence lifetime of 14.7 ns.

All these observations clearly indicate that the fluorescence spectral signatures were evidently dependent on the structure and composition of the species formed and that the complexation of U(VI) with arsenate not only occurs at acidic but also at neutral and alkaline pH ranges leading to the formation of the negatively charged species, $UO_2AsO_4^-$ that we assume in analogy to phosphate.

To determine the complexation constants Rutsch et al. applied the following equation:

$$\log \beta_{\mathrm{n,m}} = \frac{\left[UO_2(H)_{\mathrm{n}}(AsO_4)_{\mathrm{m}}^{2+\mathrm{n}-3\mathrm{m}}\right]}{\left[H^+\right]^{\mathrm{n}} \cdot \left[UO_2^{2+}\right] \cdot \left[AsO_3^{3-}\right]^{\mathrm{m}}} \tag{1}$$

with n and m as stoichiometric coefficients for hydrogen and arsenate, respectively. To determine the concentrations of the individual species within a given solution from the deconvoluted spectra they set the condition:

$$A_i = A_0 \cdot \frac{C_i}{C_{\mathrm{uranium(VI),total}}} \quad \text{and} \quad \sum_{i=1}^{n} C_i = C_{\mathrm{uranium(VI)}} \tag{2}$$

where A_0 is total fluorescence intensity at $t=0$, A_i is the fluorescence intensity of species i, C_i is concentration of species i at time t and $C_{\mathrm{uranium\ (VI),total}}$ is the total concentration of uranium (VI) ion. According to this prerequisite the fluorescence intensity of uranium is associated to its concentration. Thus, at a given concentration, the free uranyl ion in an arsenate free solution e.g. at pH 2 should bear highest fluorescence intensity to be taken as the total intensity contribution of uranium. In a solution of the same uranium concentration that contains arsenate, the resulting uranyl arsenate species as well as the remnant uncomplexed uranyl ion should individually have lower intensities. This should then allow the calculation of the concentration of the free uranyl ion and then that of the uranyl arsenate by cross-multiplication. This is only possible if the fluorescence intensity of the remnant free uranyl ion in the arsenate containing solution is known.

In our experiments, the above mentioned prerequisite did not hold. On one hand, the uranium and arsenate containing solutions, in majority of cases, showed only mono-exponential decay, thus only one species. In such a case, with the absence of the remnant uncomplexed uranium, the calculation of the concentration of the uranyl arsenate species using Eq. (1) was not possible. On the other hand, there were also cases where the remaining uncomplexed uranium bore higher florescence intensity than in an arsenate free solution. Because of these reasons we were neither able to confirm nor determine the complexation constants of the uranyl arsenate complexes. Although the complex formation constants of uranyl-arsenates are conceivably expected to be lower than those of uranyl-phosphates, there is still a need to confirm previously estimated values (Rutsch et al. 1999) using other well established detection methods.

Attempts to clarify the structure of the uranyl arsenate complexes in dilute solutions with EXAFS are bounded by two limitations. On one hand, to be detectable with EXAFS the uranium concentration has to be higher than the concentration for TRLFS measurements, but on the other hand, at concentrations beyond 5 μM there is the likeliness of precipitate formation with arsenate especially at pH ranges above 2. Of the several methods we explored to avoid uranyl arsenate precipitation, immediate shock freezing of prepared solutions at 15 K was the only method found to be suitable. Yet, even this option itself was effective only with uranium concentration of 0.05 mM at pH 2; the lowest detectable concentration possible which demanded the accumulation of several spectra. This made it possible to detect dissolved uranyl arsenate species at pH 2 for the first time ever with EXAFS. Spectral processing with EXAFSPAK and the scattering code FEFF 8.2 for this sample showed that two axial oxygen atoms (Oax) are at a distance of 1.77 Å and five equatorial oxygen atoms (Oeq) at a distance of 2.39 Å. The U-As bond distance was calculated to be 3.39 Å, which is shorter than that observed in solid phases and a significant manifestation of the dissolved uranyl-arsenate species in this sample. In comparison, a typical solid phase such as $H[UO_2AsO_4] \cdot 4H_2O$ at 15 K exhibits two axial oxygen atoms at a distance of 1.77–1.78 Å and four equatorial oxygen atoms at a distance of 2.29–2.30 Å. The monodentate U-As coordination distance lies between 3.69–3.70 Å, which is significantly longer than that observed in the uranyl arsenate solution. There also exists U-U1 bonding at a distance of 5.40 Å, which is absent in the solution species. This indicates that the solution species is monomeric.

EXAFS detection of uranyl arsenate complexes at environmentally relevant concentrations and at pH values beyond 2 still remain a challenge. Experiments that attempt to deal with these conditions will first of all have to devise a method that inhibits the precipitation of uranyl arsenates at wide pH ranges and at uranium concentration level which is detectable by EXAFS.

References

Ankudinov A.L., Bouldin C., Rehr J.J., Sims J., Hung H. (2002) Parallel calculation of electron multiple scattering using Lanczos algorithms. Phys. Rev. B 65: 104107

Barten H., Cordfunke E.H.P. (1980) The formation and stability of hydrated and anahydrous uranyl arsenates. Thermochim Acta 40: 367–375

Baumann N., Arnold T., Foerstendorf H., Read D. (2008) Spectroscopic verification of the mineralogy of an ultrathin mineral film on depleted uranium. Environ. Sci. Technol. 42: 8266–8269

Beitz J.V., Bowers D.L., Doxtader M.M., Maroni V.A., Reed D.T. (1988) Detection and speciation of transuranium elements in synthetic groundwater via pulsed-laser. Radiochim. Acta 44/45: 87–93

Bernhard G., Geipel G., Brendler V., Nitsche H. (1996) Speciation of uranium in Seepage waters of a mine tailings pile studied by time-resolved laser-induced fluorescecne spectrpscopy. Radiochim. Acta 1996, 74, 87–91

Brendler V., Geipel G., Bernhard G., Nitsche H. (1996) Complexation in the system UO22+/PO43-/OH-(aq): potentiometric and spectroscopic Investigations at very low ionic strengths. Radiochim. Acta 1996 74, 75–80

Denning R.G. (1992) Electronic structure and bonding in actinyl ions. Struct. Bond. 79: 215–276

Dieke G., Duncan A.B.F. (1949) Spectroscopic Properties of Uranium Compounds, in National Nuclear Energy Series Div. III 2 McGraw-Hill: New York

Geipel G., Brachmann A., Brendler V., Bernhard G., Nitsche H. (1996) Uranium(VI) sulfate complexation studied by time-resolved laser-induced fluorescence spectroscopy. Radiochim. Acta 75: 199–204

George G.N., Pickering I.J. (1995) EXAFSPAK – A suite of computer Programs for Analysis of X-ray Absorption Spectra, SSRL

Grenthe I., Fuger J., Konigs R.J.M., Lemire R.J., Muller A.B., Nguyen-Trung Cregu C., Wanner, H. (2003) Update on the chemical thermodynamics of uranium, neptunium, plutonium, americium and technetium. Elsevier: Amsterdam

Kato Y., Meinrath G., Kimura T., Yoshida Z., (1994) A study of U(VI) hydrolysis and carbonate complexation by time-resolved laser-induced fluorescecne spectroscopy. Radiochim. Acta 64: 107–111

Krepelova A., Brendler V., Sachs S., Baumann N., Bernhard G. (2007) U(VI)-kaolinite surface complexation in absence and presence of humic acid studied by TRLFS. Environ. Sci. Technol. 41: 6142–6147

Meinrath G. (1997) Uranium(VI) speciation by spectroscopy J. Radioanal. Nucl. Ch. 224: 119–126

Moll H., Geipel G., Brendler V., Bernhard G., Nitsche H. (1998) Interaction of uranium(VI) with silicic acid in aqueous solutions studied by time-resolved laser-induced fluorescence spectroscopy. J. Alloy Compd. 271: 765–768

Moulin C., Decambox P., Moulin V., Decaillon J.G. (1995) Uranium speciation in solution studied by TRLFS. Anal. Chem. 67: 348–353

Rutsch M., Geipel G., Brendler V., Bernhard G., Nitsche H. (1999) Interaction of uranium(VI) with arseante in aqueous solution studied by time-resolved laser-induced fluorescence sprectroscopy. Radiochim. Acta, 86: 135–141

Steinborn A., Taut S., Brendler V., Geipel G., Flach B. (2008) TRLFS: Analysing spectra with an expectation-maximization (EM) algorithm. Spectrochim Acta 71: 1425–1432

Teo B.K. (1986) EXAFS: Basic principles and data analysis. Springer: New York

Wang Z., Zachara J.M., Yantasee W., Gassman P.L., Liu, C., Joly A.G. (2004) Cryogenic laser induced fluorescence characterization of U(VI) in Hanford vadose zone pore waters. Environ. Sci. Technol. 38: 5591–5597

Effect of Temperature and Humic Acid on the U(VI) Diffusion in Compacted Opalinus Clay

C. Joseph, L.R. Van Loon, A. Jakob, K. Schmeide, S. Sachs, G. Bernhard

Abstract. The diffusion of ^{233}U(VI) in compacted Opalinus Clay from the Mont Terri test site, Switzerland was studied in the absence and presence of humic acid for two different temperatures (25, 60°C). The diffusion experiments were carried out under anaerobic conditions and a synthetic Opalinus Clay pore water (pH 7.6, $I = 0.36$ M) as background electrolyte was used. At 25°C as well as at 60°C humic acid has no influence on ^{233}U(VI) diffusion. In addition, the results showed that the interaction of ^{233}U(VI) with the clay is stronger at 60°C.

C. Joseph
Helmholtz-Zentrum Dresden-Rossendorf, Institute of Radiochemistry,
P.O. Box 510119, 01314 Dresden, Germany

L.R. Van Loon
Laboratory of Waste Management, Paul Scherrer Institut, CH-5232 Villigen PSI, Switzerland

A. Jakob
Laboratory of Waste Management, Paul Scherrer Institut, CH-5232 Villigen PSI, Switzerland

K. Schmeide
Helmholtz-Zentrum Dresden-Rossendorf, Institute of Radiochemistry,
P.O. Box 510119, 01314 Dresden, Germany

S. Sachs
Helmholtz-Zentrum Dresden-Rossendorf, Institute of Radiochemistry,
P.O. Box 510119, 01314 Dresden, Germany

G. Bernhard
Helmholtz-Zentrum Dresden-Rossendorf, Institute of Radiochemistry,
P.O. Box 510119, 01314 Dresden, Germany

B. Merkel, M. Schipek, *The New Uranium Mining Boom*
DOI 10.1007/978-3-642-22122-4

Introduction

Beside salt dome and crystalline rocks, argillaceous rocks are considered as potential host rocks for nuclear waste repositories in deep geological formations. For safety assessment purposes, a profound knowledge of the migration behavior of actinides in natural clay formations, which is governed by molecular diffusion, is required. Radioactive decay of the radionuclides will result in elevated temperatures – however, ≤ 100°C – close to the canisters emplaced in an argillaceous rock formation (Brasser et al. 2008) and such elevated temperatures can influence the migration behavior of actinides. Due to its long half-live, uranium, as a main component of spent fuel, is of special interest for diffusion studies.

Several natural clays are investigated in Europe concerning their suitability as host rock for a nuclear waste repository, e.g. Opalinus Clay (OPA) in Switzerland (Nagra 2002), Callovo-Oxfordian Clay in France (Descostes et al. 2008) or Boom Clay in Belgium (Maes et al. 2008). So far, the diffusion studies in the case of OPA are mainly focused on tracers such as HTO, $^{36}Cl^-$ or $^{125}I^-$ (Van Loon et al. 2003). Appelo et al. (2010) investigated the multicomponent diffusion in OPA for a variety of tracers (HTO, $^{22}Na^+$, $^{134}Cs^+$, $^{85}Sr^{2+}$, $^{36}Cl^-$, Br^- and I^-). However, only few studies are known, which are focusing on the diffusion properties of actinides in OPA. For instance, Wu et al. (2009) investigated the diffusion of Np(V) in OPA.

Humic acids (HA) occur ubiquitously in nature. As extraction experiments showed they can be part of the organic matter fraction of natural clays (Claret et al. 2003). Because of the variety of functional groups, HA have a pronounced ability to form complexes with metal ions. They are also known for their colloid formation and redox properties (Schmeide and Bernhard 2009). Consequently, HA may also influence the migration of metal ions such as Eu(III) (Seida et al. 2010).

In the present work, the diffusion of HTO, ^{233}U(VI) and ^{14}C-HA in OPA was studied at 25°C and 60°C under anaerobic conditions using an artificial OPA pore water. The ^{233}U(VI) diffusion was investigated in the absence and presence of ^{14}C-HA. From the HTO diffusion experiment a value for the transport porosity of the clay was determined. Such a value was needed to interpret the ^{233}U(VI) and ^{14}C-HA diffusion results.

Experimental

Samples and Solutions

Pristine OPA samples from the Mont Terri underground rock laboratory in Switzerland were used for the diffusion experiments. The bore cores were taken by the Federal Institute for Geosciences and Natural Resources (BGR). The preparation of the OPA samples (thickness: 11 mm, diameter: 25.5 mm) was performed at the Karlsruhe Institute of Technology – Institute for Nuclear Waste Disposal

(KIT-INE). Table 1 presents the average mineralogy of OPA determined by Nagra (2002).

According to Pearson (1998) a synthetic OPA pore water (pH 7.6; cf. Table 2) was prepared in Milli-Q water (18 MΩ; mod. Milli-RO/Milli-Q-System, Millipore, Schwalbach, Germany). The true ionic strength (I_t) of 0.36 M was calculated for this synthetic OPA pore water using the speciation code EQ3/6 (Wolery 1992) (cf. Table 2).

A ^{233}U(VI) stock solution (1.5×10^{-4} M $^{233}UO_2Cl_2$) was used for all experiments. For the studies with HA, a synthetic ^{14}C-labeled HA type M42 was applied (batch M180A, 23.08 MBq/g) (Pompe et al. 1998; Sachs et al. 2004). HA type M42 represents a HA like melanoidin from xylose and glutamic acid. It shows an elemental composition and functional group content (proton exchange capacity: (3.34 ± 0.18) meq/g) comparable to those of most natural HA. Consequently, its metal ion complexation and sorption behavior is comparable to that of natural HA (e.g. Pompe et al. 1998; Schmeide et al. 2003; Křepelová et al. 2006). A HA stock solution of 5 g/l was prepared by dissolving 50 mg of ^{14}C-M42 with 1890 µl of 0.1 M NaOH and filling up the volume to 10 ml with pore water.

Table 1 Average mineralogy of OPA, Error: ± 1 σ (Nagra 2002)

Mineral	wt.%
Clay minerals	66 ± 11
– Illite	23 ± 2
– Illite/smectite ML	11 ± 2
– Chlorite	10 ± 2
– Kaolinite	22 ± 2
Quartz	14 ± 4
Calcite	13 ± 8
Siderite	3 ± 1.8
Albite	1 ± 1
K-feldspar	1 ± 1.6
Pyrite	1.1 ± 0.5
Organic carbon	0.8 ± 0.5

Table 2 Composition of a generic OPA pore water according to Pearson (1998)

Element/Ion	mol/l
Na^+	2.4×10^{-1}
K^+	1.6×10^{-3}
Mg^{2+}	1.7×10^{-2}
Ca^{2+}	2.6×10^{-2}
Sr^{2+}	5.1×10^{-4}
Cl^-	3.0×10^{-1}
SO_4^{2-}	1.4×10^{-2}
CO_3^{2-}/HCO_3^-	4.8×10^{-4}
I_t	0.36

Experimental Set-Up

The details of the diffusion cells used in the experiments are described in Van Loon et al. (2003). Four identical diffusion cells were used for the experiments; two cells were conditioned at 25°C and two cells at 60°C, respectively. Each OPA core sample was placed in the cells between two stainless steel filter plates (316L, pore diameter: 0.01 mm, diameter: 25.4 mm, thickness: 1.55 mm, porosity: 0.3; MOTT industrial division, Farmington, USA). The confining pressure on each sample was 5 MPa. The experimental set-up used for the diffusion experiments at 25°C is shown in Fig. 1. The experiments were performed under anaerobic conditions (N_2, 0% CO_2). Each diffusion cell was coupled with a peristaltic pump (mod. Ecoline, Ismatec, IDEX Health & Science, Glattbrugg, Switzerland) and a high and low concentration reservoir was filled with 200 ml and 20 ml synthetic OPA pore water, respectively. The solutions were circulating through the end-plates of the cells in order to saturate the samples. The saturation time was three weeks. Subsequently, the solutions were replaced by fresh ones, whereby the respective high concentration reservoir contained the tracer and thereby tracer diffusion was started.

For the diffusion experiments at 60°C the experimental set-up was changed slightly. The diffusion cells were placed in a laboratory sand-bath (mod. ST-72, Harry Gestigkeit GmbH, Düsseldorf, Germany) and the reservoirs were stirred and heated on top of a magnetic stirrer with integrated heating function (mod. MR 3002, Heidolph, Schwabach, Germany).

At first, with all four diffusion cells HTO through- and out-diffusion experiments were performed as described by Van Loon et al. (2003) in order to determine values for the porosity (ε) of the clay samples ($[HTO]_0 = 1000$ Bq/ml). After that, the ^{233}U(VI) in-diffusion in the absence and presence of ^{14}C-HA was studied. Using two diffusion cells the ^{233}U(VI) diffusion at 25°C (cell 1) and 60°C (cell 3) was investigated. The simultaneous diffusion of ^{233}U(VI) and ^{14}C-HA was studied (cell 2 (25°C) and cell 4 (60°C)) under exclusion of light for minimizing possible light-induced HA degradation processes ($[^{233}U(VI)] = 1 \times 10^{-6}$ mol/l, $[^{14}C\text{-HA}] = 10$ mg/l). After three months, the diffusion experiments were stopped and the clay samples

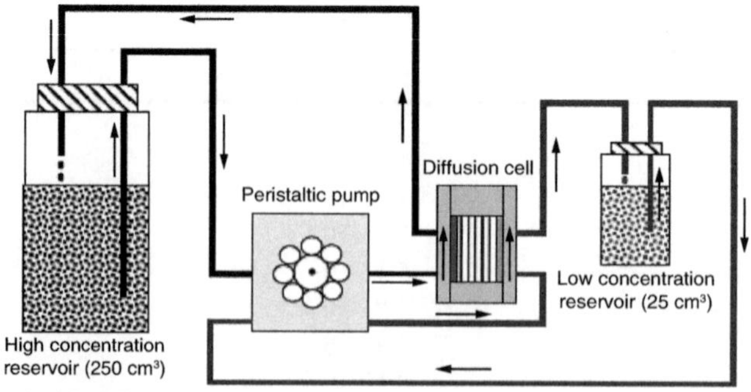

Fig. 1 Experimental set-up for the diffusion experiment at 25°C (Van Loon et al. 2003)

were removed from the cells. With the help of the abrasive peeling technique (Van Loon and Eikenberg 2005), the ^{233}U(VI) and HA diffusion profiles were determined. The peeled layers were extracted for ^{233}U(VI) content by 1 M HNO_3 (Roth, Karlsruhe, Germany) and for ^{14}C-HA content by 1 M NaOH (Merck, Darmstadt, Germany). The tracer concentrations in the extracts were determined by liquid scintillation counting (mod. TriCarb 3100 TR, Perkin Elmer, Freiburg, Germany) using Ultima Gold (Perkin Elmer) as scintillation cocktail.

Data Processing

The theoretical background of diffusion experiments has been described previously (Van Loon et al. 2003; Van Loon and Soler 2004). For HTO, the effective diffusion coefficient D_e [m^2/s] and the rock capacity factor α [–] were determined by modeling the through-diffusion flux of HTO taking into account the time-history of the HTO concentration in the low concentration reservoir. The rock capacity factor α is defined as:

$$\alpha = \varepsilon + \rho \cdot K_{\mathrm{d}} \tag{1}$$

where ε [–] is the diffusion-accessible porosity; ρ [kg/m^3] the dry bulk density and K_d [m^3/kg] is the sorption distribution coefficient. For non-sorbing tracers such as HTO where $K_d = 0$ is assumed, α is equal to ε.

For ^{233}U(VI) and ^{14}C-HA, the values for the diffusion parameters (D_e, α, ε, K_d) were determined by fitting the experimental tracer distribution profiles in the clay.

All experimental results were evaluated using the commercial software COMSOL Multiphysics 3.5a (2008), where in the model together with the clay also the adjacent filter plates were taken into account. In case of HTO, the filter K_d value was assumed to be 0. For ^{233}U(VI) and ^{14}C-HA an interaction of the tracers with the filters was considered. The respective filter K_d values were obtained by sorption (25°C; K_d(^{233}U(VI)) = 7×10^{-5} m^3/kg; K_d(^{14}C-HA) = 1×10^{-4} m^3/kg) and extraction experiments (60°C; K_d(^{233}U(VI)) = 5×10^{-3} m^3/kg; K_d(^{14}C-HA) = 3.4×10^{-3} m^3/kg) and were included in the model. A detailed discussion of the filter parameters will be given in Joseph et al. (in preparation).

Results and Discussion

Experiments at 25°C

The experiments were conducted for three months. Within this time span no ^{233}U(VI) could be detected in the low concentration reservoirs of cell 1 and 2. However, at the end of the experiment diffused HA colloids (< 1 kD) were detected in the low concentration reservoir of cell 2.

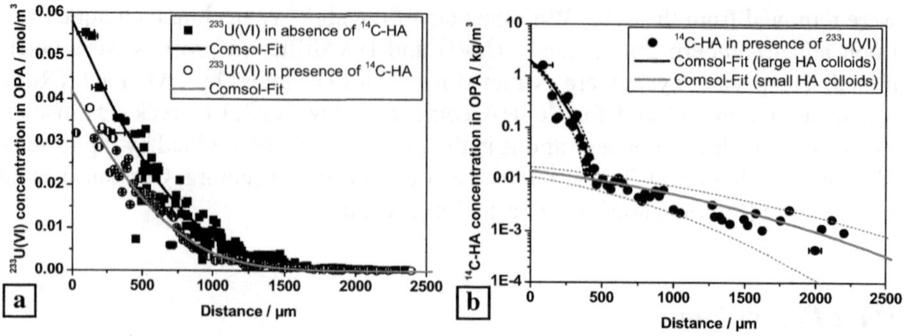

Fig. 2 Concentration profiles of ^{233}U(VI) in the absence (*filled symbols*) and presence (*open symbols*) of ^{14}C-HA (**a**) and of ^{14}C-HA in the presence of ^{233}U(VI) (**b**) in OPA at 25°C (Joseph et al. in preparation)

In Fig. 2a the ^{233}U(VI) diffusion profiles determined in the absence and presence of HA are shown. The comparison of the two data sets leads to the conclusion that in the presence of HA ^{233}U(VI) penetrates less the clay than in the absence of HA. Such a reduced diffusion was already described by Seida et al. (2010) for the Eu(III) diffusion in sedimentary rock in the presence of HA.

On the basis of the ^{233}U(VI) profiles the corresponding diffusion parameters were adjusted. Their values are summarized in Table 3. The K_d value for ^{233}U(VI) determined by batch sorption experiments using crushed OPA material (Joseph et al. 2011) is also shown in Table 3. The K_d value determined by the present diffusion studies agrees well with the former K_d value of the sorption experiments. Wu et al. (2009) determined for Np(V) in OPA a larger K_d value in diffusion experiments than in batch sorption measurements. The authors concluded that during the diffusion experiment Np(V) was partially reduced to Np(IV). Since in our study the K_d values determined from batch sorption and diffusion experiments are comparable, a reduction of U(VI) to U(IV) is excluded.

A comparison of the K_d as well as D_e values determined for ^{233}U(VI) in the absence and presence of HA shows, that the diffusion of ^{233}U(VI) is slightly hindered in the presence of HA molecules. However, based on the experimental uncertainties, it is concluded that HA does not have a significant effect on ^{233}U(VI) diffusion through compacted water-saturated OPA. In a previous study (Joseph et al. 2011) the system U(VI)/HA/OPA was investigated in terms of batch sorption experiments. There, also no influence of HA on the U(VI) interaction with OPA was observed. That result is now confirmed by diffusion experiments with intact OPA bore core samples.

In Fig. 2b the measured diffusion profile for HA in OPA at 25°C is shown. The experimental diffusion profile is indicative for the presence of two fractions of HA colloids, which are assigned to 1) a high molecular size HA colloid fraction (large HA colloids) and 2) a low molecular size HA colloid fraction (small HA colloids). Within three months the larger colloids diffused only about 500 µm into the OPA bore core due to restriction in the available pore space. However, the smaller HA

Table 3 Compilation of some further data of the experiments and of the best-fit parameter values. The data are based on the ^{233}U(VI) and ^{14}C-HA diffusion profiles and concentration data in the high and low concentration reservoir at 25°C (Joseph et al. in preparation)

	Cell 1 ^{233}U(VI)	Cell 2 ^{233}U(VI) (+ HA)	HA (large colloids)	HA (small colloids)
C_0(^{233}U(VI)) [mol/l]	1.07×10^{-6}	9.98×10^{-7}		
C_0(^{14}C-HA) [mg/l]			10.1	3
t [d]	89		87	
α [–]	61 ± 7	48 ± 7	309 ± 45	5 ± 1.2
ε [–] [a]	0.24 ± 0.02		0.20 ± 0.02	
ρ [kg/m^3]	2424		2392	
D_e [$\times0^{-12}$ m^2/s]	1.9 ± 0.4	1.2 ± 0.3	0.65 ± 0.25	0.4 ± 0.25
K_d [m^3/kg]	0.025 ± 0.003	0.020 ± 0.003	0.129 ± 0.018	0.002 ± 0.0005
K_d [m^3/kg] [b]	0.0222 ± 0.0004		0.129 ± 0.006	

[a] Determined by HTO through-diffusion
[b] Determined by sorption experiments (Joseph et al. 2011)

colloid fraction diffused through the entire OPA sample and therefore, could be detected in the low concentration reservoir. For both HA size fractions, the values for the diffusion parameters are presented in Table 3.

Joseph et al. (2011) determined the K_d value of HA with OPA by batch sorption experiments (cf. Table 3). This result is in very good agreement with the K_d value fitted for the large HA colloid fraction in the present diffusion experiment.

Experiments at 60°C

During the diffusion experiment no ^{233}U(VI) was detected in the low concentration reservoirs of cell 3 and 4. However, after one week, the solution in the low concentration reservoir of cell 4 turned yellow, which is an indication of the presence of HA. About 25 d after starting the diffusion experiment small HA colloids (< 1 kD) could clearly be detected in the low concentration reservoir.

The comparison of the ^{233}U(VI) diffusion profiles obtained at 25°C and 60°C (Fig. 3a) shows that at 60°C, the interaction of ^{233}U(VI) with OPA is stronger than at 25°C, and thus, an increased K_d value is expected for the higher temperature.

In Fig. 3b, the ^{233}U(VI) diffusion profiles at 60°C in the absence (open symbols) and presence of HA (filled symbols) are shown. The differences between both profiles are smaller compared to the analogous profiles obtained at 25°C. Thus, it is concluded that also at 60°C HA has a negligible effect on ^{233}U(VI) diffusion through OPA.

The HA diffusion through OPA was investigated at 60°C, too. The profile is very similar to that depicted in Fig. 2b and shows again the presence of at least

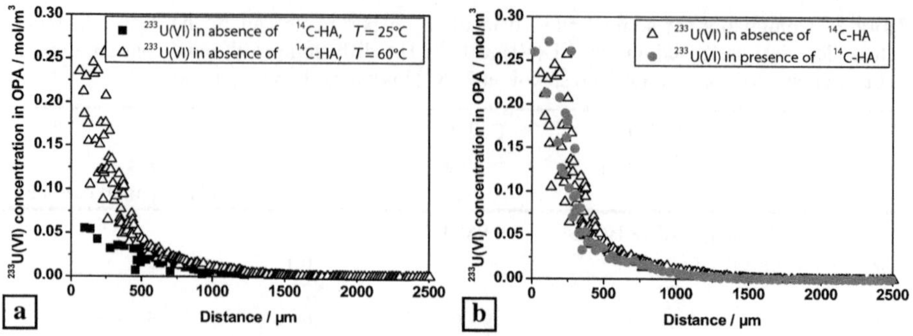

Fig. 3 Concentration profiles of ^{233}U(VI) in OPA samples at 25 and 60°C in the absence of HA (**a**) and at 60°C without and with HA in the system (**b**) (Joseph et al. in preparation)

two HA size fractions (data not shown). The profile obtained for the large HA colloid fraction is comparable to that shown for 25°C. However, the small HA colloid fraction seems to interact stronger with the clay at 60°C than at 25°C. The earlier detection of HA in the low concentration reservoir at 60°C in comparison to the experimental findings at 25°C indicates a higher D_e value for HA diffusion (small HA colloids) at 60°C.

Currently, the ^{233}U(VI) and HA profiles obtained at 60°C are being analyzed. The diffusion and distribution parameters are not available yet but will be published soon in a forthcoming manuscript (Joseph et al. in preparation).

Conclusions

Diffusion experiments with ^{233}U(VI) in pristine samples of consolidated OPA were carried out under anaerobic conditions. The influence of both HA and elevated temperature on the ^{233}U(VI) diffusion was investigated.

The experiments showed that, independent of temperature, the effect of HA on ^{233}U(VI) diffusion is negligible. The conclusion is based on the best-fit parameter values for K_d and D_e determined for both temperatures. At 60°C, a stronger interaction of ^{233}U(VI) with OPA could be observed than at 25°C.

The diffusion results for HA at 25 and 60°C showed that two distinct HA size fractions – a large- and a small-sized colloid fraction – were diffusing through the OPA samples. Within three months the high molecular size HA colloids diffused only about 500 µm in the clay, whereas the low molecular size HA colloids diffused through the OPA samples (thickness: 11 mm) and were detected in the low concentration reservoir.

In general, the results for diffusion of U(VI) and HA (large HA colloids) at 25°C in intact OPA samples show that the parameter values for (instantaneous and reversible) sorption are in good agreement with K_d values from batch sorption experiments using crushed OPA material.

Acknowledgement The authors would like to thank the BGR and KIT-INE (especially Ch. Marquardt) for providing and preparing the OPA samples, C. Nebelung for helpful discussions concerning LSC and Ch. Müller for technical support. The Federal Ministry of Economics and Technology funded this work (02E10156).

References

Appelo CAJ, Van Loon LR, Wersin P (2010) Multicomponent diffusion of a suite of tracers (HTO, Cl, Br, I, Na, Sr, Cs) in a single sample of Opalinus Clay. Geochim. Cosmochim. Acta 74: 1201–1219

Brasser T, Droste J, Müller-Lyda I, Neles JM, Sailer M, Schmidt G, Steinhoff M (2008) Endlagerung wärmeentwickelnder radioaktiver Abfälle in Deutschland, GRS – 247. Öko-Institut and Gesellschaft für Anlagen- und Reaktorsicherheit (GRS), Braunschweig

Claret F, Schäfer T, Bauer A, Buckau G (2003) Generation of humic and fulvic acid from Callovo-Oxfordian Clay under high alkaline conditions. Sci. Total Environ. 317: 189–200

COMSOL Multiphysics 3.5a (2008) Finite-element software package. http://www.comsol.com

Descostes M, Blin V, Bazer-Bachi F, Meier P, Grenut B, Radwan J, Schlegel ML, Buschaert S, Coelho D, Tevissen E (2008) Diffusion of anionic species in Callovo-Oxfordian argillites and Oxfordian limestones (Meuse/Haute-Marne, France). Appl. Geochem. 23: 655–677

Joseph C, Schmeide K, Sachs S, Brendler V, Geipel G, Bernhard G (2011) Sorption of uranium(VI) onto Opalinus Clay in the absence and presence of humic acid in Opalinus Clay pore water. Chem. Geol. 284: 240–250

Joseph C, Van Loon LR, Jakob A, Schmeide K, Sachs S, Bernhard G (in preparation) Diffusion of U(VI) in Opalinus Clay: Influence of temperature and humic acid. Geochim. Cosmochim. Acta

Křepelová A, Sachs S, Bernhard G (2006) Uranium(VI) sorption onto kaolinite in the presence and absence of humic acid. Radiochim. Acta 94: 825–833

Maes N, Salah S, Jacques D, Aertsens M, Van Gompel M, De Canniere P, Velitchkova N (2008) Retention of Cs in Boom Clay: Comparison of data from batch sorption tests and diffusion experiments on intact clay cores. Phys. Chem. Earth. 33: S149–S155

Nagra (2002) Projekt Opalinuston, Synthese der geowissenschaftlichen Untersuchungsergebnisse, Technical Report NTB 02-03. Nagra, Wettingen, Switzerland

Pearson FJ (1998) Opalinus Clay experimental water: A1Type, Version 980318, PSI Internal report TM-44-98-07. Paul Scherrer Institut, Villigen PSI, Switzerland

Pompe S, Brachmann A, Bubner M, Geipel G, Heise KH, Bernhard G, Nitsche H (1998) Determination and comparison of uranyl complexation constants with natural and model humic acids. Radiochim. Acta 82: 89–95

Sachs S, Schmeide K, Brendler V, Křepelová A, Mibus J, Geipel G, Heise KH, Bernhard G (2004) Investigation of the Complexation and the Migration of Actinides and Non-radioactive Substances with Humic Acids under Geogenic Conditions: Complexation of Humic Acids with Actinides in the Oxidation State IV Th, U, Np, Wissenschaftlich-Technische Berichte, FZR-399. Forschungszentrum Rossendorf, Dresden

Schmeide K, Bernhard G (2009) Redox stability of neptunium(V) and neptunium(IV) in the presence of humic substances of varying functionality. Radiochim. Acta 97: 603–611

Schmeide K, Sachs S, Bubner M, Reich T, Heise KH, Bernhard G (2003) Interaction of uranium(VI) with various modified and unmodified natural and synthetic humic substances studied by EXAFS and FTIR spectroscopy. Inorg. Chim. Acta 351: 133–140

Seida Y, Terashima M, Tachi Y, Iijima K, Nakazawa T, Yamada M, Yui M (2010) Sorption and diffusion of Eu in sedimentary rock in the presence of humic substance. Radiochim. Acta 98: 703–709

Van Loon LR, Eikenberg J (2005) A high-resolution abrasive method for determining diffusion profiles of sorbing radionuclides in dense argillaceous rocks. Appl. Radiat. Isot. 63: 11–21

Van Loon LR, Soler JM (2004) Diffusion of HTO, $^{36}Cl^-$, $^{125}I^-$, and $^{22}Na^+$ in Opalinus Clay: Effect of Confining Pressure, Sample Orientation, Sample Depth and Temperature, PSI-Bericht Nr. 04–03. Paul Scherrer Institut, Villigen, Switzerland

Van Loon LR, Soler JM, Bradbury MH (2003) Diffusion of HTO, $^{36}Cl^-$ and $^{125}I^-$ in Opalinus Clay samples from Mont Terri – Effect of confining pressure. J. Contam. Hydrol. 61: 73–83

Wolery TJ (1992) EQ3/6, A Software Package for the Geochemical Modeling of Aqueous Systems, UCRL-MA-110662 Part I, Lawrence Livermore National Laboratory

Wu T, Amayri S, Drebert J, Van Loon LR, Reich T (2009) Neptunium(V) sorption and diffusion in Opalinus Clay. Environ. Sci. Technol. 43: 6567–6571

Thermodynamic Data Dilemma

Broder J. Merkel

Abstract. Calculations practically are the only option to determine the uranium speciation because there is a lack of analytical techniques to analyze uranium species in natural waters. A comparison of common data sets distributed with geochemical codes shows that results differ considerable. Several important species are often not included. With respect to uranium minerals the situation is similar with differing log-*k* values for some minerals and not considering certain secondary uranium minerals that might be important for limiting uranium concentrations. Most problematic is the situation for data concerning ion exchange and surface complexation although such data is available in literature. Thus it can be stated that all available databases are not fit for purpose. Users of geochemical codes have to cope with the problem compiling or completing the data sets by deleting certain data and adding other. However, the difficulty is to define under which conditions thermodynamic data should be selected or deselected.

Introduction

Transport of Uranium in groundwater is controlled by speciation, formation of secondary minerals and sorption on solids. Modeling uranium transport thus requires thermodynamic data for species formation, mineral solubility and sorption. Uranium speciation is challenging due to the formation of negative, positive, and zero-valent species exhibiting very different sorption behavior on rocks and minerals e.g. quartz, granite, and carbonate, but as well on clay minerals and iron hydroxides.

Broder J. Merkel
Technische Universität Bergakademie Freiberg, 09599 Freiberg, Germany
E-mail: Merkel@geo.tu-freiberg.de

B. Merkel, M. Schipek, *The New Uranium Mining Boom*
DOI 10.1007/978-3-642-22122-4 © Springer 2011

The system UO_2–H_2O itself is well described. However the uranium hydroxide species are not the only decisive ones in natural groundwater. As carbon dioxide is present in any aquatic environment binary carbonate complexes or ternary complexes such as $(UO_2)_2CO_3(OH)_3^-$ are also formed. If other ligands such as sulfate (in acid mine waters), phosphate, chloride (in brines) or fluoride (in geothermal waters) are present the situation gets more complicated. In addition, ternary complexes of uranyl, carbonate and alkaline earth metals (Ca, Mg, Sr, Ba) are important and often dominant species in natural and mine-related waters (Bachmaf et al. 2008; Bernhard et al. 2001; Dong & Brooks 2006). Some databases deny the ternary $(UO_2)_2CO_3(OH)_3^-$ species and most data bases do not consider the ternary alkaline earth uranyl carbonate species and the uranyl arsenic species at all. Thus results of speciation modeling on the one hand strongly depend on the database chosen In many cases, results are even unreliable because databases do not reflect state of the art as they do not contain some of the species listed in Table 1.

On the other hand one has to consider that the criteria for accepting or denying thermodynamic data are not well defined. Minerals and their names are officially approved by the Commission on New Minerals, Nomenclature and Classifications of the International Mineralogical Association. But this is not the case for thermodynamic data for minerals. Some solubility products are calculated from experiments from historic times that were not aimed for calculating thermodynamic data and correction to ionic strength of zero is thus critical and rather difficult. With respect to aquatic species the situation is even worse because no methods are agreed on how the existence of a species is to be approved. A great advantage in comparison to older versions of thermodynamic data bases is that the log-*k* values are now mostly related to an ionic strength of zero. However, some data is still in doubt because the recalculation to zero ionic strength is not easy and depends on the availability of data and on the model used. This is in particular a problem when a buffer solution of high ionic strength has been used for the respective experiments.

Table 1 Uranium VI species that may be dominant under certain pH and redox conditions

UO_2^{2+}
UO_2OH^+, $UO_2(OH)_2^0$, $UO_2(OH)_3^-$
$UO_2CO_3^0$, $UO_2(CO_3)_2^{2-}$, $UO_2(CO_3)_3^{4-}$
$(UO_2)_2CO_3(OH)_3^-$
$UO_2SO_4^0$, $UO_2(SO_4)_2^{2-}$
$UO_2H_3PO_4^{2+}$, $UO_2H_2PO_4^+$, $UO_2HPO_4^0$, $UO_2PO_4^-$
$UO_2(H_2PO_4)_2^0$, $UO_2(HPO_4)_2^{2-}$, $UO_2(H_2PO_4)(H_3PO_4)^+$
UO_2F^+, $UO_2F_2^0$, $UO_2F_3^-$, $UO_2F_4^{2-}$
UO_2Cl^+
$UO_2H_3SiO_4^+$
$Ca_2(UO_2)(CO_3)_3^0$, $CaUO_2(CO_3)_3^{2-}$
$Mg_2(UO_2)(CO_3)_3^0$, $MgUO_2(CO_3)_3^{2-}$
$Sr_2(UO_2)(CO_3)_3^0$, $SrUO_2(CO_3)_3^{2-}$
$Ba_2(UO_2)(CO_3)_3^0$, $BaUO_2(CO_3)_3^{2-}$
$UO_2H_2AsO_4^+$, $UO_2HAsO_4^0$, $UO_2AsO_4^-$
$UO_2(H_2AsO_4)_2^0$

Species

Comparing 6 databases (NEA, LLNL, Hatches, SIT, Minteq.v4, and Wateq4f) and assuming a simple U-Ca-C system at pH 7 three databases (Minteq.v4.dat; SIT.dat; Wateq4f.dat) deny the $(UO_2)_2CO_3(OH)_3^-$ complex and assume $UO_2(CO_3)_2^{2-}$ to be the dominant species. Furthermore, the respective dominant uranium species is assumed at concentrations between 1.714e-05 and 4.000e-05 mol/l. Assuming a more realistic water composition as given in Table 2, results are even more puzzling (Fig. 1).

The most extreme results are achieved with Minteq.4.dat and Wateq4F.dat calculating that $UO_2(HPO_4)_2^{2-}$ is the only dominating species. Assuming that this water would pass through an aquifer containing some FeOOH, uranium should be sorbed readily. A completely different picture is drawn by the other databases with several dominant species, some of them anionic but some also zero-charged and thus not prone to anion exchange on FeOOH. While NEA_2007 and SIT model only a minor fraction of uranium phosphate species, LLNL assumes nearly 50% of the uranium bound with phosphate. Nevertheless, it is pointless to discuss which database gives the most realistic result in this particular case. It is fairly sure that all results are incorrect because the zero-charged $Ca_2(UO_2)(CO_3)_3$ species is not contained in any of the before mentioned data bases. Thus it is the duty of the modeler to add the reaction equation and log-*k* value in the respective database, which can be written in the following way (taken from CHESS database)

$$2\,Ca^{2+} + UO_2^{2+} + 3\,HCO_{3-} - 3\,H^+ = Ca_2(UO_2)(CO_3)_3^0 \quad \text{log-}K. \quad -1.15 \quad (1)$$

For the sake of completeness the anionic Ca-U-carbonate species $CaUO_2(CO_3)_3^{2-}$ and the respective magnesium species $Mg_2(UO_2)(CO_3)_3^0$ and $MgUO_2(CO_3)_3^{2-}$ and strontium species $Sr_2(UO_2)(CO_3)_3^0$ and $SrUO_2(CO_3)_3^{2-}$ have to be added as well. However, in this case the log-*k* values in particular for the zero-valent species of Mg and Sr are still under discussion (Bachmaf et al. 2008; Bernhard et al. 1996, 2001; Dong & Brooks 2006; Geipel et al. 2008; Nair & Merkel 2011). Figure 2 shows that the mere addition of the $Ca_2(UO_2)(CO_3)_3^0$ species according to Eq. 1 to the NEA_2007 database changes the outcome significantly, because now the zero-valent = $Ca_2(UO_2)(CO_3)_3^0$ is the most dominant species. Adding the other alkaline earth uranium carbonate species would not change the picture too much. But this depends upon the questions on the log-*k* of $Mg_2(UO_2)(CO_3)_3^0$ complex, the concentrations of Ca and Mg and the pH.

Table 2 Composition of water assumed in mg/l

Temp	pH	Ca	Mg	Na	K	HCO_3	SO_4	NO_3	Cl	PO_4	As	U
12°C	6.5	40	15	10	5	275	35	15	20	2.5	25	200

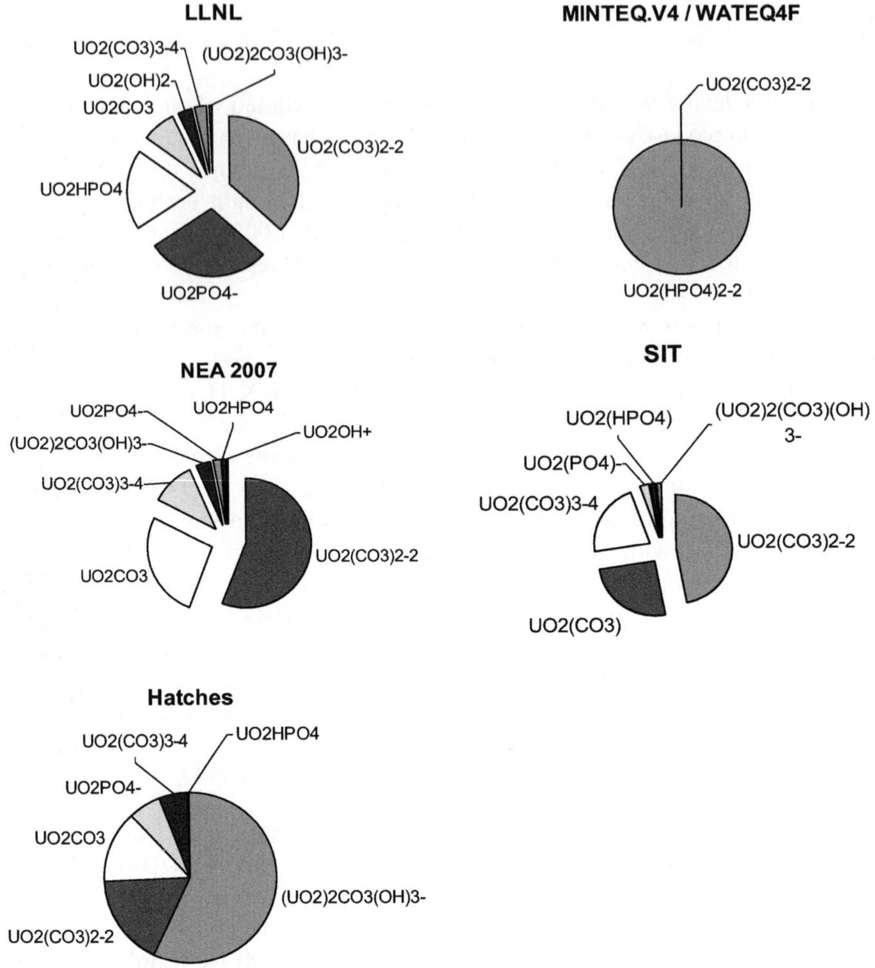

Fig. 1 Distribution of species calculated by Phreeqc (Parkhurst & Appelo 1999) using the test case given in Table 2 and six databases (only species > 1·10-9 mol/l shown)

Fig. 2 Distribution of species for the solution given in Table 2 by using NEA_2007 adding the species $Ca_2(UO_2)(CO_3)_3^0$ and $UO_2AsO_4^-$

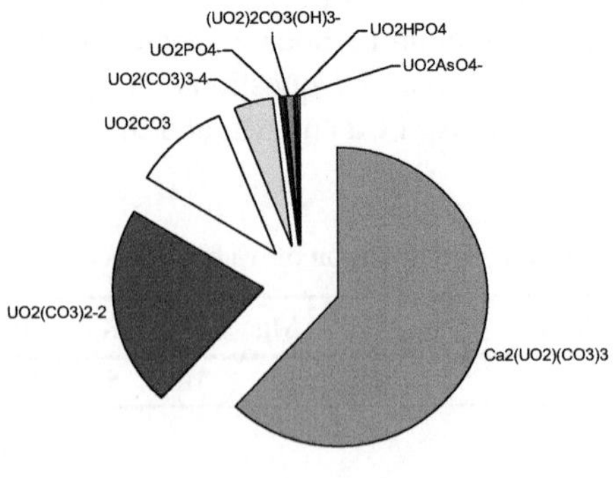

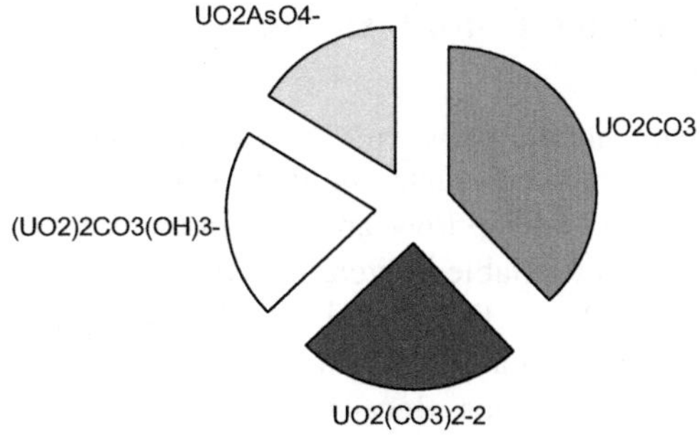

Fig. 3 Species distribution according to solution of Table 2 with the following changes: 1 μmol/l As and U, Ca 4 mg/l, TIC 1.5 mmol/l

The formation of arsenic-uranium species itself is beyond dispute. Yet, it is considered only by NEA_2007; and the most important species $UO_2AsO_4^-$ at pH 6 is not contained at all because so far no log-*k* for the $UO_2AsO_4^-$ has been published (Gezahegne et al. 2009). Based on the column experiments of (Schulze 2011) a log-k was estimated as 4.99 ± 0.2 according to Eq. 2

$$U^{4+} + AsO_4{}^{3-} + 2\,H_2O - 4\,H^+ - 2e\text{-} = UO_2AsO_4{}^- \quad \log\_k\ 4.99 \tag{2}$$

The species distribution in Fig. 2 shows that $UO_2AsO_4^-$ is not a dominant species. This is due to the concentrations of uranium (200 mg/l) and arsenic (25 mg/l). Changing the arsenic and uranium solution to equimolar concentrations of 1 μmol/l and Ca to only 4 mg/l has a significant impact on the speciation as can be seen by Fig. 3.

With respect to U-P species, the 6 above-mentioned databases present rather different species and thermodynamic data, which leads to completely diverging results. All databases are anything but consistent with respect to U-P species.

Secondary Uranium Minerals

In water containing uranium and arsenic secondary minerals such as $(UO_2)_3(AsO_4)_2 \cdot 12H_2O$ (Troegerite), $(UO_2)H(AsO_3)$ (Chadwickite), uranyl polyarsenate $K_2[(UO_2)As_2O_7]$ (Alekseev et al. 2007) or uranyl phosphate arsenates may form, but thermodynamic data is not given in the common databases.

Adding alkaline earth metal and alkali metal to a solution the number of minerals with a rather low solubility increases. Although it is known that the solubility of minerals such as $K(UO_2)_3(AsO_4)(OH)_4{\bullet}H_2O$ (Nielsbohrite) (Walenta et al. 2009), or $Ca(UO_2)[(UO_2)_3(AsO_4)_2(OH)_2] \cdot (OH)_2 \cdot 6H_2O$ (Arsenuranylite) and $U(OH)_4[(UO_2)_3\ (OH)_2(AsO_4)_2]$. $4H_2O$, (Arsenovanmeersscheite) (Frost et al. 2009) is rather low, solubility products are neither available in the above mentioned databases nor in literature. Therefore modeling of uranium and arsenic concentrations might be erroneous because of missing thermodynamic data of secondary uranium minerals with low solubility.

Surface Complexation

The situation is even worse with respect to thermodynamic data of surface complexation constants for dominant uranium species on important solids such as quartz, calcite, iron- and aluminum hydroxides, and clay minerals. Although some data is available in literature (Bachmaf et al. 2008; Catalano & Brown 2005; Davis et al. 2004; Dong et al. 2005; Fox et al. 2006; Huber & Lützenkirchen 2009; Sherman et al. 2008) this data has not jet been added to databases with only very few exceptions. Because of that, the process of creating a transport model taking into account sorption based on surface complexation, always starts from scratch and is time consuming and tedious. This might be a reason why inappropriate approaches such as sorption isotherms and Kd-concept-based models are still in use (Merkel & Nair, 2011).

Conclusions

Database integrity aims to ensure that data in a database is accurate, valid, and consistent. Certain attempts have been made in the past to compare and evaluate thermodynamic databases (Chandratillake et al. 1988; Grenthe 1992; Guillaumont et al. 2003). The major problem is that such attempts have an effect only for a while, because all projects lack a continuous financial support and reliable criteria selecting and de-selecting thermodynamic data A stringent necessity can be stated to develop and maintain a consistent database and valid database for species, minerals and surface complexation processes. This database has to be in a universal format that can be imported for the most frequently used geochemical codes. Furthermore, the database has to include literature references, uncertainties, and clues about how ionic strength has been adjusted.

References

Alekseev, E. V., Krivovichev, S. V. & Depmeier, W., 2007. K2[(UO2)As2O7] – The first uranium polyarsenate. Zeitschrift für Anorganische und Allgemeine Chemie, 633(8), 1125.

Bachmaf, S., Planer-Friedrich, B. & Merkel, B., 2008. Effect of sulfate, carbonate, and phosphate on the uranium(VI) sorption behavior onto bentonite. Radiochimica Acta, 96(6), 359–366.

Bernhard, G., Geipel, G., Brendler, V. & Nitsche, H., 1996. Speciation of Uranium in Seepage Waters of a Mine Tailing Pile Studied by Time-Resolved Laser-Induced Fluorescence Spectroscopy (TRLFS). Radiochimica Acta, 74, 87.

Bernhard, G., Geipel, G., Reich, T., Brendler, V., Amayri, S. & Nitsche, H., 2001. Uranyl(VI) carbonate complex formation: Validation of the Ca2UO2(CO3)3(aq.) species. Radiochimica Acta, 89(8), 511.

Catalano, J. G. & Brown, J. G. E., 2005. Uranyl adsorption onto montmorillonite: Evaluation of binding sites and carbonate complexation. Geochimica et Cosmochimica Acta, 69(12), 2995.

Chandratillake, M. R., Newton, G. W. A. & Robinso, V. J., 1988. Comparison of thermodynamic databases used in geochemical modelling. In: Nuclear Science and Technilogy (ed Cimmunities, C. o. t. E.), pp. 19.
Davis, J. A., Meece, D. E., Kohler, M. & Curtis, G. P., 2004. Approaches to surface complexation modeling of Uranium(VI) adsorption on aquifer sediments. Geochimica et Cosmochimica Acta, 68(18), 3621.
Dong, W., Ball, W. P., Liu, C., Wang, Z., Stone, A. T., Bai, J. & Zachara, J. M., 2005. Influence of calcite and dissolved calcium on uranium(VI) sorption to a Hanford subsurface sediment. Environmental Science and Technology, 39(20), 7949.
Dong, W. & Brooks, S. C., 2006. Determination of the formation constants of ternary complexes of uranyl and carbonate with alkaline earth metals (Mg2+, Ca2+, Sr 2+, and Ba2+) using anion exchange method. Environmental Science and Technology, 40(15), 4689.
Fox, P. M., Davis, J. A. & Zachara, J. M., 2006. The effect of calcium on aqueous uranium(VI) speciation and adsorption to ferrihydrite and quartz. Geochimica et Cosmochimica Acta, 70(6), 1379.
Frost, R. L., Cejka, J. & Dickfos, M. J., 2009. Raman spectroscopic study of the uranyl minerals vanmeersscheite U(OH)4[(UO2)3(PO4)2(OH)2]Â·4H2O and arsenouranylite Ca(UO2)[(UO2)3(AsO4)2(OH)2]Â·(OH)2Â·6H2O. Spectrochimica Acta – Part A: Molecular and Biomolecular Spectroscopy, 71(5), 1799.
Geipel, G., Amayri, S. & Bernhard, G., 2008. Mixed complexes of alkaline earth uranyl carbonates: A laser-induced time-resolved fluorescence spectroscopic study. Spectrochimica Acta Part A: Molecular and Biomolecular Spectroscopy, 71(1), 53.
Gezahegne, W., Bachmaf, S., Geipel, G., Planer-Friedrich, B. & Merkel, B., 2009. Detection of uranylarsenates in acidic and alkaline solutions with time resolved laser-induced fluorescence spectroscopy (TRLFS). Geochimica et Cosmochimica Acta, 73(13).
Grenthe, I., 1992. Chemical Thermodynamics of Uranium, Vol. 1. Elsevier, North Holland, Amsterdam, London, New York, Tokyo.
Guillaumont, R., Fanghanel, T., Fuger, J., Grenthe, I., Neck, V., Palmer, D. A., Rand, M. H., Mompean, F. J., Illemassene, M., Domenech-Orti, C. & Ben-Said, K., 2003. Update on the Chemical Thermodynamics of Uranium, Neptunium, Plutonium, Americium and Technetium (Chemical Thermodynamics). 5.
Huber, F. & Lützenkirchen, J., 2009. Uranyl Retention on Quartz-New Experimental Data and Blind Prediction Using an Existing Surface Complexation Model. Aquatic Geochemistry, 1.
Merkel, B. & Nair, S., 2011. Impact if speciation and sorption on migration of uranium in groundwater. In: Nachhaltigkeit und Langzeitaspekte bei der Sanierung von Uranbergbau- und Aufbereitungsstandorten (ed Paul, M.), Wismut GmbH.
Nair, S. & Merkel, B. J., 2011. Effect of Mg-Ca-Sr on the sorption behavior of uranium(VI) on silica. In: The new Uranium Mining boom – Challenge and lessons learned (eds Merkel, B. & Schipek, M.) Uranium Mining and Hydrogeology VI, Springer, Freiberg.
Parkhurst, D. L. & Appelo, C. A. J., 1999. User's guide to PHREEQC (version 2).
Romy Schulze, B. M., 2011. Sorption of Uranium on Iron Coated Sand in the presence of Arsenate, Selenate, and Phosphate. In: The new Uranium Mining boom – Challenge and lessons learned (eds Merkel, B. J. & Schipek, M.) Uranium Mining and Hydrogeology VI, Springer.
Sherman, D. M., Peacock, C. L. & Hubbard, C. G., 2008. Surface complexation of U(VI) on goethite (-FeOOH). Geochimica et Cosmochimica Acta, 72(2), 298.
Walenta, K., Hatert, F., Theye, T., Lissner, F. & RÃller, K., 2009. Nielsbohrite, a new potassium uranyl arsenate from the uranium deposit of Menzenschwand, southern Black Forest, Germany. European Journal of Mineralogy, 21(2), 515.

Formation of $(Ba,Ra)SO_4$ Solid Solutions – Results from Barite (Re)Precipitation and Coprecipitation Experiments

Volker Metz, Yoav O. Rosenberg, Dirk Bosbach, Melanie Böttle, Jiwchar Ganor

Abstract. Retention of $^{226}Ra^{2+}$ by barite in aqueous solution is studied in a barite (re)precipitation experiment with $^{226}Ra^{2+}$ doped barite suspensions at $I = 0.1$ mol·(kg $H_2O)^{-1}$ at ambient temperature. After about two years a steady state is achieved, demonstrating that $^{226}Ra^{2+}$(aq) concentration is controlled by the solubility of a $(Ba,Ra)SO_4$ solid solution and several orders of magnitude below the Ra^{2+} solubility with respect to a pure $RaSO_4$(s) endmember. Results of the barite (re)precipitation experiment are compared to recent $(Ba,Ra)SO_4$ coprecipitation experiments by Rosenberg and co-workers at ionic strength $I = 0.7 - 7.0$ mol·(kg $H_2O)^{-1}$.

Volker Metz
Institute for Nuclear Waste Disposal (INE), Karlsruhe Institute of Technology, Hermann-von-Helmholtz-Platz 1, 76344 Eggenstein-Leopoldshafen, Germany

Yoav O. Rosenberg
Department of Geological and Environmental Sciences, Ben-Gurion University of the Negev, P.O.B 653, Beer Sheva 84105, Israel

Dirk Bosbach
Institute of Energy and Climate Research – Safety Research and Reactor Technology (IEK-6), Research Centre Jülich, 52425 Jülich, Germany

Melanie Böttle
Institute for Nuclear Waste Disposal (INE), Karlsruhe Institute of Technology, Hermann-von-Helmholtz-Platz 1, 76344 Eggenstein-Leopoldshafen, Germany

Jiwchar Ganor
Department of Geological and Environmental Sciences, Ben-Gurion University of the Negev, P.O.B 653, Beer Sheva 84105, Israel

B. Merkel, M. Schipek, *The New Uranium Mining Boom*
DOI 10.1007/978-3-642-22122-4

Background and Motivation

The relevance of radium as the major principal radiation risk from uranium mining, milling residues and tailings has been investigated for decades (e.g. reviews of Abdelouas 2006; EPA 2008; IAEA 2003). Besides its role related to uranium mining and ore processing, radium is an abundant Naturally Occurring Radioactive Material, NORM, in many aquifers around the world (Rosenberg et al. 2011a, b and reference therein). While radium removal is technologically feasible (e.g., ion exchange, reverse osmosis desalination) it yields by-products referred to as Technologically Enhanced NORM, TENORM, which require suitable disposal (EPA 2008). The concern regarding the fate of radium TENORMs in high ionic strength environments in general and evaporitic systems in particular is expressed in two recent technical IAEA (2003) and EPA (2008) reports. Evaporation ponds for the management of residues solutions, containing high levels of Ra, may be found in the industries of geothermal energy production, oil and gas production, coal mining (settling ponds) and brackish water treatment plants. Salts precipitating in these ponds may pose a potential radiological health concern.

^{226}Ra is a critical radionuclide with respect to the long-term safety of various types of nuclear waste repositories. Hence, the fate of radium has been deserved special attention in safety case studies and performance assessment, PA, studies for planned high level waste disposal in crystalline host rocks (e.g., in Japan, Sweden, Switzerland – see Berner 1992; Berner and Curti 2002; Grandia et al. 2008; SKB 2008) and low-level waste disposal in rock salt (e.g., in Germany). In the recent safety assessment “SR Can” for a potential high level waste repository in Sweden (SKB 2008) it is shown that ^{226}Ra is one of the main contributors to radiological dose in some of the studied PA scenarios. In an exemplary scenario ^{226}Ra will become a dominant dose contributor after 50,000 years. The simulated dominance of the ^{226}Ra dose is directly dependent on the aqueous concentration of radium in the near-field of the nuclear waste.

The knowledge about the solubility behavior of radium is quite uncertain depending on the process considered. Thus, most PA calculations performed for ^{226}Ra assume that its solubility is limited by precipitation of RaSO4(s) in the range of 10^{-7} mol (kg H_2O)$^{-1}$. Abundant information from early radiochemical experimental studies and recent studies on natural systems as well as anthropogenic systems indicate that radium readily forms solid solutions with barite. It is expected that the formation of $(Ba,Ra)SO_4$ solid solutions reduces the maximal radium concentration by several orders of magnitude in comparison to the solubility with respect to a pure $RaSO_4$(s) endmember. Retention of radium by (radio)barite may occur via coprecitation or via (re)precipitation processes.

In their classical study on $(Ba,Ra)SO_4$ coprecipitation Doerner and Hoskins (1925) derived an analytical solution to calculate the concentration-based apparent partition coefficient, $K'_{\mathrm{D,barite}}$:

$$K'_{\mathrm{D,barite}} = \frac{\ln\left(\mathrm{Ra}_t^{2+}/\mathrm{Ra}_0^{2+}\right)_{\mathrm{aq}}}{\ln\left(\mathrm{Ba}_t^{2+}/\mathrm{Ba}_0^{2+}\right)_{\mathrm{aq}}},$$

where Ra_i and Ba_i are quantities (mol) of the Ra^{2+} and Ba^{2+} in the solution, respectively, and the subscripts t and 0 refer to time $= t$ and 0, respectively. The values for K'_D reported for barite are usually in the range 1.0 to 2.0 (Rosenberg et al., 2011b and references therein). Doerner and Hoskins (1925) considered the value of 1.8 as representing the thermodynamic value since it was evaluated from very slow coprecipitation experiments.

If a $(Ba,Ra)SO_4$ solid solution is formed by *coprecipitation* from an aqueous solution containing Ba, Ra and sulfate, the concept of solid solution/aqueous solution equilibria as well as the relevant activities[1] of aqueous species and species in the solid are straight forward to apply. Activities of aqueous species in concentrated solutions are accurately described by the Pitzer model (Pitzer 1973, 1979). Paige et al. (1998) measured Pitzer parameters for the $RaSO_4$-H_2SO_4-H_2O system. Recently, Pitzer parameters for the highly important $RaCl_2$ interaction were estimated by Rosenberg et al. (2011a, b) studied the $(Ba,Ra)SO_4$ coprecipitation in evaporation batch experiments at ionic strength (I) ranging from 0.7 to 7.0 mol $(kg\ H_2O)^{-1}$. In their experiments, they determined a concentration-based effective partition coefficient, $K'_{D,barite}$, of 1.04 ± 0.01, which was significantly lower than the value for relatively diluted solutions of Doerner and Hoskins (1925), i.e. $K'_{D,barite} = 1.8 \pm 0.1$.

In case of aqueous radium contacting a preexisting barite, there will be exchange occurring at the barite surface. It is less clear whether formation of a $(Ba,Ra)SO_4$ solid solution will occur via dissolution and *(re)precipitation* of the complete barite crystals or will be the solid solution process occur only in the surface regions. Curti et al. (2010) investigated the uptake of $^{226}Ra^{2+}$ by barite by means of (re)precipitation experiments in pure H_2O, 0.1 µmol·$(kg\ H_2O)^{-1}$ Na_2SO_4 solution and 0.01 mol·$(kg\ H_2O)^{-1}$ NaCl solution containing 1 µmol·$(kg\ H_2O)^{-1}$ Na_2SO_4 and 2 µmol·$(kg\ H_2O)^{-1}$ $NaHCO_3$, respectively. Unfortunately, the studied (radio)barite/solution systems did not reach steady state even in their longest experiment (120 days). The range of $K'_{D,barite}$ reported by Curti et al. (2010) is relatively low ($K'_{D,barite} = 0.12$ to 0.61) and imply a relative enrichment of radium in the liquid phase and not in the solid phase. Using barite (re)precipitation batch experiments in 0.1 mol·$(kg\ H_2O)^{-1}$ NaCl solution, Bosbach et al. (2010) showed that the aqueous radium concentration is controlled by the solubility of a $(Ba,Ra)SO_4$ solid solution and many orders of magnitude below the $^{226}Ra^{2+}$ concentration controlled by $RaSO_4(s)$. However, it remained questionable, whether a solid solution/aqueous solution equilibrium was established during the observation period (436 days) of their batch experiments.

In the present study one of the experiments of Bosbach et al. (2010) is extended in order to demonstrate unambiguously that steady state conditions were achieved. The results of the (re)precipitation experiment are compared to the recent findings in (re)precipitation experiments of Curti et al. (2010) and coprecipitation experiments of Rosenberg et al. (2011b).

---

[1] In the present communication the term activity refers solely to the product of the *molality* and *activity coefficient* (i.e., the change in the chemical potential of a solute) and not as a synonym for *radioactivity*.

Experimental Results and Discussion

Uptake of $^{226}Ra^{2+}$ by barite was determined in a static batch type experiments at room temperature (20°C). The experiments were conducted with synthetic barium sulfate "ReagentPlus®" (purity = 99%; Sigma Aldrich®) suspended in 0.1 mol·(kg $H_2O)^{-1}$ NaCl solution. Except barite, no additional solid phase was detected in the suspension by means of SEM-EDS and XRD (Figs. 1 and 2). 10 mg of the barite powder was pre-equilibrated for one week in 10 ml NaCl solution (barite surface/solution volume ratio of 0.001 m^2/ml). After the pre-equilibration 9 ml of a $^{226}Ra(NO_3)_2$ stock solution to the suspension to have a radium doping of

Fig. 1 SEM images of Sigma Aldrich barite sample used in (re)precipitation experiments (*left*) and solid phases formed in coprecipitation experiments Rosenberg et al. (2011b; *right*)

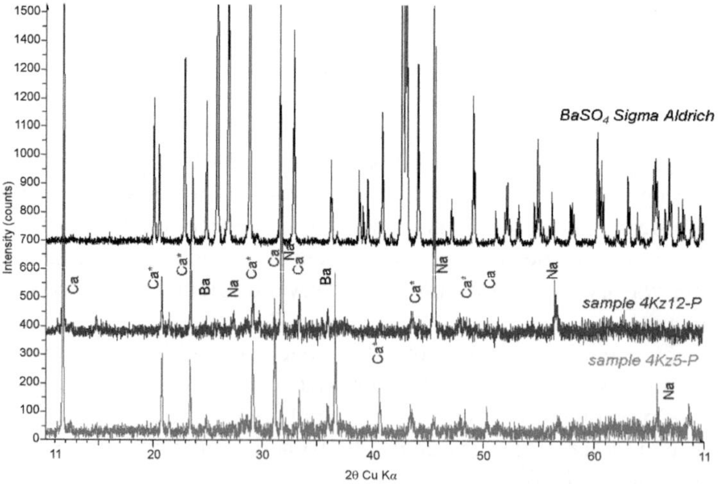

Fig. 2 XRD patterns of Sigma Aldrich barite sample used in (re)precipitation experiments and solid phases formed in exemplary coprecipitation experiments 4Kz12-P and 4Kz5-P of Rosenberg et al. (2011b), respectively. Ba denotes barite reflexes, Ca denotes gypsum reflexes, Ca* denotes gypsum reflexes superposing Ba reflexes and Na denotes halite

$2.95 \cdot 10^{-7}$ mol·(kg H_2O)$^{-1}$. Concentrations of dissolved $^{226}Ra^{2+}$ were monitored for 678 days by means of α-spectrometry using a Canberra 74/01 analysis chamber with a PIPS detector and a S100 multi channel analyzator. Detection limit of the α-spectrometry was at 10^{-2} Bq/ml and the precision $\pm 10\%$ for activities at $\sim 10^2$ Bq/ml.

The temporal evolution of the apparent $^{226}Ra^{2+}$ radioactivity concentration in solution while in contact with barite is presented in Fig. 3. A significant decrease in aqueous $^{226}Ra^{2+}$ concentration by at least one order of magnitude within the first 50 days can be observed. According to the α-spectrometry data the ^{226}Ra concentration in solution decreased from 155 Bq/ml after 92 days to 87 ± 9 Bq/ml after 678 days. Within error, a constant steady state $^{226}Ra^{2+}$ concentration was achieved between 140 and 678 days (Fig. 3).

Based on the steady state $^{226}Ra^{2+}$ molalities of $\leq 1 \cdot 10^{-8}$ mol·(kg H_2O)$^{-1}$, the stoichiometry of the $Ra_xBa_{1-x}SO_4$ solid solution was calculated. Assuming an regular solid solution, the uptake of $^{226}Ra^{2+}$ by barite was controlled by the solubility product of the end members – barite and $RaSO_4$(s) – as well as the Guggenheim interaction parameter, a_0, for the $Ra_xBa_{1-x}SO_4$ solid solution. For the $(Ba,Ra)SO_4$ solid solution Zhu (2004) estimated a Guggenheim parameter of $a_0 = 0.36$, whereas Curti et al. (2010) derived relatively high values in the range of $a_0 = 1.5$ to 2.5. Using $a_0 = 0.36$ for complete re-equilibration with the bulk of the barite crystals a solid solution composition $Ra_{0.000131}Ba_{0.999869}SO_4$ was calculated, which would limit the $^{226}Ra^{2+}$ molality one order of magnitude below the molality measured in our experiment. About the same solid solution composition and $^{226}Ra^{2+}$ molality, i.e. $2 \cdot 10^{-9}$ mol·(kg H_2O)$^{-1}$, was computed for the assumption of

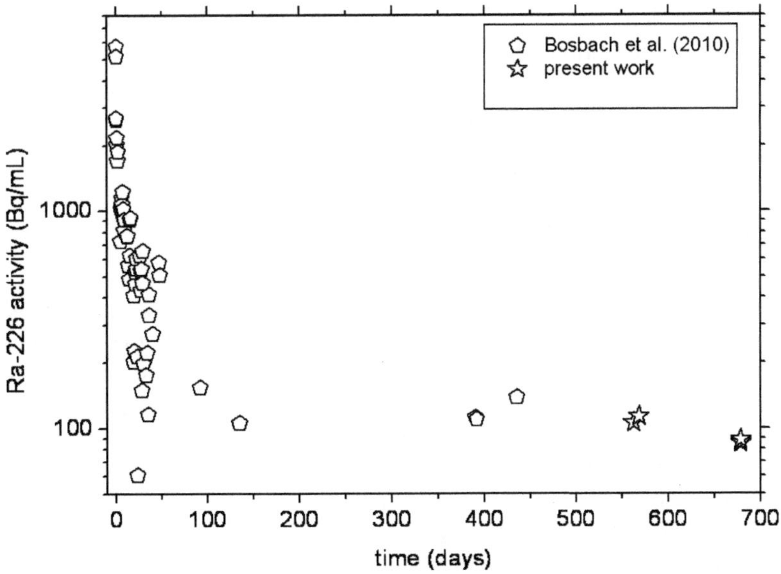

Fig. 3 Aqueous $^{226}Ra^{2+}$ concentration in Bq/ml while in contact with barite as function of time

an ideal solid solution ($a_0 = 0$). The measured $^{226}Ra^{2+}$ molality was simulated by two different approaches:

- When only 20% of the barite powder was equilibrated with the $^{226}Ra^{2+}$, $1.5 \cdot 10^{-8}$ mol·(kg H_2O)$^{-1}$ $^{226}Ra^{2+}$ in contact with $Ra_{0.000662}Ba_{0.999338}SO_4$ was calculated with respect to $a_0 = 0.36$.
- Taking into account $a_0 = 1.5$, the lowest Guggenheim parameter derived by Curti et al. (2010), a $^{226}Ra^{2+}$ molality of $1.5 \cdot 10^{-8}$ mol·(kg H_2O)$^{-1}$ in contact with $Ra_{0.000182}Ba_{0.999818}SO_4$ was computed.

In batch experiments of Rosenberg et al. (2011b), (radio)barite coprecipitation was determined under different ionic strength conditions by evaporating the fresh concentrate of a desalination plant. The NaCl rich brines were supersaturated with respect to gypsum, celestite and barite. During evaporation the ionic strength increased from 0.7 mol kg^{-1} up to halite saturation and gypsum and barite precipitation commenced within few days (Figs. 1 and 2). The decrease in dissolved $^{226}Ra^{2+}$ was concurrent with the decrease in dissolved Ba^{2+}. The linear relationship between the precipitated $^{226}Ra^{2+}$ and Ba^{2+}, as opposed to the non-concurrent precipitation of $^{226}Ra^{2+}$ both with Ca^{2+} and Sr^{2+} indicated that barite was the main host mineral phase for Ra^{2+}, while co-precipitation of Ra^{2+} with Ca^{2+} in gypsum was insignificant. Rosenberg et al. (2011b) determined a concentration-based apparent partition coefficient of $K'_{D,barite} = 1.04 \pm 0.01$. They demonstrated that the difference between their low value of $K'_{D,barite}$ compared to $K'_{D,barite}$ of Doerner and Hoskins (1925) was mainly the result of a kinetic effect but was also slightly affected by the ionic strength. Rosenberg et al. (2011b) suggested that a kinetic effect on the nucleation of $(Ra,Ba)SO_4$ reduces the value of the partition coefficient. Taking into account the $RaCl_2$ Pitzer parameters derived by Rosenberg et al. (2011a,b) calculated activities of Ra^{2+} and Ba^{2+} in the solution in order to determine an *activity*-based partition coefficient, which accounts for the ionic strength effect. Furthermore, they determined a solid activity coefficients, $\gamma_{RaSO4} = 1.1$, corresponding to a Guggenheim parameter of $a_0 \sim 0.1$.

Summary and Conclusions

Both, in the coprecipitation experiments of Rosenberg et al. (2011b) and in the present (re)precipitation experiment on the $^{226}Ra^{2+}$/barite system, the observed $^{226}Ra^{2+}$ molalities are several orders of magnitude below the radium solubility with respect to a pure $RaSO_4(s)$ endmember. With respect to the (re)precipitation experiments, equilibration between aqueous $^{226}Ra^{2+}$ and the solid over 678 days reaction time involves a substantial fraction of the barite crystals and proceeds significantly beyond pure surface adsorption processes. Aqueous molalities of $^{226}Ra^{2+}$ are controlled by the solubility of a $Ra_xBa_{1-x}SO_4$ solid solution, an equilibrium concentration of $3 \cdot 10^{-9}$ mol·(kg H_2O)$^{-1}$ is calculated both for the assumption of an ideal solid solution and for a Guggenheim parameter of $a_0 = 0.36$ taken from an approximation by Zhu (2004). The measured molalities of $^{226}Ra^{2+}$ ($\leq 1\cdot10^{-8}$ mol$\cdot$(kg $H_2O)^{-1}$) is simulated either based on the assumption that about 20% of the barite crystals are equilibrated with aqueous $^{226}Ra^{2+}$ when taking into account $a_0 = 0.36$ – or – or based on complete equilibration of a $Ra_{0.000182}Ba_{0.999818}SO_4$ solid solution with aqueous $^{226}Ra^{2+}$ taking into account a relatively high Guggenheim parameter of $a_0 = 1.5$. The latter value is in the range of a_0 determined by Curti et al. (2010). Since in the evaporation experiments of Rosenberg et al. (2011b), formation of $Ra_xBa_{1-x}SO_4$ solid solution via coprecipitation is considerably affected by kinetics, a relative strong incorporation of Ra^{2+} into the solid is observed. As a continuation of our approach to determine the equilibrium conditions of $^{226}Ra^{2+}$ retention by barite, a combination of long-term coprecipitation and (re)precipitation experiments on the formation of $Ra_xBa_{1-x}SO_4$ solid solution are conducted in the framework of the Collaborative Project SKIN.

Acknowledgments The authors would like to thank our colleagues at INE, in particular to E. Bohnert, C. Bube, A. Bluss, T. Kisely, K. Scheubeck and E. Soballa for analytical and technical assistance during the study. The research leading to these results has been financially supported in part from Svensk Kärnbränslehantering AB within the project "Experimental study on Ra^{2+} uptake by barite ($BaSO_4$) – Kinetics of solid solution formation via $BaSO_4$ dissolution and $Ra_xBa_{1-x}SO_4$ (re)precipitation" and from the European Atomic Energy Community Seventh Framework Programme, topic Fission-2010-1.1.2, Collaborative Project SKIN "Slow processes in close to equilibrium conditions of radionuclides in water/solid systems of relevance to nuclear waste management".

References

Abdelouas, A. (2006) Uranium mill tailings: Geochemistry, mineralogy and environmental impact. Elements 2, 335–341.

Bosbach, D., Böttle, M., and Metz, V. (2010) Experimental study on Ra2+ uptake by barite (BaSO4). Kinetics of solid solution formation via BaSO4 dissolution and RaxBal-xSO4 (re)precipitation. SKB TR-10-43. Svensk Kärnbränslehantering AB, Stockholm, 111 pages.

Berner, U. (1992) Project Opalinus Clay. Radionuclide concentration limits in the near-field of a repository for spent fuel and vitrified high-level waste. Technical Report 02-10, NAGRA, Wettingen, Switzerland.

Berner, U. and Curti, E. (2002) Radium solubilities from SF/HLW wastes using solid solution and co-precipitation models. Paul Scherrer Institute, Villigen, Switzerland, Internal Report TM-44-02-04.

Curti, E., Fujiwara, K., Iijima, K., Tits, J., Cuesta, C., Kitamura, A., Glaus, M. A., and Müller, W., 2010. Radium uptake during barite recrystallization at 23 ± 2 °C as a function of solution composition: An experimental 133Ba and 226Ra tracer study. Geochim. Cosmochim. Acta 74, 3553–3570.

Doerner, H. A. and Hoskins, W. M. (1925) Co-Precipitation of Radium and Barium Sulfates. J. Am. Chem. Soc. 47, 662–675.

EPA, 2008. Technologically Enhanced Naturally Occurring Radioactive Materials from Uranium Mining, Technical Report 402-R-08-005.

Grandia, F., Merino, J., and Bruno, J. (2008) Assessment of the radium-barium co-precipitation and its potential influence on the solubility of Ra in the near-field. SKB TR-08-07, Svensk Kärnbränslehantering AB, Stockholm.

IAEA (2003) Extent of Environmental Contamination by Naturally Occurring Radioactive Material (NORM) and Technological Options for Mitigation, Technical Reports Series No. 419.
Paige, C. R., Kornicker, W. A., Hileman, O. E., and Snodgrass, W. J. (1998) Solution equilibria for uranium ore processing: The BaSO4-H2SO4-H2O system and the RaSO4-H2SO4-H2O system. Geochim. Cosmochim. Acta 62, 15–23.
Pitzer, K. S. (1973) Thermodynamics of electrolytes. I. Theoretical basis and general equations. J. Phys. Chem. 77, 268–277.
Pitzer, K. S. (1979) Theory, ion interaction approach. In: Pytkowicz, R. M. (Ed.), Activity coefficients in Electrolyte Solutions. CRC Press, 157–208, Boca Raton, Florida, USA.
Rosenberg, Y. O., Metz, V., and Ganor, J. (2011a). Co-precipitation of radium in high ionic strength systems: 1. Thermodynamic Properties of the Na-Ra-Cl-SO4-H2O System – Estimating Pitzer Parameters for RaCl2. Geochim. Cosmochim. Acta (in press).
Rosenberg, Y. O., Metz, V., Oren, Y., Volkman, Y., and Ganor, J. (2011b). Co-precipitation of radium in high ionic strength systems: 2. Kinetic and ionic strength effects. Geochim. Cosmochim. Acta (in press).
SKB (2008) Long-term safety for KBS-3 repositories at Forsmark an Laxemar – a first evaluation Main Report of the SR-Can project. SKB TR-06-09, Svensk Kärnbränslehantering AB.
Zhu, C. (2004a) Coprecipitation in the Barite isostructural family: 1. Binary mixing properties. Geochim. Cosmochim. Acta 68, 3327–3337.

Study of the Speciation in the System UO_2^{2+}–SO_4^{2-}–H_2O by Means of the UV-VIS Spectrophotometry

Jakub Višňák, Aleš Vetešník, Karel Štamberg, Jiří Bok

Abstract. This work estimates speciation characteristics of the UO_2^{2+}–SO_4^{2-}–H_2O system by means of the analysis of UV-VIS absorption spectra of series of aqueous solutions of uranium(VI) (total molar concentration 0.05 mol·dm^{-3}), sodium sulfate (varying concentration from 0 to 1.3 mol·dm^{-3}), perchloric acid (to adjust pH = 2) and sodium perchlorate (to adjust ionic strength I in each series (between different series I varied from 0.7 mol·Kg^{-1} to 4.0 mol·Kg^{-1})). log β^0 for several complexes were estimated by SVD-based mathematical analysis of spectra. The existence of isomers of uranyl(VI)-sulfate complexes is discussed.

Motivation

The UO_2^{2+}–SO_4^{2-}–H_2O system is important for uranium mining studies (e.g. recovery of uranium from leach liquors), safety studies of radioactive waste reposi-

Jakub Višňák
Department of Nuclear Chemistry, Faculty of Nuclear Sciences and Physical Engineering, Czech Technical University in Prague, Břehová 7, CZ-115 19 Prague 1, Czech Republic

Aleš Vetešník
Department of Nuclear Chemistry, Faculty of Nuclear Sciences and Physical Engineering, Czech Technical University in Prague, Břehová 7, CZ-115 19 Prague 1, Czech Republic

Karel Štamberg
Department of Nuclear Chemistry, Faculty of Nuclear Sciences and Physical Engineering, Czech Technical University in Prague, Břehová 7, CZ-115 19 Prague 1, Czech Republic

Jiří Bok
Charles University in Prague, Faculty of Mathematics and Physics, Institute of Physics, Ke Karlovu 5, CZ-121 16 Prague 2, Czech Republic

B. Merkel, M. Schipek, *The New Uranium Mining Boom*
DOI 10.1007/978-3-642-22122-4

tory, geochemical modeling and several other fields (even including aqueous homogeneous reactor system modeling). Experimental study of this system can be also used for plutonium migration studies instead of PuO_2^{2+}–SO_4^{2-}–H_2O since the aquatic chemistry of U(VI) and Pu(VI) is similar and U(VI) is much less hazardous to handle than Pu(VI).

Of interest there are several thermodynamical characteristics of both individual species ($\Delta G°$, $\Delta H°$, log $\beta°$, …) and their pairs (SIT-theory interaction coefficients $\varepsilon(i, j)$ for complex-electrolyte ions, complex-ligand and complex-complex pairs).

Since we often study migration of uranium in mining area, radioactive waste repositories or homogeneous reactors under different thermodynamical conditions, the temperature (and eventually pressure) dependence of quantities mentioned above is to be studied too.

Equilibrium-chemical analysis done by means of UV-VIS absorption spectroscopy can reveal us even more – absorption spectra of individual species, number of non-negligible present species and eventually number of their isomers and hydratation number for every studied isomeric form. Spectral information when compared to the ab-initio quantum chemical computations can yield also to indirect hint for structural information for uranium complexes.

Complexation of Uranyl(VI) by Sulfate Ions in Aqueous Solutions (Theory)

Uranium(VI) forms easily trans-dioxo bonds and therefore exists as UO_2^{2+} (uranyl(VI)) stably over a wide pH-range with U-O_{ax} bond length between 176 and 182 pm (Hennig et al. 2008). The uranyl(VI) ion exists in acidic aqueous solution as pentaaqua complex – $[UO_2(H_2O)_5]^{2+}$ and with increasing pH and/or total uranium concentration is further hydrolyzed into both monomeric and polynuclear hydroxo- and oxo-complexes such as (listed only those important for the area we worked in pH <3) UO_2OH^+, $UO_2(OH)_2^0$, $(UO_2)_2OH^{3+}$, $(UO_2)_2(OH)_2^{2+}$, $(UO_2)_3(OH)_4^{2+}$, $(UO_2)_3(OH)_5^+$.

For our study is important to note that there is no convincing evidence for complexation of uranyl(VI) by perchlorate ions (ClO_4^-) in aqueous solution available (Meinrath 1998; Vdovenko et al. 1964; Johannson 1974).

Complexation of uranyl by sulfate ions in aqueous solutions under acidic conditions leads to several monomeric sulfato-uranyl complexes as described by reactions below:

$$UO_2^{2+} + SO_4^{2-} \Leftrightarrow UO_2SO_4^0, \quad (1)$$

$$UO_2^{2+} + 2\,SO_4^{2-} \Leftrightarrow UO_2(SO_4)_2^{2-}, \quad (2)$$

$$UO_2^{2+} + 3\,SO_4^{2-} \Leftrightarrow UO_2(SO_4)_3^{4-}, \quad (3)$$

There are even hints that also tetrasulfate ($UO_2(SO_4)_4^{6-}$) and pentasulfate ($UO_2(SO_4)_5^{8-}$) complexes of uranyl(VI) can be formed (Brown and Wanner 1987). Increase in pH and/or total uranium concentration leads to several ternary monouranyl-hydroxo-sulaftes such as $UO_2(OH)SO_4^-$ or $UO_2(OH)(SO_4)_2^{3-}$ and polynuclear ternary uranyl-hydroxo-sulfates such as $(UO_2)_2(OH)_2(SO_4)_2^{2-}$, $(UO_2)_3(OH)_4(SO_4)_3^{4-}$, $(UO_2)_5(OH)_8(SO_4)_n^{2-2n}$ ($n=4$, 5 or 6) (Moll et al. 2000).

Important fact is that there can exist several coordination isomers of above mentioned complexes. In the case of monouranyl sulfates $[UO_2(SO_4)_n(H_2O)_m]^{2(1-n)}$ we can distinguish isomers by the total coordination number of uranyl(VI) N^{tot} and the number of mono (N_1) and bidentate (N_2) sulfato-groups (The equation $m+N_1+2N_2=N^{tot}$ holds, so we can define a vector (N_1, N_2, m) to describe each isomer with general formula

$$\left[UO_2\left(\kappa^1-SO_4\right)_{N_1}\left(\kappa^2-SO_4\right)_{N_2}\left(H_2O\right)_m\right]^{2(1-N_1-N_2)}).$$

Both the coordination mode (mono-/bi-dentate) and N^{tot} vary with the concentration ratio $[SO_4^{2-}]/[H_2O]$ as described in Hennig et al. (2008) and may vary with ionic strength and temperature as well (Vercouter et al. 2008).

From both experimental (Hennig et al. 2007; Gál et al. 1992; Neuefeind et al. 2004) and theoretical (Hennig et al. 2007; Šebera 2009) studies there is a strong evidence for at least two isomers of $UO_2SO_4^0$ – pentacoordinated with monodentate sulfate ligand (1,0,4) and pentacoordinated with bidentate sulfate ligand (Moll et al. 2000) (0,1,3), the former being more stable (due to Šebera (2009) by 3.43 kJ/mol and due to Koch and Holthausen (2000) by 14.9 kJ/mol).

The disulfatouranyl(VI) ($UO_2(SO_4)_2^{2-}$) is assumed to exist in aqueous solutions dominantly as pentacoordinated uranyl(VI) with both sulfato-ligands as bidentate – (0,2,1) (Moll et al. 2000; Hennig et al. 2007; Nguyen-Trung et al. 1992; Hennig et al. 2008c), however, the theoretical study (Šebera 2009) speaks in favor for pentacoordinated uranyl(VI) with both mono- and bidentate SO_4^{2-} ligands – (1,1,2) as being the most stable, hexacoordinated (0,2,2) and pentacoordinated (0,2,1) lying higher by 1.34 and 5.25 kJ/mol respectively (on ΔG scale).

In the case of trisulfatouranyl(VI) ($UO_2(SO_4)_3^{4-}$) we can rely mostly only on indirect information from spectroscopic experiments (Hennig et al. 2008; Vercouter et al. 2008; Geipel et al. 1997; Vercouter et al. 2007). Notably the temperature induced bathochromic shift of $UO_2(SO_4)_3^{4-}$ fluorescence spectrum speaks in favor of the existence of at least two isomers of the complex and non-zero standard enthalpy of the isomerization reaction (Vercouter et al. 2007). Interestingly, the work (Vercouter et al. 2007) investigate also isomery of mono- and disulfate of uranyl(VI) by temperature and ionic strength (or concentration of Na^+ from indifferent electrolyte) dependence of the shape of the fluorescence spectra stating that while UO_2SO_4 may have more then one isomer in non-negligible concentrations, $UO_2(SO_4)_2^{2-}$ may exist only in one form. Quantum-chemical computations of $UO_2(SO_4)_3^{4-}$ shows us that two isomers are most likely to be detected in aqueous solutions – pentacoordinated (1,2,0) and hexacoordinated (0,3,0), the latter being more stable by 1.9 kJ/mol due to Hennig et al. (2008, 2007) or the former being more stable by 0.53 kJ/mol due to Šebera (2009).

As will be discussed in detail later, the equilibrium concentration ratio of several isomers of the complex $UO_2(SO_4)_n^{2(1-n)}$ depends on the ionic strength (or ionic composition) either through water-activity (if isomers differ by the number of coordinated water molecules, *m*) or through SIT or Pitzer interaction coefficients (this effect will be notable for charged species $UO_2(SO_4)_2^{2-}$ and $UO_2(SO_4)_3^{4-}$) (Vercouter et al. 2007).

Speciation Computation

The speciation model was based on the SIT theory. Complexing reactions of interest were precisely formulated with taking all coordinated water molecules into account with the possibility for further distinguishing isomers of sulfato-uranyl(VI) complexes differing by the number of coordinated water molecules. We used notation of mixed equilibrium constants defined by H^+ activity and concentration of other species. In this case we can write general reaction taken into account in our model as

$$aA + bB \Leftrightarrow C + d\mathrm{H}^+ + fH_2O\,, \tag{4}$$

and write for its equilibrium constant the equation

$$\log K_i = \log K_I^\circ + \Delta z_i^2 D(I) - f \log a_{H_2O} - \sum_k (\Delta\varepsilon)_{ik}\, c_k^{(m)} \tag{5}$$

where *i* is indexing different chemical reactions in the model, *I* is ionic strength in mol·Kg^{-1}, $\Delta z_i^2 = z_C^2 - az_A^2 - bz_B^2$, $(\Delta\varepsilon)_{ik} = \varepsilon_{C,k} - a\varepsilon_{A,k} - b\varepsilon_{B,k}$ and water activity a_{H_2O} is computed by slightly modified formulae (10) and (12) from Grenthe et al. (2000), $c_k^{(m)}$ is molality (mol·Kg^{-1}) of *k*-th specie. The speciation model gets pH, temperature *t*, solution density ρ and total molal concentration of U(VI), S(VI), Cl(VII) and Na(I) and give us molar concentrations (mol·dm^{-3}) for all UV-VIS absorbing species and also molal (mol·Kg^{-1}) concentration of every specie in the model. The model did consider chemical reactions in supporting electrolyte listed below (together with log $K°$ values for 25°C)

$$\mathrm{Na^+ + SO_4^{2-} \Leftrightarrow NaSO_4^-}\,, \tag{6}$$

$$\mathrm{2\,Na^+ + SO_4^{2-} \Leftrightarrow Na_2SO_4^0}\,, \tag{7}$$

$$\mathrm{HSO_4^- \Leftrightarrow SO_4^{2-} + H^+}\,, \tag{8}$$

$$\mathrm{Na^+ + HSO_4^- \Leftrightarrow NaHSO_4^0}\,, \tag{9}$$

$$\mathrm{Na^+ + ClO_4^- \Leftrightarrow NaClO_4^0}\,, \tag{10}$$

and chemical reaction for which we didn't optimize the log $K°$ values

$$UO_2^{2+} + H_2O \Leftrightarrow UO_2OH^+ + H^+ , \quad (11)$$

$$UO_2^{2+} + 2H_2O \Leftrightarrow UO_2(OH)_2^{2-} + 2H^+ , \quad (12)$$

$$2UO_2^{2+} + 2H_2O \Leftrightarrow (UO_2)_2(OH)_2^{2+} + 2H^+ , \quad (13)$$

$$3UO_2^{2+} + 4H_2O \Leftrightarrow (UO_2)_3(OH)_4^{2+} + 4H^+ , \quad (14)$$

$$3UO_2^{2+} + 5H_2O \Leftrightarrow (UO_2)_3(OH)_5^{+} + 5H^+ , \quad (15)$$

and finally, chemical reactions for which we did optimize the log $K°$ values:

$$2\left[UO_2(H_2O)_5\right]^{2+} \Leftrightarrow \left[(UO_2)_2(OH)(H_2O)_8\right]^{3+} + 2H^+ + H_2O , \quad (16)$$

$$\left[UO_2(H_2O)_5\right]^{2+} + SO_4^{2-} \Leftrightarrow \left[UO_2(SO_4)(H_2O)_{m_1}\right] + (5 - m_1)H_2O , \quad (17)$$

$$\left[UO_2(H_2O)_5\right]^{2+} + 2SO_4^{2-} \Leftrightarrow \left[UO_2(SO_4)_2(H_2O)_{m_2}\right]^{2-} + (5 - m_2)H_2O , \quad (18)$$

$$\left[UO_2(H_2O)_5\right]^{2+} + 3SO_4^{2-} \Leftrightarrow \left[UO_2(SO_4)_3\right]^{4-} + 5H_2O . \quad (19)$$

The values of numbers m_1 and m_2 in reactions (17) and (18), respectively, are: $m_1 = 3$ or 4 and $m_2 = 1$ or 2. Also there are, for reaction (18) and $m_2 = 2$, and for reaction (19) two possible isomers as products.

Experiment

The solutions for the speciation study were prepared from 0.25M UO_3 aquatic solution of pH = 2 adjusted by $HClO_4$ used to adjust total uranium concentration (0.05 mol·dm^{-3} which was then recalculated on molality for each measured solution), 5, 10, 16 mM, 0.7 and 1 M aquatic solutions of Na_2SO_4 to adjust total sulfate concentration, 1, 5 and 7 M aquatic solutions of $NaClO_4$ to adjust ionic strength to approximately $I_1 = 0.7\,mol \cdot Kg^{-1}$, $I_2 = 1.5\,mol \cdot Kg^{-1}$, $I_3 = 3.0\,mol \cdot Kg^{-1}$, $I_4 = 4.0\,mol \cdot Kg^{-1}$ for each series of solutions for UV-VIS spectra measurement (values of ionic strength were then recomputed for each solution due to the fitted values of speciation parameters and measured densities of solutions), aquatic solutions of $HClO_4$ (60% and several more diluted up to 5 mM) and demineralized water were used to adjust pH = 2 (approximately, but measured then precisely for each solution). Pure chemicals used for the experiments: Na_2SO_4 (p. a., DORAPIS), $NaClO_4 \cdot H_2O$ (p. a., Sigma-Aldrich, 101040205, ≥98%), UO_3 (p. a.,

Table 1 Coefficients a_0, a_1, a_2 in general formula $\rho = a_0 + a_1 c + a_2 c^2$ for the dependence of density of water solution of below specified substance on its molar concentration

Substance	a_0 [g·dm^{-3}]	a_1 [g·mol^{-1}]	a_2 [g·dm^3·mol^{-2}]	rReference
$HClO_4$	997.05	48.593	0	Perry and Chilton (1973)
$NaClO_4$	997.05	6.365	7.481	Janz et al. (1970)
Na_2SO_4	998.34	76.913	−0.3916	Perry and Chilton (1973)

depleted, n.p. Brno (Chemapol, Praha-Czechoslovakia), ≥98%), 60% $HClO_4$ (p. a., Fluka, 77234) and demineralized water (Milipore (Direct-Q UV 3), resistivity 18.2 MΩ·cm (25°C)).

For each solution for UV-VIS spectra measurement, pH was measured by pH-meter "Orion model 525A" and density by pycnometric method. To exactly compute total molalities of U(VI), S(VI), Cl(VII), Na(I) we used tabulated densities of Na_2SO_4 and $HClO_4$ aquatic solutions (for given concentration and 25°C, see Table 1) and both data (Janz et al. 1970) and our own pycnometric measurement of density of $NaClO_4$ aquatic solution as a function of its molar concentration. The density of 0.25 M UO_3 with pH = 2 adjusted by $HClO_4$ (Cl(VII) concentration equal to 0.42 M) was measured by pycnometric method as (1.105 ± 0.001) g·cm^{-3} for 24°C.

The UV-VIS absorption spectra measurement was done on Varian (Cary 100 Conc) spectrophotometer in spectral range 190–900 nm. Every measurement was two times repeated.

Data Analysis

Measured UV-VIS absorption spectra were corrected for the absorption of background electrolyte (solutions of the same composition but without uranium), restricted to spectral range 380–480 nm and treated by the weighted Singular Value Decomposition (wSVD). Loadings from *g* main components were fitted by the maximum likelihood method (MLM) in order to obtain both UV-VIS absorption spectra of individual species and parameters of speciation model (log β° and $\varepsilon(i,j)$) Eq. (5).

The value of *g* was estimated from a screen plot of the singular values, and, in general, *g* can be different from the expected number of individual components (UO_2^{2+}, $(UO_2)_2OH^{3+}$, UO_2SO_4, $UO_2(SO_4)_2^{2-}$, $UO_2(SO_4)_3^{4-}$) *b*.

In the case of the series of the solutions of approximate constant ionic strength $I = 1.5$ M the value of g was estimated equal to the number of studied individual components (UO_2^{2+}, $(UO_2)_2OH^{3+}$, UO_2SO_4, $UO_2(SO_4)_2^{2-}$, $UO_2(SO_4)_3^{4-}$) $b = g = 5$ but we developed methods dealing with the $g > b$ situation as well and tested them on data for $I = 3.0$ M and $I = 4.0$ M. In this moment we used the simplified speciation model where we considered all complexes to be represented by single isomer (with f (Eq. (5)) = 2, 1, 4 and 5 for complexing reactions leading to $(UO_2)_2OH^{3+}$, UO_2SO_4, $UO_2(SO_4)_2^{2-}$, $UO_2(SO_4)_3^{4-}$, respectively), but we supposed the log $\beta°$, $\varepsilon(i,j)$ and individual species UV-VIS absorption spectra to be different for different series of spectra (differing by approximate value of ionic strength I) so the dependence of those quantities on I may show the isomerism.

Results

Here, we present the results of the analysis of a series of solutions of an approximately constant ionic strength $I = 1.5$ M. Obtained values of log $\beta°$ are compared in the Table 2 to the values available in the literature.

Table 2 Decadic logarithms of stability constants extrapolated to zero ionic strength and 298.15 K (from $T = 296.5$ K with the respect to the reaction enthalpies $\Delta H°$ from NEA "Hatches18" database (program PHREEQC) (data are consistent with Grenthe et al. 1992))) log $\beta°$: obtained in this study, log $\beta°_{tab}$: data from the literature (Grenthe et al. 1992; Guillaumont et al. 2003). Errors in log $\beta°$ corresponds to 95% confidence interval

Reaction	log $\beta°$ []			$\Delta H°$ [kJ·mol^{-1}]	log $\beta°_{tab}$ []		
$2\ UO_2^{2+} + H_2O \leftrightarrow (UO_2)_2OH^{3+} + H^+$	−2.61	+	0.18	14.64	−2.70	+	1.00
		−	0.77			−	1.00
$UO_2^{2+} + SO_4^{2-} \leftrightarrow UO_2SO_4$	3.61	+	0.46	19.25	3.15	+	0.12
		−	0.43			−	0.12
$UO_2^{2+} + 2\ SO_4^{2-} \leftrightarrow UO_2(SO_4)_2^{2-}$	5.23	+	0.58	34.73	4.14	+	0.07
		−	0.58			−	0.07
$UO_2^{2+} + 3\ SO_4^{2-} \leftrightarrow UO_2(SO_4)_3^{4-}$	4.23	+	0.53	−143.93	3.02	+	0.38
		−	0.54			−	0.38

Table 3 Molar absorptivity amplitudes of individual components ε (UO_2^{2+}, $(UO_2)_2OH^{3+}$, UO_2SO_4, $UO_2(SO_4)_2^{2-}$, $UO_2(SO_4)_3^{4-}$) for spectral range 380–480 nm with dλ = 1 nm step. For each component of wSVD decomposition singular value W_n and for first $g = b = 5$ analyzed components amplitudes k_n of variance (var $V_{in} = k_n \cdot$ mean$_l(Y_{li})$) obtained by the maximum likelihood minimalization

n	W_n	k_n	Species	ε [dm^3·mol^{-1}·cm^{-1}]
1	49,6783	0,685	UO_2^{2+}	38,8
2	6,1090	0,891	$(UO_2)_2OH^{3+}$	111,3
3	0,6721	0,585	UO_2SO_4	80,2
4	0,2733	0,560	$UO_2(SO_4)_2^{2-}$	89,0
5	0,1049	0,685	$UO_2(SO_4)_3^{4-}$	94,0
6	0,0182			

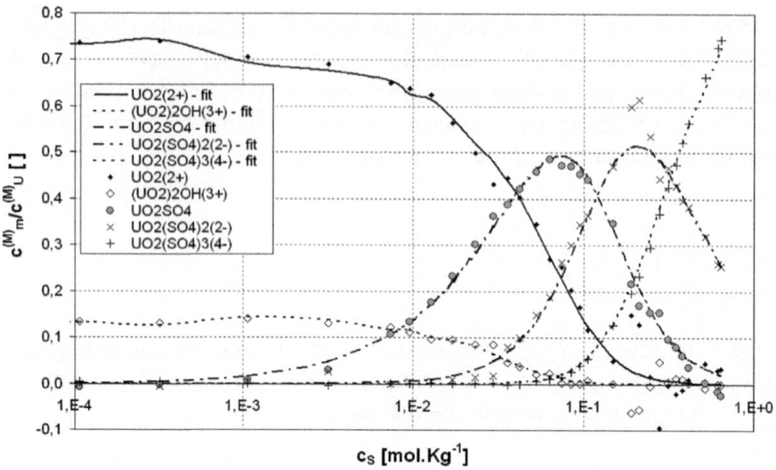

Fig. 1 Speciation diagram, lines represent ratio of molar concentrations calculated by using a speciation model with parameters: log β^o fitted by MLM (see Table 2), and $\varepsilon(i,j)$ obtained from the literature (Grenthe et al. 2000) or calculated by an approximate empiric formula for i a j ions of opposite charges $\varepsilon(i,j) = 0{,}1\ \varepsilon^D A_m z_i^2 z_j^2$, where $\varepsilon^D = 0{,}3\ \mathrm{Kg}^{1}/^{2}\ \mathrm{mol}^{-1}/^{2}$, $A_m = 0{,}5901\ \mathrm{Kg}^{1}/^{2}\ \mathrm{mol}^{-1}/^{2}$, z_i and z_j are charge-numbers of interacting ions i and j, if z_i and z_j doesn't have an opposite sign, then $\varepsilon(i,j) = 0$. $\varepsilon(UO_2^{2+}, ClO_4^-) = (0{,}46 \pm 0{,}03)$ Kg·mol^{-1} (Grenthe et al. 2000), $\varepsilon(UO_2^{2+}, SO_4^{2-}) =$ 0,12 Kg·mol^{-1} (Vercouter et al. 2008), $\varepsilon(UO_2(SO_4)_2^{2-}, Na^+) = (-0{,}12 \pm 0{,}06)$ Kg·mol^{-1} (Lemire et al. 2000) and $\varepsilon(UO_2(SO_4)_3^{4-}, Na^+) = -0{,}24$ Kg·mol^{-1} (Grenthe et al. 1992) were kept fixed for this analysis. $\varepsilon(NaSO_4^-, Na^+) = -0{,}05$ Kg·mol^{-1} was estimated from (Capewell et al. 1999) a $\varepsilon(NaSO_4^-, H^+) = 0$ Kg·mol^{-1} was estimated from (Grenthe et al. 2000). Points represent measured data

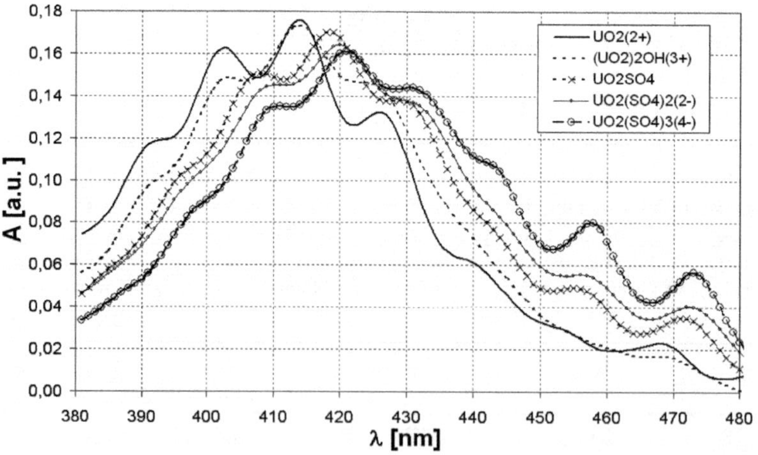

Fig. 2 Absorption spectra of individual components – UO_2^{2+}, $(UO_2)_2OH^{3+}$, UO_2SO_4, $UO_2(SO_4)_2^{2-}$, $UO_2(SO_4)_3^{4-}$ normalized due to the Euclidean norm. True molar absorptivities could be then obtained by multiplying normalized spectra by coefficients listed in Table 3

The presented values of log $\beta°$ depends, of course, on the choice of the values of log $K°$ and $\varepsilon(i,j)$ for other chemical reactions in the studied solutions. The results presented in Table 2 corresponds to $K° = 0$ for reactions (7), (9) and (10). By choosing log $K°$ equal to 1.10, 0.30 and –1.47 for reactions (7), (9) and (10), the log $\beta°$ in the Table 2 will change to –4.00 ($(UO_2)_2OH^{3+}$, almost not present in solutions), 3.73 (UO_2SO_4), 5.50 ($UO_2(SO_4)_2^{2-}$) and 4.79 ($UO_2(SO_4)_3^{4-}$).

The dependence of log $\beta°$ on background electrolyte speciation could explain discrepancy between log $\beta°$ and log $\beta°_{tab}$ values in Table 2.

Future Plans

We plan to continue in speciation studies of the system UO_2^{2+}–SO_4^{2-}–H_2O by means of UV-VIS absorption measurement. In particular, we plan to prepare solutions with both lower and higher pH values, wider range of ionic strength I values, and to study the temperature dependence of speciation parameters. TRLFS (Time Resolved Fluorescence Induced Spectroscopy) measurement can serve as the comparison.

The theoretical study of speciation parameters for the system UO_2^{2+}–SO_4^{2-}–H_2O based on the combination of relativistic quantum chemistry computations ($\Delta H°$) statistical physics (log $\beta°$, $\Delta G°$) and molecular dynamics ($\varepsilon(I)$) is also in our further field of interest.

Acknowledgement The research was supported by the Ministry of Education of the Czech Republic under contract MSM 6840770020 and RAWRA, Czech Republic.

References

Brown PL, Wanner H (1987) Predicted formation constants using the unified theory of metal ion complexation. Paris: OECD Nucl. Energy Agency 102 [5]

Capewell SG, Hefter GT, May PM (1999) Association constants for the $NaSO_4$– ion pair in concentrated caesium chloride solutions. Talanta 49: 25–30

Janz GJ, Oliver BG, Lakshminarayanan GR, Mayer GE (1970) Electrical Conductance, Diffusion, Viscosity, and Density of Sodium Nitrate, Sodium Perchlorate, and Sodium Thiocyanate in Concentrated Aqueous Solutions. J. Phys. Chem. 74 (6): 1285–1289

Johannson (1974) The Role of the Perchlorate Ion as Ligand in Solution. Coord. Chem. Rev., 12: 241–261 [4]

Gál M, Goggin PL, Mink J (1992) Vibrational spectroscopic studies of uranyl complexes in aqueous and non-aqueous solutions. Spectrochim. Acta A 48:121–132 [10]

Geipel G, Brachmann A, Brendler V, Bernhard G, Nitsche H (1996) Uranium(VI) sulfate complexation studied by time-resolved laser-induced fluorescence spectroscopy (TRLFS). Radiochim Acta 75:199–204 [16]

Grenthe I, Fuger J, Konigs RJM, Lemire RJ, Muller AB, Nguyen-Trung C, Wanner H (1992) Chemical Thermodynamics of Uranium, Vol. 1 of Chemical Thermodynamics. Amsterdam: Elsevier Science Publishers B. V.: 114, 241, 683, 715

Grenthe I, Wanner H, Oesthols E (2000) Guidelines for the extrapolation to zero ionic strength. TDB-2, NEA [18]

Guillaumont R, Fanghänel T, Fuger J, Grenthe I, Neck V, Palmer DA, Rand MH (2003) Update on the chemical thermodynamics of uranium, neptunium, plutonium, americium and technetium. Elsevier Science Publishers, Amsterdam

Hennig C, Schmeide K, Brendler V, Moll H, Tsushima S, Scheinost AC (2007) EXAFS investigation of U(VI), U(IV), and Th(IV) sulfato complexes in aqueous solution. Inorg. Chem. 46: 5882–5892 [9]

Hennig C, Tsushima S, Brendler V, Ikeda A, Scheinost AC, Bernhard G (2008) Coordination of U(IV) and U(VI) sulfate hydrate in aqueous solution. Uranium Mining and Hydrogeology V: 625–635 [1]

Hennig C, Ikeda A, Schmeide K, Brendler V, Moll H, Tsushima S, Scheinost AC, Skanthakumar S, Wilson R, Soderholm L, Servaes K, Görrler-Walrand C, Van Deun R (2008c) The relationship of monodentate and bidentate coordinated uranium(VI) sulfate in aqueous solution. Radiochim. Acta 96: 607–611 [14]

Koch W, Holthausen MC (2000) A Chemist's Guide to Density Functional Theory. WILEY-VCH, Verlag GmbH, Weinheim [15]

Lemire RJ, Fuger J, Nitsche H, Potter P, Rand MH, Rydberg J, Spahiu K, Sullivan JC, Ullman WJ, Vitorge P, Wanner H (2000) Chemical thermodynamics of neptunium and plutonium. Amsterodam, Holland: Elsevier Science B. V.

Meinrath G (1998) Aquatic Chemistry of Uranium. A Review Focusing on Aspects of Environmental Chemistry. Freiberg On-line Geoscience Vol. 1: 8 [2]

Moll H, Reich T, Hennig C, Rossberg A, Szabó Z, Grenthe I (2000) Solution coordination chemistry of uranium in the binary UO_2^{2+}–SO_4^{2-} and the ternary UO_2^{2+}–SO_4^{2-}–OH– system. Radiochim. Acta 88: 559–566 [7]

Neuefeind J, Skanthakumar S, Soderholm L (2004) Structure of the UO_2^{2+}–SO_4^{2-} ion pair in aqueous solution. Inorg. Chem. 43:2422–2426 [11]

Nguyen-Trung C, Begun, GM, Palmer DA (1992) Aqueous uranium complexes. 2. Raman spectroscopic study of the complex formation of the dioxouranium(VI) ion with a variety of inorganic and organic ligands. Inorg. Chem. 31:5280–5287 [13]

Perry RH, Chilton CH (1973) Chemical Engineers' Handbook, Fifth Edition, McGrawHill

Šebera J (2009) Quantum-chemical interpretation of UV-VIS and fluorescence spectra of uranyl(VI)-sulfate complexes. Research report for SURAO (Radioactive Waste Repository Authority (Czech Republic)): 27–42 [12]

Vercouter T, Vitorge P, Amekraz B, Moulin Ch (2008) Stoichiometries and thermodynamic stabilities for aqueous sulfate complexes of U(VI). Inorg. Chem. 47: 2180–2189 [8]

Vdovenko VM, Mashirov LG, Suglobov D N (1964) Infrared Spectra of Uranyl Perchlorate and its Crystal Hydrates: Coordination of the Perchlorate Ion. Soviet Radiochem. 6, 289–294 [3]

Characterization of the Impact of Uranium Mines on the Hydrological System in a Granitic Context: Example of the Limousin Area in France

Christian Andres, Charlotte Cazala, Emmanuel Ledoux, Jean-Michel Schmitt

Abstract. Uranium has been mined in the Limousin area (France) until the end of the 90s. Nowadays mines are closed and restored. Rehabilitation operations were devoted to assure the stability of land and infrastructures as well as to collect mine water and keep under control the radionuclide release into watersheds. Environmental monitoring is conducted by the operator (Areva-NC) under the control of the French administration. The present paper describes the methodology of investigation and the geochemical modeling approach that was developed on the Bellezane mining site where tailings are disposed of. This methodology and the corresponding modeling approach then proved to be adaptable to other tailings deposit sites.

Introduction

Uranium has been mined in the Limousin area (France) until the end of the 90s. Nowadays mines are closed and restored. Rehabilitation operations were devoted to assure the stability of land and infrastructures as well as to collect mine water and keep under control the radionuclide release into watersheds. Environmental monitoring is conducted by the operator (Areva-NC) under the control of the

---

Christian Andres
Areva NC, BG Mines

Charlotte Cazala
Institut de Radioprotection et de Sureté Nucléaire/DEI/SARG/BRN

Emmanuel Ledoux
Mines-ParisTech, Centre de Géosciences

Jean-Michel Schmitt
Areva NC, BG Mines

B. Merkel, M. Schipek, *The New Uranium Mining Boom*
DOI 10.1007/978-3-642-22122-4 © Springer 2011

French administration which defines the objectives of population and environmental protection against ionizing radiation.

In the year 2006, the French Nuclear Safety authority board (ASN) together with the Ministers of Environment, Industry and Health has set up a pluralist expert group (GEP) that enjoins experts, stakeholders and government representatives, to improve the knowledge and the management of uranium mines. A part of its work was devoted to the study of mines as radionuclide sources and to the transport of radionuclides into the environment. The group published its final report in 2010 (GEP 2010). Its recommendations are now being taken into account by the operator and the authorities. One of them consists in precisely characterizing the hydrochemical features of the different mining sites in order to predict the long term behavior of radiological and chemical potential pollutants in the hydrosystem.

The present paper describes the methodology of investigation and the geochemical modeling approach that was developed on the Bellezane mining site where tailings are disposed of. This methodology and the corresponding modeling approach then proved to be adaptable to other tailings deposit sites.

The Bellezane Uranium Mine Site

The uranium mine of Bellezane is located in a granitic geological context and was exploited from 1975 to 1993 by the mean of an open-pit and underground mining works. 3800 t of uranium were extracted and the open-pit was used as mining waste disposal since the early phase of exploitation. The open-pit is presently partially filled in with 1.5 millions tons of tailings produced by the acidic treatment of the ore.

The Hydraulic System of the Mine Site

The mine galleries and most of the tailings in the open-pit are presently under water. There is a permanent discharge of water from two adits. Superficial water and groundwater are monitored in several points of the site. The 226Ra content of mine water still exceeds the limit fixed by regulation for the release into the environment. Therefore, water is treated before being discharged into a brook.

Figure 1 displays a schematic representation of the hydraulic system of the mine and the location of monitoring facilities set up for the study. Several piezometers intercept the granitic aquifer upstream and downstream from the mine; three piezometers allow collection of water at different depths inside the tailings and one piezometer reaches the groundwater reservoir in the deep underground mining works. According to the presumed groundwater pathways, the discharge water which has been in contact with tailings and mining works is collected and treated at the outlet of galleries TB100 and BD200. The average flow rate is 50 m^3/h. In addition, the water flowing at the foot of a waste rock pile (VERSE) is also monitored.

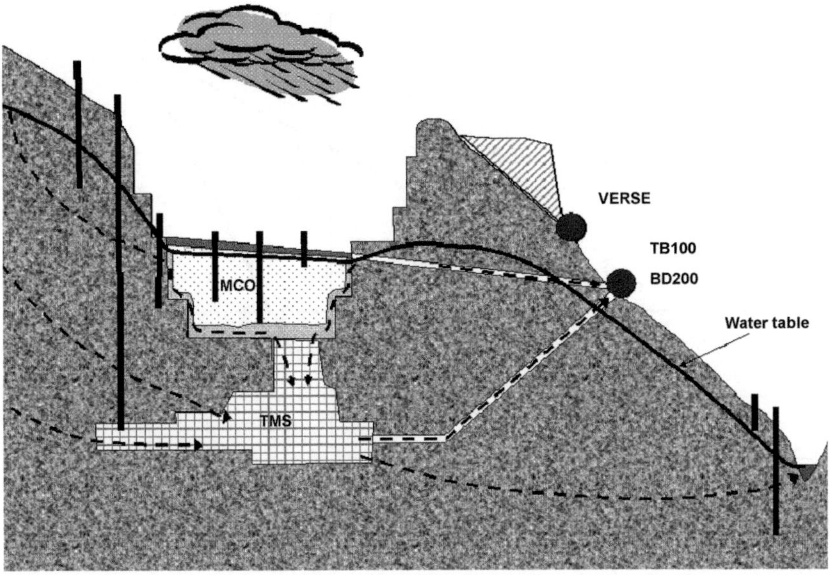

Fig. 1 Schematic representation of groundwater flow and monitoring facilities on the Bellezane site and its vicinity. Piezometers intercept the granitic aquifer, the tailings deposit (MCO) and the deep underground mining reservoir (TMS). The adits TB100 and BD200 collect the polluted discharge water. VERSE collects the waste rock lixiviates

Site Water Investigation

Superficial water and groundwater were sampled for chemical analysis. On the field, within the minutes following collection of water, temperature, pH, redox potential, electrical conductivity (EC) and alkalinity (TAC) were measured and samples were filtered and prepared for laboratory analysis (major anions and cations, heavy metals and particularly radium and uranium). Measurement campaigns were performed in April 2008 under high water conditions and in October 2008 under low water conditions.

Chemical Characterization of Water

Initial Characterization of the Chemical Poles

A first interpretation of the hydrochemical context is given by the alkalinity versus electrical conductivity diagram that can be easily drawn from field data. EC, measured in microSiemens/cm, is proportional to total dissolved solids content (TDS). Total alkalinity (TAC) indicates the aptitude of water to neutralize strong acids which usually depends on the main chemical species concerned in natural

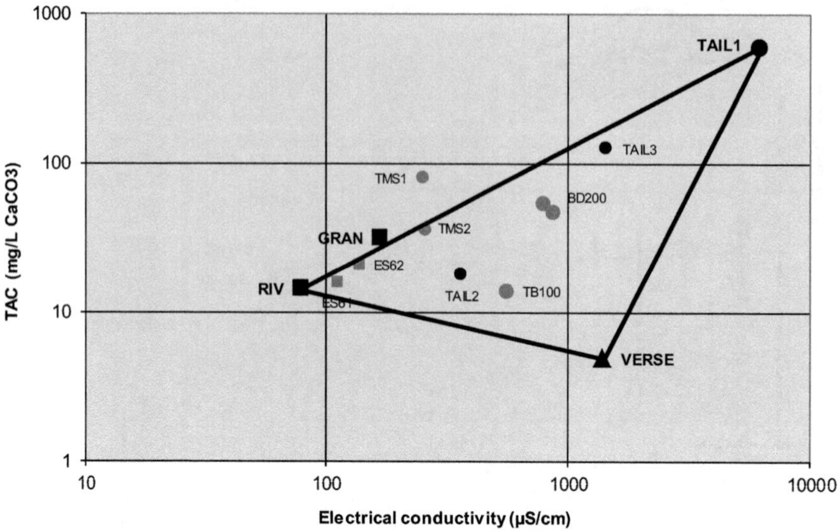

Fig. 2 Characterization of the hydrochemical poles of water in the diagram TAC/EC. TAIL1 is the 'tailings water' pole, VERSE the 'waste rock' pole and RIV and GRAN the 'natural water' respectively for surface water and groundwater

water which is hydrogeno-carbonate HCO_3^-. A water with a low dissolved content will then have at the same time a low conductivity and a low alkalinity; a water coming from acidic mining drainage which is usually rich in sulfate will show a high conductivity and a low alkalinity and a water which has reached chemical equilibrium with carbonated or silicated rocks will have medium conductivity and high alkalinity.

The diagram on Fig. 2 shows clearly several hydrochemical poles which are the consequence of water-rock interaction in the mine environment.

- Natural superficial water in river with low EC and low TAC,
- Deep granitic natural groundwater with moderate EC and TAC,
- Interstitial water from tailings with very high EC and TAC,
- Leachates from waste rock deposit with high EC but very low TAC.

Conclusions are similar for low water and high water conditions.

Other water samples plot inside the triangle determined by these hydrochemical poles. It shows that these waters may be the result of a mixture of the poles in different proportions.

Further Characterization of the Hydrochemical Features

Natural Water

Main characteristics of natural water are given in Table 1. The RIV sample corresponds to a river far from the mine site; its low pH and low TDS are typical of crystalline environment. The deep granitic water is represented by the sample GRAN, which is close to the previous one but with a higher TDS indicating a stronger water-rock interaction. Radium is present, which is a regional feature of the area.

Waste Rocks Leachates

As shown in Table 2, characteristics of waste rock water is typical of acid mine drainage (AMD) (Schmitt et al. 2004). It corresponds to acid oxidizing water with a high TDS and a calcium-magnesium facies. The uranium content is significantly higher than in natural water, the radium concentration is moderate.

Tailings Water

The interstitial water in tailings is available at three depths (see Table 3). In the middle of the tailings (TAIL1), the water is little acid with a very high TDS. It presents a reducing tendency and a sulfate-magnesium facies. Heavy metals concentrations are significant especially for iron, manganese and aluminum. The uranium content is over 2 mg/l but the radium concentration remains moderate.

Table 1 Main hydrochemical characteristics of natural water

	pH	Eh (mV)	TDS (mg/l)	Features	Metals	U (μg/l)	^{226}Ra (Bq/l)
RIV	6.5	470	55	HCO_3, Na	Fe, Mn	3	0.4
GRAN	6.4	320	125	HCO_3, Na, Ca	Fe, Mn, Zn	5	0.8

Table 2 Main hydrochemical characteristics of the waste rocks leachates

	pH	Eh (mV)	TDS (mg/l)	Features	Metals	U (μg/l)	^{226}Ra (Bq/l)
VERSE	4.7	430	1000	SO_4, Ca, Mg	Al, Mn, Fe, Zn, Ni	280	0.3

Table 3 Main hydrochemical characteristics of tailings water

	pH	Eh (mV)	TDS (mg/l)	Features	Metals	U (μg/l)	^{226}Ra (Bq/l)
TAIL1	6.3	190	8000	SO_4, Mg	Fe,Mn, Al,Ni, Co,F	2380	0.6
TAIL2	6.5	350	225	SO_4, Ca, Mg	Mn,Al, Fe	40	1.3
TAIL3	6.4	225	1215	SO_4, Ca, Mg	Fe,Mn, Al, Co	685	0.9

In the upper part of the tailings (TAIL2), the water shows globally the same characteristics but with lower ionic concentrations and higher redox potential which is probably the consequence of a dilution by rain water.

At the bottom of the tailings (TAIL3), the water seems also to be diluted compared to TAIL1. More reducing features may result of an interaction with the underlying mining reservoir.

Water from Underground Mining Works

Table 4 shows the main chemical characteristics of water from the deep mining reservoir (TMS1 and TMS2) and at the outlets of the discharge adits (TB100 and BD200).

Under high water conditions the mining reservoir water (TMS1) has a moderate TDS and plots very well on the trend of natural water (see Fig. 2). There is no indication for an abnormal metal content which indicates that the underground mining works drain essentially the deep granitic aquifer. However under low water conditions, the point TMS2 is shifted towards the pole VERSE which can be attributed to a drawdown of the piezometric surface allowing an oxidation of sulfur minerals.

Water at the outlet BD200 shows sulfate-calcium-magnesium facies fairly rich in uranium and radium. The position of the point in the diagram TAC/EC (see Fig. 2) indicates a contribution of the three hydrochemical poles "deep natural water", "waste rock" and "tailings". According to the values of concentrations in sulfate and radium, the proportion of the mixing is one third for each pole. This situation appears stable in high water and low water conditions.

Water of the outlet TB100 is gently acid with sulfate-calcium-magnesium facies and a fair content in uranium. In the TAC/EC diagram, the corresponding point plots in intermediate position between the "mining reservoir" water and "waste rock" water. This excludes a contribution of the tailings water which can be justified by the fact that this outlet is dedicated to the collection of water flowing at the upper levels of the mine.

Natural Groundwater Downstream from the Mine

Two piezometers (ES61 and ES62) are monitored in the thalweg downstream from the mine; their chemical characteristics are given in Table 5. The chemical features are calcium-sodium typical of a groundwater at moderate depth. The corresponding points plot on the trend of natural water on the TAC/EC diagram. The uranium and radium contents are not different from those of groundwater collected upstream from the mining site. It can be concluded that there is no evidence of hydraulic pathways from the mine to the granitic aquifer.

Table 4 Main hydrochemical characteristics of water from underground mining works

	pH	Eh (mV)	TDS (mg/l)	Features	Metals	U (μg/l)	^{226}Ra (Bq/l)
TMS1	6.4	435	220	HCO_3, Ca	Mn, Fe,Cu, Al	8	0.4
TMS2	6.7	460	230	HCO_3, Ca	Mn, Fe,Cu, Al	6.5	0.4
TB100	5.9	420	390	SO_4, Ca, Mg	Mn, Zn, Fe	275	0.45
BD200	6.6	440	560	SO_4, Ca	Al, Mn, Fe	320	0.2

Table 5 Main hydrochemical characteristics in piezometers downstream from the mining site

	pH	Eh (mV)	TDS (mg/l)	Features	Metals	U (μg/l)	^{226}Ra (Bq/l)
ES61	5.7	435	65	HCO_3, Ca, Na	Fe, Mn, Al	5	0.1
ES62	5.7	430	65	HCO_3, Ca, Na	Fe, Mn, Al	10	0.2

Surface Water in the Mining Site Area

A systematic study of all surface and subsurface water (streams, springs and country wells) in the mining site area was also conducted at the two sampling campaigns.

This study revealed no other evidence of pollution from the mine site, apart from the geochemical signature of the water treatment station observed downstream from the point of discharge.

Further Study of the Mine System Through Hydrodynamic and Thermodynamic Modeling

To confirm these interpretations, a detailed study and a modeling of the hydrodynamic and hydrochemical functioning of the site were also conducted (Ledoux and Schmitt 2010).

Hydrodynamic Modeling

The hydrodynamic behavior was studied from a vertical 2D model cutting the main pit and the tailings disposal. Five permeability facies were distinguished: 1) the tailings facies, 2) the sound granite facies, 3) a damaged granite facies corresponding to the pit walls, 4) a waste rock facies that corresponds to the tailings cover and to the drainage layer at the base of the tailings disposal, and 5) mining voids.

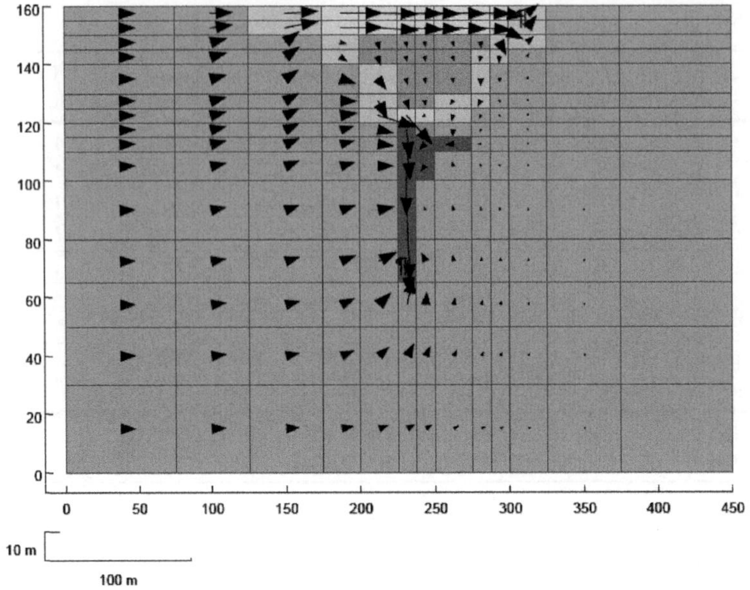

Fig. 3 2D vertical hydrodynamic model of the mine site and its tailings disposal. *Arrows* show the simulated flow field. *Shades of grey* indicate the different permeability facies (see text)

The system is fed by granitic waters from upstream and drained at two points: an outlet above the top of the tailings disposal corresponding to the opening of the TB100 gallery, and a deep outlet corresponding to the flow of the BD200 adit (see Fig. 2).

The flow field (Fig. 3) confirms that most of the water from upstream is drained by the two outlets but bypasses the tailings disposal itself. The model also shows that the water management system consisting of the two mine outlets virtually excludes any contamination downstream.

Thermodynamic Modeling

A detailed thermodynamic study of the tailings waters was conducted to clarify the hydrochemical functioning. Water in the middle of the tailings disposal is at thermodynamic equilibrium with a mineral assemblage consisting essentially of quartz, kaolinite, gypsum, dolomite, siderite, rhodochrosite, and uraninite.

The main holder of radium in this paragenesis is gypsum, as it has been shown in several other storage (Somot 1997). The heart of the tailings appears geochemically similar to what it was at the time of their disposal. This is consistent with their very low permeability and the results of the hydrodynamic modeling.

The base and the top of the tailings disposal instead show evidence of a seepage by subsurface water or by deep granite water, which is revealed by a significant drop in concentrations of major ions (sulfate, Ca, Mg) and metals (iron, manganese and uranium). The top water appears the more diluted. The fact that the radium concentration is nonetheless higher (about two times higher than in the very heart of the tailings) may be considered as an effect of the local dissolution of gypsum.

Reactive Transport Modeling

The 2D model on Fig. 3 has also been used to identify the hydrochemical functioning of the system, through the reactive transport modeling code (Bethke and Yeakel 2008). The water quality of the regional flow, entering through the left side frontier of the model was based on the deep granite water composition. The behavior of tailings was based on thermodynamic data and the mineral assemblage above, and a model for radium fixation on gypsum derived from Somot's 1997 paper.

It was verified that the 2D model allowed accounting correctly for the hydrochemical behavior of the 3D system it represents. It also predicts the main features of the evolution of the system with regard to the solid phase and the quality of discharge water.

Conclusion

Main results that can be drawn from the hydrochemical study of the mining site of Bellezane are the followings.

In the geological context of this granitic site, natural water is gently acid with low total dissolved solids content. The natural background in uranium is 5 to 10 μg/l and the activity of radium 226 is between 0.06 and 0.1 Bq/l.

The water in the deep underground mining works are almost similar of these characteristics and their chemical quality is very close to the one of natural water.

Indicators of mining activity are uranium and radium and to a lesser extent, sulfate. Tailings constitute the main source-term of those pollutants while the waste rocks are a secondary source for uranium and sulfate.

Water in the middle of tailings has a very high dissolved content and reducing features.

Chemical composition of the mine discharge water is determined by the mixing of three hydrochemical poles: 'natural granitic water', 'tailings water' and 'waste rock water'. A first assessment of the proportion of this mixing can be conducted on the basis of the alkalinity versus electrical conductivity which can be obtained easily from field data.

This hydrochemical approach successfully tested on the example of the Bellezane uranium mine and waste repository was applied to other sites in Limousin area where tailings have been stored in open-pits. Results confirm the efficiency of the method for a fast and easy characterization of the chemical impact of a mining site on the hydrosystem in a granitic environment.

References

Bethke C.M., Yeakel S. (2008) The Geochemist's Workbench® Release 7.0 Reference Manual. University of Illinois.

GEP (2010) Recommandations pour la gestion des anciens sites miniers d'uranium en France. Des sites du Limousin aux autres sites. Du court aux moyen et long termes. Septembre 2010. http://www.gep-nucleaire.org/gep

Ledoux E., Schmitt J.-M. (2010) Étude du fonctionnement hydrogéochimique de l'ancien site minier de Bellezane (Limousin, France). No R100119EL, BGM/DGS RT 10/004, Centre de Géosciences, École des Mines de Paris, Fontainebleau, France, 99 p.

Schmitt J.-M., Ledoux E., Combes P. (2004) Qualité des eaux après fermeture des mines: remplissage initial, évolution transitoire, stabilisation à long terme et gestion environnementale, Revue Française de Géotechnique, N° 106–107, 1er et 2ème trimestre 2004, pp. 95–101.

Somot S. (1997) Radium, uranium et métaux dans les résidus de traitement dynamique, acide et alcalin, de minerais d'uranium. Thèse Doctorat, Université Nancy I, 365 p.

Assessment of Uranium Waste Dump Closure Systems: Results of Long Term Test Fields at the Former Uranium Mining Site in Schlema-Alberoda, Germany

Ralf Löser, Petra Schneider, Jürgen Meyer, Andrea Schramm, Nicole Gottschalk

Abstract. Within the frame of the environmental monitoring of the former uranium mining sites in Schlema-Alberoda have been established in 1999 soil hydrological test fields on covered and revegetated waste heap dumps. The objectives of the long term investigations on the test fields are the quantification of infiltration rates as well as the evaluation of the dynamics of the soil water balance. The data, obtained from more than one decade of observation, provide the time development of the water balance in the system soil – vegetation – atmosphere. The data were used to optimize the covering and revegetation concept for the mine dumps. At the Schlema-Alberoda mine dumps the rehabilitation has been successfully carried out using of a 2-cover-layer which contains a humus surface layer.

Introduction

During the period of uranium mining in Schlema-Alberoda region (southwest Saxony, see Fig. 1) by the former Soviet-German mining company Wismut, 43

Ralf Löser
C&E Consulting und Engineering GmbH, Jagdschänkenstr. 52, D-09117 Chemnitz, Germany

Petra Schneider
C&E Consulting und Engineering GmbH, Jagdschänkenstr. 52, D-09117 Chemnitz, Germany

Jürgen Meyer
Wismut GmbH, Jagdschänkenstr. 27, D-09117 Chemnitz, Germany

Andrea Schramm
Wismut GmbH, Jagdschänkenstr. 27, D-09117 Chemnitz, Germany

Nicole Gottschalk
C&E Consulting und Engineering GmbH, Jagdschänkenstr. 52, D-09117 Chemnitz, Germany

B. Merkel, M. Schipek, *The New Uranium Mining Boom*
DOI 10.1007/978-3-642-22122-4 © Springer 2011

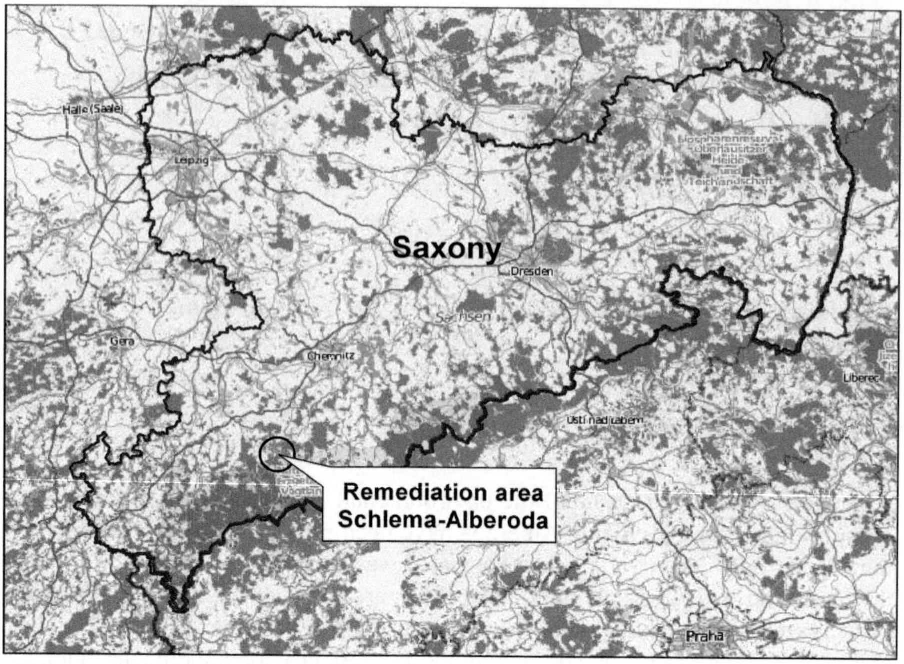

Fig. 1 Location of area under investigation

waste rock dumps have been built. The dumps have a total volume of 45 million m^3 (Runge et al. 1999) and dominate the recent landscape. In order to minimize the environmental risks and the related contaminant discharges by water and air pathways, the dumps have been profiled, reshaped, covered with inert mineral soil, and rehabilitated using engineering biological measures. The dump remediation was started after the closure of the mining activities around 1991 and is currently nearly finished.

Scope of the covering and rehabilitation works at the dumps is to reduce the water amount percolating into the dump material. Thereby also the amount of contaminants such as uranium and arsenic discharged by the percolating water will decrease. Another objective of dump remediation is the reduction of radon exhalation, which has been generated by waste rock materials and migrates on the air pathway. The discharge of contaminants from the dumps is observed continuously by a comprehensive environmental monitoring program of Wismut GmbH (see also Schneider et al., 2005).

Additional to this scheduled environmental monitoring program of Wismut GmbH special hydrological measurements have been performed by several lysimeter installations and two soil-hydrological test fields. These special measurements are aimed to quantify the precipitation water amounts percolating into the dump body throughout the soil cover system and to describe the water balance dynamics of the vegetation and soil cover (see Fig. 2). The measurements have been performed since 1999, allowing meanwhile for a detailed description of changes of the water balance of covered dump surfaces by the establishment of a vegetation cover.

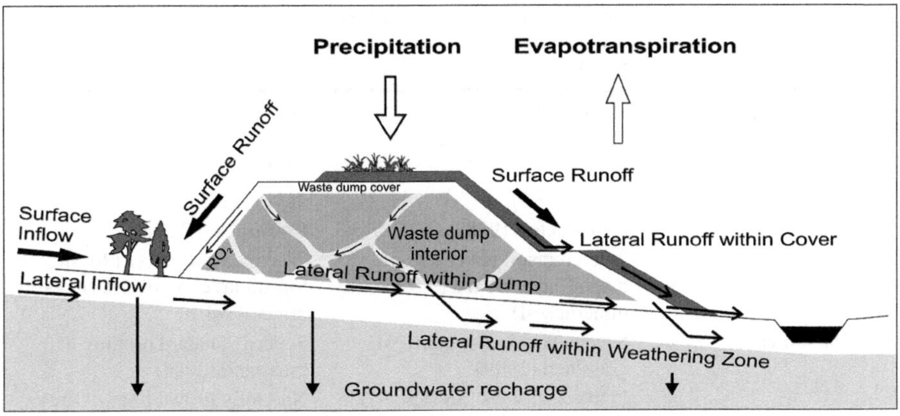

Fig. 2 Scheme of the mine dump water balance (Hähne et al. 2000)

Site Characterization

The Schlema-Alberoda region is situated in Saxony, Eastern Germany, in the lower levels of Ore Mountains, between altitudes of about 320 and 450 m NN. The climate is humid, the mean annual precipitation amount is approximately 880 mm/a, the potential evapotranspiration is between 500 and 550 mm/a. The mean air temperature is between –1.2°C in January and 16.6°C in July (LfULG 2010).

The both soil-hydrological test fields are located at the slopes of two of the larger remediated waste rock dumps covered by a soil cover system. They continuously have been operated meanwhile since 11 years (Site 1) and 7 years (Site 2).

The soil cover is a two-layered cover system consisting of a 20 cm thick organic topsoil layer and an 80 cm thick layer of an inert silty mineral soil material. The cover system does not contain any sealing or drainage component. The percolation water amount shall be reduced by an optimal soil water storage of precipitation water and, at the same time, by optimization of the water consumption by evapotranspiration. For this reason a silty soil with a high water storage capacity and a relatively low hydraulic conductivity was selected as mineral soil cover material. The 20 cm thick topsoil layer was enriched with organic material by mixing of mineral soil and humus. This was done for enhancement of water storage capacity, to provide a sufficient nutrient supply for the plant cover and to improve the ability of root penetration. Furthermore the topsoil layer reduces erosion of mineral soil as much as possible. Because of the relatively low hydraulic conductivity of the mineral soil layer, lateral runoff within the topsoil layer may occur.

The seeded grass cover has been further developed by unhindered natural succession. At present, the vegetation may be described as a fast-growing dense grassland, which is more and more dominated by planted or naturally migrated trees and shrubs. The hydrologic relevant location conditions of both test fields are listed in Table 1. As the table shows, the hydrologic conditions of both test fields are similar.

Table 1 Hydrologic conditions of the locations of test fields

Condition	Site 1	Site 2
Name of dumpsite	Halde Borbachdamm	Halde 366
Altitude	380 m NN	360 m NN
Slope direction	ENE	WSW
Inclination of slope	40% (1 : 2.5)	40% (1 : 2.5)
Cover system	Constructed 1998 2-layered soil cover (20 cm topsoil layering over 80 cm silty mineral soil)	Constructed 2004 2-layered soil cover (20 cm topsoil layering over 80 cm silty mineral soil)
Vegetation cover	Seeded standard mixture of grassland (1999) Planted willows (Salix spec.) in the surrounding of measuring installations (2000)	Seeded standard mixture of grassland (2004) Naturally grown trees of grey alder (Alnus incana) in subareas of measuring installations
Start of measurements	November 1999	November 2004
Comments	Measuring probes have been influenced more and more by the surrounding willow shrubs in a distance of approx. 2 m	The succession of plants differs between several subareas of the test field, therefore 3 different vegetation variants may be assessed

Methods

Measurements

The dynamic of soil water and soil temperature has been measured at the soil-hydrological test fields by permanently installed soil probes in two different depths of 40 cm and 80 cm. The soil probes provide hourly data, which are recorded by data loggers. The following parameters have been measured:

- Soil water tension (matrix potential) by pressure transducer tensiometers (up to 850 hPa, equal to pF 2.9) and equitensiometers (up to 15,000 hPa or pF 4.2),
- Soil water content by TDR probes (Time Domain Reflectometry) and FDR probes (Flow Domain Reflectometry).
- Soil temperature by resistance thermometers.

Additional measurements of soil water content in 6 different depths have been performed regularily by a mobile soil profile probe.

Beside the soil-hydrological test fields, lysimeter measurements have been performed at both dumpsites (Schramm et al. 2006). Two large lysimeters were installed during the construction of the cover systems with a collection area of 10 m² each. The collection basins of these lysimeters are positioned immediately underneath the lower edge of the soil cover system within the dump material. There is another large lysimeter which collects the lateral runoff flowing within the cover system. In addition to the described large lysimeters an installation of smaller slope lysimeters has been constructed. The installation consists of four slope lysimeters with a collection area of 1 m² each. Two of those lysimeters collect the amount of water percolating vertical downwards through the cover system into the dump material. The lateral runoff within the cover system is collected additional by the other two slope lysimeters.

Data of 3 meteorological stations were used for the evaluation of the measurement results. The stations are operated within the frame of environmental monitoring program of Wismut GmbH. They are located in a short distance of a few kilometers to the test fields and lysimeters. There is an additional opportunity to use official climate data of meteorological station Aue of German Meteorological Service (DWD). At soil-hydrological test field Site 2 another rain gauge is operated.

The soil physical characteristics of the cover and dump materials have been determined by in-situ and laboratory investigations within the frame of applicability tests prior to the application of the materials, during the construction works of the cover system as well as several years after construction of the cover system. The soils and materials may be described in detail concerning the relationship between soil water content and soil water tension and concerning to the unsaturated and saturated hydraulic conductivity. The stage of vegetation development at the test fields was documented regularly. The composition of plant species was surveyed and evaluated by plant sociological investigations due to the method of Braun-Blanquet (1964). Additionally the root distribution in the cover soil was determined.

Data Analysis

The measured data were assessed annually. The time series of water tension, water content and soil temperature were used for calibration of the water balance model CoupModel (Jansson and Karlberg 2011). Then the calibrated model was used to calculate water balances of the plant and soil cover of both dumpsites. All major parts of the water balance like precipitation, interception evaporation, soil water evaporation, soil water transpiration, surface runoff, infiltration, lateral runoff inside the soil cover and percolation through the soil cover were calculated. Additionally, the calibrated model was verified by the results of lysimeter measurements.

CoupModel (Coupled Heat and Mass Transfer Model for Soil-Plant-Atmosphere Systems, (Jansson and Karlberg 2011) initially was developed as a water balance model of agricultural and silvicultural sites. It is a physically based model. Due to the application of well-known physical equations the model may applied to a wide range of site conditions and ecosystems independent from geographical situation. The basic structure of the model is a depth profile of the soil. Vegetation characteristics may be described by a variety of parameters and subroutines. All relevant hydrological processes like snow accumulation and melt, soil heat flow and frost, interception, potential and real evapotranspiration, surface runoff, lateral runoff and vertical water flow are considered. Potential evapotranspiration is modeled by using the Penman–Monteith approach (FAO 1998). Calculation of saturated and unsaturated water flow may be performed using the Richards equation in one-dimensional direction.

For the model application CoupModel initially was calibrated by the standard potential evapotranspiration of a short grass layer using the Penman–Monteith approach described by FAO (1998). Then the model was composed step by step by activating the necessary subroutines. The actual calibration of CoupModel is performed by adjustment of relevant model parameters as long as the measured water tension, water content and soil temperature of both investigated depths of soil profile are reproduced sufficiently. For each period of model calibration the most important parameters describing the vegetation cover by the approach of Penman–Monteith had to be adjusted. Those parameters are roughness length, displacement height and surface resistance. Additional parameters like leaf area index, root depth, root distribution and the parameters of the reduction function of transpiration regarding to oxygen stress and drought stress have to be adjusted for each model application.

Results

Measured soil water tension, soil water content and soil temperature of each profile depth could be simulated by the calibrated model without remaining substancial discrepancies over the entire investigation period. The same applies to the calculated runoff amounts, which could be validated by the results of lysimeter measurements. Hence the model has the ability to simulate the dynamics of hydrological processes and soil water balance. An exemplary comparison of measured values and model results is given by Figs. 3 and 4. The figures illustrate the currently investigated conditions during the vegetation period of year 2010. As the second half of this year was very wet, an unusual large amount of precipitation was measured.

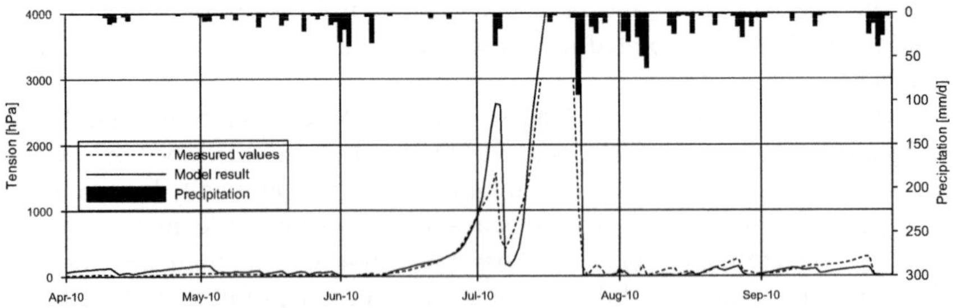

Fig. 3 Exemplary comparison of soil water tension, measured values and calculated by calibrated water balance model (Site 2, grey alder)

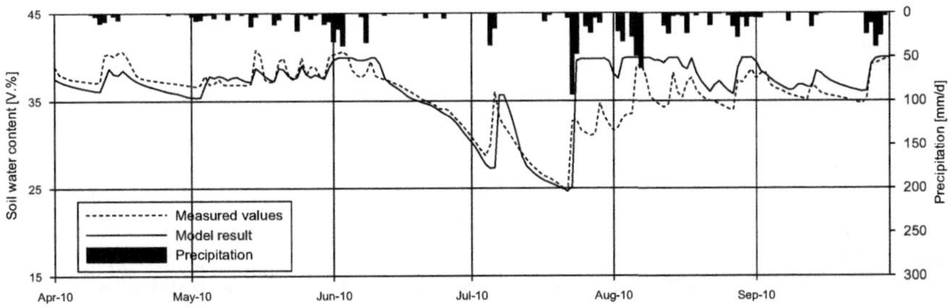

Fig. 4 Exemplary comparison of soil water content, measured values and calculated by calibrated water balance model (Site 2, grey alder)

Present Results of Site 1

Site 1 is characterized by a mineral cover soil with a relatively low hydraulic conductivity of 1E-07 m/s. Therefore there is a possibility for occurrence of saturated zones within the cover system during periods with high precipitation amounts in winter and springtime. So, positive pressures caused by backwater in the mineral soil may be measured in each winter, especially in a depth of 80 cm. Furthermore the air capacity of the soil is relatively small. Although these characteristics are rather unfavorable for the establishment of vegetation, the cover system was penetrated by plant roots over its entire profile within two or three years only. During the second year after seed the soil was covered by a dense layer of grass and herbs. This vegetation realized a water consumption of 460 mm/a, compared to an evaporation of only 300 mm/a of bare soil during construction of cover system in 1999. In the following years from 2002 to 2006 the grassland realized an evapotranspiration amount in the range of approx. 450 … 500 mm/a, this is approximately 80% of potential evapotranspiration of a short grass layer. Roughly around the years 2006 and 2007 the willow shrubs in the surrounding of the installation of the soil probes started to influence the soil water consumption obviously. The total evapotranspiration ranges now around 650 mm/a or between 110 and 130% of potential evapotranspiration of a short grass layer, respectively. The change of total evapotranspiration and its three main components are shown in Fig. 5. Transpiration of the plant cover becomes the dominant component of evapotranspiration.

An increase of water consumption by the vegetation cover leads to a reduction of the runoff amounts (see Fig. 6). So the water percolation through the soil cover system could be reduced from approximately 500 mm/a at bare soil conditions to less than 200 mm/a in the years 2007 to 2009. The percolation nearly falls to zero in summer beginning in the third year after seed. However if the precipitation amounts are very high like in 2010 a percolation of soil cover system occurs permanently. The mean amount of lateral runoff within the organic topsoil layer did not considerable change, it normally stays below 100 mm/a.

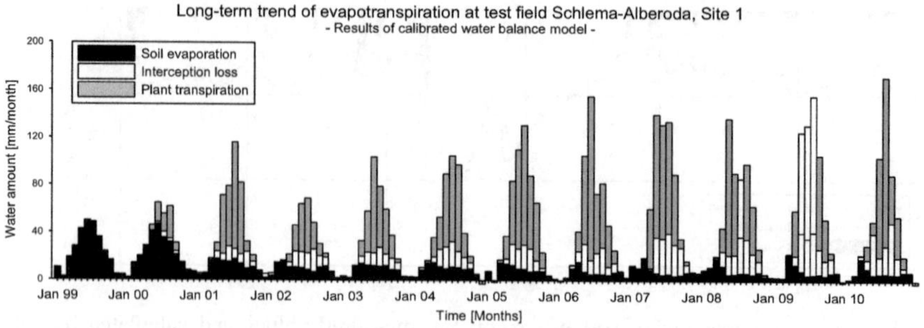

Fig. 5 Trend of total evapotranspiration at test field Site 1 and its components during the entire observation period, calculated by the calibrated water balance model

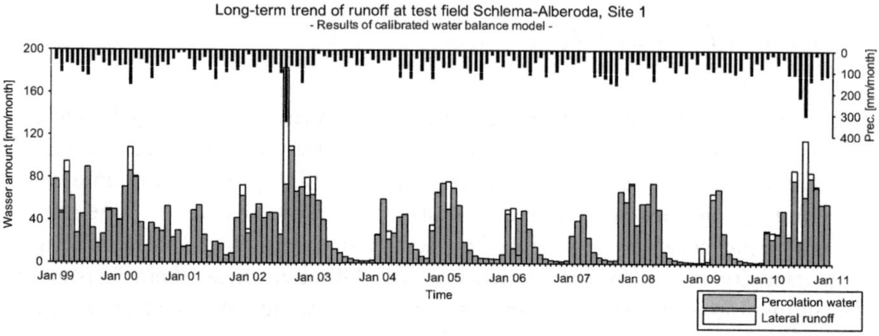

Fig. 6 Trend of monthly runoff amounts at test field Site 1 during the entire observation period, calculated by the calibrated water balance model

Present Results of Site 2

At soil-hydrological test field Site 2 there are different subareas with various vegetation types, so it is possible to consider three model variants starting at year 2009. The first variant is unmowed grassland vegetation without trees or shrubs. Here the water consumption caused by evapotranspiration amounts to about 450 ... 500 mm/a. This amount is comparable to the situation at test field Site 1. At another subarea the grass layer has been mowed periodically, the vegetation type may be described as a short grass layer or lawn. Here the evapotranspiration only lies around 60...70% of the potential evapotranspiration of a short grass layer, it ranges around the small amount of 350 mm/a. However, in a subarea with young trees of grey allow the water consumption has been rised to approximately 120% of potential evapotranspiration of a short grass layer (approx. 650 mm/a). The described change of total evapotranspiration of both test fields and subareas is shown in relation to the potential evapotranspiration of a short grass layer in Fig. 7.

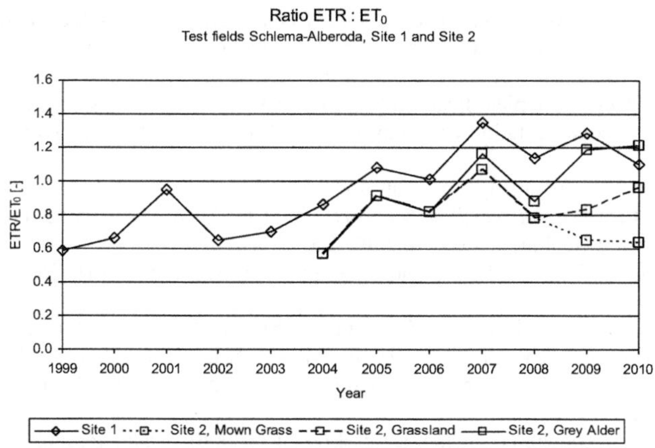

Fig. 7 Trend of real evapotranspiration (ETR) in relation to potential evapotranspiration of a short grass layer (ET_0)

Conclusions

The results of the performed investigations by test fields and lysimeter measurements and of the data evaluations verify the efficiency of the soil cover system of waste rock dumps consisting of a 20 cm thick layer of organic topsoil and a 80 cm thick layer of silty mineral soil in regard to the reduction of percolation water amount. It is shown that the plant cover grows fast and dense under these conditions. The plants considerably increase the evapotranspiration of soil water. Trees and shrubs which are planted or naturally grown may realize a water consumption higher than predicted by water balance models in the past. The soil of the constructed dump cover system in Schlema-Alberoda provides very good conditions for plant growth. The hydrologic characteristic of the used soil materials (water storage characteristics, hydraulic conductivity) did not change significantly during the present observation period compared to the dynamic and fast developments of plant cover. Hence the unchanged low hydraulic conductivity is a contribution to the effective reduction of percolation water amount in combination with the water consumption of the dense vegetation cover of the remediated waste rock dumps.

References

Braun-Blanquet (1964): Pflanzensoziologie. Grundzüge der Vegetationskunde. Springer-Verlag Berlin 1928, 3. renewed ed. Springer-Verlag Wien u. New York 1964

FAO (1998): Crop evapotranspiration – Guidelines for computing crop water requirements. Edited by Food and Agriculture Organization of the United Nations (FAO), FAO Irrigation and drainage paper 56, Rome, 1998

Hähne R., Löser R., Schöpe M. (2000): Wasserhaushaltliche Renaturierung von Halden und Deponien. Freiberger Forschungshefte C 482, Berg- und Hüttentag Freiberg 2000, S. 311–321

Jansson P.-E., Karlberg L. (2011): Coupled heat and mass transfer model for soil-plant-atmosphere systems. Official documentation of CoupModel, KTH Mark- och Vattenteknik Stockholm (http://www2.lwr.kth.se/Vara%20 Datorprogram/CoupModel/index.htm, 30.05.2011)

LfULG (2010): Gewässerkundliche Monatsberichte. Monthly edited by Sächsiches Landesamt für Umwelt, Landwirtschaft und Geologie, Dresden, 2002–2011

Runge W. et al. (1999): Chronik der Wismut. Edited and copyright © by Wismut GmbH, Chemnitz, 2738 p.

Schneider P., Löser R., Meyer J., Kreyßig E. (2005): Concept of a Surface Water Monitoring at the Former Uranium Mining Site Schlema-Alberoda, in: Merkel B.J., Hasche-Berger, A.: Uranium in the Environment – Mining Impact and Consequences, pp. 779–787, Springer Verlag ISBN 3-540-28363-3

Schramm A., Nindel K., Meyer J., Knoche D., Schneider P., Löser R. (2006): Lysimeteruntersuchungen der Wismut GmbH im Bereich der Niederlassungen Aue (Standort I) und Königstein (Standort II). In: Langzeitverhalten von Deponieabdichtungen, Wissenschaftliche Berichte der Hochschule Zittau/Görlitz, Heft 91 (2006), Nr. 2234–2257, S. 123–136 (ISBN-Nr. 3-9811021-3-4)

Implementation of a Modeling Concept to Predict Hydraulic and Geochemical Conditions During Flooding of a Deep Mine

Thomas Metschies, Ulf Jenk

Abstract. More than four decades of intensive uranium mining in Germany resulted in a legacy of surface and underground mining objects requiring remediation in a densely populated area. The remediation concepts have to take into account the predictions of future environmental impacts for various remediation activities. Controlled flooding of deep uranium mines to a natural flooding or overflow level is the general remediation solution selected. Thereby the local and regional hydraulic and geochemical conditions of lager areas at and around the former mining site are influenced over a long period. For the optimization of the operations as well as the application for the necessary permissions model predictions are regularly used. In the presented case of the Königstein deep mine flooding the modeling concept followed combines commercially available groundwater modeling software to account for the relevant hydrogeological and geochemical processes in the mine and the influenced surroundings. It was found that there are no off-the-shelf solutions available allowing to model both the hydraulic and geochemical conditions on a local and regional scale. Furthermore a flexible modeling concept is required to include additional modeling tools if necessary and to allow better traceable model predictions. Based on monitoring of the ongoing flooding process the concept is further developed and the models are continuously checked, adopted and extended. Thereby the system understanding is improved allowing to further narrow the prediction uncertainties.

In addition to a description of the applied modeling tools and their interaction the paper presents the experiences gained during this process and their implications for the further improvement of the model predictions.

Thomas Metschies
WISMUT GmbH, Jagdschänkenstraße 29,D-09117 Chemnitz

Ulf Jenk
WISMUT GmbH, Jagdschänkenstraße 29,D-09117 Chemnitz

B. Merkel, M. Schipek, *The New Uranium Mining Boom*
DOI 10.1007/978-3-642-22122-4 © Springer 2011

Introduction

Flooding of former underground deep mines is necessary to achieve long-term sustainable conditions as consequence of the uranium mining legacy remediation in the Eastern Part of Germany. Restoring the natural flow conditions a release of contaminants from the former mines into the surrounding environment could appear. Based on the national and European environmental legislation an optimization of the flooding process is necessary to avoid or at least minimize any adverse effects on the neighboring ground and surface water resources.

The approach followed at Wismut consists of a controlled, stepwise increase of the mine water level with simultaneous monitoring of the hydrochemical, hydraulic and geomechanical response in the mine and the surrounding environment resulting in an extensive hydraulic and hydrochemical data set (Gatzweiler et al. 1998).

The Königstein mine site is situated in the densely populated Elbe river valley southeast of the Saxon capital Dresden. From the early sixties through 1990 approx. 19,000 t of uranium were produced (Jenk et al. 2008). The ore body situated in the deepest of four sandstone aquifers was first mined conventionally and later leached using sulfuric acid (2 to 3 g/l H_2SO_4). The leaching operation took place from underground mine workings which were used for injection and drainage of the lixiviant. As the result the leached rocks remained hydraulically isolated from the surrounding aquifer. Due to the leaching operation and secondary effects as pyrite oxidation within the leached sandstone blocks about 2 mil. m³ highly concentrated pore water solutions remained (Table 1).

Flooding of the mine is connected with the mobilization of this contaminant potential resulting in a possible spreading of contaminants into the surrounding aquifers. It is the aim of the remediation concept to wash out the main part of the mobile contaminants from the mine in a controlled way to avoid contamination of the surrounding aquifers. For the approval of an underground mine flooding an assessment of environmental impacts has to be provided to the licensing authorities. Basis for such an assessment are predictions of the expected future conditions using hydraulic and geochemical models.

Table 1 Pore water concentrations in leached sandstone blocks (Jenk et al. 2008)

pH	EC	SO_4^{2-}	Fe	Zn	U	Al
–	mV	mg/l	mg/l	mg/l	mg/l	mg/l
2.0	700	10,000	2–3	200	200	300

Modeling Concept

Modeling the environmental impacts of mine flooding on ground and surface waters is a complex problem. The identification of relevant processes which are necessary to be taken into account shows that different temporal and spatial scales have to be considered. While hydraulic flow conditions usually have to be seen in a more regional context contaminant releases are determined by processes on the local down to the microscopic scale. There is no commercial modeling tool available appropriately reflecting all these aspects. It was therefore necessary to develop a consistent modeling approach based on a number of commercially available but also on purpose made modeling tools which have clearly defined interfaces to ensure traceable predictions.

Precondition for any modeling activity is a consistent conceptual understanding of the site and the relevant processes. At the Königstein mine hydrogeological and hydraulic understanding is based on the knowledge collected during the period of prospecting, mining and remediation planning. The available information is further extended during the stepwise mine flooding requiring a continuous revision of the conceptual understanding. The resulting conceptual understanding of the natural and anthropogenic conditions is summarized as Conceptual Site Model (Fig. 1) being the basis for all analytical and numerical analyses.

Based on a review of the existing knowledge base and additional measurements a well founded understanding of the regional groundwater flow and a great number of level and flow measurements as a basis for a regional flow model exist. A vast number of program codes for regional flow models are commercially available. Together with the licensing authorities the FE code SPRING was selected based on the experience gathered by implementing the flow conditions in MODFLOW in the early 1990s. The flexibility in defining geometric objects and the ability to describe unsaturated flow conditions within the chosen FE code that time was a main reason for this decision.

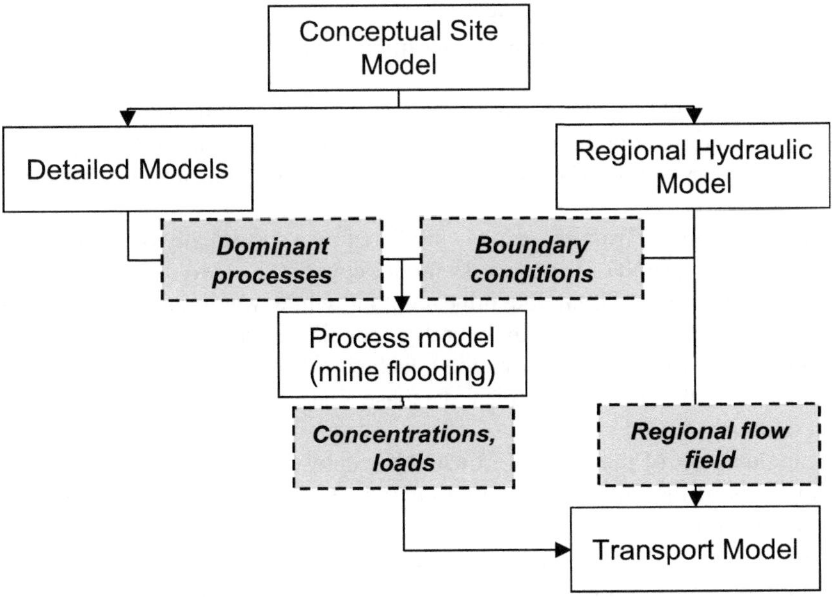

Fig. 1 Modeling approach for the Königstein mine

The regional flow model is a combined saturated/unsaturated model applying the Van-Genuchten-functions (van Genuchten 1980) allowing to describe the respective free water table in the relevant aquifers of the stratified sandstone aquifer system. In total the model includes about 400,000 knots on 19 layers with a finer vertical discretization of the relevant 2 deep aquifers and the aquiclude in between. Relevant tectonical elements such as fracture zones and anthropogenic hydraulic connections between the aquifers are separately implemented.

In addition a specific mine boundary condition was implemented in the modeling code SPRING allowing an appropriate representation of the water level changes in the mine based on water balance calculation for all mine cells in each modeling time step. During and after flooding the mine water level is an important boundary condition determining the development of the regional groundwater flow field.

The regional hydraulic model was calibrated based on the water level and the mine water inflow measurements conducted during and after the active mining phase. Having both detailed water level and groundwater inflow measurements for a complex flow system the model calibration became very time-consuming. First model calibrations were done using the measurement results of the mining phase which was characterized by continuous drawdown of the water table in the adjacent aquifers. Detailed monitoring measurements during the stepwise flooding starting in 2001 were used for continuous model verification. Comparing modeling results for the ongoing flooding process adjustments of model parameters were required due to a different hydraulic behavior during the drawdown and flooding phase but also because of additional monitoring points installed meanwhile. In contrast to active mining when water table depressions were generated in the adjacent aquifers the period of stepwise flooding resulted in a backfilling of the groundwater depressions. In the upper unsaturated aquifers this turned out to be a long-lasting process.

Using the hydraulic model the development of the regional flow field can be predicted for various flooding options providing the necessary information on hydraulic boundary conditions such as the mine water in- and outflows for other models (Fig. 1).

Contaminant release from the mine is predicted by a specifically developed box model called FLOODING. This process model covers the relevant hydraulic and geochemical processes within the mine contour while water in- and outflows through this contour are defined as boundary conditions taken from the regional flow model. The process model uses the geochemical modeling code PhreeqC in combination with the thermodynamical database Wateq4f which was amended by relevant uranium species.

The single boxes of the model FLOODING represent the hydraulic properties of certain parts of the mine and have an internal structure. They include the volume of the open mine workings as well as the saturated and unsaturated pore volume of the rocks. Boxes representing the base mining levels in impermeable bedrock contain only open mine working volumes while boxes of the mining areas in the sandstone consist of separate volumes for mine workings and pore spaces.

Within the hydraulic model the open mine workings of neighboring boxes are connected allowing the water flow through the mine. Within the single boxes the pore volume is coupled with the open mine workings based on the Darcy equation using effective hydraulic parameters.

Pore water concentrations and secondary mineral contents are defined based on field and lab measurements as geochemical boundary conditions representing the source of contaminants. Flooding of the mine leads to a dissolution of the secondary minerals and dilution of the pore water. At the beginning the contaminant release from the flooded mine is dominated by the advective transport of the dissolved contaminants in the open mine workings. Flushing out of the hydraulically well connected mine openings results in a peak of released contaminant concentrations and loads. The long-term release of contaminants depends on slower processes. It was found in detailed model and field tests in separated mine parts that in the Königstein mine density driven flow will be the relevant process for medium to long-term contaminant release from the sandstone blocks. Diffusive processes connecting stagnant parts are less important.

A main requirement for calibration of the model is that all boxes have a similar parametrization as long as no differences could be justified e.g. by different mining technologies or remediation activities such as contaminant immobilization were applied in certain parts.

The process model was calibrated based on field experiments in isolated mine parts described with separate boxes and is continuously tested based on the monitoring results of the pumped mine water as integral response of the flooding process. The conditions within the single boxes are checked for plausibility. However the model concept and the available parameters does not allow to describe local effects as e.g. reported in single boreholes within specific mine spaces (Fig. 2).

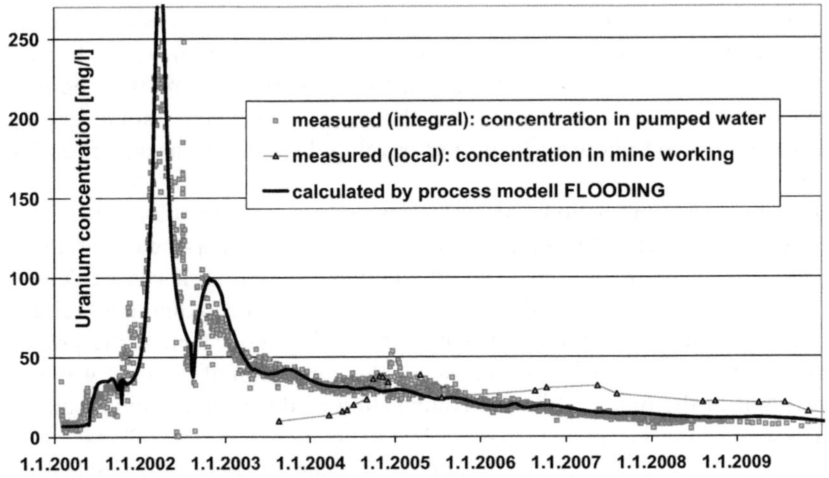

Fig. 2 Measured and calculated uranium concentrations in the pumped mine water exemplary compared to local conditions within the flooded mine

Based on the results of the geochemical process model and the predicted regional flow field the effect of the contaminant release can be estimated for the downstream aquifer areas. At present several 1D-flow models are prepared based on the Transport option of PHREEQC to exemplary predict the hydrochemical conditions within the affected aquifers along the flow paths determined in the regional flow model. The results of the regional flow and geochemical process model define the boundary conditions for these reactive transport models. The results of the transport modeling in the surrounding aquifers is a precondition for the approval of the final remediation steps by the authorities.

Experiences Gained Following the Modeling Concept

In the middle of the 1990s the expectations concerning model predictions were high. Three dimensional models including all relevant hydraulic and geochemical processes seemed to be possible due to the development of the computer sciences and modeling codes. On the other hand it was required that models have to be flexible, precise and the model results traceable. Technically such models could be realized. However the necessary conceptual approximations concerning the relevant processes and the natural variability of main model parameters result in a wide range of model uncertainties. Tracing the uncertainties in a complex model is a difficult task. The effect of single processes and parameters might be superposed and therefore lead to faulty conclusions.

The modeling concept followed at the Königstein site allows to study the influence of the process approximations as well as the parameter variability using the respective modeling tools representing the relevant temporal and spatial scales. Scenario analyses can be run for the variation of certain parameters without getting superposed effects. Hydraulic effects of the mine flooding on the regional flow are determined with the appropriate modeling tool while the contaminant release is modeled on the scale of the flooded mine. Using this approach the understanding of the complex system, the major influencing parameters and processes were refined allowing to further develop the conceptual understanding of the site.

Concerning exact predictions of water levels and concentrations at certain points in the flow system this approach shows the effect of parameter variability which can not be significantly reduced by additional sampling. It can be demonstrated that instead of precise predictions for a single reference point more reliable predictions are achieved for integral units. As for the flooded mine the process model does not allow to predict the geochemical composition of flooding waters at a certain point in the mine, while the composition of the pumped mine water as integral system response can be well described by leveling out local effects.

Using this modeling concept clearly defined interfaces between the single models concerning water levels, flow volumes or concentrations are necessary. Based on the continuous model development the effect of changes in one model have regularly to be transferred to the other models requiring a permanent maintenance of the used modeling tools. This can be best ensured in remediation activities with a longer project period.

Conclusion

Basis for a sound modeling is a conceptual model of the site documenting the understanding of relevant processes and parameters. In course of the project this conceptual model is further refined using ongoing research and monitoring results. Process identification and the assessment of relevant parameters and their data range is essential for a reliable model prediction.

Furthermore appropriate modeling tools have to be selected reflecting the temporal and special scale of the relevant processes. In case of the Königstein mine it was found that there are no off-the-shelf solutions available allowing to model both the hydraulic and geochemical conditions on a local and regional scale. Therefore the combination of single modeling tools based on commercially available and on purpose prepared models was considered. Such a modeling system requires a continuous maintenance of the models and the model data which is transferred from one model to the other at the defined interfaces.

This additional effort on the other hand provides a higher traceability of parameter and process uncertainties in the respective model and reduces the risk of misinterpretations.

The presented modeling concept of the Königstein deep mine allows to easily include additional technical and geochemical enhancement measures to be implemented to reduce the contaminant release. Different flooding scenarios can be calculated to estimate the resulting environmental effects. Furthermore this modeling concept allows to define mine water treatment requirements and helps to optimize the flooding process.

Acknowledgements The authors want to acknowledge the contributions to the presented modeling concept made by Prof. Christoph König and Martin Becker (deltah, Witten) concerning the regional flow model, Dr. Harald Kalka (UIT GmbH, Dresden) implementing the concept of the process model FLOODING and Prof. Wolfgang van Berk (Technical University Clausthal-Zellerfeld) developing the geochemical model of the influenced aquifers downstream the Königstein mine.

References

Gatzweiler R, Jakubick AT (1998) Hydrogeological predictions in the remediation of the decommissioned WISMUT facilities. In: Proc. Int. Workshop "Uranium Mining and Hydrogeology II" (ed. by B Merkel and C Helling), Freiberg, September 1998, Verlag Sven von Loga, Köln 1998, p. 1

Jenk U, Nindel K, Zimmermann U (2008) Underground in-situ mine water treatment in a flooded uranium mine at the WISMUT Königstein site – motivation activities and outlook. In: Merkel B, Hasche-Berger A (Eds), 2008 Uranium Mining and Hydrogeology, Springer Heidelberg, pp 431–436
Van Genuchten T (1980) A closed form equation for predicting the hydraulic conductivity of unsaturated soils. Soil Sci. Soc. Am. J., pp. 892–898

Radiological Impact Assessment of the Uranium Tailings Pond at Turamdih in India

R.N. Nair, Faby Sunny, Manish Chopra, V.D. Puranik

Abstract. Groundwater flow and contaminant transport modeling has been carried out for the uranium tailings pond at Turamdih in Singhbhum district of Jharkhand, India using the finite element based FEFLOW software with a view to assess its radiological impact on human and the environment. Hydrogeological investigations; including laboratory and field based studies; have been carried out to collect site-specific data on geological settings, cross sections and aquifer characteristics of the site. Results indicate that the concentrations of U-238 and its progenies and the corresponding annual effective dose rates to members of the public through groundwater drinking pathway are less than the corresponding standards even at a distance of 500 m from the boundary of the tailings pond. The radiological impact in groundwater at this distance is practically nil up to a period of 4000 years and trivial beyond this period.

Introduction

Uranium tailings ponds by the virtue of the longevity of its contents may pose as a long-term radiological hazard if not designed and monitored properly. Hence,

---

R.N. Nair
Environmental Assessment Division, Bhabha Atomic Research Centre, Mumbai-400 085, India

Faby Sunny
Environmental Assessment Division, Bhabha Atomic Research Centre, Mumbai-400 085, India

Manish Chopra
Environmental Assessment Division, Bhabha Atomic Research Centre, Mumbai-400 085, India

V.D. Puranik
Environmental Assessment Division, Bhabha Atomic Research Centre, Mumbai-400 085, India

B. Merkel, M. Schipek, *The New Uranium Mining Boom*
DOI 10.1007/978-3-642-22122-4

the long-term radiological impact assessment of uranium tailings ponds becomes a regulatory mandate. Such an exercise has been carried out for the uranium tailings pond at Turamdih in the Singhbhum district of Jharkhand, India. Hydrogeological investigations; including laboratory and field based studies; have been carried out to collect site-specific data on geological settings, cross sections and aquifer characteristics of the site. Groundwater flow in the Turamdih watershed is modeled using the finite element based software, FEFLOW using these site-specific hydrogeological data. The transport of U-238 and its important progenies from the tailings pond is also modeled using the same software and a three-dimensional numerical decay chain transport model.

Hydrogeological Data

Hydrogeological investigations have been carried out at the site to collect site-specific data on geological settings, cross sections and aquifer characteristics (UCIL 2003; Tewary et al. 2008). The resistivity survey shows that the site consists of three distinct vertical layers (top layer of soil, middle weathered rock zone and bottom fractured hard rock zone) forming a single unconfined aquifer with varying thicknesses. Pumping tests are carried out at Talsa, Bada Talsa and Kherva Dungri to characterize the unconfined aquifer in the site. The hydraulic conductivities at Talsa and Bada Talsa lie in between 0.11 and 0.16 m/day. Very low hydraulic conductivity (0.017 m/day) is observed at Kherva Dungri which is about 3 km from Bada Talsa. The domain of study including the uranium tailings pond is shown in Fig. 1.

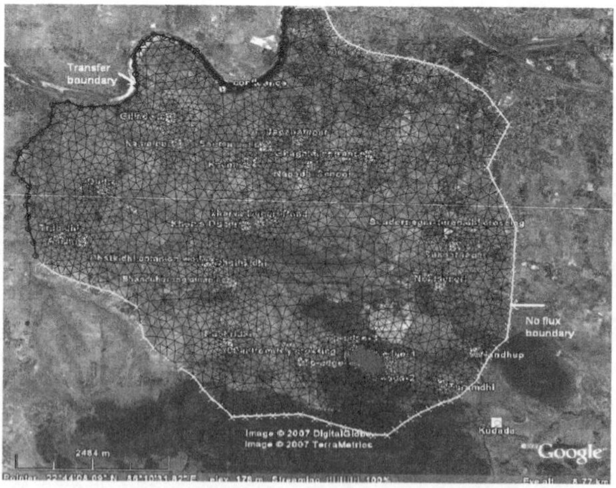

Fig. 1 Domain of study including the uranium tailings pond

Contaminant Transport Data

The amount of U-238 in the tailings pond is calculated by determining the un-recovered U-238 in the U_3O_8 ore (processed for an operational life of 20 years) that would be dumped in the tailings pond. U-238 is the parent radionuclide of a long natural radioactive decay series. Since the parent radionuclide is very long-lived, its daughter products are in equilibrium with their parent. The series has radon gas in the chain and the equilibrium will be disturbed from radon gas onwards. Short-lived radionuclides like Th-234 and Pa-234 are not considered in the transport model. The inventory of radionuclides in the tailings waste corresponds to 3000 t/day ore having a grade of 0.03%. As 90% recovery is expected for natural uranium, the inventory of its daughters will be higher by a factor of 10 in the tailings waste. Thus, the initial activities of U-238, U-234, Th-230 and Ra-226 are 6.04×10^{12} Bq, 6.04×10^{12} Bq, 6.04×10^{13} Bq and 6.04×10^{13} Bq respectively. The distribution coefficients (K_d values) of these radionuclides for soil are reported as 500, 500, 3000 and 500 ml/g respectively (IAEA 2003).

FEFLOW Model Software

Groundwater flow in the Turamdih watershed is modeled using the finite element based software, FEFLOW (WASY FEFLOW 2002). The FEFLOW is a finite element simulation package for two-dimensional and three-dimensional density-dependent flow, mass and heat transport in groundwater. It solves the advection-dispersion equations for contaminant mass and heat. The model allows the analysis of both saturated and unsaturated aquifers. Adsorption effects can be modeled via Henry, Freundlich or Langmuir isotherms.

Groundwater Flow Modeling

The mathematical treatment of groundwater flow through a porous medium depends upon an equation that captures the essence of the physics of flow. The basis for developing such an equation is a conservation statement that balances the inflow, outflow, and change in water mass within a representative volume of porous medium (continuity equation). A combination of continuity equation and Darcy's law can be used to formulate the groundwater flow equations. There are different formulations of these differential equations. For example, Eq. (1) states the rate of change of hydraulic head, h, in a two-dimensional domain of an unconfined aquifer.

$$S_y \frac{\partial h}{\partial t} = \frac{\partial}{\partial x}\left(K_x h \frac{\partial h}{\partial x} \right) + \frac{\partial}{\partial y}\left[K_y h \frac{\partial h}{\partial y} \right] \pm q \qquad (1)$$

where S_y is the specific yield; K_x, K_y are the hydraulic conductivities in x and y directions (m/s); and q represents the pumping and/or recharge rates (m/s).

Contaminant Transport Modeling

The equation describing the transport and dispersion of a dissolved chemical in groundwater can be derived from the principle of conservation of mass of the solute. The principle of conservation of mass requires that the net mass of solute entering or leaving a specified volume of aquifer during a given time interval must equal to the accumulation or loss of mass stored in that volume during the interval. This relationship may then be expressed mathematically by considering all fluxes into and out of a representative elementary volume. A generalized form of the solute transport equation for a porous medium is given by:

$$\frac{\partial}{\partial t}(\varphi \mathrm{C}) = -\nabla(v\mathrm{C} - \varphi \mathrm{D}\nabla \mathrm{C}) \pm Q \tag{2}$$

with porosity φ, combined coefficient of diffusion and dispersion tensor D, specific discharge *v* and source/sink term *Q*.

Boundary Conditions

The conceptualization of the flow model is carried out by considering the drainage pattern of the site. Drainage of the site measuring 9.1 × 7.3 km (Fig. 1) is controlled through a network of small streamlets with a general northward flow direction ultimately draining into Kharkai River. The tailings pond which is enclosed by the Banduhurang Ridge in the south is drained by a number of small drainage channels, which flow into small steams ultimately joining the Kharkai River. The most prominent stream of the area, the Seema Nala originates from the western end of the Banduhurang Ridge which flows northward into the Kharkai River located at about 3.5 km from the tailings pond. The Kharkai River is a perennial river and a major tributary of the Subarnarekha River on the south-western fringes of Jamshedpur city near Kagalnagar. The terrain elevation is higher (400 m) in the southern side (Banduhurang ridges) compared to the northern side (130 m).

The site is bounded by natural hydrogeological boundaries such as the Banduhurang Ridge on the south and the Kharkai River on the north. In the flow model, the north region bounded by the Kharkai River is given as transfer boundary condition, with specified heads. Amount and direction of flux between river and aquifer depends on the water head in the river and is being controlled by the collimation layer having leakance of 10 m/day. Leakance is the conductance of the interface joining the Kharkai River and the unconfined aquifer and is the ratio (K/d) of the hydraulic conductivity of the river bed (25 m/day) to the distance through which the groundwater enters/leaves the river bed (2.5 m). The south boundary of the hill ridge forms a water divide leading to a zero flux boundary condition. The western and eastern sides of the site are bounded by the prominent streams joining the Banduhurang Ridge to the Kharkai River and the groundwater flow is parallel to river, so no flux boundary condition is assumed on this side

Table 1 Features of numerical model used for the Turamdih watershed

Problem class	Combined flow and mass transport model	Vertical exaggeration	1 : 1
Time class	Transient (unsteady) flow, transient (unsteady) mass transport	Problem measure	1100 : 0
Time stepping scheme	Automatic using predictor-corrector scheme	Number of layers	3
Constraint in time step	Upper limit of 10,000 days	Number of Slices	4
Upwinding	Full upwinding	Type	Saturated
Maximum number of iterations per time step	13	Dimension	Three-dimensional
Form of transport equation	Convective	Mesh elements	28,593
Solver	Iterative equation solver	Element type	6-nodded triangular prism
Adsorption law	Henry isotherm	Aquifer	Unconfined with top surface free and moving and the other layers as unspecified

also. A three-dimensional model of the site is developed with the top slice having 100 injection wells representing the infiltration from the tailings slurry and the bottom slice having a zero flux boundary condition. The features of the numerical model used are presented in Table 1.

Results and Discussion

Three-dimensional numerical groundwater flow and contaminant transport modeling has been carried out for the uranium tailings pond at Turamdih. The approximate surface area of the tailing pond is 320,000 m^2. This study has been carried out to estimate the contamination migration pattern within the watershed area of 9.1×7.3 km. The groundwater flow simulation shows that the groundwater flow is towards the north-west direction. The groundwater velocity varies between 0.008 and 0.4 m/day (Fig. 2).

Among the long-lived nuclides, the maximum concentrations are shown by Ra-226 because of its ingrowths from the parent radionuclides and higher inventory. The minimum concentrations are shown by Th-230 because of its high K_d value. U-238 and U-234 show same concentrations as they are having similar inventories and K_d values. The concentration of Ra-226 at 500 m from the centre of the tailings pond at 1000 y is about 0.1 Bq/l and it increases to about 1 Bq/l at 10,000 y. The concentrations of U-238, U-234 and Th-230 are about 0.1 Bq/l at 10,000 y. The plume migration of U-238 and Ra-226 at 1000 y after disposal is depicted in Figs. 3 and 4 respectively.

Fig. 2 Groundwater flow velocities and directions after 1 y of model simulation

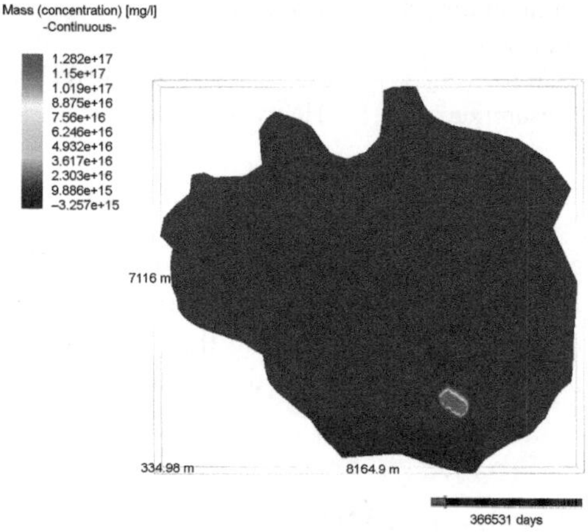

Fig. 3 Migration of U-238 from the tailings pond after 1000 y of disposal

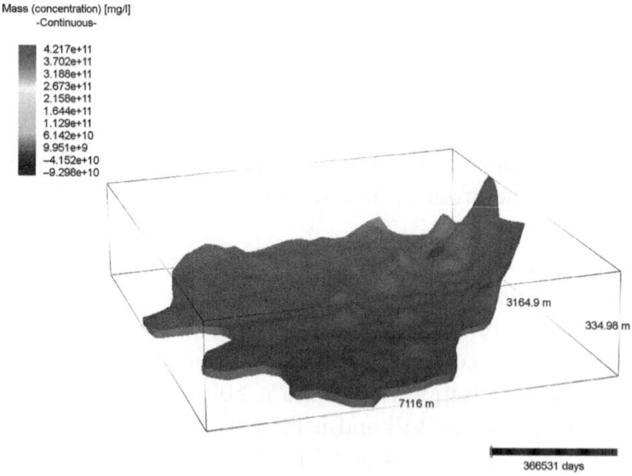

Fig. 4 Migration of Ra-226 from the tailings pond after 1000 y of disposal

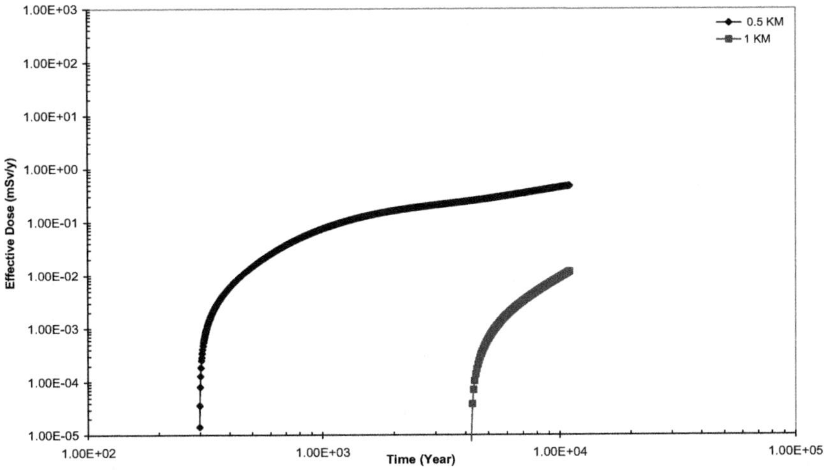

Fig. 5 Total annual effective doses to members of the public through groundwater drinking pathway at different distances from the uranium tailings pond

The concentrations of these radionuclides are translated into annual committed effective doses through groundwater drinking pathway using the ingestion dose coefficients reported by IAEA (IAEA 1996). The effect of build-up of progenies is included in the calculations. The results are depicted in Fig. 5. The total annual effective dose to members of the public at 1 km from the centre of the tailings pond (500 m from the boundary of the tailings pond) through groundwater drinking pathway is trivial up to a period of 4000 y. The dose is estimated to be 0.01 mSv/y after 10,000 years and it is 10 times lower than 0.1 mSv/y, which is considered as a safe dose limit for drinking water pathway (WHO 2004).

It is observed that Rn-222, Po-210, Pb-210 and Ra-226 together contribute about 99.75% of the total dose. All other radionuclides including U-238 contribute only about 0.25% of the total dose (Nair et al. 2010). This indicates the necessity of monitoring these radionuclides in groundwater in the vicinity of uranium tailings ponds. The doses with the decay chain transport are about 100 times higher than that without the decay chain transport. Thus the ingrowths of progenies are very important for the radiological impact assessment of uranium tailings ponds.

Conclusions

In summary, an estimation of concentrations of uranium and its daughter products in groundwater and the resultant annual effective dose rates to members of the public confidently shows that the radiological impact due to the uranium tailings pond at Turamdih is far below any concern with respect to groundwater transport scenario. The concentrations of these radionuclides and the corresponding annual effective dose rates to members of the public are found to be less than the corresponding standards at a distance of 1 km from the centre of the tailings pond. The radiological impact in groundwater at this distance is practically nil up to a period of 4000 years and trivial beyond this period.

References

IAEA (1966) International Basic Safety Standards for Protection against Ionizing Radiation and for the Safety of Radiation Sources. Safety Series No. 115, International Atomic Energy Agency, Vienna

IAEA (2003) Derivation of Activity Limits for the Disposal of Radioactive Waste in Near Surface Disposal Facilities. IAEA TECDOC-1380, International Atomic Energy Agency, Vienna

Nair R N, Faby Sunny, Manikandan S T (2010) Modelling of decay chain transport in groundwater from uranium tailings ponds. Applied Mathematical Modelling 34: 2300–2311

Tewary B K et al. (2008) Groundwater Modelling for Banduhurang Opencast Uranium Mine and Proposed Tailing Pond Area of UCIL. BRNS Project Report. Central Institute of Mining & Fuel Research, Dhanbad, India

UCIL (2003) Uranium Ore Processing Project at Turamdih: A Design Basis Report. Uranium Corporation of India Ltd, Jaduguda

WASY FEFLOW (2002) Finite Element Subsurface Flow and Transport Simulation System. Institute for Water Resource Planning and Systems Research Ltd. Germany

WHO (2004) Guidelines for Drinking-Water Quality Third Edition Volume 1 Recommendations. World Health Organization, Geneva

The Mean Hydraulic Residence Time and Its Use for Assessing the Longevity of Mine Water Pollution from Flooded Underground Mines

Michael Paul, Thomas Metschies, Marcus Frenzel, Jürgen Meyer

Abstract. The prediction of water quality trends, and the forecast of the longevity of mine water pollution are of tremendous importance for strategic decision making and long-term budget planning for water management and treatment in the wake of mine flooding. Numerical simulation tools are able to consider the influence of a huge variety of processes and parameters, however, simple and straightforward methods are sometimes necessary to apply, either for a quick estimate or due to the lack of resources and/or input data. The paper describes a simple approach for assessing the long term water quality of flooded underground mines based on very generic hydraulic data namely the mean hydraulic residence time as a key parameter. The comparison with long term monitoring data for some of Wismut's flooded underground uranium mines will be used to demonstrate the usefulness, but also the constraints of this approach.

Introduction

Flooding of underground mines is a key element of mine closure programs. At the end of the flooding process, after groundwater rebound is complete and steady-

Michael Paul
Wismut GmbH, 09117 Chemnitz, Jagdschänkenstraße 29

Thomas Metschies
Wismut GmbH, 09117 Chemnitz, Jagdschänkenstraße 29

Marcus Frenzel
Wismut GmbH, 09117 Chemnitz, Jagdschänkenstraße 29

Jürgen Meyer
Wismut GmbH, 09117 Chemnitz, Jagdschänkenstraße 29

B. Merkel, M. Schipek, *The New Uranium Mining Boom*
DOI 10.1007/978-3-642-22122-4

state flow conditions are emerging, the mine waters start to discharge into downstream water reservoirs, either of surface or groundwaters. At most flooded uranium mines the mine waters need to be treated before discharge to the environment, as a consequence of a low pH and/or unacceptable concentrations of heavy metals, arsenic and radionuclides. In many cases, water treatment turns out to be the biggest burden in the frame of long term management of remediated mine sites. Therefore, the prediction of water quality trends and the forecast of the longevity of mine water pollution are of crucial importance for strategic decision making and long-term budget planning for water management and water treatment. The same is true, if the recovery of uranium is a profitable option.

Under a post-flooding flow regime the quality of mine waters is controlled by a number of geochemical and transport processes, including the following key mechanisms: i) the flushing of pore waters in conjunction with the flood water rise, including the dissolution of readily available secondary minerals stored in formerly dewatered parts of the mine, ii) the dissolution of primary minerals, both in the saturated and the unsaturated zone of the mine, and iii) precipitation and sorption reactions of dissolved species as a consequence of changes in the hydrogeochemical milieu in the flood water column.

Concentration vs. time curves for mine waters are very often characterized by a concentration peak during or soon after completion of the flooding process, the so-called "*first flush*" as a consequence of process i) (Younger et al. 2002). On the other hand, the contribution of processes ii) and iii) very much determine the long-term trend of the element concentrations in the mine water.

To identify general rules of the hydro geochemical behavior of flooded mines it is necessary to analyze and compare key data from different mines with regard to the water quality measurements vs. time. Key factors influencing the characteristics of contaminant release from abandoned mines include, besides original source strength and hydraulic residence time, many more like mine geometry, degree of internal convection, ore composition, general hydro-geochemical milieu, and the availability of key reactants like oxygen.

In order to gain a conceptual understanding of the governing processes for an individual mine site, to assess long-term water quality trends, but also to compare the behavior of different mines, a straightforward analysis of the concentration vs. time curves based on the hydraulic residence time can be a valuable tool. The paper describes a simple approach for assessing the long term water quality of flooded underground mines based on the mean hydraulic residence time as a key parameter. The comparison with long term monitoring data for some of Wismut's flooded underground uranium mines will be used to demonstrate the usefulness, but also the constraints of this approach.

Methodology

The proposed methodology is focused on the analytical assessment of water quality trends for flooded underground mines, based on the so called "perfectly mixed flow reactor" (PMFR) approach (see Brusseau 1996). Flooded underground mines are assessed as well mixed defined water reservoirs, which solute contents will be diluted slowly by inflow. Preconditions for the applicability of this approach include the following:

- steady-state (post-)flooding conditions of the mine(s) to be assessed,
- homogeneity of the mine water reservoir,
- knowledge of the volume of the flooded mine voids and of the mean water inflow rate of the mine, in order to calculate the mean hydraulic residence time (HRT),
- existence of water quality data including peak concentration at a monitoring station representative for the entire mine water pool.

The approach includes the followings steps:

1. Identification of the element peak concentration (c_0) after setting up steady-state flooding conditions,
2. Calculation of the mean hydraulic residence time T of the mine, by dividing the flooded mine volume through the mean inflow rate of the mine,
3. Calculation of the theoretical dilution curve (TDC), which forms a asymptotic curve, subtending the element peak concentration (c_0), and comparison with the measured concentration data.

The TDC for a given mine can be calculated for any time t after observation of the peak concentration c_0 as follows

$$c(t) = c_0 \cdot \mathrm{e}^{\frac{-t}{T}} \tag{1}$$

with

c(t) concentration at time t after the peak concentration,
c_0 peak concentration,
t time after the peak concentration,
T hydraulic residence time.

This very simple and straightforward approach assumes that the concentration of the component of concern in the diluting agent (fresh groundwater) is equal to zero or at least very small in comparison to the concentration in the mine water. Most ore deposits, however, are surrounded by geochemical anomalies with a higher and specific content of mineralisation than the outer unaffected environment. Furthermore the upper parts of the underground mines are mostly non flooded, caused by their position above the ground water level. Mineral weathering processes, mainly in the unsaturated zone, lead to a slow but long term washout of pollutants. Hence the inflow in the water reservoir of the mines is mostly mineralized and not negligible. Therefore, the formula can be completed with a summand reflecting the inflow concentration (c_1):

$$c(t) = c_0 \cdot \mathrm{e}^{\frac{-t}{T}} + c_1 \tag{2}$$

Each TDC is site specific and strictly defined by only three variables: (a) the initial peak concentration c_0, which defines the starting point of the curve, (b) the hydraulic residence time T, which defines the slope gradient of the curve and (c) the inflow concentration c_1 which limits the lower concentration level.

Comparing the measured data with the TDC the following cases can be generally distinguished: If the concentration vs. time curve is governed by dilution alone and no other process contributes significantly, the measured data should plot more or less close to the TDC (Type A). If the measured data lie above the TDC (Type B), additional dissolution takes place or a more slowly delivering from a pore water source is present. Dissolution can be either from the saturated mine water pool, or the unsaturated zone above the flood water level. If the measured data lie below the TDC (Type C), precipitation reactions from the liquid phase must be taken into account, or the water flow might be controlled by a more rapid process (e.g. piston flow model).

Analysis of *c* vs. *t* Curves for Example Mines

In order to demonstrate the usefulness, but also the constraints of the approach it will be applied in the following for three flooded example mines. In this respect theoretical dilution curves (TDCs) will be compared with long-term monitoring data for Wismut's Saxon uranium mines at Schlema-Alberoda, Pöhla, and Königstein. The analysis will focus on uranium and arsenic as key contaminants. Table 1 summarizes key variables of the three mines. The hydraulic residence times were calculated by dividing the flooded mine volume through the average water inflow over the entire period of steady-state conditions.

Schlema Mine

The Schlema mine being the largest uranium mine in Saxony is situated in the Western Ore Mountains (Erzgebirge). The mine contains voids at over 60 levels from close to the surface down to a depth of 1,800 m bsl. The uranium ore occurs in hydrothermal veins, which were formed at faults in a quasi non-permeable Palaeozoic rock-assemblage. After termination of mining activities and with the step by step underground remediation of the voids mine flooding began in 1991. In 2001 the major part of mine workings was flooded, while the mine water level had been risen up to 110 m below the Markus-Semmler drainage adit. Nearly steady-state conditions were reached with just small changes of the flooding level afterwards (Fig. 1). The flooded volume of mine workings determined by mine surveying is about 36.5 Mm^3. The mean natural inflow was estimated at 6.0 Mm^3/a for steady-state conditions.

Table 1 Key variables including the hydraulic residence time (HRT) for the three example mines

Parameter		Schlema	Pöhla	Königstein
Flooding status		Quasi fully flooded	Fully flooded	Steady-state flooding level
Monitoring data used		2000–2010	1995–2010	2005–2010
Volume flooded (V)	[Mm^3]	36.5	1.0	4.4
Mean mine water inflow (Q)	[Mm^3/a]	6.0	0.14	3.6
Mean HRT (T)	[a]	6.1	7.1	1.2
Inflow concentration (c_I)	U [mg/l]	0.3	negligible	1.02*
	As [µg/l]	100	100	–

* including technical inflow of untreated flooding water

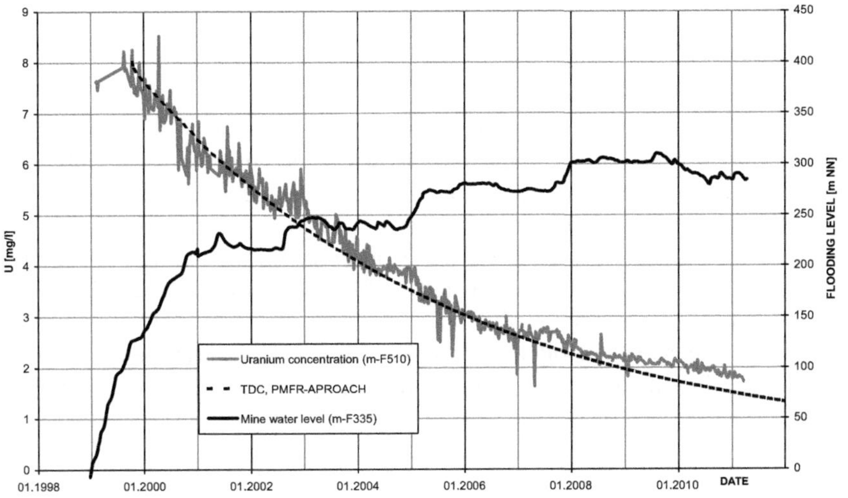

Fig. 1 Comparison between the decrease of the measured uranium concentration and the TDC with respect to the mine water level in the Schlema mine

For the inflow into the mine, including seepage water from mine dumps and drainage of mineralized host rock and galleries located above the mine water level, a medium uranium concentration of 0.3 mg/l in the mine inflow was estimated.

The dilution curve calculated by application of the above described TDC-concept using the parameters given in Table 1 compared to the measured uranium concentrations is shown in Fig. 1. The uranium concentrations are measured in the observation well m-F510. The mine water quality can be considered as homogeneous due to strong thermal convection in the flooded mine voids. In 1999 at nearly steady-state flooding conditions the uranium peak concentration was reached at about 8 mg/l. Since then the concentrations have been decreasing asymptotic basically following the TDC (case Type A). Except temporal fluctuations due to operational reasons the uranium concentration curve fits well the TDC until 2007. In

2007 the flooding level was further increased inundating new mine areas which were formerly situated in the unsaturated zone. An additional flushing of mobile oxidized minerals occurred from these areas resulting in a slight increase of the uranium concentration.

The good fit of the TDC with the measured uranium concentration in the mine water allows to conclude, that dilution by meteoric water is the key process for the uranium evolution in the Schlema mine, and that the mine works like a PMFR. There are no signs for a significant uranium dissolution. Reasons for this behavior are stable hydrochemical conditions, with intermediate Eh conditions and neutral pH-values in the mine water. The huge mine volume in conjunction with the well mixed mine water reservoir due to strong convection lead to a homogenization effect overcompensating temporal changes of the water inflow.

Pöhla Mine

The Pöhla mine is a medium size uranium mine located in the western ore mountains close to the Czech border. The combined scarn and vein deposit is embedded in a sequence of quasi non-permeable cambro-ordivician schists. The deposit is of polymetallic nature. The mine contains mine openings down to 500 m below the main drainage adit as well as above it. After termination of mining the flooding process began in 1992 and was completed in 1995. At this time the mine water level reached the main drainage adit. Since then the water discharges at steady-state flow conditions. The flooded mine volume is much smaller than that of the Schlema mine while the retention time is a little longer. The inflow concentration

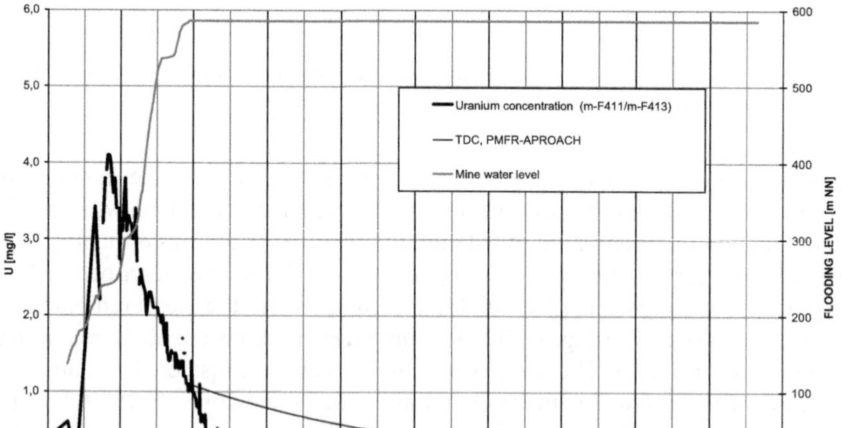

Fig. 2 Comparison between the decrease of the measured uranium concentration and the TDC with respect to the mine water level in the Pöhla mine

for uranium is negligible. The observation wells m-F411 and m-F413, which were used for the analysis, are situated in blind shaft 2 and in the main drainage adit.

After the start of mine flooding the highest uranium concentrations were reached at a peak of about 4 mg/l (Fig. 3). Shortly after, the uranium concentration decreased rapidly. At the end of the flooding process at steady-state flow conditions a uranium concentration of about 1 mg/l was measured, which is used as the start concentration for the TDC application. The main model parameters are given in Table 1. The rapid decrease of the uranium concentration proceeded until 1996, it has been slowing down until present. The latest measured uranium concentrations in February 2011 were as low as 14 µg/l. As indicated in Fig. 2, the measured uranium concentrations lay significantly below the TDC. This is a clear indication for uranium precipitation in the mine water body caused by a change of the hydrochemical conditions from an oxidizing into a reducing milieu.

Both at Schlema and Pöhla, arsenic has also an enormous importance as an environmental pollutant. The high arsenic concentrations are mainly caused by oxidation of arsenic minerals in ore veins. In both mines the concentrations are relatively high. The inflow concentrations in the Schlema mine are about 100 µg/l, which are 10% of the present arsenic concentration in the mine water. A higher arsenic content occurs in the host rocks of the deeper parts of the Schlema mine. As a result flooding of the deeper parts led to a first flush peak in 1999 followed by a typical dilution curve until 2000 (Fig. 3). After reaching quasi steady-state flow conditions in 2001 the dilution process started again from a changed initial level. This process was interrupted by a massive water inflow into the mine caused by the big flood of August, 2002. At a limited extend the effect of the flood could also be seen in the uranium concentrations (Fig. 1). For Schlema the TDC in Fig. 3

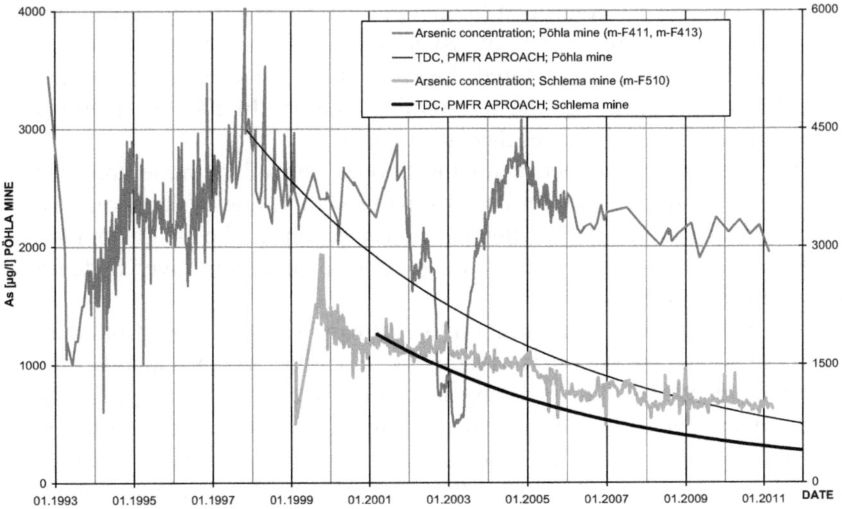

Fig. 3 Comparison between the decrease of the measured arsenic concentration and the TDC in the Schlema and Pöhla mines

fits relatively good with the arsenic data, if the effect of the 2002 flood is considered. Since the end of 2006, however, the arsenic concentrations are at a constant level significantly higher than the TDC. Such higher concentrations are a hint at dominant dissolution processes of arsenic minerals, which were left in the mine, over-compensating the dilution process. It is to suppose that kinetic effects of the dissolution process determine the arsenic concentration level.

For Pöhla the arsenic concentration curve is also shown in Fig. 3. The curve deviates significantly from the uranium curve in Fig. 2. The TDC plots significantly below the measured arsenic data. The arsenic peak was reached in 1997 with about 3000 µg/l. The concentration decreased until 2002 down to 2200 µg/l. After this the measured concentrations vary between 2000 and 3000 µg/l, with the exception of lower values in 2002/2003 as the result of an in-situ dilution experiment. In contrast to uranium, arsenic is obviously mobilized under reducing conditions, which have been developed in Pöhla for years. Therefore, the dilution process was superposed by the dissolution of arsenic minerals present in the mine.

Königstein Mine

The medium sized Königstein mine is situated southeast of Dresden close to the Elbe river. Uranium mineralisation occurs in a roll-front type deposit in upper cretaceous sandstones. The uranium ore is disseminated at reduction zones in the lowest of four permeable cretaceous aquifers. The main difference to other Saxon uranium mines was the mining technology. Due to the low grade uranium mineralisation a controlled underground acid leach process was established where mine workings were used for acid injection and drainage. Flooding of the mine started in 2001. For the stepwise controlled flooding process the natural inflow is amended by a technical input of water from groundwater sources. Through active pump and treat the mine water is captured downstream of the contaminant source to avoid contaminant release into the surrounding aquifers. In 2005 an intermediate steady-state flooding level was reached at 110 m asl.

The main hydraulic parameters for the flooded part of the mine are given in Table 1. The hydraulic residence time is relatively short due to the additional injection of technical waters. The estimated concentrations of the natural and technical inflow to the mine consider effects of operational conditions given by temporal recycling of the pumped waters.

The measured uranium concentrations are shown in Fig. 4. Apart from the general decreasing trend the data show a temporal variability which is mainly effected by operational conditions like pumping and inflow rates. A uranium peak concentration of about 35 mg/l was measured when the intermediate flooding level was reached in early 2005. Until 2009 the concentration decreased down to a level of 7.6 mg/l. Compared to the TDC-estimations according to the simple mixing model the measured concentration decreased much slower. This indicates the influence of an additional source. In the case of the Königstein mine the hydraulic system is

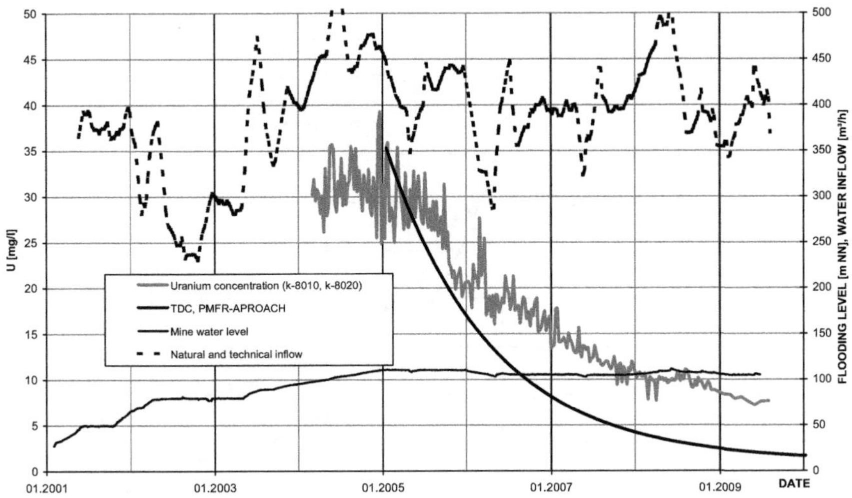

Fig. 4 Comparison of the measured uranium concentration and the TDC mixing model in conjunction with the hydraulic conditions in the Königstein mine

characterized by two flow sub-systems which are i) the open mine workings and ii) the sandstone blocks. Even with backfilled and dammed parts the open mine workings are hydraulically well connected, channeling the main water flow through the flooded parts. The advective flow through the sandstone blocks is less significant and mainly driven by density differences resulting from salinity contrasts between the pore water and the water in the mine workings. As the mixing model describes the flushing of the mine workings only, the model underestimates the real contaminant concentrations. The flushing of the mine galleries is, however, superposed by the release of contaminants from the more stagnant part of the sandstone blocks. Because of the mining technology the pore waters in these blocks have a tremendous contaminant potential and will therefore influence the long-term contaminant release.

In case of such a coupled system a simple mixing approach fails. Conceptually the simple approach can be extended by using two coupled separate mixing cells for the hydraulically well connected mine workings on the one hand and the stagnant bedrock parts on the other. However, this approach faces the problem of too many free parameters, so it usually does not result in a unique parameterization. While an estimation of the relevant mining and pore volumes is possible based on the mine survey, the other parameters (water entering the compartments, initial concentrations in the pore water) remain variable in wide ranges, since they cannot easily be deduced from the integral parameters which are measured in the mine waters at the mine overflow or the pump. In Fig. 5 this approach is shown as an example for two different ratios assumed for the flow through the stagnant part compared to the total throughput. This approach allows a better fit but remains nevertheless uncertain concerning the ability to predict long-term conditions because of the difficulty to determine reliable effective parameters.

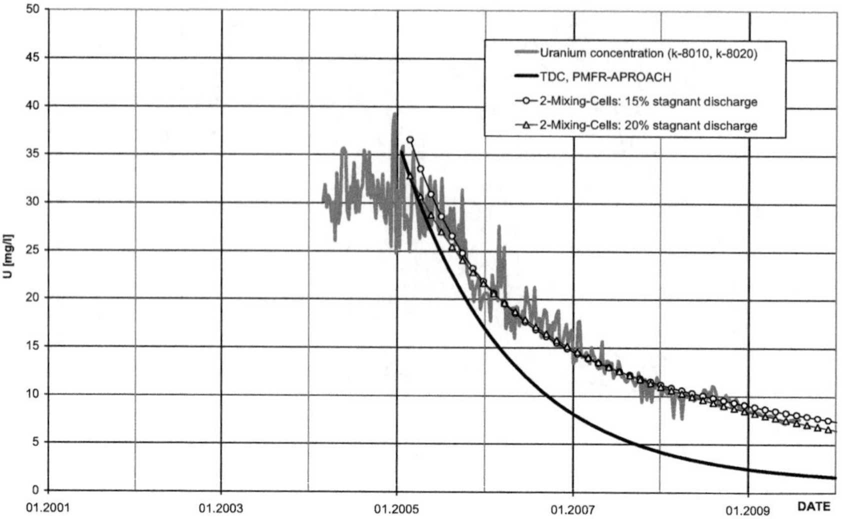

Fig. 5 Comparison of the measured uranium concentration, the TDC mixing model and a mixing cell model (with two parameter sets) taking into account the different hydraulic characteristics of the mine workings and the pore volume in the Königstein mine

For the prediction of the future contaminant release from the Königstein mine a more sophisticated box model named FLOODING was implemented which considers both parts as an internal structure of the box (Metschies et al. 2010, not shown in the figures). The volume of the open mine workings as well as the pore space are considered as mixing cells. The geochemical composition of the separate parts was determined based on a wide range of sampling and field investigations. Using a flooding experiment the parameters were fixed for the boxes representing the experiment areas and later up-scaled on the total mine area. During the flooding process this model was continually checked and if necessary amended. However, the original geochemical composition had generally not to be changed. Only in the case of hotspots and a part of conventional mining with backfilling of a concrete paste the parameters needed some justified corrections to reflect the resulting variations.

A comparable behavior is to observe in the case of zinc, which is also a typical element with high concentrations in the acidic oxidizing Königstein mine water. The peak was reached in June 2005 with 21,000 µg/l. The zinc concentration curve followed the TDC until December 2005, than it remained static until July 2006. A level of 3,800 µg/l was reached in August 2009. The measured zinc data lay above the TDC, which is also caused by the effect of zinc mobilization from pore waters in the permeable sandstone.

Conclusions

Prediction of contaminant release from flooded mines is of interest for the estimation of environmental effects which are to be expected, and to plan for the necessary technical measures to avoid adverse consequences. The comparison of a simple concept for prediction based on a mixing cell concept using physically based parameters with measured concentrations in overflowing or pumped mine water showed that in the case of the well mixed Schlema mine a reasonable prediction quality can be achieved for a parameter like uranium. In mines with non permeable boundary conditions, without additional dissolution or precipitation, the simple TDC-concept using the hydraulic residence time as a key parameter and assuming dilution as the main process works sufficiently to estimate future concentration trends in mine waters of flooded underground mines.

As far as geochemical (Pöhla) or hydraulic processes (Königstein) influence the contaminant release significantly the simple approach fails and more detailed models taking into account these relevant processes are necessary to apply.

Nevertheless, the comparison of the measured data from mine flooding with the simple mixing approach allows to identify the effect of additional release or immobilization/retardation processes and might justify the applicability of a simple extrapolation of the measured results based on an exponential relationship based on parameters derived from the mines geometric and hydraulic conditions.

References

Brusseau ML (1996) Evaluation of Simple Methods for Estimation Contaminant Removal by Flushing. Ground Water, Vol. 34, No. 1: 19–22.

Metschies T et al. (2010) Modelling of the hydraulic behaviour of deep mines in regional aquifer systems. Proceedings of the International Mine Water Association Symposium 2010, Sydney/Canada. CBU Press, Cape Breton.

Younger PL, Banwart SA, Hedin RS (2002) Mine Water – Hydrology, Pollution, Remediation. Kluwer Academic Publishers, Dordrecht.

Uranium Mineralization in Fractured Welded Tuffs of the Krasnokamensk Area: Transfer from Ancient to Modern Oxidizing Conditions

Vladislav Petrov, Valery Poluektov, Jörg Hammer, Sergey Schukin

Abstract. The main aim of this contribution is to describe the primary controls of the hydrothermal mineralization, the preferential pathways for ancient and recent meteoric water infiltration, mineral-chemical modification of the wall rocks, and transformation of uranium mineralization in the context of redox front propagation through unsaturated fractured porous welded tuffs. The data on the vein-type Tulukuevskoe uranium deposit in SE Transbaikalia, Russia are applied for modeling of uranium migration and deposition of secondary concentrations using quasi-stationary state approximation (QSSA) approach.

Introduction

Transformation of primary to secondary uranium mineralization in oxidizing environment is a phenomenon, which is necessary to be understood in terms of survey and exploration of mineral resources, remediation of territories contaminated by radionuclides, and as natural analogues in the context of safety assessment of the nuclear waste facilities (Smellie et al. 1997; Bruno et al. 2002).

Vladislav Petrov
IGEM RAS, Moscow, Russia

Valery Poluektov
IGEM RAS, Moscow, Russia

Jörg Hammer
BGR, Hannover, Germany

Sergey Schukin
PPGKhO, Krasnokamensk, Russia

B. Merkel, M. Schipek, *The New Uranium Mining Boom*
DOI 10.1007/978-3-642-22122-4 © Springer 2011

Conceptual approach to the spent nuclear fuel (SNF) isolation within the vadose zone of crystalline massifs is based on the assumption that oxidized meteoric water will inevitably destroy fuel rods and assemblages (Vadose Zone 2000). As a result the major part of the actinides and fission products will be leached from the UO_2 crystal lattice and will be transferred to the environment in unacceptable content. The main pathways for transferring of the radionuclides are hydraulically active fault zones. However, this probabilistic scenario does not mainly take into account the fact that different types of permeable reactive barriers where retention of U(VI) and its reduction again to insoluble U(IV) form could be created and preserved inside the fault zones over a long period of time (Petrov et al. 2008).

In addition the study of physicochemical processes in the fracture–wall rock system has shown that, along with the morphogenetic features and spatial combinations of different fluid conduits (macro- and microfractures, pore channels, intergranular boundaries, etc.), the duration of interaction between solid, liquid, and gas phases is also important for reconstructing filtration and transport processes. The necessity to account the factor of time resulted in the development of reactive transport modeling to natural systems. A model of quasi-stationary state approximation (Lichtner 1988) is applicable to porous-fractured media. The approximation is based on the assumption that hydrothermal solutions or meteoric water that penetrates into the fracture pore space interacts with rocks during the period that is necessary for reaching equilibrium in the water–rock system with formation of proper mineral assemblage. For this purpose, a rock column is divided into elementary volumes (blocks i, $i+1$ ..., $i+n$). Hot gas-saturated fluids gradually ascend through the system of blocks or oxygen-bearing meteoric water filtrates down to the column. The calculation of rates of chemical reactions and mineral formation and the thermodynamic state of the system in each block is implemented with specialized databases and software packages.

Processes of transformation of primary to secondary uranium mineralization were indicated and studied in the vadose zone of the Mesozoic (140 Ma) welded tuff unit of volcanic strata, which have hosted the large (~35 k tons U) Tulukuevskoe uranium deposit in the Krasnokamensk Area, SE Transbaikalia, Russia. The vein-type deposit contains primary mineralization occurring as pitchblende subjected to secondary transformations within the upper part of the deposit, mined by the Tulukuevsky open pit (TOP) during 1972–1998. There are two types of ores such as pitchblende controlling by steeply dipping faults and pitchblende-molybdenite developed along flat faults at the bottom of the TOP. The deposit provides remarkable examples of processes governing uranium migration and accumulation in oxidizing conditions (Petrov et al. 2005).

The main objective of this contribution is to indicate the conditions for transformation of primary to secondary uranium mineralization in the oxidizing environment of the volcanic strata. We suppose that these processes have to be studied in detail and considered during conceptual and numerical filtration-transport modeling as well as in the context of the total system performance assessment of the SNF underground facilities.

Results

Investigations were realized at seven levels of the open pit NW wall (block ~200 × 200 × 200 m). Investigated block includes relatively fresh and differently altered rock varieties, the Fault 1 A controlling ore-body with primary pitchblende at depth ~60 m from the day surface and distinct occurrences of secondary uranium mineralization, and well developed fracture network.

At each level the principal Sample Lines (SL-A from the top to SL-G at the bottom) with length from 120 to 200 m have been studied (Fig. 1).

Sample points were positioned every other 10 m at the lines. Besides, transversely to the block plane gallery 250 m long at the bottom was inspected. Reasoning from the requirements of the uranium migration observations the following fieldwork and lab test activities during 2000–2010 have been performed:

- Geological-structural survey and digital photo mapping of fracture families with positioning (X, Y, Z coordinates) and subsequent statistical data analysis (dip, aperture, morphology, spacing, extension, mineral filling);
- Investigation of composition, nature and degree of hydrothermal-metasomatic alteration and oxidation of the rocks (optical microscopy, wet chemistry, XR fluorescent analysis, NAA, XRD diagnostics of clay minerals, TEM for studying of oxyhydroxides, and coulometer titration of organic matter content);
- Studies of primary and secondary minerals (optical microscopy, UV lighting, SEM and EDS, fission-track radiography analysis).

The main aim of the activity is to draw up the data for uranium transport modeling using QSSA approach.

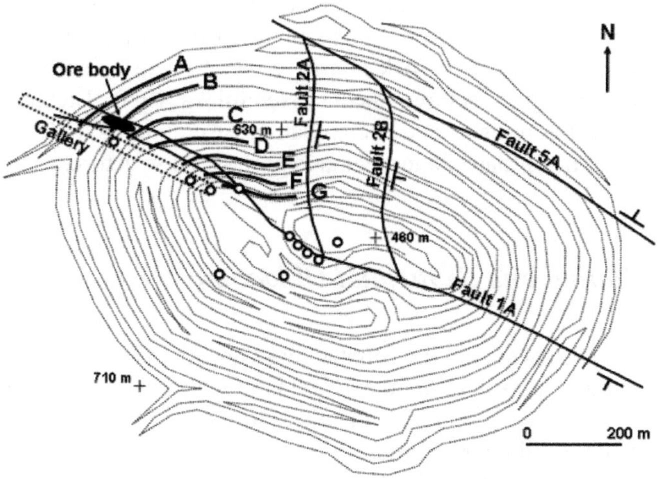

Fig.1 Surveying plan of the TOP showing location of the Sample Lines (*from A to G*), prospecting gallery, fault zones, exposed ore body, and fracture water sources (*circles*)

Fracture Network

A total of 3717 fractures were studied at the TOP with laser theodolite positioning of the main discontinuities such as NW-striking Fault 1A and its branch.

Obtained data show that two steep fracture families are most widely developed (direction/dip): N140–175°/70–85°SW and N10–20°/70–85°NW. All the fractures are classified as non-mineralized (open) and mineralized, and by the aperture (A) they form three groups of respective percentage: A1 – subveinlets (< 1 mm) (55%), A2 – veinlets (1–10 mm) (35%) and A3 – veins (> 10 mm) (10%).

The mineral filling is mainly carbonate, chlorite, hematite, goethite and rarely fluorite. In the vertical section hematite, goethite, carbonate and secondary uranium minerals filled fractures decreasing and density of the primary uranium mineralization filled fractures increasing with depth are observed. Large fractures (aperture > 10 mm) are strongly affected by the exogenous processes. The fracture planes have residual gouge, gliding striation, and crusts of Fe-Mn oxyhydroxides.

Obtained data were used for the development of 2D density and aperture models for all fracture families. The models show pathways for the meteoric water flow down to the bottom of the welded tuff block. The main fast water-gas flow goes along the Fault 1A as well as along high fracture density zone located at the distance and stretched out from top to bottom of the TOP. These discontinuities form V-shape structure that appears for the preferential palaeo- and current fluid flows. High fracture density zone consists of two main steeply dipping fracture families, which constitute a rhombic network. Inside the rhombic blocks the nets of small sub-parallel discontinuities occur. They may serve as fingers for the flow of liquid and gaseous phases.

Obtained data show necessity to superpose fracture filtration parameters with filling mineralogy and sorption capacity of the minerals concerning uranium. In addition, oxidizing solutions may penetrate into the stratum also in the process of molecular diffusion into the near-fracture rock matrix. As matrix diffusion is an important transport mechanism in fractured media (Neretnieks 1980; Lichtner et al. 1996), its influence on the transfer and accumulation of uranium should be also taken into account in terms of matrix alteration and pore space structure.

Mineral-Chemical Zoning and Sequential Oxidation

Late Mesozoic hydrothermal processes in the country rocks are strongly connected with fracture network and are classified (Andreeva and Golovin 1998) as pre-ore, ore-related and post-ore mineral associations (Table 1) organized in vertically oriented external, intermediate and internal zones.

Relatively fresh rocks of gray-violet color form external zone where biotite is well preserved or partially replaced by ankerite and hematite. Intermediate zone consists of subzone of preservation of fine-grained hematite, and subzone of re-

Table 1 Mineral composition of welded tuffs and veins of the TOP

State-event	Mineral composition
Fresh rock	Quartz, feldspar, biotite, hornblende, hematite, accessory minerals (zircon, sphene, magnetite, ilmenite, apatite, etc.)
Pre-ore events	*Rocks*: quartz, hydromica (illite, mixed-layered illite- smectite), carbonate (ankerite, calcite, siderite, breunnerite), relict minerals (K-feldspar, albite, quartz); *Veins*: quartz, siderite, ankerite, pyrite, sphalerite, galena, chalcopyrite
U and ore-related mineralization	Hypogene (pitchblende and *tucholite*). *Veinlet-metasomatic mineralization*: berthierine, siderite, ankerite, quartz, hematite, fluorite (dark), organic matter
Post-ore events	*Veinlet-metasomatic mineralization*: kaolinite, smectite, berthierine; *Veinlets*: calcite, fluorite (light), quartz, sulfides
Transformed U mineralization	Ancient oxidation zone (hydropitchblende, urhyte, uranophane I) and secondary minerals (uranophane II, heyviite, calcurmolite, uranylcarbonates-liebigite?)

placement of feldspar by metasomatic carbonates, hydromica and quartz. Internal zone is clearly close to the Fault 1A with intensive development of fine-flake light-color micas, carbonates and quartz as well as cataclasis and substantial fracturing. Post-ore argillization is superposed predominantly on this particular zone.

Original and hydrothermally altered rocks have undergone ancient and modern oxidation. Ancient processes of oxidation form three main stratified subzones (from top to bottom): leaching (ancient vadose zone, SL-A, B, C), complete oxidation (SL-D, E) and incomplete oxidation (SL-F, G). A horizon of secondary uranium enrichment (ancient transition zone) at ~+640 masl is located.

The ancient oxidation belongs to a hydroxide-silicate type with gradual transition of primary ores to uranium hydroxides and silicates and unchanged morphology of primary segregations (Belova 2000). This process can be described as the chain: $U^{IV}O_2 + O_2 + H_2O$ (pitchblende)→ $U^{VI}O_3nH_2O + Me$ (hydropitchblende)→ $MeU^{VI}{}_2O_7{\cdot}nH_2O + Si$ (hydroxides-velsendorphite, urhyte)→ $Me(UO_2)_2[SiO_4]_2nH_2O$ (silicates-uranophane), where Me = Ca, Pb, Ba, Sr, K, Na.

Hence uranium minerals formed in conditions of the ancient oxidation zone (hydropitchblende, urhyte) characterize the beginning of this process, while uranophane indicates its completion. In case of high-grade ore replacement with massive and veinlet pitchblende in form of pseudomorphoses the complex uranium hydroxides prevail, while in case of base veinlet-impregnated ores the primary uranium oxides are replaced mainly by uranium silicates. Uranium hydroxides and silicates occur directly at the place of the primary minerals forming semi-amorphous, in some cases gel-like, "gummite" rims.

The whole hydrogeological and hydrogeochemical system was disturbed, and redox conditions were changed in connection with the deposit mining by the open pit and recession of the groundwater table. In this case, the ancient oxidation zone including subzone of leaching, and those of complete and incomplete oxidation and the secondary enrichment horizon turned out to be exposed to the effect of modern oxidation processes.

The modern oxidation is controlled by zone of Fault 1 A and large fracture families and is caused by the lowering of groundwater level during the TOP mining. Supergene alteration is accompanied by the formation of Fe-Mn oxyhydroxides such as goethite, Fe-vernadite, hematite, ferrihydrite and protoferrihydrite. Goethite and Fe-vernadite have predominantly stratified allocation (SL-A, B). Hematite is characterized by sheetlike distribution in the upper parts and linear occurrence in the lower parts of the section while clustered aggregates of ferrihydrite and protoferrihydrite occur at the bottom of the TOP (SL-G).

Secondary uranium minerals on pitchblende ores are presented mainly by uranophane, heyviite, calcurmolite, umochoite, mourite, and liebigite. In case of pitchblende-molybdenite ores the pattern of hypergene processes becomes more complicated. Acid molybdenum-bearing solutions formed in the process of jordisite oxidation reacted with primary uranium minerals, such as pitchblende, coffinite, tucholite. The solution mobilized uranium, which later included in the composition of soluble uranium molybdates and sulfates.

Fluid Flow Pathways and Transport Properties of Rock

The fieldwork and statistical data were used as a basis for the development of 3D density and aperture (Fig. 2) models for all fracture families as well as fracture hydraulic conductivity computation model with the help of the GoCAD software.

The 3D model of the allocation of the fracture aperture values (A1, A2 and A3) proves the conclusion on the wide distribution (dispersion) of filtration channels (with the exception of the main drainage channel of the Fault 1A) at the upper

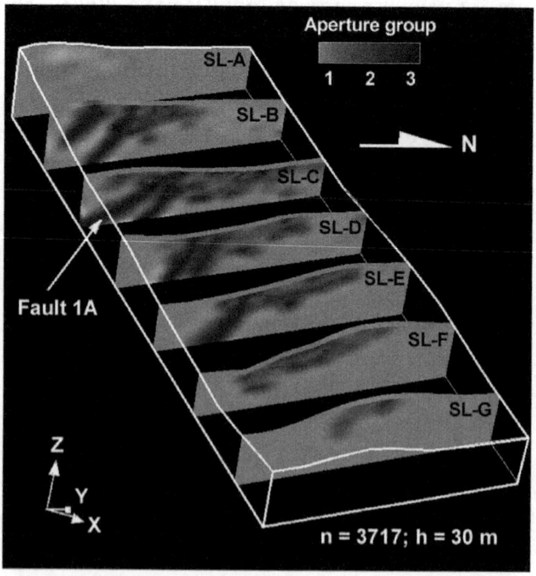

Fig. 2 3D aperture model for all fracture families of the TOP created in GoCAD. SL – sample line, n – number of fractures involved in calculation for space of 30 m height

horizons of the section and their gradual tightening towards lower parts of the rock section to form a single channel. The SL-C (location of ancient transition zone) is characterized by the highest degree of non-uniformity in terms of channel size.

Available data provide an opportunity for evaluating hydraulic properties of the fracture families as function of sorption capacity of mineral fillings and rock matrix molecular diffusion in relation to uranium. One of the main factors affecting the modern redox front development and uranium accumulation is the presence of permeable reactive barriers with mainly reducing conditions within the block of oxidizing rocks (Petrov et al. 2008).

Retention of U(VI) and its reduction again to insoluble U(IV) form is connected with the reactive ability of the Fe-Mn oxyhydroxides, organic matter and modern microbial activity (ferrihydrite and protoferrihydrite). In addition, U(VI) contained in easily soluble carbonate compounds, is able to migrate in colloidal form under oxidizing conditions whereas in the reducing environment of the reactive barriers the redistribution of uranium occurs with gradual augmentation of secondary mineralization and sorptive forms.

Data required for the calculation of the time for achieving the transport properties and equilibrium in the system "water-rock" is as follows: mineral composition and grain-size distribution of enclosing rocks and uranium ores, as well as kinetics

Table 2 Average mineral composition, size of mineral grains and their content in fresh and altered welded tuffs of the vadose zone of the Tulukuevskoe deposit

Mineral	Fresh rock		Altered rock	
	Volume, %	Grain size, mm	Volume, %	Grain size, mm
Oligoclase	37.5	1.5		
Albite			10.05[a]	1.5
Quartz	18.8	1	10.05[a]	1
K-feldspar	33	2	13.7[a]	2
Biotite	3.8	0.5		
Calcite			15.5	0.2
Ankerite			7.3	0.05
Siderite			7.3	0.05
Illite			6.3	0.005
Illite-smectite			9	0.002
Chlorite			3.7	0.01
Smectite			1.4	0.001
Kaolinite			1.8	0.005
Hematite + LA[b]	0.9	0.001	2.7	0.001
Goethite			0.9	0.0005
Fluorite			0.41	0.1
Pitchblende			0.8	0.5
Uranophane			0.09	0.002
Porosity	9.7		10.5	

[a] relic minerals, [b] leucoxene-like aggregate

of mineral transformations. Specific characteristics of the mineral composition of welded tuffs differing in the degree of alteration are determined by the zoning of hydrothermal-metasomatic transformations. Data on grain-size distribution of rocks and ores required for the assessment of the reaction surface area of minerals have been obtained from the detailed microscopic analysis of numerous thin sections (Table 2). The data on the rates of mineral formation may be retrieved from the available computer bases of thermodynamic and mineral-chemical data.

Discussion and Conclusion

The analysis and synthesis of the data allow compiling three main stages of the formation, modification and re-deposition of uranium concentrations at the Tulukuevskoe deposit: 1) hydrothermal ore formation, 2) original state of the geological environment before the deposit mining, 3) deposit mining by an open pit, recession of water table and re-deposition of ore concentrations (Fig. 3).

After hydrothermal ore formation, mineral phases of U(VI) prevailed in upper parts of ore bodies, which for a long period of time had occurred above the level of underground waters in oxidizing conditions, while in lower parts mineral phases of U(IV) prevailed in water-saturated environment and relatively reducing conditions. With the progress of open pit mining and recession of the underground water level the lower parts of ore bodies also occurred in water-unsaturated oxidizing environment, and it resulted in the transition of the residual part of U(IV) to U(VI), and accordingly, to the replacement of primary pitchblende by secondary minerals. Due to this fact, today the secondary uranium mineralization is predominantly met at the TOP.

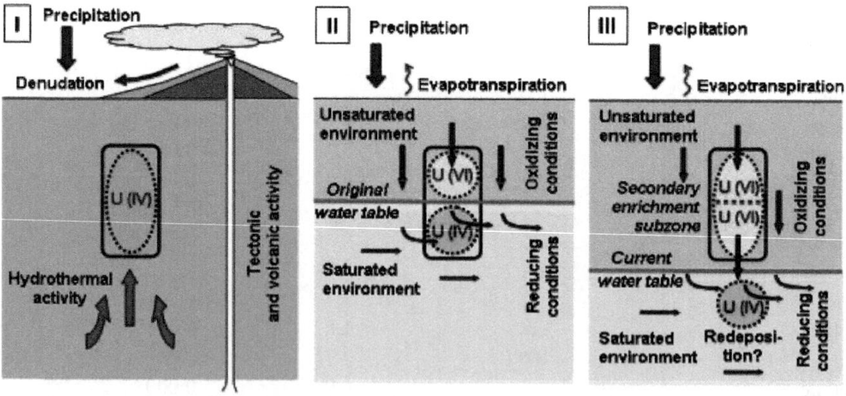

Fig. 3 Simplified model of uranium ore formation, modification and re-deposition in connection with temporal and spatial changes of oxidizing/reducing conditions in the Tulukuevskoe deposit. I – stage of hydrothermal ore formation, II – stage of the original state of the geological environment before the deposit mining, III – stage of the deposit mining by an open pit, recession of water table, uranium transport and re-deposition (?) of ore concentrations

It seems feasible to take into account the above-described peculiarities for developing conceptual model of uranium transfer in the vadose zone of the Tulukuevskoe deposit. In many aspects it affects the difference in oxidizing/reducing conditions of the occurrence of uranium ores in the process and after open pit mining of the deposit due to the lowering of the level of ground waters. In addition fractured porous environment such as welded tuffs requires specific conceptual and numerical treatments because both the fractures and porous matrix are active parts of the flow and transport regime as well as development of secondary uranium mineralization.

In the context of detection of spatiotemporal relations between redox front propagation and geochemical events, which have occurred in the vadose zone of the deposit mined by TOP, the QSSA approach is the most convenient. This approach is based on the postulation that meteoric waters gradually penetrating the rocks react with the latter during the period required for the adjustment of equilibrium into the water-rock system with development of the adequate rockforming and uranium mineral association. It correlates with the stationary state in the elementary volume in column (from block i to block $i+6$) (Fig. 4). At the model all the blocks are characterized by equivalent fracture network geometry and specific mineral-chemical composition organized with depth.

It is evident that QSSA approach will allow more reliable and detailed description of the dynamics of the redox front propagation and thus the identification of the conditions of uranium transfer in oxidizing fractured unit as well as deposition of secondary uranium concentrations at the TOP. Unfortunately, this model will not be completely adequate at least by two reasons. Firstly, the model does not

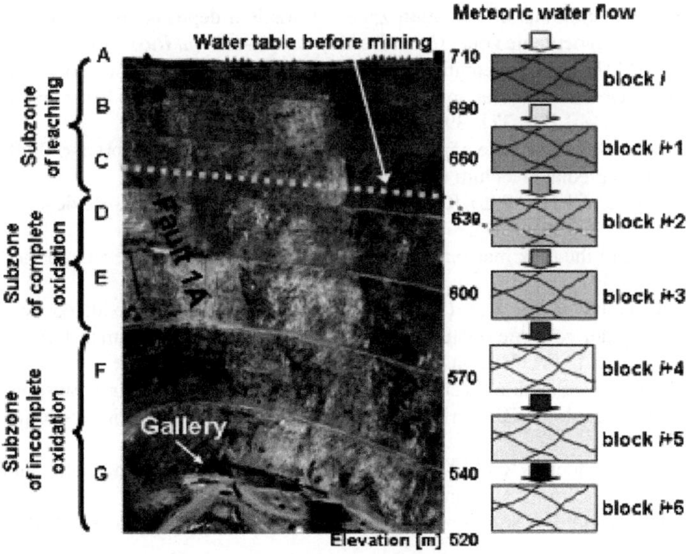

Fig. 4 Conceptual model of the oxidizing front penetration into the welded tuff unit with QSSA approach at the *right* and parts of the ancient oxidizing zone at the *left*

take into account that Fault 1A forms the main pathway for the meteoric water descend and simultaneously has permeable reactive barriers for retention of U(VI) and its reduction again to insoluble U(IV) (Petrov et al. 2008). Secondly, the effect of the stressed-strained state of the environment on filtration and transport processes is disregarded. The analysis of the post-ore stress field shows that Fault 1A (main pathway) was oriented relative to the effective normal (σ_n) and shear (τ) stresses so that $\mu = \tau/\sigma_n$ is about 0.6. Precisely such μ values are characteristic of the most hydraulically active faults (Zoback and Townend 2001). Current hydraulic activity of Fault 1A is confirmed by the fact that eight of eleven fracture water sources are localized in the zone affected by this fault.

Regardless of these circumstances we believe that obtained data could be applied for developing numerical reactive transport model which would be useful for deeper insight into the processes of actinide migration in the natural environment.

Acknowledgements We almighty appreciate Marc Lespinasse and Judith Sausse from UMR G2R (7566) University Nancy-1, France for their expert assistance for modeling of the rock filtration properties using GoCAD. We extend our thanks to Peter C. Lichtner from Los-Alamos National Laboratory, USA for fruitful discussions during the progress of the work. This study was partially supported by the Presidium of RAS program No. 4, project 02.740.11.0049, and was carried out in the frame of the scientific and technical cooperation agreement between IGEM RAS and BGR.

References

Andreeva OV, Golovin VA (1998) Metasomatic processes at uranium deposits of Tulukuevskaya caldera, Eastern Transbaikalia, Russia. Geol Ore Dep 3: 205–220 (*in Russian*).

Belova LN (2000) Formation conditions of oxidation zones of uranium deposits and accumulation of uranium minerals in hypergenic zone. Geol Ore Dep 2: 113–121 (*in Russian*).

Bruno J, Duro L, Grive M (2002) The applicability and limitations of thermodynamic geochemical models to simulate trace element behaviour in natural waters. Lessons learned from natural analogue studies. Chem Geol 190: 371–393.

Lichtner PC (1988) The quasi-stationary state approximation to coupled mass transport and fluid-rock interaction in a porous media. Geochim Cosmochim Acta 52: 143–165.

Lichtner PC, Steefel CI, Oelkers EH (1996) Reactive transport in porous media. Reviews in Mineralogy 34.

Neretnieks I (1980) Diffusion in the rock matrix: an important factor in radionuclide retardation? J Geophys Res, 85: 4379–4397.

Petrov VA, Poluektov VV, Golubev VN et al. (2005) Uranium mineralization in oxidized fractured environment of the giant volcanic related uranium field from the Krasnokamensk Area. Proc Int Symp Uranium Prod. IAEA, Vienna, Austria: 260–264.

Petrov V, Poluektov V, Hammer J, Schukin S (2008) Fault-related barriers for uranium transport. Uranium Mining and Hydrogeology. BJ Merkel and A Hasche-Berger (eds.) Springer-Verlag Berlin Heidelberg: 779–789.

Smellie JAT, Karlsson F, Alexander WR (1997) Natural analogue studies: present status and performance assessment implications. Contam Hydrol 26: 3–17.

Vadose Zone: Science and Technology Solutions (2000) BB Looney and RW Falta (eds.). Battelle Press, Columbus, U.S.A.

Zoback MD, Townend J (2001) Implication of hydrostatic pore pressures and high crustal strength for the deformation of intraplate lithosphere. Tectonophysics 336: 19–30.

Synthesis and Research of Uranium Minerals That Form in the Disposals of Waste Radioactive Products and Under Natural Conditions

Anna Shiryaeva, Maria Gorbunova

Abstract. Two types of uranium minerals (silicates and molybdates) have been synthesized for modeling experiments in order to research the dynamics of uranium and its fission products in the geochemical environment. The attention has been paid to soddyite, Na-boltwoodite, cesium uranyl oxonium silicate hydrate and cesium uranyl molybdenum oxide, which typically form under natural conditions and during temporary storage of spent nuclear fuel. The synthesized minerals have been characterized and the conditions under which they form have been determined.

Introduction

Research of the processes that occur in places of nuclear waste disposal and temporary storage of spent nuclear fuel contributes to the safety of the nuclear-fuel cycle. Secondary phases that form while corrosion of radiating installations or in zones of contact between radioactive waste and geological environment are often difficult to investigate due to their inaccessibility. In this connection, obtaining laboratory-synthesized analogues of radionuclide-containing minerals is used for modeling the processes that take place under different geochemical conditions. Furthermore, carrying out such experiments make it possible to determine the conditions under which uranium can be extracted or the mobility of the radionuclides can be limited.

Anna Shiryaeva
Moscow State University, Leninskie Gory, 119991 Moscow, Russian Federation

Maria Gorbunova
Moscow State University, Leninskie Gory, 119991 Moscow, Russian Federation

B. Merkel, M. Schipek, *The New Uranium Mining Boom*
DOI 10.1007/978-3-642-22122-4

In this work specimen of two groups of uranium minerals: uranyl molybdates and silicates have been examined.

Uranyl Molybdates

Uranyl molybdates tend to form under natural conditions as well as in places of temporary storage of spent nuclear fuel. This group of minerals is one of the most spread in zones of oxidation of uranium-molybdenum deposits. Moreover, formation of some molybdenum isotopes such as ^{95}Mo, ^{97}Mo, ^{98}Mo and ^{100}Mo takes place in the nuclear reactor; these isotopes occur to be abundant and are usually presented in spent nuclear fuel.

Cesium is one of the products of uranium fission, usually presented in uranium deposits, and its isotope ^{137}Cs sufficiently contributes to the total radioactivity of uranium deposits and spent nuclear fuel, so, investigation of cesium compounds makes it possible to find out the conditions of such substances formation and to determine dynamics of Cs (as well as U) migration.

This part of work is devoted to the development of methodology of cesium uranyl molybdate synthesis. There are two topologically related structures of $Cs_2(UO_2)_2(MoO_4)_3$ (Krivovichev S. et al. 2002): α-$Cs_2(UO_2)_2(MoO_4)_3$, that forms under hydrothermal conditions at the temperature about 90°C and β-$Cs_2(UO_2)_2(MoO_4)_3$ that crystallizes from the melt of α-phase at the temperature 650°C.

This work is devoted to the development of methodology of β-$Cs_2(UO_2)_2(MoO_4)_3$ synthesis from $CsNO_3$, $UO_2(NO_3)_2{\cdot}6H_2O$, H_2MoO_4, $NH_4Mo_7O_{24}{\cdot}H_2O$ by crystallization from the melt. Three experiments have been carried out (Table 1). Initial substances were dissolved in water and intermixed under heating by magnetic stirrer in order to obtain homogeneous mixture. After the water had been vaporized, the mixture was grinded in porcelain crucible and then heated in muffle fur-

Table 1 Terms of experiments

Experiment	1	2	3
Reagents	$UO_2(NO_3)_2{\cdot}6H_2O$ H2MoO4 $CsNO_3$	$UO_2(NO_3)_2{\cdot}6H_2O$ $NH_4Mo_7O_{24}{\cdot}H_2O$ $CsNO_3$	$UO_2(NO_3)_2{\cdot}6H_2O$ $NH_4Mo_7O_{24}{\cdot}H_2O$ $CsNO_3$
Proportion of elements	Mo, U, Cs in stechiometric proportion: 3 : 2 : 2	Mo, U, Cs in stechiometric proportion: 3 : 2 : 2	Mo in 10% shortage
Temperature conditions	Heating up to 600°C for 1.5 h, keeping at the pointed temperature for 20 h, sharp cooling	Heating up to 750°C for 2 h, keeping at the pointed temperature for 2.5 h, cooling down to 600°C, keeping at 600°C for 22 h, slow cooling	Heating up to 750°C for 2 h, keeping at the pointed temperature for 2.5 h, cooling to 600°C, keeping at 600°C for 22 h, slow cooling

nace up to the temperature of melting. β-$Cs_2(UO_2)_2(MoO_4)_3$ was crystallized by cooling down to 600°C keeping at this temperature during 20–22 h and further cooling down to the room temperature.

Three samples, synthesized under various conditions (t°, Mo%), have been characterized by powder X-ray diffraction (XRD). Comparison of the analysis results with data from database (Serezhkin V. et al. 1987) showed the presence of β-$Cs_2(UO_2)_2(MoO_4)_3$ in all the samples (Figs. 1–3). Parameters of the tetragonal crystal cell have been determined: a = 10.13(6), c = 16.28(3).

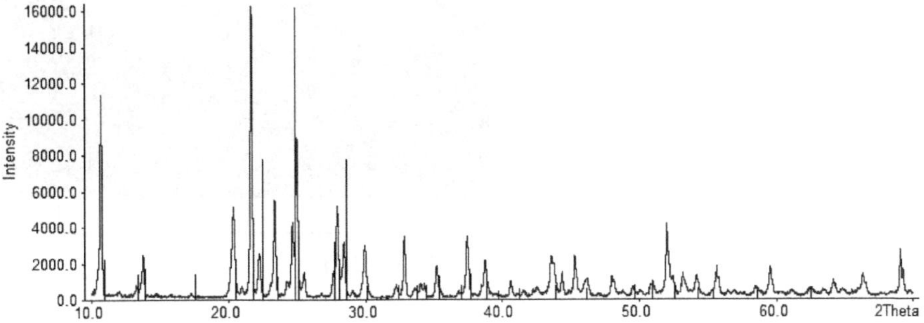

Fig. 1 Comparison of XRD – results of the first sample with data from database

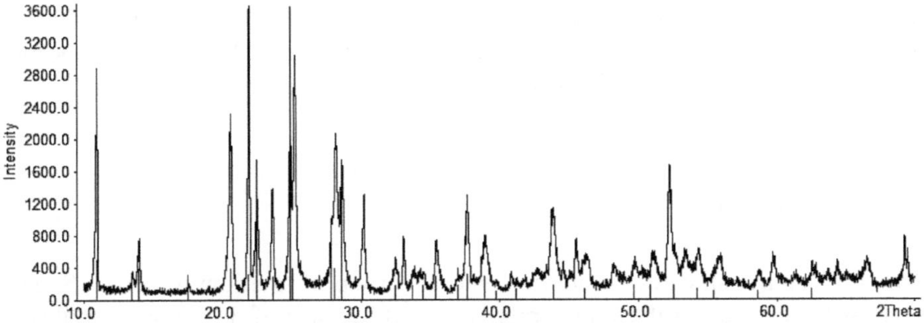

Fig. 2 Comparison of XRD – results of the second sample with data from database

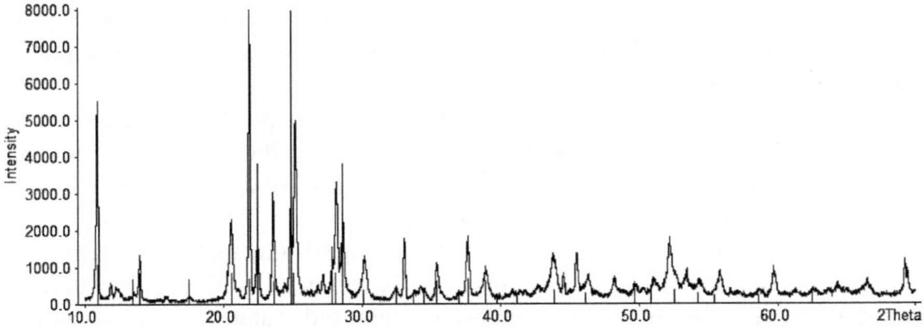

Fig. 3 Comparison of XRD – results of the third sample with data from database

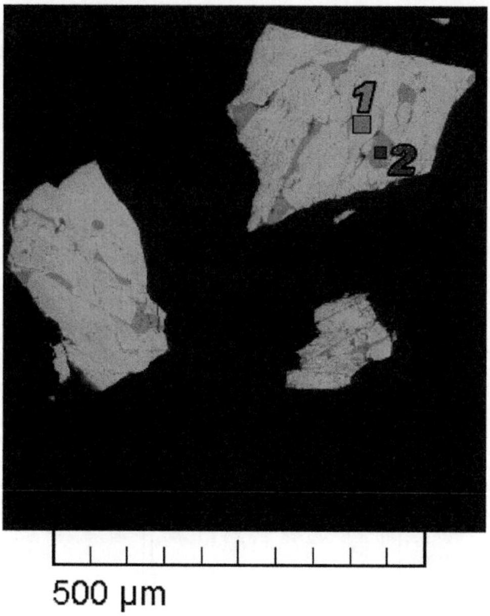

Fig. 4 REM-image of the synthesized sample

The samples have been grounded and analyzed by raster electron microscope (REM) with electron microprobe analysis (EMPA). Domination of the phase $Cs_2(UO_2)_2(MoO_4)_3$ (Fig. 4, point 1) has been determined, and the admixture of the second phase (Fig. 4, point 2) that consisted of Cs : U : Mo : O in proportion 1.5 : 0.5 : 4.5 : 16.0 has been indicated.

Uranyl Silicates

Group of uranyl silicates is the most various group of uranium minerals. Special interest to research of these minerals is caused because of their low solubility in ground and surface water and ability of crystallizing from water solutions what restricts uranium migration in environment. As group of silicates is one of the most abundant groups of uranium minerals, its constituents are tend to form under oxidation conditions in deposits of uraninite (UO_2) – one of the most important raw materials for fuel energetics.

In this work three representatives of uranosilicate group (uranyl silicate hydrate – soddyite ($(UO_2)_2SiO_4{\cdot}2H_2O$), sodium uranyl oxonium silicate hydrate – Na-boltwoodite ($NaUO_2SiO_3OH{\cdot}1.5H_2O$) and cesium uranyl oxonium silicate hydrate ($CsUO_2SiO_3OH{\cdot}1.5H_2O$)) were studied precisely.

According to the published works on this group of minerals (N. Chernorukov et al. 2007), uranium silicates of alkaline metals had been synthesized by reaction between crumbled up granules of siliceous glass and water solution of alkaline metal nitrate under the following conditions: 200°C, 16 h, pH = 11.5. Crystallized substances were washed by water and dried out. Uranosilicates of alkaline-earth metals had been obtained by reaction of ionic exchange under hydrothermal conditions.

For synthesis of the studied uranyl silicates: soddyite ($(UO_2)_2SiO_4{\cdot}2H_2O$), Na-boltwoodite ($NaUO_2SiO_3OH{\cdot}1.5H_2O$) and cesium uranyl oxonium silicate hydrate ($CsUO_2SiO_3OH{\cdot}1.5H_2O$) – have been tested the following methodology: initial substances have been dissolved in water and poured together in stechiometric proportion according to the conjectured reactions of three experiments:

1) $2UO_2(NO_3)_2{\cdot}6H_2O + Na_2SiO_3{\cdot}9H_2O = (UO_2)_2SiO_4{\cdot}2H_2O\downarrow + 2NaNO_3 + 2HNO_3 + 18H_2O$;
2) $UO_2(NO_3)_2{\cdot}6H_2O + Na_2SiO_3{\cdot}9H_2O = NaUO_2SiO_3OH{\cdot}1.5H_2O\downarrow + NaNO_3 + HNO_3 + 12.5H_2O$;
3) $UO_2(NO_3)_2{\cdot}6H_2O + Na_2SiO_3{\cdot}9H_2O + CsCl = CsUO_2SiO_3OH{\cdot}1.5H_2O\downarrow + NaCl + NaNO_3 + HNO_3 + 12.5H_2O$.

In experiments on synthesis of Na-boltwoodite and cesium uranyl oxonium silicate hydrate pH of mixture had been brought up to 12 by addition of NaOH solution. The mixtures had been heated in autoclave at 200°C for 16 h after what the obtained compounds were filtered, washed and dried out.

Analyses have been made by raster electron microscope with electron microprobe analysis and X-ray difraction. REM and EMPA confirmed the expected composition of the substances. At the REM-image of soddyite and cesium uranyl oxonium silicate hydrate needle-shaped crystals are observed (Figs. 5 and 6).

Comparison of X-ray pictures with data from mineral database (Stohl and Smith 1981; Honea 1961) showed that synthesized compounds are presented by mixtures of several phases with researched substances among them (Figs. 7–9). Full determination of other phases is to be carried out. Exact data of cesium uranyl oxonium silicate hydrate were not found in database, thus, this specimen was compared with K-boltwoodite ($KUO_2SiO_3OH{\cdot}1.5H_2O$). Correlation peak/background for the last substance indicates presence of amorphous phase.

Further investigations to optimize terms of synthesis are to be carried out.

Fig. 5 REM-image of soddyite

Fig. 6 REM-image of cesium uranyl oxonium silicate hydrate

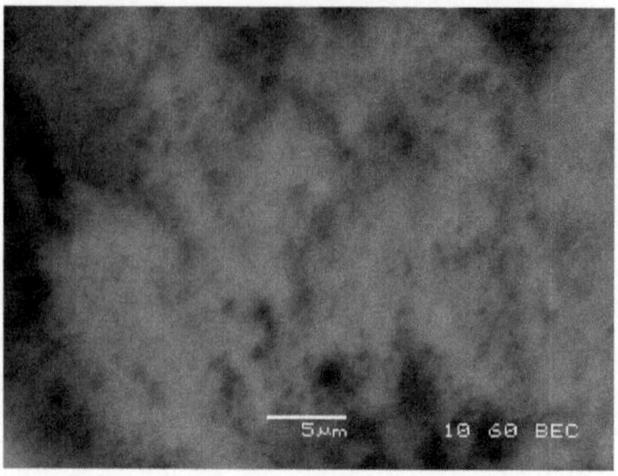

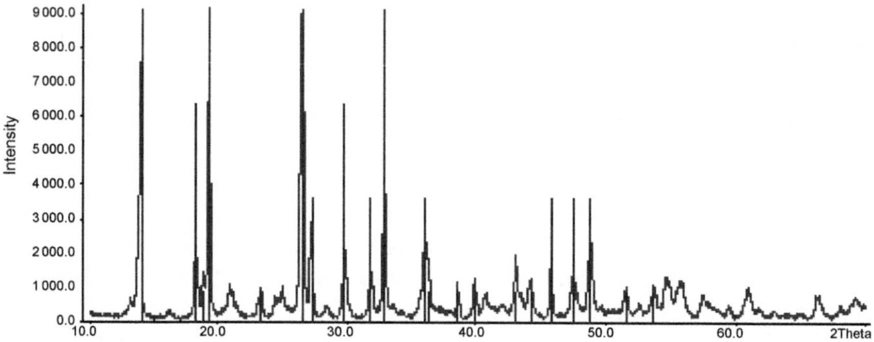

Fig. 7 Comparison of XRD – results of the first synthesized substance with soddyite from database

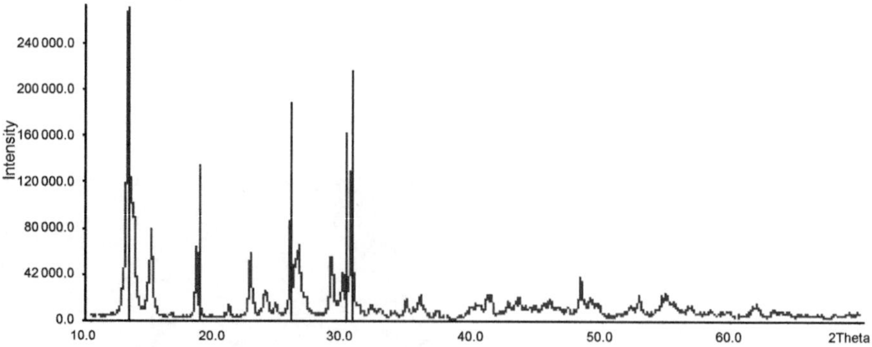

Fig. 8 Comparison of XRD – results of the second synthesized substance with Na-boltwoodite from database

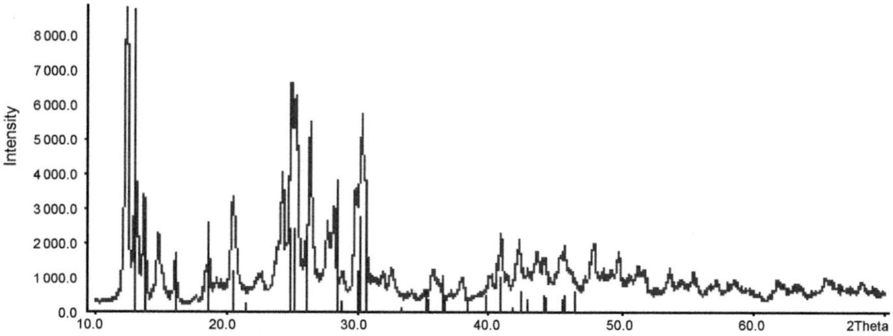

Fig. 9 Comparison of XRD – results of the third synthesized substance with K-boltwoodite from database

Conclusion

New methodology of synthesis of β-$Cs_2(UO_2)_2(MoO_4)_3$ by solid-state method was developed, the synthesized compound was characterized by REM with EMPA and XRD.

Attempts to obtain uranyl silicates under hydrothermal condition were made. The proposed methodology is to be optimized, in order the selectivity of the synthesized substance be achieved.

The obtained data on conditions of formation of the synthesized compounds are useful for determination of the processes that occur in spent nuclear fuel temporary storages. Synthesized compounds may be used in experiments on examination of migratory qualities of uranium and products of its fission.

References

Chernorukov N. et al. (2007) Synthesis and Study of Uranosilicates of the Uranophane-Kasolite Group. Radiochemistry. Vol. 49, №4: 300–304

Honea., (1961), Am. Mineral., 46, 12

Krivovichev S. et al. (2002) Syntheses and Crystal Structures of two Topologically related Modifications of $Cs_2[(UO_2)_2(MoO_4)_3]$. Inorg Chem. Vol. 41: 34–39

Serezhkin V. et al. (1987). Inorg. Chem. 32

Stohl, Smith (1981), Am. Mineral, 66, 610

Modeling the Groundwater Flow of a ^{90}Sr Plume Through a Permeable Reactive Barrier Installed at the Chalk River Laboratories, Chalk River, Ontario, Canada

Jutta Hoppe, Jeff Bain, David Lee, Dale Hartwig, Sung-Wook Jeen, David Blowes

Abstract. In 1998, a passive remediation system known as the wall and curtain was installed at the Chalk River Laboratories in Ontario, Canada, to prevent a ^{90}Sr plume present in the lower part of a 12 m thick aquifer from discharging into a nearby swamp. To provide a better understanding of the groundwater flow-through (or throughput) and the capture zone of the system, a model was constructed using the three-dimensional finite element numerical computer code HydroGeoSphere. The model was calibrated with observed hydraulic head values. The results show a good agreement between the simulated and observed heads and the capture zone width and depth determined for two data sets from different dates. Further physical and geochemical data will be collected in order to refine the current model.

Jutta Hoppe
Department of Earth and Environmental Sciences, University of Waterloo, Ontario, Canada

Jeff Bain
Department of Earth and Environmental Sciences, University of Waterloo, Ontario, Canada

David Lee
Atomic Energy of Canada Limited, Chalk River Laboratories, Chalk River, Ontario, Canada
Department of Earth and Environmental Sciences, University of Waterloo, Ontario, Canada

Dale Hartwig
Atomic Energy of Canada Limited, Chalk River Laboratories, Chalk River, Ontario, Canada

Sung-Wook Jeen
Atomic Energy of Canada Limited, Chalk River Laboratories, Chalk River, Ontario, Canada
Department of Earth and Environmental Sciences, University of Waterloo, Ontario, Canada

David Blowes
Department of Earth and Environmental Sciences, University of Waterloo, Ontario, Canada

B. Merkel, M. Schipek, *The New Uranium Mining Boom*
DOI 10.1007/978-3-642-22122-4 © Springer 2011

Introduction

Strontium-90 (^{90}Sr) is a mobile, radioactive isotope. It is found at low concentrations in the environment as a result of nuclear testing that occurred in the 1950s and 1960s and in nuclear reactor waste (Department of Health and Human Services 2005). Strontium-90 decays with a half-life of 28.8 years to yttrium-90 (^{90}Y) by emitting a beta particle, yttrium-90 decays by emitting a more energetic beta particle with a half-life of 64 h to zirconium-90 (^{90}Zr). The main health concerns for strontium-90 are related to the energetic beta particle from yttrium-90 (Argonne National Laboratory 2006).

Permeable Reactive Barriers (PRBs) are passive remediation systems that are installed in the path of a contaminant plume, which can be applied for many different contaminants. Depending on the reactive material used in a PRB, it either promotes chemical reactions with the contaminant that lead to products which do not cause a health concern or the reactive material sorbs the contaminant. PRBs composed of zero-valent iron have been used successfully for the treatment of Cr, U and Tc (Blowes et al. 2000), as well as TCE. Organic carbon PRBs composed of sawdust or woodchips have been used to treat constituents of acid-mine drainage, including SO_4, Fe, Ni, Co and Zn (Blowes et al. 2000). Zeolites are aluminosilicates with a high sorption capacity and have been used in PRBs for sorption of heavy metals and radionuclides (Lee and Hartwig 2005).

Site Description

The Chalk River Laboratories (CRL) are located about 193 km West of Ottawa in Ontario, Canada. The surficial geology has been described by King and Killey (1994). In the early 1950s, a pilot plant was operated at the site for the purpose of decomposing and reducing the volumes of ammonium nitrate solutions containing mixed fission products. Some of these solutions, including solutions containing ^{90}Sr, were released into a pit lined with crushed limestone (Parsons 1963). These solutions subsequently infiltrated into an underlying aquifer. The aquifer is composed of 5 to 13 m of saturated sand units, including very fine sand, fine sand, fine-medium sand, minor interstratified silty sand and in some areas, is underlain by a basal till unit (Killey and Munch 1987). A fine layer of gravel was detected in some boreholes at or near the top of the till. Hydraulic conductivity values range from 6.6×10^{-6} m/s to 4.0×10^{-5} m/s. Groundwater velocities are in the range of 100 to 150 m/a (Klukas and Moltyaner 1995). Due to geochemical interactions in the aquifer, the leading edge of the ^{90}Sr plume moves slower and the travel time from the source area to Duke Swamp 440 m downgradient exceeds 40 years. In 1994 the plume was approaching Duke Swamp and was about 7 m wide and was located in the bottom half of the aquifer (Lee and Hartwig 2005). A large number of monitoring wells is installed on the site. The horizontal hydraulic gradient ranges from 0.01 to 0.016. The groundwater flow direction is to the Southwest.

In December 1998, a permeable reactive barrier system was installed using, as the reactive medium, granular clinoptilolite. This mineral had been suggested for control of Sr-90 groundwater plumes (Cantrell et al. 1995; Fuhrmann et al. 1996). The installed reactive barrier system is known as the wall and curtain (Fig. 1). It consists of a 30 m long sealed-joint, steel-sheet pile cut-off wall (Waterloo Barrier®), located on the down-gradient side of a granular zeolite "curtain" or barrier (Lee and Hartwig 2005). The cut-off wall ranges in depth from 9.5 to 12 m and is in contact with the till or bedrock. Two sheet piling walls were installed perpendicular to the cut-off wall to act as the sides of the box containing the curtain of reactive material. The sides and wall were grouted with a cement-based grout to create an impermeable barrier (Lee and Hartwig 2005). The reactive curtain is a 2 m thick, 11 m wide and 6 m deep volume of granular clinoptilolite. As Ca concentrations are about 10 mg/l, conditions for ^{90}Sr sorption on the clinoptilolite are favorable (Lee et al. 1998; Lee and Hartwig 2005). The hydraulic conductivity of the clinoptilolite is about 2.6×10^{-4} m/s. Ten fully penetrating wells were installed in the reactive material directly in front the sheet piling wall, which are connected to a manhole by a horizontal pipe. The outflow of this drain can be adjusted, making it possible to control the flow rate and capture zone of the wall and curtain. The outflow of a horizontal drain 60 m upgradient of the wall and curtain is also adjustable. The function of this upgradient horizontal drain is to capture uncontaminated groundwater from the shallow parts of the aquifer and thus ensure less loading of the clinoptilolite and a longer time to replacement of the reactive material. The deeper aquifer has a road salt label, and so the electric conductivity of the water from the upgradient drain is monitored to ensure that capture does not include the lower contaminated groundwater. Another feature of the wall and curtain

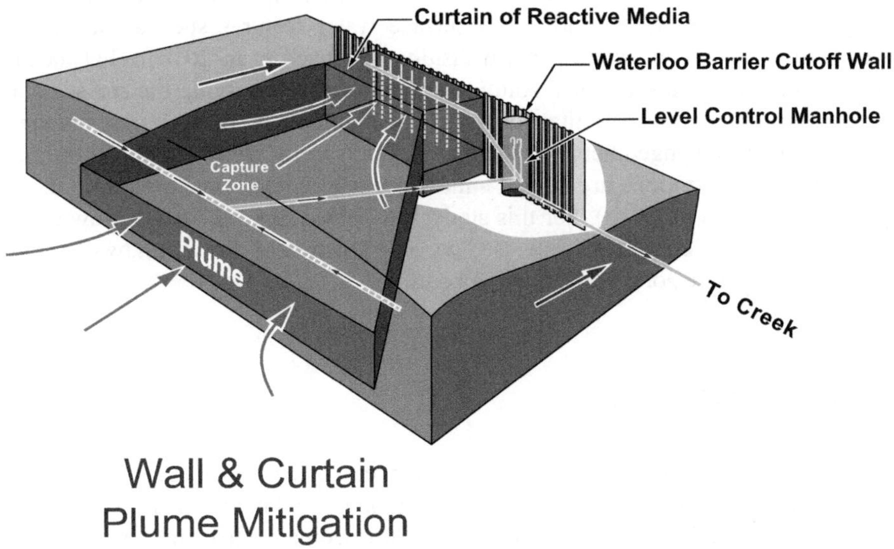

Fig. 1 Schematic of AECL's Wall and Curtain PRB (from Lee and Hartwig 2005)

design is that the flow of both the PRB drain and the upgradient drain are measured with direct, manual measurement (Lee and Hartwig, in prep.).

Objective

The objectives of this study are to provide a better understanding of the hydraulic characteristics of the wall and curtain and the groundwater diversion drain and provide quantitative estimates of the distribution of flow within the upgradient aquifer and flow through the wall and curtain system.

Model

For this study, the three-dimensional finite element numerical computer code HydroGeoSphere was used. This code was developed at the University of Waterloo, Ontario, Canada (Therrien et al. 2010). HydroGeoSphere provides a rigorous simulation capability that combines fully-integrated hydrologic/water quality/subsurface flow and transport capabilities with a well-tested set of user interface tools (Therrien et al. 2010). Flow in the saturated and unsaturated zone as well as through fractures can be simulated using the control volume finite element approach. Pumping and injection wells, streams and tile drains can be included in the simulation. A unique feature is that when the flow of water is simulated in a fully-integrated mode, water derived from rainfall inputs is allowed to partition into components such as overland and stream flow, evaporation, infiltration, recharge and subsurface discharge into surface water features such as lakes and streams in a natural, physically-based fashion (Therrien et al. 2010). That means that the fully-coupled numerical solution approach allows solving the surface and variably saturated flow simultaneously at each time step. In a similar manner, solutes can be exchanged between the surface and subsurface in a realistic fashion and solute concentrations are solved simultaneously at each time step in both regimes (Therrien et al. 2010). For this study, only physical groundwater flow capabilities of HydroGeoSphere were used in order to model the groundwater flow through and capture zone of the wall and curtain.

Simulation Setup

The domain for the flow model was set up as a rectangular block measuring 200 m in the direction of flow, the x-dimension, 150 m in the y-dimension and 12 m in the z-dimension. The large length and width of this domain have the advantage of lowering the possible effects on the simulated flow by the lateral domain boundaries. As the aquifer thickness varies little across the site, a uniform depth of 12 m is reasonable. The spacing of the finite element grid ranges from 1 to 3 m in every dimension for elements far from the wall and curtain in the domain; however, the spacing was refined in certain parts of the model domain to achieve a more detailed view of the groundwater flow in the vicinity of the wall and curtain. Thus, the elements close to the reactive barrier were assigned a length of 0.25 to 0.5 m. The sheet piling walls were modeled with a thickness of 0.005 m.

Hydraulic conductivity values were assigned to zones in the grid based on the stratigraphy of the aquifer (see Table 1). The gravel layer at the bottom of aquifer was included with a hydraulic conductivity of 1.3×10^{-4} m/s. This value was incorporated into a 0.5 m thick layer in the lower third of the aquifer, which has a hydraulic conductivity of 3.6×10^{-5} m/s. A weighted average of 0.01 m of gravel and 0.49 m of the sandy lower part of the aquifer results in a hydraulic conductivity of 5.5×10^{-5} m/s for this layer.

The wall and curtain was positioned in the center of the grid in the y-dimension and 115 m from the upgradient boundary in the x-dimension. The dimensions of the reactive material are the same as in the field. The clinoptilolite was assigned a hydraulic conductivity of 2.6×10^{-4} m/s. The hydraulic conductivity of the sheet piling walls was set to 1×10^{-9} m/s, to make it nearly impermeable. The ten wells behind the curtain, which act as outflow drains, were included in the model as ten wells with a diameter of 0.054 m, spaced 1 m apart from each other in the y-dimension. Each of these wells is simulated to remove one tenth of the discharge of the wall and curtain. The horizontal upgradient drain was included in the model as a horizontal tile drain. Flow from a small pond downgradient of the wall and curtain is recorded by readings of a weir. These recordings were represented in the model by a 10 m long (in x-direction) and 30 m wide (in y-direction) area on the surface of the model through which a specified discharge, which was varied depending on the reading on a particular date.

Constant head boundaries were assigned to the inflow (x = 0 m) and outflow (x = 200 m) boundaries of the model based on observed heads close to the bounda-

Table 1 Hydraulic conductivity values based on the aquifer stratigraphy used in the model

Conductive Unit	Description	Material	Elevation (m)	Hydraulic Conductivity (m/s), isotropi
K1	Upper 2/3 of aquifer	Silty fine and very fine sand	6–12	1.3×10^{-5}
K2	Lower 1/3 of aquifer	Fine and medium sand	1.5–6	3.6×10^{-5}
K3	Gravel above till	Fine gravel/washed till	1–1.5	5.5×10^{-5}
K4	Till above bedrock	Stony till	0–1.0	1.4×10^{-6}
K5	Curtain	Clinoptilolite	6–12	2.6×10^{-4}
K6	Wall and base curtain	Sheet piling; bentonite + plastic		1.0×10^{-9}

Table 2 Input parameters for June 2006 and November 8, 2010 used for the simulations

Parameter	Simulation Jun-06	8-Nov 10	Units
Curtain discharge	2.9E-04	3.9E-04	m^3/s
Drainage tile discharge	4.2E-05	2.2E-04	m^3/s
Seepage area discharge	9.2E-05	7.0E-05	m^3/s
Area of downgradient seepage	300	300	m^2
Seepage area (specific discharge)	3.1E-07	2.3E-07	m/s
Upgradient boundary head	11.4	11.1	m
Downgradient boundary head	8.9	9.7	m
Aquifer porosity (all units)	33	33	%

ries of the model domain. The observed heads were adjusted to a different datum to fit simulations in the 12 m thick aquifer. The boundaries in the y-dimension were considered to be no flow boundaries. A constant precipitation recharge flux rate of 0.3 m/a was assigned over the entire top surface domain (Klukas and Moltyaner, 1995). Flow through the wall and curtain was simulated in the saturated zone under steady-state conditions.

The model was calibrated by comparing observed heads to simulated heads over a large area of the model domain. Only the influent and effluent constant head boundaries were adjusted to achieve a good fit between observed and simulated heads over the model domain. Two data sets from different dates were analyzed, which included head measurements and flow rates of the wall and curtain and the upgradient drain as well as the flow from the pond. The data sets from June 2006 and November 2010 were chosen as the flow rates differ significantly from each other, thus a comparison of the capture zone is made possible. The input parameters of both dates are summarized in Table 2.

Summary and Discussion of Results

June 2006

For the June 2006 data set, heads of 11.4 m and 8.9 m were assigned to the influent and effluent boundaries in the x-dimension, respectively. The simulated heads across the model domain coincide well with the observed heads, with R^2 values of near 0.9 for observation points along the centerline of the domain (see Fig. 2). In general, the difference between observed and simulated heads is less than 0.5 m. The observed heads are distributed equally below and above the line, which suggests that the comparison is reasonable.

The width and depth of the zone of influence of the wall and curtain can be estimated. For June 2006, the zone of influence is 70 m wide at the upgradient boundary of the model, and 12 m deep. The upgradient drain deflects the water

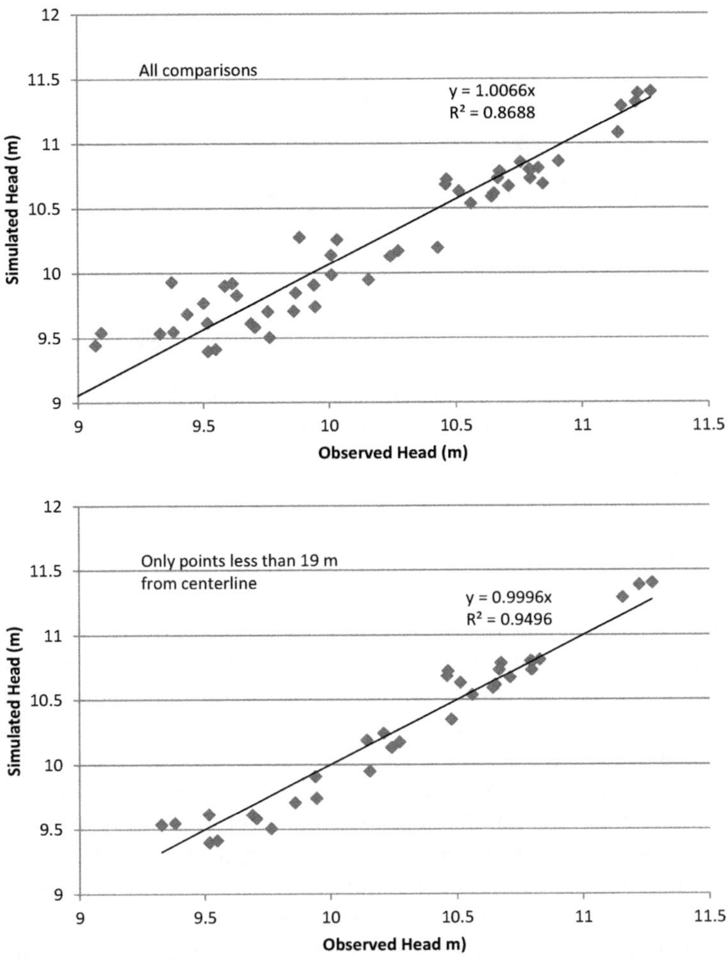

Fig. 2 Comparison of simulated heads and heads observed in June 2006, with a reference line showing a 1 : 1 ratio of the values. The upper plot includes all of the observation points. The lower plot includes only data less is than 20 m from the ($y = 75$ m) center line

table slightly and intercepts shallow groundwater, but also draws deeper flow lines to shallower depths. Flow lines from the deeper, contaminated parts of the aquifer are intercepted by the wall and curtain, as is intended by design (see Figs. 1 and 3). Discharge into the pond downgradient of the wall and curtain occurs from flow lines outside of the zone of influence only. Modeled groundwater velocities range from 20 to 40 m/a in the upper part of the aquifer to 40 to 60 m/a in the lower parts of the aquifer and the washed till layer. Velocities downgradient of the wall and curtain are low (<30 m) due to a low hydraulic gradient, but high near the wall and curtain due to a greater hydraulic gradient.

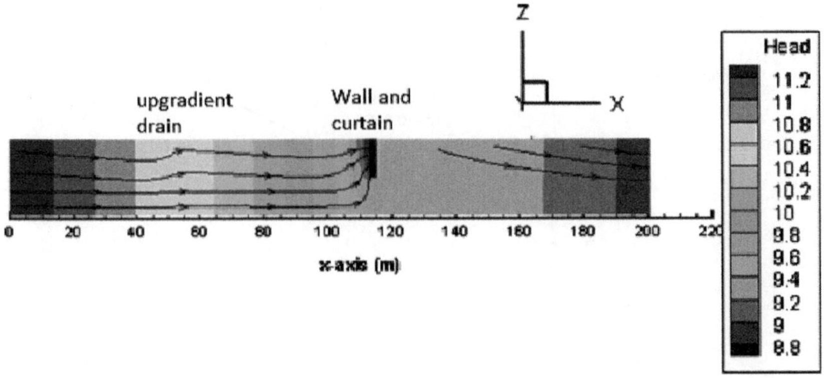

Fig. 3 Vertical-longitudinal cross-section of simulated heads (m) calibrated to June 2006 head data and streamlines, plotted on the centerline ($y=75$ m) of the domain

November 8, 2010

Influent and effluent boundary heads of 11.1 and 9.7 m, respectively, were assigned for the data set of November 8, 2010. The simulated and observed heads across the site show a good agreement, with R^2 values of 0.92 for observation points along the centerline of the domain (see Fig. 4). Differences between observed and simulated heads were generally less than 0.2 m. Although there is reasonable agreement between the measured and calculated hydraulic head values, the zone of influence estimated from the simulations is significantly larger than is indicated from monitoring-well measurements. This discrepancy may result from the simplifying assumptions inherent in the simulations, including the homogeneity and isotropy of the aquifer, assumptions regarding the extent and permeability of the coarse-grained layer at the base of the aquifer, and assumptions regarding the hydraulic head levels down gradient of the wall and curtain. Additional field measurements and modeling are planned to resolve these uncertainties.

The estimated discharge of the wall and curtain is one third higher and the discharge from the upgradient drain is five times higher than in June 2006. The pond and swamp receive 25% less water than in June 2006. Due to the different flow rates of the upgradient drain and the wall and curtain, the zone of influence width nearly doubles to 130 m. This means that the plume is captured by the wall and curtain, however, a large portion of the zone of influence lies within the uncontaminated parts of the aquifer. If the head of the upgradient drain is too low due to a high drainage rate, the drain has the potential to remove contaminated water from considerable depths in the upgradient aquifer (Shikaze and Sudicky 1994). In November 2010, the discharge rate of the upgradient drain is 56% of the wall and curtain discharge. In the simulation, the upgradient drain draws more water from the deeper parts of the aquifer into it than it did in June 2006, mainly from the

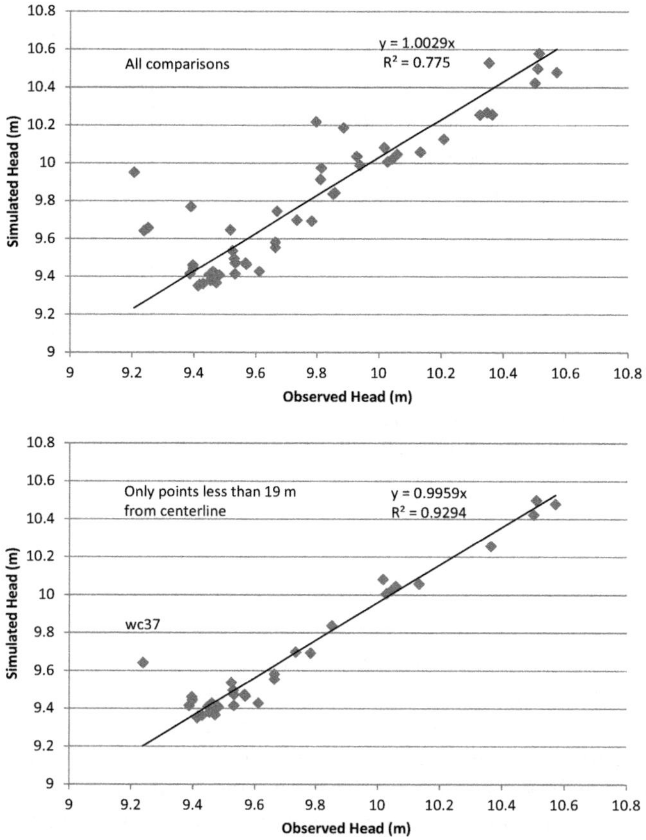

Fig. 4 Comparison of simulated heads and heads observed on Nov 8, 2010, with a reference line showing a 1 : 1 ratio of the values. The upper plot includes data that is more than 20 m away from the $y = 75$ centerline of the domain. The lower plot includes only data within 20 m from the center line. The isolated high simulated head value on this plot is for WC37, at the west end of the wall

upper half of the aquifer, but partly also from the contaminated lower third of the aquifer (see Fig. 5). If this situation persisted for a longer time, there would be a potential of upward movement of the plume and the contaminated groundwater would be captured by the upgradient drain. Regular electrical conductivity measurements of the outflow of the upgradient drain are used to prevent this scenario (Lee and Hartwig, in prep). The discharge of the wall and curtain was increased by 50% on November 10, 2010 and the discharge of the upgradient drain was decreased to $1.6 \times 10^{-4}\,m^3/s$ by December 3, 2010. Simulations of these flow rates show that a smaller portion of deeper water is drawn into the upgradient drain, ensuring the capture of the plume by the wall and curtain and preventing inadvertent diversion of contaminated groundwater.

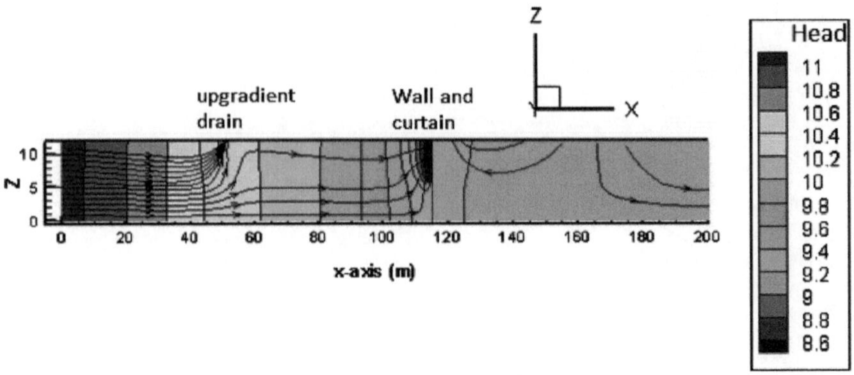

Fig. 5 Vertical-longitudinal cross-section of simulated heads (m) calibrated to Nov 8, 2010 head data and streamlines, plotted on the centerline ($y = 75$ m) of the domain. Streamlines as deep as 3.3 m above bedrock are predicted to enter the upgradient drain

Flow Balance

As the simulations of the data set from November 8, 2010 show that there is a potential of the upgradient drain to draw in water from the contaminated, deeper parts of the aquifer, both data sets were further analyzed to estimate a ratio of the upgradient flow rate to the discharge of the wall and curtain required to prevent this scenario from occurring. In June 2006, the flow rate of the upgradient drain was 14% of the wall and curtain discharge and only intercepted shallow, uncontaminated groundwater. The plume is located in the lower half of the aquifer, or below depths of 5.5 m. In further simulations, the discharge of the upgradient drain was incrementally increased until flow lines from depths of 5.5 m were directed into the upgradient drain, keeping all other parameters of the model constant. These simulations show that the flow rate of the upgradient drain for the June 2006 data set could be increased from 4.2×10^{-5} m^3/s to 1.3×10^{-4} m^3/s, or from 14 to 45% of the wall and curtain discharge rate, without drawing contaminated groundwater. Increasing the upgradient drain flow rate to this value would have the advantage of the upgradient drain intercepting a larger amount of uncontaminated groundwater, thus prolonging the time-to-replacement of the reactive materials.

A similar set of simulations was performed with the November 8, 2010 data set. The flow rate of the upgradient drain in this case was 56% of the wall and curtain discharge. As the upgradient drain already intercepted water from the lower half of the aquifer, the flow rate was incrementally decreased while keeping other parameters constant until only water from depths of less than 5 m was drawn into the upgradient drain. As a result of the simulations, the flow rate of the upgradient drain would have to be decreased from 2.2×10^{-4} m^3/s to less than 1.1×10^{-4} m^3/s in order to prevent the upgradient drain to intercept water from depths greater than 5.5 m. In other words, the upgradient drain discharge would have to be reduced from 56 to 28% of the flow rate of the wall and curtain.

These findings are based on two data sets only and due to the different results for June 2006 and November 8, 2010, one single value will not be appropriate to all groundwater flow situations of the site. However, the simulations suggest that keeping the flow rate of the upgradient drain at a value of 25 to 30% of the wall and curtain discharge will ensure that only uncontaminated groundwater is intercepted by the upgradient drain, while the contaminant plume is captured by the wall and curtain. This will ensure remediation of the plume while preventing unnecessary overloading of the clinoptilolite due to the treatment of uncontaminated groundwater, thus leading to a greater longevity of the wall and curtain. The current practice of monitoring the electric conductivity of the outflow of the upgradient drain is a practical approach for ensuring that only uncontaminated groundwater is captured.

Conclusions

Two data sets were analyzed with the constructed model. Both data sets show a good agreement between simulated and observed hydraulic heads. Although reasonable agreement between the measured and simulated hydraulic head levels was observed, a discrepancy remains with respect to the extent of the zone of influence of the wall and curtain. Additional field and modeling studies are planned to resolve this discrepancy.

A ratio of flow rates through the wall and curtain and the upgradient drain was estimated to ensure that no contaminated groundwater is pulled into the upgradient drain. However, a single ratio cannot provide optimal conditions for ensuring that the uncontaminated, shallow groundwater is intercepted by the upgradient drain while the plume is captured by the wall and curtain. In general, the flow rate of the upgradient drain should be 25 to 30% less than that of the wall and curtain.

The current model has limitations in the accuracy of describing the flow through the wall and curtain. A major factor is a lack of head data downgradient of the wall and curtain, as well as inconsistencies in the observed head measurements. Moreover, simulated heads for November 2010 at opposite ends of the sheet piling were 0.6 m higher than the observed heads, which may suggest inaccuracies in the hydraulic conductivity values used for the simulation. In order to minimize the limitations mentioned above, field work will be performed at the site to collect more physical and geochemical data. The current flow model will then be refined to show the flow through the wall and curtain more accurately. A geochemical transport model is planned to be constructed after the geochemical characterization of the site, which will be used to provide estimates on the longevity of the wall and curtain.

References

Argonne National Laboratory, Environmental Science Division, 2006. Human Health Fact Sheet on Strontium.

Blowes, D.W., Ptacek, C.J., Benner, S.G., McRae, C.W.T., Bennett, T.A., Puls, R.W., 2000. Treatment of inorganic contaminants using permeable reactive barriers. Journal of Contaminant Hydrology, Volume 45. pp. 123–137.

Cantrell, K.J., P.F. Martin, and J.E. Szecsody, 1995. "Clinoptilolite as an In-Situ Permeable Barrier to Strontium Migration in Ground Water," Pacific Northwest National Laboratory, Richland, Washington, April 1995.

Department of Health and Human Services, Centers for Disease Control and Prevention, 2005. Radiation Emergencies. Radioisotope Brief Strontium-90 (Sr-90).

Fuhrmann, M., Aloysius, D., Zhou, H., 1996. Permeable, subsurface sorbent barrier for ^{90}Sr: Laboratory studies of natural and synthetic materials. Waste Management, Vol. 15, 7, p. 485–493.

Killey, R.W.D. and J.H. Munch, 1987. Radiostrontium migration from a 1953–54 liquid release to a sand aquifer. Water Poll. Res. J. Canada, Vol. 22, No. 1, pp. 107–128.

King, K.J. and R.W.D. Killey. 1994 Quaternary geology of the AECL Chalk River Laboratories Property: A review. Siting Task Force, Tech. Bib No. 345 Oct 1994.

Klukas, M.H. and G.L. Moltyaner. 1995. Numerical simulations of groundwater flow and solute transport in the Lake 233 Aquifer. Environmental Research Branch, Chalk River Laboratories, Chalk River, AECL. 54 p.

Lee, D.R. and D.S. Hartwig, 2005. Zeolite prevents discharge of strontium-90-contaminated groundwater. Canadian Nuclear Society Waste Management, Decommissioning and Environmental Restoration for Canada's Nuclear Activities: Current Practices and Future Needs Ottawa, Ontario Canada May 8–11, 2005.

Lee, D.R., D.J.A. Smyth, S.G. Shikaze, R. Jowett, D.S. Hartwig and C. Milloy. 1998. Wall and curtain passive collection/treatment of contaminant plumes. In Proceedings of the First International Conference on Remediation of Chlorinated and Recalcitrant Compounds. Designing and Applying Treatment Technologies, pp. 77–84. Battelle Press, Columbus, OH.

Lee, D.R. and D.S. Hartwig, in preparation. Performance assessment of a zeolite PRB that has prevented release of 90Sr for 13 years.

Parsons, P.J. 1963. Migration from a disposal of radioactive liquid in sands. Health Physics, Vol. 9, pp. 333–343.

Shikaze, S.G. and E.A. Sudicky, 1994. Quantitative modelling for selection and design of a passive gravity-driven system to collect the nitrate-plant plume: Final simulations. Report to: David Lee, Atomic Energy of Canada Ltd., Chalk River, Ontario. Waterloo Centre for Groundwater Research, University of Waterloo.

Therrien, R., McLaren, R.G., Sudicky, E.A. and S.M. Panday, 2010. HydroGeoSphere: A Three-dimensional Numerical Model Describing Fully-integrated Subsurface and Surface Flow and Solute Transport. Groundwater Simulations Group, University of Waterloo. 443 p.

Part 6
Miscellaneous

Cleaning of NORM Contaminated Pipes from Dismantling of Oil/Gas Production Facilities at a North African Site

Rainer Barthel

Abstract. At a North African site of oil/gas production, a large volume of NORM contaminated tubings, pipes and fittings resulting from dismantling of production wells and other facilities was accumulated and stockpiled over years. To manage this scrap showing different types and levels of radioactive contamination, a comprehensive project was initiated for radiological survey of the scrap, the comparison of feasible cleaning options, and scrap cleaning. Dry abrasive grit blasting was selected as the most efficient and reliable technique for cleaning.

Introduction

Surface contamination of pipes and other facility components by naturally occurring radionuclides is a common issue in oil/gas industry (IAEA 2003). Usually, the reservoir water contains Group II (Periodic Table) cations of calcium, strontium, barium and radium dissolved from the reservoir rock. As a consequence, the produced water contains the long-lived radium isotopes Ra-226 ($T_{1/2} = 1620$ a) from the U-238 series and Ra-228 ($T_{1/2} = 5.75$ a) from the Th-232 series. The parent nuclides U-238 and Th-232 are not mobilized with the formation water of the reservoir rocks. Due to operation history, pipes, valves and further components of the production facilities are more or less contaminated with Ra-226, Ra-228, and their daughter nuclides. In some oil/gas fields, also Pb-210 ($T_{1/2} = 22.3$ a) is dissolved in considerable amounts in the reservoir water, resulting in elevated surface contaminations with Pb-210. Another mechanism that may lead to high Pb-210

Rainer Barthel
Brenk Systemplanung GmbH, Heider-Hof-Weg 23, D-52080 Aachen

B. Merkel, M. Schipek, *The New Uranium Mining Boom*
DOI 10.1007/978-3-642-22122-4 © Springer 2011

contamination in components of oil/gas production facilities is the decay of Rn-222 ($T_{1}/_{2}=3.8$ d), which is highly soluble in oily reservoir waters.

At a North African oil/gas production site, large volumes of contaminated tubings, pipes, fittings etc. resulting from dismantling of production wells and other facilities were accumulated and stockpiled over years. To manage this scrap in compliance with national regulations and international recommendations for the clearance of steel scrap, a comprehensive project was initiated comprising initial radiological survey and assessment of the stockpiled scrap, the development and test of a clearance measurement method straightforwardly applicable under field conditions with an acceptable expenditure of time, the study and comparison of technically feasible cleaning techniques, subsequent specification and construction of a semi-mobile cleaning facility at the site, and last but not least, execution of the cleaning operations and clearance of decontaminated scrap for unrestricted reuse. In the following, some results and experiences gained in the framework of this project, conducted between 2006 and 2010, are summarized.

Radiological survey of stockpiled steel scrap

Some stockpiles of tubings, pipes and so-called "scrap of complicated geometry" (SCG, valves, fittings, bended pipes etc.) are shown in Fig. 1.

In total, a volume of about 1270 t of steel scrap was pertinent to the project, comprising about 730 t of tubings from production wells (diameters of $2^{7}/_{8}$", $3^{1}/_{2}$" and $4^{1}/_{2}$"), about 390 t of pipes from the dismantling of various facilities with different diameters (mainly 6" to 10"), and about 150 t of SCG.

For the project planning, radiological survey measurements concerning surface contaminations were performed to

- assess the fraction of the stockpiled scrap with radioactive contamination above the clearance level, therefore necessary to be cleaned,
- to take scratch samples of different scale types for a detailed radiological, chemical and mineralogical analysis aimed at calibrating the measurement devices, and at assessing the suitability of different cleaning techniques.

Figure 2 qualitatively illustrates the variability of surface contamination levels for a certain stockpile with $3^{1}/_{2}$" tubings measured with a surface contamination monitor of type MicroCont II, probe Norm1 (Rados GmbH). Table 1 exemplifies the diversity of scale types.

Fig. 1 Stockpiles of tubings with diameters of $3^1/_2$" (**a**) and $4^1/_2$" (**b**), and of SCG (**c**)

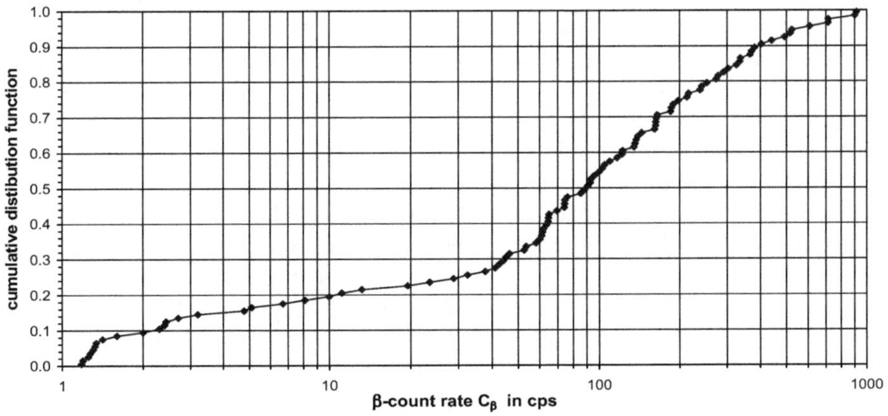

Fig. 2 Cumulative distribution function of the β-count rates C_β measured for 100 randomly selected tubings of a stockpile (cf. Fig. 1a)

Table 1 Results of γ-spectrometric analyses of specific activities (A_r) of scale samples

Scrap type	Scale type	A_{Ra-226} in Bq/g	A_{Pb-210} in Bq/g	A_{Ra-228} in Bq/g	A_{Th-228} in Bq/g
Tubings 3$^1/_2$"	A	20	260	8	10
Tubings 2$^7/_8$"	C	3.7	51	0.4	0.4
Tubings 4$^1/_2$"	D	2.0	108	1.5	1.2
Pipes 10"	F	350	240	140	180
Tubings 4$^1/_2$"	P	0.13	220	0.05	0.07
Tubings 3$^1/_2$"	R	170	630	50	60
Pipes 6"	S	290	155	37	50
SCG	W	350	70	250	280

Method for Clearance Measurements

The national regulations for clearance of steel scrap that had to be complied with define radioactive surface contamination as "the presence of radioactivity on a surface in excess of 3.0 Bq/cm^2 (β) or 0.3 Bq/cm^2 (α)", i.e.

$$S_\beta \le L_\beta = 3\,\text{Bq/cm}^2\text{, and} \tag{1}$$

$$S_\alpha \le L_\alpha = 0.3\,\text{Bq/cm}^2, \tag{2}$$

where S_β and S_α denote the area specific β- and α-activity of surface contamination, L_β and L_α the respective limits for clearance. For any surface contaminated with NORM, Eq. (2) is obviously more restrictive than Eq. (1).

However, measurement of the area specific α-activity of NORM-contaminated surfaces is technically very challenging, and it is virtually impossible in the field with required accuracy and acceptable effort. To overcome this difficulty in compliance with the regulatory requirements, a method for the measurement of surface contamination in large scrap volumes was developed, which is straightforwardly applicable under field conditions with acceptable effort. The method is based on the measurement of area specific β-activity that can be related to the area specific α-activity, if the activity ratios between relevant long-lived radionuclides are known (Barthel and Bunzmann 2007).

For a given type of radioactive contamination with known activity ratios, the requirements (1) and (2) can be transformed into each other taking into account the numbers of α- and β-emissions, which are associated with the relevant long-lived radionuclides Ra-226, Pb-210, R-228 and Th-228, given in Table 2.

Based on Table 2 and the laboratory analyses of scale samples determining the specific activities A_{Ra-226}, A_{Pb-210}, A_{Ra-228} and A_{Th-228}, limits $L_\beta(L_\alpha)$ were derived for the various scale types occurring in the stockpiled scrap, which correspond to the limit L_α of requirement Eq. (2):

$$L_\beta\left(L_\alpha\right) = L_\alpha \cdot \frac{2\cdot\left(A_{Ra-226} + A_{Pb-210} + A_{Ra-228} + A_{Th-228}\right)}{4\cdot A_{Ra-226} + A_{Pb-210} + 5\cdot A_{Th-228}} \tag{3}$$

Table 2 α- and β-emitters in the sub-chains of relevant long-lived radionuclides

Sub-chain	α-emitters		β-emitters	
	Radionuclides	Number	Radionuclides	Number
Ra-226+	Ra-226, Rn-220, Po-218, Po-214	4	Pb-214, Bi-214	2
Pb-210+	Po-210	1	Pb-210, Bi-210	2
Ra-228+	–	0	Ra-228, Ac-228	2
Th-228+	Th-228, Ra-224, Rn-220, Po-216, Bi-212 (36%), Po-212 (64%)	5	Pb-212, Bi-212 (64%), Tl-208 (36%)	2

The next step conducted for various scale types concerned the calibration of the β-sensitive surface contamination probe Norm1, also depending on the ratios between the activities of the long-lived radionuclides because of different energy distributions of emitted β-particles, which influence the detector response – the β-count rate C_β for a given radionuclide composition of the surface contamination (Barthel and Bunzmann 2007). By means of calibration factors derived for the probe Norm1, derived limits LC_β were specified for each scale type, such that the following condition for clearance corresponds to Eq. (2):

$$C_\beta \leq LC_\beta \tag{4}$$

For clearance, the calculated values of LC_β were conservatively rounded to the next smaller multiples of 0.5 cps, yielding derived limits from 4 cps for scales with high activity ratio Ra-226 to Pb-210 (e.g. scale types F, S and W, cf. Table 1) to 6.5 cps for scales with very low activity ratio Ra-226 to Pb-210 (scale type P).

The devices for the measurement of surface contamination are also sensitive to γ-radiation. Thus, the output of a usual measurement is a count rate $C_{\beta+\gamma}$, representing the number of detector counts per second (cps) resulting from β- and γ-radiation in total. To determine the pure β-count rate C_β, a second measurement of the same surface area is required while shielding the β-particles (but not influencing the γ-rays significantly) to measure the portion C_γ of $C_{\beta+\gamma}$, which is caused by γ-rays only (i.e. the γ-count rate). This can be done by shielding the detector window of the Norm1 probe against β-particles by an aluminum plate of 3 mm thickness. The β-count rate C_β, i. e. the quantity applied for clearance according to Eq. (4), then results by calculating the difference

$$C_\beta = C_{\beta+\gamma} - C_\gamma \tag{5}$$

Unwanted contributions to the directly measurable count rates from natural background radiation and from contamination of surrounding scrap areas are canceled down in C_β. For achieving an accuracy for the measurement of C_β of about 10%, the rates $C_{\beta+\gamma}$ and C_γ were measured over 1 minute each, in general.

Comparison of Cleaning Techniques

Five technologically feasible options for managing the radioactively contaminated steel scrap were considered:

- *a*brasive *g*rit *b*lasting (AGB),
- *h*igh *p*ressure *w*ater *j*etting (HPWJ),
- *s*cale *d*issolving (SD),
- *s*cale *m*illing (SM), and
- *s*crap *s*melting (SS).

For the options SM and SS, an export of the contaminated scrap to a licensed company (e. g. to Germany) would have been required. For this reason, they were excluded from further considerations. The cleaning options AGB, HPWJ and SD were assessed as techniques that could be carried out at the site, in principle.

Application of chemical methods by the injection of so-called inhibitors is a well-known practice for scale prevention in production wells. Chemical methods can also be applied for removal of scale using mixtures of acids or combinations of acids and complexing agents (IAEA 2003). Depending on specific chemical, mineralogical and physical conditions, chemical scale removal is often the approach with the lowest costs, especially when scale is not easily accessible and the mechanical removal methods are ineffective or expensive to deploy. However, the scales in tubings have often too little porosity for an effective chemical treatment within a reasonable time frame. The effectiveness of scale removal by SD depends on its chemical/mineralogical composition and on physical properties like scale porosity, which determines the ratio of the reactant surface to the scale mass. Hard scale types such as Galena (PbS) and Celestine ($Sr(Ba)SO_4$), which are dominating at the site, are extremely resistant to chemical removal. The nonporous sheets of these scales are slow to react with any but the strongest chemical reactants. A further critical issue of option SD was the disposal of radioactive liquid produced. For these reasons, option SD was excluded from further considerations.

Dry abrasive grid blasting (AGB) is used in a wide range of industries for many different purposes including the removal of rust, scale, paint etc., and for various forms of surface preparation. An abrasive material (grit), the choice of which depends on specific circumstances, is propelled on the surface at high speed. The most common method of abrasive grit blasting uses compressed air to propel the grit material from a blast pot through a blasting hose to a nozzle that is manually controlled by the operator. In automatic systems, the hose is fitted to a lance with blast head that is maneuvered by a lance trolley. For the cleaning of inner pipe surfaces, the lance is inserted into the tube up to the rear end, which is connected and tightly sealed with the collecting adapter of the abrasive recovery box. Due to the vacuum prevailing in the collecting adapter, the abrasive material is recovered and conveyed to the reclaim unit. The produced waste (scale debris, smashed grit) is collected in a disposal drum.

High pressure water jetting (HPWJ) is a commonly used technique for cleaning surfaces. In the oil and gas industry, HPWJ has been applied to remove scales from tubings and other types of surface contaminated equipment. In general, this technique can provide effective scale removal. Generation of contaminated dust is reduced by keeping the material wet. By HPWJ, scale is removed by cavitation, whereby small bubbles form in the fluid jet stream. These bubbles are created by the large pressure release as the fluid passes through the jet nozzle. The bubbles collapse on impact with scale, causing a forceful (explosive) erosive effect. The ability of HPWJ to remove tight paint, rust, scale etc. from surfaces can be much improved by injecting small concentration of an abrasive into the water stream at the nozzle. HPWJ utilizes a pressure pump (pressures from about 300 to 3,000 bar are applied), a large volume of water, HP-hoses, a pressure lance and a nozzle assembly. The used water should be collected, filtered (to remove scale), reused by circulation, and finally disposed of.

The comparison and evaluation of tenders for executing the cleaning project by AGB or HPWJ was conducted by means of 4 basic criteria (completeness and conformity with tender specifications; likelihood of successful and safe cleaning with the offered technique; practical experiences in radiation protection applicable to the cleaning task; costs), 8 technical criteria (time schedule including the completion and mobilization of the equipment; exhaustive description of the working procedure; potential for process optimization concerning flexibility/effort for different conditions; immobilization of contaminated waste; required support concerning the shipment, installation, and supply of resources; documentation of clearance measurements; experiences/references concerning the technology and the items cleaned in the past; qualification of employees), and of 4 radiation protection criteria (radiological risks for employees and appropriateness of the Radiation Protection Procedure; prevention of radioactive emissions and spreading of contaminated material; monitoring/supervision of possible emissions and contaminations; final decontamination of the facility equipment). After intensive evaluation of all criteria, it was decided to execute the cleaning project by means of AGB, because of the robustness of this technique and likelihood for successful, efficient, and safe performance.

Construction and Operation of the Cleaning Facility

Company sat. Kerntechnik GmbH was awarded to conduct the cleaning project by means of AGB. Brenk Systemplanung GmbH was ordered by the client for the management and supervision of the project execution on site. In addition, a local company was charged with service provision.

The cleaning facility (CF) was designed and constructed in co-operation of sat. Kerntechnik and company Munk & Schmitz Oberflächentechnik GmbH. After an acceptance test conducted in June 2009, the facility components were shipped to the site. The construction of the facility at the site of operation was finished in October 2009. Thereafter it was operated until December 2010, with an extended break in summer 2010, because of the climatic conditions in the desert.

The CF presents a semi-mobile facility. Its components are installed in six 40' containers and one 20' container. Four 40' containers for cutting, lance traversing, blasting and measurement are placed in a row. The "Measurement Container" was originally envisaged for executing clearance measurements. However, conduction of clearance measurements inside the container caused logistic problems, which have been solved by using the output-rack and an additionally erected rack for clearance measurements. Directly connected alongside the "Blasting Container", the "Blasting Machine Container" (40') was placed, which contains the Problast machine, the control system, and the equipment for collection of blasting waste into 200 l drums. From here, a ramp was constructed for discharging filled blasting waste drums. The sixth 40' container – the so-called "Deduster Container" – was placed in second floor above the "Blasting Machine Container". It contains the deduster cyclone with grit recycle box, and the filter system for exhaust air. The CF is completed by a "Personnel Access Lock Container" (length of 20'), which is sub-divided into a "white area" and a "black area". It is used for changing clothes and contains sanitary equipment for regular use, a shower for incidental use, and the hand-foot-monitor. The "Personnel Access Lock Container" allows for access into the CF via the "Blasting Machine Container". Ancillary containers for storage of spare parts and expendable items, for staff recreation, office work and sanitary use were provided by the client.

Tubings and pipes with diameters from $2^7/_8$" to 10" with original lengths of about 9 to 12 m were cut into pieces between 4.2 and 6.2 m of length, appropriate for blasting with the lance traversing unit (see Fig. 3). SCG, including pipes with diameters above 10", were dismantled and cut into pieces suitable for blasting manually in a so-called blasting box.

In total, about 1270 t of scrap were managed, including about 500 t that were cleared as uncontaminated on the basis of respective measurements. About 720 t were cleared after successful blasting, and about 50 t had to be re-stockpiled in a supervised area for interim storage, because of their oily or tar-like contamination, which could not be removed by grit blasting with acceptable effort.

Fig. 3 Cutting and blasting of tubings in the cleaning facility. *Left:* Cutting of tubings with band saw. *Right:* blasting of tubings with lance unit

Conclusions

The abrasive grit blasting cleaning facility and the methods developed for proving compliance with regulatory requirements for the clearance of clean/cleaned steel scrap by means of accurate and effective field measurements turned out to be appropriate.

References

Barthel R, Bunzmann C (2007) Measurement of surface contamination according to legal requirements, Proceedings of the 5th International Symposium on Naturally Occurring Radioactive Material (NORM V), Seville, Spain, 19–22 March 2007: 455–466

IAEA (2003) Radiation Protection and the Management of Radioactive Waste in the Oil and Gas Industry, Safety Reports Series No. 34, Vienna, 2003

Direct and Indirect Effects of Uranium on Microstructure of Sedimentary Phosphate: Fission Tracks and Radon Diffusion

Fatima Zahra Boujrhal, Rajaâ Cherkaoui El Moursli

Abstract. Moroccan sedimentary phosphates are essentially constituted by apatite and contain some tens ppm of uranium. This uranium undergoes spontaneous fission and radioactive decay in chain, since its geological formation. The spontaneous fission causes the radiation damage in the structure, by recoil of the fission products, whereas the radioactive decay products various radioactive daughters, where one of them is gas; Radon. This radioactive gas causes microstructure defects (nanopores) by diffusion. The present abstract focuses on the exam of the micro/nanostructure due to these two phenomena. The fission tracks of several grains of phosphate are studied by microscopy observations, whereas the radon emanation is analyzed by gamma spectrometry analysis of bulk of phosphate. Some interesting and correlated results are obtained.

Introduction

Phosphate refers to rock or ore containing phosphate ions and is the occurring of the element phosphorus. It is used in everything from fertilizer to various industries.

Morocco houses approximately 2/3 of the world's phosphate reserves. The main component of these materials is apatite, with a general chemical formula

Fatima Zahra Boujrhal
Université Sultan Moulay Slimane, Faculté des Sciences et Techniques, Laboratoire de Gestion et de Valorisation des ressources naturelle (LGVEN), B.P. 523, Béni Mellal, Morocco

Rajaâ Cherkaoui El Moursli
Université Mohammed V, Faculté des Sciences, Laboratoire de Physique Nucléaire (LPNR), Av. Ibn Battouta, B.P. 1014, Rabat, Morocco

B. Merkel, M. Schipek, *The New Uranium Mining Boom*
DOI 10.1007/978-3-642-22122-4

$Me_{10}(XO_4)Z_2$, Me represents mainly a divalent cation (Ca^{2+}, Ca^{2+}, Ca^{2+}, Ca^{2+}, U^{4+}, ...), XO_4 represents mainly a trivalent anion (PO_4^{3-}, SO_4^{3-}, VO_4^{3-}, CO_3^{2-}, ...) and Z a monovalent anion like Cl^-, Br^-, F^-, OH^-, ... This mineral which generally crystallize in the hexagonal system (space group P63/m), finds application in all domains; like medicine, nuclear wastes, biology and geology.

Sedimentary phosphate contains uranium (in traces) which undergoes spontaneous fission since its geological formation and causes radiation damage (fission tracks) in the structure, by recoil of the fission products. Indeed, each single atom of this radionuclide creates fission tracks and all of fission track accumulated during geological time can be selectively dissolved and enlarged by etching in dilute nitric acid. The number of tracks depends in uranium content and the time, that's why, fission tracks analysis can be used as a technique for age determination (Fission tack age).

On the other hand, the Uranium-238 is attempt to become stable element, it will undergo a series of 14 radioactive decays until it eventually becomes lead-206, a stable non-radioactive element. This Uranium-238 is more or less in radioactive equilibrium with its radioactive daughters, principally radium-226 and radon-222.

This last radionuclide is a noble gas can easily emanate, at normal conditions from such natural materials in various rates depending on the micro/nanostructure: (Boujrhal et al.1999, 2004, 2005, 2007). With a half-live of 3.8 day. It is useful used as a natural tracer in physical science. Its diffusion can cause some microstructure defects (micropores).

A previous work to determine these three radionuclide contents is presented in: Berrada et al. (1992, 1995).

Material and methods

Materials

Several sedimentary phosphates of various Moroccan deposits are studied at natural state as like after annealing in a furnace for one and two hours at 50 to 950°C.

For fission track analysis, some apatite grains are selected from heated and unheated samples of phosphate, and are assembled on resin to form corresponding pastilles. Each pastille is polished in order to obtain a plane internal surface in the grains and then etched in dilute nitric acid in order to facilitate the counting of fission tracks density by optic microscopy.

For radon analysis, only a bulk of phosphate is needed to analyze uranium-238, radium-226 and radon-222 contents by gamma spectrometry.

Techniques

Scattering Electron Microscopy and Optical Microscopy

The characterization of the microstructure of internal surface in apatite's grains, Scattering Electron Microscopy (SEM) and optical Microscopy (OP) are used. With the first microscopy, external and internal morphologies, the chemical composition and the structure of the mineral are largely study in: (Boujrhal and Cherkaoui 2009). Indeed, the technique of spectrometry by energy dispersion (EDX) gives a qualitative and quantitative analysis whereas the technique of spectrometry by wavelength dispersion (WDX) makes X-cartographies of the elements constituting the grains of the samples. But with this microscopy it is impossible to distinguish between the fission tack and nanopores. In order to raise this problem, an optical microscope is used and the nanopores density (D_P) is counted in the non-etched grains (pastilles), whereas the total tracks density (D_T) is counted in the etched grains for each sample. The fission tracks density (D_{TF}) results from substation between the total tracks density and micropores density for each sample (heated and unheated one).

$$D_{TF} = D_T + D_P$$

Spectrometry Gamma

Method of regions for Gamma spectrometry is put in evidence to determine the presence of natural radionuclides, especially Uranium-238, Radium-226 and Radon emanation. The measuring equipment consists of NaI (Tl) scintillation detector 3″ × 3″, with the well of 1″ × 2″ and electronic chain. The measured spectrum is divided in three regions. A system of only two equations to two unknowns is developed in the two first regions. The resolution of this system follows to the determination of uranium and radium activity. This method is largely described in: Berrada et al. (1995). For radon emanation, two spectra are measured for each sample, one after radioactive equilibrium between radon-222 and its immediate daughters and the other between rafium-226 and eadon-222. Then the radon emanation rate is determined from following substation between the radium activity and radon activity for each sample.

$$T(\%) = 100(Ra - Rn)/Ra$$

Rn: Activity of radon-222 in the sample (1st measurement)
Ra: Activity of radium-226 in the sample (2nd measurement)

Note that, the total radon activity is equal to radium activity at radioactive equilibrium between radium and radon (after 10 time of half-live of Rn-222 (3.8 j)).

Results and Discussions

Apatite Fission Tracks Analysis (AFTA)

The characterization of these sedimentary apatite was carried out by various techniques (Electron microscopy and X-ray diffraction) giving the similar and/or complementary results. Microanalysis by electron microscopy (SEM or TEM) gives an idea on external and internal morphology, the chemical composition and the structure of the mineral. The technique of spectrometry by energy dispersion (EDX) gives a qualitative and quantitative analysis whereas the technique of spectrometry by wavelength dispersion (WDX) makes X-cartographies of the elements constituting the grains of the samples., the visualization of the structure and the microstructure of apatite by TEM, allows to measure the size of crystallites: Boujrhal and Cherkaoui (2009). These results obtained by electron microscopy were confirmed by the diffraction of the rays.

Distribution of fission tracks and pores in sedimentary apatite is reported in Fig. 1. Based in analysis of radiation damage, each fission track is progressively repaired, at a rate which increases with temperature. A study of the annealing characteristic for 1 h of fission tracks in this sedimentary apatite is reported in Fig. 2. Fission track is progressively repaired, at a rate which increases with temperature. At 600°C, all fission tracks are annealed out. Similar results are obtained for 2 h of heat treatment and are reported in: Boujrhal and Cherkaoui (2009, 2011).

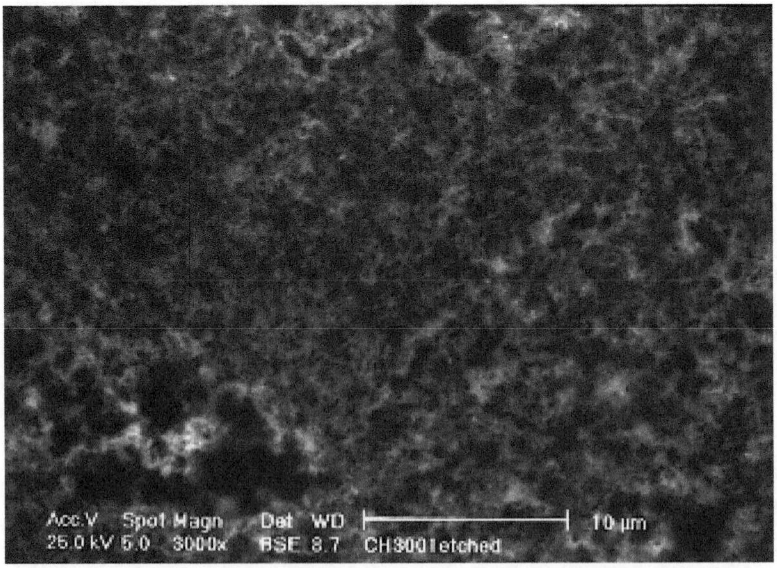

Fig. 1 Nonopores and etched spontaneous fission tracks (confined tracks) in apatite grain obtained by SEM. Etching conditions: HNO_3, 7%; room temperature, time 5 s. These nanostructure defects can't be distinguish by this microscopy because thy have the same form and dimension

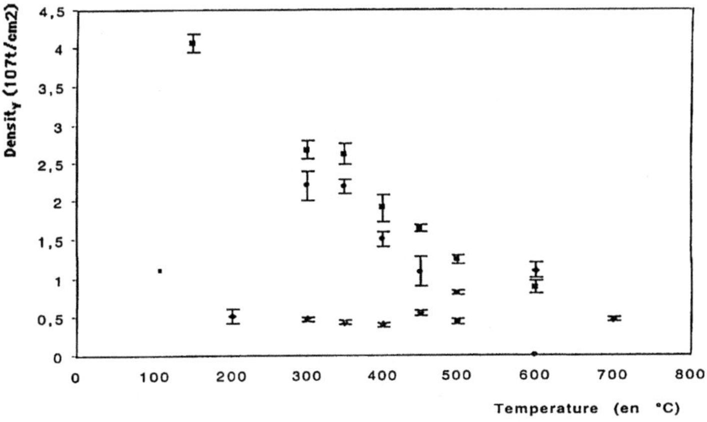

Fig. 2 Variation of total tracks density (■), of nanopores density (♦) and of fission track density (•) with annealing temperature. Annealing time is 1 h

Radon Diffusion Analysis

The emanation rate is generally affected by the nature of the sample and the exterior conditions (temperature, pressure …). In order to exam the effect of the radon diffusion on macro/nanostructure of apatite and the try to related it to fission tracks results, radon emanation rates measured at temperature ranging from room temperature to 950°C are done and reported in Fig. 3.

The radon emanation rate of the phosphate presented here is about 8.4% at room temperature, and it increases with increasing annealing temperature. At 600°C, the radon emanation rates began to increase brutally in order to reach maximum at 800°C.

The same result is obtained of all of sample which has various natural radon emanations.

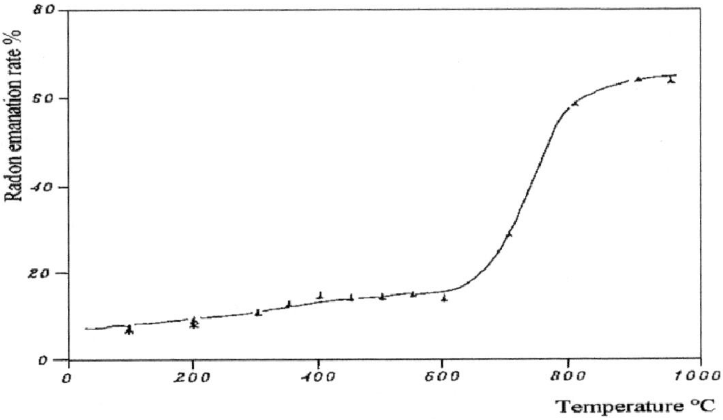

Fig. 3 Effect of annealing on radon emanation in phosphate 1

Conclusion

The microstructure of apatite is affected by radiation damage (fission tracks) since its geological formation so as by radon diffusion; both are results of Uranium transformations (spontaneous fission and radioactive decay respectively). Heat treatment can repair the radiation damage and accelerate radon diffusion progressively.

The fission tracks are clearly observed by scattering electron microscopy (SEM), but confusion takes place because since the fission track and the nanopores are the same form and size. Optical microscope resolves this problem and fission tracks density is deduced for heated and unheated samples.

The evolution of fission tracks with annealing temperature follows their total reparation at 600°C.

On the other hand, the immediate effect of heat on radon emanation provokes a change of it velocity at 600°C until a maximum at 800°C.

Acknowledgements Fission track study has been supported by the collaboration between the Centre Nationalde l'Energie, des Sciences et des TEchniques Nucleaires (CNESTEN, Maroc) and The Commissariat à l'Energie Atomique (CEA, France). I thank Professor/Doctor Joëlle Carpéna.

References

Berrada M, Boujrhal F.Z, Choukri A, El Khoukhi T, and Iraqi M.R (1992) Radon emanation from sedimentary phosphate and corresponding phosphogypsum, in: Radon and inert Gaz in the Earth Sciences and Environment, Toelicht. Verhand. Geologische en mijnkaarten van belgie (Geological and mining chart Mem. Expl. of Belgum(Mém. Expl. Cartes Géologique et Minière de la Belgique), N°32, 322p/b, pp. 253–258

Berrada M, Boujrhal F.Z, Couchot P, Chambaudet A, Mercier R (1995) Effet de la température de cuisson sur le potentiel d'émanation et le taux de dégazage en radon de phosphates sédimentaires, Dans: Gaz Geochimestry (C. Dubois, D. Clein, A. Chambaudet, M. Rebetez). Science Reviews 335–358

Boujrhal F.Z, Berrada M, Cherkaoui El Moursli R (1999) Retention of radon by apatite structure: the case for sedimentary phosphate, Phosphorus Bulletin 10: 274–282

Boujrhal F.Z, Carpena J, Cherkaoui El Moursli R (2001) A study of radon retention and fission tracks annealing with temperature in natural apatite. Radiation Physics and Chemistry 61: 645–647

Boujrhal F.Z, Hlil E.K, Cherkaoui El Moursli R (2004) Study of apatite behaviour in the presence of radionuclides U and Rn and local modification of their crystalline and electronic structure. Radiation Physics and Chemistry 69: 1–6

Boujrhal F.Z, Hlil E.K, Cherkaoui El Moursli R, El Khoukhi T, Sghir B (2005) A comparative study of radon retention ability of crystalline apatite and amorphous oxide materials. Materials Science Forum 480–481:169–174

Boujrhal F.Z, Sghir B, Ossama S, Cherkaoui El Moursli R, (2007) Investigation of the micro/nanostructure and the structure defect of sedimentary phosphates by electron microscopy. Acta Cryst A63: s148

Boujrhal F.Z Cherkaoui El Moursli R (2009) Morphologie externe et interne des phosphates sédimentaires marocains: effet de recuit. International Journal of Environmental Studies 66 Issue 2: 229–249

Uranium in German Mineral Water – Occurrence and Origins

Friedhart Knolle, Ewald Schnug, Manfred Birke, Rula Hassoun, Frank Jacobs

Abstract. The U content in 1216 mineral water samples from Germany collected over the last decade is presented and discussed. Nearly all waters containing >2 µg/l U come from Permo-Triassic aquifers in Bavaria, Hesse, Lower Saxony, Saxony-Anhalt and Southwest Germany. Most important U source rocks are Bunter and Keuper strata, only secondarily Permian and crystalline aquifers. Anthropogenic loads are not significant, but show up in principal component analyses.

Introduction

The degree of U contamination in humans is majorly caused by the consumption of drinking water containing U (De Kok and Schnug 2008). U is known for its

Friedhart Knolle
GeoPark Harz, Braunschweiger Land, Ostfalen Network, Grummetwiese 16, D-38640 Goslar,
E-mail: fknolle@t-online.de

Ewald Schnug
Technical University Braunschweig, Faculty 2 Life Sciences, Pockelsstraße 14,
D-38106 Braunschweig

Manfred Birke
Federal Institute for Geosciences and Natural Resources, Wilhelmstraße 25–30, D-13593 Berlin

Rula Hassoun
Technical University Braunschweig, Faculty 2 Life Sciences, Pockelsstraße 14,
D-38106 Braunschweig

Frank Jacobs
GeoPark Harz, Braunschweiger Land, Ostfalen Network, Grummetwiese 16, D-38640 Goslar,
E-mail: fknolle@t-online.de

B. Merkel, M. Schipek, *The New Uranium Mining Boom*
DOI 10.1007/978-3-642-22122-4 © Springer 2011

extreme variability in concentration and distribution – so the geological setting and possible anthropogenic sources must be taken into account. In order to obtain an overall view of the nationwide occurrence and sources of U in German mineral, curative and table waters (summarily called mineral waters), the quantity of U and 65 further elements as well as the radioactivity in 1216 mineral water samples of German origin were analyzed.

Material and Methods

Samples derived from a worldwide survey conducted by the former Institute of Plant Nutrition and Soil Science of the Federal Agricultural Research Centre (FAL) in Braunschweig, since 2008 the Institute for Crop and Soil Science (PB) of the Julius Kühn Institute in Braunschweig, Germany. Concentrations of U and other elements in those waters have been analyzed by PB or collected from published sources. Data sets retrieved from literature have been validated through U analyses at PB (Knolle 2008).

The database is continuously updated, latest with data derived from Reimann and Birke (2010). It contains actually 2879 brands from all over the world, of which are 1216 brands true "German mineral waters" which source is located in Germany and 510 brands from "German and neighboring countries" with sources located in the direct neighbor countries of Germany (Austria, Belgium, Czech Republic, Denmark, France, Hungary, Poland, Switzerland and the UK). The authors would have been glad to align this database with the data of 1456 mineral samples of the Federal Institute for Risk Assessment (Bundesinstitut für Risikobewertung 2005), but access to the data was denied to the public as well as the science community. Also the Federal Office for Radiation Protection published data on the radioactivity of 366 specified mineral waters of Germany plus 35 from European countries (Bundesamt für Strahlenschutz 2006), but even on an official request for cooperation between two federal research organizations denied access to information of the equivalent doses of individual nuclides in the waters.

To correlate the hydrogeological parameters to the mineral waters, a quality classification geocoding of the springs and wells was carried out. In order to combine the locations with the geological, hydrogeological and hydrogeochemical maps, data were converted to a uniform coordinate system using the World Geodetic System 1984 (WGS 84). The transformation of coordinates and preparation of the maps were carried out using ArcGIS. For the statistical calculations and analyses the statistic program SPSS 12.0 was employed.

Results

In a comparison of the content of 66 elements as well as HCO_3 and dose rate of the mineral waters tested, by far the highest coefficient of variation was determined with 820% for U. In the mineral waters analyzed, the following 30 out of a total of 65 elements correlated weakly but significantly ($p < 0.05$) with U: As, B, Be, Ca, Ce, Cl, Cu, Dy, Er, F, Fe, Ge, Gd, Ho, K, Li, Mg, Mn, Na, Ni, Mo, Nb, S, Se, Ta, Ti, V, W, Y und Zn as well as the dose rate and the TDS. The concentration of Ra-226 and Th also correlated only weakly. The result of principle component and multiple regression analyses was that the U content of mineral waters is only to a minor degree related to the content of other chemical elements (Knolle 2008).

In the dataset of Knolle (2008), the mean mineral water U content was 3.08 µg/l U, the median 0.13 µg/l U. 82% of the samples showed a content of <2 µg/l U. The lowest U contents below the proven limit of ICP-QMS of 0.015 µg/l U were found in quaternary granular aquifers, the highest in the curative water of the Nürtinger Heinrichsquelle in Nürtingen, Baden-Württemberg (474 µg/l U), the Purio Aqua Römer Mainhard water of the Römerquelle in Mainhardt-Baad, Baden-Württemberg (27.4 µg/l U) and the water of the Alwa Bonalwa-Quelle in Bad Peterstal-Griesbach, Baden-Württemberg (24.5 µg/l U). As the Nürtinger Heinrichsquelle has been closed and individual mineral water bottlers took countermeasures, these values were a market snapshot before 2007.

Table 1 summarizes the mean and P 95 occurrence data for U and selected other elements in the most recent version of the mineral water database (Hassoun in prep.).

Compared to the mean of the world collective and also the sample collective from Germany plus its neighbor countries the trace element concentrations in German mineral waters are now the lowest. The highest mean concentrations for elements are usually found in mineral waters with a high mineralisation (TDS > 1000 mg/l). Exception is U which concentration is more or less unaffected by the mineralisation in a particular water (Hassoun in prep.; Jacobs in prep.).

Table 1 Mean and P 95 concentration of As, B, Cu, Li, Ni, Mo, Pb, U and Zn concentration in mineral waters in µg/l

Element	Mineral Waters					
	World		Germany and neighboring countries		Germany	
	Mean	P 95	Mean	P 95	Mean	P 95
n	2879		1598		1216	
As	2.36	6.23	1.60	4.68	0.98	4.18
B	500	1401	290	839	178	688
Cu	2.11	6.67	1.70	5.01	1.59	3.37
Li	165	701	159	632	131	568
Mo	1.06	3.68	0.86	3.31	0.75	2.77
Ni	1.78	8.36	1.87	8.97	1.74	8.81
Pb	0.19	0.89	0.16	0.56	0.14	0.45
U	1.78	6.06	1.92	5.54	1.48	5.49
Zn	7.16	22.3	5.84	18.8	5.09	16.8

Regional Distribution and Geological Setting

Following Knolle (2008), the U content of the German mineral waters tested with >2 µg/l U can be localized in the following regional clusters of springs (from North to South): Fulda – Upper Weser, Southwest Harz foreland, Northeast Harz foreland (eastern Subhercynian Cretaceous Basin), Saale-Unstrut, Thüringer Forest und forelands, Rhön, Bad Kissingen, South German Keuper und Northern Black Forest (Fig. 1).

In the framework of a German-wide regional geological and hydrogenetic general view, nearly all mineral waters with a content of >2 µg/l U could be classified as resulting from geogenic influences. The mineral waters with contents >2 µg/l U can almost all be found within or on the immediate border of the Permo-Triassic sediment complexes of Bavaria, Hesse, Lower Saxony and Saxony-Anhalt as well as the Southwest German cuesta landscape.

The main source aquifers are rocks of the Bunter sandstone and Keuper, only secondarily Permian and crystalline aquifers:

- **Middle Keuper**: Northern Germany: Steinmergel-Keuper = Arnstadt Formation and/or Gypsum Keuper = Grabfeld Formation (Extal); South Germany:

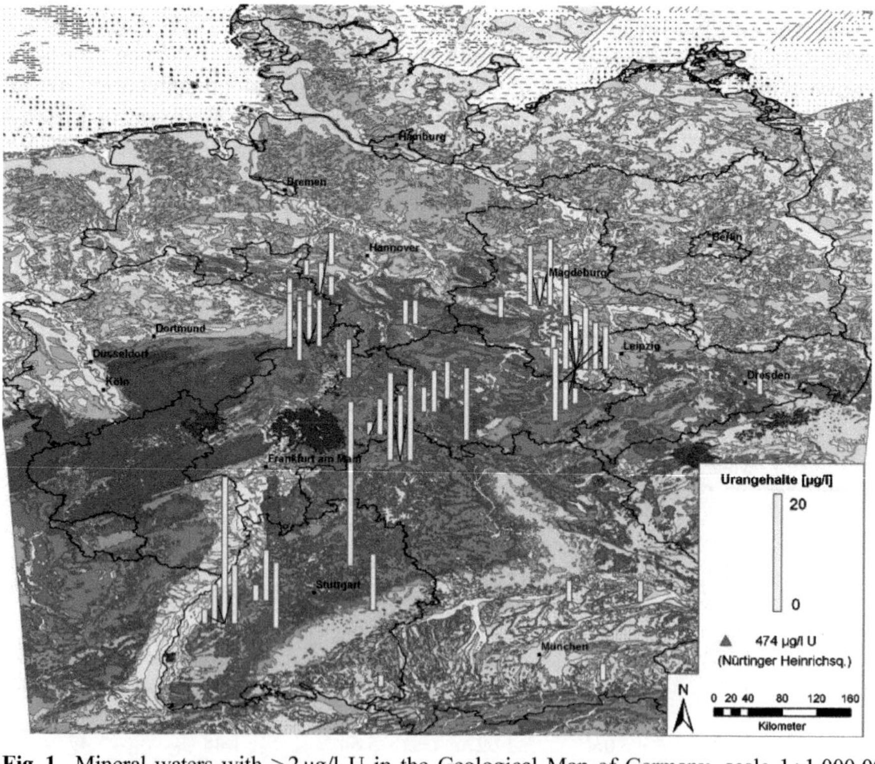

Fig. 1 Mineral waters with >2 µg/l U in the Geological Map of Germany, scale 1 : 1,000,000 (from Knolle 2008)

Löwenstein Formation, i.e. Stubensandstein in Baden-Württemberg and Burgsandstein in Bavaria (Bad Überkingen, Mainhardt-Baad und Nürtingen),

- **Bunter**, primarily Middle, partially Upper Bunter (Bad Driburg, Bad Pyrmont, Fuldatal, Hecklingen-Gaensefurth, Katlenburg, Hessberg, Hessian Rhön, Saale-Unstrut/Leisslingen, Steinheim-Vinsebeck, Volkmarsen and Warburg),
- **Permian** (Bad Kissingen, Friedrichroda, Saale-Unstrut/Leisslingen und Schmalkalden),
- **Paragneisses** of Rench area, Black Forest (Bad Peterstal-Griesbach).

This geological assignment of the U contents in the analyzed mineral waters was also found by Birke and Demetriades (2010) as well as Reimann and Birke (2010). Most of the U contents in German mineral waters derive from sandstone-hosted uranium enrichments in the Germanic Permo-Triassic (Knolle 2008).

Discussion

The mineral waters analyzed clearly show a dominating geogenic U content originating from the German Permo-Triassic. Following the scientific and public discussion in this matter, individual spring-water bottlers have already taken counteractive measures and sunk the U content in their bottled water. Anthropogenic U loads in the mineral waters investigated to present play no significant role, but principal component analyses show signals from mainly phosphate fertilizer effects (Birke et al. 2010). This indicates that the surface/groundwater cycle is getting influenced by U from fertilizers.

This investigative study is to be seen as a geochemical reconnaissance, postulating further and more detailed investigation into the infiltration passage, in particular in regard to the anthropogenic U contamination along the fertilization path (De Kok and Schnug 2008; Schnug and Haneklaus 2008). Particular attention should be paid to the interconnected zones in which >2 μg/l U-loaded bodies of surface water lie, which appear to have been caused by decades of fertilization (Birke et al. 2009, 2010).

Acknowledgement The authors wish to express their most sincere gratitude to Dr. Holger Lilienthal and Dr. Kerstin Panten (Institute for Crop and Soil Science, Julius Kühn Institute, Braunschweig, Germany) who contributed significantly to the present work, their right as co-author, however, being denied.

References

Birke M, Rauch U, Lorenz H (2009) Uranium in stream and mineral water of the Federal Republic of Germany. Environ Geochem Health 31: 693–706

Birke M, Demetriades A, De Vivo B, Eds. (2010) Special Issue: Mineral Waters of Europe. J Geochem Explor 107: i–viii + 217–422

Bundesinstitut für Risikobewertung (2005) Uran in Mineralwasser: Bei Erwachsenen geringe Mengen tolerierbar, Wasser für Säuglingsnahrung sollte uranfrei sein. Stellungnahme 24/2005 vom 13.5.2005. www.bfr.bund.de/cms5w/sixcms/detail.php/6488; www.bfr.bund.de/cm/208/uran_in_mineralwasser.pdf; Pressemitteilung 22/2005 vom 30.6.2005; www.bfr.bund.de/cd/6488

Bundesamt für Strahlenschutz (2010) Natürliche Radionuklide in Mineralwässern. www.bfs.de/ion/nahrungsmittel/mineralwasser.html, Stand 2010

De Kok LJ, Schnug E (2008) Loads and fate of fertilizer derived uranium. Backhuys, Leiden, The Netherlands, 229 pp.

Hassoun R (in prep.) A statistical evaluation of the contribution of mineral and tap water to the dietary intake of As, B, Cu, Li, Mo, Ni, Pb, U and Zn by humans. Diss. Fak. f. Lebenswissenschaften TU Carolo-Wilhelmina zu Braunschweig

Jacobs F (in prep.) Ein Beitrag zu Vorkommen und Herkunft von As, B, Cu, Li, Mo, Ni, Pb und Zn in deutschen Mineralwässern. Diss. Fak. f. Architektur, Bauingenieurwesen und Umweltwissenschaften TU Carolo-Wilhelmina zu Braunschweig

Knolle F (2008) Ein Beitrag zu Vorkommen und Herkunft von Uran in deutschen Mineral- und Leitungswässern. Diss. Fak. f. Lebenswissenschaften TU Carolo-Wilhelmina zu Braunschweig, Digital Library Braunschweig 2009: www.digibib.tu-bs.de/?docid=00027200

Knolle F, Haneklaus S, Schnug E (2009) Indications for Contamination of Water Bodies with Uranium through Fertilization. In: Plant Nutrient Management Under Stress Conditions. Proc. 17th Int Symp CIEC, Nov 24–27, 2008, Cairo, Egypt, pp. 311–318

Knolle F, Schnug E (2009) Vorkommen und Herkunft von Uran in deutschen Mineral- und Trinkwässern. bbr 60: 56–59

Knolle F (2010) Neue Befunde zur Herkunft von Uran in Grund- und Leitungswässern aus Phosphatdüngern. Julius-Kühn-Archiv 424 (2009): 70–73

Reimann C, Birke M, Eds. (2010) Geochemistry of European Bottled Water. Borntraeger Science Publ, Stuttgart, XI + 268 pp.

Schnug E, Haneklaus S (2008) Dispersion of uranium in the environment by fertilization. In: Merkel, B.J., Hasche-Berger, A.: Uranium, Mining and Hydrogeology, Springer, Berlin, 958 pp.

Heavy Metal Loads to Agricultural Soils in Germany from the Application of Commercial Phosphorus Fertilizers and Their Contribution to Background Concentration in Soils

Sylvia Kratz, Frauke Godlinski, Ewald Schnug

Abstract. Statistical data on the annual sale of phosphorus (P) containing mineral fertilizers in Germany combined with data on toxic elements in mineral fertilizers allow an estimation of heavy metal loads to agricultural land. For the time period from 1950/51 to 2009/2010, the mean annual loads of the elements As, Cd, Cu, Ni, Pb, U and Zn to agricultural land in Germany exclusively from the application of P fertilizers amounted to: As 40, Cd 22, Cu 95, Ni 54, Pb 11, Zn 431 and U 114 t/yr, while maximum values reached: As 73, Cd 42, Cu 146, Ni 90, Pb 20, Zn 764 and U 228 t/yr. Depending on the soil group looked at, the contribution over the last sixty years of mineral fertilizer bound heavy metal inputs to average background concentrations of agricultural in Germany ranges between 0.4–1.4% for As, 3.5–12.3% for Cd, 0.2–1.1% for Cu, 0.03–1.5% for Ni, 0.03–0.1% for Pb, 0.3–1.8% for Zn and 4.4–13.7% for U. It can be concluded that there is an important urgent need to limit the concentration of these elements in mineral fertilizers given that these heavy metals are toxic, causing harm to the environment and public health and for example the preferable pathway for Cd into the food chain is the soil-plant system, and for U intake by drinking water.

Sylvia Kratz
Institute for Crop and Soil Science, Julius Kühn-Institute (JKI), Federal Research Centre for Cultivated Plants, Bundesallee 50, 38116 Braunschweig, Germany

Frauke Godlinski
Institute for Crop and Soil Science, Julius Kühn-Institute (JKI), Federal Research Centre for Cultivated Plants, Bundesallee 50, 38116 Braunschweig, Germany

Ewald Schnug
Institute for Crop and Soil Science, Julius Kühn-Institute (JKI), Federal Research Centre for Cultivated Plants, Bundesallee 50, 38116 Braunschweig, Germany

B. Merkel, M. Schipek, *The New Uranium Mining Boom*
DOI 10.1007/978-3-642-22122-4 © Springer 2011

Introduction

Commercial mineral fertilizers are one of the main contributing factors to heavy metal accumulation in agricultural soils. As a background for a reliable estimate of their relevance, solid statistical data is needed. At the JKI-Institute for Crop and Soil Science statistical data on the sale of phosphorus (P) containing mineral fertilizers dating back to the year 1950/1951 have been collected. Combined with the institute's database on heavy metals in commercial fertilizers, annual heavy metal loads to agricultural land can be estimated. This article describes the material and methods used for calculation and the resulting estimated loads of the heavy metals arsenic (As), cadmium (Cd), copper (Cu), nickel (Ni), lead (Pb), zinc (Zn) and uranium (U) to agriculture soils. These values can be used to estimate the relative contribution of fertilizer bound heavy metal inputs to soil background concentrations.

Material and Methods

Data on the yearly sale of P containing mineral fertilizers was taken from the official statistical yearbook for Germany (Statistisches Jahrbuch) or since 2003 from the on-line statistical tables (Destatis). According to Destatis (2010), various fertilizer groups were defined (Table 1).

Data on agricultural or cultivated area was also taken from official statistics. While this data is based on calendar years (January to December), data on fertilizer sales is given for crop years (July to June). In order to combine them, fertilizer sales data is always referred to by the second named year (e.g. 2009 for 2008/2009), assuming that the majority of mineral fertilizer bought in the first named year is used to achieve the yield of the second named year.

Table 1 Fertilizer groups used for calculation according to Destatis (2010)

Fertilizer group	Fertilizer types included:
Superphosphate	Single super phosphate, concentrated super phosphate, triple super phosphate
Single nutrient fertilizers	Soft (ground) rock phosphate[a], partially acidulated rock phosphate, rock phosphate with water soluble fraction, rock phosphate with calcium carbonate, dicalcium phosphate, Thomas meal (=basic slag)[b]
PK fertilizers	
NP and NPK fertilizers	
Fused phosphate (data until 1963/64)	

[a] Only since 1964/65, before that time soft (ground) rock phosphate was listed as a separate group.
[b] Only since 1998/99, before that time Thomas meal was listed as a separate group.

The official statistics do not contain total amounts of sold fertilizers as such, but refer to nutrient amounts sold (i.e. 1000 t P_2O_5). In order to estimate the related amounts of heavy metals, the total amounts of sold fertilizers need to be recalculated. To do this, average P_2O_5 contents must be assumed for each fertilizer group. This was done based on the legal definitions of minimum contents given in the German Fertilizer Ordinance (DüMV, 2008), or the European Fertilizer Regulation (EU, 2003). For PK, NP and NPK fertilizers, the most frequently sold types (based on the internet offers by www.agravis.de) were used for calculation (Table 2).

Table 2 Average P_2O_5 contents of fertilizer groups assumed for calculation

Fertilizer group	Fertilizer type	P_2O_5 content (in %)
Superphosphate	Single	16
(EU)	Concentrated	25
	Triple	38
	Average (mean)	*23*
Single nutrient fertilizers	Partially digested rock phosphate (with Mg)	16
(DüMV)	Rock phosphate with water soluble fraction	23
	Soft (ground) rock phosphate	23
	Soft (ground) rock phosphate with Mg	16
	Dicalcium phosphate with Tricalcium phosphate	8
	Dicalcium phosphate (with Mg)	20
	Thomas meal (EU)	12
	Average (mean)	*17*
Fused phosphate (EU)		25
NP	Monoammonium	52
(www.agravis.de)	Diammonium phophate	46
	NP 20/20	20
	NP 10/34	34
	NP 17/17 + 0,1 Bor	17
	NP average (mean)	*34*
NPK	13/9/16/4/7	9
(www.agravis.de)	14/6/16/8	6
	14/5/20/6	5
	8/10/18/6/10	10
	12/7/15/2/13	7
	15/15/15/2	15
	NPK average (mean)	*9*
NP + NPK average		*21*
PK	PK 12/24	12
(www.agravis.de)	PK 12/24 + MgO	12
	PK 11/22/4/6	11
	PK 10/20/3/5	10
	PK 20/30	20
	PK average (mean)	*13*

Table 3 Heavy metal contents (mg/kg) of phosphorus fertilizer groups used for calculation

mg/kg	Super-phosphate	Single nutrient fertilizers[a]	Thomas meal	Fused phosphate	Soft ground rock phosphate	PK	NP + NPK
As	8.9	8.9	3.9	3.2	7.9	19.3	13.9
Cd	25	16.3	No data	8.9	14.1	6.9	6.5
Cu	53.7	20	22.5	32.0	14.4	15.4	31.8
Ni	27.4	25.9	10.1	29.2	24.0	17.7	15.1
Pb	5.4	5	1.3	3.5	6.8	5.5	3
Zn	280	232	5	280	246	133	189
U	97.5	92.6	0.8	58.2	93.6	61.1	32.5
n	16	27	4	5	8	32	99

[a] Including Thomas meal and soft ground rock phosphate.

Fertilizer amounts for the different fertilizer groups were calculated using the following equation:

$$\text{Fertilizer amount }(\text{in }1000\,\text{t}) = \text{total amount of } P_2O_5\ (1000\,\text{t})*100/\text{average } P_2O_5 \text{ content } (\%).$$

To estimate the fertilizer bound heavy metal inputs, data from the institute's mineral fertilizer database was used. Since the estimate was for German agriculture, only samples from the German market were included in the calculations. The collection dates back to samples from 1970 onwards, the most recent samples coming from the governmental fertilizer control of the crop year 2008/2009. Since the database is not large enough to allow for a grouping of samples by year or decade, an average heavy metal concentration was calculated from all samples (Table 3).

Annual fertilizer bound heavy metal loads for each fertilizer group were calculated using the following equations:

$$\text{Total amount of heavy metal }(\text{t}) = \text{amount of fertilizer }(\text{kg})*\text{heavy metal content }(\text{mg/kg})*10^{9}.$$

After summing up the amounts from all fertilizer groups, the total annual heavy metal load was estimated.

$$\text{Heavy metal load }(\text{g/ha}) = \text{total amount of heavy metal }(\text{t})*1000/\text{agricultural area }(1000\,\text{ha}).$$

In order to estimate the contribution of fertilizer bound heavy metal inputs to background heavy metal concentrations in agricultural soils, official data on background concentrations (Table 4) in agricultural topsoils was used and converted from concentrations in mg/kg to contents in g/ha, assuming a soil depth (ploughing depth) of 30 cm and a soil bulk density of 1.5 g/cm³. In order to illustrate the influence of soil parent material and soil texture, calculations were carried out for 5 different soil groups.

Table 4 Average (median) heavy metal background values for agricultural topsoils (mg/kg) (Utermann et al. 2008)

mg/kg	As	Cd	Cu	Ni	Pb	Zn	U
Periglacial till on intermediate or basic[a] magmatites and metamorphites	4	0.53	42	148	37	126	2.4
Periglacial till on clay[b]	8.8	0.3	23	37	39	99	2.5
Marl/glacial till	3.4	0.15	8.4	8.8	9.4	31	1.2
Silt	9.2	0.36	15	20	31	54	2
Sand	2.7	0.16	8.5	3.4	15	23	0.8

[a] For U: acid magmatites and metamorphites.
[b] For As and U: claystones.

Results and Discussion

For the time period from 1950/51 to 2009/2010, the mean annual loads of the elements As, Cd, Cu, Ni, Pb, Zn and U to agricultural land in Germany exclusively from the application of P fertilizers amounted to 40 t/yr As, 22 t/yr Cd, 95 t/yr Cu, 54 t/yr Ni, 11 t/yr Pb, 431 t/yr Zn and 114 t/yr U. Over this 60 year period, annual loads varied greatly (Table 5). Estimated annual loads and summarized total heavy metal loads per hectare agricultural land are shown in Table 6.

The wide range in heavy metals and loads per hectare can be explained by a change in the types and amounts of fertilizers used over this time period. Figure 1 illustrates how heavy metal loads change over time, using Cd and U as an example. It should be pointed out that the observed variations are not the result of decreasing or increasing heavy metal concentrations, since the average concentration values used in the calculations for each fertilizer group were constant over the whole time period (see material and methods). Rather, the relative proportions of the various fertilizer types used changed considerably over the years, as did the total amount of P containing mineral fertilizers applied to the soil (Fig. 2).

As the estimates in Table 7 show, the contribution of mineral fertilization to soil background concentrations of heavy metals ranges from less than 1 up to 14%. The relative proportion of fertilizer bound heavy metals in agricultural topsoils is largest for the elements Cd (3.5–11.5%) and U (4.4–13.7%). Both elements are known to have a high bio-toxicity as well as a high mobility in environmental systems, entering the food chain most commonly via the soil-plant system (Cd) or drinking water (U).

Fertilizer derived heavy metals are primarily a risk to the quality of groundwater and drinking water, depending on their mobility in soils. While for instance Cd and Zn are relative immobile, U complexes are highly mobile in the presence of carbonate ions (Utermann and Fuchs 2008). The German fertilizer ordinance (DüMV) already limits the concentration of most heavy metals, whereas no regulation currently exists for U. It is vital that the levels of U input to soils from fertilizer application do not exceed the harvested crop products uptake of U to avoid an accumulation in agricultural soils and to protect water bodies (Ekardt und Schnug 2008). It seems reasonable to assume an off-take of 1 g/ha * yr U for agricultural crops and so critical values for U in fertilizers should be set as follows: an obligation to label at concentrations of > 20 mg/kg U per kg P_2O_5 and a limit value of 50 mg/kg U per kg P_2O_5, (following the regulation for Cd). Furthermore it needs to be determined whether more stringent measures are required in water protection areas.

Table 5 Range, average and total heavy metal amounts coming from P containing mineral fertilizers (see Table 1 for definition) between 1951 and 2010

	As	Cd	Cu	Ni	Pb	Zn	U
Minimum	12.0	5.6	26.4	13.5	2.8	118	28.2
Maximum	73.1	42.1	146	90.5	20.2	764	228
Average	40.3	22.1	95.0	54.1	11.2	431	114
Total 1951–2010	2,419	1,325	5,702	3,243	670	25,849	6,839

Table 6 Range, average and total heavy metal loads coming from P containing mineral fertilizers (see Table 1 for definition) between 1951 and 2010

	As	Cd	Cu	Ni	Pb	Zn	U
Minimum	0.71	0.42	1.61	0.81	0.20	8.32	1.9
Maximum	6.03	2.92	11.2	7.21	1.61	62.0	18.1
Average	2.92	1.41	6.81	3.91	0.80	30.8	8.21
Total 1951–2010	174	83.0	409	233	48.3	1850	493

Table 7 Background concentrations (converted into g/ha, see material and methods) for agricultural topsoils and relative contribution to soil background values of estimated heavy metal loads from P containing mineral fertilizers over the past sixty years

	As	Cd	Cu	Ni	Pb	Zn	U
Background concentrations (g/ha)							
Periglacial till on intermediate or basic[a] magmatites and metamorphites	18,000	2385	189,000	666,000	166,500	567,000	10,800
Periglacial till on clay[b]	39,600	1350	103,500	166,500	175,500	445,500	11,250
Marl/glacial till	15,300	675	37,800	39,600	42,300	139,500	5400
Silt	41,400	1620	67,500	90,000	139,500	243,000	9000
Sand	12,150	720	38,250	15,300	67,500	103,500	3600
Estimated total heavy metal load from P containing mineral fertilizers from 1951–2010 (g/ha)							
	174.1	82.8	408.7	233.4	48.3	1850	493
Relative contribution of fertilizer bound heavy metal loads to background concentrations of agricultural topsoils (%)							
Periglacial till on intermediate or basic[a] magmatites and metamorphites	0.97	3.47	0.22	0.04	0.03	0.34	4.56
Periglacial till on clay[b]	0.44	6.13	0.39	0.14	0.03	0.43	4.38
Marl/glacial till	1.14	12.3	1.08	0.59	0.11	1.37	9.12
Silt	0.42	5.11	0.61	0.26	0.03	0.79	5.47
Sand	1.43	11.5	1.07	1.53	0.07	1.85	13.7

[a] For U: acid magmatites and metamorphites.
[b] For As and U: claystones.

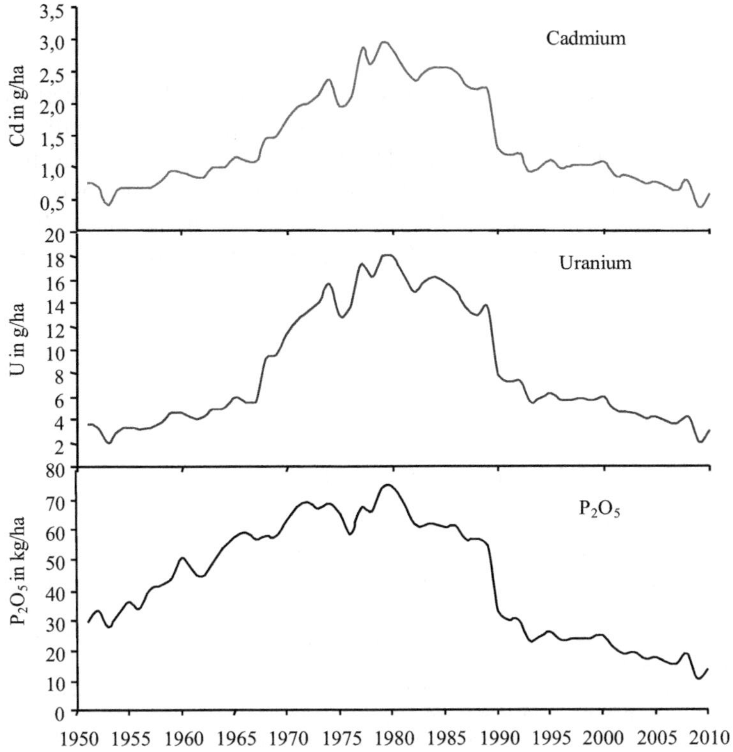

Fig. 1 Applied P mineral fertilizers and estimated heavy metal loads of cadmium (Cd) and uranium (U) to agricultural soils in Germany from 1950 to 2010

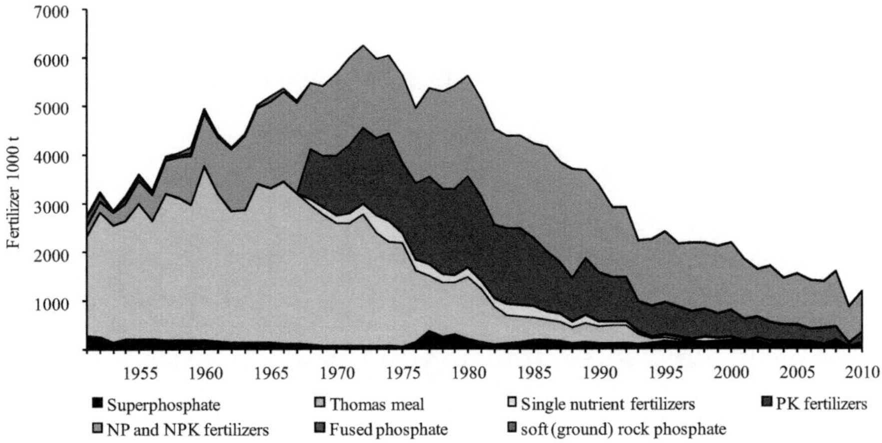

Fig. 2 Various P containing mineral fertilizers types used in Germany from 1950 to 2010

Conclusion

It can be concluded that there is an urgent need to set limits for heavy metals in mineral fertilizers on an EU wide legal basis. With the exception of U, fertilizers underlying the German fertilizer ordinance (DüMV) are already limited in their heavy metal concentrations, however, in order to effectively diminish fertilizer bound heavy metal loads to agricultural land, spatially related limit values ("load limits") in g/(ha * yr) should also be defined based on ecotoxicological considerations (including uranium).

References

Destatis (2010) Fachserie 4, Reihe 8.2, Düngemittelversorgung 2009/2010, (https://www-genesis.destatis.de/genesis)

Düngemittelverordnung, DüMV, from Dec 12, 2008, BGBl. I 60, pp. 2524

Ekardt, F. and Schnug, E. (2008) Legal aspects of uranium in environmental compartments. In: de Kok, L.J. and Schnug, E. (eds.) (2008) Loads and Fate of fertlizer-derived uranium. Backhuys Publishers, Leiden. ISBN/EAN 978-90-5782-193-6, 209–217

Regulation (EC) No 2003/2003 of the European Parliament and of the Council of 13 October 2003 relating to fertilizers

Statistisches Jahrbuch (2005) Statistisches Jahrbuch über Ernährung, Landwirtschaft und Forsten. Münster-Hiltrup: Landwirtschaftsverlag, 562 pp.

Utermann J., Düwel O., Fuchs M. (2008) Flächenrepräsentative Hintergrundwerte für As, Be, Co, Mo, Sb, Se, Tl, U, und V in Böden Deutschlands aus länderübergreifender Sicht. BGR-Bericht, Archiv Nr. 10040/08, 71 pp.

Utermann, J. and Fuchs, M. (2008) Uranium in German soils. In: de Kok, L.J. and Schnug, E. (eds.) (2008) Loads and Fate of fertlizer-derived uranium. Backhuys Publishers, Leiden. ISBN/EAN 978-90-5782-193-6, 33–46

Effect of Mg-Ca-Sr on the Sorption Behavior of Uranium(VI) on Silica

Sreejesh Nair, Broder J. Merkel

Abstract. The effect of alkaline earth uranyl carbonate complexes on the sorption behavior of U(VI) on quartz was studied by means of well defined batch experiments. In the pure U(VI) – silica system, the pH dependency was minor and about 90% of the uranium sorbed onto quartz. However, in the presence of Ca, Sr, or Mg the sorption of uranium decreased to 10, 30, and 50% respectively. The reduction in sorption behavior is due to the presence of zero valent and anionic complexes of alkaline earth uranyl carbonate species and show the importance to include these species in thermodynamic data bases. It is assumed that, in particular, the zero-valent species are responsible for less sorption in the presence of alkaline earth metals.

Introduction

Uranium is a significant contaminant to the environment. The mobility of uranium in the aquatic system is mainly dependent on the speciation and the major controlling factors of uranium speciation are pH, ionic strength, redox potential, availability of organic and inorganic ligands etc. Ternary complexes of alkaline earth uranyl carbonates, $M_2UO_2(CO_3)_3^0$ (M = Ca, Sr) and $MUO_2(CO_3)_3^{2-}$ (M = Mg, Ca,

Sreejesh Nair
Department of Hydrogeology, Technische Universität Bergakademie Freiberg, Gustav-Zeuner Str. 12, 09599 Freiberg, Germany

Broder J. Merkel
Department of Hydrogeology, Technische Universität Bergakademie Freiberg, Gustav-Zeuner Str. 12, 09599 Freiberg, Germany

B. Merkel, M. Schipek, *The New Uranium Mining Boom*
DOI 10.1007/978-3-642-22122-4

Sr), may play a key role in the environmental chemistry of uranium under neutral to alkaline pH conditions. The formation of $Ca_2UO_2(CO_3)_3{}^0$ was first reported by Bernhard et al. (1996) and further studies on the alkaline earth uranyl carbonates revealed that they can play a significant role in the aqueous speciation of U(VI) at neutral to alkaline pH range (Kalmykov and Choppin 2000; Zheng et al. 2003). Thus it is important to know the major uranium species in order to predict the distribution and migration behavior. The migration of radionuclide in groundwater is often controlled by the sorption on minerals present along the flow path. Hence the sorption behavior in the subsurface is important to estimate the suitability of geologic repositories for nuclear waste materials, the long-term behavior of uranium mining and milling sites, and migration of uranium in groundwater with slightly elevated uranium concentrations. The objective of this study was to understand the importance of alkaline earth uranyl carbonates in the sorption behavior of U(VI) on quartz as a rock forming mineral (e.g. granites, sandstones).

Materials and Methods

The quartz sand used for this work, termed as F32, was obtained from Quarzwerke Frechen, Germany. The properties of F32 and the purification process are explained in detail in (Nair and Merkel 2011). U(VI) stock solution (0.126×10^{-6} M) was prepared from $UO_2(NO_3)_2 \cdot 6H_2O$ (Chemapol, Czech Republic). Solutions of 1×10^{-3} M magnesium chloride, calcium chloride, strontium chloride, and sodium hydrogen carbonate were prepared by dissolving $MgCl_2 \cdot 6H_2O$, $CaCl_2 \cdot 6H_2O$, $SrCl_2 \cdot 6H_2O$, and $NaHCO_3$ in deionized and purified water. All chemicals used were of ACS reagent grade or better.

Four sets of sorption experiments were conducted with uniform amount of uranium concentration, quartz (10 g) and solution (100 ml) under normal laboratory conditions (temp. 23 ± 1°C) in 150 ml teflon (PTFE) beakers:

1. only uranium and quartz,
2. uranium, calcium (1×10^{-3} M) and quartz,
3. uranium, strontium (1×10^{-3} M) and quartz,
4. uranium, magnesium (1×10^{-3} M) and quartz.

All experiments were conducted at a pH ranging from 6.5 to 9. Saturation indices were modeled in advance using the geochemical speciation code PHREEQC (Parkhurst and Appelo 1999) and the Nuclear Energy Agency thermodynamic database NEA_2007 (Grenthe et al. 2007) to avoid over saturation and thus precipitation of mineral phases. The experiments were performed for a period of 48 h and the samples were collected at a time interval of 1, 3, 6, 12, 24 and 48 h. Solution was analyzed for uranium with differential pulse adsorptive cathodic stripping voltammetry (797 Va Computrace, Metrohm, Switzerland) using a hanging mer-

Table 1 Aqueous speciation of U(VI) with alkaline earth metals and stability constants

Aqueous Reactions	log K
1. $2Ca^{2+} + UO_2^{2+} + 3CO_3^{2-} = Ca_2UO_2(CO_3)_3^0$	30.7[a] 30.79[b]
2. $Ca^{2+} + UO_2^{2+} + 3CO_3^{2-} = CaUO_2(CO_3)_3^{2-}$	27.18[a]
3. $Mg^{2+} + UO_2^{2+} + 3CO_3^{2-} = MgUO_2(CO_3)_3^{2-}$	26.11[a]
4. $Sr^{2+} + UO_2^{2+} + 3CO_3^{2-} = SrUO_2(CO_3)_3^{2-}$	26.86[a]
5. $2Sr^{2+} + UO_2^{2+} + 3CO_3^{2-} = Sr_2UO_2(CO_3)_3^0$	29.73[c]

[a] Dong and Brooks (2006).
[b] Bernhard et al. (2001).
[c] Geipel et al. (2008).

cury drop electrode (HMDE) as reference electrode (Benedikt 2007). The percentage of uranium sorption on quartz was calculated using Eq. 1:

$$\text{Sorption } (\%) = \left(\frac{C_\mathrm{i} - C_\mathrm{f}}{C_\mathrm{i}} \right) \times 100 \tag{1}$$

where C_i and C_f are the initial and the final concentration of uranium in the solution.

The aqueous speciation calculation of anionic or zero-valent complexes of Mg, Ca, and Sr were done using PHREEQC with NEA_2007 database adding log K values from Dong and Brooks (2006) for alkaline earth uranyl carnonates. Different results would show by using the data from Bernhard et al. (2001) and Geipel et al. (2008), because these authors also stated zero valent species for Ca as well as Sr, and discussed elsewhere (Merkel and Nair 2011). Aqueous speciation and log K values of alkaline earth uranyl carbonates are summarized in Table 1.

Results and Discussion

Aqueous Speciation

Figures 1 to 4 shows the aqueous speciation calculation of U(VI) with respect to pH in the presence as well as in the absence of Mg, Ca and Sr. In the absence of alkaline earth elements, uranyl carbonate complexes – $(UO_2)_2CO_3(OH)^{3-}$, $UO_2(CO_3)_2^{2-}$ and $UO_2(CO_3)_3^{4-}$ – are the most dominant species under neutral to alkaline pH conditions (Fig. 1).

The effect of Mg on the speciation of U(VI) is illustrated in Fig. 2. Here the most dominant species in alkaline pH range is $MgUO_2(CO_3)_3^{2-}$. Figure 3 depicts the importance of Ca in the aqueous speciation of U(VI). The most predominant species under alkaline pH conditions are $Ca_2UO_2(CO_3)_3^0$(aq.) and $CaUO_2(CO_3)_3^{2-}$.

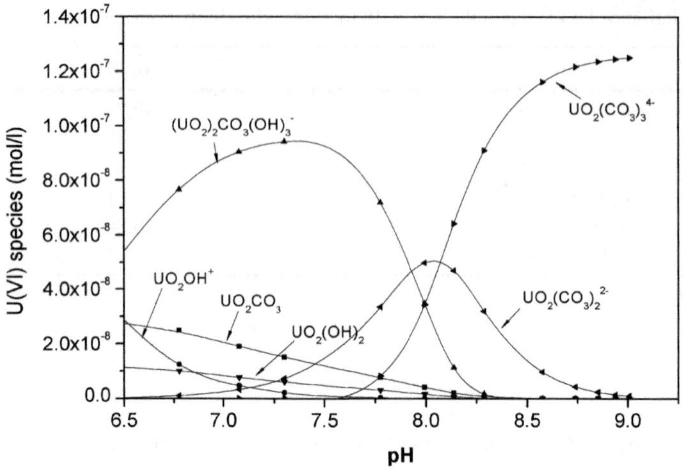

Fig. 1 Calculated U(VI) species distribution (Phreeqc with NEA_2007 data base) in relation to pH (0.126×10^{-6} M U, 1.5×10^{-3} M NaCl, 1×10^{-3} M $NaHCO_3$, temp: 23°C, p_{CO2}: $10^{-3.5}$ hPa)

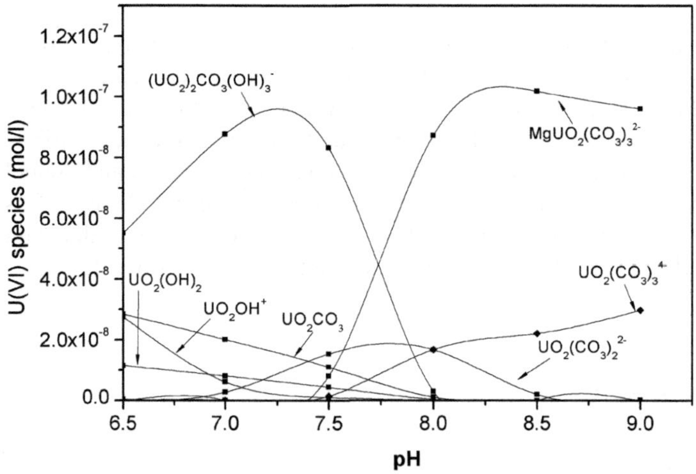

Fig. 2 Aqueous speciation calculation of Magnesium uranyl carbonate species in relation to pH using Phreeqc with NEA_2007 data base and log *K* from Dong et al. (2006). (0.126×10^{-6} M U, 1×10^{-3} M Mg, 1×10^{-3} M $NaHCO_3$, 1.5×10^{-3} M NaCl, temp: 23°C, p_{CO2}: $10^{-3.5}$ hPa)

The change in U(VI) speciation by Sr is shown in Fig. 4. The $SrUO_2(CO_3)_3^{2-}$ is significantly dominant in the alkaline pH range. Ternary complexes of alkaline earth uranyl carbonates dominate in the alkaline pH conditions and other uranyl carbonate complexes are less important within the given experimental conditions.

Due to the uncertainty of the existence of $Mg_2UO_2(CO_3)_3^0$ species, it was not taken into account for the speciation calculation.

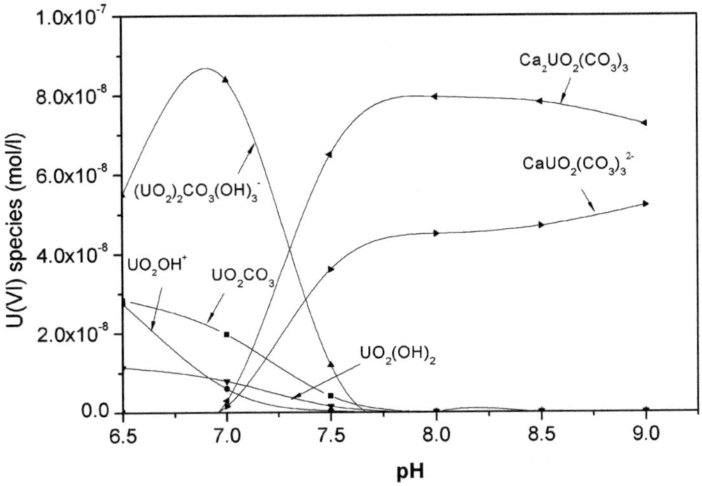

Fig. 3 Aqueous speciation calculation of Calcium uranyl carbonate species in relation to pH using Phreeqc with NEA_2007 data base and log *K* from Dong et al. (2006). (0.126×10^{-6} M U, 1×10^{-3} M Ca, 1×10^{-3} M $NaHCO_3$, 1.5×10^{-3} M NaCl, temp: 23°C, p_{CO2}: $10^{-3.5}$ hPa)

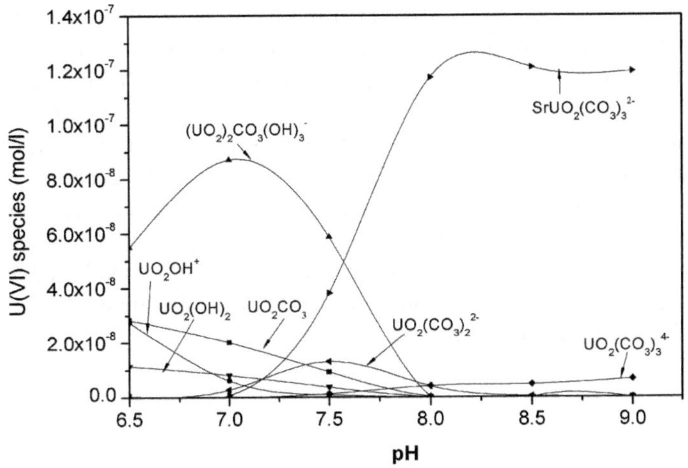

Fig. 4 Aqueous speciation calculation of Strontium uranyl carbonate species in relation to pH using Phreeqc with NEA_2007 data base and log *K* from Dong et al. (2006). (0.126×10^{-6} M U, 1×10^{-3} M Sr, 1×10^{-3} M $NaHCO_3$, 1.5×10^{-3} M NaCl, temp: 23°C, p_{CO2}: $10^{-3.5}$ hPa)

Sorption

U(VI) sorption on quartz in the pure U-CO_3 system is independent of pH and about 90% of uranium sorbed onto quartz. This high sorption behavior is due to the very low uranium concentration and the higher solid to solution ratio in the

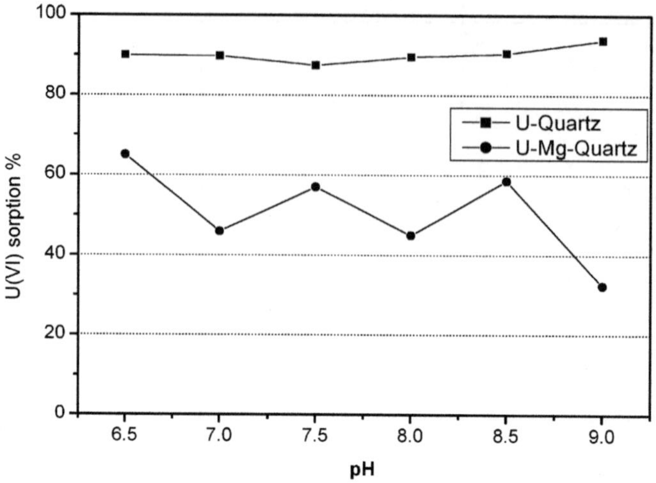

Fig. 5 U(VI) sorption onto quartz as a function of pH, with and without Mg. (0.126×10^{-6} M U, 1×10^{-3} M Mg, 1.5×10^{-3} M NaCl, 1×10^{-3} M $NaHCO_3$, period: 48 h, 23°C, p_{CO2}: $10^{-3.5}$ hPa)

current study. Similar sorption behavior was reported for uranium-silica system under alkaline pH range (Stamberg et al. 2003). In the presence of alkaline earth elements, U(VI) sorption is significantly less with an average of 50% for Mg, 30% for Sr and 10% for Ca in the solution (Figs. 5–7). The sorption of U(VI) on quartz was decreased from 65% at pH 6.5 to 32% at pH 9 in the presence of Mg and the sorption rate was variable throughout the pH range. (Fig. 5). The sorption retardation is due to the formation of $MgUO_2(CO_3)_3^{2-}$ (Fig. 2). Moreover, the decrease in U(VI) sorption, in the presence of Mg, on anion exchange resin was reported elsewhere (Dong and Brooks 2008). Instead of Mg, the introduction pH range. Competitive sorption of U(VI) on quartz with Mg, Ca and Sr can be excluded as proved by sorption experiments with solutions containing only alkaline earth metals in the absence of uranium (data not shown).

Conclusion

The presence of alkaline earth elements has a significant impact on the sorption behavior of U(VI) on quartz. In the presence of Mg, Ca and Sr, the sorption of U(VI) on quartz was decreased considerably. This sorption retardation is due to the formation of ternary complexes of alkaline earth uranyl carbonates. The U(VI) speciation calculation using PHREEQC with NEA_2007 database enhanced with log *K* values from Dong and Brooks (2006), Bernhard et al. (2001) and Geipel et al. (2008) revealed that the alkaline earth uranyl carbonate complexes are dominant in the presence of Mg, Ca and Sr at neutral to alkaline pH conditions. Moreover, the log *K* values from the literature are sensitive to the speciation calculation.

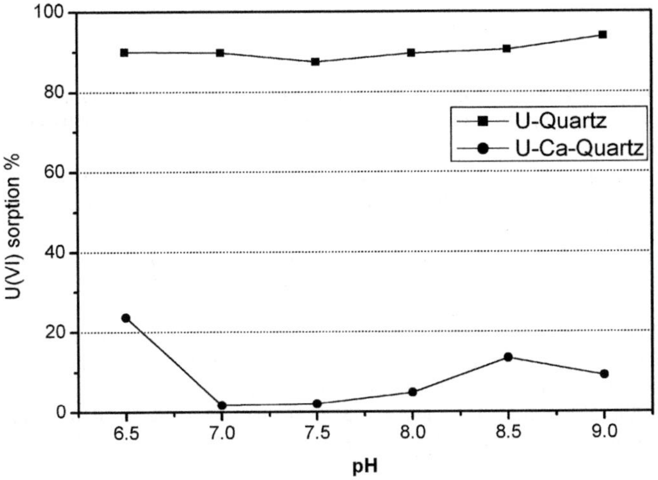

Fig. 6 U(VI) sorption onto quartz as a function of pH, with and without Ca. (0.126×10^{-6} M U, 1×10^{-3} M Ca, 1.5×10^{-3} M NaCl, 1×10^{-3} M $NaHCO_3$, period: 48 h, 23°C, p_{CO2}: $10^{-3.5}$ hPa)

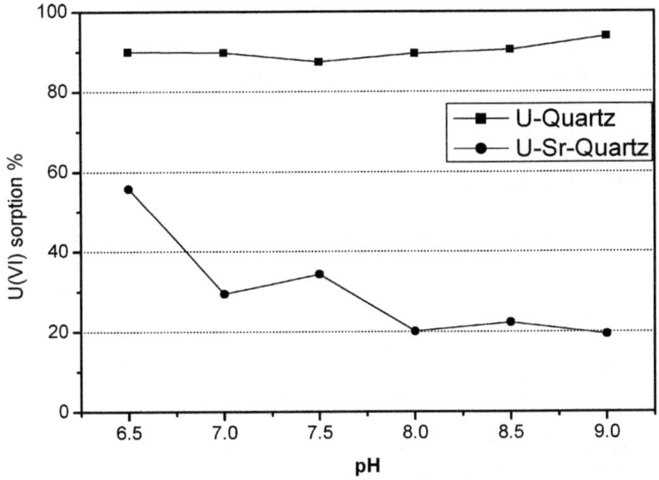

Fig. 7 U(VI) sorption onto quartz as a function of pH, with and without Sr. (0.126×10^{-6} M U, 1×10^{-3} M Sr, 1.5×10^{-3} M NaCl, 1×10^{-3} M $NaHCO_3$, period: 48 h, 23°C, p_{CO2}: $10^{-3.5}$ hPa)

From this study, it is evident that Mg-, Ca-, and Sr-Uranyl-Carbonato complexes play a major role in the sorption behavior of U(VI) at neutral to alkaline pH conditions. Therefore these log-*K* values have to be added by the user to existing data bases if they do not contain the alkaline earth uranyl carbonate species. Further studies are recommended to evaluate the formation and existence of alkaline earth uranyl carbonate species, especially $Mg_2UO_2(CO_3)_3^0$ and $Sr_2UO_2(CO_3)_3^0$.

References

Benedikt G (2007) 797 VA Computrace – voltammetric trace determination of uranium(VI) in drinking and mineral water. Metrohm Information Issue 2/2007. Metrohm Ltd.,CH-9101 Herisau, Switzerland. 36

Bernhard G, Geipel G, Brendler V, Nitsche H (1996) Speciation of uranium in seepage waters of a mine tailing pile studied by time-resolved laser-induced fluorescence spectroscopy (TRLFS). Radiochimica Acta 74:87–91

Bernhard G, Geipel G, Reich T, Brendler V, Amayri S, Nitsche H (2001) Uranyl(VI) carbonate complex formation: Validation of the $Ca_2UO_2(CO3)_{(3)}$(aq.) species. Radiochimica Acta 89:511–518

Dong WM, Brooks SC (2006) Determination of the formation constants of ternary complexes of uranyl and carbonate with alkaline earth metals (Mg^{2+}, Ca^{2+}, Sr^{2+}, and Ba^{2+}) using anion exchange method. Environmental Science & Technology 40:4689–4695

Dong WM, Brooks SC (2008) Formation of aqueous $MgUO_2(CO_3)_3^{2-}$ complex and uranium anion exchange mechanism onto an exchange resin. Environmental Science & Technology 42:1979–1983

Fox PM, Davis JA, Zachara JM (2006) The effect of calcium on aqueous uranium(VI) speciation and adsorption to ferrihydrite and quartz. Geochimica Et Cosmochimica Acta 70:1379–1387

Geipel G, Amayri S, Bernhard G (2008) Mixed complexes of alkaline earth uranyl carbonates: A laser-induced time-resolved fluorescence spectroscopic study. Spectrochimica Acta Part a-Molecular and Biomolecular Spectroscopy 71:53–58

Grenthe I, Fuger J, Konings R, Lemire RJ, Muller AB, Wanner J (2007) The Chemical Thermodynamics of Uranium. Elsevier: New York

Kalmykov SN, Choppin GR (2000) Mixed $Ca^{2+}/UO_2^{2+}/CO_3^{2-}$ complex formation at different ionic strengths. Radiochimica Acta 88:603–606

Merkel BJ, Nair S (2011) Impact of speciation and Sorption on migration of uranium in groundwater. Nachhaltigkeit und Langzeitaspekte bei der Sanierung von Uranbergbau und Aufbereitungsstandorten. Proceedings des Internationalen Bergbausymposiums WISSYM_2011, Ronneburg, Germany, p 269–274

Nair S, Merkel BJ (2011) Impact of Alkaline Earth Metals on Aqueous Speciation of Uranium(VI) and Sorption on Quartz. Aquatic Geochemistry 17:209–219

Parkhurst DL, Appelo CA (1999) User's Guide to PHREEQC (version 2). A Computer Program for Speciation, Batch-Reaction, One-Dimensional Transport, and Inverse Geochemical Calculation. U.S.G.S., Water Resources Investigation Report 99–4259

Stamberg K, Venkatesan KA, Rao PRV (2003) Surface complexation modeling of uranyl ion sorption on mesoporous silica. Colloids and Surfaces a-Physicochemical and Engineering Aspects 221:149–162

Zheng ZP, Tokunaga TK, Wan JM (2003) Influence of calcium carbonate on U(VI) sorption to soils. Environmental Science & Technology 37:5603–5608

Radiological Hazard of Mine Water from Polymetallic and Uranium Deposits in the Karkonosze Mountains, South-West Poland

Nguyen Dinh Chau, Nowak Jakub, Bialic Marcin, Rajchel Lucyna, Czop Mariusz, Wróblewski Jerzy

Abstract. Surface and ground waters associated with abandoned uranium mines in the Kowary region, south-west Poland, has been investigated for their natural radioactivity levels. Concentrations of ^{238}U, ^{234}U, ^{226}Ra, ^{228}Ra and ^{222}Rn isotopes were analyzed in 15 water samples collected from the old mining galleries, from Jedlica river and its inflow streams, as well as from several water supply plants in Kowary and Jelenia Góra towns. The results showed that although the protection measures against leaching of radioactive elements from the old galleries in the Kowary region have been undertaken and cultivation of the mining area had been performed, the levels of radioactivity in surface and ground waters on the study area are still relatively high.

Dinh Chau Nguyen
AGH University of Science and Technology, al. Mickiewicza 30, 30-059 Kraków, Poland

Jakub Nowak
AGH University of Science and Technology, al. Mickiewicza 30, 30-059 Kraków, Poland

Marcin Bialic
AGH University of Science and Technology, al. Mickiewicza 30, 30-059 Kraków, Poland

Lucyna Rajchel
AGH University of Science and Technology, al. Mickiewicza 30, 30-059 Kraków, Poland

Mariusz Czop
AGH University of Science and Technology, al. Mickiewicza 30, 30-059 Kraków, Poland

Jerzy, Wróblewski
National Atomic Energy Agency, ul. Sudecka 38, 58-500 Jelenia Góra, Poland

B. Merkel, M. Schipek, *The New Uranium Mining Boom*
DOI 10.1007/978-3-642-22122-4

Study Area

The study area is located in the vicinity of Kowary and Jelenia Góra towns belonging to the Sudeten hydrological province situated in south-western Poland (Paczyński and Płochniewski 1996). This region is built up of Precambrian metamorphic formations folded and formed as a semisphere covering the younger intrusive granite (Fig. 1). Outer metamorphic formations are composed of gneiss, mica-schist and amphibolite. The inner intrusive granite is represented mainly by grey red porphyritic granite of small and medium grains, consisting of orthoclase, oligoclase, quartz and biotite minerals. Apart from the red granite, in this region there is also other type of granite differs from the first not only in isomorphic grains but also in significantly larger contribution of quartz minerals than biotite ones. Both types of granites were formed as dikes intersected by numerous younger intrusive veins (Adamski and Gawor 1986; Mochnacka and Banaś 2000; Zdulski 2000).

The uranium deposits are concentrated mostly in the metamorphic zones adhering to the NW-SE faults adjoining with the intrusive veins. Within the study area there are seven locations, where uranium deposits occurred: Karpacz, Wołowa Góra, Budniki, Kowary, Ogorzelec-Victoria, Ruebezahl and Podgórze. The Podgórze ores were the most valuable. Here the uranium concentrations varied in a range of 0.1% (12,200 Bq/kg) to 0.3% (33,600 Bq/kg) (Mochnacka and Banaś 2000). The uranium ores are composed predominantly of black blende associated with sulfur and silicon minerals. The uranium concentrations in intrusive rocks range from 8 to 18.5 ppm (98–226 Bq/kg) and are up to four times larger than aver-

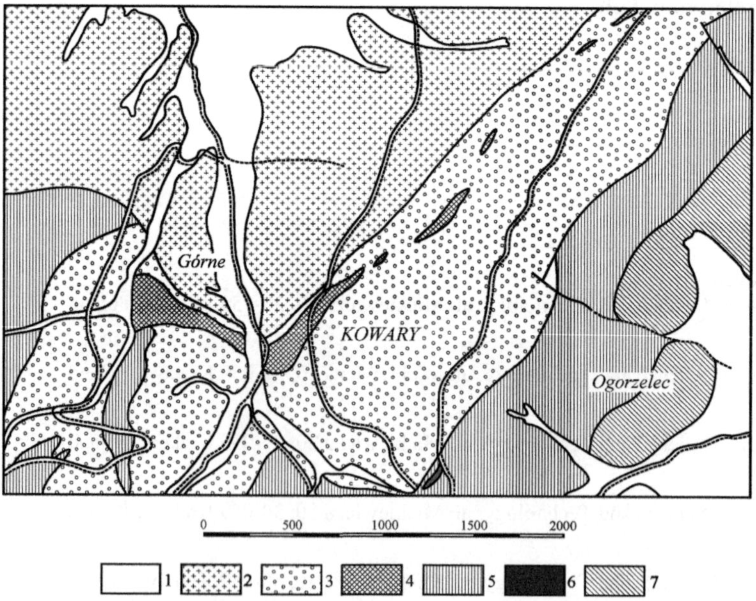

Fig. 1 Geological sketch of the study area (Mochnacka 1965) 1-alluvium, 2-granite, 3-gneiss, 4-ore-bearing complex, 5-mica schists, 6-marbles, 7-amphibolites

age content of this element in typical granite (4.7 ppm–57 Bq/kg) (Nieć 2009). The uranium contents in the waste materials from the exploitation of the Podgórze uranium deposits are very high and range from 24 to 402 ppm (293–4900 Bq/kg), with the geometric mean of 54.2 ppm (660 Bq/kg).

The hydrological conditions in the study area are similar to the general conditions occurring in the Sudeten mountains. Groundwater occurs mostly in Quaternary sandstone and gravel aquifers as well as in catchment area of Jedlica river and its streams. The permeability of Quaternary formations in the region varies in the range of $2.8 \cdot 10^{-5}$ to $1.9 \cdot 10^{-3}$ m/s. Typical water discharge of the wells varies between 20 and 70 m^3/h (Staśko and Michniewicz 2007). In the Kowary region, groundwater occurs also in fissures and cracks of granite and metamorphic formations. The permeability of these aquifers is usually very low and varies from 10^{-6} to 5.7×10^{-4} m/s. Water discharge of springs in the metamorphic formations ranges from 0.2 to 3.5 m^3/h (Staśko and Michniewicz 2007).

Groundwater in the study area is of meteoric origin. The mean annual precipitation amounts to approximately 870 mm. The flow rate of the Jedlica river is about 0.37 m^3/s (Dubicki 2002). Quaternary, Paleozoic and Pre-Cambrian aquifers in the Kowary region are connected by different fissure systems.

Mining activities were present in the region from the beginning of 17th century. Till the end of the 19th century mainly iron, copper and other metal ores were exploited. The exploitation of uranium deposits began in the 20s and end up in the 60s of the last century, with several breaks due to economic crisis and the Second World War. During prospecting and exploitation many galleries and drillings were created. Nowadays, most of them are destroyed or abandoned and their entrances are buried. Although, in some cases the adits cutting the groundwater table and also lowering the groundwater table in the larger area. As a result of that, water from some adits constantly outflows into the surface collectors.

Gallery No 17 is placed near the pathway "Żółta Droga" and its outflow is connected with Jedlica river. The entrance to the gallery is closed by the iron grids. Inside the gallery water stays up to 20 cm above the bottom. The galleries No 19 and 19a were being used for radon therapy. Water from these galleries flows into Jedlica river. Gallery No 10 is placed on the level of 615 m a.s.l. connecting Wolność mine with Wulkan mine. Water in this gallery is characterized by high concentration of sulfur and organic compounds probably originating from decomposed organic materials in interior of this gallery. It flows into Piszczak stream, which discharges into Jedlica river. During periods of strong rainfall, the contribution of water from galleries No 17, 19 and 19a can reach 40% of the total discharge of Jedlica river (Adamski and Gawor 1986). Potentially, this water can leach heavy and radioactive elements from the rocks and bear highly radioactive wastes from old mines in the area.

Due to the uranium and other metallic ore mining excavations, the drainages area of the Jedlica river significantly enlarged and strongly influence on the groundwater table.

In 1985 four water samples were collected and analyzed for the uranium, radium and radon isotopes. The radon concentrations were measured by emanome-

Table 1 Concentrations of uranium, radium and radon in the water samples collected from four locations in the study area measured in 1985 (Adamski and Gawor 1986)

Sampling site	U		Ra		Rn	
	[g/l]	[mBq/l][1]	[g/l]	[mBq/l][2]	[eman/l]	[Bq/l][3]
Upstr. gallery No 17	$2.0 \cdot 10^{-6}$	24	$5 \cdot 10^{-12}$	185		
Gallery No 17	$4.8 \cdot 10^{-5}$	576	$5 \cdot 10^{-12}$	185	108/107	400/650
Gallery No 19a	$2.1 \cdot 10^{-4}$	2520	$9.6 \cdot 10^{-11}$	355		
Gallery No 19	$4.6 \cdot 10^{-5}$	552	$5 \cdot 10^{-12}$	185		

[1] 1 μg of uranium = 12.2 mBq.
[2] 1 ng of ^{226}Ra = 37 mBq.
[3] 1 eman = 3.7 Bq.

ter, the uranium and radium isotope concentrations were determined by the fluorescence and radiochemical methods respectively (Adamski and Gawor 1986). Results of these analyses are summarized in Table 1. The original results were expressed in g/l and eman/l units for uranium, radium and radon respectively. To facilitate comparison with the results of the present study the concentration and activity data in Table 1 were translated into specific activity concentration [Bq/l].

Sampling Sites and Measurement Methods

In the framework of the present study 15 water samples have been collected. Location of sampling sites is shown in Fig. 2: Samples 1 and 2 (Table 2) represent Jedlica river. The first sample was collected upstream of the gallery No 17, while the second one was collected below the outflow from gallery No 19, in front of the "First Building". Samples 3, 4, 5 and 6 were collected at outflows from galleries No 17, 19a, 19 and 10. Sample 7 represents outflow from Gallery "Jedlica". Samples 8 and 9 represent stream "Piszczak". Sample 8 was collected 5 m upstream and sample 9 5 m downstream of the water outflow from gallery No. 10. Sample 10 was collected from private well in the possession No. 49. Sample 11 represents the spring located at granite hills named "Parking". The last four samples (No. 12, 13, 14, 15) were collected from water supply plants in Kowary and Jelenia Góra.

Analyses of uranium (^{238}U, ^{234}U), radium (^{228}Ra, ^{226}Ra) and radon (^{222}Rn) contents have been carried out for each sample. For radon, 200 ml glass bottle was carefully filled up with water and transferred to the laboratory. The time of sampling was recorded. For the radium and uranium isotopes, the analyzed water was collected in plastic canister (ca. 5 l). To avoid adsorption on the canister wall, 5 ml of HNO_3 8M was added.

To determine the radon concentration, 10 ml of water was transferred to counting vial containing 10 ml of HiSafe 3™ Wallac scintillator. Then the sample was measured sequentially in liquid scintillation spectrometer with α/β discrimination.

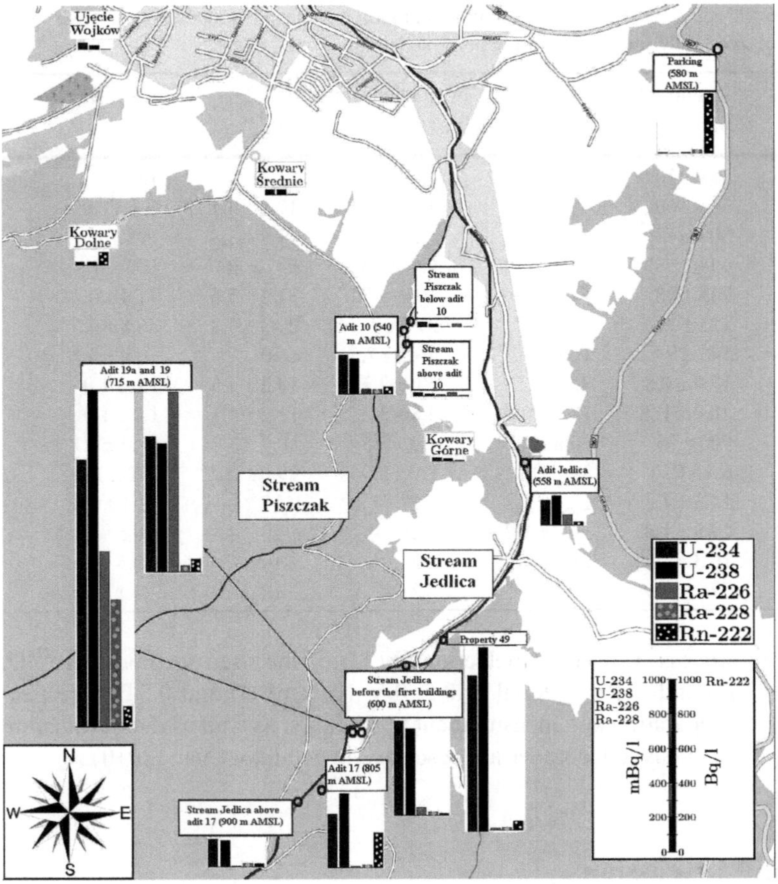

Fig.2 Specific activities of uranium, radium and radon isotopes in water samples collected in the study area

Radium isotopes were determined in 21 water samples. Radium was co-precipitated together with barium as a sulfate. Then the precipitate was purified using alkaline EDTA. The radium sample was again precipitated at pH 4.5 using acetic acid. The sample was placed into a counting vial and mixed with 12 ml of Instagel™ Packard scintillator. Activities of ^{226}Ra and ^{228}Ra isotopes were measured using liquid scintillation counter with α/β discrimination.

Uranium was co-precipitated from 31 water sample together with magnesium hydroxide at pH 9–10 using ammonia solution. Then the precipitate was purified using a DOWEX column and the source for alpha counting was prepared by micro-coprecipitation with $NdCl_3$ and placed onto 0.1 μm Eichrom™ membrane filter. After drying the filter was placed in alpha spectrometer coupled with PIPS detector. The measurement of alpha particles emitted from ^{234}U and ^{238}U isotope lasted typically 50 h. To determine the chemical efficiency of sample preparation procedure, the spike solution of ^{232}U was added to the water sample at the beginning of

Table 2 Specific activities of uranium, radium and radon isotopes in water sales collected in the study area analyzed in the framework of the presented study

Sample No	Uranium [mBq/l]		Radium [mBq/l]		^{222}Rn [Bq/l]
	^{238}U	^{234}U	^{226}Ra	^{228}Ra	
1	158 ± 10	152 ± 10	4.4 ± 2.1	21.3 ± 4.6	16 ± 3
2	544 ± 60	498 ± 60	40.8 ± 6.4	17.1 ± 4.1	6.3 ± 1.9
3	309 ± 18	429 ± 25	6.2 ± 2.5	14.7 ± 3.8	198 ± 10
4	1446 ± 87	1930 ± 120	949 ± 30	687 ± 26	103 ± 8
5	783 ± 48	744 ± 46	1039 ± 32	31.4 ± 5.6	71.4 ± 6.3
6	223 ± 17	201 ± 17	27.5 ± 5.2	23.3 ± 4.8	36.8 ± 4.5
7	143 ± 9	166 ± 11	56.2 ± 7.5	≤ 10	16.9 ± 3.1
8	17.0 ± 0.8	7.7 ± 0.5	5.0 ± 2.2	18.2 ± 4.5	1.6 ± 1.0
9	28.4 ± 1.2	18.1 ± 0.9	4.8 ± 2.2	17.2 ± 4.1	1.7 ± 1.0
10	906 ± 53	1066 ± 62	6.2 ± 2.5	10.4 ± 3.2	47.4 ± 5.1
11	6.4 ± 0.2	3.6 ± 0.1	8.9 ± 3.0	19.0 ± 0.2	352 ± 14
12	14.3 ± 1.3	12.6 ± 1.2	≤2	≤ 10	72.1 ± 6.3
13	29.8 ± 1.8	30.2 ± 1.8	≤2	≤ 10	≤0.5
14	16.3 ± 1.5	11.2 ± 1.2	≤2	≤ 10	≤0.5
15	37.2 ± 1.9	25.5 ± 1.5	≤2	≤ 10	2.4 ± 1.1

the preparing process. Low limit detection (LLD) of the used methods for ^{238}U, ^{234}U, ^{228}Ra, ^{226}Ra and ^{222}Rn is equal to 0.5, 0.5, 10, 2 mBq/l and 0.5 Bq/l respectively. Sample preparation and measurement procedures, as well as data evaluation algorithms for the analyzed isotopes are described in detail by Chau (2010).

Results and Discussion

The measured specific activities of uranium, radium and radon isotopes are presented in Table 2 and in Fig. 2. Specific activities of uranium isotopes in the analyzed samples range from a few mBq/l to about 2000 mBq/l. Similar range is observed for radium isotopes (from few mBq/l to 1000 mBq/l). The specific activity of radon varies from less than 1 Bq/l in the samples collected from Kowary water supply plants to above 350 Bq/l for sample No. 11 (Parking place).

The measured specific activities of ^{238}U in the collected water samples are similar to those analyzed in 1985. Concentration of ^{238}U in outflows from mine galleries and in the water from private well (sample No. 10) exceed the maximum contamination level (MCL = 180 mBq/l) for drinking water (BSS 1996). Large concentrations of uranium in water from the galleries can be linked to oxidizing conditions prevailing there, which enable uranium to be leached from the weathered rocks. The specific activities of ^{226}Ra presented in this work are very different from those published in the work of Adamski and Gawor (1986). The differences may result from poor sensitivity of the methods used in that time.

In water sample from gallery No 10 the uranium specific activities are near 200 mBq/l, whereas specific activities of radium isotopes are below 20 mBq/l. This could result from partial precipitation of radium ions with SO_4^{2-} ions in the gallery waters (Chau et al. 2011).

The concentrations of natural radionuclides in the water seeping from crystalline granite (sample No. 11, "Parking" site) are worth noting. The specific activities of uranium and radium isotopes are very low (from a few mBq/l to a several tens Bq/l), but the radon activity reaches high values (ca. 350 Bq/l). This interesting phenomenon can be explained by the fact, that in the interior of the granite hills reducing conditions prevail and the rock is weekly weathered. Consequently, transfer of both uranium and radium from the rock matrix into the water is very limited (Chau et al. 2011). On the other hand, high radon concentration most probably results from the emanation process (Przylibski 2000; Asikainen 1981). The concentrations of radioactive isotopes in Jedlica river are rather high, resulting mainly from discharges of water from old galleries. The highest concentrations were observed at outflows from galleries No 17, 19 and 19a. The specific activities of uranium isotopes in the analyzed tap water samples (No. 12, 13, 14, 15) vary from ca. 14 to near 40 mBq/l.

Conclusions

Although the protection measures against leaching of radioactive elements from the old galleries of abandoned uranium mines in the Kowary region has been undertaken and cultivation of the mining area had been performed, the levels of radioactivity in surface and groundwater on this area are still large. The maximum measured uranium concentrations in outflows from old galleries reach the values up to 2000 mBq/l. Measures should be taken to reduce the radon content in tap water from Kowary Dolne well.

Acknowledgements Financial support of the work through statutory funds of the AGH University of Science and Technology (projects No. 11.11.220.01) is kindly acknowledged.

References

Adamski W, Gawor F (1986) Underground mining galleries in Kowary region, their detriments and influence on the hydrological conditions. Wrocław Techniques University, Wrocław (in Polish)

Asikainen M (1981) State of disequilibrium between ^{238}U, ^{234}U, ^{226}Ra and ^{222}Rn in groundwater from bedrock. Geochem Cosmochem 45: 201–206

BSS (1996) International Basic Safety Standards for Protection against Ionizing Radiation. IAEA Vienna

Chau N (2010) Natural radioactive elements in the mineral waters occurring in the Polish Carpathians. Wydawnictwo Jak Kraków (in Polish)

Chau N, Nowak J, Bialic M, Rajchel L, Czop M and Wróblewski J (2011) New data of the natural radionuclide concentrations of the environment water in the vicinity of Kowary town (Sudetes Mountain, West Poland). Biuletyn Panstwowego Instytutu Geologicznego, Warszawa (in press)

Dubicki A (red) (2002) The water resources in upper and medium discharge area of the Odra river in the dry weather conditions. Inst Meteor Wat Manag in Warsaw (in Polish)

Mochnacka K (1965) Ore minerals of the polymetallic deposit at Kowary (Lower Silesia). Prace Mineral., Kom. Nauk Mineral. PAN (in Polish)

Mochnacka K, Banaś M (2000) Occurrence and genetic relationship of uraniom and thorium mineralization in the Karkonosze Izera Block (the Sudeten Mts, SW Poland). Ann Soc Geol Pol 70(2): 137–150

Nieć M (2009) Uranium deposit occurring and prospecting in future in Poland. Polityka Energetyczna 12(2/2): 435–451 (in Polish)

Przylibski T (2000) Estimating the radon emanation coefficient from crystalline rocks into groundwaters. Appl Radiat Isot 53: 473–479

Staśko S, Michniewicz M (2007) Sudeten Subregion. in Paczyński B, Sadurski A (red.) Regional Hydrogeology of Poland. Part I – Common Water. Wyd Geol Warsaw (in Polish)

WHO (2008) Guidelines for drinking water quality, 3rd edit. Vol 1. Recommendations, Geneva

Zdulski M (2000) The sources of uranium mining in Poland. Wydawnictwo DiG, Warsaw (in Polish)

Uranium – a Problem in the Elbe Catchment Area?

Petra Schneider, Heinrich Reincke, Sylvia Rohde, Uwe Engelmann

Abstract. Data from the Elbe River and its tributaries indicate that the river water quality has been improved since reunification due to the closure of mining and industrial sites. Despite those actions uranium concentrations which exceed the environmental quality norms are still measured in some parts of the Elbe catchment. Former mining areas are still the main uranium sources. The contributing loads from Zwickauer Mulde and Saale tributaries are still relevant for the total uranium concentrations in the Elbe catchment. Further, there is a relevant amount of heavy metals still available from the sediments, which acted as metal sinks. As in the Hamburg Harbor uranium values were analyzed in the sediment in the range of the natural background, the relevant geochemical sink for uranium is situated between Magdeburg and Hamburg harbor and should be subjected to further investigations.

Petra Schneider
C&E Consulting und Engineering Chemnitz, Jagdschänkenstr. 52, D-09117 Chemnitz, Germany

Heinrich Reincke
Senatskanzlei der Stadt Hamburg, Senatskanzlei-Planungsstab,
Poststr.11 D-20354 Hamburg, Germany

Sylvia Rohde
Sächsisches Landesamt für Umwelt, Landwirtschaft und Geologie, Pillnitzer Platz 3,
D-01326 Dresden, Germany

Uwe Engelmann
Sächsisches Landesamt für Umwelt, Landwirtschaft und Geologie, Pillnitzer Platz 3,
D-01326 Dresden, Germany

B. Merkel, M. Schipek, *The New Uranium Mining Boom*
DOI 10.1007/978-3-642-22122-4 © Springer 2011

Introduction

The former GDR was the largest uranium producing country in Eastern Europe in the past. After the closure of mining activities in the period after the reunification of Germany, the transport and deposition of contaminated material as a result of flood events as well as the discharge of waste water from mining and milling facilities into rivers were the main paths of sediment and soil contamination in the 1990s. Beside uranium input from the discharge of former uranium sites in Saxony, the Elbe River receives other heavy metals from its tributaries Vereinigte Mulde and Saale which as well dewater catchment areas where former mining sites (uranium, hard coal, copper) are situated.

The Elbe River is one of the major rivers in Western Europe. From its spring in the Giant Mountains (Czech Republic) to its mouth at the North Sea near Cuxhaven (Germany) it covers a distance of 1091 km and a catchment area of 148,268 km^2 (see Fig. 1), one third of it is located in the Czech Republic and two thirds are in the Federal Republic of Germany, smaller areas belong to Poland and Austria. The river length in Germany is 728 km. Along its way the catchment drains some of Northern and Central Europe's major cities including Prague, Dresden, Berlin and Hamburg. With respect to the EC Water Framework Directive the Elbe river basin consists of three different types of water: river, estuary and coastal water. In Germany the Elbe catchment was divided into 2775 water bodies.

The Elbe's major tributaries include the Vltava, Bilina and Ohře rivers in Czech Republic as well as Saale, Havel and Mulde rivers in Germany.

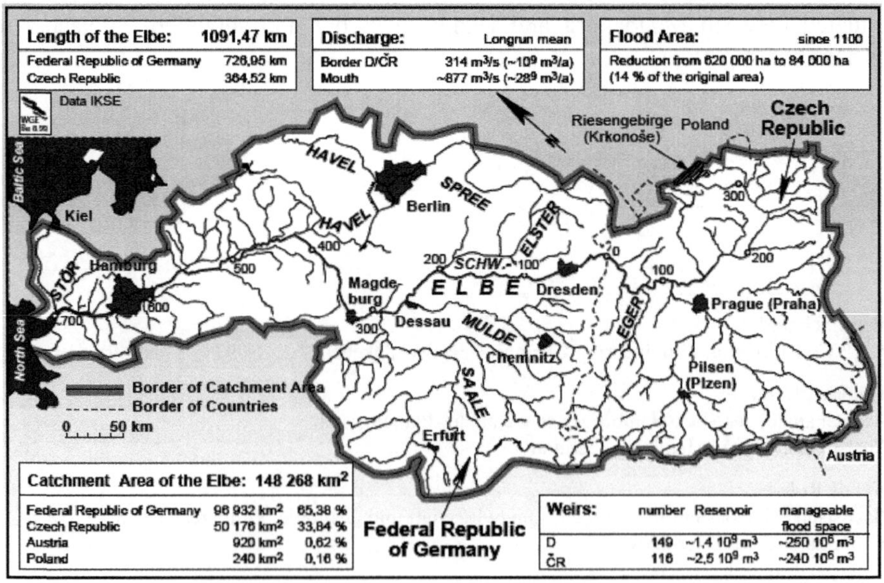

Fig. 1 Catchment area of the Elbe river

The *Vltava catchment* has a size of about 28,090 km², the river runs north from its source in Šumava until Prague (Czech Republic), merging with the Elbe at Mělník (Czech Republic). The *Bilina catchment* amounts to 1071 km². The Bilina rises on the slopes of Ore Mountains in the Czech Republic, north of Chomutov. The river flows between the Czech Central Mountains and the Ore Mountains to the north-east and runs into the Elbe in Ústí nad Labem (Czech Republic). The river has no major tributaries. The Bilina River is considered to be one of the most important pollution sources in the Elbe River catchment area. Despite gradual improvement of water quality, the sediments are still contaminated (Havel et al. 2007). The *Ohře River* originates as Eger, later becomes Czech Ohře, and has a catchment size of 6255 km², being a left tributary of the Elbe. It's the third longest river in the Czech Republic.

The *Saale River* with a length of 427 km and a catchment area of 24,079 km² is one of the largest tributaries of the River Elbe (Zerling et al. 2003). Some of the main tributaries to the Saale River are Weiße Elster, Unstrut, Schwarza, and Wipper. The Saale catchment, mainly located in Saxony-Anhalt (Germany) drains former mining areas, namely copper mining (through the Schlenze tributary) and lignite as well as uranium mining (through the Weiße Elster tributary). The copper mining activities in the Mansfeld mining district were closed in 1990. Nevertheless the old subsurface mines, adits and dumps are still emitting metals into the environment. According to Haack and Plimer (1998) the dumps of the Mansfeld copper area contain from 40 to 150 mg/kg uranium.

The *Havel* is a river in north-eastern Germany, flowing through the German states of Mecklenburg-Vorpommern, Brandenburg, Berlin and Saxony-Anhalt. It is a right tributary of the Elbe River with a catchment area of 24,096 km².

The *Mulde River* is a left side tributary of the Elbe River and is situated in Saxony and Saxony-Anhalt, Germany. The river system consists of the Freiberger Mulde River and the Zwickauer Mulde River, which merge to form the Vereinigte Mulde River. The Zwickauer Mulde River drains the former uranium mining and milling areas and the former hard coal mining areas in Saxony.

With respect to uranium concentrations in surface waters, the proposal for the environmental quality norm is 1 µg/l in the filtered sample, considering the natural background value as 90th percentile. Uranium is considered in the current draft of the Surface Water Regulation OGewV, which is still under discussion in the Federal Council of Germany.

Uranium Concentrations in the Elbe Catchment

Historical Data

By the end of the 1980s the Elbe was ranked among the most polluted water courses in Europe (ICPER 2009). However, since the International Commission for the Protection of the Elbe River was established in October 1990, the Czech

Republic and Germany worked together in an intense cooperation to improve the status of the Elbe and its catchment area.

Historical data measured in 1987 in the Elbe River near Brunsbüttel close to the estuary into North Sea showed values of around 2 μg/l in the river water. The major part of the uranium contamination was bound to suspended matter and sediments. The average uranium concentrations in the Elbe stream after the closure of uranium mining in the period between 1993 and 1998 reached values up to 1.3 μ/l in the river water, 3.9 μg/l in suspended matter and 5.8 mg/kg in sediments (internal data Hamburg City). For instance, in Magdeburg uranium concentrations were measured in the Elbe sediment between 2.5 and 5.1 mg/kg in 1996 (12 samples) and around 1.4 μg/l in the filtrated water (GKSS Forschungszentrum 1997). In 2002 the sediment concentrations in the region of Hamburg Harbor ranged between 1.9 and 3.3 mg/kg (location Seemannshöft, internal data Hamburg City).

Historical values from the Czech part of the catchment ranged between 1 μg/l (Vltava), 2 μg/l (Eger) and 0.8 μg/l (Bilina). The corresponding sediment values ranged between 2.3 mg/kg (Vltava), 7.2 mg/kg (Eger) and 4.5 mg/kg (Bilina) in 1994, and between 5.9 mg/kg (Vltava), 4.5 mg/kg (Eger) and 6.2 mg/kg (Bilina) in 1996 (GKSS Forschungszentrum 1997). The dilution in the Czech part of the Elbe catchment resulted in lower concentrations at the border to Germany, showing uranium concentrations in Schmilka at the border of around 1.4 μg/l in whole the period between 1996 and 2004.

Already in that time period the uranium contributions of the tributaries Bilina, Saale and Mulde were significant. After the millennium flood in 2002, the Saale River including Weiße Elster catchment area contributed with up to 4 μg U/l to the concentrations in the Elbe River. In the Mulde River system concentrations up to 4.5 μg U/l were observed (all values 2002 to 2003). An old adit from the Mansfeld mining district flowing into the Schlenze River with mean uranium concentrations of about 60 μg/l increased the uranium concentration of the River Saale by 0.5 μg/l (Baborowski and Bozau 2006). Uranium concentrations in the tributaries of the Saale measured near their mouths ranged from about 1 to 65 μg/l (Baborowski and Bozau 2006). Locally, also the former lignite mining areas emit uranium, particularly the flooded lignite open cast mines in Witznitz and Bockwitz as well as the Gladegraben, a tributary to the Weiße Elster River (Delakowitz et al. 2003). The measured values spread between 10 and 70 μg/l (for comparison: the values in the Lusatian lignite mine area are lower than 2 μg/l, having only a few slight exceptions in the data base; see also Delakowitz et al. 2003).

Current Data

Nowadays, even though the major emission sources are closed, there are still diffuse uranium emissions in the Elbe catchment area which are mainly caused from old mining sites. The uranium concentrations of the River Elbe near Magdeburg decreased from 1 to 4 μg/l (2002 to 2006) to 0.8 to 2.7 μg/l (2007 to 2009). In

Saxony a reduction of the average concentration from 9.2 µg/l in 2003 to 4.2 µg/l in 2009 in the River Zwickauer Mulde was found. Uranium concentrations > 0.2 µg/l, resulting from Wismut site discharges were observed in the period from 2005 to 2008 in the Crossen area only. Since 2009 the uranium concentrations in the Elbe River which were measured at the border to Czech Republic (Schmilka) have been < 1 µg/l. The sediment concentrations in Schmilka ranged between 2.9 and 4.9 mg/kg (11 values in 2009). The 2009 concentrations in Magdeburg ranged between 0.8 and 2.1 µg/l (12 values). In 2010 the sediment concentrations in the region of Hamburg Harbor ranged between 0.9 and 1.8 mg/kg (location Seemannshöft).

In Saxony-Anhalt the following average values of uranium concentrations were analyzed in the period between 2005 and 2007: 1.8 µg/l (Bode in Nienburg), 1.9 µg/l (Saale in Bad Dürrenberg), 2.0 µg/l (Elbe in Magdeburg), 2.5 µg/l (Saale in Halle-Trotha), 2.7 µg/l (Unstrut in Freyburg), 3.6 µg/l (Weiße Elster in Halle-Ammendorf), 16.8 µg/l (Laucha in Schkopau), up to 40.7 µg/l (Schlenze before entering Saale River), (Schneider et al. 2010). The data show clearly that the largest contribution of uranium loads to the Saale tributary results from the Schlenze River.

Sources of Uranium Concentrations in the Elbe Catchment

Natural Background

Natural background values were determined for all rivers in Europe as screening within the FOREGS mapping (Salminen et al. 2005), see Fig. 2. Further, more detailed investigations have been carried out by the environmental authorities of the German counties. The natural background of uranium concentrations in the main tributaries of the Elbe catchment in Saxony is given in Table 1 (Greif and Klemm 2010).

The natural background values for uranium in the Elbe catchment in Saxony-Anhalt show a range between 0.25 and 3.3 µg/l in the filtered samples, and 1.2 mg/kg in the sediments (Schneider et al. 2010).

Table 1 Natural background of uranium concentrations in the contributing rivers of the Elbe catchment in Saxony (values measured by BfUL, detection limit 0.1 µg/l), all values in µg/l (P50 – 50th percentile, P90 – 90th percentile) (Greif and Klemm 2010)

Environm. Quality Norm	Freiberger Mulde		Bobritzsch		Zwickauer Mulde		Schwarzwasser		Rote Weißeritz		Wilde Weißeritz	
	P50	P90	P50	P90	P50	P90	P50	P90	P50	P90	P50	P90
1	<0.5	<0.5	<0.5	<0.5	1.9	3.7	0.15	1.3	0.36	1.4	0.16	1

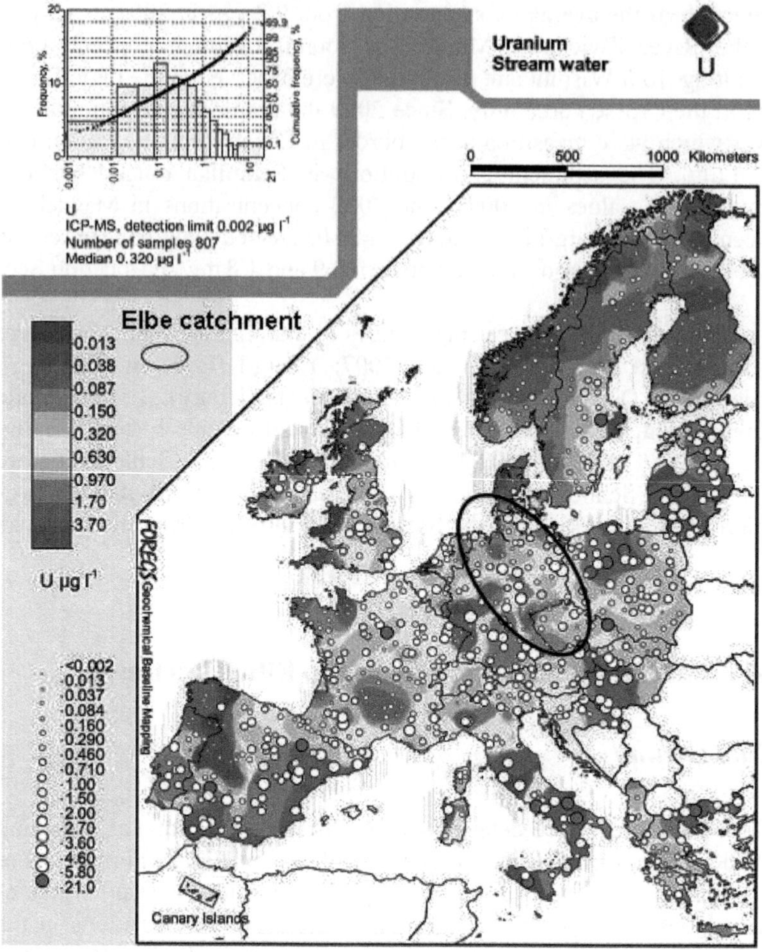

Fig. 2 Natural background of uranium concentrations in stream water in Europe – FOREGS map (Salminen et al. 2005)

The natural uranium background in creeks arriving to the Elbe River in the area of Hamburg from the north show uranium contents between 0.5 and 1 µg/l, while creeks arriving from the south to the Elbe river like Moorburger Landscheide and Seeve show lower uranium concentrations. Generally, the values correspond with the range given in the FOREGS map due to Salminen et al. (2005).

Uranium Concentrations from Former and Current Mining Sites

Since the termination of active uranium mining operations in Germany, Wismut GmbH conducts the rehabilitation of its former mining and milling sites. Site-

Table 2 Uranium discharge from former Wismut mining sites in the Elbe catchment area (after Paul 2007)

Water Treatment Plant	Flow rate [m³/h]	Type of feed water	uranium concentration in mg/l	U_{nat} (mg/l) Permitted discharge	Average discharge	U_{nat} (t/a)
Schlema	1150	Mine water	5	0.5	0.2	4.4
Helmsdorf	250	Supernatant water	10	0.5	0.3	0.88
Königstein	650	Mine water	20	0.3	0.006	1.7
Ronneburg*	450	Mine water	2	0.3		
Seelingstädt	300	Supernatant water, seepage	2	0.3	0.12	0.63

* The Ronneburg WWTP was designed for a treatment capacity of 450 m³/h with an option to expand the capacity to 600 m³/h by converting the sludge lines (Paul 2009).

specific rehabilitation concepts are complemented by strategies outlining individual flooding approaches for the large underground mine workings at the Königstein, Ronneburg and Aue as well as Pöhla and Freital-Gittersee sites (Gatzweiler et al. 2002). The uranium discharge from former Wismut mining sites in the Elbe catchment area is given in Table 2.

Uranium Concentrations from Communal Waste Water Treatment Plants

Slightly increased uranium concentrations even in the effluent of urban wastewater treatment plants were measured in regions of former mining activities and may be attributed to rain water and extraneous water entering the connected sewerage system. In 56 out of 83 wastewater treatment plant effluents uranium values higher than the detection limit of 0.2 µg/l were found in Saxony in the investigation period 2004–2010. The median value was below the detection limit, the 90th percentile reached 1.3 µg/l (data from the Saxon Environmental Authority, 2011). In the north of the Elbe catchment area, close to the North Sea, uranium concentrations up to 0.3 µg/l were observed (sample from WWTP Köhlbrandhöft/Dradenau 2011).

Uranium Concentrations in Sediments

When evaluating the status of surface water bodies in the Elbe it was found out that some substances still do not meet the objectives defined for the good status (ICPER 2009), among them uranium. The analysis of causes showed that the pollutant load originates especially from sediments. The major role in this process is played by stored pollutants originating from the previous industrial inputs (especially in the period prior to 1990; ICPER 2009). Despite extensive improvement in water quality since reunification data from the Elbe River and its tributaries indicate that the sediment situation of uranium has locally not reached an acceptable level.

Recent sediment investigations in the river meadow sediments of the Zwickauer Mulde River showed values up to 450 Bq/kg U-238 downstream the former mining areas, between 120 and 400 Bq/kg U-238 in the Saxonian Lowlands, between 60 and 200 Bq/kg U-238 in the United Mulde River (after inflow of Freiberger Mulde), and at the border to Saxony-Anhalt values < 100 Bq/kg U-238 (Ritzel 2010). In the river sections, investigated between the towns Grimma and Eilenburg values between 220 and 360 Bq/kg U-238 were analyzed. The investigated areas were affected by the Millennium Flood in 2002 (Ritzel 2010).

In 2000, the distribution of uranium in the river sediments of the whole Saale catchment was investigated by Zerling et al. (2003), later by Barborowski and Bozau (2006). Barborowski and Bozau (2006) reported sediment values from the Mansfeld area in the Saale catchment between 3 and 7 mg/kg. The authors state that there are two mechanisms which are responsible for the accumulation of uranium in sediments: the uptake of uranium by organisms (phytoplankton and bacteria), which die and sink to the bottom of the water column (so called biosorption), and the interaction of uranium with colloidal materials and organic acids. This was confirmed by Kinzel (2004) for the Mulde River. Generally, in Saxony-Anhalt average uranium concentrations were analyzed in the period between 2005 and 2007 of 1.2 mg/kg (Unstrut in Freyburg), 1.4 mg/kg (Bode in Neugattersleben), 4.1 mg/kg (Saale in Halle-Trotha), 8.9 mg/kg (Weiße Elster in Halle-Ammendorf) up to 11.3 mg/kg (Mulde in Jeßnitz), (Schneider et al. 2010). Recent sediment investigations in Magdeburg showed uranium concentrations around 3 mg/kg. In sampling locations Bunthaus and Seemannshöft near Hamburg Harbor uranium has been measured since 2010. The values range from 2 mg/kg in Bunthaus to 1 mg U/kg in Seemannshöft.

Conclusions

The geochemical investigations of the last 20 years still show an impact from former mining sites in terms of the uranium concentrations. The main input comes from the Saale catchment (rivers Weiße Elster and Schlenze) and the Mulde catchment (Zwickauer Mulde). The mines, tailings and sedimentation ponds of hard coal, metal and uranium mining along the Mulde River present long-term pollutant sources. Despite the fact that the concentrations in the Mulde catchment lowered significantly, the influence from the former mining activities can still be seen. This is due to the long term relevant amounts of heavy metals which are bound to the sediment phase. In this regard, the conclusions from Schneider and Reincke (2005) are still valid regarding the need of a river basin-wide sediment assessment which should start with the assessment of inventories of interim depots in the catchment area.

Within a typical river section of 1 km length in the lower middle Elbe, the estimated nutrient and pollutant loads deposited on the floodplains and in the river course clearly demonstrate the specific sink function of both sites. However, at the same time the results suggest that in contrast to the deposits in the floodplains sediments within the river course may partly be remobilized.

As in the Hamburg Harbor the uranium concentrations are in the range of the natural background the resulting open questions are: Where is the geochemical sink between Magdeburg and Hamburg Harbor? Which risks are to be expected from the sediment bound uranium in the Elbe catchment area? An investigation program on the mentioned issues in the Elbe catchment area should be established as soon as possible. The results should be included in the Elbe River Basin Managements Plan due to EC Water Framework Directive 2000/60/EC.

References

Baborowski M., Bozau E. (2006): Impact of former mining activities on the uranium distribution in the River Saale (Germany), Applied Geochemistry 21, 1073–1082.

Delakowitz B., Ender V., Mehner C., Wetzel C., Heidenreich H., Heidrich R., Heidrich U., Schneider P., Neitzel P.L., Kupsch H., Franke K., Rößler D., Lippold H., Mansel A., Burckhardt M., Crustewitz C., Schößler C. (2003): Verhalten von Radionukliden bei der Flutung von Braunkohlentagebauen; Forschungsbericht AZ 13.8802.3528/47 im Auftrag des Sächsischen Landesamtes für Umwelt und Geologie, August 2003.

European Community/Europäische Komission (2000): Richtlinie 2000/60/EG des Europäischen Parlaments und des Rates vom 23.10.2000 zur Schaffung eines Ordnungsrahmens für Maßnahmen der Gemeinschaft im Bereich der Wasserpolitik. in: Amtsblatt der Europäischen Gemeinschaften, L 327 vom 22.12.2000, Luxemburg.

Gatzweiler R., Jakubick A.T., Meyer J., Paul M., Schreyer J. (2002): Flooding of the WISMUT mines – learning by doing or applying a comprehensive systematic approach? – In: Merkel, B.J., Planer-Friedrich, B., Wolkersdorfer, Ch.: Uranium in the Aquatic Environment. – pp. 745–754; Heidelberg (Springer).

GKSS Forschungszentrum (1997): Erfassung und Beurteiluing der Elbe mit Schadstoffen, Teilprojekt 2: Schwermetalle – Schwermetallspezies. Zusammenfassende Aus- und Bewertung der Längsprofiluntersuchungen in der Elbe, BMBF-Forschungsprojekt.

Greif A., Klemm W. (2010): Geogene Hintergrundbelastungen, Schriftenreihe des Sächsischen Landesamtes für Umwelt, Landwirtschaft und Geologie, 04/2010.

Haack U., Plimer I. (1998): Zum Stoffbestand der Kupferschieferschlacken im Raume Mansfeld-Hettstedt-Eisleben. Mitt Geol Sachsen-Anhalt 4, 153–162.

Havel L., Vlasak P., Aronova K. (2007): Temporal and Spacial Changes of the Bilina River Ecosystem (North West Bohemia, Czech Republic), SESF 5 Symposium for European Freshwater Sciences, Palermo 2007.

ICPER (2009): Water Framework Directive in the Elbe catchment area – Structure of the River Basin Management Plan – Further Information, Information Sheet published by the ICPER No. 3 – December 2009.

Kinzel T. (2004): Untersuchungen zur Belastungssituation und zu Bindungsformen von Schadelementen in industriell geprägten Fließgewässern, Dissertation University Hamburg in 2004.

Paul M., Baacke D., Metschies T., Kuhn W. (2009): Post-Flooding Water Management at the Ronneburg Uranium Mine: Lessons Learned and Remaining Challenges. – In: Water Institute of Southern Africa & International Mine Water Association: Proceedings, International Mine Water Conference. – pp. 952–957.

Paul M. (2007): The WISMUT experience in remediation of Uranium Mining & Milling Legacies, IAEA Technical Meeting, Swakopmund/Namibia, 1–5 October 2007.

Reincke H., Hurst S., Schneider P. (2002): Strategy Concept Elbe, in: Merkel B.J.; Planer-Friedrich, B.; Wolkersdorfer, C: Uranium in the Aquatic Environment, pp. 41–48, Springer Verlag ISBN 3-540-43927-7.

Reincke H., Schneider P. (2004): Strategy Concept Elbe – Passive Water Treatment Methods for the Minimisation of Impacts on Water Bodies by Ore Mining Activities; In: Geller, W. et al: Proceedings of the 11th Magdeburg Seminar on Waters in Central and Eastern Europe: Assessment, Protection, Management; ISSN 0948-9452.

Ritzel S. (2010): Ermittlung und Bewertung von radiologisch relevanten Sedimentablagerungen in den landwirtschaftlich genutzten Auengebieten der Zwickauer und Vereinigten Mulde – Ergebnisse des Fremdleistungsvorhabens, Mitteilung des Landesamte für Umwelt, Landwirtschaft und Geologie vom 07.01.2010.

Schneider P., Neitzel P., Schaffrath M., Schlumprecht H. (2003): Physico-chemical Assessment of the Reference Status in German Surface Waters: A Contribution to the Establishment of the EC Water Framework Directive 2000/60/EG in Germany, In: Hydrochimica et hydrobiologica Acta. 31 (1), pp. 49–63.

Schneider P., Reincke H. (2005): Contaminated Sediments in Elbe Basin and its Tributary Mulde, in: Merkel B.J., Hasche-Berger, A.: Uranium in the Environment – Mining Impact and Consequences, pp. 655–662, Springer Verlag ISBN 3-540-28363-3.

Schneider P., Süß A., Gottschalk N., Löser R., Schaffrath M., Stüwer B, Ackermann S, Tröger K. (2010): Ermittlung geogener Hintergrundbelastungen durch Schwermetalle in Oberflächengewässern des Landes Sachsen – Anhalt. Unpublished report for Landesamt für Hochwasserschutz und Wasserwirtschaft Sachsen-Anhalt.

Salminen R., Batista M.J., Bidovec M., Demetriades A., De Vivo B., De Vos W., Duris M., Gilucis A., Gregorauskiene V., Halamic J., Heitzmann P., Lima A., Jordan G., Klaver G., Klein P., Lis J., Locutura J., Marsina K., Mazreku A., O'Connor P.J., Olsson S.Å., Ottesen R.-T., Petersell V., Plant J.A., Reeder S., Salpeteur I., Sandström H., Siewers U., Steenfelt A., Tarvainen T. (2005): Geochemical Atlas of Europe. Part 1 – Background Information, Methodology and Maps.

Verordnung zum Schutz der Oberflächengewässer (Oberflächengewässerverordnung – OGewV), draft of the Surface Water Protection Regulation, May 2011.

Zerling L., Hanisch C., Junge F.W., Müller A., (2003). Heavy metals in Saale sediment – changes in the contamination since 1991. Acta Hydrochim hydrobiol 31, 368–377.

Energetic and Economic Significance of Uranium in Mineral Phosphorous Fertilizers

Ewald Schnug, Nils Haneklaus

Abstract. During the past decade up to 167 t/yr U has been applied to farmland by P fertilization in Germany and this amount would have been sufficient to satisfy the energy demand of 2.4·106 households if U had been extracted previously. This energy source equals the heating value of timber from 5.6·106 ha of forest land. The use of U from P fertilizers is an uncommon contribution of agriculture to climate protection, which has been valued to 1.8 €/kg P when an equivalent value to CO_2 emission car tax bands in Germany is assumed.

Contribution of Uranium in Rock Phosphates to Environmentally-Sound and Cost-Efficient Energy Production

Uranium (U) is an entirely indispensable raw material for producing nuclear power. The basic idea of using U from rock phosphates for energetic purposes at competitive expenses in order to satisfy the U demand for future nuclear power plants whilst minimizing adverse environmental effects by fertilizer-derived U loads to agricultural soils has been proposed by Hu et al. (2008). The extraction of

Ewald Schnug
Technical University Braunschweig – Faculty 2 Life Sciences, Pockelsstraße 14,
D-38106 Braunschweig, Germany,
E-mail: e.schnug@tu-braunschweig.de

Nils Haneklaus
Institute of Nuclear Technology and Energy Systems (IKE), University of Stuttgart,
Pfaffenwaldring 31, D-70569 Stuttgart, Germany,
E-mail: nils.haneklaus@ike.uni-stuttgart.de

B. Merkel, M. Schipek, *The New Uranium Mining Boom*
DOI 10.1007/978-3-642-22122-4 © Springer 2011

U during fertilizer processing is technically no problem (Imphos 2009). It has been common practice until the end of the 1990s in Europe and the USA, and plants are still operating worldwide. Several methods exist for the extraction of U from rock phosphate such as precipitation of U from phosphoric acid, ion exchange separation with a chelating resin, membrane separation of U, froth flotation and solvent extraction of U (Kratz and Schnug 2006). Gupta and Singh (2003) provide a detailed description of these procedures. The technology experiences a revival as the contract between France and Maroc in 2009 shows, which comprises the extraction of U from rock phosphates (Ellersick 2007).

U is an accompanying element of phosphate rock, particularly that of sedimentary origin. Phosphate rock of sedimentary and igneous origin can be distinguished. Depending on the geographical and biogenic origin of the rock phosphate U concentrations may be as high as 150 mg kg^{-1} in sedimentary and 220 mg kg^{-1} in igneous rock phosphates (Kratz and Schnug 2006). About 6% of the world's known P deposits exhibit U in recoverable concentrations (Orris and Chernoff 2002). Countries possessing such P/U deposits are Afghanistan, Angola, Australia, Belgium, Brazil, Canada, Central-Africa, Ecuador, Finland, Greenland, Hungary, India, Israel, Jordan, Mauritania, Morocco, Mozambique, New Zealand, Saudi Arabia, Senegal, South Africa, Spain, Sweden, Syria, Togo, Turkey and United States (Orris and Chernoff 2002; Ragheb 2008).

Rock phosphates and processed phosphorous (P) fertilizers are a main source for U contamination and accumulation in agricultural soils (Rogasik et al. 2008; Kratz and Schnug 2006). Even more critical is the translocation of U into water bodies with yet unknown effects on human health (Knolle et al. 2011; Smidt et al. 2011). IAEA (2001) speculated that extraction of U from phosphates in the future depends on the development of the world market price for U, or environmental regulations which request re-installation of the process. In addition, the argument that a lower U content of the ore in U mines increase the price costs for fuel production and the life-cycle contribution to the CO_2 balance is no longer valid if the U source is a side-product of fertilizer production. The unconventional world U reserves in P deposits are estimated to be as high as $22 \cdot 10^6$ t and could deliver U for another 440 years for the same price as conventional U resources, which is about five times longer than the time span calculated for conventional U reserves (Comby 2008; OECD 2005; Ragheb 2008).

On a long-term basis it can be expected that the prices for P fertilizers will remain unaffected after global implementation of the technology for the extraction of U as the demand of U for nuclear power plants is worldwide rapidly increasing. At the same time the U supply from conversion of nuclear weapons is declining so that the price of U in the world market can be expected to remain on a constantly high level. Other U secondary recovery sources are bottom and fly ashes from coal combustion and sea water. Hereby, the energetic use of coal is in a similar dilemma as is the agricultural utilization of phosphates: Coal and fly ashes show U concentrations, which are in a similar range like those in rock phosphates. The extraction of U from fly ashes is explored with respect to limited U reserves.

Table 1 Energetic and ecological characteristics of various energy sources. Sources: Anonymous (2007, 2008a,b); Deal (2010)

Energy source	Energy density (MJ/kg)	Electricity, produced (kWh/kg)	CO_2-emission (g/kWh)	Land use (ha/1000 MW)
1 kg Firewood[1]	12	1	1851	5,333,333
1 kg Coal	33	3	1000	
1 kg Oil	46	4	814	
1 kg Natural gas	54	5	480	
1 kg U (0.7% ^{235}U)	600,000	50,000	32	768
1 kg LEU[2] (3.5% ^{235}U)	3,456,000	288,000		
Solar			27	12,961
Wind			24	51,842
1 kg Water[3]	0.00008	0.001	22	125,000

[1] Based on 1 kg firewood = 1 kW/h and a yield potential of 1.5 t/ha.
[2] LEU – Low Enriched Uranium.
[3] Potential at 100 m dam height.

U in rock phosphates is a precious metal with a view to energetic and economic aspects as a simple calculation reveals (Table 1). During the past decade up to 167 t/yr U has been applied in Germany exclusively by P fertilization. This amount would have been sufficient to satisfy the energy demand of 2.4 million households of average size. Thus this energy source equals the heating value of timber from 5.6 million ha of forest land. It can be concluded that the use of U-depleted P fertilizers is an uncommon contribution of agriculture to climate protection, which has been valued to 1.8 €/kg P when supposing an equivalent value to CO_2 emission car tax bands in Germany. A detailed description of the calculation basis is provided below. Either a CO_2 tax on mineral P fertilizer products or a monetary reward for farmers in the form of a CO_2 bonus are supposed to provide a strong motivation support for the re-vitalisation of the existing technology for U extraction from rock phosphates.

The Potential of Fertilizer Practice to Reduce Adverse Effects of Climate Change

"Nothing in life is to be feared, it is only to be understood. Now is the time to understand more, so that we may fear less" (Marie Curie, 1867–1934). The relevance of this quotation of Marie Curie is most likely more important than ever if it comes to the civil use of nuclear energy after the nuclear accidents in Chernobyl and Fukushima. Nuclear power is the only large-scale, non-greenhouse-gas emitting electricity source with only a small environmental footprint if latest technology is employed. Several small and medium reactors with advanced technology have been developed in different countries (Anonymous 2011a). For instance a Small & Modular Reactor (SMR) developed by Hyperion which stabilizes itself through a prompt negative temperature coefficient of reactivity. In other words basic physics prevent the reactor from going supercritical (Anonymous 2011b). It produces 70 MWt or 25 MWe which can supply 35,000 average German households with electric energy. It is a small and sealed unit which remains invisible in the surrounding landscape and which needs to be refueled only after 7–10 years. Nonetheless U is required to run the reactor and secondary recovery sources for U such as U from rock phosphates or coal ashes would complete a self-contained "green" cycle for energy production.

Here, U from rock phosphates is undoubtedly a privileged source as it otherwise enters the food chain. Fertilizer-derived U has been shown to be discharged into water bodies whereby rate and velocity are closely linked to fertilizer practices and soil characteristics (Smidt et al. 2011). A preliminary, conservative estimate of the U concentration in 369 water samples from areas in the plains of northern Germany where a higher, geogenically originating, U content can be excluded, revealed that ¼ up to ⅔ (worst case scenario with concentrations above 0,1 µg/L) of all specimens may be contaminated by fertilizer-derived U by now (11% > 2 µg/L; 0,5% > 10 µg/L; 0,3% > 20 µg/L) (Smidt et al. 2011).

A CO_2 tax on P fertilizer products in relation to the U content or a CO_2 bonus for farmers seem opportune political instruments to improve environmental quality in terms of contamination of surface water with U and reduction of CO_2 emissions. Such approach makes a compulsory declaration of the U content of mineral P fertilizers a prerequisite. The following basic calculations provide an estimate on the CO_2 savings potential in Germany. A P fertilizer rate of 22 kg/ha P is recommended when following the codes of good agricultural practice (GAP). On an average 10 g/ha U are applied to agricultural soils following this recommendation. 10 g U deliver 500 kW and this saves 500 kg CO_2 when compared to yielding the same energy from coal. A CO_2 tax on basis of prevailing emission car tax bands charges 2 € per g CO_2 above a base rate of 100 g/km CO_2 so that the holder of a medium class car with an exhaust of 250 g/km CO_2 and an annual mileage of 15,000 km will have to pay 300 € per year. Thus 0.08 € are charged per kg CO_2. Accordingly the extraction of U from rock phosphates yields a CO_2 tax/bonus of 40 €/ha. This means that the fictive CO_2 bonus and commodity price are higher than the value of fertilizer P if a spot price of 1.11 € for 10 g U (as U_3O_8) is assumed as it was on January 28, 2010 and costs for 22 kg P of 32 € (including tax) as published on 28.1.2010 (Anonymous 2010).

Conclusions

Recovery of U from rock phosphates is a challenging, yet intriguing concept to combine the allocation of U for nuclear power with an efficient measure of environmental protection. And it adds value to agricultural production if U-free fertilizers are applied which will not be covered by product prices, but which is highly regarded by the society as it preserves the quality of drinking water (Smidt et al. 2011) and contributes to a significant reduction of CO_2 emissions. A CO_2 tax/bonus based on emission car tax bands seems an appropriate approach to promote the production and use of U-free mineral fertilizer products. The environmental benefits of U extraction from rock phosphates are immanent. Last but not least it can be stated that the utilization of U in worldwide P-resources vouches for cleaner fertilizers, cleaner soils, cleaner waters, cleaner atmosphere and a clairvoyant, profitable fertilizer manufacturing.

Acknowledgement The authors wish to express their most sincere gratitude to Dir. & Prof. Dr. Silvia Haneklaus (Institute for Crop and Soil Science, Julius Kühn-Institute, Braunschweig, Germany) who contributed significantly to the presented work, her right as co-author, however, being denied.

References

Anonymous (2007) http://www.physik.uni-muenchen.de/lehre/vorlesungen/wise_06_07/ep1/vorlesung/skript26_5_2.pdf

Anonymous (2008a) http://www.xemplar.ca/de/about_uranium.php

Anonymous (2008b) http://www.co2-emissionen-vergleichen.de/Stromerzeugung/CO2-Vergleich-Stromerzeugung.html

Anonymous (2010) http://www.agrarmarkt-nrw.de/duengermarkt.shtm

Anonymous (2011a) http://www.world-nuclear.org/info/inf33.html

Anonymous (2011b) http://www.hyperionpowergeneration.com/product.html

Comby B (2008) Uranium resources worldwide. http://www.ecolo.org/documents/documents_in_english/uranium_resources_BC-04.htm

Deal J (2010) "Night with a futurist". Webinar of the The DaVinci Institute. PO Box 270315. Louisville. CO 80027. USA. January 10, 2010

Ellersick S (2007) Klimawandel à la francaise. http://www.cl-netz.de/read.php?id=26361

Gupta CK, Singh H (2003) Uranium resource processing: Secondary resources. Springer, Berlin

Hu Z, Zhang H, Wang Y, Haneklaus S, Schnug E (2008) Combining energy and fertilizer production – vision for China's future. In: Loads and fate of fertilizer-derived uranium. LJ De Kok, E Schnug (Eds), Backhys Publishers, Leiden, 127–134

IMPHOS (2009) Newsletter Uranium recovery from phosphoric acid. Phosphate Newsletter 26, 11 http://www.imphos.org/download/imphos_news_special26_web.pdf

Knolle F, Birke M, Hassoun R, Jacobs F, Schnug E (2011) Uranium in German mineral water – occurrence and origins. This volume

Kratz S, Schnug E (2006) Rock phosphates and P fertilizers as sources of U contamination in agricultural soils. In: Merkel, BJ and Hasche-Berger, A (eds.), Uranium in the Environment, pp. 57–67. Springer, Berlin

OECD (2005) Uranium 2005 Resources, Production and Demand. OECD, International Atomic Energy Agency (IAEA). Published by OECD Publishing. ISBN 9789264024267

Orris GJ, Chernoff CB (2002) Data set of world phosphate mines, deposits, and occurrences-Part B. Location and mineral Economic data: U.S. Geological Survey Open-File Report 02-156A, 328 p. http://geopubs.wr.usgs.gov/open-file/of02-156/OF02-156B.PDF

Ragheb M (2008) Uranium resources in phosphate rocks. https://netfiles.uiuc.edu/mragheb/www/NPRE%20402%20ME%20405%20Nuclear%20Power%20Engineering/Uranium%20Resources%20in%20Phosphate%20Rocks.pdf

Rogasik J, Kratz S, Funder U, Panten K, Schnug E, Barkusky D, Baumecker M, Gutser R, Lausen P, Scherer HW, Schmidt L (2008) Uranium in soils of German long-term fertilizer trials. In: Loads and fate of fertilizer-derived uranium. LJ De Kok, E Schnug (eds.), Backhys Publishers, Leiden, 135–146

Smidt GA, Hassoun R, Birke M, Erdinger L, Schäf M, Knolle F, Utermann J, Duijnisveld WHM, Schnug E (2011) Uranium in German tap and groundwater – occurrence and origins. This volume

Contribution of Mineral and Tap Water to the Dietary Intake of As, B, Cu, Li, Mo, Ni, Pb, U and Zn by Humans

Rula Hassoun, Ewald Schnug

Abstract. This paper reports on the contribution of mineral and tap water to the dietary intake of As, B, Cu, Li, Mo, Ni, Pb, U and Zn by humans in order to identify potential hazards from contaminations with these elements through fertilizer use in agriculture. Part of the research work presented is the development of standardized diet types as bias-free data background for the human exposure to As, B, Cu, Li, Mo, Ni, Pb, U and Zn through solid food. The average contribution of drinking water to the total daily intake was highest for U (64.7%) followed by Li (24.3%), Cu (4.65%) Zn (2.40%), B (2.20%), Pb (1,77%), Ni (1.72%), As (1.00%) and was lowest for Mo (0.23%).

Introduction

Agriculture is a main contributor to environmental loads of nearly all elements of the periodic system. Not only waste-based fertilizer materials such as sewage sludge, but also mineral fertilizers, particularly mineral phosphorous fertilizers, contain significant amounts of elements which affect the quality of the environment and food plants. For the time period from 1950/51 to 2009/2010, the mean annual loads of the elements As, Cd, Cu, Ni, Pb, Zn and U to agricultural land in Germany exclusively from the application of P fertilizers amounted to (T/a) As 40 (73), Cd 22.1 (42.1), Cu 95 (146), Ni 54 (91), Pb 11 (20), U 114 (228) and Zn 431

Rula Hassoun
Technical University Braunschweig, Department of Life Sciences, Pockelstr. 14, D-38106 Braunschweig, Germany

Ewald Schnug
Technical University Braunschweig, Department of Life Sciences, Pockelstr. 14, D-38106 Braunschweig, Germany
E-mail: e.schnug@tu-braunschweig.de

B. Merkel, M. Schipek, *The New Uranium Mining Boom*
DOI 10.1007/978-3-642-22122-4 © Springer 2011

(764) tons (values in brackets are maxima values; Kratz et al. 2011). Of this elements Cd the whereabouts of Cd in the food chain is well investigated, but much less is known in this context for As, B, Cu, Li, Mo, Ni, Pb, U and Zn.

There are two major pathways through which elements enter the food chain: either by uptake of agricultural crops or by leaching into potable ground and surface water bodies (Smidt et al. 2011). A closer look to the transfer pathways for such elements is of great importance when it comes to a risk assessment through which pathway elements that are applied together with fertilization may enter the human body and which measures are suitable to influence the intake of these elements by humans, either with view to the prevention of intake negatively affecting health, or with the objective to compensate deficiency of an essential micro nutrient.

Finally, for a correct assessment of risks arising from a dispersal of these elements with fertilizers in the environment it is necessary to know to which extent each of the two general pathways, solid or liquid food, contributes to their total daily intake. This information is particularly required in order to assess efficient abatement strategies for contaminations in the food chain.

Material and Methods

Meta data on occurrence concentrations of various elements and their daily intake with solid foods: The main objective of this study was to evaluate the relative significance of water to the total daily intake of As, B, Cu, Li, Mo, Ni, Pb, U and Zn. This evaluation requires information about the amount of intake deriving from solid foods. EFSA (2009a,b) provides a so-called "Concise European Food Consumption Database" which holds food consumption data reported from individual EU countries. EFSA (2009a,b) itself mentions the different methodologies employed for data collection as the main limitation of the data base. This results in sometimes very different consumption patterns between different countries. According to these data a Dane would for instance consume nearly 6 times more liquids than a Bulgarian citizen. Another limitation highlighted by EFSA (2009) is the "broad food categories used". In comparison, in this study the considerably variable food sources in the human diet were assembled in a few broader categories as a means to achieve a low bias from element intake with solid food.

Consequently in this study a standardized healthy diet has been designed based on an energy requirement of 2000 kcal/day (Eastwood 2003) and according to the rules of the well-known "nutrition pyramid" (Anonymous 2010). Four food categories were defined: "cereal and cereal products", "meat/fish and products of them", "milk/eggs and products of them" and "vegetables/fruits and products of them". For the standard diet it was adopted that 55% of the total energy demand derive from carbohydrates, 15% from protein (5% milk and milk products, 10% meat/fish/shellfish and meat/fish/shellfish products) and 30% from fats (15% from milk and milk products, 15% meat/fish/shellfish and meat/fish/shellfish products). This is more or less equivalent to a daily diet including 314 g cereals and cereal products, 122 g meat/fish/shellfish and products of them ((meat : fish = 6 : 1) includes 2% offal), 120 g milk/egg and products of them, 320 g vegetables and fruits (of which 44% leaf vegetables 38% root vegetables and 18% fruits). The assumed average energy content of these food categories was: cereal and cereal products: 350 kcal/100 g, fish/shellfish and fish/shellfish and products of them (10% fat): 100 kcal/100 g, meat and meat products including offal and products of them (20% fat): 250 kcal/100 g (at a ratio of meat/fish = 6 : 1 the average energy content of this source would be 228 kcal/100 g), milk, eggs and products of them (20% fat): 250 kcal/100 g, vegetables and fruits: 60 kcal/100 g.

The standard diet has been diversified in three additional diet types: an ovo-lacto vegetarian, a vegan and a carnivore type. The average mass consumption of the different food categories to maintain an energy input of 2000 kcal/day is shown in Table 1.

Meta data for the concentration of As, B, Cu, Li, Mo, Ni, Pb, U and Zn in different food categories mentioned in Table 1 were collected through a comprehensive literature research by Hassoun (2011). In statistics a meta-data analysis combines the results of several studies that address a set of related research hypotheses. The main results of the meta-analysis for elemental concentrations in different food categories are presented in Table 2.

Table 1 Average consumption (g/day) of food assigned to different categories in order to maintain an energy input of 2000 kcal/day in each diet type

Diet type	Daily consumption (g/day) in food category			
	Cereals and cereal products	Meat/fish/shellfish and products of them	Milk/egg and products of them	Vegetables and fruits
Standard	314	123	120	320
Vegetarian (ovo-lacto)	200	–	200	833
Vegan	343	–	–	1333
Carnivore	171	439	80	333

Table 2 Mean occurrence concentrations (mg/kg) of As, B, Cu, Li, Mo, Ni, Pb, U and Zn in the food categories "cereal and cereal products", "meat/fish and products of them", "milk/eggs and products of them" and "vegetables/fruits and products of them"

Element	Cereals and cereal products	Meat/fish and products of them	Milk/eggs and products of them	Vegetables/fruits and products
As	0.09	2.32	0.01	0.03
B	2.5	0.63	0.43	5.06
Cu	1.92	11.3	0.69	1.95
Li	0.21	0.07	0.07	0.1
Mo	0.41	0.38	0.14	0.46
Ni	0.38	0.18	0.05	0.37
Pb	0.07	0.11	0.03	0.15
U	0.002	0.002	0.001	0.001
Zn	15.5	25.7	10.5	5.67

Details about the *origin and analysis of mineral and tap water samples* are presented by Knolle et al. (2011) and Smidt et al. (2011). Occurrence means and 95th percentile concentrations are presented in Tables 3 and 4. Note that data given there may differ from those provided by Knolle et al. (2011) because of differences in the underlying database.

The mineral water data are presented as regional collectives (Table 3) and in addition for three different classes of total dissolved substance content (TDS). Low mineralized waters with a TDS < 500, medium mineralized waters with a TDS between 500 and 1000 and high mineralized waters with a TDS > 1000.

Table 3 Occurrence mean and 95th percentile concentration (P95) of As, B, Cu, Li, Ni, Mo, Pb, U and Zn in mineral waters from (world, Germany and neighboring countries, Germany) and German tap water (µg/l) (Date of database: July 11, 2010; Hassoun 2011)

	Origin of mineral waters						German tap water	
Element	World		Germany and Neighboring Countries		Germany			
	Mean	P95	Mean	P95	Mean	P95	Mean	P95
As	6.52	16.2	4.09	12.2	1.92	11.1	1.27	3.21
B	672	2024	326	1599	227	1297	40.7	102
Cu	6.24	23.6	4.22	16.9	3.45	10.4	78.8	356
Li	328	1115	415	1030	263	910	20.3	15.2
Mo	1.29	4.02	1.15	3.70	0.69	2.52	0.32	1.11
Ni	4.10	13.3	4.45	14.1	3.69	11.9	2.74	5.56
Pb	0.91	1.68	0.88	1.79	0.69	1.73	1.07	3.06
U	2.54	8.73	2.92	10	3.08	9.68	1.67	7.21
Zn	14.4	38.3	9.11	32.4	0.73	2.66	151	639

Table 4 Occurrence mean and 95th percentile concentration (P95) of As, B, Cu, Li, Ni, Mo, Pb, U and Zn in mineral waters from (world, Germany and neighboring countries, Germany) and German tap water (µg/l) (Date of database: July 11, 2010; Hassoun 2011)

	Mineral waters of the world					
Element	Low mineralization		Medium mineralization		High mineralization	
	Mean	P95	Mean	P95	Mean	P95
As	2.03	9.27	2.48	12.2	23.4	155
B	55.2	168	368	740	1719	9611
Cu	0.18	0.15	0.03	0.1	0.03	0.12
Li	5.01	19.6	4.77	17.3	10.5	33.8
Mo	17.4	72.9	65.2	310	1060	5000
Ni	0.7	2.99	1.10	3.85	2.24	6.83
Pb	2.9	10.9	3.51	12.2	7.13	26.6
U	0.95	1.38	0.94	1.85	0.86	2.09
Zn	1.45	6.3	1.23	6.24	7.54	15.7

Examples of well-known brand names for the three TDS classes are: low mineralized waters: SPA Barisart (49), Volvic (109), Wittenseer Quelle (425); medium mineralized waters: Ramloesa (817), Vittel Bonne Source (841), highly mineralized waters: Appolinaris (2767), Heppinger (4566), Westerwald Quelle (8500).

Results and Discussion

Based on the nine elements investigated in this research work U shows the smallest range of variation in daily intake within the four dietary group types. An ovo-lacto vegetarian diet supplies the smallest amount of U with solid food to the human body (1.46 μg/day) whereas a carnivore diet delivers 1.4 times more U than an ovo-lacto vegetarian diet. The element U takes the seventh place among the nine elements in terms of the concentration range in drinking waters. A consumer of German tap or low mineralized bottled waters would have the lowest (2.9 μg/day), but a consumer of high mineralized bottled waters a 5.2 times higher intake of U (Table 5).

Except for Li, where in the standard scenario a quarter of the daily intake comes from tap water, the contribution of tap water to the total daily intake was always below 5% (Table 5).

These figures change dramatically when the scenario with the highest daily element intake is taken into account: for U and Li nearly the entire total daily intake (>95%) can be attributed to highly mineralized bottled water. In comparison, the corresponding values are for As and B>75%, for Cu, Mo, Ni and Pb about 50%, and for Zn 33%. However, for the elements Cu, Pb and Zn (famously known for a strong anthropogenic background through plumbing materials) tap water is a more powerful contributor than even highly mineralized bottled waters (Table 6). Consumers with a carnivore habit would take in more As, Cu, U and Zn, but vegans have a higher intake of B, Li, Mo, Ni and Pb with their solid food.

The standard diet together with German tap water exceeds the limit for As twice if the "tolerable daily intake (TDI)" for toxic elements is considered (Table 6). But even the mean value of As intake reported in literature (Hassoun 2011) meets already the TDI of 146 μg/day As. It needs to be mentioned here that the data given here do not differentiate between inorganic and organic As. According to EFSA (2009b), however, inorganic As compounds are by far more toxic than organic As compounds which make up to the majority of the total As concentration. Considering that As in waters is almost completely inorganic, consumers of high mineralized bottled waters are especially at risk for dangerously high As intake (Table 4).

In the high input scenario around two third of the TDI with Ni and Pb are met and more than three quarters with U (Table 6). Also six times more Li is supplied than tolerable but with the same element the standard diet with German tap water seems to deliver too small amounts of Li. Li, however is not considered as an essential element but with a very small bandwidth between beneficial and adverse effects (Anonymous 2011).

Table 5 Occurrence mean and 95th percentile intake estimates (μg/day) for uranium in different water consumer types according to different exposure scenarios compared to the average daily intake of U (μg/day) through dietary group types (maintained at an energy input of 2000 kcal/day)

	Scenario	
	A: Occurrence mean scenario	B: Occurrence 95th percentile scenario
Water consumer type (μg/day)		
1 Tap water consumer	2.90	16.0
2 German bottled water consumer	6.16	19.4
3 German and neighbors bottled water consumer	5.84	20.0
4 World bottled water consumer	5.08	17.5
5 Low mineralized bottled water consumer	2.90	12.6
6 Medium mineralized bottled water consumer	2.46	12.5
7 High mineralized bottled water consumer	15.1	31.4
Dietary group type * (μg/day)		
Standard	1.58	
Vegetarian (ovo-lacto)	1.46	
Vegan	1.96	
Carnivore	2.09	
Intake		
Lowest daily U intake (water consumer + dietary group type combination) (μg/day)	1 + Vegetarian (ovo-lacto)	
Lowest daily U intake (μg/day)	4.36	
Contribution of water (%) to the daily intake of U at lowest intake scenario	66.6	
Maximum daily U intake (water consumer + dietary group type combination) (μg/day)	7 + Carnivore	7 + Carnivore
Maximum daily U intake (μg/day)	17.7	33.5
Maximum contribution of water (%) to the daily intake of U at type combination	87.8	95.3

Remarks: Scenario A: U exposure obtained by multiplying 2 l/day consumption by occurrence means. Scenario B: U exposure obtained by multiplying 2 l/day consumption by 95th percentile concentration.

* Values for dietary intake by different dietary group types are calculated from mean U occurrence concentration (see Tables 1–5).

The standard diet with German tap water supplies already four times more Cu, two times more Mo and Ni and nearly fully the recommended amount of Zn than the recommended daily intake if the average values for recommended intake of minerals in the literature are taken into account (Table 6; Hassoun 2011).

Table 6 Mean values of As, B, Cu, Li, Mo, Ni, Pb, U and Zn intake estimates (μg/day) in different diet scenarios types and the contribution of waters to the daily intake of these elements (%) (dietary group types maintained at an energy input of 2000 kcal/day; water intake 2 l/day)

Intake (μg/day)	As	B	Cu	Li	Mo	Ni	Pb	U	Zn
Reported *	142	1650	1600	586	220	213	71	2.10	10,200
Standard diet** (solid only)	296	3610	3240	128	429	343	111	1.58	12,300
Standard diet + German tap water ***	299	3691	3398	169	430	349	113	4.48	12,602
Contribution of water at standard scenario (%)	1.00	2.20	4.65	24.3	0.23	1.72	1.77	64.7	2.40
High input scenario ****	1235 HMBWC carnivore	26,802 HMBWC vegan	7101 GTWC carnivore	10,194 HMBWC vegan	720 HMBWC vegan	678 HMBWC vegan	208 GTWC vegan	33.5 HMBWC carnivore	17,877 GTWC carnivore
Contribution of water at high input scenario (%)	76.0	86.5	54.4	98.7	40.4	49.4	46.6	95.3	31.2
TDI*/*****	146	–	10,000	1800	–	1000	309	42	–
% of TDI at high input scenario	846	–	71.0	566		67.8	67.3	79.7	
Recommended* or required (r/r)	–	–	900	1000	200	250	–	–	11,000
% of r/r at standard diet + German tap water scenario	–	–	378	16.9	215	140	–	–	115

Remarks:
* Meta data analysis see Hassoun (2011).
** See tables.
*** At occurrence means scenario.
**** At occurrence 95th percentile scenario.
***** TDI: Tolerable Daily Intake for a 70 kg person; GTWC: German tap water consumer; HMBWC: High mineralized bottled water consumer.

Conclusions

Among the elements investigated in this study As and Li show the most susceptible bandwidth in daily intake through dietary habits: the highest input scenario for As (a highly mineralized bottled water consuming carnivore) supplies nearly 20 times more As than the lowest one (a German tap water consuming ovo-lacto vegetarian). In comparison, Zn is the element which daily intake varies only 1.5 times between lowest and highest input scenario. The ranking of the other elements from highest to lowest intake through dietary habits is As = Li > B > Cu = U = Ni > Pb > Mo > Zn. Among these elements U is that with the highest contribution of waters to the dietary daily intake.

Considering the significant amounts charged annually to agricultural land with mineral P fertilizers (according to Kratz et al. 2011: (t/a) for U: 114 mean, 228 maximum and 40 mean, 73 maximum for As and the fact that As as well as U are easily leached from the soil matrix (Smidt et al. 2011) requests for immediate action of agricultural politics to limit further pollution of soils and water resources by these elements. In case of U, where agriculture is the only significant source of input to the environment and because recent research indicates leaching of U from arable soils and presence of fertilizer-derived U in ground and drinking water (Smidt et al. 2011), it is suggested that there should be a legit limitation in the fertilizer ordinance (Ekardt and Schnug 2008) as it would be the most promising measure to protect drinking water in Germany from further U contamination.

References

Anonym (2010) The Nutrition Source, Healthy Eating Pyramid. http://www.hsph.harvard.edu/nutritionsource/what-should-you-eat/pyramid/ and http://www.dietbites.com/FDApyramid.gif.

Anonymous (2011) Lithium dosage. http://bipolar-disorder.emedtv.com/lithium/lithium-dosage.html.

Eastwood M (2003) Principles of human nutrition. 2. ed. Wiley-Blackwell. ISBN: 978-0-632-05811-2.

Ekardt F, Schnug E (2009) Legal aspects of uranium in environmental compartments. In: De Kok LJ, Schnug E (2008) Loads and fate of fertilizer derived uranium. Backhuys, Leiden, The Netherlands, 209–216.

EFSA (2009a) Scientific opinion of the panel on contaminants in the food chain on a request from German Federal Institute for Risk Assessment (BfR) on uranium in food stuff, in particular mineral water. The EFSA Journal 1018, 1–59. Question number: EFSA-Q-2007-135.

EFSA (2009b) Scientific Opinion on Arsenic in Food. The EFSA Journal 1351, 77–99. Question number: EFSA-Q-2008-425.

Hassoun R (in prep.) A statistical evaluation of the contribution of mineral and tap water to the dietary intake of As, B, Cu, Li, Mo, Ni, Pb, U and Zn by humans. Diss. Fak. f. Lebenswissenschaften TU Carolo-Wilhelmina zu Braunschweig.

Knolle F, Birke M, Hassoun R, Schnug E (2011) Uranium in German mineral waters – occurrence and origins. This volume.

Kratz S, Godlinski F, Schnug E (2011) Heavy metal loads to agricultural soils from the application of commercial phosphorus fertilizers in Germany and their contribution to background concentration in soils. This volume.

Smidt GA, Landes FC, de Carvalho LM, Koschnisky A, Schnug E, (2011) Cadmium and uranium in German and Brazilian phosphorous fertilizers. This volume.

Smidt GA, Hassoun R, Birke M, Erdinger L, Schäf M, Knolle F, Utermann J, Duijnisveld WHM, Schnug E (2011) Uranium in German tap and groundwater – occurrence and origins. This volume.

Uranium in Secondary Phosphate Fertilizers and Base Substrates from Water Treatment Plants

Dorit Julich, Christine Waida, Stefan Gäth

Abstract. As a limiting factor for plant growth the demand for phosphorous (P) fertilizers is huge while the global phosphate reserves are strongly limited. Thus, the recycling of P from secondary raw materials such as sewage sludges and ashes or other organic waste materials will be a more and more important resource. In fertilizers, both primary and secondary, not only the nutrients but also the content of pollutants is important for an application in agriculture. Mineral P fertilizers may contain remarkable concentrations of uranium (U), cadmium (Cd) and other trace elements depending on their origin and chemical treatment.

In our study, we analyzed the P and U contents of secondary phosphate fertilizers and theirs base substrates (sewage sludges and ashes), which were treated with different P recovery methods. The aim was to evaluate several P recovery methods processes (thermal, thermochemical, acid hydrolysis, etc.) with regard to the P enrichment on the one hand and the heavy metals exclusion on other hand. We measured total P and U contents (aqua regia) as well as the mobile fractions (EDTA, $Ca(NO_3)_2$) in the base substrates and products. These element contents were compared to common mineral P fertilizer (triple super phosphate and rock phosphate).

Dorit Julich
Institute of Landscape Ecology and Resources Management, Justus-Liebig-University Giessen, Heinrich-Buff-Ring 26C, 35392 Giessen, Germany,
E-mail: dorit.julich@umwelt.uni-giessen.de

Christine Waida
Institute of Landscape Ecology and Resources Management, Justus-Liebig-University Giessen, Heinrich-Buff-Ring 26C, 35392 Giessen, Germany

Stefan Gäth
Institute of Landscape Ecology and Resources Management, Justus-Liebig-University Giessen, Heinrich-Buff-Ring 26C, 35392 Giessen, Germany

B. Merkel, M. Schipek, *The New Uranium Mining Boom*
DOI 10.1007/978-3-642-22122-4 © Springer 2011

The results show that the different recovery processes considerably influence the P enrichment in the products, though the P contents could not reach the contents in the mineral P fertilizers. In most P recovery processes, the heavy metals (including U) contents were decreased clearly, whereas the chemical treatment methods were more effective than thermal procedures. The plant available metal fraction as well was decreased clearly in most. In the case of U, the contents in the recovery products seemed to be stronger dependent on U concentrations of the base substrates than on the treatment process. However, the total U contents in all secondary fertilizer products were by far lower (<6 mg kg^{-1}) than contents in commonly used mineral phosphate. Compared to triple super and rock phosphate, the fertilization with the analyzed secondary phosphate products therefore would lead to much lower U inputs into agricultural soils (2–50 mg U per kg P) and the plant available fractions here are negligible. Only one product, an Al-containing sewage sludge ash, reached higher loads of 130 mg U per kg P caused by very low P contents. For triple super phosphate, we measured an uranium load of 645 mg U per kg P, for rock phosphate a load of 310 mg U per kg P. Nevertheless, it has to consider that the investigated base substrates already had comparable low U contents resulting in non-problematic products, at least regarding U. Thus, higher U pollutions in sewages sludge may result in more contaminated products.

Uranium in German Tap and Groundwater – Occurrence and Origins

Geerd A. Smidt, Rula Hassoun, Lothar Erdinger, Mathias Schäf, Friedhart Knolle, Jens Utermann, Wilhelmus H.M. Duijnisveld, Manfred Birke, Ewald Schnug

---

Geerd A. Smidt
Jacobs University Bremen, Integrated Environmental Studies program, Campus Ring 1, D-28759 Bremen, Germany
E-mail: g.smidt@jacobs-university.de

Rula Hassoun
Technical University Braunschweig – Faculty 2 Life Sciences, Pockelsstraße 14, D-38106 Braunschweig, Germany

Lothar Erdinger
University Hospital Heidelberg, Dept. of Medical Microbiology and Hygiene, Im Neuenheimer Feld 324, D-69120 Heidelberg, Germany

Mathias Schäf
University Hospital Heidelberg, Dept. of Medical Microbiology and Hygiene, Im Neuenheimer Feld 324, D-69120 Heidelberg, Germany

Friedhart Knolle
GeoPark Harz. Braunschweiger Land. Ostfalen Network, Grummetwiese 16, D-38640 Goslar, Germany

Jens Utermann
Federal Environmental Agency, Wörlitzer Platz 1, D-06844 Dessau-Roßlau, Germany

Wilhelmus H.M. Duijnisveld
Federal Institute for Geosciences and Natural Resources, Stilleweg 2, D-30655 Hannover, Germany

Manfred Birke
Federal Institute for Geosciences and Natural Resources, Stilleweg 2, D-30655 Hannover, Germany

Ewald Schnug
Technical University Braunschweig – Faculty 2 Life Sciences, Pockelsstraße 14, D-38106 Braunschweig, Germany

B. Merkel, M. Schipek, *The New Uranium Mining Boom*
DOI 10.1007/978-3-642-22122-4

Abstract. This study presents data of uranium (U) concentrations in groundwater ($N=155$) and tap water ($N=4092$) to which 76% of the entire German population has access. The mean U concentration was 0.68 µg/l, the median 0.50 µg/l. 3.7% of all samples had U concentrations below the detection limits, which accounts for water to which 11.7% of the entire population has access. 14.3% of samples were above 2 µg/l U, 3.3% above 10 µg/l U, representing an exposed population of 10.5 and 1.31% respectively. The regional distribution of U concentrations largely agrees with the geological setting reported for mineral waters, however, in addition clear evidence for anthropogenic influence through agricultural activities were found in certain areas.

Introduction

Uranium (U) is an element of considerable environmental concern, due to its radioactive and toxic nature. Initiated by studies and publications on drinking water quality by European and German state authorities as well as private institutes (Foodwatch 2008; Schnug et al. 2008; Schulz et al. 2008; Alexander et al. 2009) U contamination in mineral and drinking water and the discussion of thresholds alignment has made the headlines in several newspapers and magazines since 2008 in Germany. A study by the European Food Safety Authority (Alexander et al. 2009) pointed out that the exposure of infants to U from drinking water is 3-times higher than for adults on the body weight basis. Uranium concentrations below 2 µg/l in drinking water already induce 50% of the tolerable daily intake of 0.6 µg per kg body weight per day (WHO 2005) for infants.

The long-term diffuse input by agricultural activity has long been ignored as a possible way of U transfer to drinking water. In 2008, the Committee for Drinking Water (TWK 2008) of the German Environmental Agency (UBA) published a report focusing on U in drinking water, which points out the influence of agriculture on the introduction of U into the ecosystem. The TWK recommends an investigation of the possible impacts of phosphorous (P) fertilizers on U concentrations in shallow groundwater which is used as drinking water. P fertilizers originating from marine sedimentary phosphate rock deposits are naturally enriched with U in concentrations up to 220 mg/kg (Kratz et al. 2008). German fertilizers were found to contain U in range of 0.73–206 mg/kg, with a mean concentration of 61.3 mg U/kg (Smidt et al. 2011).

In addition to reports from literature (Schnug et al. 2008), meanwhile there exists further solid evidence that fertilizer-derived U contaminates groundwater and surface waters in Germany. The U concentration in German watercourses of agricultural areas proved to be about 10 times higher than in forest regions (0.08 and 0.8 µg/l U, respectively); the anthropogenic origin of U was reflected in higher factor values (principle component analyses) of the U-Se-Mg factor (Birke et al. 2008). A U content of >2 µg/l U was measured for stream waters of the loess areas in the eastern and northern Harz foreland, the Thuringian valley and the northern parts of the upper Rhine valley, south of Mainz and the regions north of Stuttgart, and partly in Mecklenburg-Western Pomerania (Schnug et al. 2008). In groundwater under arable land the U content was 3 to 17-fold higher than under forest soils (Huhle et al. 2008; Utermann and Fuchs 2008; Zielinski et al. 2000). The anthropogenic factor is underlined by elevated concentrations of other highly mobile elements applied regularly by fertilization such as B, Mg and K. These elements deliver a high contribution in the multiple regression analysis conducted for the evaluation of origins of the U content in tap water (Schnug et al. 2008). In addition, it has been shown that a close relationship exists between the omnipresent anthropogenic groundwater pollutant nitrate and U in groundwater and drinking water (Schäf et al. 2007; Popit et al. 2004).

This paper reports on the U content in 4092 tap waters and 155 groundwaters from Germany collected by several research groups over the last five years and discusses the possible origin of the U measured in the waters. This database is by far the largest available in the country, and has the additional distinction that each of the taps have been geocoded and the number of people with potential access to the water supplied from each tap was estimated. The database represents data on drinking water to which 76% of the entire German population has access. Besides the regional distribution of the U concentrations which largely agrees with the geological setting reported for mineral waters, clear evidence for anthropogenic influence on U concentrations in groundwater resulting from agricultural activities are discussed.

Materials and Methods

Tap and Groundwater Sampling and Analyses

The initial tap water sampling campaign of this study was conducted by the former Institute of Plant Nutrition and Soil Science of the Federal Agricultural Research Center (FAL-PB, now the Institute for Crop and Soil Science of the Federal Research Centre for Cultivated Plants (JKI)). With a view to maximum efficiency of the survey (fast, low cost, reliable) the samples were collected by individuals who responded to a chain letter action launched by FAL-PB in August 2006. A preliminary selection aimed primarily at collecting samples from German communities with more than 50,000 inhabitants. 266 individuals (97% of the entire group involved) delivered 471 samples representing 458 geocoded locations. In order to minimize the sampling error the participants were asked to employ a simple, but standardized sampling procedure: As sampling containers freshly emptied 500 ml polyethylene Diet Coke® bottles were stipulated, because their inward walls had been continuously in contact with an acidic, heavy metal free buffer solution (Schnug et al. 1996) and thus any sorption sites could be assumed to have been saturated with protons. Additionally, these bottles would contain no sugars which may otherwise have promoted the growth of microorganisms during the time elapsed from sample collection to laboratory delivery. The suitability of the sampling method has been checked in the laboratory through the recovery of U from aqueous U standard solutions found to be within the combined averaged repeatability of the methods used for U analysis (ICP-QMS and alpha spectroscopy) which amounts to ± 12.3% deviation of means (calculated from 95% confidence interval for the regression of 208 sample pairs from 17 different sources, $r^2 = 89.6\%$, Knolle 2008). The theoretical lower limit of detection (LLD) for ^{238}U by ICP-QMS is 2 ng/l (El-Himri et al. 2000), but practically the LLD was found to be 15 ng/l, which fits well with the 13 ng/l reported by UNEP (2001).

The entire data set has been enlarged with 299 data sets from Schäf et al. (2007) plus 2911 data sets collected from official federal states authorities by Foodwatch in 2009, 65 from the so called "German Environmental Survey on Children" (Schulz et al. 2008) and 168 from a survey by the Federal Office for Radiation Protection BfS (Beyermann et al. 2009). For U in total 4092 non redundant data sets were available. Of the 4092 data sets for U 9.5% originated from the FAL survey (Knolle 2008), 7.2% from Schäf et al. (2007), 1.7% Schulz et al. (2008), 77.5% from the Foodwatch (2009) and 4.3% from the BfS survey (Beyermann et al. 2009).

Groundwater samples ($N = 155$) were taken in the south-eastern part of Lower Saxony between Göttingen, Uelzen and Hannover from groundwater wells monitored regularly by the Lower Saxony State Authority for Water Resource Management (NLWKN) for the European Water Framework Directive (WRRL). The sampling area represents a region of intense cropping production, implying an intense use of mineral fertilizers. The pH, electric conductivity (EC) and the oxygen (O_2) content of groundwater was determined in a flow-through chamber after the stagnation water was discarded and the water volume of the well was exchanged three times. All water samples were collected in acid cleaned PE bottles, 0.2 μm pore size filtered prior to acidification to pH 2 with suprapure concentrated HCl for the U determination by ICP-QMS analyses. Data on major cation and anion contents were determined by the NLWKN using DIN-certified analytical methods.

Geocoding of Samples

A geological evaluation of the origin of U in the German mineral waters is given by Knolle (2008) and Knolle et al. (2011). In order to assess the population with potential access to a particular quality of tap water, the tap water data were geocoded by means of the 5 digit German postcode. The actual population number for each post code was retrieved from the most recent entry to the internet. However, the German postcode system has cases where the same town name has more than one postcode (e.g. Aachen: 52064, 52068, 5270, 5070) or the same postcode is attributed to more than one town (e.g. postcode 01651 stands for: Lampertswald, Schoenfeld, Tauscha, Thiendorf, Weissig). In the case where the same town name has more than one postcode and if for the subunits of the postcode no individual population numbers were available, the total population number for the town found on the internet (in this example for Aachen = 258,772) was divided by the number of data sets for the town (in this example 4). In the case where the same postcode was attributed to more than one town the most recent population number for the particular town was taken from the online available information.

If for the same postcode and the same town name more than one U measurement was available (e.g. two entries for post code 01665 "Kaelbschuetztal") and if the difference from the combined mean (value A: 0.15 µg/l U/value B: 6.20 µg/l U: combined mean = 3.23 µg/l U) was smaller than the 95% confidence interval given for this mean in table 2 (10% = ±0.32 µg/l U) they were entered as separate measurements in the data. Otherwise the values were averaged and entered as a single entry under either the same postcode or the same town name. The available dataset (N = 4092) covers 60,354,408 German inhabitants, which is 76% of the German population.

Results and Discussion

Population Exposure Assessment and Regional Distribution of U Concentrations in Tap Water

A new and unique approach of this research work is that it links the U concentration data within drinking waters to the number of individuals who consume them (see Material and Methods). Table 1 displays the exposure of German population to U in tap water. This value is designated in the following text as the "percentage of population exposed". This provides far more exact information on exposures than just a simple frequency analysis of the numbers of samples to concentrations. In addition to the "percentage of population exposed" for mean and P 95 occurrence also the "population weighted mean" is given in Table 2. The "population weighted mean" is the sum of all individual measurements multiplied with "percentage of population covered" * 0.01. It is de facto the average concentration of U to which the majority of the entire population covered is exposed. It is similar to the median value although that only indicates the most frequently occurring U concentration value in a data set with only a limited value in estimating exposures. In contrast the P 95 value gives the concentration which covers 95% of all samples. In consideration of the intensive and diverse discussion about permissible concentrations for U in drinking waters this table has a more detailed division of the concentration range. Table 1 shows that considering the larger data set with 4092 entries 3.3% of all samples covering 1.31% of the population observed is exposed to U concentrations above the anticipated critical value of 10 µg/l U (Vigelahn 2010).

Analysis of the smaller data set with "only" 750 entries indicates that 4.3% of the samples contained U concentration values above 10 µg/l, which is a higher

Table 1 Exposure of the German population to uranium in tap water (% of samples taken and % of total population exposed) (Citations for critical values: BfR 2005: 0.2 µg/l U; WHO 1988, EFSA 2008: 2 µg/l U; UBA 2005: 10 µg/l U, action level by UBA 2010: 20 µg/l U)

Concentration of U (µg/l)	Percentage of			
	Samples	Population exposed	Samples	Population exposed
	N=750	N=29,551,132	N=4092	N=60,354,408
<LLD*	25.2	31.9	3.70	11.7
LLD–0.2	19.1	20.6	20.5	27.1
0.2–2	31.7	37.4	58.2	49.4
2–5	10.0	5.29	10.1	7.64
5–10	9.60	4.58	4.20	2.85
10–15**	2.80	0.24	1.80	0.71
15–20**	0.50	0.02	0.60	0.22
>20**	1.10	0.02	0.90	0.38

* LLD = Lower limit of detection.
** Above maximum permissible value according to TrinkwV (2010).

Table 2 Mean, population weighted mean (PWM) concentrations related to the population coverage (PC = and % of the population exposed of the total population observed) for U in German tap waters

Element	Median		Mean		PWM		P95	
	µg/l	% PC	µg/l	% PC	µg/l	% PC	µg/l	% PC
U*	0.280	57.8	0.51	70.9	1.45	92.8	8.00	99.6
U**	0.500	53.4	0.68	67.1	1.67	86.6	7.21	97.7

* N=750.
** N=4092.

figure, but the percentage of the population exposed to this concentration is lower. Extreme exposure to concentrations above 20 µg/l U were observed in around 1% of the samples covering less than 0.4% of the population covered by the data set. In comparison UBA (2009) underestimates the figures for the number of cases with high U exposure, reporting that less than 0.6% of the population is exposed to concentrations higher than 10 µg/l U, with less than 0.1% exposed to extreme concentrations above 20 µg/l U.

The map displayed in Fig. 1 presents the regional distribution of the tap water samples (N=4097) in Germany. The six different symbols represent different U concentration ranges in µg/l (<0.5, 0.5–2, 2–5, 5–10, 10–20, >20), reflecting the different critical values for U in drinking water discussed since 2005 (BfR 2005: 0.2 µg/l U; WHO 1988, EFSA 2008: 2 µg/l U; UBA 2005: 10 µg/l U, action level by UBA 2010: 20 µg/l U). Uranium concentrations in waters show a distinctive regional distribution in Germany, originating from the strong local influence of the geological background. Information on the geological sources of this variation in

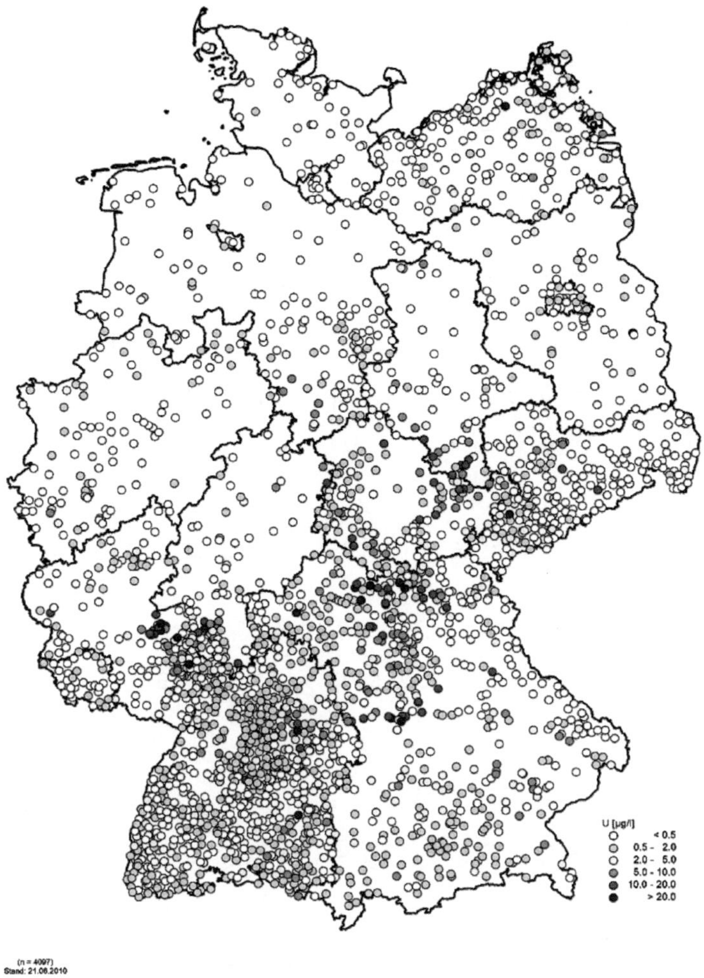

Fig. 1 Regional distribution of uranium concentrations in German tap water ($N = 4092$)

distribution were not the task of this research work, but details can be found in Birke et al. (2008), Birke et al. (2010) and Knolle et al. (2011).

Table 3 presents a breakdown of the 4092 data sets for uranium by the 16 individual federal states and displays the percentage of the total population exposed to water with different U concentrations. The table also allows an evaluation of the extent to which the individual states are represented in the entire survey. For all except two federal states the population covered in % of the total population investigated reflects more or less the % of the federal states contribution to the total German population of 80.6 million persons. Baden-Württemberg has a 21.4% coverage by the survey, compared with a contribution of only 12.4% to the German population, therefore is overrepresented by the survey when compared with North Rhine-Westphalia (16.6% survey coverage, 21.9% of the German population).

Table 3 Uranium concentrations in tap water from German federal states (Länder) sorted ascending with increasing population (pop.) access to waters with < 2 µg/l U. (Citations for critical values: BfR 2005: 0.2 µg/l U; EFSA 2008, WHO 1988: 2 µg/l U; UBA 2005: 10 µg/l U, action level by UBA 2010: 20 µg/l U)

							% of population covered				
Federal State	Area [km²]	Area [%]	Total pop. (*10⁶)	% of total pop.	No. of samples	Pop. covered in % of total pop. investigated	<0.2 µg/l U	0.2–2 µg/l U	225 µg/l U	5–10 µg/l U	10–20 µg/l U
Saxony-Anhalt	20443	5.7	2.80	3.5	93	2.5	36.6	28.0	11.8	15.1	8.6
Thuringia	16251	4.5	2.54	3.2	330	2.8	2.1	65.8	20	10.6	1.5
Hesse	21114	5.9	5.90	7.3	126	5.3	23.8	44.4	17.5	7.1	7.1
Bavaria	70553	19.7	11.60	14.4	630	11.7	21.1	49.4	14.5	7.6	5.7
Rhineland-Palatinate	19486	5.4	3.88	4.8	298	4.6	34.9	39.6	9.7	5.0	7.4
Baden-Württemberg	35751	10.0	10.00	12.4	1333	21.4	8.5	75.4	11.2	3.0	1.2
Schleswig-Holstein	15731	4.4	2.70	3.3	52	5.9	75.0	11.5	13.5	0.0	0.0
Saxony	18338	5.1	4.60	5.7	344	6.0	62.5	26.5	7.6	1.7	1.5
Mecklenburg-West Pomerania	23170	6.5	1.85	2.3	485	2.9	1.7	89.5	5.4	2.3	0.6
Lower Saxony	47343	13.2	7.48	9.3	150	6.8	62.0	29.3	6.0	2.7	0.0
North Rhine West-phalia	37070	10.3	17.69	21.9	109	16.6	45.0	48.6	5.5	0.9	0.0
Brandenburg	29053	8.1	2.67	3.3	74	2.1	68.9	25.7	7.4	0.0	0.0
Saarland	2570	0.7	1.08	1.3	38	1.5	43.6	51.3	5.1	0.0	0.0
Berlin	889	0.2	3.45	4.3	18	6.2	44.4	55.6	0.0	0.0	0.0
Bremen	404	0.1	0.68	0.8	6	0.9	50.0	50.0	0.0	0.0	0.0
Hamburg	755	0.2	1.69	2.1	8	2.9	100	0.0	0.0	0.0	0.0
Germany	358921	100	80.61	100	4095	74.9	22.3	59.0	10.9	4.5	2.5

In contrast to the downscaling overall evaluation by UBA (Vigelahn et al. 2010) the exposure situation in the individual German counties is quite different. This is of concern especially in regions where exposure can be to concentrations exceeding the anticipated critical value of 10 μg/l U and, or with exposure to extremes with more than 20 μg/l U. The largest population exposed to high U content in tap water and above 10 μg/l for Germany (3.3%) was found in Rhineland-Palatinate with 10.8% followed by Saxony-Anhalt with 8.6%, Bavaria with 7.4% and Hesse with 7.1%. The largest populations with an exposure below 0.2 μg/l were found in Hamburg (100%), Schleswig-Holstein (75%) and Brandenburg (68.9%). In all states other than Lower Saxony, North Rhine-Westphalia, Brandenburg, Berlin, Bremen, Saarland and Hamburg samples with extreme U concentrations exceeding 20 μg/l were found. The highest number of samples containing such high exposures were found in Rhineland-Palatinate (3.4%), followed by Bavaria (1.7%), Baden-Württemberg (0.7%), Saxony (0.2%) and Hesse (0.1%). These samples represent exposures to only 0.38% of the entire population covered by this survey, but this represents in absolute figures 229,347 persons in the population observed, or if transposed to the entire German population 306,318 persons for which a higher health risk due to extreme U exposure can be assumed.

Figure 2 shows a distinctive pattern for the relationship between the size of supply systems and the exposure to U in tap water. The risk of higher exposure to U in tap water appears to increase with decreasing size of a supply unit. A more detailed look into the frequency of the size of supply systems delivering water with more than 10 μg/l U in Fig. 2 shows that such supply units usually serve a population of up to 5,000 consumers, but high U concentrations may also occur

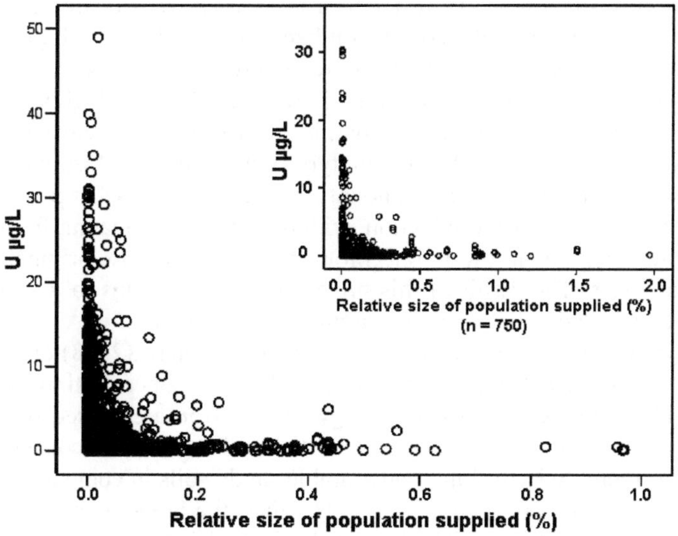

Fig. 2 U concentration in 4092 German tap water samples and % of the population exposed of the total population observed ($N = 60{,}354{,}408$) (Inlay graph: samples $N = 750$ and eposes population $N = 29{,}551{,}132$)

in supply units serving more than 60,000 consumers. The use of shallow groundwater for the supply of drinking water is common in small water supply systems, which are often located in rural areas. The distinct distribution pattern shown in Fig. 2 gives further indications for the impact of fertilizer-derived U on ground and tap water.

Indications for Fertilizer-Derived U in Groundwater

In Germany 73.3% of the drinking water abstraction is from groundwater. In areas where shallow groundwater resources are used for drinking water supply the hazard of contamination with the diffuse anthropogenic pollutants such as nitrate has been known for several decades. The nitrate content of groundwater and drinking water correlates directly with fertilization intensities including that of P loads in areas with high livestock density. In areas of intense cropping production both nutrients, N and P (and thus U), are favorably applied as mineral NP and NPK fertilizers. The proportion of P applied as NP and NPK fertilizers in relation to the total P input from mineral fertilizers increased from 42% in 1979/80 to 76% in 2007/08 (BMELV 1979–2007). In this study U was found in a concentration range of 0.002–16.9 μg/l with a mean concentration of 0.76 μg/l in groundwater in south-eastern Lower Saxony. Nitrate was found in concentrations ranged between 0.44 and 159 mg/l with a mean concentration of 14.5 mg/l. The results of the groundwater analysis ($N = 155$) could corroborate the hypothesis of the interrelation between U and NO_3 ($R = 0.285$, $p < 0.01$) in groundwater (3.5–56 m) in the area with intensive cropping production in the south-eastern part of Lower Saxony. A better correlation is indicated, if only shallow wells of depth of 0–15 m are statistically analyzed ($R = 0.331$, $p < 0.01$). Although the correlation between U and NO_3 is highly significant, the Pearson correlation coefficients are relatively low compared to the results found by Schäf et al. (2007). In their study, however, an area of highly intensive arable agriculture and horticulture predominates, with a near exclusively use of mineral fertilizers (mostly NPK, PK and NP types). In contrast to this, in the south-eastern part of Lower Saxony large amounts of nitrate are discharged to groundwater from intensive livestock production, with this nitrate originating from organic fertilizers containing much lower concentrations of U than those originating from mineral fertilizers. The result of this mixed fertilization leads to a still significant but weaker correlation between the parameters NO_3 and U.

These results are confirmed by studies from Utermann and Fuchs (2008) on U contents in soils ($N = 1000$) and Duijnisveld et al. (2008) on U concentrations in percolation water and shallow groundwater (pw/sgw) ($N = 50$) in areas used predominantly for agricultural production in Germany. It was shown that U concentrations in pw/sgw are found twice as high under arable sandy soils in comparison to sandy soils under forest. In dependency to the pH values the predominant species in pw/sgw under arable soils U carbonate complexes, which are known to be highly mobile in soils, while U in pw/sgw under forest consists mainly of the relatively immobile bivalent uranyl species (Utermann 2009).

Table 4 U and nitrate content in water of shallow and deep wells from two waterworks in southern Germany in 2009 (data reported to the authors by the local waterworks)

Location	Well type	U (μg/l)	NO_3 (mg/l)
Straubing	Shallow	2.8	40.0
	Deep	<0.2	2.8
Rehlingen	Shallow	10	22.0
	Deep	1.6	8.2

Increased U concentrations in surface waters were already measured in south-eastern Lower Saxony, which correlate with phosphate and therefore the use of P fertilizers (Fauth et al. 1985; Birke et al. 2006). This hypothesis is strengthened by results from shallow (7–9 m) and deep wells (70–100 m) of two waterworks in southern Germany (Table 4), which are located in regions without geogenically elevated U concentrations in ground and surface waters.

Particularly on arable soils with high carbonate contents, fertilizer-derived U has a high potential to be leached into deeper soil layers as a uranyl-carbonate complex. Laboratory batch experiments could show that additionally spiked U was not adsorbed by a leptosol soil (Smidt in prep.) and higher U concentrations in the easily mobilized fraction (Zielinski et al. 2006) of arable soils in comparison to unfertilized sandy and loess soils.

The mobility of U in soils and the leaching behavior on arable soils is much higher than that of other heavy metals and thus strongly resembles the transloca-tion dynamics of nitrate. Another possibility for accelerated U leaching from agri-cultural soils is suggested by Wu et al. (2011) who claim that immobile U (IV) might be oxidized through nitrate to easily leachable U (VI).

Model calculations estimated that the breakthrough of U in deeper soil layers and into groundwater can be expected after 50 years under climatic and soil condi-tions of northern Belgium if an annual load of 9 g/ha U is assumed. The equilib-rium U concentration for this combination of factors would be about 22 μg/l after that period of time (Jacques et al. 2008). This would indicate in return that a U concentration in this order of magnitude may be caused exclusively by fertilization.

Using data from tap waters from a geogenically U-poor area in the northern German lowland enabled us to estimate the percentage of tap waters which might be potentially contaminated with fertilizer-derived U. A conservative estimation regards 25% of drinking waters to be contaminated with U above 0.5 μg/l with a worst case scenario of tap water contamination above 0.1 μg/l indicated in 67% of the drinking waters in regions that are contaminated with fertilizer-derived U.

Conclusion

The presented tap water study, which covers 76% of the entire German popula-tion, showed that 1.3% or 1 million of the 80.6 million inhabitants in Germany are exposed to U concentrations in tap water which are higher than the anticipated threshold limit of 10 µg/l (Vigelahn et al. 2010). Noteworthy is that if the lower threshold limit of 2 µg/l recommended for nursing children diet (Vigelahn et al. 2010) would be considered, the percentage of population exposed to tap water at or above this limit would increase to 11.8% of the population (9.62 Million inhabitants). The data revealed also an increasing risk for exposure to elevated concentrations of U through tap water with decreasing size of the supply unit.

Besides the unavoidable geogenic contribution of U in tap water in some German regions, evidences for a diffuse anthropogenic contamination of tap and groundwater in areas of intense agricultural cropping production were found especially in shallow groundwater.

Uranium would be the one element for which transport in soils could be tracked most efficiently by using nuclear tracer techniques: but the labeling of fertilizers with even small amounts of LEU (low enriched U) for lab experiments would go against current legal regulations associated with the use of nuclear materials.

Acknowledgement The authors wish to express their most sincere gratitude to Dir. & Prof. Dr. Silvia Haneklaus, Dr. Sylvia Kratz and Dr. Holger Lilienthal (Institute for Crop and Soil Science, Julius Kühn-Institute, Braunschweig, Germany) who contributed significantly to the presented work, their right as co-author, however, being denied. The authors express their gratitude to the German Environmental Agency in Dessau for granting access to the water data of the "German Environmental Survey on Children".

References

Alexander J, Benford D, Cockburn A, Cravedi JP, Dogliotti E, Di Domenico A, Férnandez Cruz ML, Fürst P, Fink-Gremmels J, Galli C, Grandjean P, Gzyl J, Heinemeyer G, Johansson N, Mutti A, Schlatter J, van Leeuwen R, Van Peteghem C, Verger P (2009) Scientific Opinion of the Panel on Contaminants in the Food Chain. EFSA Journal 1018: 1–59

Beyermann M, Bünger T, Gehrcke K, Obrikat D (2009) Strahlenexposition durch natürliche Radionuklide im Trinkwasser in der Bundesrepublik Deutschland. BfS, Salzgitter, Dezember 2009; urn:nbn:de:0221-20100319945

Birke M, Rauch U (2008) Uranium in stream water of Germany. In: de Kok L J, Schnug E (eds.) Loads and Fate of fertilizer-derived uranium. Backhuys Publishers, Leiden. ISBN/EAN 978-90-5782-193-6, 79–91

Birke M (2006) Geochemischer Atlas der Bundesrepublik Deutschland. Vorabinformationen Uran in http://www.bgr.de (18.2.2008)

Birke M, Rauch U, Lorenz H (2009) Uranium in stream and mineral water of the Federal Republic of Germany Environ Geochem Health 31: 693–706

Birke M, Demetriades A, De Vivo B, Eds. (2010) Special Issue: Mineral Waters of Europe. J Geochem Explor 107: i–viii + 217–422

BMELV – Bundesministerium für Ernährung, Landwirtschaft und Verbraucherschutz (1979/2007) Statistisches Jahrbuch über Ernährung, Landwirtschaft und Forsten. Landwirtschaftsverlag GmbH, Münster-Hiltrup

BMU – Bundesministerium für Umwelt (2011) http://www.bmu.de/binnengewaesser/trinkwasser/doc/3259.php

DIN (2004) Die Trinkwasserprobenahme (DIN 38402; Teil 14, 38411-1, EN 25667-1,2,3, BGesundhBl 3/2004: 296–300

Duijnisveld WHM, Godbersen L, Dilling J, Gäbler H E, Utermann J, Klump J and Scheeder G (2009) Ermittlung flächenrepräsentativer Hintergrundkonzentrationen prioritärer Schadstoffe in Bodensickerwasser. UBA-Forschungsbericht FKZ 20472264

El-Himri M, Pastor A, de la Guardia M (2000) Determination of uranium in tap water by ICP-MS. Fresenius J Anal Chem 367: 151–156

Fauth H, Hindel R, Siewers U, Zinner J (1985) Geochemischer Atlas Bundesrepublik Deutschland. Verteilung in Wässern und Bachsedimenten. Bundesanstalt für Geowissenschaften und Rohstoffe, Komm. E. Schweizerbart'sche Verlagsbuchhandlung, Stuttgart

Foodwatch (2008) Uran im Leitungswasser – gefährlich für Säuglinge, 4.8.08, http://www.foodwatch.de/kampagnen__themen/mineralwasser/trinkwasser/index_ger.html

Foodwatch (2009) http://www.foodwatch.de/kampagnen__themen/mineralwasser/index_ger.html

Huhle B, Kummer S, Merkel B (2008) Mobility of uranium from phosphate fertilizers in sandy soils. In: de Kok LJ, Schnug E (eds.) Loads and Fate of fertilizer-derived uranium. Backhuys Publishers, Leiden. ISBN/EAN 978-90-5782-193-6, 47–56

Jaques D, Mallants D, Šimůnek J, Van Genuchten MT (2008) Modelling the fate of uranium from inorganic phosphorus fertilizer applications in agriculture. In: de Kok LJ, Schnug E (eds.) (2008) Loads and Fate of fertlizer-derived uranium. Backhuys Publishers, Leiden. ISBN/EAN 978-90-5782-193-6, 57–64

Knolle F (2008) Ein Beitrag zu Vorkommen und Herkunft von Uran in deutschen Mineral- und Leitungswässern, Dissertation, Technischen Universität Carolo-Wilhelmina zu Braunschweig, 161 pp., Digital Library Braunschweig 2009: www.digibib.tu-bs.de/?docid=00027200

Knolle F, Birke M, Hassoun R, Schnug E (2011) Uranium in German mineral waters – occurrence and origins. This volume

Popit A, Vaupotic J, Kukar N (2004) Systematic radium survey in spring waters of Slovenia. J Environm. Radioactivity 76: 337–347

Schäf M (2007a) Uran in Trinkwasserproben im Rhein-Neckar Gebiet, Diplomarbeit, Institut für Physikalische Chemie der Universität Heidelberg

Schäf M, Daumann L, Erdinger L (2007b) Uran in Trinkwasserproben im Rhein-Neckar Gebiet. Umweltmed. Forsch. Praxis 12: 315

Schnug E, Fleckenstein J, Haneklaus S (1996) Coca Cola Is It! The Ubiquitous Extractant for Micronutrients in Soil. Comm. Soil Sci. Plant Anal. 27: 1721–1730

Schnug E, Birke M, Costa N, Knolle F, Fleckenstein J, Panten K, Lilienthal H, Haneklaus S (2008) Uranium in German mineral and tap waters. In: de Kok LJ, Schnug E (eds.) Loads and Fate of fertilizer-derived uranium. Backhuys Publishers, Leiden. ISBN/EAN 978-90-5782-193-6, 91–110

Schulz C, Rapp T, Conrad A, Hünken A, Seiffert L, Becker K, Seiwert M, Kolossa-Gehring M (2008) Elementgehalte im häuslichen Trinkwasser aus Haushalten mit Kindern in Deutschland. Forschungsbericht 202 62 219 UBA-FB 001026. WaBoLu Hefte 04/08. ISSN 1862–4340, Dessau-Roßlau

Smidt GA, Landes FC, de Carvalho LM, Koschnisky A, Schnug E (2011) Cadmium and uranium in German and Brazilian phosphorous fertilisers. This volume

TWK (2008) Trinkwasserkommission des Bundesgesundheitsministeriums beim Umweltbundesamt, Stellungnahme der TWK zu einer Reihe häufig gestellter Fragen zu Uran im Trinkwasser, Online-Release: 3.11.08, http://www.umweltbundesamt.de/wasser/themen/trinkwasser/empfehlungen.htm

UBA – Umweltbundesamt (2008) Kinder-Umwelt-Survey 2003/06 Trinkwasser Elementgehalte im häuslichen Trinkwasser aus Haushalten mit Kindern in Deutschland. WaBoLu-Hefte 04/08, Forschungsbericht 202 62 219, UBA-FB 001026, Dessau-Roßlau

UBA – Umweltbundesamt (2009) Uran (U) im Trinkwasser: Kurzbegründung des gesundheitlichen UBA-Leitwertes (10 µg/l U) und des Grenzwertes für „säuglingsgeeignete" abgepackte Wässer (2 ng/l U) http://www.umweltdaten.de/wasser/themen/trinkwassertoxikologie/kurzbegruendunguran_leitwert.pdf

UNEP – United Nations Environmental Programme (2001) Depleted uranium in Kosovo – post conflict environmental assessment. United Nations Environmental Programme, Nairobi, Kenya

Utermann J, Fuchs M (2008) Uranium in German soils. In: de Kok LJ, Schnug E (eds.) (2008) Loads and Fate of fertilizer-derived uranium. Backhuys Publishers, Leiden. ISBN/EAN 978-90-5782-193-6: 33–46

Utermann J, Duijnsveld WHM, Godbersen L (2009) Uran in Böden und Sickerwässern – gibt es Indizien für eine Phosphordüngerbürtige U-Anreicherung. Mitt. Dt. Bodenkdl. Ges. 19: 4 pp.

Vigelahn L, Schmoll O, Hermann H-D, Chorus I (2010) Rund um das Trinkwasser. 88 pp., Dessau-Roßlau

Wu WM, Carley J, Green SJ, Luo J, Kelly SD, van Nostrand J, Lowe K, Mehlhorn T, Caroll S, Boonchayanant B, Löffler FE, Watson D, Kemner KM, Zhou J, Kitanidis PK, Kostka JE, Jardine PM, Criddle C (2010) Effects of nitrate on the stability of uranium in a bioreduced region of the subsoil. Environ Sci Technol 2010, 44, 5104–5111

WHO (2005) Uranium in drinking-water, Background document for development of WHO guidelines for drinking-water quality, WHO/SDE/WSH/03.04/118

Zielinski RA, Simmons KR, Oremn WH (2000) Use of 234U and 238U Isotopes to identify fertilizer derived uranium in the Florida Everglades Appl Geochem 15: 369–383

Zielinski RA, Oremand WH, Simmons KR, and Bohlen PJ (2006) Fertilizer-derived uranium and sulfur in rangeland soil and runoff: A case study in central florida. Water Air Soil Poll. 176: 163–183

Speciation of Uranium – from the Environment to Living Cells

Gerhard Geipel, Katrin Viehweger, Gert Bernhard

Abstract. Depleted uranium used a ammunition corrodes in the environment forming mineral phases and then dissolved uranium species like uranium carbonates (Schimmack et al. 2007) and hydroxides. These hydroxide species were contacted with plant cells (canula). After 24 h contact time the cells were fractionated and the uranium speciation in the fraction was determined by TRLFS (time resolved laser-induced fluorescence spectroscopy) at room temperature as well at 120 K. It could be shown that the uranium speciation in the fractions is different to that in the nutrient solution. Comparison of the emission bands with literature data allows assignment of the uranium binding forms.

Introduction

Uranium is an ubiquitous element. Besides this, depleted uranium ammunition as well as uranium mining and milling and manifold other use of uranium leads to an increase of uranium contamination in the environment. Application of laser-induced and time-resolved methods allow the direct determination of uranium speci-

Gerhard Geipel
Helmholtz-Zentrum Dresden Rossendorf, Institute of Radiochemistry, Bautzener Landstr. 400, D-01328 Dresden

Katrin Viehweger
Helmholtz-Zentrum Dresden Rossendorf, Institute of Radiochemistry, Bautzener Landstr. 400, D-01328 Dresden

Gert Bernhard
Helmholtz-Zentrum Dresden Rossendorf, Institute of Radiochemistry, Bautzener Landstr. 400, D-01328 Dresden

B. Merkel, M. Schipek, *The New Uranium Mining Boom*
DOI 10.1007/978-3-642-22122-4

ation at extremely low concentrations. This behavior can be directly observed due to the extraordinary luminescence properties of uranium-(VI).

Application of Cryo-TRLFS allows to determine additionally species which are strongly quenched at room temperature (Wang et al. 2004; Geipel 2006).

The uptake of uranium by plants is mainly described by transfer factors. Nevertheless, it is of common interest in which binding forms uranium exits within plants (Günther et al. 2003) and in which forms uranium may be bond inside plant cells.

Experimental

Canola cells (*brassica nappus*) were grown in a standard nutrition solution. 48h before the disruption the nutrition solution was exchanged by a half concentrated nutrition solution containing no phosphate ions and 24 h before disruption to the nutrition solution uranium(VI) was added to a concentration of 1×10^{-5} M. The pH of the solution was not changed in order to guaranty optimize living conditions for the cells. At the beginning of the experiment the canola cells were disrupted by ultrasonic and high pressure homogenization. Afterwards a differential centrifugation was performed with 15 min at 1000 g, 15 min at 15,000 g and 60 min at 50,000 g. The supernatant contains the cytoplasma and the content of the vacuole. The resulting pellet was dissolved and by two phase partitioning divided into the plasma membrane vesicle fraction (upper fraction, polyethylene glycol 3550) and a fraction containing the other membranes (lower fraction, dextran T-550). The residual solid fraction contains all non-soluble cell compartments like the cell nuclei (Viehweger 2010).

Results and Discussion

Dissolved uranium under natural, oxidizing conditions exists as hydroxide (Sachs unpublished), carbonate (Wang 2004; Geipel 2006) or phosphate (Brendler 1996; Scapolan 1998) species. In Fig. 1 for example the spectrum of $(UO_2)_3(OH)_5^-$ is shown. As uranium carbonate species do emit very little luminescence at room temperature often cryo-techniques were used to increase the luminescence yield (Fig. 2).

In previous experiments in order to study the corrosion of depleted uranium ammunition under nearly natural conditions (Schimmack 2007) besides the formation of mineral phases (i.e. sabugalite) in the seepage water dissolved species were assigned to be uranium(VI)-carbonate and uranium(VI)-hydroxide. The spectrum of the hydroxide containing water is shown in Fig. 3.

To estimate the uranium speciation in the nutrition solution luminescence spectra were obtained from the phosphate free nutrition solution, where uranium(VI)

Fig. 1 Luminescence spectrum of $(UO_2)_3(OH)_5^-$ at room temperature

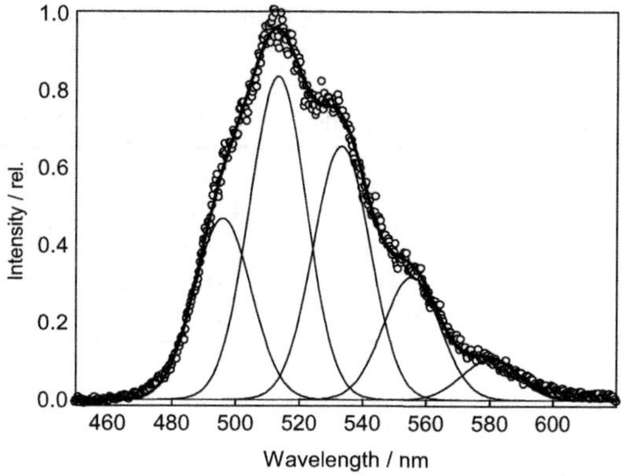

Fig. 2 Luminescence spectrum of $UO_2(CO_3)_3^{4-}$ at 150 K

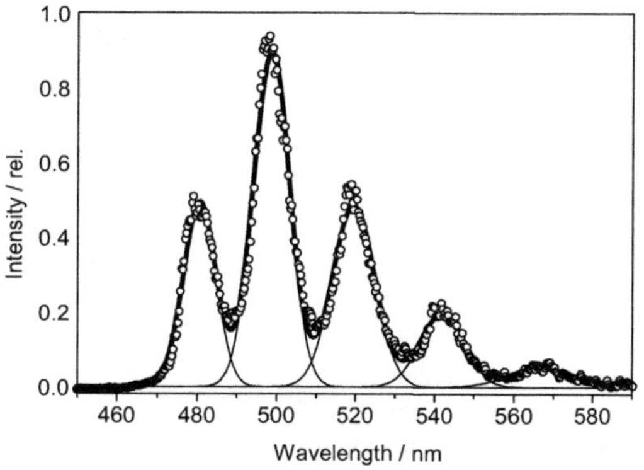

Fig. 3 Luminescence spectrum of $(UO_2)_3(OH)_5^-$ in seepagewater (room temperature)

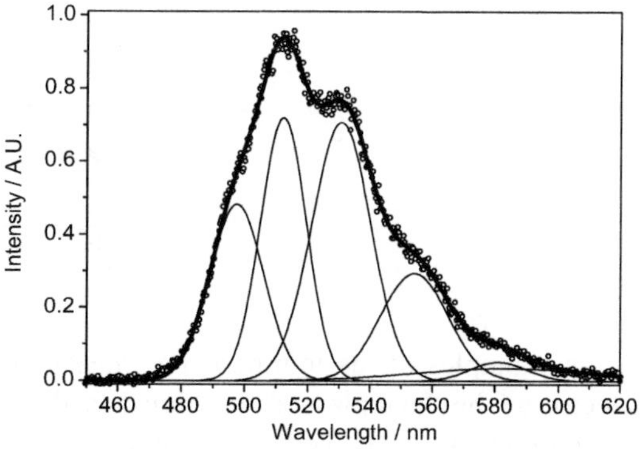

was added to a concentration of 1 x 10^{-5} M. The resulting spectrum is shown in Fig. 4, reveal clearly the uranium speciation to be assigned as $(UO_2)_3(OH)_5^-$.

The emission wavelengths of the several spectra are summarized in Table 1. The luminescence lifetimes were not estimated in this study due to strong and mainly unknown quench effects of other components in the several solutions.

Table 1 Emission wavelength of several synthetic and natural uranium species

Synthetic species			Natural species	
Hydroxide	Phosphate	Carbonate	Seepage water	Nutrition solution
495.9	498.9	480.4	497.5	498.8
513.2	520.2	498.6	512.2	515.0
533.3	543.8	519.3	530.7	535.9
555.5	568.9	541.6	554.2	559.2
580.3	597.8	567.3	581.1	574.2

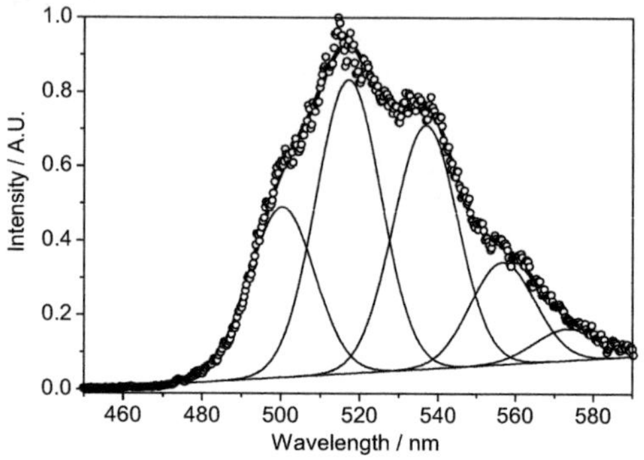

Fig. 4 Luminescence spectrum of the nutrition solution at room temperature

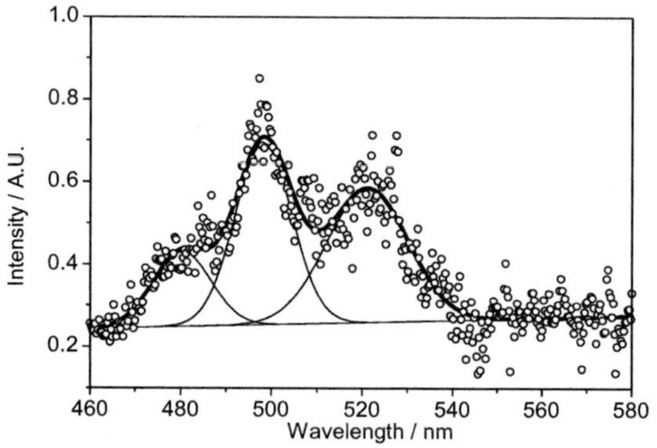

Fig. 5 Luminescence spectrum of the "cytoplasma" fraction at room temperature

As examples of the luminescence measurements of the cell compartments the spectra of the "cytoplasma" fraction and the "cell nuclei" fraction at room temperature and at 150 K were shown in the Figs. 5–8. The estimated emission wavelength were summarized together with data obtained from synthetic species given in the literature in Table 2.

The spectra show clearly the change of the uranium speciation from the nutrition solution, which can be seen as the pore water in soil under natural conditions to the cell compartments.

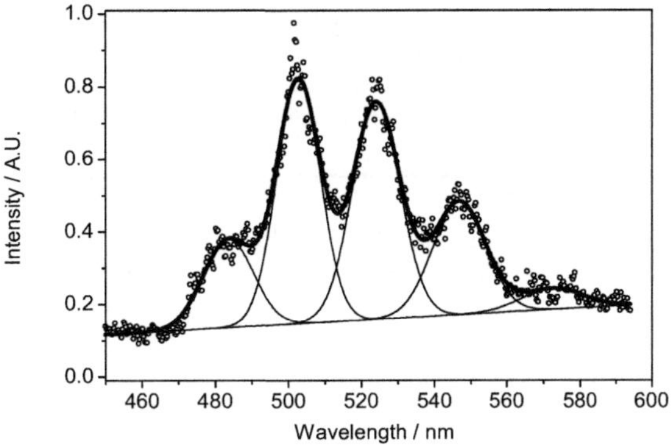

Fig. 6 Luminescence spectrum of the "cytoplasma" fraction at 150 K

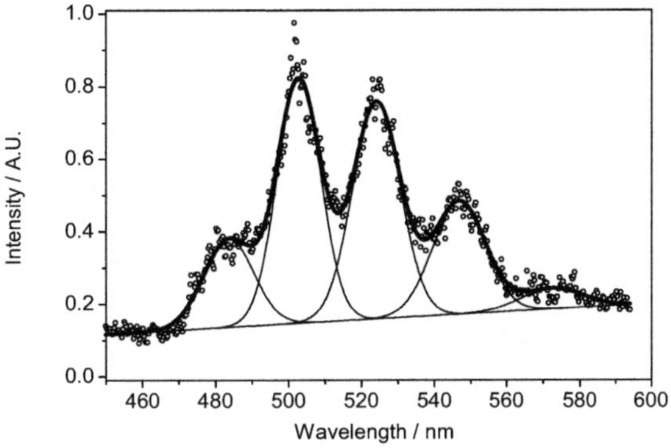

Fig. 7 Luminescence spectrum of the "cell nuclei" fraction at room temperature

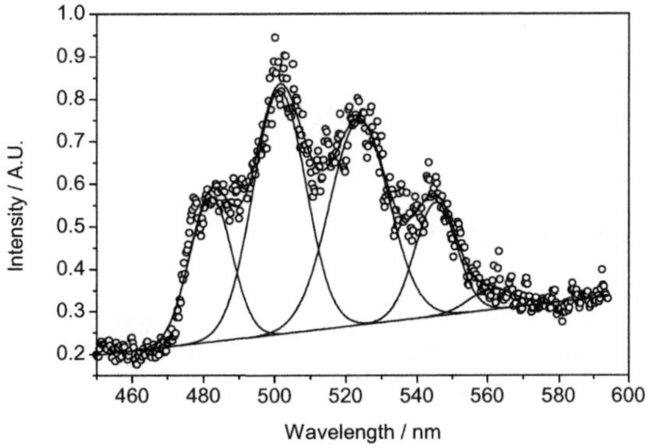

Fig. 8 Luminescence spectrum of the "cell nuclei" fraction at 150 K

At 150 K the spectra show nearly the same mission bands for both fractions. As for the cell nuclei the binding form can be assigned to the carboxyl group, we cannot exclude that under freezing conditions additionally the NH_2-group may be involved. Additionally phosphate is also involved in binding of the UO_2-moiety.

Table 2 Emission wavelength of uranium species in cell compartments and comparison with synthetic species in nm

Cytoplasma		Cell nuclei		Lipopoly-saccharides	O-Phospho-L-Threonine	Malonate
293 K	150 K	293 K	150 K			
480.3	483.6	479.4	482.0	483.6	484.0	479.0
498.2	502.8	497.2	501.3	499.5	502.0	496.0
521.1	524.0	517.4	523.7	520.6	523.0	517.0
	546.9	538.9	545.3	543.9	547.0	542.0
	572.4	567.1	561.2	568.9	573.0	566.0

Summary

Plant cells are able to incorporate uranium. The speciation of uranium after contact with plant cells is different to that in the culture medium.

In the cell nuclei and in the cytoplasm fraction different uranium species were detected. The uranium is bond probably COO^- groups (Brachmann 2002) in the cell nuclei fraction and to ligands comparable to lipopolysaccharide (Barkleit 2008) in the cytoplasm.

Cryo-TRLFS results in slightly shifted emission spectra if compared to room temperature measurements. This may be due to reduced quench effects, so that species may be detected which do emit much less luminescence at room temperature. The detected species show luminescence properties with ligands comparable to O-Phospho-L-Threonine (Günther 2006). It cannot be excluded that under these conditions besides carboxylic groups also the NH_2-group participates in the interaction with uranium. However, due to contact to the atmosphere expected carbonato species were not observed.

References

Barkleit A, et al. (2008) Dalton Trans 2879–2886
Brachmann A, et al. (2002) Radiochim Acta 90:147–149
Brendler V, et al. (1996) Radiochim Acta 74:75–79
Geipel G (2006) in Vij D R Handbook of appl solid state spect. 577–593
Günther A, et al. (2003) Radiochim Acta 91:319–328
Günther A, et al. (2006) Radiochim Acta 94:845–851
Sachs S, unpublished results
Scapolan S, et al. (1998) J Alloys Compd 271–273:106–111
Schimmack W, et al. (2007) Radiat Environ Biophys 46:221–227
Viehweger K, Geipel G (2010) Environ Exp Botany 69:39–46
Wang et al. (2004) Environ Sci Technol 38:5591–5597

The Implications of New Legislation on NORM-Generating Industries

David Read

Abstract. This paper provides a brief overview of industrial NORM (Naturally Occurring Radioactive Material) sources and attempts to place the volumes and activities generated in context by comparison to nuclear arisings. Greater regulatory scrutiny and new legislation has the potential to substantially increase the costs of treatment and disposal. It is argued that waste management strategies for both nuclear and NORM wastes should be based on the hazards the materials pose rather than origin and that some Public provision could be made to ensure the continued competitiveness of NORM-affected industries.

Introduction

The volume of radioactive waste created each year by the nuclear industry is dwarfed by that from other sources, notably the production and combustion of fossil fuels and exploitation of industrial minerals. Some of the scales affecting production equipment contain activity levels corresponding to those of intermediate level nuclear waste (ILW), defined in the United Kingdom as >4 GBq/t alpha and/or >12 GBq/t beta-gamma. These materials require specialized treatment and are usually dealt with on a case by case basis (e.g. Read et al. 2004). The majority of industrial waste is less active but, nevertheless, requires careful handling and disposal to limit exposures.

Formal recognition of the problem by legislators has taken a long time and it is unlikely that the Basic Safety Standards now being re-drafted (IAEA 2010; CEC 2009) will be able to address all eventualities. Similar difficulties have been en-

---

David Read
Chemistry Department, Loughborough University, Leics., LE11 3TU, UK

B. Merkel, M. Schipek, *The New Uranium Mining Boom*
DOI 10.1007/978-3-642-22122-4

countered by national regulators owing to the diversity of sources and hence, types of waste arising. However, the main concern, at least in the UK, has been gross under-estimation of the extent of the problem by Government. In compiling its estimate of the country's low level waste inventory for the next 100 years, Defra (2005) assigned a volume of just 410,000 m^3 for non-nuclear wastes as opposed to around two million m^3 for nuclear arisings (excluding Sellafield). One industrial process in the north east of England generates this quantity of radioactive waste each year.

This paper provides a brief overview of industrial NORM (Naturally Occurring Radioactive Material) sources and attempts to place the volumes and activities generated in context by comparison to nuclear arisings. The types of waste found in the oil & gas and mineral processing industries are discussed and methods used for treatment and disposal of residues are outlined.

NORM Industries

The international framework for radiological protection was developed primarily to meet the demands of the nuclear and medical sectors. It focuses on the use of artificial radionuclides; exposure to natural sources is considered mainly in terms of uranium mining and radon. This omission was partly addressed by Directive 96/29 Euratom (Basic Safety Standards, Title VII; CEC 1996), which for the first time placed an obligation on Member States to identify work activities that could cause significant radiation exposures caused by enhanced concentrations of natural radioactivity.

The new draft Basic Safety Standards (IAEA 2010; CEC 2009) takes the above further by identifying a number of "NORM industries" with defined thresholds for notification, clearance and exemption. The industries currently listed are shown in Table 1. Member States are free to include additional industries depending on local needs, for example the extraction of kaolinite (China Clay) has been added to the list in draft UK legislation. Similarly, the national authorities in Sweden and Finland may decide that their large paper and pulp industries require regulatory control.

Table 1 Sectors designated as "NORM Industries"

Extraction of REE from monazite	Production of phosphate fertilizers
Production of Th compounds	Cement production, maintenance of clinker ovens
Processing of Nb/Ta ore	Coal-fired power plants, maintenance of boilers
Oil and gas production	Phosphoric acid production,
Geothermal energy production	Primary iron production,
TiO_2 pigment production	Tin/lead/copper smelting,
Thermal phosphorus production	Groundwater filtration facilities
Zircon and zirconium industry	China Clay refining (UK)

The term NORM or TENORM (Technologically Enhanced NORM) is defined somewhat loosely in the literature and the above list is provisional. Often it is taken to include all mining residues and some military wastes, for example radium contamination from luminizing operations. Strictly, depleted uranium (DU) could also be considered though much of the DU used in weapons is derived from reprocessing and contains artificial nuclides (Trueman et al. 2004).

Volumes and Activity Levels

The range of industrial processes that can lead to NORM being generated (Table 1) accounts for the wide variety of phases encountered. A detailed description of them all is beyond the scope of this short paper; however, the most common scale minerals and associated nuclide enrichments are common to a number of the key sectors, as the following examples, taken from the extractive industries, show.

There are more than 40,000 oil & gas fields worldwide; the majority of which are believed to be affected by NORM contamination (e.g. IAEA 2003a,b). Workers in the United States have estimated that 100t of scale is generated on average per well each year (American Petroleum Institute 1992). Though this figure will vary substantially with geology, production methods and age of the installation, the potential inventory is enormous. In Western Europe, production is dominated by the North Sea with around 500 platforms, each fed by several wells. The residues removed from contaminated equipment fall into two main categories. Those from the northern North Sea (NNS) comprise deposits of crystalline, sulfate-rich scale; mostly barite-celestine, sometimes inter-grown with carbonates and/or iron oxy-hydroxides (Fig. 1a). Radium isotopes dominate; the relative proportion of ^{226}Ra and ^{228}Ra depending on the source rocks. Published data from a number of North Sea operators indicate mean activities of 20 and 10 Bq/g for the two isotopes, respectively, though the range is large (Sniffer 2005). The corresponding annual arisings are in the region of 400 GBq ^{226}Ra and 200 GBq ^{228}Ra.

Scales from the southern sector comprise thinner and less persistent metallic deposits containing lead, often of a high purity (Fig. 1b). Radioactive ^{210}Pb occurs by isotopic substitution reaching kBq/g levels in some locations (Worden et al. 2000; Ceccarello et al. 2004). The lead may be oxidized or altered by reaction with circulating fluids to form, for example, hydroxy-carbonates such as hydrocerussite or arsenates; arsenic may be present at levels up to 10 wt%. Several toxic heavy metals including mercury are more abundant further east, particularly in German onshore gas wells where an unusual lead-mercury amalgam, termed altmarkite, has been recorded (Ceccarello et al. 2004).

Radioactive wastes from the central area contain signatures characteristic of both the northern and southern sectors of the North Sea fields. Individual samples differ in the relative proportion of barite-celestine and lead/litharge or galena, if conditions are highly reducing. This accounts for the activities of the radium iso-

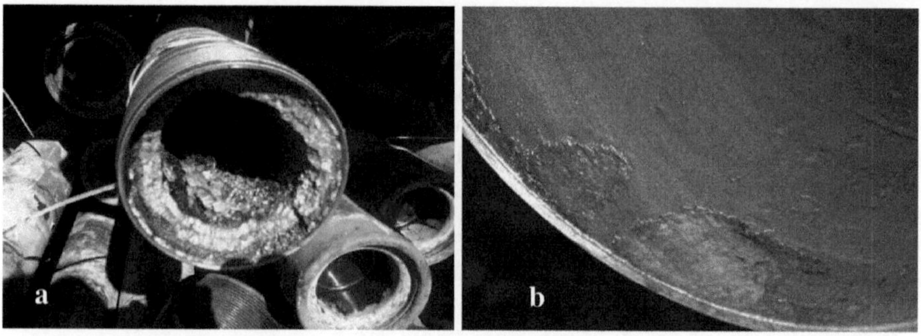

Fig. 1 Scale on pipe work from the northern (**a**) and southern (**b**) sectors of the North Sea

topes and ^{210}Pb, respectively. The above phases are found in other oil & gas facilities elsewhere in the world (e.g. Al Masri and Aba 2005).

Barite is also the principal scale mineral in metalliferous (Read et al. 2002; IAEA 2003b) and non-metalliferous (e.g. Read et al. 2004; Beddow et al. 2006) mineral ore processing. Almost all metal smelting and refinement processes have the potential to generate NORM, including iron and base metal production (IAEA 2003a,b; Paschoa and Steinhäusler 2010). However, the issue is most acute with rare earth, titanium and zirconium ores owing to the naturally high uranium and/or thorium contents in the source materials. Monazite may contain up to 15% thorium and was exploited for this metal before the rare earth elements themselves became such a valuable resource (Read et al. 2002). The more soluble bastnasite, which is the main commercial ore today, also contains substantial uranium and thorium. Concentrations are lower in the case of rutile and ilmenite (titanium) or zircon, but significant enrichment occurs in process residues. Again, barite is an important phase, often containing >1000 Bq/g radium isotopes and, in extreme situations, >100 kBq/g. Titania plants in eastern England between them generate in excess of one million tons of low activity (<10 Bq/g) NORM residues each year together with a much smaller quantity of high activity scales.

Two key industries in non-metalliferous mineral ore processing are production of phosphate (e.g. Beddow et al. 2006) and China Clay (Read et al. 2004). A typical 1,000 t/day phosphate plant can generate >200,000 t phosphogypsum per annum, as evidenced by the huge disposal heaps found, for example, in Florida (Hilton 2007). It is estimated that 30 billion t exist worldwide much of it re-used in building materials. These deposits are enriched in radium isotopes (≤20 Bq/g) whereas uranium itself is mainly found in the final products, such as fertilizers. Workers in Germany have estimated that industrial fertilizers typically add 3 kg/ha to intensively farmed land per annum (Schmitz-Feuerhake and Bertell 2008).

China Clay (kaolinite) is mined in several European countries from primary, igneous sources. Production in one area, south west England, is currently around one million ton per annum, though volumes were much higher in the past. As the yield is low (~10%), 10 million t of source rock need to be excavated to produce this amount of useable clay. Such a large volume of raw material means that there may be a substantial inventory of even ultra-trace components passing through the plants. Each stage of refining has the potential to accumulate enhanced activities of selected radioelements; radium by co-precipitation in barite, uranium by co-precipitation with iron oxy-hydroxides, lead and polonium by volatilization. Activity levels in scale deposits frequently reach several thousand Bq/g, requiring bespoke methods to decontaminate equipment and deal with active residues (Read et al. 2004).

Waste Management

Much of the scale arising from decontamination operations in the offshore oil & gas industry is disposed of at sea, in a practice that is attracting increasing scrutiny owing to the OSPAR Convention on sea dumping (OSPAR 1998). Cessation of sea disposal would substantially increase the already large volumes of scaled equipment (tubulars, vessels etc.) and contaminated sludge that requires treatment onshore, the former usually by ultra high pressure (UHP) water and/or grit blasting. Two commercial operators carry out the bulk of this work in the UK and there are similar facilities in Norway, Germany and the Netherlands. A major increase in capacity is foreseen in connection with decommissioning North Sea platforms. UHP jetting is also the preferred decontamination method employed in the China Clay industry (Read et al. 2004).

Chemical dissolution has been used where there is potential for re-injection of slurry, such as at the onshore Wytch Farm asset in southern England (Worden et al. 2000). The disadvantages of this method stem from the low solubility of sulfate and lead-rich scales, necessitating the use of aggressive reagents. In addition to problems of handling on site, there are concerns that re-injecting strongly acidic solutions into the subsurface could lead to the mobilization of other contaminants and impact on local water resources.

Where re-use of equipment is not practicable, metals can be recycled following smelting. Suitable facilities exist, for example in Sweden and Germany, where Siempelkamp operates an induction furnace in Krefeld. The GERTA plant can treat up to 2000 t of metal scrap per annum, including NORM contaminated pipe work from oil and gas producing facilities. Slag is released for use in road construction, subject to limits on activity levels and consequential dose constraints, or landfill. This contrasts with the approach in Sweden where waste residues are returned to origin.

So-called “soft wastes” can be dealt with by incineration or thermal desorption. As with smelting, there is a still a need to dispose of slag and fly ash residues and this is normally the most intractable part of the NORM waste management process. Few countries have bespoke disposal facilities for NORM. Historically, decommissioning waste has either been sent to nuclear repositories, incurring disproportionate costs and expending a valuable national resource (e.g. Capper Pass, Brent Spar) or consigned to landfill following special dispensation by the regulators. Three countries that do possess well engineered disposal routes for NORM are Germany, Canada and Norway. The WETRO site, near Bautzen, approximately 50 miles east of Dresden is a former open-cast mine which has been developed as a waste disposal site servicing not only the oil and gas sector but the iron and steel, power generation and building industries. It is a significant operation with an annual capacity of more than 400,000 t. In addition to accepting German wastes, consignments have been received from several other EU countries including Ireland, Austria and Poland. The NORMCAN operation in Alberta, Canada was specifically designed for the oil & gas industry as is the Stangeneset site in Norway, which opened in 2008. Plans for similar facilities at Stoneyhill in Scotland are well advanced.

Financial Provisions and Concluding Remarks

In the past, less than optimal practices have been adopted whereby organizations have disposed of NORM wastes to the marine environment, terrestrial water courses or unlined, municipal landfills. Increasing awareness of the health risks posed by such materials has led to greater regulatory scrutiny and new legislation. This should improve standards but, undoubtedly, has the potential to substantially increase the costs of treatment and disposal.

As noted above, the volume of NORM generated each year by industry is enormous and far exceeds that produced by nuclear activities. Accumulated total disposals at the only operating nuclear repository in the UK, the Low Level Waste Repository (LLWR) in Cumbria, amount to only 1 million m^3 and projected arisings of LLW from decommissioning all sites is estimated at only $4.5 \times 10^6 \, m^3$ (NDA 2010).

However, unlike the nuclear sector, where the costs are borne by the taxpayer, the burden of remediating NORM contamination invariably falls on private concerns. Several countries fund nuclear clean-up via a levy on electricity prices but no such provision is made for NORM. In the UK, the cost of nuclear decommissioning is estimated at approximately £75 billion, the bulk of which is earmarked for disposal of relatively innocuous soil and rubble. A case could be made for diverting the least active nuclear wastes away from expensive, highly engineered facilities to inert landfills. Conversely, it is argued that some of the Public funds available could be re-allocated for dealing with the more hazardous NORM wastes. A strategy in which decisions for treatment and disposal are based on the actual hazards the material pose rather than origin would offer both enhanced protection for the Public while ensuring the continued competitiveness of key industries.

Acknowledgements The author would like to thank N. Terry (Reclaym Ltd), P. Reid (Scotoil Ltd), L. Howden (Conoco-Phillips plc), A. Parvin (BP plc), G. McNulty (Huntsman plc), G. Beatty (MEL Chemicals) and S. Chandler (DECC) for their valuable input at various stages of this work.

References

Al Masri M, Aba A (2005). Distribution of scales containing NORM in different oilfield equipment (2005) App Rad Isotope 63: 457–463

American Petroleum Institute (1992) Bulletin on management of Naturally Occurring Radioactive Materials (NORM) in oil and gas production. API Bulletin E2

Beddow H, Black S, Read D (2006) Characterisation of scale from a former phosphoric acid processing plant. J Env Rad 86: 289–312

CEC (1996) Council Directive 96/29/Euratom. Laying down Basic Safety Standards for the protection of the health of workers and the general Public against the dangers arising from ionizing radiation. OJ EC L159, 39

CEC (2009) European Commission services considerations with regard to natural radiation sources in BSS Directive. Draft

Ceccarello S, Black S, Read D, Weiss H, Schubert M, Kunze C, Grossmann J (2004) Radioactive scales from a natural gas production facility in the Altmark region, Germany. App Mineral, Special Ed. Pecchio et al. (eds.): 395–398

Defra (2005) The low level radioactive waste inventory. Proc. LLW management policy workshop, London

Hilton J (2007) Dealing with a phosphate industry NORM safety concern at source: Regulating phosphoric acid and phosphogypsum as co-products. Proc. NORM V, Seville

IAEA (2003a) Radiation protection and the management of radioactive waste in the oil and gas industry. Safety Reports Series No. 34

IAEA (2003b) Extent of environmental contamination by naturally occurring radioactive material (NORM) and technological options for mitigation. Technical Reports Series No. 419

IAEA (2010) International Basic Safety Standards for protection against ionising radiation and for the safety of radiation sources. Draft Safety Requirement DS370, draft 3.0

NDA (2011) The 2010 UK Radioactive Waste Inventory. Report URN 10D/985 NDA/ST/STY(11)0004

OSPAR (1998) Convention for the Protection of the Marine Environment of the NE Atlantic. Options for the objectives and timeframe of OSPAR's strategy with regard to radioactive substances. Ministerial meeting of the OSPAR Commission, Sintra

Paschoa A, Steinhäusler F (2010) TENR – Technologically Enhanced Natural Radiation (Radioactivity in the Environment) Elsevier, 244p

Read D, Andreoli M, Knoper M, Williams C, Jarvis, N (2002) The degradation of monazite: Implications for the mobility of rare earth and actinide elements during low temperature alteration. Eur J Mineral 14: 487–498

Read D, Rabey B, Black S, Glasser F, Grigg C, Street A (2004) Implementation of a strategy for managing radioactive scale in the China Clay industry. Min Eng 17: 293–304

Schmitz-Feuerhake I, Bertell R, (2008) Radiological aspects of uranium contamination. In Loads and Fate of Fertilizer-derived Uranium. LJ De Kok and E Schnug (Eds.), Backhuys

Sniffer (2005) A review of the application of "Best Practicable Means" within a regulatory framework for managing radioactive wastes. Final Report Project UKRSR05

Trueman E, Black S, Read D (2004) Characterisation of depleted uranium (DU) from an unfired CHARM-3 penetrator. Sci Tot Env 327: 337–340

Worden R, Manning D, Lythgoe P (2000) The origin and production geochemistry of radioactive lead (^{210}Pb) in NORM contaminated formation waters. J Geochem Explor 69–70: 605–699

References

Low-Lying States for the $^{103,\,105}$Mo Isotopes

O. Jisar, J. Inchaouh, M.K. Jammari, A. Morsad, H. Chakir

Abstract. Low-lying levels of the isotope ^{103,105}Mo are investigated in the framework of the quasiparticle-phonon coupling plus rotor method. The intrinsic and collective states are determined by using the deformed mean field of Nilsson, the monopole-pairing interaction (BCS) and the quadrupole-quadrupole force. Microscopic structure of quadrupole phonon is given from the Tamm-Dancoff Approximation. The two effects of recoil and Coriolis forces are included with the hypothesis of a symmetric rotational motion. The levels schemes of $^{103,\,105}$Mo are discussed and compared with the experimental data.

O. Jisar
Laboratoire de recherche Subatomique et applications, Université Hassan II, Faculté des sciences Ben M'Sik, Département de Physique, BP. 7955, Casablanca, Maroc

J. Inchaouh
Laboratoire de recherche Subatomique et applications, Université Hassan II, Faculté des sciences Ben M'Sik, Département de Physique, BP. 7955, Casablanca, Maroc

M.K. Jammari
Université Hassan II, Faculté des sciences Ain chock, Département de Physique, BP. 5366, Casablanca, Maroc
E-mail: Omar05jdair@yahoo.fr

A. Morsad
Laboratoire de recherche Subatomique et applications, Université Hassan II, Faculté des sciences Ben M'Sik, Département de Physique, BP. 7955, Casablanca, Maroc

H. Chakir
Laboratoire de recherche Subatomique et applications, Université Hassan II, Faculté des sciences Ben M'Sik, Département de Physique, BP. 7955, Casablanca, Maroc

B. Merkel, M. Schipek, *The New Uranium Mining Boom*
DOI 10.1007/978-3-642-22122-4

Introduction

Several experimental investigations have been made for low-lying states in A ~ 100 mass of transitional region. So this has given important information about the characterization of both collective and intrinsic excitation for odd-A nuclei. A large variety of shapes may be observed in Mo and Zr nuclei of this region when the number of neutrons increases from $N = 58$ to $N = 64$. The lighter isotopes are rather spherical. It is also well established that three shapes co-exist in the transitional odd-A, $N = 59$, Sr and Zr nuclei. For $N > 59$, strongly deformed axially symmetric bands are observed (Genevey et al. 2006). They were populated as fission fragments produced by the ^{238}U(α,*f*) fusion-fission reaction (Hua et al. 2004) and by the spontaneous fissions fragments of ^{252}Cf (Ding et al. 2006). The main part of these nuclei have been described with success by different theoretical approaches, the RTRP model is described in detail by Larsson, Leander, and Ragnarsson (Orlandi et al. 2006), the ASYRMO model, this code diagonalizes the particle plus triaxial rotor Hamiltonian in the strong coupling basis (Pinston et al. 2006) and the self-consistent total Routhian Surface (TRS) model shows the shape coexistence of the nuclei (Prolate/Oblate) (Ding et al. 2006). In this work, the structure in $^{103,\,105}$Mo was studied at low-excitation energy by using another theoretical approach based on the quasiparticle-phonon coupling. For giving a description of the structure of the level scheme in a transitional nucleus, the study is confined to the case where a microscopic picture is considered for the quadrupole phonon by means of Tamm-Dancoff Approximation (TDA) (Ring and Schuck 1980). To obtain intrinsic states we have used the deformed average field of Nilsson, a monopole pairing, a quadrupole-quadrupole interaction and the recoil force. The states of rotational bands are determined by inclusion of the Coriolis force. The levels schemes of ^{103}Mo are discussed and compared with the experimental data (Hua et al. 2004).

The Theoretical Formalism

In this section, we make a brief reminder of the quasiparticle-phonon coupling plus rotor method (QPRM) developed in (Boulal et al. 2000; Jdair et al. 2011). The excited state for an odd-nucleus is then reproduced as been combination of rotational state.

$$|I_{a=}\rangle = \sum_{k} b_k^{I_\alpha} \,|K\rangle. \tag{1}$$

The index α labels the states with the same angular momentum and b_k^I represents the amplitudes of Coriolis mixing determined after diagonalization of the total Hamiltonien which is separated into three terms: the intrinsic Hamiltonien H_{int}, the rotational terms H_I, and the Coriolis force H_C.

$$H = H_{\text{int}} + H_I + H_C \tag{2}$$

where

$$H_I = A_R\left(I^2 - I_3^2\right)$$

$$H_C = -A_R\left(I_+J_- + I_-J_+\right).$$

The intrinsic motion is essentially represented by quasiparticle system (BCS approximation) by considering a Nilsson average deformed field H_{sp} plus a monopole pairing interaction. As residual interaction, the quadrupole-quadrupole interaction H_Q and the recoil force H_J are also added to H_{int}.

$$H_{\text{int}} = H_{sp} + H_P + H_Q + H_J \tag{3}$$

where

$$H_{sp} = \sum_{\nu\tau} e_{\nu\tau} a_{\nu\tau}^{+} a_{\nu\tau}$$

$$H_p = -\sum_{\nu\mu\tau} G_\tau a_{\nu\tau}^{+} a_{-\nu\tau}^{+} a_{-\mu\tau} a_{\mu\tau}$$

$$H_Q = -\frac{1}{2}\chi \sum_{\tau\tau'} \left\{ Q_{22}^{+}(\tau) Q_{22} + (\tau') + Q_{2-2}^{+}(\tau) Q_{2-2}(\tau') \right\}$$

$$H_J = A_R\left(J^2 - J_3^2\right) \tag{4}$$

with $I_\pm = I_1 \pm iI_2$, $J_\pm = J_1 \pm iJ_2$ and $A_R = \hbar^2/2\jmath$ is the rotational coefficient. Where the quadrupole moment of mass with $\gamma = \pm 2$ is given as one-body interaction

$$Q_{2\gamma} = \sum_{\nu\tau\mu} \langle \nu\tau | r^2 Y_{2\gamma} | \mu\tau \rangle a_{\nu\tau}^{+} a_{\mu\tau}. \tag{5}$$

The inclusion of the Coriolis force H_C requires the matrix of the model Hamiltonian H to be constructed and diagonalized within the space of symmetrized functions (Bohr and Mottelson 1975)

$$\left| IMK_\rho \right\rangle = \sqrt{\frac{2I+1}{16\pi^2}} \left\{ D_{MK}^{I} \left| K_\rho \right\rangle + (-)^{I+K} D_{M-K}^{I} \left| \bar{K}_\rho \right] \right\}. \tag{6}$$

Here ρ is the quantum number of a given intrinsic states with a projection K of the intrinsic angular momentum along the symmetric axis. $|K_\rho\rangle$ can be obtained by resolution of the secular problem

$$H_{\text{int}} \left| K_\rho \right\rangle = \left(H_{sp} + H_P + H_Q + H_J \right) \left| K_\rho \right\rangle = E_{K_\rho}^{\text{int}} \left| K_\rho \right\rangle. \tag{7}$$

In the form of the QPRM method, configuration for wave functions of an intrinsic state $|K_\rho\rangle$ must renferms contributions of both one-quasiparticle and quasiparticle-phonon components.

$$|K\rangle = \left(\sum_{\nu} C_{\nu}^{\rho} \delta_{K\Omega_\nu} \alpha_{\nu\tau}^{+} + \sum_{\nu\gamma} D_{\nu\gamma}^{\rho} \delta_{K=\Omega_\nu+\gamma} \alpha_{\nu\tau}^{+} B_{\gamma}^{+} \right) |BCS\rangle . \tag{8}$$

With $|BCS\rangle$ is the BCS ground state. C_ν is the amplitude of each one-quasiparticle component and D_ν represents the amplitude of the quasiparticle-phonon component. The BCS method is an approximate approach to treat pairing correlation by using the Bogoliubov-Valatin transformation which makes change from particle to quasiparticle operators

$$a_{\sigma\nu\tau}^{+} = U_{\nu\tau} \alpha_{\sigma\nu\tau}^{+} + \sigma V_{\nu\tau} \alpha_{-\sigma\nu\tau} . \tag{9}$$

Here the operator $\alpha_{\sigma\nu\tau}^{+}(\alpha_{\sigma\nu\tau})$ creates (destroys) a quasiparticle for a nucleon type τ, with a σ-sign depending to time reversal symmetry and where the occupation (non-occupation) probability is expressed by $U_{\nu\tau}(V_{\nu\tau})$.

The quadrupole-phonon operator is defined in the frame of the Tamm-Dancoff Approximation (TDA) (Ring and Schuck 1980)

$$B_{\gamma}^{+} = \frac{1}{2} \sum_{\nu\mu\tau} \left(X_{\gamma}^{\tau} \right)_{\nu\mu} \alpha_{\nu\tau}^{+} \alpha_{\mu\tau}^{+} . \tag{10}$$

This expression permits a microscopic structure description for the quadrupole vibrational core (γ-phonon state) by showing the X-amplitudes which are related to two-quasiparticle excitations.

Results and Discussion

The quasiparticle-phonon plus rotor method describes the odd-A nuclei by coupling the simple quasiparticle to the γ-vibration of the core. We have used this method in order to give a description of the positive/negative low lying levels of the ^{105}Mo nucleus by coupling a proton to the ^{104}Mo even-even core. The theoretical calculations are performed for ^{105}Mo, which have been investigated at low and high-spin (Boulal et al. 2000). For the Nilsson deformed average field, the parametrization (Zhang et al. 1989) adopted for the transitional region around the mass A = 100 is used. A deformation parameter for the even-even core ^{104}Mo is taken equal to $\varepsilon_2 = 0.320$ using Moller data (Moller et al. 1995).

The pairing gap is fixed for proton and neutron by the well-known phenomenological relation $\Delta_p = \Delta_n = 12/A^{1/2}$ (Moller and Nix 1992). The inertia parameters are determined by using the energy of first excited state ($A_R = E(2^+)/6$). The parameter of quadrupole force is fitted so as to reproduce the experimental energy of quadrupole vibrational core, ^{104}Mo ($E(2^+) = 200$ keV) (Guessous et al. 1996).

In the present calculation, values Kappa = 0,068 and Mu = 0,35 were taken for the strength parameters of the ls and l^2 terms [14]. Results provided by our method of calculating by using the average field of Nilsson for neutron treated in the (Fig. 1). To minimize l' energy means, one introduces method BCS. The resolution of equations BCS was made with a relative precision of 0.00000010 at the end of one iteration. Results provided by this method, treated in Table 1. In Table 2, the correlations between ground and excited states are presented for ^{104}Mo. The same for the nucleus of ^{103}Mo, our objective is to study the low-lying states of ^{103}Mo, which is a system of even-even core of ^{102}Mo plus an extra nucleon. The correlation of this last with the core system makes the treatment more complex in the sense of the number of combinations. However, to simplify our study we use the Tamm-Dancoff Approximation (TDA) in which the system is treated in simple way as a two-body interaction. The combination of both methods (BCS and TDA) leads to our QPRM method in which the different possible correlations (states) are presented in Fig. 2. Our theoretical calculation, shows in the case of the nucleus ^{103}Mo the existence of three bands of positive parity and two bands of negative parity at low-lying excitation energy in agreement with the experimental data (see Fig. 3) (Hua et al. 2004). The intrinsic state $3/2^+[411]$ is considered as a ground state for ^{103}Mo nucleus with a spherical orbital $g_{7/2}$, and the negative parity has a bandhead $5/2^-[532]$ coming from the spherical shell $h_{11/2}$. We have been able to propose wave function configurations for the most of the observed low-lying states and bands built upon some of them. Thus, there is a band structure $3/2^+$, $5/2^+$, $7/2^+$, … which was observed for the first time and which has been assigned from our results by the intrinsic states $3/2^+[411]$. For this intrinsic state, the vibrational contribution has caused an energy shift so to approach nearly the ground state $3/2^+$ and reproduce the experimental scheme. The other two positive parity bands provide level structures with various mixing between the two intrinsic states

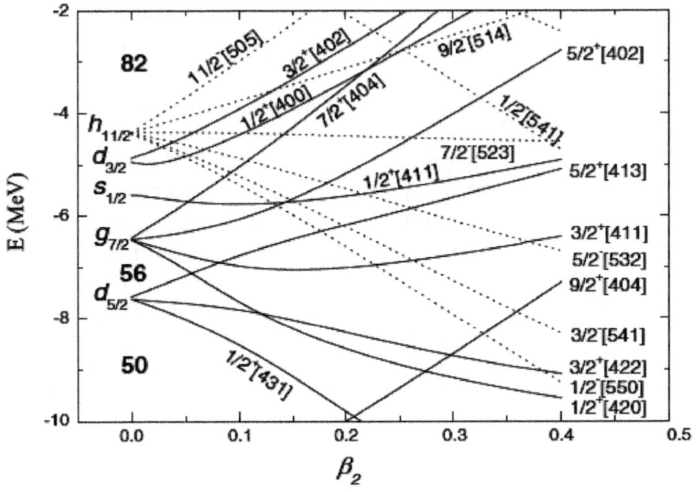

Fig. 1 Partial Nilsson diagram for neutrons

$5/2^{+}[411]$ and $7/2^{+}[411]$. For the configuration of the negative parity band, it is dominated by the intrinsic state $5/2^{-}[532]$. The calculated energy position reproduces the observed positive parity band head. Also, an important contribution from the quasiparticle-phonon coupling is established for the configuration of positive parity wave functions. However, for the negative parity wave functions, the configuration indicates a large component related to only a one-quasiparticle excitation. One can then consider the negative parity intrinsic states in ^{103}Mo as been a pure one-quasiparticle excitation without any contribution from the vibrational excitation mode. The construction and diagonalization of the total Hamiltonian matrix are performed by using the basis function. The energies and wave functions of rotational bands based upon the dominant intrinsic states are determined for each value of the total angular momentum and parity. Although, the structure of intrinsic states is given by two components- the dominant one-quasiparticle component and the quasiparticle-phonon component- it is the first of these that provides a direct matching to the experimental bands. This is a particularly simple process since bands in ^{103}Mo have already been assigned by the dominant Nilsson configuration on the basis of experimental properties.

Table 1 Energy of 20 levels of quasiparticle calculated for ^{104}Mo (case of the neutrons)

Number of states	Identification	EQ(Mev)
22	1/2+(431)	5.021890
23	5/2–(303)	4.925972
24	1/2–(301)	4.908247
25	7/2+(413)	4.039942
26	1/2+(420)	2.783517
27	1/2–(550)	2.605045
28	3/2+(422)	2.284541
29	3/2–(541)	1.599449
30	9/2+(404)	1.289890
31	3/2+(411)	0.030631
Level of Fermi		
32	5/2–(532)	0.030631
33	5/2+(413)	0.748420
34	1/2+(411)	1.564369
35	7/2–(523)	2.127608
36	1/2–(541)	2.216676
37	5/2+(402)	3.143135
38	1/2–(530)	4.037612
39	7/2+(404)	4.062152
40	1/2+(660)	4.386178
41	3/2–(532)	4.444315

Table 2 Element of matrices of the moment d'inertie obtained between Nilsson orbitals

Couple of states			
MU	NU	(MU *I* (*J*+) + (*J*–) *I* NU)	EQP(MU) + EQP(NU)
–1/2+(431)	1/2+(431)	0.598264	10.043780
–1/2+(431)	1/2+(420)	2.981715	7.805407
1/2+(431)	3/2+(422)	–3.023857	7.306431
1/2+(431)	3/2+(411)	0.176612	5.052521
–1/2+(431)	1/2+(411)	–0.254003	6.586259
–1/2+(431)	1/2+(660)	0.032004	9.408069
5/2–(303)	3/2–(541)	–0.001039	6.525421
5/2–(303)	7/2–(523)	–0.000679	7.053580
5/2–(303)	3/2–(532)	–0.052025	9.370287
–1/2–(301)	1/2–(301)	–1.004599	9.816493
–1/2–(301)	1/2–(550)	–0.001864	7.513291
1/2–(301)	3/2–(541)	0.001741	6.507695
–1/2–(301)	1/2–(541)	–0.008214	7.124923
1/2–(301)	1/2–(530)	–0.046345	8.945859
1/2–(301)	3/2–(532)	–0.057582	9.352562
7/2+(413)	9/2+(404)	–3.021134	5.329833
7/2+(413)	5/2+(413)	0.512361	4.788362
7/2+(413)	5/2+(402)	–0.718673	7.183077
–1/2+(420)	1/2+(420)	0.040721	5.567033
1/2+(420)	3/2+(422)	–0.967840	5.068057
1/2+(420)	3/2+(411)	–3.080726	2.814148
–1/2+(420)	1/2+(411)	1.997832	4.347885
–1/2+(420)	1/2+(660)	–0.007181	7.169695
–1/2–(550)	1/2–(550)	5.339473	5.210090
1/2–(550)	3/2–(541)	–5.402582	4.204493
–1/2–(550)	1/2–(541)	–1.371685	4.821721
–1/2–(550)	1/2–(530)	0.316304	6.642657
1/2–(550)	3/2–(532)	0.522606	7.049360
3/2+(422)	5/2+(413)	–3.153995	3.032960
3/2+(422)	1/2+(411)	–1.839107	3.848909
3/2+(422)	5/2+(402)	0.103450	5.427675
3/2+(422)	1/2+(660)	0.005122	6.670719
3/2–(541)	5/2–(532)	–5.363432	1.630080
3/2–(541)	1/2–(541)	1.018619	3.816125
3/2–(541)	1/2–(530)	–1.294144	5.637061
9/2+(404)	7/2+(404)	0.329963	5.352043
3/2+(411)	5/2+(413)	–0.929644	0.779051

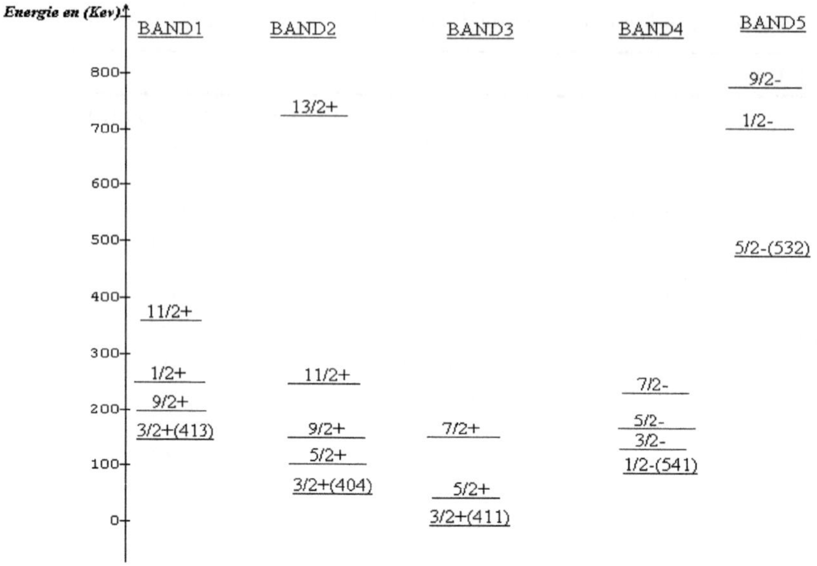

Fig. 2 The partial level scheme of ^{103}Mo nucleus obtained by using the quasiparticle-phonon coupling plus rotor method

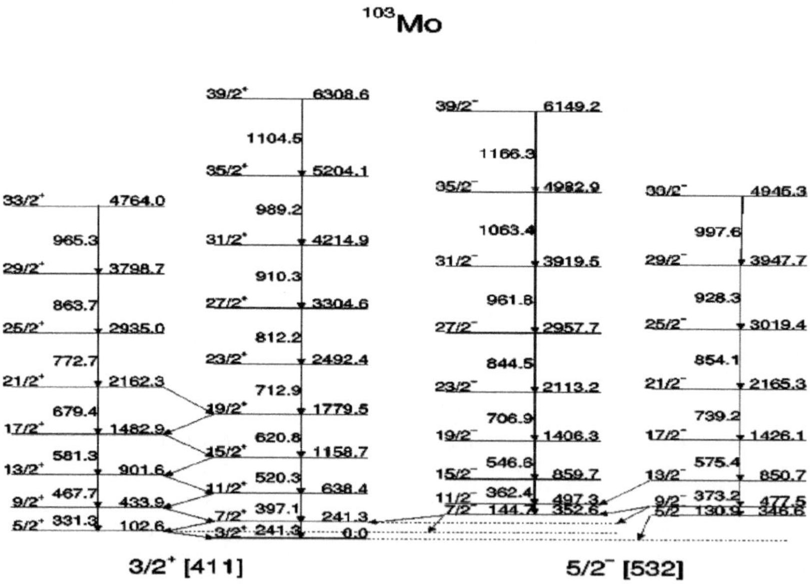

Fig. 3 Partial level scheme of ^{103}Mo. Energies are in KeV (Hua et al. 2004)

To review the contribution of interaction terms from the quadrupole interaction H_{QQ} are the recoil force H_J, and a pairing interaction term H^P, is plotted in Fig. 4 the development of clean energy intrinsic states, which lie closest to the Fermi

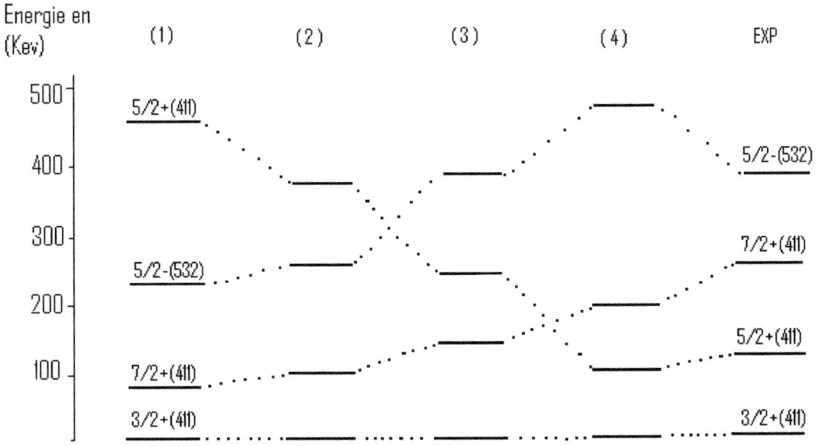

Fig. 4 Energy evolution of intrinsic states in ^{103}Mo caused by including successive interaction terms of quadrupole, recoil forces, and pairing interaction with: (*1*): H_{BCS}; (*2*): $H_{BCS} + H_{QQ}$; (*3*): $H_{BCS} + H_{QQ} + H_J$; (*4*): $H_{BCS} + H_{QQ} + H_J + H^P$

level, following the successive addition of different interaction terms. We accept the following comments:

- The lowest state 3/2+(411) given by Nilsson + BCS calculation does not change following the inclusion of other interactions. It gives a perfect ground state 3/2+ of ^{103}Mo.
- Lowering the state 5/2+(411) is due exclusively to the quadrupole interaction. It goes well for the experimental energy of the head band 5/2+ (to 102 Kev).
- The position of the state 7/2+(411) has experienced a sequential increase following the introduction of the three interaction terms H_{QQ}, H_J, and H^P. His final position close to that found experimentally for the head band (to 397 Kev).
- The effect of these terms for the negative parity states is somewhat negligible.
- It therefore holds that the positive parity states are very sensitive to the quadrupole interaction and recoil force. For cons, the negative parity states were influenced only by the terms of the one-body of recoil force.

Conclusion

In this contribution, we have study the low-lying states of the $^{103,\,105}$Mo nucleus within the quasiparticle-phonon plus rotor method. We have shown for $^{103,\,105}$Mo that the contribution of quadrupole phonon coupling plays an important role for positive parity than negative parity states. The low-lying states have been obtained under the effect of the Coriolis mixing force. The results of our calculation show that the quasiparticle-vibration coupling is a good theoretical approach to produce the main feature of the experimental level scheme. Some discrepancies between theory and experiment might be due to the choice of the residual interaction (addition of the β-vibration), or necessity to replace the Tamn-Dancoff Approximation (TDA) phonon by the Random Phase Approximation (RPA) one.

References

Bohr A., Mottelson B.R. (1975) Nuclear Structure. Vol. 2, Benjamin
Boulal A., Inchaouh J., Jammari M.K. (2000) Eur Phys J A7: 317
Ding H.B. et al. (2006) Phys Rev C74: 054301
Genevey J. et al. (14 Nov 2006) in2p3-00113688, version 1-DANF06
Guessous A. et al. (1996) Phys Rev C53: 1191
Hua H. et al. (2004) Phys Rev C69: 014317
Jdair O., Inchaouh J., Jammari M.K., Chakir H. (March 2011) International Journal of Academic Research Vol. 3.N°2, Part IV
Jing-Ye Zhang et al. (1989) Phys Rev C39: 714
Meyer R.A. et al. (1985) Nucl Phys A439: 510
Moller P. et al. (1995) Atomic data and Nuclear Data tables. Vol. 59, N°. 2, 185
Moller P., Nix J. R. (1992) Nucl Phys A536: 20
Orlandi R. et al. (2006) Phys Rev C73: 054310
Pinston J.A., Urban W., Droste Ch., Genevey J., Rzaca-Urban T. (2006) Phys Rev C74: 064304
Ring P., Schuck P. (1980) The nuclear Many-body Problem. Springer, New York

Index

GPSR Compliance

The European Union's (EU) General Product Safety Regulation (GPSR) is a set of rules that requires consumer products to be safe and our obligations to ensure this.

If you have any concerns about our products, you can contact us on ProductSafety@springernature.com

In case Publisher is established outside the EU, the EU authorized representative is:

Springer Nature Customer Service Center GmbH
Europaplatz 3
69115 Heidelberg, Germany

Batch number: 10755308

Printed by Printforce, the Netherlands